Packed with  the 1993 NE...

The Code

Word for word, you get the entire text of the 1993 *National Electrical Code!*

Official Commentary

Commentary and Formal Interpretations explain the intent of *NEC* requirements and help you apply the Code to real situations. You get the official word—directly from the NFPA experts who publish the *NEC!*

Real-Life Examples

Hundreds of charts, diagrams, and graphs clarify requirements and help you solve real problems you encounter on the job.

Valuable Supplements

Six information-packed supplements (including three that are all new) bring you valuable, time-saving facts that make working with the *NEC* an easier task!

New Expanded Format

The Handbook has been completely redesigned in a large 8$\frac{1}{2}$" x 11" size to be easier to read and use!

Get up to speed on the new Code with the *NEC Changes Video!*

210-7(d), Exception: Receptacle Replacements

GFCI receptacle

NATIONAL ELECTRICAL CODE 1993 CHANGES

Here's a fast, easy way to get a handle on the latest *NEC* changes. Created by NFPA, the *NEC* publisher, this video carefully examines the new 1993 Code. It provides helpful background on the code-making process, as well as the purpose, intent, and application of major changes.

Get first-hand guidance on these vital *NEC* changes...and more!

- Ground-fault circuit-interruptor protection for receptacle outlets in dwellings
- GFCI protection for receptacles in bathrooms of commercial, industrial, and other non-dwelling occupancies
- Phase converters (Article 455)
- Harmonization of communications cable types to recognize designations used in Canada (Article 800)

Learn at your own pace...

Perfect for staff training, the *NEC Changes Video* is an informative, self-paced course. It helps you quickly understand the 1993 Code changes, so you are better prepared to interpret and apply the new *NEC* on the job. Reserve your video today!

Item No. KJ-VCNEC93* $179.50 (Members: $161.55)

*Specify 3/4", VHS, or Betamax.

NFPA 70

National Electrical Code®

1993 Edition

This 1993 edition of the *National Electrical Code* (NFPA 70-1993) was adopted by the National Fire Protection Association, Inc. on May 21, at its 1992 Annual Meeting, in New Orleans, LA, and was released by the Standards Council on July 17, 1992. It was approved by the American National Standards Institute on August 14, 1992. This 1993 edition supersedes all other previous editions, supplements, and printings dated 1897, 1899, 1901, 1903, 1904, 1905, 1907, 1909, 1911, 1913, 1915, 1918, 1920, 1923, 1925, 1926, 1928, 1930, 1931, 1933, 1935, 1937, 1940, 1942, 1943, 1947, 1949, 1951, 1953, 1954, 1955, 1956, 1957, 1958, 1959, 1962, 1965, 1968, 1971, 1975, 1978, 1981, 1984, 1987, and 1990.

Changes in this 1993 edition of the *National Electrical Code* (as compared with the 1990 edition) are indicated by vertical lines in the margin.

The location (in the 1990 edition) of material not appearing in the 1993 edition, and not identified as a change by a vertical line, is identified by a bullet (·) in the margin. Changes in section and table numbers are not identified.

Material identified by the superscript letter "x" includes text extracted from other NFPA documents as identified in Appendix A.

This Code is purely advisory as far as the NFPA and ANSI are concerned, but is offered for use in law and for regulatory purposes in the interest of life and property protection. Anyone noticing any errors should notify the Secretary of the National Electrical Code Committee at the NFPA Executive Office.

History and Development of the National Electrical Code

The National Fire Protection Association has acted as sponsor of the *National Electrical Code* since 1911. The original Code document was developed in 1897 as a result of the united efforts of various insurance, electrical, architectural, and allied interests.

In accordance with the provisions of the NFPA Regulations Governing Committee Projects, a National Electrical Code Technical Committee Report containing proposed amendments to the 1990 *National Electrical Code* was published by the NFPA in June 1991. This report recorded the actions of the various Code-Making Panels and the Correlating Committee of the National Electrical Code Committee on each proposal that had been made to revise the 1990 Code. The report was circulated to all members of the National Electrical Code Committee, and was made available to other interested NFPA members and to the public for review and comment. Following the close of the public comment period, the Code-Making Panels met, acted on each comment, and reported their action to the Correlating Committee. The NFPA published the National Electrical Code Technical Committee Documentation in April 1992, which recorded the actions of the Code-Making Panels and the Correlating Committee on each public comment to the National Electrical Code Technical Committee Report. The NFPA also published the Advance Printing of the Proposed 1993 *National Electrical Code* in April 1992, to permit the study and evaluation by those interested, prior to formal action on the Committee Report by the 1992 NFPA Annual Meeting. The National Electrical Code Technical Committee Report and the National Electrical Code Technical Committee Documentation were presented to the 1992 NFPA Annual Meeting for adoption. The proceedings of that adoption are published in the September 1992 issue of the NFPA *Journal*.

The NFPA has an Electrical Section that provides particular opportunity for NFPA members interested in electrical safety to become better informed and to contribute to the development of the *National Electrical Code* and other NFPA electrical standards. Each of the Code-Making Panels and the Chairman of the Correlating Committee reported their recommendations to meetings of the Electrical Section at the 1992 NFPA Annual Meeting. The Electrical Section thus had opportunity to discuss and review the report of the National Electrical Code Committee prior to the adoption of this edition of the Code by the Association.

The time schedule for processing the 1996 edition of the *National Electrical Code* is as follows:

1996 NATIONAL ELECTRICAL CODE SCHEDULE
(1995 ANNUAL MEETING)

Date	Event
Nov. 5, 1993 (5:00 p.m. EST)	Receipt of Proposals
Jan. 10-22, 1994	Code-Making Panel Meetings
May 9-13, 1994	Correlating Committee Meeting
June 17, 1994	NEC-TCR to Mailing House
Oct. 21, 1994 (5:00 p.m. EST)	Closing Date for Comments
Dec. 5-17, 1994	Code-Making Panel Meetings (TCD)
March 6-10, 1995	Correlating Committee Meeting
April 14, 1995	NEC-TCD to Mailing House
May 22-25, 1995	NFPA Annual Meeting (Denver, CO)
July 19-21, 1995	Standards Council Issuance

Anyone may submit proposals to amend the 1993 Code. Sample forms for this purpose may be obtained from the Secretary of the Standards Council at NFPA Headquarters.

The NFPA *Electrical Code for One- and Two-Family Dwellings*, NFPA 70A, 1993 edition, is an abridged version of the 1993 text, edited only as dictated to eliminate extraneous material not of concern to this type occupancy, and to place in the text only the more popular types of wiring methods, not to exclude any other type authorized by the complete Code.

The *National Electrical Code Handbook* is published by the National Fire Protection Association. This text was prepared by the NFPA Chief Electrical Engineer and the NFPA Electrical Field Service Specialists.

Method of Submitting a Proposal to Revise the National Electrical Code

The following is based on the NFPA Regulations Governing Committee Projects, adopted by the Board of Directors on December 3, 1977 (last amended on June 16, 1992).

A proposal to revise the 1993 edition of the *National Electrical Code* must be submitted so that the proposal is received at NFPA Headquarters by November

5, 1993, as indicated in the time schedule for the 1996 *National Electrical Code*. A proposal received after this date will be returned to the submitter. The proposal is to be sent to the Secretary of the Standards Council at NFPA Headquarters, 1 Batterymarch Park, P.O. Box 9101, Quincy MA 02269-9101.

Each proposal shall include:

(a) identification of the submitter (the person's name) and his or her affiliation (i.e., committee, organization, company), where appropriate, and

(b) an indication that the proposal is for revision of the 1993 *National Electrical Code* and identification of the specific section number, table number (or equivalent identification) of the section, etc., to be revised, and

(c) a statement of the problem and substantiation for proposal, and

(d) the proposed text of the proposal, including the wording to be added, revised (and how revised) or deleted.

Proposals that do not include all of the above information may not be acted on by the National Electrical Code Committee.

It is preferred that the forms available from NFPA for submittal of proposals be used. A separate proposal form should be used for revision of each section of the Code.

An example of a properly submitted (although not necessarily technically correct or acceptable) proposal appears on the following page.

NOTICE

Following issuance of this edition of NFPA 70, The *National Electrical Code*, an appeal was filed with the NFPA Board of Directors.

The appeal requests that Section 250-45 be revised to require redundant grounding of exposed noncurrent-carrying metal parts of cord- and plug-connected equipment likely to become energized.

NFPA will announce the disposition of the appeal when it has been determined. Anyone wishing to receive automatically the disposition of the appeal should notify in writing the Secretary, Standards Council, NFPA Headquarters.

FORM FOR PROPOSALS ON NFPA TECHNICAL COMMITTEE DOCUMENTS

Mail to: Secretary, Standards Council
    National Fire Protection Association, 1 Batterymarch Park, Quincy, Massachusetts 02269-9101
    Fax No.: 617-770-3500

Note: All proposals must be received by 5:00 p.m. E.S.T./E.D.S.T. on the published proposal closing date.

Date Oct. 2, 1992 Name   John Doe   Tel. No.   (617) 770-3000

Address 1234 Main St., Anyplace, MA 02269

Representing (Please indicate organization, company or self)   ABC Manufacturing Co.

1. a) Document Title: National Elec. Code   NFPA No. & Year 70-1993

   b) Section/Paragraph: 663-23(a), Exception

2. Proposal recommends: (Check one) ☒ new text
                       ☐ revised text
                       ☐ deleted text.

3. Proposal (include proposed new or revised wording,
   or identification of wording to be deleted):

   Add a new exception as follows:

   Exception: In industrial occupancies where all of the following
   conditions are met:
     a. The equipment is listed;
     b. Conditions of maintenance and supervision assure that only
     qualified persons will service the equipment;
     c. The supply voltage does not exceed 600 volts, nominal.

> **FOR OFFICE USE ONLY**
>
> Log #: _____
>
> Date Rec'd: _____
>
> Proposal #: _____

4. Statement of Problem and Substantiation for Proposal:
   We have had over 20 years of experience using such equipment in 18 plants
   throughout the United Sates and Canada, and have found that it is often
   necessary to service the equipment while it is energized. Such servicing
   includes measurement of voltages and currents between various parts, and
   examination of the voltage and current waveforms. The use of interlocks on
   the equipment doors prevents such needed servicing. The equipment we use is
   listed by UL, FM, and CSA, and is not provided with interlocks because
   they were not judged necessary, and until the 1993 NEC, there was no Code rule
   requiring interlocks. The limitations in the proposed new exception should
   provide the necessary safeguards against electric shock without interlocks.

5. ☒ This Proposal is original material.
   ☐ This Proposal is not original material; its source (if known) is as follows:

   _____

(Note: Original material is considered to be the submitter's own idea based on or as a result of his own experience, thought, or research and, to the best of his knowledge, is not copied from another source.)

I hereby grant NFPA the non-exclusive, royalty-free rights, including non-exclusive, royalty-free rights in copyright, in this proposal and I understand that I acquire no rights in any publication of NFPA in which this proposal in this or another similar or analogous form is used.

John Doe
Signature (Required)

PLEASE USE SEPARATE FORM FOR EACH PROPOSAL

CONTENTS

First Printing, September 1992
Printed in U.S.A.

NATIONAL ELECTRICAL CODE COMMITTEE

This list represents the membership at the time the Committee was balloted on the text of this edition. Since that time, changes in the membership may have occurred.

Correlating Committee

R. G. Biermann, *Chairman*
Des Moines, IA

Mark W. Earley, *Secretary*
National Fire Protection Association
(Nonvoting)

Jean A. O'Connor, *Recording Secretary*
National Fire Protection Association
(Nonvoting)

M. F. Borleis, Baltimore, MD
Milton E. Cox, Northbrook, IL
William R. Drake, Novato, CA
Kenneth R. Edwards, Washington, DC
Edward L. Eldridge, Bridgewater, NJ
Joseph E. Pipkin, Washington, DC
L. H. Sessler, Cary, NC
J. Philip Simmons, Richardson, TX
Jay A. Stewart, Marquette, MI
Wilford I. Summers, Colorado Springs, CO

Alternates

Melvin J. Anna, Morristown, NJ
(Alt. to L. H. Sessler)
Clyde H. Craig, Toledo, OH
(Alt. to R. G. Biermann)
Paul Duks, Northbrook, IL
(Alt. to M. E. Cox)
Robert Fagotti, Washington, DC
(Alt. to K. R. Edwards)
Earl W. Roberts, Mystic, CT
(Alt. to E. L. Eldridge)
John W. Troglia, Milwaukee, WI
(Alt. to M. F. Borleis)

PANEL NO. 1—Articles 90, 100, 110

Frank K. Kitzantides, *Chairman*
National Electrical Manufacturer's Association

Randall A. Ambuehl, IBEW Local 291
Rep. International Brotherhood of Electrical Workers
Paul Duks, Underwriters Laboratories Inc.
Jerald M. Gerber, Cleveland Electric Illuminating Co.
Rep. Electric Light Power Group/Edison Electric Institute
Charles J. Hart, Bethesda, MD
Rep. National Electrical Contractors Association
Vincent J. Hession, Bellcore
Rep. Exchange Carriers Standards Association
E. Palko, Plant Engineering Magazine
Rep. Institute of Electrical & Electronics Engineers, Inc.
Charles L. Roach, Tennessee Eastman Company
Rep. Chemical Manufacturers Association
Steven G. Roll, ETL Testing Laboratories

J. Philip Simmons, International Association of Electrical Inspectors
Wilford I. Summers, Colorado Springs, CO

Alternates

M. F. Borleis, Baltimore Gas & Electric
(Alt. to J. M. Gerber)
William F. Clark, City of Sacramento, CA
(Alt. to J. P. Simmons)
Philip H. Cox, National Electrical Manufacturers Association
(Alt. to F. K. Kitzantides)
H. Landis Floyd, E. I. du Pont de Nemours & Co.
(Alt. to E. Palko)
Darwin E. McCurry, IBEW - Local 12
(Alt. to R. A. Ambuehl)
Charles H. Williams, National Electrical Contractors Association
(Alt. to C. J. Hart)

PANEL NO. 2—Articles 210, 215, 220, Chapter 9, Examples 1 through 6

William I. Summers, *Acting Chairman*
Colorado Springs, CO

Raffic Ali, Underwriters Laboratories Inc.
James W. Carpenter, North Carolina Dept. of Insurance
Rep. International Association of Electrical Inspectors
Robert J. Cunningham, Duquesne Light Co.
Rep. Electric Light Power Group/Edison Electric Institute
Winfield E. Haggard, Win-Electric, Inc.
Rep. IEC
Thomas L. Harman, University of Houston Clear Lake
Billy H. McClendon, PPG Industries, Inc.
Rep. Chemical Manufacturers Association

Bernard Mericle, IBEW
Rep. International Brotherhood of Electrical Workers
Earl W. Roberts, REPTEC
Rep. National Electrical Manufacturers Association
Thomas E. Sparling, Tom Sparling Associates
Rep. Institute of Electrical & Electronics Engineers, Inc.
H. Brooke Stauffer, SMART HOUSE Limited Partnership
Rep. National Association of Home Builders
Stephen J. Thorwegen, Fisk Electric Company
Rep. National Electrical Contractors Association

Alternates

J. F. Abegglen, San Diego Gas & Electric
  (Alt. to R. J. Cunningham)
Richard W. Becker, Engineered Electrical Systems, Inc.
  (Alt. to T. E. Sparling)
Ernest S. Broome, City of Knoxville, TN
  (Alt. to J. W. Carpenter)

George F. Hawkes, Crouse-Hinds Co.
  (Alt. to E. W. Roberts)
J. Morris Trimmer, University of Florida
  (Alt. to T. L. Harman)
Michael P. Wissman, Henderson Electric Co. of Ohio, Inc.
  (Alt. to W. E. Haggard)
William Wusinich, IBEW Local 98
  (Alt. to IBEW Rep.)

Panel No. 3—Articles 300, 305, 690

E. C. Lawry, *Chairman*
WI Dept. Industrial Labor & Human Relations
Rep. International Association of Electrical Inspectors

Michael F. Auth, Auth Electric
  Rep. Associated Builders and Contractors, Inc.
C. John Beck, Pacific Gas and Electric Co.
  Rep. Electric Light Power Group/Edison Electric Institute
Charles W. Beile, Allied Tube & Conduit
  Rep. National Electrical Manufacturers Association
Ronald R. Bishop, Carlon
  Rep. SPI/VI
Richard G. Gewain, Hughes Associates Inc.
Charles E. Jackson, Celanese Chemical Co. Inc.
  Rep. Chemical Manufacturers Association
Donald A. Mader, Underwriters Laboratories Inc.
Robert D. Nicholson, Solar Cells Inc.
  Rep. Solar Energy Industries Association
  (Vote Limited to Article 690)
Michael Palivoda, LTV Steel Tubular Co.
  Rep. American Iron and Steel Institute
B W. Whittington, Whittington Engineering Inc.
  Rep. Institute of Electrical & Electronics Engineers, Inc.

Thomas H. Wood, Cecil B. Wood Inc.
  Rep. National Electrical Contractors Association

Alternates

Ward I. Bower, Sandia National
  (Alt. to R. D. Nicholson)
Edward A. Hamilton, Niagara Mohawk Power Corp.
  (Alt. to C. J. Beck)
Daleep C. Mohla, Union Carbide Corporation
  (Alt. to B. W. Whittington)
George A. Straniero, Triangle PWC Inc.
  (Alt. to C. W. Beile)
Raymond W. Weber, City of Wisconsin Rapids
  (Alt. to E. C. Lawry)
Craig M. Wellman, E. I. du Pont de Nemours & Co.
  (Alt. to C. E. Jackson)

PANEL NO. 4—Articles 225, 230

Glen W. Cock, *Chairman*
Tampa Electric Co.
Rep. Electric Light Power Group/Edison Electric Institute

K. W. Carrick, B. P. Oil Company
  Rep. Institute of Electrical & Electronics Engineers, Inc.
Robert T. Davis, Knoxville Utilities Board
  Rep. American Public Power Association
Howard D. Hughes, Hughes Electric Co. Inc.
  Rep. National Electrical Contractors Association
James Lee Hunter, Southwire Co.
  Rep. The Aluminum Association
John H. Kassebaum, Eli Lilly & Co.
  Rep. Chemical Manufacturers Association
Ray C. Mullin, Bussmann, Cooper Industries
Robert J. Pollock, Underwriters Laboratories Inc.
Albert J. Reed, New York Board of Fire Underwriters
Donald R. Strassburg, State of Washington
  Rep. International Association of Electrical Inspectors
Jack K. Veazey, IBEW, Local 850
John W. Young, Siemens Energy & Automation, Inc.
  Rep. National Electrical Manufacturers Association

Alternates

Hugh D. Butler, Southwire
  (Alt. to J. L. Hunter)
Dennis Darling, Molan Engineering
  (Alt. to K. W. Carrick)
Junior L. Owings, State of Oregon
  (Alt. to D. R. Strassburg)
Philip M. Piqueira, General Electric Company
  (Alt. to J. W. Young)
Vincent J. Saporita, Bussmann, Cooper Insustries
  (Alt. to R. C. Mullin)
Robert Skinner, Northeast Utilities Service
  (Alt. to G. W. Cock)
Jack K. Veazey, IBEW - Local 850
  (Alt. to IBEW Rep.)

Nonvoting

M. R. Holman, Square D. Canada

PANEL NO. 5—Articles 200, 250, 280

L. H. Sessler, *Chairman*
Cary, NC
Rep. Exchange Carriers Standards Association

Robert E. Bernd, Underwriters Laboratories Inc.
James E. Bessine, IBEW Local 13
  Rep. International Brotherhood of Electrical Workers
Edward L. Eldridge, Thomas & Betts Co.
  Rep. National Electrical Manufacturers Association

Richard A. Haman, ND State Electrical Board
  Rep. International Association of Electrical Inspectors
Richard Lloyd, Alflex Corp. Rep. The Aluminum Association
D. H. McIntosh, Newark, DE
  Rep. Chemical Manufacturers Association

William J. Neiswender, Orlando, FL
   Rep. Institute of Electrical & Electronics Engineers, Inc.
William F. Saffell, Black & Decker (US) Inc.
   Rep. Power Tool Institute, Inc.
Ronald J. Toomer, Toomer Electrical Co. Inc.
   Rep. National Electrical Contractors Association

Alternates

Melvin J. Anna, Bell Communications Research
   (Alt. to L. H. Sessler)
Paul M. Bowers, City of Iowa City, IA
   (Alt. to R. A. Haman)
Norman H. Davis, Underwriters Laboratories Inc.
   (Alt. to R. E. Bernd)

Donald J. Frost, IBEW Local 288
   (Alt. to J. Bessine)
Abjul Q. Haideri, E I. du Pont de Nemours & Co.
   (Alt. to D. H. McIntosh)
Robert C. Oldham, Reynolds Metals Co.
   (Alt. to R. Lloyd)
David Peot, The Singer Co.
   (Alt. to W. F. Saffell)
Elliot Rappaport, Electro Technology Consultants Inc.
   (Alt. to W. J. Neiswender)
Jerome W. Seigel, General Electric Co.
   (Alt. to E. L. Eldridge)
Douglas White, Houston Lighting & Power Co.
   (Alt. to EEI Rep.)

PANEL NO. 6—Articles 310, 400, 402, Chapter 9, Tables 5 through 9

Clyde H. Craig, *Chairman*
Craig Electric Co.
Rep. National Electrical Contractors Association

Harold Baggett, IBEW, Local 1701
   Rep. International Brotherhood of Electrical Workers
David P. Brown, Baltimore Gas & Electric Co.
   Rep. Electric Light Power Group/Edison Electric Institute
Dwight Durham, Southwire Co.
   Rep. The Aluminum Association
L. Bruce McClung, Union Carbide Corp.
   Rep. Institute of Electrical & Electronics Engineers, Inc.
Robert R. Sallaz, City of Akron, OH
   Rep. International Association of Electrical Inspectors
George A. Straniero, Triangle PWC Inc.
   Rep. National Electrical Manufacturers Association
Donald A. Voltz, Dow Chemical USA
   Rep. Chemical Manufacturers Association
Leonard B. Zafonte, Underwriters Laboratories Inc.

Alternates

Joseph W. Aitken, Hercules Inc.
   (Alt. to D. A. Voltz)
William C. Ferrell, IBEW
   (Alt. to H. Baggett)
Barry N. Hornberger, Philadelphia Electric Co.
   (Alt. to D. P. Brown)
W Terry Lindsay, Duncan Electric Company Inc.
   (Alt. to C. H. Craig)
Patricia McCune, Alcan Cable
   (Alt. to D. Durham)
Dann N. Strube, Dept. of Housing, Building & Construction
   (Alt. to R. R. Sallaz)

Nonvoting

Austin C. Tingley, Canada Wire & Cable Co. Ltd.

PANEL NO. 7—Articles 320, 321, 324, 325, 326, 328, 330, 333, 334, 336, 337, 338, 339, 340, 342, 363

Constantine S. Golovko, *Chairman*
Building & Safety Division
Rep. International Association of Electrical Inspectors

Bobby C. Gentry, Southwire Co.
   Rep. The Aluminum Association
Thomas J. Guida, Underwriters Laboratories Inc.
A P. Haggerty, Scott Paper Co.
   Rep. Institute of Electrical & Electronics Engineers, Inc.
Charles J. Hart, Bethesda, MD
   Rep. National Electrical Contractors Association
Russell W. Higginbottom, Triangle Conduit & Cable Co.
   Rep. National Electrical Manufacturers Association
John A. Kroiss, Shell Oil Co.
   Rep. Chemical Manufacturers Association
Robert J. Eugene, City of Spokane,WA
Peter F. V. Lloyd, B. F. Goodrich Co.
Don E. Morgan, IBEW - Local 756
   Rep. International Brotherhood of Electrical Workers
D. C. Roberts, American Electric Power Service Corp.
   Rep. Electric Light Power Group/Edison Electric Institute

David E. Schumacher, Schumacher Electric Company Inc.
   Rep. Associated Builders and Contractors, Inc.

Alternates

J. Randy Bottomlee, B'ham Electrical JATC
   (Alt. to D. E. Morgan)
James M. Daly, Cablec Corporation
   (Alt. to R. W. Higginbottom)
Patricia McCune, Alcan Cable
   (Alt. to B. C. Gentry)
Eugene M. Rubin, City of Los Angeles, CA
   (Alt. to C. S. Golovko)
Lynn F. Saunders, General Motors Corp.
   (Alt. to A. P. Haggerty)
Ronald L. Spees, Consumers Power Co.
   (Alt. to D. C. Roberts)
D F. Tulloh, Union Carbide Corp.
   (Alt. to J. A. Kroiss)
Charles H. Williams, National Electrical Contractors Association
   (Alt. to C. J. Hart)

PANEL NO. 8—Articles 318, 331, 345, 346, 347, 348, 349, 350, 351, 352, 353, 354, 356, 358, 362, 364, 365, 374, Chapter 9, Tables 1 through 4

Armond C. Webb, *Chairman*
Pacific Gas & Electric Co.
Rep. Electric Light Power Group/Edison Electric Institute

John E. Bond, IBEW Local 948
Rep. International Brotherhood of Electrical Workers
Wayne Brinkmeyer, Biddle Electric Corp.
Rep. National Electrical Contractors Association
John S. Corry, Corry Electric Inc.
Rep. Associated Builders and Contractors, Inc.
Dwight Durham, Southwire Co.
Rep. The Aluminum Association
Charles W. Forsberg, Carlon
Rep. SPI/VI
James L. Griffin, Wheatland Tube Co.
Rep. American Iron and Steel Institute
Kenneth E. Haas, Underwriters Laboratories Inc.
J. Earl Kreuzer, American Electric Co.
Rep. National Electrical Manufacturers Association
Wayne A. Lilly, City of Harrisburg
Robert B. Phillips, Monsanto Co.
Rep. Chemical Manufacturers Association
Dennis L. Rowe, NY Board of Fire Underwriters
James E. Tyson, US Department of Veterans Affairs

Alternates

Richard J. Baltes, Baltes Electric Corporation
(Alt. to J. S. Corry)
Khimchand H. Chudasama, US Department of Veterans Affairs
(Alt. to J. E. Tyson)
Robert W. Cox, ICI Americas, Inc.
(Alt. to R. B. Phillips)
Wyman H. Hawley, Athens-Clarke County
(Alt. to W. A. Lilly)
Richard Lloyd, Alflex Corp.
(Alt. to D. Durham)
Gary J. O'Neil, Southern California Edison Co.
(Alt. to A. C. Webb)
James E. Robertson, Dow Chemical Co.
(Alt. to IEEE Rep.)
D. Ray Rink, IBEW Local 153
(Alt. to J. E. Bond)
Ronald J. Toomer, Toomer Electrical Co. Inc.
(Alt. to W. Brinkmeyer)
L. Vincent Siemens, Energy & Automation Inc.
(Alt. to J. E. Kreuzer)

PANEL NO. 9—Articles 370, 373, 380, 384

Dale R. Deming, *Chairman*
American Electric Co.

Artie O. Barker, Boise, ID
Joseph A. Cannatelli, Arco Chemical Co.
Rep. Chemical Manufacturers Association
Frederic P. Hartwell, Electrical Construction & Maintenance Magazine
Orlen G. Holien, City Electric
Rep. Associated Builders and Contractors, Inc.
Philip Lyle Johnson, Siemens Energy & Automation Inc.
Rep. National Electrical Manufacturers Association
Robert Kaemmerlen, Kaemmerlen Electric Co.
Rep. National Electrical Contractors Association
Wayne Menuz, Underwriters Laboratories Inc.
Anthony J. Mussi, NY Board of Fire Underwriters
Rep. International Association of Electrical Inspectors
David K. Norton, US Department of Veterans Affairs
Sukanta Sengupta, Rhone-Poulenc Inc.
Rep. Institute of Electrical & Electronics Engineers, Inc.
Fred J. Smith, IBEW Local 117
Rep. International Brotherhood of Electrical Workers

William R. Stalker, Commonwealth Edison Co.
Rep. Electric Light Power Group/Edison Electric Institute

Alternates

Michael A. Chetney, Milwaukee Area Electrical JATC
(Alt. to F. J. Smith)
Kermit B. Duncan, Pacific Power & Light Co.
(Alt. to W. R. Stalker)
Anthony Montuori, New York Board of Fire Underwriters
(Alt. to A. J. Mussi)
Wallace Salshutz, Pfizer Inc.
(Alt. to J. A. Cannatelli)
William E. Slater, RACO Inc.
(Alt. to P. L. Johnson)
Constantinos V. Vasiliades, US Department of Veterans Affairs
(Alt. to D. K. Norton)

PANEL NO. 10—Articles 240, 780

Stanley D. Kahn, *Chairman*
Forest Electric Corporation
Rep. National Electrical Contractors Association

Robert J. Deaton, Union Carbide Corp.
Richard L. Doughty, E I. de Pont de Nemours & Co.
Rep. Chemical Manufacturers Association
Donald G. Fischer, Square D. Co.
Rep. National Electrical Manufacturers Association
Donald E. Fritz, Pennsylvania Power & Light
Rep. Electric Light Power Group/Edison Electric Institute
Patrick R. Herbert, High Voltage Maintenance Corp.
Rep. International Electrical Testing Association Inc.
John J. Mahal, Underwriters Laboratories Inc.
Leo F. Martin, City of Boston, MA
Rep. International Association of Electrical Inspectors

Arden L. Munson, Hussmann Corp.
Rep. Air Conditioning and Refrigeration Institute
George J. Ockuly, Bussmann, Cooper Industries
Clarence Whelan, IBEW Local 242
Rep. International Brotherhood of Electrical Workers

Alternates

Joseph J. Andrews, Westinghouse Savannah River Co.
(Alt. to R. J. Deaton)
Douglas A. Fisher, IBEW Local 474
(Alt. to C. Whelan)
Carl J. Fredericks, Dow Chemical USA
(Alt. to R. L. Doughty)

Paul P. Gubany, Bussmann, Cooper Industries
  (Alt. to G. J. Ockuly)
A. F. KnicKrehm, The KnicKrehm Co.
  (Alt. to S. D. Kahn)
James D. Mars, TU Electric - TP&L Division
  (Alt. to D. E. Fritz)
William F. Nagle, Town of Danvers, MA
  (Alt. to L. F. Martin)

R. O. D. Whitt, Westinghouse Electric Corp.
  (Alt. to D. G. Fischer)
Kevin L. Young, White-Rodgers
  (Alt. to A. Munson)

Nonvoting

Rick C. Gilmour, Canadian Standards Association

PANEL NO. 11—Articles 430, 440, 670, Chapter 9, Example No. 8

Richard W. Osborn, *Chairman*
Rep. National Electrical Contractors Association

David S. Baker, Brown & Root USA Inc.
  Rep. Institute of Electrical & Electronics Engineers, Inc.
Milton E. Cox, Underwriters Laboratories Inc.
Thomas E. Dye, Olin Corp.
  Rep. Chemical Manufacturers Association
Francis Finnegan, Texas Instruments Inc.
William T. Fiske, ETL Testing Laboratories Inc.
Paul P. Gubany, Bussmann, Cooper Industries
T. A. Jacoby, Tecumseh Products Co.
  Rep. Air Conditioning and Refrigeration Institute
Miguel G. Lopez, US Department of Veterans Affairs
Don P. Ruedi, Square D. Co.
  Rep. National Electrical Manufacturers Association
Charles B. Schram, Wilmette, IL
Robert Schultz, A. D. S. Electric Company Inc.
  Rep. Associated Builders and Contractors, Inc.
Mark N. Shapiro, City of Madison Heights, MI
  Rep. International Association of Electrical Inspectors
Russell F. Sheehan, IBEW Local 103
  Rep. International Brotherhood of Electrical Workers
David W. Trudeau, Factory Mutual Research
James A. Wilson, Black & Veatch

Alternates

Joe David Cox, Tennessee Eastman Co.
  (Alt. to T. Dye)
Michael D. Landolfi, Landolfi Electric Co. Inc.
  (Alt. to R. Schultz)
Joseph E. Martin, Emerson Electric Co.
  (Alt. to D. P. Ruedi)
Philip W. Mason, JATC of Greater Boston
  (Alt. to R. F. Sheehan)
George J. Ockuly, Bussmann, Cooper Industries
  (Alt. to P. P Gubany)
Tim Pinnick, City of Lawrence, KS
  (Alt. to M. N. Shapiro)
Richard A. Rasmussen, Underwriters Laboratories, Inc.
  (Alt. to M. Cox)
Lynn F. Saunders, General Motors Corp.
  (Alt. to D. S. Baker)
Jeffrey L. Steplowski, Department of Veterans Affairs
  (Alt. to M. Lopez)
M. Edward Thomas, Alabama Power Company
  (Alt. to ELPG/EEI Rep.)
Leonard K. VanTassel, Carrier Air Conditioning
  (Alt. to T. A. Jacoby)
Thomas H. Wood, Cecil B. Wood Inc.
  (Alt. to R. W. Osborn)

NONVOTING

Robert Colvin, Allen-Bradley Canada Ltd.

PANEL NO. 12—Articles 427, 610, 620, 630, 645, 660, 665, 668, 669, 685

B. W. Whittington, *Chairman*
Whittington Engineering Inc.

A. R. Cartal, Middle Dept. Inspection Agency, Inc.
  Rep. International Association of Electrical Inspectors
James F. Cook, Eagle Electric Manufacturing
  Rep. National Electrical Manufacturers Association
Michael J. DeMartini, Underwriters Laboratories Inc.
Ralph E. Droste, United Technologies/Otis Elevator Co.
  Rep. National Elevator Industry Inc.
  (Vote Limited to Articles 610, 620, 630)
C. James Erickson, Newark, DE
  Rep. Chemical Manufacturer's Association
  (Vote Limited to Articles 427, 610, 645, 660, 665, 668, 669, 685)
Arthur Fine, AT&T Bell Laboratories
  Rep. Insulated Cable Engineers Association Inc.
William J. Kelly, Eastman Kodak Co.
  Rep. Institute of Electrical & Electronics Engineers, Inc.
B. J. Lowery, IBEW Local 175
  Rep. International Brotherhood of Electrical Workers
David G. Nutt, International Business Machines
  Rep. Computer & Business Equipment Manufacturers Association
  (Vote Limited to Article 645)
C. L. Pittman, Maryville, TN
  Rep. The Aluminum Association
  (Vote Limited to Articles 645, 660, 665, 668, 669, 685)

Ronald L. Purvis, Georgia Power Company
  Rep. Electric Light Power Group/Edison Electric Institute
Robert H. Reuss, Harnischfeger Corp.
  Rep. Crane Manufacturers Association of America Inc.
  (Vote Limited to Article 610)
T. Neil Thorla, Inland Steel Co.
  Rep. Association of Iron & Steel Engineers
  (Vote Limited to Articles 610, 620, 630)
Charles M. Trout, Main Electric Company Inc.
  Rep. National Electrical Contractors Association

Alternates

Thomas C. Feindel, Digital Equipment Corporation
  (Alt. to D. G. Nutt)
  (Vote Limited to Article 645)
Wayne Hoffman, ESAB Welding Products, Inc.
  (Alt. to J. F. Cook)
R H. Keis, First State Inspection Agency, Inc.
  (Alt. to A. R. Cartal)
Richard H. Laney, Siecor Corporation
  (Alt. to A. Fine)
Nick Marchitto, Otis Elevator Co.
  (Alt. to R. E. Droste)
  (Vote Limited to Articles 610, 620, 630)

S L. Ralston, PPG Industries Inc.
 (Alt. to C. J. Erickson)
 (Vote Limited to Articles 427, 610, 645, 660, 665, 668, 669, 685)
Merritt D. Redick, BHP-UTAH Int'l Inc.
 (Alt. to W. J. Kelly)
Leo R. Wasilewski, Reynolds Metals Co
 (Alt. to C. L. Pittman)
 (Vote Limited to Articles 645, 660, 665, 668, 669, 685)

John N. Wright, IBEW Local 270
 (Alt. to B. J. Lowery)

Nonvoting

K. McCullough, National Elevator & Escalator Association

PANEL NO. 13—Articles 450, 460, 470, 710

Kernan R. Dennis, *Chairman*
Northwest Electric Co.
Rep. National Electrical Contractors Association

William A. Brunner, IBEW
Hugh D. Butler, Southwire Carrollton, GA
 Rep. The Aluminum Association
E E. Carlton, Menlo Park, CA
Walter Kroboth, Exxon Chemical Co.
 Rep. Chemical Manufacturers Association
Hiram J. Lamb, City of Charlottesville, VA
 Rep. International Association of Electrical Inspectors
William T. O'Grady, Underwriters Laboratories Inc.
Milton D. Robinson, Milt Robinson Engineering Co.
 Rep. Institute of Electrical & Electronics Engineers, Inc.
Truman C. Surbrook, Michigan State University
 Rep. American Society of Agricultural Engineers
Walter E. Thomas, Westinghouse Electric Corp.
 Rep. National Electrical Manufacturers Association
John W. Troglia, Wisconsin Electric Power Co.
 Rep. Electric Light Power Group/Edison Electric Institute

Alternates

David A. Barnard, ACME Electrical Corp.
 (Alt. to W. E. Thomas)
J. Alan Barringer, North Carolina Department of Insurance
 (Alt. to H. J. Lamb)
O. L. Davis, Manzano Western, Inc.
 (Alt. to K. R. Dennis)
Richard P. Fogarty, Consolidated Edison of New York
 (Alt. to J. W. Troglia)
Alfred D. Y. Pong, US Department of Veterans Affairs
 (Alt. to VA Rep.)
Robert L. Simpson, Simpson Electrical Engineering Co.
 (Alt. to M. D. Robinson)
Samuel Solomon, Alameda County Electrical
 (Alt. to W. A. Brunner)
Ivan L. Winsett, Ronk Electrical Industries
 (Alt. to T. C. Surbrook)
Ralph H. Young, Tennessee Eastman Co. (TEC)
 (Alt. to W. Kroboth)

PANEL NO. 14—Articles 500, 501, 502, 503, 504, 510, 511, 513, 514, 515, 516

Richard J. Buschart, *Chairman*
Chesterfield, MO

Albert A. Bartkus, Underwriters Laboratories Inc.
Joseph H. Kuczka, Killark Electric Mfg. Co.
 Rep. National Electrical Manufacturers Association
Francis J. McGowan, Factory Mutual Research Corp.
Donald T. Murphy, Ralston Purina Co.
 Rep. American Feed Manufacturers Association, Inc.
 (Vote Limited to Articles 500 & 502)
Elliot M. Nesvig, Erdco Engineering Corporation
 Rep. Instrument Society of America
W. L. Raines, St Louis County Dept. of Public Works
 Rep. International Association of Electrical Inspectors
J H. Rannells, Amoco Oil Co.
 Rep. American Petroleum Institute
Carl R. Roselli, ANR Pipeline Co.
Mark G. Saban, Modern Electric Company of Illinois
 Rep. National Electrical Contractors Association
Peter J. Schram, Braintree, MA
Harold B. Smith, Wunderlich-Malec Ing, Inc.
 Rep. GEPS
Lawrence C. Strachota, Iowa Power & Light Co.
 Rep. Electric Light Power Group/Edison Electric Institute
James A. Weldon, IBEW Local 728
 Rep. International Brotherhood of Electrical Workers
Donald W. Zipse, FMC Corp.
 Rep. Institute of Electrical & Electronics Engineers, Inc.

Alternates

Lawrence J. Adamcik, The Dow Chemical Company
 (Alt. to D. W. Zipse)
Alonza W. Ballard, Crouse-Hinds ECM
 (Alt. to J. Kuczka)
David N. Bishop, Chevron USA Inc.
 (Alt. to J. H. Rannells, Jr)
Marion A. Cabler, Phillips Petroleum Co.
 (Alt. to CMA Rep)
James D. Cospolich, Waldemar S. Nelson & Co. Inc.
 (Alt. to E. M. Nesvig)
Milton L Foster, IBEW
 (Alt. to J. A. Weldon)
Pedro M. Garcia, Florida Power & Light Co.
 (Alt. to L. C. Strachota)
Gerald W. Lemberg, Lemberg Electric Co. Inc.
 (Alt. to M. G. Saban)
James M. Lynch, City of Chicago
 (Alt. to W. L. Raines)
David F. Moorhead, Blue Seal Feeds Inc.
 (Alt. to D. T. Murphy)
Ronald C. Vaickauski, Underwriters Laboratories Inc.
 (Alt. to A. A. Bartkus)

Nonvoting

John A. Bossert, Hazloc, Inc.
Vincent Rowe, Shell Canada Ltd.
 (Alt. to J. Bossert)
Fred K. Walker, HQ United States Air Force

PANEL NO. 15—Articles 445, 480, 518, 520, 530, 540, 700, 701, 702, 705

J. A. Stewart, *Acting Chairman*
Jay Stewart Assoc. Inc.

Joseph S. Dudor, Joseph S. Dudor, Consulting Elec. Eng.
Rep. Institute of Electrical & Electronics Engineers, Inc.
Robert C. Duncan, Reedy Creek Improvement District
George W. Flach, Flach Consultants
George T. Howard, George T. Howard Associates
LaVell J. Jensen, Utah Power & Light Co.
Rep. Electric Light Power Group/Edison Electric Institute
Gordon S. Johnson, Dundee, FL
Rep. Electrical Generating Systems Association
(Vote Limited to Articles 445, 480, 700, 701, 702, 705)
Jack W. Kalbfeld, Kalico Technology Inc.
(Vote Limited to Articles 518, 520, 530, 540)
Michael B. Klein, Labyrinth Electrical, Inc.
Rep. Illuminating Engineering Society of North America
J. J. Kubisz, Underwriters Laboratories Inc.
Michael A. Lanni, Universal City Studios
Rep. Motion Picture Association of America, Inc.
Keku M. Mistry, Bell Communications Research
Rep. Exchange Carriers Standards Association
(Vote Limited to Articles 445, 480, 700, 701, 702, 705)

Michael Mowrey, IBEW
Edward D. Rheaume, IBEW Local 970
Rep. International Brotherhood of Electrical Workers
Jay A. Stewart, Jay Stewart Assoc. Inc.
Rep. Chemical Manufacturers Association
David G. Strasser, Marathon Electric Mfr. Corp.
Rep. National Electrical Manufacturers Association
Kenneth E. Vannice, Lee Colortran, Inc.
Rep. US Institute of Theatre Technology
(Vote Limited to Articles 518, 520, 530, 540)

Alternate

Anthony J. DeLuca, IBEW Local 313
(Alt. to M. Mowrey)
Michael V. Glenn, Longview Fibre Co.
(Alt. to J. Dudor)
James R. Iverson, Onan Corp.
(Alt. to G. Johnson)
Bobby J. Love, City of Houston, TX
(Alt. to IAEI Rep - Voting Alternate)
Randall Perdue, Arkansas Power & Light Company
(Alt. to L. J. Jensen)
Charles G. Pounds, Dow Chemical USA
(Alt. to J. Stewart)
Michael D. Skinner, CBS/MTM Studios
(Alt. to M. A. Lanni)

PANEL NO. 16—Articles 640, 650, 720, 725, 760, 770, 800, 810, 820

D. Harold Ware, *Chairman*
Libra Electric Co.
Rep. National Electrical Contractors Association

Melvin J. Anna, Bell Communications Research
Rep. Exchange Carriers Standards Association
George J. Bagnall, Rural Electrification Admin
Charles D. Hansell, C. D. Hansell Consultants Inc.
Rep. Institute of Electrical & Electronics Engineers, Inc.
David A. Heywood, US Testing Co.
Harry Katz, South Texas Electrical JATC
Rep. International Brotherhood of Electrical Workers
Stanley Kaufman, AT&T Bell Laboratories
Michael A. Lanni, Universal City Studios
Rep. Motion Picture Association of America, Inc.
(Vote Limited to Articles 640, 720, 725, 760, 810, 820)
Raymond I. Lease, Middle Dept. Inspection Agency
Rep. International Association of Electrical Inspectors
Irving Mande, Edwards Co. Inc.
Rep. National Electrical Manufacturers Association
Vieney B. Mascarenhas, Canada Wire & Cable Limited
Rep. Insulated Cable Engineers Association Inc.
William Pike, L & B. Enterprises
Rep. Building Industry Consulting Service International
William L. Schallhammer, Underwriters Laboratories Inc.
Arthur E. Schlueter Jr, Pipe Organ Sales & Service Inc.
Rep. American Institute of Organ Builders/Associated Pipe Organ Builders of America
(Vote Limited to Articles 640, 650, 720, 725)
James W. Stilwell, Tele Services R&D
Rep. National Cable Television Association
(Vote Limited to Articles 770, 800, 810, 820)

Kyle E. Todd, Gulf States Utilities Co.
Rep. Electric Light Power Group/Edison Electric Institute
Inder L. Wadehra, IBM Corporation
Rep. Computer & Business Equipment Manufacturers Association
David Wechsler, Union Carbide Corp.
Rep. Chemical Manufacturers Association
Dean K. Wilson, Industrial Risk Insurers

Alternates

David K. Baker, Superior Cable Corp.
(Alt. to V. B. Mascarenhas)
Ronald P. Cantrell, IBEW Local 72
(Alt. to H. F. Katz)
S E. Egesdal, Honeywell Inc.
(Alt. to I. Mande)
Elias Eduardo, Gil Houston Lighting & Power Co.
(Alt. to K. Todd)
Ray A. Jones, E I. du Pont de Nemours & Co.
(Alt. to D. Wechsler)
Ronald G. Jones, Ronald G. Jones, PE
(Alt. to C. D. Hansell)
John Mangan, Medford City Hall, MA
(Alt. to R. I. Lease Jr.)
Jan Rowland, Houston, TX
(Alt. to A. E. Schlueter)
(Vote Limited to Articles 640, 650, 720, 725)
L H. Sessler, Cary, NC
(Alt. to M. J. Anna)
Gary W. Victorine, Hewlett-Packard Co.
(Alt. to I. Wadehra)

PANEL NO. 17—Article 517

George Schuck, *Chairman*
Rep. IBEW Local 3

D. R. Borden, Tri-City Electric Company Inc.
Rep. National Electrical Contractors Association
Timothy M. Croushore, Allegheny Power System, Inc.
Rep. Electric Light Power Group/Edison Electric Institute
Floyd J. DeVore, General Electric Medical
Alvin Friendlich, Crown Electric
Thomas H. Gilliam, Riverside Methodist Hospital
Rep. Association for the Advancement of Medical Instrumentation
Glenn T. Keates, Harley, Ellington, Pierce, Yee
Rep. Institute of Electrical & Electronics Engineers, Inc.
Ode Richard Keil, Joint Commission on Accreditation of Healthcare Orgs
Phillip W. Knight, Knight Electric Co. Inc.
Rep. Associated Builders and Contractors, Inc.
M. T. Merrigan, Underwriters Laboratories Inc.
James A. Meyer, Pettis Memorial VA Hospital
Rep. American Society of Anesthesiologists
Hugh O. Nash, Smith Seckman Reid Inc.
Gary D. Slack, Healthcare Engineering Consultants
Rep. American Hospital Association
Jeffrey L. Steplowski, Department of Veterans Affairs
Duane J. Telecky, Washoe Medical Center
Rep. NFPA Health Care Section

Bradley Zempel, City of Seattle, WA
Rep. International Association of Electrical Inspectors

Alternates

Steven Benesh, Square D. Co.
(Alt. to F. J. DeVore)
Thomas J. Castor, American Electric Power
(Alt. to T. M. Croushore)
James R. Duncan, Sparling & Associates, Inc.
(Alt. to G. T. Keates)
Douglas S. Erickson, American Hospital Association
(Alt. to G. D. Slack)
Stanley D. Kahn, Forest Electric Corporation
(Alt. to D. R. Borden)
Gerald L. Kavanaugh, IBEW Local 145
(Alt. to G. Schuck Jr)
Elver L. Madsen, Montana Dept. Commerce/ Building Codes Bureau
(Alt. to B. Zempel)
Bryan Parker, Montefiore Hospital & Medical Center
(Alt. to T. H. Gilliam)
James E. Tyson, US Department of Veterans Affairs
(Alt. to J. L. Steplowski)

PANEL NO. 18—Articles 410, 600, 605

A. F. KnicKrehm, *Chairman*
The KnicKrehm Co.
Rep. National Electrical Contractors Association

Kenneth F. Kempel, Underwriters Laboratories Inc.
E G. Kiener, Solar Turbine Inc.
Rep. Institute of Electrical & Electronics Engineers, Inc.
Bernard J. Mezger, Bostrom Management Corp.
Rep. American Home Lighting Institute
(Vote Limited to Article 410)
Saul Rosenbaum, Leviton Mfg. Co. Inc.
Rep. National Electrical Manufacturers Association
James P. Suttle, Ohio Edison Co.
Rep. Electric Light Power Group/Edison Electric Institute
Thomas E. Trainor, City of San Diego, CA
Rep. International Association of Electrical Inspectors
Jack Wells, Pass & Seymour/LeGrand
Randall K. Wright, Wright Sign Company
Rep. Electric Sign Association
(Vote Limited to Article 600)

Alternates

Rudy T. Elam, Calspan Corp.
(Alt. to E. G. Kiener)
Howard D. Hughes, Hughes Electric Co. Inc.
(Alt. to A. F. KnicKrehm)
Stephen G. Kieffer, Kieffer & Co. Inc.
(Alt. to R. K. Wright)
Robert M. Milatovich, Clark County
(Alt. to T. E. Trainor)
Joe Mogan, Lithonia Lighting
(Alt. to S. Rosenbaum)
John L. Revil, IBEW Local 223
(Alt. to IBEW Rep.)
T M. Von Sprecken, Mississippi Power Co.
(Alt. to J. P. Suttle)
Sidney Wolkin, Lightolier Inc.
(Alt. to B. J. Mezger)
(Vote Limited to Article 410)

PANEL NO. 19—Articles 545, 547, 550, 551, 553, 555, 604, 675

John H. Stricklin, *Chairman*
State of Idaho Mountain Home, ID
Rep. International Association of Electrical Inspectors

Barry Bauman, Wisconsin Power & Light Co.
Rep. American Society of Agricultural Engineers
(Vote Limited to Articles 545, 547, 604, 675)
Douglas R. Betts, Fleetwood Enterprises Inc.
Rep. Manufactured Housing Institute
(Vote Limited to Articles 550, 551)
A. L. Buxton, Burndy Corp.
Rep. National Electrical Manufacturers Association
Richard A. Clarey, IBEW Local 551
Rep. International Brotherhood of Electrical Workers
Randol O. Farlow, Georgia Power Co.
Rep. Electric Light Power Group/Edison Electric Institute

James W. Finch, Kampgrounds of America, Inc.
(Vote Limited to Articles 550, 551)
Bruce A. Hopkins, Recreation Vehicle Industry Association
(Vote Limited to Articles 550 & 551)
Melvin Kallenbach, Mel-Kay Electric Co. Inc.
Rep. National Electrical Contractors Association
Richard P. McEnroe, Mill Mutuals Itasca, IL
(Vote Limited to Article 547)
Tug Miller, California Travel Parks
Donald A. Nissen, Underwriters Laboratories, Inc.
Henry Omson, US Department of Housing & Urban Development
(Vote Limited to Articles 545, 550, 551)

Glenn C. Safford, Kirby Risk Supply Co Inc.
  Rep. National Association of Electrical Distributors
  (Vote Limited to Articles 545, 547, 604, 675)
Ed Starostovic, PFS Corp.
  Rep. National Association of Home Builders
  (Vote Limited to Articles 545, 547, 604, 675)
Michael L. Zieman, Radco

Alternates

William C. Boteler, Hubbell Incorporated
  (Alt. to A. L. Buxton)
Paul Duks, Underwriters Laboratories Inc.
  (Alt. to D. A. Nissen)
Peter F. Locke, Northeast Utilities
  (Alt. to R. O. Farlow)

Kent Perkins, Recreation Vehicle Industry Assn.
  (Alt. to B. A. Hopkins VL 550, 551)
Frank J. Potucek, Mill Mutuals
  (Alt. to R. P. McEnroe VL 547)
Howard Slemmer, State of Washington
  (Alt. to J. H. Stricklin)
Homer Staves, Kampgrounds of America, Inc.
  (Alt. to J. Finch VL 550, 551)
LaVerne E. Stetson, US Dept. of Agriculture
  (Alt. to B. Bauman)
Raymond F. Tucker, RADCO
  (Alt. to M. L. Zieman)

PANEL NO. 20—Articles 422, 424, 426, 680

Russell J. Helmick, *Chairman*
City of Irvine, CA
Rep. International Association of Electrical Inspectors

Art Buxbaum, City of San Diego, CA
Edward S. Charkey, American Insurance Services Group, Inc.
Ray Franklin, IBEW
  Rep. International Brotherhood of Electrical Workers
Robert C. Lester, New York Board of Fire Underwriters
Morris L. Markel, Mor-Mar Products, Inc.
  Rep. National Electrical Manufacturers Association
James R. McNicol, Brett Aqualine Inc.
  Rep. National Spa and Pool Institute
  (Vote Limited to Article 680)
Herbert P. Spiegel, Corona Industrial Electric, Inc.
  Rep. National Electrical Contractors Association
Leo Stambaugh, Texas Utilities Electric Co.
  Rep. Electric Light Power Group/Edison Electric Institute
Donald J. Talka, Underwriters Laboratories Inc.
Leon T. Uhl, Leon T. Uhl, PE
  Rep. Association of Home Appliance Manufacturers
  (Vote Limited to Articles 422, 424, 426)
Leonard K. VanTassel, Carrier Air Conditioning
  Rep. Air Conditioning and Refrigeration Institute
  (Volume Limited to Articles 422, 424, 426)

Robert M. Yurkanin, Electran Process Intl Inc.
  Rep. Institute of Electrical & Electronics Engineers, Inc.

Alternates

R. L. Bunch, Tecumseh Products Company
  (Alt. to L. Van Tassel)
Kenneth J. Byrne, Virginia Power
  (Alt. to L. Stambaugh)
Richard E. King, IBEW
  (Alt. to R. Franklin)
James N. Pearse, Leviton Mfg. Co. Inc.
  (Alt. to M. Markel)
Joel A. Rencsok, City of Phoenix, AZ
  (Alt. to R. J. Helmick)

Nonvoting

William H. King, US Consumer Product Safety Comm.
G W. Lawrence, Canadian Standards Association

NOTE: Membership on a Committee shall not in and of itself constitute an endorsement of the Association or any document developed by the Committee on which the member serves.

National Electrical Code

NFPA 70

ARTICLE 90 — INTRODUCTION

90-1. Purpose.

(a) **Practical Safeguarding.** The purpose of this Code is the practical safeguarding of persons and property from hazards arising from the use of electricity.

(b) **Adequacy.** This Code contains provisions considered necessary for safety. Compliance therewith and proper maintenance will result in an installation essentially free from hazard but not necessarily efficient, convenient, or adequate for good service or future expansion of electrical use.

(FPN): Hazards often occur because of overloading of wiring systems by methods or usage not in conformity with this Code. This occurs because initial wiring did not provide for increases in the use of electricity. An initial adequate installation and reasonable provisions for system changes will provide for future increases in the use of electricity.

(c) **Intention.** This Code is not intended as a design specification nor an instruction manual for untrained persons.

90-2. Scope.

(a) **Covered.** This Code covers:

(1) Installations of electric conductors and equipment within or on public and private buildings or other structures, including mobile homes, recreational vehicles, and floating buildings; and other premises such as yards, carnival, parking, and other lots, and industrial substations.

(FPN): For additional information concerning such installations in an industrial or multibuilding complex, see the National Electrical Safety Code, ANSI C2-1990.

(2) Installations of conductors and equipment that connect to the supply of electricity.

(3) Installations of other outside conductors and equipment on the premises.

(4) Installations of optical fiber cable.

(b) **Not Covered.** This Code does not cover:

(1) Installations in ships, watercraft other than floating buildings, railway rolling stock, aircraft, or automotive vehicles other than mobile homes and recreational vehicles.

(2) Installations underground in mines and self-propelled mobile surface mining machinery and its attendant electrical trailing cable.

(3) Installations of railways for generation, transformation, transmission, or distribution of power used exclusively for operation of rolling stock or installations used exclusively for signaling and communications purposes.

(4) Installations of communications equipment under the exclusive control of communications utilities located outdoors or in building spaces used exclusively for such installations.

(5) Installations under the exclusive control of electric utilities for the purpose of communications or metering; or for the generation, control, transformation, transmission, and distribution of electric energy located in buildings used exclusively by utilities for such purposes or located outdoors on property owned or leased by the utility or on public highways, streets, roads, etc., or outdoors by established rights on private property.

(FPN): It is the intent of this section that this Code covers all premises wiring or wiring other than utility-owned metering equipment, on the load side of the service point of buildings, structures, or any other premises not owned or leased by the utility. Also, it is the intent that this Code cover installations in buildings used by the utility for purposes other than listed in Section 90-2(b)(5) above, such as office buildings, warehouses, garages, machine shops, and recreational buildings that are not an integral part of a generating plant, substation, or control center.

(c) Special Permission. The authority having jurisdiction for enforcing this Code may grant exception for the installation of conductors and equipment that are not under the exclusive control of the electric utilities and are used to connect the electric utility supply system to the service-entrance conductors of the premises served, provided such installations are outside a building or terminate immediately inside a building wall.

90-3. Code Arrangement. This Code is divided into the Introduction and nine chapters. Chapters 1, 2, 3, and 4 apply generally; Chapters 5, 6, and 7 apply to special occupancies, special equipment, or other special conditions. These latter chapters supplement or modify the general rules. Chapters 1 through 4 apply except as amended by Chapters 5, 6, and 7 for the particular conditions.

Chapter 8 covers communications systems and is independent of the other chapters except where they are specifically referenced therein.

Chapter 9 consists of tables and examples.

Material identified by the superscript letter "x" includes text extracted from other NFPA documents as identified in Appendix A.

90-4. Enforcement. This Code is intended to be suitable for mandatory application by governmental bodies exercising legal jurisdiction over electrical installations and for use by insurance inspectors. The authority having jurisdiction for enforcement of the Code will have the responsibility for making interpretations of the rules, for deciding upon the approval of equipment and materials, and for granting the special permission contemplated in a number of the rules.

The authority having jurisdiction may waive specific requirements in this Code or permit alternate methods where it is assured that equivalent objectives can be achieved by establishing and maintaining effective safety.

This Code may require new products, constructions, or materials that may not yet be available at the time the Code is adopted. In such event, the authority having jurisdiction may permit the use of the products, constructions, or materials that comply with the most recent previous edition of this Code adopted by the jurisdiction.

90-5. Mandatory Rules and Explanatory Material. Mandatory rules of this Code are characterized by the use of the word "shall." Explanatory material is in the form of Fine Print Notes (FPN).

90-6. Formal Interpretations. To promote uniformity of interpretation and application of the provisions of this Code, Formal Interpretation procedures have been established.

(FPN): These procedures may be found in the "NFPA Regulations Governing Committee Projects."

90-7. Examination of Equipment for Safety. For specific items of equipment and materials referred to in this Code, examinations for safety made under standard conditions will provide a basis for approval where the record is made generally available through promulgation by organizations properly equipped and qualified for experimental testing, inspections of the run of goods at factories, and service-value determination through field inspections. This avoids the necessity for repetition of examinations by different examiners, frequently with inadequate facilities for such work, and the confusion that would result from conflicting reports as to the suitability of devices and materials examined for a given purpose.

It is the intent of this Code that factory-installed internal wiring or the construction of equipment need not be inspected at the time of installation of the equipment, except to detect alterations or damage, if the equipment has been listed by a qualified electrical testing laboratory that is recognized as having the facilities described above and that requires suitability for installation in accordance with this Code.

(FPN No. 1): See Examination, Identification, Installation, and Use of Equipment, Section 110-3.

(FPN No. 2): See definition of "Listed," Article 100.

90-8. Wiring Planning.

(a) Future Expansion and Convenience. Plans and specifications that provide ample space in raceways, spare raceways, and additional spaces will allow for future increases in the use of electricity. Distribution centers located in readily accessible locations will provide convenience and safety of operation. See Sections 110-16 and 240-24 for clearances and accessibility.

(b) Number of Circuits in Enclosures. It is elsewhere provided in this Code that the number of wires and circuits confined in a single enclosure be varyingly restricted. Limiting the number of circuits in a single enclosure will minimize the effects from a short-circuit or ground fault in one circuit.

90-9. Metric Units of Measurement. For the purpose of this Code, metric units of measurement are in accordance with the modernized metric system known as the International System of Units (SI).

Values of measurement in the Code text will be followed by an approximate equivalent value in SI units. Tables will have a footnote for SI conversion units used in the table.

Conduit size, wire size, horsepower designation for motors, and trade sizes that do not reflect actual measurements, e.g., box sizes, will not be assigned dual designation SI units.

(FPN): For metric conversion practices, see Standard for Metric Practice, ANSI/ASTM E380-1984.

Chapter 1.   General

ARTICLE 100 — DEFINITIONS

Scope.   This article contains only those definitions essential to the proper application of this Code. It is not intended to include commonly defined general terms or commonly defined technical terms from related codes and standards. In general, only those terms used in two or more articles are defined in Article 100. Other definitions are included in the article in which they are used but may be referenced in Article 100.

Part A of this article contains definitions intended to apply wherever the terms are used throughout this Code. Part B contains definitions applicable only to the parts of articles covering specifically installations and equipment operating at over 600 volts, nominal.

A. General

Accessible: (As applied to wiring methods.) Capable of being removed or exposed without damaging the building structure or finish, or not permanently closed in by the structure or finish of the building. (See "Concealed" and "Exposed.")

Accessible: (As applied to equipment.) Admitting close approach: not guarded by locked doors, elevation, or other effective means. (See "Accessible, Readily.")

Accessible, Readily: (Readily Accessible.) Capable of being reached quickly for operation, renewal, or inspections, without requiring those to whom ready access is requisite to climb over or remove obstacles or to resort to portable ladders, chairs, etc. (See "Accessible.")

Ampacity: The current in amperes that a conductor can carry continuously under the conditions of use without exceeding its temperature rating.

Appliance: Utilization equipment, generally other than industrial, normally built in standardized sizes or types, that is installed or connected as a unit to perform one or more functions such as clothes washing, air conditioning, food mixing, deep frying, etc.

Appliance Branch Circuit: See "Branch Circuit, Appliance."

Approved: Acceptable to the authority having jurisdiction.

Askarel: A generic term for a group of nonflammable synthetic chlorinated hydrocarbons used as electrical insulating media. Askarels of various compositional types are used. Under arcing conditions, the gases produced, while consisting predominantly of noncombustible hydrogen chloride, can include varying amounts of combustible gases depending upon the askarel type.

Attachment Plug (Plug Cap) (Cap):   A device that, by insertion in a receptacle, establishes connection between the conductors of the attached flexible cord and the conductors connected permanently to the receptacle.

Automatic: Self-acting, operating by its own mechanism when actuated by some impersonal influence, as for example, a change in current strength, pressure, temperature, or mechanical configuration. (See "Nonautomatic.")

Bare Conductor: See under "Conductor."

Bonding: The permanent joining of metallic parts to form an electrically conductive path that will assure electrical continuity and the capacity to conduct safely any current likely to be imposed.

Bonding Jumper: A reliable conductor to assure the required electrical conductivity between metal parts required to be electrically connected.

Bonding Jumper, Circuit: The connection between portions of a conductor in a circuit to maintain required ampacity of the circuit.

Bonding Jumper, Equipment: The connection between two or more portions of the equipment grounding conductor.

Bonding Jumper, Main: The connection between the grounded circuit conductor and the equipment grounding conductor at the service.

Branch Circuit: The circuit conductors between the final overcurrent device protecting the circuit and the outlet(s).

Branch Circuit, Appliance: A branch circuit supplying energy to one or more outlets to which appliances are to be connected; such circuits to have no permanently connected lighting fixtures not a part of an appliance.

Branch Circuit, General Purpose: A branch circuit that supplies a number of outlets for lighting and appliances.

Branch Circuit, Individual: A branch circuit that supplies only one utilization equipment.

Branch Circuit, Multiwire: A branch circuit consisting of two or more ungrounded conductors having a potential difference between them, and a grounded conductor having equal potential difference between it and each ungrounded conductor of the circuit and that is connected to the neutral or grounded conductor of the system.

Building: A structure that stands alone or that is cut off from adjoining structures by fire walls with all openings therein protected by approved fire doors.

Cabinet: An enclosure designed either for surface or flush mounting and provided with a frame, mat, or trim in which a swinging door or doors are or may be hung.

Circuit Breaker: A device designed to open and close a circuit by nonautomatic means and to open the circuit automatically on a predetermined overcurrent without damage to itself when properly applied within its rating.

(FPN): The automatic opening means can be integral, direct acting with the circuit breaker, or remote from the circuit breaker.

Adjustable: (As applied to circuit breakers.) A qualifying term indicating that the circuit breaker can be set to trip at various values of current, time, or both, within a predetermined range.

Instantaneous Trip: (As applied to circuit breakers.) A qualifying term indicating that no delay is purposely introduced in the tripping action of the circuit breaker.

Inverse Time: (As applied to circuit breakers.) A qualifying term indicating that there is purposely introduced a delay in the tripping action of the circuit breaker, which delay decreases as the magnitude of the current increases.

Nonadjustable: (As applied to circuit breakers.) A qualifying term indicating that the circuit breaker does not have any adjustment to alter the value of current at which it will trip or the time required for its operation.

Setting: (of circuit breaker.) The value of current, time, or both at which an adjustable circuit breaker is set to trip.

Concealed: Rendered inaccessible by the structure or finish of the building. Wires in concealed raceways are considered concealed, even though they may become accessible by withdrawing them. [See "Accessible — (As applied to wiring methods)."]

Conductor:

Bare: A conductor having no covering or electrical insulation whatsoever. (See "Conductor, Covered.")

Covered: A conductor encased within material of composition or thickness that is not recognized by this Code as electrical insulation. (See "Conductor, Bare.")

Insulated: A conductor encased within material of composition and thickness that is recognized by this Code as electrical insulation.

Conduit Body: A separate portion of a conduit or tubing system that provides access through a removable cover(s) to the interior of the system at a junction of two or more sections of the system or at a terminal point of the system.

Boxes such as FS and FD or larger cast or sheet metal boxes are not classified as conduit bodies.

Connector, Pressure (Solderless): A device that establishes a connection between two or more conductors or between one or more conductors and a terminal by means of mechanical pressure and without the use of solder.

Continuous Duty: See under "Duty."

Continuous Load: A load where the maximum current is expected to continue for three hours or more.

Controller: A device or group of devices that serves to govern, in some predetermined manner, the electric power delivered to the apparatus to which it is connected.

Cooking Unit, Counter-Mounted: A cooking appliance designed for mounting in or on a counter and consisting of one or more heating elements, internal wiring, and built-in or separately mountable controls. (See "Oven, Wall-Mounted.")

Copper-Clad Aluminum Conductors: Conductors drawn from a copper-clad aluminum rod with the copper metallurgically bonded to an aluminum core. The copper forms a minimum of 10 percent of the cross-sectional area of a solid conductor or each strand of a stranded conductor.

Covered Conductor: See under "Conductor."

Cutout Box: An enclosure designed for surface mounting and having swinging doors or covers secured directly to and telescoping with the walls of the box proper. (See "Cabinet.")

Damp Location: See under "Location."

Dead Front: Without live parts exposed to a person on the operating side of the equipment.

Demand Factor: The ratio of the maximum demand of a system, or part of a system, to the total connected load of a system or the part of the system under consideration.

Device: A unit of an electrical system that is intended to carry but not utilize electric energy.

Disconnecting Means: A device, or group of devices, or other means by which the conductors of a circuit can be disconnected from their source of supply.

Dry Location: See under "Location."

Dustproof: So constructed or protected that dust will not interfere with its successful operation.

Dusttight: So constructed that dust will not enter the enclosing case under specified test conditions.

(FPN):   For test conditions other than for rotating equipment, see Enclosures for Electrical Equipment (1000 Volts Maximum), Clause 6.5, ANSI/NEMA 250-1985.

Duty:

Continuous Duty: Operation at a substantially constant load for an indefinitely long time.

Intermittent Duty: Operation for alternate intervals of (1) load and no load; or (2) load and rest; or (3) load, no load, and rest.

Periodic Duty: Intermittent operation in which the load conditions are regularly recurrent.

Short-Time Duty: Operation at a substantially constant load for a short and definitely specified time.

Varying Duty: Operation at loads, and for intervals of time, both of which may be subject to wide variation.

Dwelling:

Dwelling Unit: One or more rooms for the use of one or more persons as a housekeeping unit with space for eating, living, and sleeping, and permanent provisions for cooking and sanitation.

Multifamily Dwelling: A building containing three or more dwelling units.

One-Family Dwelling: A building consisting solely of one dwelling unit.

Two-Family Dwelling: A building consisting solely of two dwelling units.

Electric Sign: A fixed, stationary, or portable self-contained, electrically illuminated utilization equipment with words or symbols designed to convey information or attract attention.

Enclosed: Surrounded by a case, housing, fence, or walls that will prevent persons from accidentally contacting energized parts.

Enclosure: The case or housing of apparatus, or the fence or walls surrounding an installation to prevent personnel from accidentally contacting energized parts, or to protect the equipment from physical damage.

(FPN):   For enclosure types, see Enclosures for Electrical Equipment (1000 Volts Maximum), ANSI/NEMA 250-1985.

Equipment: A general term including material, fittings, devices, appliances, fixtures, apparatus, and the like used as a part of, or in connection with, an electrical installation.

Equipment Grounding Conductor: See "Grounding Conductor, Equipment."

Explosionproof Apparatus: Apparatus enclosed in a case that is capable of withstanding an explosion of a specified gas or vapor that may occur within it and of preventing the ignition of a specified gas or vapor surrounding the enclosure by sparks, flashes, or explosion of the gas or vapor within, and that operates at such an external temperature that a surrounding flammable atmosphere will not be ignited thereby.

(FPN):   For further information, see Explosion-Proof and Dust-Ignition-Proof Electrical Equipment for Use in Hazardous (Classified) Locations, ANSI/UL 1203-1988.

Exposed: (As applied to live parts.) Capable of being inadvertently touched or approached nearer than a safe distance by a person. It is applied to parts not suitably guarded, isolated, or insulated. (See "Accessible" and "Concealed.")

Exposed: (As applied to wiring methods.) On or attached to the surface or behind panels designed to allow access. [See "Accessible — (As applied to wiring methods)."]

Externally Operable: Capable of being operated without exposing the operator to contact with live parts.

Feeder: All circuit conductors between the service equipment or the source of a separately derived system and the final branch-circuit overcurrent device.

Fitting: An accessory such as a locknut, bushing, or other part of a wiring system that is intended primarily to perform a mechanical rather than an electrical function.

Garage: A building or portion of a building in which one or more self-propelled vehicles carrying volatile flammable liquid for fuel or power are kept for use, sale, storage, rental, repair, exhibition, or demonstrating purposes, and all that portion of a building that is on or below the floor or floors in which such vehicles are kept and that is not separated therefrom by suitable cutoffs.

(FPN): For commercial garages, repair, and storage, see Section 511-1.

General-Purpose Branch Circuit: See "Branch Circuit, General Purpose."

General-Use Snap Switch: See under "Switches."

General-Use Switch: See under "Switches."

Ground: A conducting connection, whether intentional or accidental, between an electrical circuit or equipment and the earth, or to some conducting body that serves in place of the earth.

Grounded: Connected to earth or to some conducting body that serves in place of the earth.

Grounded, Effectively: Intentionally connected to earth through a ground connection or connections of sufficiently low impedance and having sufficient current-carrying capacity to prevent the buildup of voltages that may result in undue hazards to connected equipment or to persons.

Grounded Conductor: A system or circuit conductor that is intentionally grounded.

Grounding Conductor: A conductor used to connect equipment or the grounded circuit of a wiring system to a grounding electrode or electrodes.

Grounding Conductor, Equipment: The conductor used to connect the noncurrent-carrying metal parts of equipment, raceways, and other enclosures to the system grounded conductor, the grounding electrode conductor, or both, at the service equipment or at the source of a separately derived system.

Grounding Electrode Conductor: The conductor used to connect the grounding electrode to the equipment grounding conductor, to the grounded conductor, or to both, of the circuit at the service equipment or at the source of a separately derived system.

Ground-Fault Circuit-Interrupter: A device intended for the protection of personnel that functions to deenergize a circuit or portion thereof within an established period of time when a current to ground exceeds some predetermined value that is less than that required to operate the overcurrent protective device of the supply circuit.

Ground-Fault Protection of Equipment: A system intended to provide protection of equipment from damaging line-to-ground fault currents by operating to cause a disconnecting means to open all ungrounded conductors of the faulted circuit. This protection is provided at current levels less than those required to protect conductors from damage through the operation of a supply circuit overcurrent device.

Guarded: Covered, shielded, fenced, enclosed, or otherwise protected by means of suitable covers, casings, barriers, rails, screens, mats, or platforms to remove the likelihood of approach or contact by persons or objects to a point of danger.

Hoistway: Any shaftway, hatchway, well hole, or other vertical opening or space in which an elevator or dumbwaiter is designed to operate.

Identified: (As applied to Equipment.) Recognizable as suitable for the specific purpose, function, use, environment, application, etc., where described in a particular Code requirement. (See "Equipment.")

(FPN): Suitability of equipment for a specific purpose, environment, or application may be determined by a qualified testing laboratory, inspection agency, or other organization concerned with product evaluation. Such identification may include labeling or listing. (See "Labeled" and "Listed.")

Individual Branch Circuit: See "Branch Circuit, Individual."

In Sight From (Within Sight From, Within Sight): Where this Code specifies that one equipment shall be "in sight from," "within sight from," or "within sight," etc., of another equipment, one of the equipments specified is to be visible and not more than 50 feet (15.24 m) distant from the other.

Insulated Conductor: See under "Conductor."

Intermittent Duty: See under "Duty."

Interrupting Rating: The highest current at rated voltage that a device is intended to interrupt under standard test conditions.

(FPN): Equipment intended to break current at other than fault levels may have its interrupting rating implied in other ratings, such as horsepower or locked rotor current.

Isolated: Not readily accessible to persons unless special means for access are used.

Labeled: Equipment or materials to which has been attached a label, symbol, or other identifying mark of an organization acceptable to the authority having jurisdiction and concerned with product evaluation, that maintains periodic inspection of production of labeled equipment or materials and by whose labeling the manufacturer indicates compliance with appropriate standards or performance in a specified manner.

Lighting Outlet: An outlet intended for the direct connection of a lampholder, a lighting fixture, or a pendant cord terminating in a lampholder.

Listed: Equipment or materials included in a list published by an organization acceptable to the authority having jurisdiction and concerned with product evaluation, that maintains periodic inspection of production of listed equipment or materials, and whose listing states either that the equipment or material meets appropriate designated standards or has been tested and found suitable for use in a specified manner.

(FPN): The means for identifying listed equipment may vary for each organization concerned with product evaluation, some of which do not recognize equipment as listed unless it is also labeled. The authority having jurisdiction should utilize the system employed by the listing organization to identify a listed product.

Location:

Damp Location: Partially protected locations under canopies, marquees, roofed open porches, and like locations, and interior locations subject to moderate degrees of moisture, such as some basements, some barns, and some cold-storage warehouses.

Dry Location: A location not normally subject to dampness or wetness. A location classified as dry may be temporarily subject to dampness or wetness, as in the case of a building under construction.

Wet Location: Installations underground or in concrete slabs or masonry in direct contact with the earth, and locations subject to saturation with water or other liquids, such as vehicle washing areas, and locations exposed to weather and unprotected.

Multioutlet Assembly: A type of surface or flush raceway designed to hold conductors and receptacles, assembled in the field or at the factory.

Multiwire Branch Circuit: See "Branch Circuit, Multiwire."

Nonautomatic: Action requiring personal intervention for its control. (See "Automatic.")

(FPN): As applied to an electric controller, nonautomatic control does not necessarily imply a manual controller, but only that personal intervention is necessary.

Nonincendive Circuit: A circuit in which any arc or thermal effect produced, under intended operating conditions of the equipment or due to opening, shorting, or grounding of field wiring, is not capable, under specified test conditions, of igniting the flammable gas, vapor, or dust-air mixture.

(FPN): For test conditions, see Electrical Equipment for Use in Class I, Division 2 Hazardous (Classified) Locations, ANSI/ISA-S12.12.

Outlet: A point on the wiring system at which current is taken to supply utilization equipment.

Outline Lighting: An arrangement of incandescent lamps or electric-discharge lighting to outline or call attention to certain features such as the shape of a building or the decoration of a window.

Oven, Wall-Mounted: An oven for cooking purposes designed for mounting in or on a wall or other surface and consisting of one or more heating elements, internal wiring, and built-in or separately mountable controls. (See "Cooking Unit, Counter-Mounted.")

Overcurrent: Any current in excess of the rated current of equipment or the ampacity of a conductor. It may result from overload (see definition), short circuit, or ground fault.

(FPN): A current in excess of rating may be accommodated by certain equipment and conductors for a given set of conditions. Hence the rules for overcurrent protection are specific for particular situations.

Overload: Operation of equipment in excess of normal, full-load rating, or of a conductor in excess of rated ampacity that, when it persists for a sufficient length of time, would cause damage or dangerous overheating. A fault, such as a short circuit or ground fault, is not an overload. (See "Overcurrent.")

Panelboard: A single panel or group of panel units designed for assembly in the form of a single panel; including buses, automatic overcurrent devices, and equipped with or without switches for the control of light, heat, or power circuits; designed to be placed in a cabinet or cutout box placed in or against a wall or partition and accessible only from the front. (See "Switchboard.")

Periodic Duty: See under "Duty."

Plenum: A compartment or chamber to which one or more air ducts are connected and that forms part of the air distribution system.

Power Outlet: An enclosed assembly that may include receptacles, circuit breakers, fuseholders, fused switches, buses, and watt-hour meter mounting means; intended to supply and control power to mobile homes, recreational vehicles, or boats; or to serve as a means for distributing power required to operate mobile or temporarily installed equipment.

Premises Wiring (System): That interior and exterior wiring, including power, lighting, control, and signal circuit wiring together with all of their associated hardware, fittings, and wiring devices, both permanently and temporarily installed, that extends from the service point of utility conductors or source of a separately derived system to the outlet(s). Such wiring does not include wiring internal to appliances, fixtures, motors, controllers, motor control centers, and similar equipment.

Qualified Person: One familiar with the construction and operation of the equipment and the hazards involved.

Raceway: An enclosed channel designed expressly for holding wires, cables, or busbars, with additional functions as permitted in this Code.

(FPN): Raceways may be of metal or insulating material, and the term includes rigid metal conduit, rigid nonmetallic conduit, intermediate metal conduit, liquidtight flexible conduit, flexible metallic tubing, flexible metal conduit, electrical nonmetallic tubing, electrical metallic tubing, underfloor raceways, cellular concrete floor raceways, cellular metal floor raceways, surface raceways, wireways, cablebus, and busways.

Rainproof: So constructed, protected, or treated as to prevent rain from interfering with the successful operation of the apparatus under specified test conditions.

Raintight: So constructed or protected that exposure to a beating rain will not result in the entrance of water under specified test conditions.

Readily Accessible: (See "Accessible, Readily.")

Receptacle: A receptacle is a contact device installed at the outlet for the connection of a single attachment plug.

(FPN): A single receptacle is a single contact device with no other contact device on the same yoke. A multiple receptacle is a single device containing two or more receptacles.

Receptacle Outlet: An outlet where one or more receptacles are installed.

Remote-Control Circuit: Any electric circuit that controls any other circuit through a relay or an equivalent device.

Sealable Equipment: Equipment enclosed in a case or cabinet that is provided with a means of sealing or locking so that live parts cannot be made accessible without opening the enclosure. The equipment may or may not be operable without opening the enclosure.

Separately Derived System: A premises wiring system whose power is derived from generator, transformer, or converter windings and that has no direct electrical connection, including a solidly connected grounded circuit conductor, to supply conductors originating in another system.

Service: The conductors and equipment for delivering energy from the electricity supply system to the wiring system of the premises served.

Service Cable: Service conductors made up in the form of a cable.

Service Conductors: The supply conductors that extend from the street main or from transformers to the service equipment of the premises supplied.

Service Drop: The overhead service conductors from the last pole or other aerial support to and including the splices, if any, connecting to the service-entrance conductors at the building or other structure.

Service-Entrance Conductors, Overhead System: The service conductors between the terminals of the service equipment and a point usually outside the building, clear of building walls, where joined by tap or splice to the service drop.

Service-Entrance Conductors, Underground System: The service conductors between the terminals of the service equipment and the point of connection to the service lateral.

(FPN):   Where service equipment is located outside the building walls, there may be no service-entrance conductors, or they may be entirely outside the building.

Service Equipment: The necessary equipment, usually consisting of a circuit breaker or switch and fuses, and their accessories, located near the point of entrance of supply conductors to a building or other structure, or an otherwise defined area, and intended to constitute the main control and means of cutoff of the supply.

Service Lateral: The underground service conductors between the street main, including any risers at a pole or other structure or from transformers, and the first point of connection to the service-entrance conductors in a terminal box or meter or other enclosure with adequate space, inside or outside the building wall. Where there is no terminal box, meter, or other enclosure with adequate space, the point of connection shall be considered to be the point of entrance of the service conductors into the building.

Service Point: Service point is the point of connection between the facilities of the serving utility and the premises wiring.

Setting (of Circuit Breaker):   See under "Circuit Breaker."

Short-Time Duty: See under "Duty."

Show Window: Any window used or designed to be used for the display of goods or advertising material, whether it is fully or partly enclosed or entirely open at the rear and whether or not it has a platform raised higher than the street floor level.

Sign: See "Electric Sign."

Signaling Circuit: Any electric circuit that energizes signaling equipment.

Solar Photovoltaic System: The total components and subsystems that, in combination, convert solar energy into electrical energy suitable for connection to a utilization load.

Special Permission: The written consent of the authority having jurisdiction.

Switchboard: A large single panel, frame, or assembly of panels on which are mounted, on the face or back, or both, switches, overcurrent and other protective devices, buses, and usually instruments. Switchboards are generally accessible from the rear as well as from the front and are not intended to be installed in cabinets. (See "Panelboard.")

Switches:

Bypass Isolation Switch: A bypass isolation switch is a manually operated device used in conjunction with a transfer switch to provide a means of directly connecting load conductors to a power source, and of disconnecting the transfer switch.

General-Use Switch: A switch intended for use in general distribution and branch circuits. It is rated in amperes, and it is capable of interrupting its rated current at its rated voltage.

General-Use Snap Switch: A form of general-use switch so constructed that it can be installed in flush device boxes or on outlet box covers, or otherwise used in conjunction with wiring systems recognized by this Code.

Isolating Switch: A switch intended for isolating an electric circuit from the source of power. It has no interrupting rating, and it is intended to be operated only after the circuit has been opened by some other means.

Motor-Circuit Switch: A switch, rated in horsepower, capable of interrupting the maximum operating overload current of a motor of the same horsepower rating as the switch at the rated voltage.

Transfer Switch: A transfer switch is a device for transferring one or more load conductor connections from one power source to another.

(FPN): A transfer switch may be either automatic or nonautomatic.

Thermally Protected: (As applied to motors.) The words "Thermally Protected" appearing on the nameplate of a motor or motor-compressor indicate that the motor is provided with a thermal protector.

Thermal Protector: (As applied to motors.) A protective device for assembly as an integral part of a motor or motor-compressor that, when properly applied, protects the motor against dangerous overheating due to overload and failure to start.

(FPN): The thermal protector may consist of one or more sensing elements integral with the motor or motor-compressor and an external control device.

Utilization Equipment: Equipment that utilizes electric energy for electronic, electromechanical, chemical, heating, lighting, or similar purposes.

Varying Duty: See under "Duty."

Ventilated: Provided with a means to permit circulation of air sufficient to remove an excess of heat, fumes, or vapors.

Volatile Flammable Liquid: A flammable liquid having a flash point below 38°C (100°F), or a flammable liquid whose temperature is above its flash point, or a Class II combustible liquid having a vapor pressure not exceeding 40 psia (276 kPa) at 38°C (100°F) whose temperature is above its flash point.

Voltage (of a Circuit): The greatest root-mean-square (effective) difference of potential between any two conductors of the circuit concerned.

(FPN): Some systems, such as 3-phase 4-wire, single-phase 3-wire, and 3-wire direct-current may have various circuits of various voltages.

Voltage, Nominal: A nominal value assigned to a circuit or system for the purpose of conveniently designating its voltage class (e.g., 120/240, 480Y/277, 600).

The actual voltage at which a circuit operates can vary from the nominal within a range that permits satisfactory operation of equipment.

(FPN): See Voltage Ratings for Electric Power Systems and Equipment (60 Hz), ANSI C84.1-1989.

Voltage to Ground: For grounded circuits, the voltage between the given conductor and that point or conductor of the circuit that is grounded; for ungrounded circuits, the greatest voltage between the given conductor and any other conductor of the circuit.

Watertight: So constructed that moisture will not enter the enclosure under specified test conditions.

(FPN): For test conditions other than for rotating equipment, see Enclosure for Electrical Equipment (1000 Volts Maximum), Section 250-5.07, Hosedown Test, NEMA Standards Publication No. 250-1979.

Weatherproof: So constructed or protected that exposure to the weather will not interfere with successful operation.

(FPN): Rainproof, raintight, or watertight equipment can fulfill the requirements for weatherproof where varying weather conditions other than wetness, such as snow, ice, dust, or temperature extremes, are not a factor.

Wet Location: See under "Location."

B. Over 600 Volts, Nominal

Whereas the preceding definitions are intended to apply wherever the terms are used throughout this Code, the following definitions are applicable only to parts of the article specifically covering installations and equipment operating at over 600 volts, nominal.

Circuit Breaker: See under "Switching Devices."

Cutout: See under "Switching Devices."

Disconnect (Isolator): See under "Switching Devices."

Disconnecting Means: See under "Switching Devices."

Fuse: An overcurrent protective device with a circuit opening fusible part that is heated and severed by the passage of overcurrent through it.

(FPN): A fuse comprises all the parts that form a unit capable of performing the prescribed functions. It may or may not be the complete device necessary to connect it into an electrical circuit.

Expulsion Fuse Unit (Expulsion Fuse): A vented fuse unit in which the expulsion effect of gases produced by the arc and lining of the fuseholder, either alone or aided by a spring, extinguishes the arc.

Power Fuse Unit: A vented, nonvented, or controlled vented fuse unit in which the arc is extinguished by being drawn through solid material, granular material, or liquid, either alone or aided by a spring.

Vented Power Fuse: A fuse with provision for the escape of arc gases, liquids, or solid particles to the surrounding atmosphere during circuit interruption.

Nonvented Power Fuse: A fuse without intentional provision for the escape of arc gases, liquids, or solid particles to the atmosphere during circuit interruption.

Controlled Vented Power Fuse: A fuse with provision for controlling discharge circuit interruption such that no solid material may be exhausted into the surrounding atmosphere.

(FPN): The fuse is designed so that discharged gases will not ignite or damage insulation in the path of the discharge or propagate a flashover to or between grounded members or conduction members in the path of the discharge where the distance between the vent and such insulation or conduction members conforms to manufacturer's recommendations.

Interrupter Switch: See under "Switching Devices."

Multiple Fuse: An assembly of two or more single-pole fuses.

Oil (Filled) Cutout: See under "Switching Devices."

Power Fuse: See under "Fuse."

Regulator Bypass Switch: See under "Switching Devices."

Switching Device: A device designed to close, open, or both, one or more electric circuits.

Switching Devices:

Circuit Breaker: A switching device capable of making, carrying, and breaking currents under normal circuit conditions, and also making, carrying for a specified time, and breaking currents under specified abnormal circuit conditions, such as those of short circuit.

Cutout: An assembly of a fuse support with either a fuseholder, fuse carrier, or disconnecting blade. The fuseholder or fuse carrier may include a conducting element (fuse link), or may act as the disconnecting blade by the inclusion of a nonfusible member.

Disconnecting (or Isolating) Switch (Disconnector, Isolator): A mechanical switching device used for isolating a circuit or equipment from a source of power.

Disconnecting Means: A device, group of devices, or other means whereby the conductors of a circuit can be disconnected from their source of supply.

Interrupter Switch: A switch capable of making, carrying, and interrupting specified currents.

Oil Cutout (Oil-Filled Cutout): A cutout in which all or part of the fuse support and its fuse link or disconnecting blade is mounted in oil with complete immersion of the contacts and the fusible portion of the conducting element (fuse link), so that arc interruption by severing of the fuse link or by opening of the contacts will occur under oil.

Oil Switch: An oil switch is a switch having contacts that operate under oil (or askarel or other suitable liquid).

Regulator Bypass Switch: A specific device or combination of devices designed to bypass a regulator.

ARTICLE 110 — REQUIREMENTS FOR ELECTRICAL INSTALLATIONS

A. General

110-2. Approval. The conductors and equipment required or permitted by this Code shall be acceptable only if approved.

(FPN): See Examination of Equipment for Safety, Section 90-7 and Examination, Identification, Installation, and Use of Equipment, Section 110-3. See definitions of "Approved," "Identified," "Labeled," and "Listed."

110-3. Examination, Identification, Installation, and Use of Equipment.

(a) Examination. In judging equipment, considerations such as the following shall be evaluated:

(1) Suitability for installation and use in conformity with the provisions of this Code.

(FPN): Suitability of equipment use may be identified by a description marked on or provided with a product to identify the suitability of the product for a specific purpose, environment, or application. Suitability of equipment may be evidenced by listing or labeling.

(2) Mechanical strength and durability, including, for parts designed to enclose and protect other equipment, the adequacy of the protection thus provided.

(3) Wire-bending and connection space.

(4) Electrical insulation.

(5) Heating effects under normal conditions of use and also under abnormal conditions likely to arise in service.

(6) Arcing effects.

(7) Classification by type, size, voltage, current capacity, and specific use.

(8) Other factors that contribute to the practical safeguarding of persons using or likely to come in contact with the equipment.

(b) Installation and Use. Listed or labeled equipment shall be used or installed in accordance with any instructions included in the listing or labeling.

110-4. Voltages. Throughout this Code, the voltage considered shall be that at which the circuit operates.

110-5. Conductors. Conductors normally used to carry current shall be of copper unless otherwise provided in this Code. Where the conductor material is not specified, the material and the sizes given in this Code shall apply to copper conductors. Where other materials are used, the size shall be changed accordingly.

(FPN): For aluminum and copper-clad aluminum conductors, see Section 310-15.

110-6. Conductor Sizes. Conductor sizes are expressed in American Wire Gage (AWG) or in circular mils.

110-7. Insulation Integrity. All wiring shall be so installed that, when completed, the system will be free from short circuits and from grounds other than as required or permitted in Article 250.

110-8. Wiring Methods. Only wiring methods recognized as suitable are included in this Code. The recognized methods of wiring shall be permitted to be installed in any type of building or occupancy, except as otherwise provided in this Code.

110-9. Interrupting Rating. Equipment intended to break current at fault levels shall have an interrupting rating sufficient for the nominal circuit voltage and the current that is available at the line terminals of the equipment.

Equipment intended to break current at other than fault levels shall have an interrupting rating at nominal circuit voltage sufficient for the current that must be interrupted.

110-10. Circuit Impedance and Other Characteristics. The overcurrent protective devices, the total impedance, the component short-circuit withstand ratings, and other characteristics of the circuit to be protected shall be so selected and coordinated as to permit the circuit protective devices that are used to clear a fault without the occurrence of extensive damage to the electrical components of the circuit. This fault shall be assumed to be either between two or more of the circuit conductors, or between any circuit conductor and the grounding conductor or enclosing metal raceway.

(FPN): For circuit breakers, see Section 240-83(c).

110-11. Deteriorating Agents. Unless identified for use in the operating environment, no conductors or equipment shall be located in damp or wet locations; where exposed to gases, fumes, vapors, liquids, or other agents having a deteriorating effect on the conductors or equipment; nor where exposed to excessive temperatures.

(FPN No. 1): See Section 300-6 for protection against corrosion.

(FPN No. 2): Some cleaning and lubricating compounds can cause severe deterioration of many plastic materials used for insulating and structural applications in equipment.

Equipment approved for use in dry locations only shall be protected against permanent damage from the weather during building construction.

110-12. Mechanical Execution of Work. Electric equipment shall be installed in a neat and workmanlike manner.

(a) Unused Openings. Unused openings in boxes, raceways, auxiliary gutters, cabinets, equipment cases, or housings shall be effectively closed to afford protection substantially equivalent to the wall of the equipment.

(b) Subsurface Enclosures. Conductors shall be racked to provide ready and safe access in underground and subsurface enclosures, into which persons enter for installation and maintenance.

(c) Integrity of Electrical Equipment and Connections. Internal parts of electrical equipment, including busbars, wiring terminals, insulators, and other surfaces shall not be damaged or contaminated by foreign materials such as paint, plaster, cleaners, or abrasives.

110-13. Mounting and Cooling of Equipment.

(a) Mounting. Electric equipment shall be firmly secured to the surface on which it is mounted. Wooden plugs driven into holes in masonry, concrete, plaster, or similar materials shall not be used.

(b) Cooling. Electrical equipment that depends upon the natural circulation of air and convection principles for cooling of exposed surfaces shall be installed so that room airflow over such surfaces is not prevented by walls or by adjacent installed equipment. For equipment designed for floor mounting, clearance between top surfaces and adjacent surfaces shall be provided to dissipate rising warm air.

Electrical equipment provided with ventilating openings shall be installed so that walls or other obstructions do not prevent the free circulation of air through the equipment.

110-14. Electrical Connections.

Because of different characteristics of copper and aluminum, devices such as pressure terminal or pressure splicing connectors and soldering lugs shall be identified for the material of the conductor and shall be properly installed and used. Conductors of dissimilar metals shall not be intermixed in a terminal or splicing connector where physical contact occurs between dissimilar conductors (such as copper and aluminum, copper and copper-clad aluminum, or aluminum and copper-clad aluminum), unless the device is identified for the purpose and conditions of use. Materials such as solder, fluxes, inhibitors, and compounds, where employed, shall be suitable for the use and shall be of a type that will not adversely affect the conductors, installation, or equipment.

(FPN): Many terminations and equipment are marked with a tightening torque.

(a) Terminals. Connection of conductors to terminal parts shall ensure a thoroughly good connection without damaging the conductors and shall be made by means of pressure connectors (including set-screw type), solder lugs, or splices to flexible leads.

Exception: Connection by means of wire binding screws or studs and nuts having upturned lugs or equivalent shall be permitted for No. 10 or smaller conductors.

Terminals for more than one conductor and terminals used to connect aluminum shall be so identified.

(b) Splices. Conductors shall be spliced or joined with splicing devices identified for the use or by brazing, welding, or soldering with a fusible metal or alloy. Soldered splices shall first be so spliced or joined as to be mechanically and electrically secure without solder and then soldered. All splices and joints and the free ends of conductors shall be covered with an insulation equivalent to that of the conductors or with an insulating device identified for the purpose.

(c) Temperature Limitations. The temperature rating associated with the ampacity of a conductor shall be so selected and coordinated as to not exceed the lowest temperature rating of any connected termination, conductor, or device.

(1) Termination provisions of equipment for circuits rated 100 amperes or less, or marked for Nos. 14 through 1 conductors, shall be used only for conductors rated 60°C (140°F).

Exception No. 1: Conductors with higher temperature ratings shall be permitted to be used, provided the ampacity of such conductors is determined based on the 60°C (140°F) ampacity of the conductor size used.

Exception No. 2: Equipment termination provisions shall be permitted to be used with higher rated conductors at the ampacity of the higher rated conductors, provided the equipment is listed and identified for use with the higher rated conductors.

(2) Termination provisions of equipment for circuits rated over 100 amperes, or marked for conductors larger than No. 1, shall be used only with conductors rated 75°C (167°F).

Exception No. 1: Conductors with higher temperature ratings shall be permitted to be used, provided the ampacity of such conductors is determined based on the 75°C (167°F) ampacity of the conductor size used.

Exception No. 2: Equipment termination provisions shall be permitted to be used with the higher rated conductors at the ampacity of the higher rated conductors, provided the equipment is listed and identified for use with the higher rated conductors.

(3) Separately installed pressure connectors shall be used with conductors at the ampacities not exceeding the ampacity at the listed and identified temperature rating of the connector.

(FPN): With respect to Sections 110-14(c)(1), (2), and (3), equipment markings or listing information may additionally restrict the sizing and temperature ratings of connected conductors.

110-16. Working Space About Electric Equipment (600 Volts, Nominal, or Less). Sufficient access and working space shall be provided and maintained about all electric equipment to permit ready and safe operation and maintenance of such equipment.

(a) Working Clearances. Except as elsewhere required or permitted in this Code, the dimension of the working space in the direction of access to live parts operating at 600 volts, nominal, or less to ground and likely to require examination, adjustment, servicing, or maintenance while energized shall not be less than indicated in Table 110-16(a). Distances shall be measured from the live parts if such are exposed or from the enclosure front or opening if such are enclosed. Concrete, brick, or tile walls shall be considered as grounded.

In addition to the dimensions shown in Table 110-16(a), the work space shall not be less than 30 inches (762 mm) wide in front of the electric equipment. The work space shall be clear and extend from the floor or platform to the height required by this section. In all cases, the work space shall permit at least a 90-degree opening of equipment doors or hinged panels.

Table 110-16 (a). Working Clearances

Voltage to Ground, Nominal		Minimum Clear Distance (feet)		
	Condition:	1	2	3
0-150		3	3	3
151-600		3	3½	4

For SI units: one inch = 25.4 millimeters; one foot = 0.3048 meter.

Where the "Conditions" are as follows:

1. Exposed live parts on one side and no live or grounded parts on the other side of the working space, or exposed live parts on both sides effectively guarded by suitable wood or other insulating materials. Insulated wire or insulated busbars operating at not over 300 volts shall not be considered live parts.

2. Exposed live parts on one side and grounded parts on the other side.

3. Exposed live parts on both sides of the work space (not guarded as provided in Condition 1) with the operator between.

Exception No. 1: Working space shall not be required in back of assemblies such as dead-front switchboards, or motor control centers where there are no renewable or adjustable parts such as fuses or switches on the back and where all connections are accessible from locations other than the back.

Exception No. 2: By special permission, smaller spaces may be permitted (1) where it is judged that the particular arrangement of the installation will provide adequate accessibility or (2) where all uninsulated parts are at a voltage no greater than 30 volts RMS, 42 volts peak, or 60 volts dc.

(b) Clear Spaces. Working space required by this section shall not be used for storage. When normally enclosed live parts are exposed for inspection or servicing, the working space, if in a passageway or general open space, shall be suitably guarded.

(c) Access and Entrance to Working Space. At least one entrance of sufficient area shall be provided to give access to the working space about electric equipment.

For equipment rated 1200 amperes or more and over 6 ft (1.83 m) wide, containing overcurrent devices, switching devices, or control devices, there shall be one entrance not less than 24 inches (610 mm) wide and 6½ feet (1.98 m) high at each end.

Exception No. 1: Where the location permits a continuous and unobstructed way of exit travel.

Exception No. 2: Where the work space required by Section 110-16(a) is doubled, only one entrance to the working space is required.

Working space with one entrance provided shall be so located that the edge of the entrance nearest the equipment is the minimum clear distance given in Table 110-16(a) away from such equipment.

(d) Front Working Space. In all cases where there are live parts normally exposed on the front of switchboards or motor control centers, the working space in front of such equipment shall not be less than 3 feet (914 mm).

(e) Illumination. Illumination shall be provided for all working spaces about service equipment, switchboards, panelboards, or motor control centers installed indoors.

(FPN): Additional lighting fixtures are not intended where the workspace is illuminated by an adjacent light source.

(f) Headroom. The minimum headroom of working spaces about service equipment, switchboards, panelboards, or motor control centers shall be 6½ feet (1.98 m).

Exception: Service equipment or panelboards, in existing dwelling units, that do not exceed 200 amperes.

(FPN No. 1): For higher voltages, see Article 710.

(FPN No. 2): As used in this section, a motor control center is an assembly of one or more enclosed sections having a common power bus and principally containing motor control units.

110-17. Guarding of Live Parts (600 Volts, Nominal, or Less).

(a) Live Parts Guarded Against Accidental Contact. Except as elsewhere required or permitted by this Code, live parts of electric equipment operating at 50 volts or more shall be guarded against accidental contact by approved enclosures or by any of the following:

(1) By location in a room, vault, or similar enclosure that is accessible only to qualified persons.

(2) By suitable permanent, substantial partitions or screens so arranged that only qualified persons will have access to the space within reach of the live parts. Any openings in such partitions or screens shall be so sized and located that persons are not likely to come into accidental contact with the live parts or to bring conducting objects into contact with them.

(3) By location on a suitable balcony, gallery, or platform so elevated and arranged as to exclude unqualified persons.

(4) By elevation of 8 feet (2.44 m) or more above the floor or other working surface.

(b) Prevent Physical Damage. In locations where electric equipment would be exposed to physical damage, enclosures or guards shall be so arranged and of such strength as to prevent such damage.

(c) Warning Signs. Entrances to rooms and other guarded locations containing exposed live parts shall be marked with conspicuous warning signs forbidding unqualified persons to enter.

(FPN): For motors, see Sections 430-132 and 430-133. For over 600 volts, see Section 110-34.

110-18. Arcing Parts. Parts of electric equipment which in ordinary operation produce arcs, sparks, flames, or molten metal shall be enclosed or separated and isolated from all combustible material.

(FPN): For hazardous (classified) locations, see Articles 500 through 517. For motors, see Section 430-14.

110-19. Light and Power from Railway Conductors. Circuits for lighting and power shall not be connected to any system containing trolley wires with a ground return.

Exception: Car houses, power houses, or passenger and freight stations operated in connection with electric railways.

110-21. Marking. The manufacturer's name, trademark, or other descriptive marking by which the organization responsible for the product may be identified shall be placed on all electric equipment. Other markings shall be provided indicating voltage, current, wattage, or other ratings as are specified elsewhere in this Code. The marking shall be of sufficient durability to withstand the environment involved.

110-22. Identification of Disconnecting Means. Each disconnecting means required by this Code for motors and appliances, and each service, feeder, or branch circuit at the point where it originates, shall be legibly marked to indicate its purpose unless located and arranged so the purpose is evident. The marking shall be of sufficient durability to withstand the environment involved.

Where circuit breakers or fuses are applied in compliance with the series combination ratings marked on the equipment by the manufacturer, the equipment enclosure(s) shall be legibly marked in the field to indicate the equipment has been applied with a series combination rating. The marking shall be readily visible and state "Caution — Series Rated System _____ A Available. Identified Replacement Component Required."

(FPN): See Section 240-83(c) for interrupting rating marking for end-use equipment.

B. Over 600 Volts, Nominal

110-30. General. Conductors and equipment used on circuits over 600 volts, nominal, shall comply with all applicable provisions of the preceding sections of this article and with the following sections, which supplement or modify the preceding sections. In no case shall the provisions of this part apply to equipment on the supply side of the service conductors.

110-31. Enclosure for Electrical Installations. Electrical installations in a vault, room, or closet or in an area surrounded by a wall, screen, or fence, access to which is controlled by lock and key or other approved means, shall be considered to be accessible to qualified persons only. The type of enclosure used in a given case shall be designed and constructed according to the nature and degree of the hazard(s) associated with the installation.

A wall, screen, or fence shall be used to enclose an outdoor electrical installation to deter access by persons who are not qualified. A fence shall not be less than 7 feet (2.13 m) in height.

(FPN No. 1): Article 450 covers minimum construction requirements for transformer vaults.

(FPN No. 2): Isolation by elevation is covered in paragraph (b) of this section and in Section 110-34.

(FPN No. 3): A fence made of a combination of 6 feet (1.80 m) or more of fence fabric and a 1-foot (300-mm) or more extension utilizing three or more strands of barbed wire is equivalent to a 7-foot (2.13-m) fence.

(a) Indoor Installations.

(1) In Places Accessible to Unqualified Persons. Indoor electrical installations that are open to unqualified persons shall be made with metal-enclosed equipment or shall be enclosed in a vault or in an area access to which is controlled by a lock. Metal-enclosed switchgear, unit substations, transformers, pull boxes, connection boxes, and other similar associated equipment shall be marked with appropriate caution signs. Openings in ventilated dry-type transformers or similar openings in other equipment shall be designed so that foreign objects inserted through these openings will be deflected from energized parts.

(2) In Places Accessible to Qualified Persons Only. Indoor electrical installations considered accessible to qualified persons only in accordance with this section shall comply with Sections 110-34, 710-32, and 710-33.

(b) Outdoor Installations.

(1) In Places Accessible to Unqualified Persons. Outdoor electrical installations that are open to unqualified persons shall comply with Article 225.

(FPN): For clearances of conductors for system voltages over 600 volts, nominal, see National Electrical Safety Code, ANSI C2-1990.

(2) In Places Accessible to Qualified Persons Only. Outdoor electrical installations having exposed live parts shall be accessible to qualified persons only in accordance with the first paragraph of this section and shall comply with Sections 110-34, 710-32, and 710-33.

(c) Metal-Enclosed Equipment Accessible to Unqualified Persons. Ventilating or similar openings in equipment shall be so designed that foreign objects inserted through these openings will be deflected from energized parts. Where exposed to physical damage from vehicular traffic, suitable guards shall be provided. Metal-enclosed equipment located outdoors and accessible to the general public shall be designed so that exposed nuts or bolts cannot be readily removed, permitting access to live parts. Where metal-enclosed equipment is accessible to the general public and the bottom of the enclosure is less than 8 feet (2.44 m) above the floor or grade level, the enclosure door or hinged cover shall be kept locked.

110-32. Work Space about Equipment. Sufficient space shall be provided and maintained about electric equipment to permit ready and safe operation and maintenance of such equipment. Where energized parts are exposed, the minimum clear work space shall not be less than 6½ feet (1.98 m) high (measured vertically from the floor or platform), or less than 3 feet (914 mm) wide (measured parallel to the equipment). The depth shall be as required in Section 110-34(a). In all cases, the work space shall be adequate to permit at least a 90-degree opening of doors or hinged panels.

110-33. Entrance and Access to Work Space.

(a) Entrance. At least one entrance not less than 24 inches (610 mm) wide and 6½ feet (1.98 m) high shall be provided to give access to the working space about electric equipment.

On switchboard and control panels exceeding 6 feet (1.83 m) in width, there shall be one entrance at each end of such board.

Exception No. 1: Where the switchboards and control panels location permits a continuous and unobstructed way of exit travel.

Exception No. 2: Where the work space required in Section 110-34(a) is doubled.

Working space with one entrance provided shall be so located that the edge of the entrance nearest the switchboards and control panels is the minimum clear distance given in Table 110-34(a) away from such equipment.

Where bare energized parts at any voltage or insulated energized parts above 600 volts, nominal, to ground are located adjacent to such entrance, they shall be suitably guarded.

(b) Access. Permanent ladders or stairways shall be provided to give safe access to the working space around electric equipment installed on platforms, balconies, mezzanine floors, or in attic or roof rooms or spaces.

110-34. Work Space and Guarding.

(a) Working Space. The minimum clear working space in front of electric equipment such as switchboards, control panels, switches, circuit breakers, motor controllers, relays, and similar equipment shall not be less than specified in Table 110-34(a) unless otherwise specified in this Code. Distances shall be measured from the live parts, if such are exposed, or from the enclosure front or opening if such are enclosed.

Table 110-34(a).
Minimum Depth of Clear Working Space in Front of Electric Equipment

Nominal	Conditions		
Voltage to Ground	1	2	3
	(Feet)	(Feet)	(Feet)
601-2500	3	4	5
2501-9000	4	5	6
9001-25,000	5	6	9
25,001-75 kV	6	8	10
Above 75 kV	8	10	12

For SI units: one foot = 0.3048 meter.

Where the "Conditions" are as follows:

1. Exposed live parts on one side and no live or grounded parts on the other side of the working space or exposed live parts on both sides effectively guarded by suitable wood or other insulating materials. Insulated wire or insulated busbars operating at not over 300 volts shall not be considered live parts.

2. Exposed live parts on one side and grounded parts on the other side. Concrete, brick, or tile walls will be considered as grounded surfaces.

3. Exposed live parts on both sides of the work space (not guarded as provided in Condition 1) with the operator between.

Exception: Working space is not required in back of equipment such as dead-front switchboards or control assemblies where there are no renewable or adjustable parts (such as fuses or switches) on the back and where all connections are accessible from locations other than the back. Where rear access is required to work on deenergized parts on the back of enclosed equipment, a minimum working space of 30 inches (762 mm) horizontally shall be provided.

(b) Separation from Low-Voltage Equipment. Where switches, cutouts, or other equipment operating at 600 volts, nominal, or less, are installed in a room or enclosure where there are exposed live parts or exposed wiring operating at over 600 volts, nominal, the high-voltage equipment shall be effectively separated from the space occupied by the low-voltage equipment by a suitable partition, fence, or screen.

Exception: Switches or other equipment operating at 600 volts, nominal, or less, and serving only equipment within the high-voltage vault, room, or enclosure shall be permitted to be installed in the high-voltage enclosure, room, or vault if accessible to qualified persons only.

(c) Locked Rooms or Enclosures. The entrances to all buildings, rooms, or enclosures containing exposed live parts or exposed conductors operating at over 600 volts, nominal, shall be kept locked.

Exception: Where such entrances are under the observation of a qualified person at all times.

Where the voltage exceeds 600 volts, nominal, permanent and conspicuous warning signs shall be provided, reading substantially as follows: "Danger—High Voltage—Keep Out."

(d) Illumination. Adequate illumination shall be provided for all working spaces about electrical equipment. The lighting outlets shall be so arranged that persons changing lamps or making repairs on the lighting system will not be endangered by live parts or other equipment.

The points of control shall be so located that persons are not likely to come in contact with any live part or moving part of the equipment while turning on the lights.

(e) Elevation of Unguarded Live Parts. Unguarded live parts above working space shall be maintained at elevations not less than required by Table 110-34(e).

Table 110-34(e).
Elevation of Unguarded Live Parts above Working Space

Nominal Voltage Between Phases	Elevation
1001-7500	8'6"
7501-35000	9'
Over 35kV	9' + 0.37" per kV above 35

For SI units: one inch = 25.4 millimeters; one foot = 0.3048 meter.

Chapter 2. Wiring and Protection

ARTICLE 200 — USE AND IDENTIFICATION OF GROUNDED CONDUCTORS

200-1. Scope. This article provides requirements for (1) identification of terminals; (2) grounded conductors in premises wiring systems; and (3) identification of grounded conductors.

(FPN): See Article 100 for definitions of "Grounded Conductor" and "Grounding Conductor."

200-2. General. All premises wiring systems shall have a grounded conductor that is identified in accordance with Section 200-6.

Exception: Circuits and systems exempted or prohibited by Sections 210-10, 215-7, 250-3, 250-5, 250-7, 503-13, 517-63, 668-11, 668-21, and 690-41, Exception.

The grounded conductor, where insulated, shall have insulation (1) that is suitable, other than color, for any ungrounded conductor of the same circuit on circuits of less than 1000 volts, or (2) rated not less than 600 volts for solidly grounded neutral systems of 1 kV and over as described in Section 250-152(a).

200-3. Connection to Grounded System. Premises wiring shall not be electrically connected to a supply system unless the latter contains, for any grounded conductor of the interior system, a corresponding conductor that is grounded.

For the purpose of this section, "electrically connected" shall mean connected so as to be capable of carrying current as distinguished from connection through electromagnetic induction.

200-6. Means of Identifying Grounded Conductors.

(a) Sizes No. 6 or Smaller. An insulated grounded conductor of No. 6 or smaller shall be identified by a continuous white or natural gray outer finish along its entire length.

Exception No. 1: Multiconductor varnished-cloth-insulated cables.

Exception No. 2: Fixture wires as outlined in Article 402.

Exception No. 3: Where the conditions of maintenance and supervision assure that only qualified persons will service the installation, grounded conductors in multiconductor cables shall be permitted to be permanently identified at their terminations at the time of installation by a distinctive white marking or other equally effective means.

Exception No. 4: The grounded conductor of a mineral-insulated, metal-sheathed cable shall be identified at the time of installation by distinctive marking at its terminations.

For aerial cable, the identification shall be as above, or by means of a ridge so located on the exterior of the cable as to identify it.

Wires having their outer covering finished to show a white or natural gray color but having colored tracer threads in the braid, identifying the source of manufacture, shall be considered as meeting the provisions of this section.

(b) Sizes Larger than No. 6. An insulated grounded conductor larger than No. 6 shall be identified either by a continuous white or natural gray outer finish along its entire length or at the time of installation by a distinctive white marking at its terminations. Multiconductor flat cable No. 4 or larger shall be permitted to employ an external ridge on the grounded conductor.

Exception: Where the conditions of maintenance and supervision assure that only qualified persons will service the installation, grounded conductors in multiconductor cables shall be permitted to be permanently identified at their terminations at the time of installation by a distinctive white marking or other equally effective means.

(c) Flexible Cords. An insulated conductor intended for use as a grounded conductor, where contained within a flexible cord, shall be identified by a white or natural gray outer finish or by methods permitted by Section 400-22.

(d) Grounded Conductors of Different Systems. Where conductors of different systems are installed in the same raceway, box, auxiliary gutter, or other type of enclosure, one system grounded conductor, if required, shall have an outer covering conforming to Section 200-6(a) or (b). Each other system grounded conductor, if required, shall have an outer covering of white with an identifiable colored stripe (not green) running along the insulation, or other and different means of identification as allowed by Section 200-6(a) or (b).

200-7. Use of White or Natural Gray Color. A continuous white or natural gray covering on a conductor or a termination marking of white or natural gray color shall be used only for the grounded conductor.

Exception No. 1: An insulated conductor with a white or natural gray finish shall be permitted as an ungrounded conductor where permanently reidentified to indicate its use, by painting or other effective means at its termination, and at each location where the conductor is visible and accessible.

Exception No. 2: A cable containing an insulated conductor with a white or natural gray outer finish shall be permitted for single-pole, 3-way, or 4-way switch loops where the white or natural gray conductor is used for the supply to the switch, but not as a return conductor from the switch to the switched outlet. In these applications, reidentification of the white or natural gray conductor shall not be required.

Exception No. 3: A flexible cord for connecting an appliance, having one conductor identified by a white or natural gray outer finish or by any other means permitted by Section 400-22, shall be permitted whether or not the outlet to which it is connected is supplied by a circuit having a grounded conductor.

Exception No. 4: A white or natural gray conductor of circuits of less than 50 volts shall be required to be grounded only as required by Section 250-5(a).

200-9. Means of Identification of Terminals. The identification of terminals to which a grounded conductor is to be connected shall be substantially white in color. The identification of other terminals shall be of a readily distinguishable different color.

Exception: Where the conditions of maintenance and supervision assure that only qualified persons will service the installations, terminals for grounded conductors shall be permitted to be permanently identified at the time of installation by a distinctive white marking or other equally effective means.

200-10. Identification of Terminals.

(a) Device Terminals. All devices provided with terminals for the attachment of conductors and intended for connection to more than one side of the circuit shall have terminals properly marked for identification.

Exception No. 1: Where the electrical connection of a terminal intended to be connected to the grounded conductor is clearly evident.

Exception No. 2: The terminals of lighting and appliance branch-circuit panelboards.

Exception No. 3: Devices having a normal current rating of over 30 amperes other than polarized attachment plugs and polarized receptacles for attachment plugs as required in Section 200-10(b).

(b) Receptacles, Plugs, and Connectors. Receptacles, polarized attachment plugs and cord connectors for plugs and polarized plugs shall have the terminal intended for connection to the grounded (white) conductor identified.

Identification shall be by a metal or metal coating substantially white in color or the word "white" located adjacent to the identified terminal.

If the terminal is not visible, the conductor entrance hole for the connection shall be colored white or marked with the word "white."

Exception: Terminal identification shall not be required for 2-wire nonpolarized attachment plugs.

(FPN): See Section 250-119 for identification of wiring device equipment grounding conductor terminals.

(c) Screw Shells. For devices with screw shells, the terminal for the grounded conductor shall be the one connected to the screw shell.

(d) Screw-Shell Devices with Leads. For screw-shell devices with attached leads, the conductor attached to the screw shell shall have a white or natural gray finish. The outer finish of the other conductor shall be of a solid color that will not be confused with the white or natural gray finish used to identify the grounded conductor.

(e) Appliances. Appliances that have a single-pole switch or a single-pole overcurrent device in the line or any line-connected screw-shell lampholders, and that are to be connected (1) by permanent wiring methods; or (2) by field-installed attachment plugs and cords with three or more wires (including the equipment grounding conductor) shall have means to identify the terminal for the grounded circuit conductor (if any).

200-11. Polarity of Connections. No grounded conductor shall be attached to any terminal or lead so as to reverse designated polarity.

ARTICLE 210 — BRANCH CIRCUITS

A. General Provisions

210-1. Scope. This article covers branch circuits except for branch circuits that supply only motor loads, which are covered in Article 430. Provisions of this article and Article 430 apply to branch circuits with combination loads.

Exception: Branch circuits for electrolytic cells as covered in Section 668-3(c), Exception Nos. 1 and 4.

210-2. Other Articles for Specific-Purpose Branch Circuits. Branch circuits shall comply with this article and also with the applicable provisions of other articles of this Code. The provisions for branch circuits supplying equipment in the following list amend or supplement the provisions in this article and shall apply to branch circuits referred to therein:

	Article	Section
Air-Conditioning and Refrigerating Equipment		440-6
		440-31
		440-32
Busways		364-9
Circuits and Equipment Operating at Less than 50 Volts...	720	
Central Heating Equipment Other than Fixed Electric Space Heating Equipment....................................		422-7
Class 1, Class 2, and Class 3 Remote-Control, Signaling, and Power-Limited Circuits........................	725	
Closed-Loop and Programmed Power Distribution	780	
Cranes and Hoists		610-42
Electronic Computer/Data Processing Equipment...........		645-5
Electric Signs and Outline Lighting...........................		600-6
Electric Welders	630	
Elevators, Dumbwaiters, Escalators, Moving Walks, Wheelchair Lifts, and Stairway Chair Lifts................		620-61

210-3. Rating. Branch circuits recognized by this article shall be rated in accordance with the maximum permitted ampere rating or setting of the overcurrent device. The rating for other than individual branch circuits shall be: 15, 20, 30, 40, and 50 amperes. Where conductors of higher ampacity are used for any reason, the ampere rating or setting of the specified overcurrent device shall determine the circuit rating.

Exception: Multioutlet branch circuits greater than 50 amperes shall be permitted to supply nonlighting outlet loads on industrial premises where maintenance and supervision indicate that qualified persons will service equipment.

210-4. Multiwire Branch Circuits.

(a) General. Branch circuits recognized by this article shall be permitted as multiwire circuits. A multiwire branch circuit shall be permitted to be considered as multiple circuits. All conductors shall originate from the same panelboard.

(FPN): A 3-phase, 4-wire power system used to supply power to computer systems or other similar electronic loads may necessitate that the power system design allow for the possibility of high harmonic neutral currents.

(b) Dwelling Units. In dwelling units, a multiwire branch circuit supplying more than one device or equipment on the same yoke shall be provided with a means to disconnect simultaneously all ungrounded conductors at the panelboard where the branch circuit originated.

(c) Line to Neutral Load. Multiwire branch circuits shall supply only line to neutral load.

Exception No. 1: A multiwire branch circuit that supplies only one utilization equipment.

Exception No. 2: Where all ungrounded conductors of the multiwire branch circuit are opened simultaneously by the branch-circuit overcurrent device.

(FPN): See Section 300-13(b) for continuity of grounded conductor on multiwire circuits.

(d) Identification of Ungrounded Conductors. Where more than one nominal voltage system exists in a building, each ungrounded system conductor shall be identified by phase and system.

The means of identification shall be permanently posted at each branch-circuit panelboard.

(FPN): The means of identification of each system phase conductor, wherever accessible, may be by separate color coding, marking tape, tagging, or other equally effective means. See Sections 215-8, 230-56, and 384-3(e) for high leg marking.

210-5. Color Code for Branch Circuits.

(a) Grounded Conductor. The grounded conductor of a branch circuit shall be identified by a continuous white or natural gray color. Where conductors of different systems are installed in the same raceway, box, auxiliary gutter, or other types of enclosures, one system grounded conductor, if required, shall have an outer covering of white or natural gray. Each other system grounded conductor, if required, shall have an outer covering of white with an identifiable colored stripe (not green) running along the insulation or other and different means of identification.

Exception No. 1: The grounded conductors of mineral-insulated, metal-sheathed cable shall be identified by distinctive marking at the terminals during the process of installation.

Exception No. 2: As permitted in Exception No. 3 of Section 200-6(a) and the Exception to Section 200-6(b).

(b) Equipment Grounding Conductor. The equipment grounding conductor of a branch circuit shall be identified by a continuous green color or a continuous green color with one or more yellow stripes unless it is bare.

Exception No. 1: As permitted in Section 250-57(b), Exception Nos. 1 and 4 and Section 310-12(b), Exception Nos. 1 and 2.

Exception No. 2: The use of conductor insulation having a continuous green color or a continuous green color with one or more yellow stripes shall be permitted for internal wiring of equipment if such wiring does not serve as the lead wires for connection to branch-circuit conductors.

210-6. Branch Circuit Voltage Limitations.

(a) Occupancy Limitation. In dwelling units and guest rooms of hotels, motels, and similar occupancies, the voltage shall not exceed 120 volts, nominal, between conductors that supply the terminals of:

(1) Lighting fixtures.

(2) Cord- and plug-connected loads 1440 volt-amperes, nominal, or less, or less than 1/4 horsepower.

(b) 120 Volts Between Conductors. Circuits not exceeding 120 volts, nominal, between conductors shall be permitted to supply:

(1) The terminals of medium-base screw-shell lampholders or lampholders of other types applied within their voltage ratings.

(2) Auxiliary equipment of electric-discharge lamps.

(3) Cord- and plug-connected or permanently connected utilization equipment.

(c) 277 Volts to Ground. Circuits exceeding 120 volts, nominal, between conductors and not exceeding 277 volts, nominal, to ground shall be permitted to supply:

(1) Listed electric-discharge lighting fixtures equipped with medium-base screw-shell lampholder.

(2) Listed incandescent lighting fixtures equipped with medium-base screw-shell lampholder, where supplied at 120 volts or less from the output of a stepdown autotransformer that is an integral component of the fixture, and the outer screw-shell terminal is electrically connected to a grounded conductor of the branch circuit.

(3) Lighting fixtures equipped with mogul-base screw-shell lampholders.

(4) Lampholders other than the screw-shell type applied within their voltage ratings.

(5) Auxiliary equipment of electric discharge lamps.

(6) Cord- and plug-connected or permanently connected utilization equipment.

(d) 600 Volts Between Conductors. Circuits exceeding 277 volts, nominal, to ground and not exceeding 600 volts, nominal, between conductors shall be permitted to supply:

(1) The auxiliary equipment of electric-discharge lamps mounted in permanently installed fixtures where the fixtures are mounted in accordance with one of the following:

a. Not less than a height of 22 feet (6.71 m) on poles or similar structures for the illumination of outdoor areas, such as highways, roads, bridges, athletic fields, or parking lots.

b. Not less than a height of 18 feet (5.49 m) on other structures, such as tunnels.

(2) Cord- and plug-connected or permanently connected utilization equipment.

(FPN):   See Section 410-78 for auxiliary equipment limitations.

Exception No. 1 to (b), (c), and (d) above: For lampholders of infrared industrial heating appliances as provided in Section 422-15(c).

Exception No. 2 to (b), (c), and (d) above: For railway properties as described in Section 110-19.

210-7. Receptacles and Cord Connectors.

(a) Grounding Type. Receptacles installed on 15- and 20-ampere branch circuits shall be of the grounding type. Grounding-type receptacles shall be installed only on circuits of the voltage class and current for which they are rated, except as provided in Tables 210-21(b)(2) and (b)(3).

Exception: Nongrounding-type receptacles installed in accordance with Section 210-7(d), Exception.

(b) To Be Grounded. Receptacles and cord connectors having grounding contacts shall have those contacts effectively grounded.

Exception No. 1: Receptacles mounted on portable and vehicle-mounted generators in accordance with Section 250-6.

Exception No. 2: Ground-fault circuit-interrupter replacement receptacles as permitted by Section 210-7(d), Exception.

(c) Methods of Grounding. The grounding contacts of receptacles and cord connectors shall be grounded by connection to the equipment grounding conductor of the circuit supplying the receptacle or cord connector.

(FPN):　For installation requirements for the reduction of electrical noise, see Section 250-74, Exception No. 4.

The branch circuit wiring method shall include or provide an equipment grounding conductor to which the grounding contacts of the receptacle or cord connector shall be connected.

(FPN No. 1):　Section 250-91(b) describes acceptable grounding means.

(FPN No. 2):　For extensions of existing branch circuits, see Section 250-50.

(d) Replacements. Grounding-type receptacles shall be used as replacements for existing nongrounding types and shall be connected to a grounding conductor installed in accordance with Section 210-7(c).

Ground-fault circuit-interrupter protected receptacles shall be provided where replacements are made at receptacle outlets that are required to be so protected elsewhere in this Code.

Exception: Where a grounding means does not exist in the receptacle enclosure, either a nongrounding or a ground-fault circuit-interrupter-type of receptacle shall be used. A grounding conductor shall not be connected from the ground-fault circuit-interrupter-type receptacle to any outlet supplied from the ground-fault circuit-interrupter-type receptacle. Existing nongrounding-type receptacles shall be permitted to be replaced with grounding-type receptacles where supplied through a ground-fault circuit-interrupter. These receptacle locations shall be marked "GFCI protected."

(e) Cord- and Plug-Connected Equipment. The installation of grounding-type receptacles shall not be used as a requirement that all cord- and plug-connected equipment be of the grounded type.

(FPN):　See Section 250-45 for type of cord- and plug-connected equipment to be grounded.

(f) Noninterchangeable Types. Receptacles connected to circuits having different voltages, frequencies, or types of current (ac or dc) on the same premises shall be of such design that the attachment plugs used on these circuits are not interchangeable.

210-8. Ground-Fault Circuit-Interrupter Protection for Personnel.

(a) Dwelling Units.

(1) All 125-volt, single-phase, 15- and 20-ampere receptacles installed in bathrooms shall have ground-fault circuit-interrupter protection for personnel.

(2) All 125-volt, single-phase, 15- or 20-ampere receptacles installed in garages shall have ground-fault circuit-interrupter protection for personnel.

Exception No. 1 to (a)(2): Receptacles that are not readily accessible.

Exception No. 2 to (a)(2): A single receptacle or a duplex receptacle for two appliances located within dedicated space for each appliance that in normal use is not easily moved from one place to another, and that is cord- and plug-connected in accordance with Section 400-7(a)(6), (a)(7), or (a)(8).

Receptacles installed under exceptions to Section 210-8(a)(2) shall not be considered as meeting the requirements of Section 210-52(g).

(3) All 125-volt, single-phase, 15- and 20-ampere receptacles installed outdoors where there is direct grade level access to the receptacles shall have ground-fault circuit-interrupter protection for personnel.

(FPN): See Section 215-9 for feeder protection.

For the purposes of this section, "direct grade level access" is defined as being located not more than 6 feet, 6 inches (1.98 m) above grade level and being readily accessible.

(4) All 125-volt, single-phase, 15- and 20-ampere receptacles installed in crawl spaces at or below grade level and in unfinished basements shall have ground-fault circuit-interrupter protection for personnel.

For purposes of this section, unfinished basements are defined as portions or areas of the basement not intended as habitable rooms and limited to storage areas, work areas, and the like.

Exception No. 1: A single receptacle supplied by a dedicated branch circuit that is located and identified for specific use by a cord- and plug-connected appliance, such as a refrigerator or freezer.

Exception No. 2: The laundry circuit as required by Sections 210-52(f) and 220-4(c).

Exception No. 3: A single receptacle supplying a permanently installed sump pump.

(5) All 125-volt, single-phase, 15- and 20-ampere receptacles to serve counter top surfaces, installed within 6 feet (1.83 m) of a wet bar sink or kitchen sink, shall have ground-fault circuit-interrupter protection for personnel.

(FPN): The intent of this subsection is to permit the exemption of receptacles that are located specifically for appliances such as refrigerators and freezers from ground-fault circuit-interrupter protection for personnel.

(6) All 125-volt, single-phase, 15- or 20-ampere receptacles installed in boathouses shall have ground-fault circuit-interrupter protection for personnel.

(b) Other than Dwelling Units.

(1) All 125-volt, 15- and 20-ampere receptacles installed in bathrooms of commercial, industrial, and all other nondwelling occupancies shall have ground-fault circuit-interrupter protection for personnel.

Bathroom: As used in this article, a bathroom is an area including a basin with one or more of the following: a toilet, a tub, or a shower.

(2) All 125-volt, single-phase, 15- and 20-ampere receptacles installed on roofs shall have ground-fault circuit-interrupter protection for personnel.

210-9. Circuits Derived from Autotransformers. Branch circuits shall not be derived from autotransformers unless the circuit supplied has a grounded conductor that is electrically connected to a grounded conductor of the system supplying the autotransformer.

Exception No. 1: An autotransformer shall be permitted to extend or add a branch circuit for an equipment load without the connection to a similar grounded conductor when transforming from a nominal 208 volts to a nominal 240-volt supply or similarly from 240 volts to 208 volts.

Exception No. 2: In industrial occupancies, where conditions of maintenance and supervision ensure that only qualified persons will service the installation, autotransformers shall be permitted to supply nominal 600-volt loads from nominal 480-volt systems, and 480-volt loads from nominal 600-volt systems, without the connection to a similar grounded conductor.

210-10. Ungrounded Conductors Tapped from Grounded Systems. Two-wire dc circuits and ac circuits of two or more ungrounded conductors shall be permitted to be tapped from the ungrounded conductors of circuits having a grounded neutral conductor. Switching devices in each tapped circuit shall have a pole in each ungrounded conductor. All poles of multipole switching devices shall manually switch together where such switching devices also serve as a disconnecting means as required by Section 410-48 for double-pole switched lampholders; Section 410-54(b) for electric-discharge lamp auxiliary equipment switching devices; Section 422-21(b) for an appliance; Section 424-20 for a fixed electric space heating unit; Section 426-51 for electric de-icing and snow-melting equipment; Section 430-85 for a motor controller; and Section 430-103 for a motor.

B. Branch-Circuit Ratings

210-19. Conductors — Minimum Ampacity and Size.

(a) General. Branch-circuit conductors shall have an ampacity not less than the maximum load to be served. In addition, conductors of multioutlet branch circuits supplying receptacles for cord- and plug-connected portable loads shall have an ampacity of not less than the rating of the branch circuit. Cable assemblies where the neutral conductor is smaller than the ungrounded conductors shall be so marked.

(FPN No. 1):   See Section 310-15 for ampacity ratings of conductors.

(FPN No. 2):   See Part B of Article 430 for minimum rating of motor branch-circuit conductors.

(FPN No. 3): See Section 310-10 for temperature limitation of conductors.

(FPN No. 4): Conductors for branch circuits as defined in Article 100, sized to prevent a voltage drop exceeding 3 percent at the farthest outlet of power, heating, and lighting loads, or combinations of such loads and where the maximum total voltage drop on both feeders and branch circuits to the farthest outlet does not exceed 5 percent, will provide reasonable efficiency of operation. See Section 215-2 for voltage drop on feeder conductors.

(b) Household Ranges and Cooking Appliances. Branch-circuit conductors supplying household ranges, wall-mounted ovens, counter-mounted cooking units, and other household cooking appliances shall have an ampacity not less than the rating of the branch circuit and not less than the maximum load to be served. For ranges of 8¾ kW or more rating, the minimum branch-circuit rating shall be 40 amperes.

Exception No. 1: Tap conductors supplying electric ranges, wall-mounted electric ovens, and counter-mounted electric cooking units from a 50-ampere branch circuit shall have an ampacity of not less than 20 and shall be sufficient for the load to be served. The taps shall not be longer than necessary for servicing the appliance.

Exception No. 2: The neutral conductor of a 3-wire branch circuit supplying a household electric range, a wall-mounted oven, or a counter-mounted cooking unit shall be permitted to be smaller than the ungrounded conductors where the maximum demand of a range of 8¾ kW or more rating has been computed according to Column A of Table 220-19, but shall have an ampacity of not less than 70 percent of the branch-circuit rating and shall not be smaller than No. 10.

(c) Other Loads. Branch-circuit conductors supplying loads other than cooking appliances as covered in (b) above and as listed in Section 210-2 shall have an ampacity sufficient for the loads served and shall not be smaller than No. 14.

Exception No. 1: Tap conductors for such loads shall have an ampacity not less than 15 for circuits rated less than 40 amperes and not less than 20 for circuits rated at 40 or 50 amperes and only where these tap conductors supply any of the following loads:

a. Individual lampholders or fixtures with taps extending not longer than 18 inches (457 mm) beyond any portion of the lampholder or fixture.

b. A fixture having tap conductors as provided in Section 410-67.

c. Individual outlets with taps not over 18 inches (457 mm) long.

d. Infrared lamp industrial heating appliances.

e. Nonheating leads of de-icing and snow-melting cables and mats.

Exception No. 2: Fixture wires and cords as permitted in Section 240-4.

210-20. Overcurrent Protection. Branch-circuit conductors and equipment shall be protected by overcurrent protective devices having a rating or setting (1) not exceeding that specified in Section 240-3 for conductors; (2) not exceeding that specified in the applicable articles referenced in Section 240-2 for equipment; and (3) as provided for outlet devices in Section 210-21.

Exception No. 1: Tap conductors as permitted in Section 210-19(c) shall be permitted to be protected by the branch-circuit overcurrent device.

Exception No. 2: Fixture wires and cords as permitted in Section 240-4.

(FPN): See Section 240-1 for the purpose of overcurrent protection and Sections 210-22 and 220-3 for continuous loads.

210-21. Outlet Devices. Outlet devices shall have an ampere rating not less than the load to be served and shall comply with (a) and (b) below.

(a) Lampholders. Where connected to a branch circuit having a rating in excess of 20 amperes, lampholders shall be of the heavy-duty type. A heavy-duty lampholder shall have a rating of not less than 660 watts if of the admedium type and not less than 750 watts if of any other type.

(b) Receptacles.

(1) A single receptacle installed on an individual branch circuit shall have an ampere rating of not less than that of the branch circuit.

Exception: Where installed in accordance with Section 430-81(c).

(FPN): See definition of "Receptacle" in Article 100.

(2) Where connected to a branch circuit supplying two or more receptacles or outlets, a receptacle shall not supply a total cord- and plug-connected load in excess of the maximum specified in Table 210-21(b)(2).

(3) Where connected to a branch circuit supplying two or more receptacles or outlets, receptacle ratings shall conform to the values listed in Table 210-21(b)(3), or where larger than 50 amperes, the receptacle rating shall not be less than the branch-circuit rating.

(4) It shall be permitted to base the ampere rating of a range receptacle on a single range demand load specified in Table 220-19.

Table 210-21(b)(2).
Maximum Cord- and Plug-Connected Load to Receptacle

Circuit Rating Amperes	Receptacle Rating Amperes	Maximum Load Amperes
15 or 20	15	12
20	20	16
30	30	24

Table 210-21(b)(3).
Receptacle Ratings for Various Size Circuits

Circuit Rating Amperes	Receptacle Rating Amperes
15	Not over 15
20	15 or 20
30	30
40	40 or 50
50	50

210-22. Maximum Loads. The total load shall not exceed the rating of the branch circuit, and it shall not exceed the maximum loads specified in Sections 210-22(a) through (c) under the conditions specified therein.

(a) Motor-Operated and Combination Loads. Where a circuit supplies only motor-operated loads, Article 430 shall apply. Where a circuit supplies only air-conditioning equipment, refrigerating equipment, or both, Article 440 shall apply. For circuits supplying loads consisting of motor-operated utilization equipment that is fastened in place and that has a motor larger than $1/8$ horsepower in combination with other loads, the total computed load shall be based on 125 percent of the largest motor load plus the sum of the other loads.

(b) Inductive Lighting Loads. For circuits supplying lighting units having ballasts, transformers, or autotransformers, the computed load shall be based on the total ampere ratings of such units and not on the total watts of the lamps.

(c) Other Loads. The rating of the branch-circuit overcurrent device serving continuous loads, such as store lighting and similar loads, shall be not less than the noncontinuous load plus 125 percent of the continuous load.

Exception: Circuits supplied by an assembly together with its over-current devices that is listed for continuous operation at 100 percent of its rating.

It shall be acceptable to apply demand factors for range loads in accordance with Table 220-19, including Note 4.

210-23. Permissible Loads. In no case shall the load exceed the branch-circuit ampere rating. An individual branch circuit shall be permitted to supply any load for which it is rated. A branch circuit supplying two or more outlets shall supply only the loads specified according to its size in (a) through (d) below and summarized in Section 210-24 and Table 210-24.

(a) 15- and 20-Ampere Branch Circuits. A 15- or 20-ampere branch circuit shall be permitted to supply lighting units, other utilization equipment, or a combination of both. The rating of any one cord- and plug-connected utilization equipment shall not exceed 80 percent of the branch-circuit ampere rating. The total rating of utilization equipment fastened in place shall not exceed 50 percent of the branch-circuit ampere rating where lighting units, cord- and plug-connected utilization equipment not fastened in place, or both, are also supplied.

Exception: The small appliance branch circuits required in a dwelling unit(s) by Section 220-4(b) shall supply only the receptacle outlets specified in that section.

(b) 30-Ampere Branch Circuits. A 30-ampere branch circuit shall be permitted to supply fixed lighting units with heavy-duty lampholders in other than dwelling unit(s) or utilization equipment in any occupancy. A rating of any one cord- and plug-connected utilization equipment shall not exceed 80 percent of the branch-circuit ampere rating.

(c) 40- and 50-Ampere Branch Circuits. A 40- or 50-ampere branch circuit shall be permitted to supply cooking appliances that are fastened in place in any occupancy. In other than dwelling units, such circuits shall be permitted to supply fixed lighting units with heavy-duty lampholders, infrared heating units, or other utilization equipment.

(d) Branch Circuits Larger than 50 Amperes. Branch circuits larger than 50 amperes shall supply only nonlighting outlet loads.

210-24. Branch-Circuit Requirements — Summary. The requirements for circuits having two or more outlets, other than the receptacle circuits of Section 220-4(b) as specifically provided for above, are summarized in Table 210-24. Branch circuits in dwelling units shall supply only loads within that dwelling unit or loads associated only with that dwelling unit. Branch circuits required for the purpose of lighting, central alarm, signal, communications, or other needs for public or common areas shall not be supplied from a dwelling unit panelboard.

Table 210-24.
Summary of Branch-Circuit Requirements

CIRCUIT RATING	15 Amp	20 Amp	30 Amp	40 Amp	50 Amp
CONDUCTORS: (Min. Size)					
Circuit Wires*	14	12	10	8	6
Taps	14	14	14	12	12
Fixture Wires and Cords			Refer to Section 240-4		
OVERCURRENT PROTECTION	15 Amp	20 Amp	30 Amp	40 Amp	50 Amp
OUTLET DEVICES:					
Lampholders Permitted	Any Type	Any Type	Heavy Duty	Heavy Duty	Heavy Duty
Receptacle Rating**	15 Max. Amp	15 or 20 Amp	30 Amp	40 or 50 Amp	50 Amp
MAXIMUM LOAD	15 Amp	20 Amp	30 Amp	40 Amp	50 Amp
PERMISSIBLE LOAD	Refer to Section 210-23(a)	Refer to Section 210-23(a)	Refer to Section 210-23(b)	Refer to Section 210-23(c)	Refer to Section 210-23(c)

*These gauges are for copper conductors.
**For receptacle rating of cord-connected electric-discharge lighting fixtures, see Section 410-30(c).

C. Required Outlets

210-50. General. Receptacle outlets shall be installed as specified in Sections 210-52 through 210-63.

(a) Cord Pendants. A cord connector that is supported by a permanently installed cord pendant shall be considered a receptacle outlet.

(b) Cord Connections. A receptacle outlet shall be installed wherever flexible cords with attachment plugs are used. Where flexible cords are permitted to be permanently connected, it shall be permitted to omit receptacles for such cords.

(c) Appliance Outlets. Appliance receptacle outlets installed in a dwelling unit for specific appliances, such as laundry equipment, shall be installed within 6 feet (1.83 m) of the intended location of the appliance.

210-52. Dwelling Unit Receptacle Outlets.

(a) General Provisions. In every kitchen, family room, dining room, living room, parlor, library, den, sun room, bedroom, recreation room, or similar room or area of dwelling units, receptacle outlets shall be installed so that no point along the floor line in any wall space is more than 6 feet (1.83 m), measured horizontally, from an outlet in that space, including any wall space 2 feet (610 mm) or more in width and the wall space occupied by fixed panels in exterior walls, but excluding sliding panels in exterior walls. The wall space afforded by fixed room dividers, such as freestanding bar-type counters or railings, shall be included in the 6-foot (1.83-m) measurement.

As used in this section, a "wall space" shall be considered a wall unbroken along the floor line by doorways, fireplaces, and similar openings. Each wall space 2 feet or more (610 mm or more) wide shall be treated individually and separately from other wall spaces within the room. A wall space shall be permitted to include two or more walls of a room (around corners) where unbroken at the floor line.

(FPN):   The purpose of this requirement is to minimize the use of cords across doorways, fireplaces, and similar openings.

Receptacle outlets shall, insofar as practicable, be spaced equal distances apart. Receptacle outlets in floors shall not be counted as part of the required number of receptacle outlets unless located close to the wall.

The receptacle outlets required by this section shall be in addition to any receptacle that is part of any lighting fixture or appliance, located within cabinets or cupboards, or located over 5½ feet (1.68 m) above the floor.

Exception: Permanently installed electric baseboard heaters equipped with factory-installed receptacle outlets or outlets provided as a separate assembly by the manufacturer shall be permitted as the required outlet or outlets for the wall space utilized by such permanently installed heaters. Such receptacle outlets shall not be connected to the heater circuits.

(FPN):   Listed baseboard heaters include instructions that may not permit their installation below receptacle outlets.

(b) Small Appliances.

(1) The two or more 20-ampere small appliance branch circuits required by Section 220-4(b) shall serve all receptacle outlets, including refrigeration equipment, in the kitchen, pantry, breakfast room, dining room, or similar area of a dwelling unit. Such circuits, whether two or more are used, shall have no other outlets.

Exception No. 1: A receptacle installed solely for the electrical supply to and support of an electric clock in any of the rooms specified above.

Exception No. 2: Outdoor receptacles.

Exception No. 3: In addition to the required receptacles specified by Section 210-52, switched receptacles supplied from a general-purpose branch circuit as defined in Section 210-70(a), Exception No. 1 shall be permitted.

Exception No. 4: A receptacle served by a circuit supplying only motor loads.

Exception No. 5: Receptacles installed to provide power for electric ignition systems or clock timers for gas-fired ranges, ovens, or counter-mounted cooking units.

(2) Receptacles installed in the kitchen to serve counter top surfaces shall be supplied by not less than two small appliance branch circuits, either or both of which shall also be permitted to supply receptacle outlets in the kitchen and other rooms specified in Section 210-52(b)(1). Additional small appliance branch circuits shall be permitted to supply receptacle outlets in the kitchen and other rooms specified in Section 210-52(b)(1).

(c) Counter Tops. In kitchens and dining areas of dwelling units, a receptacle outlet shall be installed at each wall counter space 12 inches (305 mm) or wider. Receptacle outlets shall be installed so that no point along the wall line is more than 24 inches (610 mm), measured horizontally from a receptacle outlet in that space.

A receptacle outlet shall be installed at each island or peninsular counter top with a long dimension of 24 inches (610 mm) or greater and a short dimension of 12 inches (305 mm) or greater. Receptacle outlets to serve island or peninsular counter tops shall be installed above, or within 12 inches (305 mm) below the counter top. Receptacle outlets shall be installed so that no point along the centerline of the long dimension is more than 24 inches (610 mm), measured horizontally from a receptacle outlet in that space. A peninsular counter top is measured from the connecting edge.

Counter top spaces separated by range tops, refrigerators, or sinks shall be considered as separate counter top spaces. Receptacle outlets rendered not readily accessible by appliances fastened in place or appliances occupying dedicated space shall not be considered as these required outlets.

Receptacle outlets shall not be installed in a face-up position in the work surfaces or counter tops in a kitchen or dining area.

(FPN): The 24-inch (610-mm) dimension is measured along the wall line or centerline, and the intent is that there be a receptacle outlet for every 4 linear feet (1.2 m) or fraction thereof of counter length.

(d) Bathrooms. In dwelling units, at least one wall receptacle outlet shall be installed in the bathroom adjacent to each basin location. See Section 210-8(a)(1).

(e) Outdoor Outlets. For a one-family dwelling and each unit of a two-family dwelling that is at grade level, at least one receptacle outlet accessible at grade level shall be installed at the front and back of the dwelling. See Section 210-8(a)(3).

(f) Laundry Areas. In dwelling units, at least one receptacle outlet shall be installed for the laundry.

Exception No. 1: In a dwelling unit that is an apartment or living area in a multifamily building where laundry facilities are provided on the premises that are available to all building occupants, a laundry receptacle shall not be required.

Exception No. 2: In other than one-family dwellings where laundry facilities are not to be installed or permitted, a laundry receptacle shall not be required.

(g) Basements and Garages. For a one-family dwelling, at least one receptacle outlet, in addition to any provided for laundry equipment, shall be installed in each basement and in each attached garage, and in each detached garage with electric power. See Sections 210-8(a)(2) and (a)(4).

(h) Hallways. In dwelling units, hallways of 10 feet (3.05 m) or more in length shall have at least one receptacle outlet.

As used in this subsection, the hall length shall be considered the length along the centerline of the hall without passing through a doorway.

210-60. Guest Rooms. Guest rooms in hotels, motels, and similar occupancies shall have receptacle outlets installed in accordance with Section 210-52. See Section 210-8(b)(1).

Exception: In rooms of hotels and motels, the required number of receptacle outlets determined by Section 210-52(a) shall be permitted to be located convenient for permanent furniture layout.

210-62. Show Windows. At least one receptacle outlet shall be installed directly above a show window for each 12 linear feet (3.66 m) or major fraction thereof of show window area measured horizontally at its maximum width.

210-63. Rooftop Heating, Air-Conditioning, and Refrigeration Equipment Outlet. A 125-volt, single-phase, 15- or 20-ampere-rated receptacle outlet shall be installed at an accessible location for the servicing of heating, air-conditioning, and refrigeration equipment on rooftops and in attics and crawl spaces. The receptacle shall be located on the same level and within 25 feet (7.62 m) of the heating, air-conditioning, and refrigeration equipment. The receptacle outlet shall not be connected to the load side of the equipment disconnecting means.

Exception: Rooftop equipment on one- and two-family dwellings.

210-70. Lighting Outlets Required. Lighting outlets shall be installed where specified in Sections 210-70(a), (b), and (c) below.

(a) Dwelling Unit(s). At least one wall switch-controlled lighting outlet shall be installed in every habitable room; in bathrooms, hallways, stairways, attached garages, and detached garages with electric power; and at outdoor entrances or exits.

(FPN): A vehicle door in a garage is not considered as an outdoor entrance.

At least one lighting outlet controlled by a light switch located at the point of entry to the attic, underfloor space, utility room, and basement shall be installed where these spaces are used for storage or contain equipment requiring servicing. The lighting outlet shall be provided at or near the equipment requiring servicing.

Where lighting outlets are installed according to (a) above in interior stairways, there shall be a wall switch at each floor level to control the lighting outlet where the difference between floor levels is six steps or more.

Exception No. 1: In habitable rooms, other than kitchens and bathrooms, one or more receptacles controlled by a wall switch shall be permitted in lieu of lighting outlets.

Exception No. 2: In hallways, stairways, and at outdoor entrances, remote, central, or automatic control of lighting shall be permitted.

(b) Guest Rooms. At least one wall switch-controlled lighting outlet or wall switch-controlled receptacle shall be installed in guest rooms in hotels, motels, or similar occupancies.

(c) Other Locations. At least one wall switch-controlled lighting outlet shall be installed at or near equipment requiring servicing such as heating, air-conditioning, and refrigeration equipment in attics or underfloor spaces. The wall switch shall be located at the point of entry to the attic or underfloor space.

ARTICLE 215 — FEEDERS

215-1. Scope. This article covers the installation requirements and minimum size and ampacity of conductors for feeders supplying branch-circuit loads as computed in accordance with Article 220.

Exception: Feeders for electrolytic cells as covered in Section 668-3(c), Exception Nos. 1 and 4.

215-2. Minimum Rating and Size. Feeder conductors shall have an ampacity not less than required to supply the load as computed in Parts B, C, and D of Article 220. The minimum sizes shall be as specified in (a) and (b) below under the conditions stipulated. Feeder conductors for a dwelling unit or a mobile home need not be larger than service-entrance conductors. Article 310, Note 3, Notes to Ampacity Tables of 0 to 2000 Volts shall be permitted to be used for conductor size.

(a) For Specified Circuits. The ampacity of feeder conductors shall not be less than 30 amperes where the load supplied consists of any of the following number and types of circuits: (1) two or more 2-wire branch circuits supplied by a 2-wire feeder; (2) more than two 2-wire branch circuits supplied by a 3-wire feeder; (3) two or more 3-wire branch circuits supplied by a 3-wire feeder; or (4) two or more 4-wire branch circuits supplied by a 3-phase 4-wire feeder.

(b) Ampacity Relative to Service-Entrance Conductors. The feeder conductor ampacity shall not be less than that of the service-entrance conductors where the feeder conductors carry the total load supplied by service-entrance conductors with an ampacity of 55 amperes or less.

(FPN No. 1): See Examples 1 through 10 in Chapter 9.

(FPN No. 2): Conductors for feeders as defined in Article 100, sized to prevent a voltage drop exceeding 3 percent at the farthest outlet of power, heating, and lighting loads, or combinations of such loads, and where the maximum total voltage drop on both feeders and branch circuits to the farthest outlet does not exceed 5 percent, will provide reasonable efficiency of operation.

(FPN No. 3): See Section 210-19(a) for voltage drop for branch circuits.

215-3. Overcurrent Protection. Feeders shall be protected against overcurrent in accordance with the provisions of Part A of Article 240.

215-4. Feeders with Common Neutral.

(a) Feeders with Common Neutral. Feeders containing a common neutral shall be permitted to supply two or three sets of 3-wire feeders, or two sets of 4-wire or 5-wire feeders.

(b) In Metal Raceway or Enclosure. Where installed in a metal raceway or other metal enclosure, all conductors of all feeders using a common neutral shall be enclosed within the same raceway or other enclosure as required in Section 300-20.

215-5. Diagrams of Feeders. If required by the authority having jurisdiction, a diagram showing feeder details shall be provided prior to the installation of the feeders. Such a diagram shall show the area in square feet of the building or other structure supplied by each feeder, the total connected load before applying demand factors, the demand factors used, the computed load after applying demand factors, and the size and type of conductors to be used.

215-6. Feeder Conductor Grounding Means. Where a feeder supplies branch circuits in which equipment grounding conductors are required, the feeder shall include or provide a grounding means in accordance with the provisions of Section 250-57 to which the equipment grounding conductors of the branch circuits shall be connected.

215-7. Ungrounded Conductors Tapped from Grounded Systems. Two-wire dc circuits and ac circuits of two or more ungrounded conductors shall be permitted to be tapped from the ungrounded conductors of circuits having a grounded neutral conductor. Switching devices in each tapped circuit shall have a pole in each ungrounded conductor.

215-8. Means of Identifying Conductor with the Higher Voltage to Ground. On a 4-wire, delta-connected secondary where the midpoint of one phase is grounded to supply lighting and similar loads, the phase conductor having the higher voltage to ground shall be identified by an outer finish that is orange in color or by tagging or other effective means. Such identification shall be placed at each point where a connection is made if the grounded conductor is also present.

215-9. Ground-Fault Protection for Personnel. Feeders supplying 15- and 20-ampere receptacle branch circuits shall be permitted to be protected by a ground-fault circuit-interrupter in lieu of the provisions for such interrupters as specified in Section 210-8 and Article 305.

215-10. Ground-Fault Protection of Equipment. Ground-fault protection of equipment as specified by Section 230-95 shall be provided for a feeder disconnect rated 1000 amperes or more in a solidly grounded wye system with greater than 150 volts to ground, but not exceeding 600 volts phase-to-phase.

Exception: Feeder ground-fault protection of equipment shall not be required where ground-fault protection of equipment is provided on the supply side of the feeder.

ARTICLE 220 — BRANCH-CIRCUIT, FEEDER, AND SERVICE CALCULATIONS

A. General

220-1. Scope. This article provides requirements for determining the number of branch circuits required and for computing branch-circuit, feeder, and service loads.

Exception: Branch-circuit and feeder calculations for electrolytic cells as covered in Section 668-3(c), Exception Nos. 1 and 4.

220-2. Voltages. Unless other voltages are specified, for purposes of computing branch-circuit and feeder loads, nominal system voltages of 120, 120/240, 208Y/120, 240, 480Y/277, 480, and 600 volts shall be used.

220-3. Computation of Branch Circuits. Branch-circuit loads shall be computed as shown in (a) through (d) below.

(FPN): See Section 600-6(c) for exterior signs and outline lighting.

(a) Continuous and Noncontinuous Loads. The branch-circuit rating shall not be less than the noncontinuous load plus 125 percent of the continuous load.

Exception: Where the assembly, including overcurrent devices, is listed for continuous operation of 100 percent of its rating.

(b) Lighting Load for Listed Occupancies. A unit load of not less than that specified in Table 220-3(b) for occupancies listed therein shall constitute the minimum lighting load for each square foot (0.093 sq m) of floor area. The floor area for each floor shall be computed from the outside dimensions of the building, apartment, or other area involved. For dwelling unit(s), the computed floor area shall not include open porches, garages, or unused or unfinished spaces not adaptable for future use.

(FPN): The unit values herein are based on minimum load conditions and 100 percent power factor, and may not provide sufficient capacity for the installation contemplated.

Table 220-3(b). General Lighting Loads By Occupancies

Type of Occupancy	Unit Load per Square Foot (Volt-Amperes)
Armories and Auditoriums	1
Banks	$3\frac{1}{2}$**
Barber Shops and Beauty Parlors	3
Churches	1
Clubs	2
Court Rooms	2
Dwelling Units*	3
Garages — Commercial (storage)	$\frac{1}{2}$
Hospitals	2
Hotels and Motels, including apartment houses without provision for cooking by tenants*	2
Industrial Commercial (Loft) Buildings	2
Lodge Rooms	$1\frac{1}{2}$
Office Buildings	$3\frac{1}{2}$**
Restaurants	2
Schools	3
Stores	3
Warehouses (storage)	$\frac{1}{4}$
In any of the above occupancies except one-family dwellings and individual dwelling units of two-family and multifamily dwellings:	
Assembly Halls and Auditoriums	1
Halls, Corridors, Closets, Stairways	$\frac{1}{2}$
Storage Spaces	$\frac{1}{4}$

For SI units: one square foot = 0.093 square meter.

*All general-use receptacle outlets of 20-ampere or less rating in one-family, two-family, and multifamily dwellings and in guest rooms of hotels and motels [except those connected to the receptacle circuits specified in Sections 220-4(b) and (c)] shall be considered as outlets for general illumination, and no additional load calculations shall be required for such outlets.

**In addition, a unit load of 1 volt-ampere per square foot shall be included for general-purpose receptacle outlets where the actual number of general-purpose receptacle outlets is unknown.

(c) Other Loads — All Occupancies. In all occupancies, the minimum load for each outlet for general-use receptacles and outlets not used for general illumination shall be not less than the following, the loads shown being based on nominal branch-circuit voltages.

(1) Outlet for a specific appliance or other load except for a motor load .... ampere rating of appliance or load served.

(2) Outlet for motor load .... See Sections 430-22 and 430-24 and Article 440.

(3) An outlet supplying recessed lighting fixture(s) shall be the maximum volt-ampere rating of the equipment and lamps for which the fixture(s) is rated.

(4) Outlet for heavy-duty lampholder .... 600 volt-amperes.

(5) Track Lighting. See Section 410-102.

(6) Other outlets* ....180 volt-amperes per outlet.

For receptacle outlets, each single or each multiple receptacle on one strap shall be considered at not less than 180 volt-amperes.

*This provision shall not be applicable to receptacle outlets connected to the circuit specified in Section 220-4(b).

Exception No. 1: Where fixed multioutlet assemblies are employed, each 5 feet (1.52 m) or fraction thereof of each separate and continuous length shall be considered as one outlet of not less than 180 volt-amperes capacity; except in locations where a number of appliances are likely to be used simultaneously, each 1 foot (305 mm) or fraction thereof shall be considered as an outlet of not less than 180 volt-amperes. The requirements of this exception shall not apply to dwelling unit(s) or the guest rooms of hotels or motels.

Exception No. 2: Table 220-19 shall be permitted for computing the load of household electric ranges.

Exception No. 3: A load of not less than 200 volt-amperes per linear foot (305 mm) of show window, measured horizontally along its base, shall be permitted instead of the specified unit load per outlet.

Exception No. 4: The loads of outlets serving switchboards and switching frames in telephone exchanges shall be waived from the computations.

Exception No. 5: Section 220-18 shall be considered as a permitted method of computing the load for a household electric clothes dryer.

(d) Loads for Additions to Existing Installations.

(1) Dwelling Units. Loads for structural additions to an existing dwelling unit or to a previously unwired portion of an existing dwelling unit, either of which exceeds 500 square feet (46.5 sq m), shall be computed in accordance with (b) above. Loads for new circuits or extended circuits in previously wired dwelling units shall be computed in accordance with either (b) or (c) above.

(2) Other than Dwelling Units. Loads for new circuits or extended circuits in other than dwelling units shall be computed in accordance with either (b) or (c) above.

220-4. Branch Circuits Required. Branch circuits for lighting and for appliances, including motor-operated appliances, shall be provided to supply the loads computed in accordance with Section 220-3. In addition, branch circuits shall be provided for specific loads not covered by Section 220-3 where required elsewhere in this Code; for small appliance loads as specified in (b) below; and for laundry loads as specified in (c) below.

(a) Number of Branch Circuits. The minimum number of branch circuits shall be determined from the total computed load and the size or rating of the circuits used. In all installations, the number of circuits shall be sufficient to supply the load served. In no case shall the load on any circuit exceed the maximum specified by Section 210-22.

(b) Small Appliance Branch Circuits — Dwelling Unit. In addition to the number of branch circuits determined in accordance with (a) above, two or more 20-ampere small appliance branch circuits shall be provided for all receptacle outlets specified by Section 210-52 for the small appliance loads.

(c) Laundry Branch Circuits — Dwelling Unit. In addition to the number of branch circuits determined in accordance with (a) and (b) above, at least one additional 20-ampere branch circuit shall be provided to supply the laundry receptacle outlet(s) required by Section 210-52(f). This circuit shall have no other outlets.

(d) Load Evenly Proportioned Among Branch Circuits. Where the load is computed on a volt-amperes-per-square-foot (0.093-sq m) basis, the wiring system up to and including the branch-circuit panelboard(s) shall be provided to serve not less than the calculated load. This load shall be evenly proportioned among multioutlet branch circuits within the panelboard(s). Branch-circuit overcurrent devices and circuits need only be installed to serve the connected load.

(FPN): See Examples 1(a), 1(b), 2(b), and 4(a), Chapter 9.

B. Feeders

220-10. General.

(a) Ampacity and Computed Loads. Feeder conductors shall have sufficient ampacity to supply the load served. In no case shall the computed load of a feeder be less than the sum of the loads on the branch circuits supplied as determined by Part A of this article after any applicable demand factors permitted by Parts B, C, or D have been applied.

(FPN): See Examples 1 through 10, Chapter 9. See Section 210-22(b) for maximum load in amperes permitted for lighting units operating at less than 100 percent power factor.

(b) Continuous and Noncontinuous Loads. Where a feeder supplies continuous loads or any combination of continuous and noncontinuous loads, the rating of the overcurrent device shall not be less than the noncontinuous load plus 125 percent of the continuous load.

Exception: Where the assembly including the overcurrent devices protecting the feeder(s) are listed for operation at 100 percent of their rating, neither the ampere rating of the overcurrent device nor the ampacity of the feeder conductors shall be less than the sum of the continuous load plus the noncontinuous load.

220-11. General Lighting. The demand factors listed in Table 220-11 shall apply to that portion of the total branch-circuit load computed for general illumination. They shall not be applied in determining the number of branch circuits for general illumination.

(FPN): See Section 220-16 for application of demand factors to small appliance and laundry loads in dwellings.

Table 220-11.   Lighting Load Feeder Demand Factors

Type of Occupancy	Portion of Lighting Load to which Demand Factor Applies (volt-amperes)	Demand Factor Percent
Dwelling Units	First 3000 or less at	100
	From 3001 to 120,000 at	35
	Remainder over 120,000 at	25
Hospitals*	First 50,000 or less at	40
	Remainder over 50,000 at	20
Hotels and Motels — Including Apartment Houses without Provision for Cooking by Tenants*	First 20,000 or less at	50
	From 20,001 to 100,000 at	40
	Remainder over 100,000 at	30
Warehouses	First 12,500 or less at	100
(Storage)	Remainder over 12,500 at	50
All Others	Total volt-amperes	100

*The demand factors of this table shall not apply to the computed load of feeders to areas in hospitals, hotels, and motels where the entire lighting is likely to be used at one time, as in operating rooms, ballrooms, or dining rooms.

220-12. Show-Window Lighting. For show-window lighting, a load of not less than 200 volt-amperes shall be included for each linear foot (305 mm) of show window, measured horizontally along its base.

(FPN): See Section 220-3(c), Exception No. 3, for branch circuits supplying show windows.

220-13. Receptacle Loads — Nondwelling Units. In other than dwelling units, receptacle loads computed at not more than 180 volt-amperes per outlet in accordance with Section 220-3(c)(6) shall be permitted to be added to the lighting loads and made subject to the demand factors given in Table 220-11, or they shall be permitted to be made subject to the demand factors given in Table 220-13.

Table 220-13.
Demand Factors for Nondwelling Receptacle Loads

Portion of Receptacle Load to which Demand Factor Applies (volt-amperes)	Demand Factor Percent
First 10 kVA or less	100
Remainder over 10 kVA at	50

220-14. Motors. Motor loads shall be computed in accordance with Sections 430-24, 430-25, and 430-26.

220-15. Fixed Electric Space Heating. Fixed electric space heating loads shall be computed at 100 percent of the total connected load; however, in no case shall a feeder load current rating be less than the rating of the largest branch circuit supplied.

Exception No. 1: Where reduced loading of the conductors results from units operating on duty-cycle, intermittently, or from all units not operating at one time, the authority having jurisdiction may grant permission for feeder conductors to have an ampacity less than 100 percent, provided the conductors have an ampacity for the load so determined.

Exception No. 2: The use of the optional calculations in Sections 220-30 and 220-31 shall be permitted for fixed electric space heating loads in a dwelling unit. In a multifamily dwelling, the use of the optional calculation in Section 220-32 shall be permitted.

220-16. Small Appliance and Laundry Loads — Dwelling Unit.

(a) Small Appliance Circuit Load. In each dwelling unit, the feeder load shall be computed at 1500 volt-amperes for each 2-wire small appliance branch circuit required by Section 220-4(b) for small appliances supplied by 15- or 20-ampere receptacles on 20-ampere branch circuits in the kitchen, pantry, dining room, and breakfast room. Where the load is subdivided through two or more feeders, the computed load for each shall include not less than 1500 volt-amperes for each 2-wire branch circuit for small appliances. These loads shall be permitted to be included with the general lighting load and subjected to the demand factors permitted in Table 220-11 for the general lighting load.

(b) Laundry Circuit Load. A feeder load of not less than 1500 volt-amperes shall be included for each 2-wire laundry branch circuit installed as required by Section 220-4(c). It shall be permissible to include this load with the general lighting load and subject it to the demand factors provided in Section 220-11.

220-17. Appliance Load — Dwelling Unit(s). It shall be permissible to apply a demand factor of 75 percent to the nameplate-rating load of four or more appliances fastened in place that are served by the same feeder in a | one-family, two-family, or multifamily dwelling.

Exception: This demand factor shall not be applied to electric ranges, clothes dryers, space heating equipment, or air-conditioning equipment.

220-18. Electric Clothes Dryers — Dwelling Unit(s). The load for household electric clothes dryers in a dwelling unit(s) shall be 5000 watts (volt-amperes) or the nameplate rating, whichever is larger, for each dryer served. The use of the demand factors in Table 220-18 shall be permitted.

Table 220-18.
Demand Factors for Household Electric Clothes Dryers

Number of Dryers	Demand Factor Percent
1	100
2	100
3	100
4	100
5	80
6	70
7	65
8	60
9	55
10	50
11-13	45
14-19	40
20-24	35
25-29	32.5
30-34	30
35-39	27.5
40 and over	25

220-19. Electric Ranges and Other Cooking Appliances — Dwelling Unit(s). The feeder demand load for household electric ranges, wall-mounted ovens, counter-mounted cooking units, and other household cooking appliances individually rated in excess of 1¾ kW shall be permitted to be computed in accordance with Table 220-19. Where two or more single-phase ranges are supplied by a 3-phase, 4-wire feeder, the total load shall be computed on the basis of twice the maximum number connected between any two phases. kVA shall be considered equivalent to kW for loads computed under this section.

(FPN): See Example 5(a), Chapter 9.

220-20. Kitchen Equipment — Other than Dwelling Unit(s). It shall be permissible to compute the load for commercial electric cooking equipment, dishwasher booster heaters, water heaters, and other kitchen equipment in accordance with Table 220-20. These demand factors shall be applied to all equipment that has either thermostatic control or intermittent use as kitchen equipment. They shall not apply to space heating, ventilating, or air-conditioning equipment.

However, in no case shall the feeder demand be less than the sum of the largest two kitchen equipment loads.

Table 220-19. Demand Loads for Household Electric Ranges, Wall-Mounted Ovens, Counter-Mounted Cooking Units, and Other Household Cooking Appliances over 1¾ kW Rating. Column A to be used in all cases except as otherwise permitted in Note 3 below

Number of Appliances	Maximum Demand (See Notes)	Demand Factors Percent (See Note 3)	
	Column A (Not over 12 kW Rating)	Column B (Less than 3½ kW Rating)	Column C (3½ kW to 8¾ kW Rating)
1	8 kW	80%	80%
2	11 kW	75%	65%
3	14 kW	70%	55%
4	17 kW	66%	50%
5	20 kW	62%	45%
6	21 kW	59%	43%
7	22 kW	56%	40%
8	23 kW	53%	36%
9	24 kW	51%	35%
10	25 kW	49%	34%
11	26 kW	47%	32%
12	27 kW	45%	32%
13	28 kW	43%	32%
14	29 kW	41%	32%
15	30 kW	40%	32%
16	31 kW	39%	28%
17	32 kW	38%	28%
18	33 kW	37%	28%
19	34 kW	36%	28%
20	35 kW	35%	28%
21	36 kW	34%	26%
22	37 kW	33%	26%
23	38 kW	32%	26%
24	38 kW	31%	26%
25	40 kW	30%	26%
26-30	15 kW plus 1 kW	30%	24%
31-40	for each range	30%	22%
41-50	25 kW plus ¾	30%	20%
51-60	kW for each	30%	18%
61 and over	range	30%	16%

Note 1. Over 12 kW through 27 kW ranges all of same rating. For ranges individually rated more than 12 kW but not more than 27 kW, the maximum demand in Column A shall be increased 5 percent for each additional kW of rating or major fraction thereof by which the rating of individual ranges exceeds 12 kW.

Note 2. Over 8¾ kW through 27 kW ranges of unequal ratings. For ranges individually rated more than 8¾ kW and of different ratings, but none exceeding 27 kW, an average value of rating shall be computed by adding together the ratings of all ranges to obtain the total connected load (using 12 kW for any range rated less than 12 kW) and dividing by the total number of ranges; and then the maximum demand in Column A shall be increased 5 percent for each kW or major fraction thereof by which this average value exceeds 12 kW.

Note 3. Over 1¾ kW through 8¾ kW. In lieu of the method provided in Column A, it shall be permissible to add the nameplate ratings of all ranges rated more than 1¾ kW but not more than 8¾ kW and multiply the sum by the demand factors specified in Column B or C for the given number of appliances.

Note 4. Branch-Circuit Load. It shall be permissible to compute the branch-circuit load for one range in accordance with Table 220-19. The branch-circuit load for one wall-mounted oven or one counter-mounted cooking unit shall be the nameplate rating of the appliance. The branch-circuit load for a counter-mounted cooking unit and not more than two wall-mounted ovens, all supplied from a single branch circuit and located in the same room, shall be computed by adding the nameplate rating of the individual appliances and treating this total as equivalent to one range.

Note 5. This table also applies to household cooking appliances rated over 1¾ kW and used in instructional programs.

(FPN No. 1):   See Table 220-20 for commercial cooking equipment.

(FPN No. 2):   See Examples, Chapter 9.

Table 220-20.
Feeder Demand Factors for Kitchen Equipment—
Other than Dwelling Unit(s)

Number of Units of Equipment	Demand Factors Percent
1	100
2	100
3	90
4	80
5	70
6 and over	65

220-21. Noncoincident Loads. Where it is unlikely that two dissimilar loads will be in use simultaneously, it shall be permissible to omit the smaller of the two in computing the total load of a feeder.

220-22. Feeder Neutral Load. The feeder neutral load shall be the maximum unbalance of the load determined by this article. The maximum unbalanced load shall be the maximum net computed load between the neutral and any one ungrounded conductor, except that the load thus obtained shall be multiplied by 140 percent for 3-wire, 2-phase or 5-wire, 2-phase systems. For a feeder supplying household electric ranges, wall-mounted ovens, counter-mounted cooking units, and electric dryers, the maximum unbalanced load shall be considered as 70 percent of the load on the ungrounded conductors, as determined in accordance with Table 220-19 for ranges and Table 220-18 for dryers. For 3-wire dc or single-phase ac, 4-wire, 3-phase, 3-wire, 2-phase, or 5-wire, 2-phase systems, a further demand factor of 70 percent shall be permitted for that portion of the unbalanced load in excess of 200 amperes. There shall be no reduction of the neutral capacity for that portion of the load that consists of electric-discharge lighting, electronic computer/data processing, or similar equipment, and supplied from a 4-wire, wye-connected, 3-phase system nor the grounded conductor of a 3-wire circuit consisting of two phase wires and the neutral of a 4-wire, 3-phase wye-connected system.

(FPN No. 1):   See Examples 1(a), 1(b), 2(b), 4(a), and 5(a), Chapter 9.

(FPN No. 2):   A 3-phase, 4-wire power system used to supply power to computer systems or other similar electronic loads may necessitate that the power system design allow for the possibility of high harmonic neutral currents.

C. Optional Calculations for Computing Feeder and Service Loads

220-30. Optional Calculation — Dwelling Unit.

(a) Feeder and Service Load. For a dwelling unit having the total connected load served by a single 3-wire, 120/240-volt or 208Y/120-volt set of service-entrance or feeder conductors with an ampacity of 100 or greater, it shall be permissible to compute the feeder and service loads in accordance with Table 220-30 instead of the method specified in Part B of this article. Feeder and service-entrance conductors whose demand load is determined by this optional calculation shall be permitted to have the neutral load determined by Section 220-22.

(b) Loads. The loads identified in Table 220-30 as "other load" and as "remainder of other load" shall include the following:

(1) 1500 volt-amperes for each 2-wire, 20-ampere small appliance branch circuit and each laundry branch circuit specified in Section 220-16.

(2) 3 volt-amperes per square foot (0.093 sq m) for general lighting and general-use receptacles.

(3) The nameplate rating of all appliances that are fastened in place, permanently connected, or located to be on a specific circuit, ranges, wall-mounted ovens, counter-mounted cooking units, clothes dryers, and water heaters.

(4) The nameplate ampere or kVA rating of all motors and of all low-power-factor loads.

Table 220-30.
Optional Calculation for Dwelling Unit
Load in kVA

Largest of the following five selections.

(1) 100 percent of the nameplate rating(s) of the air conditioning and cooling, including heat pump compressors.

(2) 100 percent of the nameplate ratings of electric thermal storage and other heating systems where the usual load is expected to be continuous at the full nameplate value. Systems qualifying under this selection shall not be figured under any other selection in this table.

(3) 65 percent of the nameplate rating(s) of the central electric space heating, including integral supplemental heating in heat pumps.

(4) 65 percent of the nameplate rating(s) of electric space heating if less than four separately controlled units.

(5) 40 percent of the nameplate rating(s) of electric space heating of four or more separately controlled units.

Plus: 100 percent of the first 10 kVA of all other load.

40 percent of the remainder of all other load.

220-31. Optional Calculation for Additional Loads in Existing Dwelling Unit. For an existing dwelling unit presently being served by an existing 120/240-volt or 208Y/120, 3-wire service, it shall be permissible to compute load calculations as follows:

Load (in kVA)	Percent of Load
First 8 kVA of load at	100%
Remainder of load at	40%

Load calculation shall include lighting at 3 volt-amperes per square foot (0.093 sq m); 1500 volt-amperes for each 20-ampere appliance circuit; range or wall-mounted oven and counter-mounted cooking unit; and other appliances that are permanently connected or fastened in place, at nameplate rating.

If air-conditioning equipment or electric space heating equipment is to be installed, the following formula shall be applied to determine if the existing service is of sufficient size.

Air-conditioning equipment*	100%
Central electric space heating*	100%
Less than four separately controlled space heating units*	100%
First 8 kVA of all other load	100%
Remainder of all other load	40%

Other loads shall include:

1500 volt-amperes for each 20-ampere appliance circuit.

Lighting and portable appliances at 3 volt-amperes per square foot (0.093 sq m).

Household range or wall-mounted oven and counter-mounted cooking unit.

All other appliances fastened in place, including four or more separately controlled space heating units, at nameplate rating.

*Use larger connected load of air conditioning and space heating, but not both.

220-32. Optional Calculation — Multifamily Dwelling.

(a) Feeder or Service Load. It shall be permissible to compute the feeder or service load of a multifamily dwelling in accordance with Table 220-32 instead of Part B of this article where all the following conditions are met:

(1) No dwelling unit is supplied by more than one feeder.

(2) Each dwelling unit is equipped with electric cooking equipment.

Exception: When the computed load for multifamily dwellings without electric cooking in Part B of this article exceeds that computed under Part C for the identical load plus electric cooking (based on 8 kW per unit), the lesser of the two loads shall be permitted to be used.

(3) Each dwelling unit is equipped with either electric space heating or air conditioning or both.

Feeders and service-entrance conductors whose demand load is determined by this optional calculation shall be permitted to have the neutral load determined by Section 220-22.

(b) House Loads. House loads shall be computed in accordance with Part B of this article and shall be in addition to the dwelling unit loads computed in accordance with Table 220-32.

(c) Connected Loads. The connected load to which the demand factors of Table 220-32 apply shall include the following:

(1) 1500 volt-amperes for each 2-wire, 20-ampere small appliance branch circuit and each laundry branch circuit specified in Section 220-16.

(2) 3 volt-amperes per square foot (0.093 sq m) for general lighting and general-use receptacles.

(3) The nameplate rating of all appliances that are fastened in place, permanently connected or located to be on a specific circuit, ranges, wall-mounted ovens, counter-mounted cooking units, clothes dryers, water heaters, and space heaters.

Table 220-32.
Optional Calculation—Demand Factors for Three or More
Multifamily Dwelling Units

Number of Dwelling Units	Demand Factor Percent
3–5	45
6–7	44
8–10	43
11	42
12–13	41
14–15	40
16–17	39
18–20	38
21	37
22–23	36
24–25	35
26–27	34
28–30	33
31	32
32–33	31
34–36	30
37–38	29
39–42	28
43–45	27
46–50	26
51–55	25
56–61	24
62 and over	23

If water heater elements are so interlocked that all elements cannot be used at the same time, the maximum possible load shall be considered the nameplate load.

(4) The nameplate ampere or kVA rating of all motors and of all low-power-factor loads.

(5) The larger of the air-conditioning load or the space heating load.

220-33. Optional Calculation — Two Dwelling Units. Where two dwelling units are supplied by a single feeder and the computed load under Part B of this article exceeds that for three identical units computed under Section 220-32, the lesser of the two loads shall be permitted to be used.

220-34. Optional Method — Schools. The calculation of a feeder or service load for schools shall be permitted in accordance with Table 220-34 in lieu of Part B of this article where equipped with electric space heating or air conditioning, or both. The connected load to which the demand factors of Table 220-34 apply shall include all of the interior and exterior lighting, power, water heating, cooking, other loads, and the larger of the air-conditioning load or space heating load within the building or structure.

Feeders and service-entrance conductors whose demand load is determined by this optional calculation shall be permitted to have the neutral load determined by Section 220-22. Where the building or structure load is calculated by this optional method, feeders within the building or structure shall have ampacity as permitted in Part B of this article; however, the ampacity of an individual feeder need not be larger than the ampacity for the entire building.

This section shall not apply to portable classroom buildings.

Table 220-34.
Optional Method—Demand Factors for Feeders
and Service-Entrance Conductors for Schools

Connected Load Volt-Amperes per Square Foot	Demand Factors Percent
Connected load up to and including 3, plus	100
Conected load over 3 and including 20, plus	75
Connected load over 20 at	25

For SI units: one square foot = 0.093 square meter.

220-35. Optional Calculations for Additional Loads to Existing Installations. For the purpose of allowing additional loads to be connected to existing feeders and services, it shall be permitted to use actual maximum kVA demand figures to determine the existing load on a service or feeder when all the following conditions are met:

(1) The maximum demand data is available in kVA for a minimum of a one-year period.

(2) The existing demand at 125 percent plus the new load does not exceed the ampacity of the feeder or rating of the service.

(3) The feeder has overcurrent protection in accordance with Sections 240-3, and the service has overload protection in accordance with Section 230-90.

220-36. Optional Calculation — New Restaurants. Calculation of a service load or feeder, where the feeder serves the total load, for a new restaurant shall be permitted in accordance with Table 220-36 in lieu of Part B of this article.

The overload protection of the service-entrance conductors shall be in accordance with Sections 230-90 and 240-3.

Feeder conductors shall not be required to be of greater ampacity than | the service-entrance conductors.

Service-entrance or feeder conductors whose demand load is determined by this optional calculation shall be permitted to have the neutral load determined by Section 220-22.

Table 220-36.
Optional Method—Demand Factors for
Service-Entrance and Feeder Conductors for New Restaurants

Connected Load (kVA)	All Electric Demand Factor (Percent)	Not All Electric Demand Factor (Percent)
0-325	80	100
Remainder over 325	50	70

D. Method for Computing Farm Loads

220-40. Farm Loads — Buildings and Other Loads.

(a) Dwelling Unit. The feeder or service load of a farm dwelling unit shall be computed in accordance with the provisions for dwellings in Part B or C of this article. Where the dwelling has electric heat and the farm has electric grain drying systems, Part C of this article shall not be used to compute the dwelling load.

(b) Other than Dwelling Unit. For each farm building or load supplied by two or more branch circuits, the load for feeders, service-entrance conductors, and service equipment shall be computed in accordance with demand factors not less than indicated in Table 220-40.

(FPN): See Section 230-21 for overhead conductors from a pole to a building or other structure.

220-41. Farm Loads — Total. The total load of the farm for service-entrance conductors and service equipment shall be computed in accordance with the farm dwelling unit load and demand factors specified in

Table 220-40.
Method for Computing Farm Loads for Other than Dwelling Unit

Ampere Load at 240 Volts	Demand Factor Percent
Loads expected to operate without diversity, but not less than 125 percent full-load current of the largest motor and not less than the first 60 amperes of load	100
Next 60 amperes of all other loads	50
Remainder of other load	25

Table 220-41. Where there is equipment in two or more farm equipment buildings or for loads having the same function, such loads shall be computed in accordance with Table 220-40 and shall be permitted to be combined as a single load in Table 220-41 for computing the total load.

(FPN): See Section 230-21 for overhead conductors from a pole to a building or other structure.

Table 220-41.
Method for Computing Total Farm Load

Individual Loads Computed in Accordance with Table 220-40	Demand Factor Percent
Largest load	100
Second largest load	75
Third largest load	65
Remaining loads	50

To this total load, add the load of the farm dwelling unit computed in accordance with Part B or C of this article. Where the dwelling has electric heat and the farm has electric grain drying systems, Part C of this article shall not be used to compute the dwelling load.

ARTICLE 225 — OUTSIDE BRANCH CIRCUITS AND FEEDERS

225-1. Scope. This article covers requirements for outside branch circuits and feeders run between buildings, structures, or poles on the premises; and electric equipment and wiring for the supply of utilization equipment that is located on or attached to the outside of buildings, structures, or poles.

Exception: Outside branch circuits and feeders for electrolytic cells as covered in Section 668-3(c), Exception Nos. 1 and 4.

(FPN): For additional information on wiring over 600 volts, see National Electrical Safety Code, ANSI C2-1990.

225-2. Other Articles. Application of other articles, including additional requirements to specific cases of equipment and conductors, is as follows:

225-3. Calculation of Load.

(a) Branch Circuits. The load on outdoor branch circuits shall be as determined by Section 220-3.

(b) Feeders. The load on outdoor feeders shall be as determined by Part B of Article 220.

225-4. Conductor Covering.
Where within 10 feet (3.05 m) of any building or other structure, open wiring on insulators shall be insulated or covered. Conductors in cables or raceways, except Type MI cable, shall be of the rubber-covered type or thermoplastic type and in wet locations shall comply with Section 310-8. Conductors for festoon lighting shall be of the rubber-covered or thermoplastic type.

Exception: Equipment grounding conductors and grounded circuit conductors where permitted to be bare or covered elsewhere in this Code.

225-5. Size of Conductors.
The ampacity of outdoor branch-circuit and feeder conductors shall be in accordance with Section 310-15 based on loads as determined under Section 220-3 and Part B of Article 220.

225-6. Minimum Size of Conductor.

(a) Overhead Spans. Open individual conductors shall not be smaller than the following:

(1) For 600 volts, nominal, or less, No. 10 copper or No. 8 aluminum for spans up to 50 feet (15.2 m) in length and No. 8 copper or No. 6 aluminum for a longer span.

Exception: Where supported by a messenger wire.

(2) For over 600 volts, nominal, No. 6 copper or No. 4 aluminum where open individual conductors and No. 8 copper or No. 6 aluminum where in cable.

(b) Festoon Lighting. Overhead conductors for festoon lighting shall not be smaller than No. 12.

Exception: Where supported by messenger wires.

(FPN): See Section 225-24 for outdoor lampholders.

Definition. Festoon lighting is a string of outdoor lights suspended between two points more than 15 feet (4.57 m) apart.

225-7. Lighting Equipment Installed Outdoors.

(a) General. For the supply of lighting equipment installed outdoors, the branch circuits shall comply with Article 210 and (b) through (d) below.

(b) Common Neutral. The ampacity of the neutral conductor shall not be less than the maximum net computed load current between the neutral and all ungrounded conductors connected to any one phase of the circuit.

(c) 277 Volts to Ground. Circuits exceeding 120 volts, nominal, between conductors and not exceeding 277 volts, nominal, to ground shall be permitted to supply lighting fixtures for illumination of outdoor areas of industrial establishments, office buildings, schools, stores, and other commercial or public buildings where the fixtures are not less than 3 feet (914 mm) from windows, platforms, fire escapes, and the like.

(d) 600 Volts Between Conductors. Circuits exceeding 277 volts, nominal, to ground and not exceeding 600 volts, nominal, between conductors shall be permitted to supply the auxiliary equipment of electric-discharge lamps in accordance with Section 210-6(d)(1).

225-8. Disconnection.

(a) Disconnecting Means. The disconnecting means for branch-circuit and feeder fuses shall be in accordance with Section 240-40.

(b) Disconnect Required for Each. Where more than one building or other structure is on the same property and under single management, each building or other structure served shall be provided with means for disconnecting all ungrounded conductors.

The disconnecting means shall be installed either inside or outside of a building or structure at a readily accessible location nearest the point of entrance of the supply conductors.

Disconnects shall be installed in accordance with the requirements of Sections 230-71 and 230-72.

Exception No. 1: For large capacity multibuilding industrial installations under single management, where it is assured that the disconnecting can be accomplished by establishing and maintaining safe switching procedures, the disconnecting means shall be permitted to be located elsewhere on the premises.

Exception No. 2: Buildings or other structures qualifying under the provisions of Article 685.

(c) Suitable for Service Equipment. The disconnecting means specified in (b) above shall be suitable for use as service equipment.

Exception: For garages and outbuildings on residential property, a snap switch or a set of 3-way or 4-way snap switches suitable for use on branch circuits shall be permitted as the disconnecting means.

225-9. Overcurrent Protection.   Overcurrent protection shall be in accordance with Section 210-20 for branch circuits and Article 240 for feeders.

(FPN):   See Section 240-24 for requirements for access to overcurrent devices.

225-10. Wiring on Buildings.   The installation of outside wiring on surfaces of buildings shall be permitted for circuits of not over 600 volts, nominal, as open wiring on insulators, as multiconductor cable, as Type MC cable, as Type MI cable, as messenger supported wiring, in rigid metal conduit, in intermediate metal conduit, in rigid nonmetallic conduit, in cable trays, as cablebus, in wireways, in auxiliary gutters, in electrical metallic tubing, in flexible metal conduit, in liquidtight flexible metal conduit, liquidtight flexible nonmetallic conduit, and in busways. Circuits of over 600 volts, nominal, shall be installed as provided in Section 710-4. Circuits for signs and outline lighting shall be installed in accordance with Article 600.

225-11. Circuit Exits and Entrances.   Where outside branch and feeder circuits leave or enter a building, the requirements of Sections 230-52 and 230-54 shall apply.

225-12. Open-Conductor Supports.   Open conductors shall be supported on glass or porcelain knobs, racks, brackets, or strain insulators.

225-13. Festoon Supports.   In spans exceeding 40 feet (12.2 m), the conductors shall be supported by a messenger wire; and the messenger wire shall be supported by strain insulators. Conductors or messenger wires shall not be attached to any fire escape, downspout, or plumbing equipment.

225-14. Open-Conductor Spacings.

(a) 600 Volts, Nominal, or Less. Conductors of 600 volts, nominal, or less, shall comply with the spacings provided in Table 230-51(c).

(b) Over 600 Volts, Nominal. Conductors of over 600 volts, nominal, shall comply with the spacings provided in Part D of Article 710.

(c) Separation from Other Circuits. Open conductors shall be separated from open conductors of other circuits or systems by not less than 4 inches (102 mm).

(d) Conductors on Poles. Conductors on poles shall have a separation of not less than 1 foot (305 mm) where not placed on racks or brackets. Conductors supported on poles shall provide a horizontal climbing space not less than the following:

Power conductors, below communication conductors	30 inches (762 mm)
Power conductors alone or above communication conductors:	
300 volts or less	24 inches (610 mm)
Over 300 volts..	30 inches (762 mm)
Communication conductors below power conductors..	same as power conductors
Communication conductors alone	no requirement

225-15. Supports Over Buildings. Supports over a building shall be in accordance with Section 230-29.

225-16. Point of Attachment to Buildings. The point of attachment to a building shall be in accordance with Section 230-26.

225-17. Means of Attachment to Buildings. The means of attachment to a building shall be in accordance with Section 230-27.

225-18. Clearance from Ground. Overhead spans of open conductors and open multiconductor cables of not over 600 volts, nominal, shall conform to the following:

10 feet (3.05 m) — above finished grade, sidewalks, or from any platform or projection from which they might be reached where the supply conductors are limited to 150 volts to ground and accessible to pedestrians only.

12 feet (3.66 m) — over residential property and driveways, and those commercial areas not subject to truck traffic where the voltage is limited to 300 volts to ground.

15 feet (4.57 m) — for those areas listed in the 12-foot (3.66-m) classification where the voltage exceeds 300 volts to ground.

18 feet (5.49 m) — over public streets, alleys, roads, parking areas subject to truck traffic, driveways on other than residential property, and other land traversed by vehicles such as cultivated, grazing, forest, and orchard.

(FPN): Note: For clearances of conductors of over 600 volts, see National Electrical Safety Code, ANSI C2-1990.

225-19. Clearances from Buildings for Conductors of Not Over 600 Volts, Nominal.

(a) Above Roofs. Overhead spans of open conductors and open multiconductor cables shall have a vertical clearance of not less than 8 feet

(2.44 m) above the roof surface. The vertical clearance above the roof level shall be maintained for a distance not less than 3 feet (914 mm) in all directions from the edge of the roof.

Exception No. 1: The area above a roof surface subject to pedestrian or vehicular traffic shall have a vertical clearance from the roof surface in accordance with the clearance requirements of Section 225-18.

Exception No. 2: Where the voltage between conductors does not exceed 300 and the roof has a slope of not less than 4 inches (102 mm) in 12 inches (305 mm), a reduction in clearance to 3 feet (914 mm) shall be permitted.

Exception No. 3: Where the voltage between conductors does not exceed 300, a reduction in clearance above only the overhanging portion of the roof to not less than 18 inches (457 mm) shall be permitted if (1) not more than 4 feet (1.22 m) of the conductors pass above the roof overhang, and (2) they are terminated at a through-the-roof raceway or approved support.

Exception No. 4: The requirement for maintaining the vertical clearance 3 feet (914 mm) from the edge of the roof shall not apply to the final conductor span where the conductors are attached to the side of a building.

(b) From Nonbuilding or Nonbridge Structures. From signs, chimneys, radio and television antennas, tanks, other nonbuilding or nonbridge structures, clearances, vertical, diagonal and horizontal, shall not be less than 3 feet (914 mm).

(c) Horizontal Clearances. Clearances shall not be less than 3 feet (914 mm).

(d) Final Spans. Final spans of feeders or branch circuits to a building they supply or from which they are fed shall be permitted to be attached to the building, but they shall be kept 3 feet (914 mm) from windows, doors, porches, fire escapes, or similar locations.

Exception: Conductors run above the top level of a window shall be permitted to be less than the 3 feet (914 mm) requirement above.

Overhead branch-circuit and feeder conductors shall not be installed beneath openings through which materials may be moved, such as openings in farm and commercial buildings, and shall not be installed where they will obstruct entrance to these building openings.

(e) Zone for Fire Ladders. Where buildings exceed three stories or 50 feet (15.2 m) in height, overhead lines shall be arranged, where practicable, so that a clear space (or zone) at least 6 feet (1.83 m) wide will be left either adjacent to the buildings or beginning not over 8 feet (2.44 m) from them to facilitate the raising of ladders when necessary for fire fighting.

(FPN): Note: For clearance of conductors over 600 volts, see National Electrical Safety Code, ANSI C2-1990.

225-20. Mechanical Protection of Conductors. Mechanical protection of conductors on buildings, structures, or poles shall be as provided for services in Section 230-50.

225-21. Multiconductor Cables on Exterior Surfaces of Buildings. Supports for multiconductor cables on exterior surfaces of buildings shall be as provided in Section 230-51.

225-22. Raceways on Exterior Surfaces of Buildings. Raceways on exterior surfaces of buildings shall be raintight and arranged to drain.

Exception: As permitted in Section 350-2.

225-23. Underground Circuits. Underground circuits shall meet the requirements of Section 300-5.

225-24. Outdoor Lampholders. Where outdoor lampholders are attached as pendants, the connections to the circuit wires shall be staggered. Where such lampholders have terminals of a type that puncture the insulation and make contact with the conductors, they shall be attached only to conductors of the stranded type.

225-25. Location of Outdoor Lamps. Locations of lamps for outdoor lighting shall be below all energized conductors, transformers, or other electric utilization equipment.

Exception No. 1: Where clearances or other safeguards are provided for relamping operations.

Exception No. 2: Where equipment is controlled by a disconnecting means that can be locked in the open position.

| **225-26. Vegetation.** Vegetation such as trees shall not be used for support of overhead conductor spans.

Exception: For temporary wiring in accordance with Article 305.

ARTICLE 230 — SERVICES

230-1. Scope. This article covers service conductors and equipment for control and protection of services and their installation requirements.

(FPN):   See Diagram 230-1.

A. General.

230-2. Number of Services. A building or other structure served shall be supplied by only one service.

Where more than one service is permitted by any of the following exceptions, a permanent plaque or directory shall be installed at each service drop or lateral or at each service-equipment location denoting all other services on or in that building or structure and the area served by each.

Exception No. 1: For fire pumps where a separate service is required.

Exception No. 2: For emergency, legally required standby, optional standby, or parallel power production systems where a separate service is required.

General .... Part A
Overhead Service Drop Conductors .... Part B
Underground Service-Lateral Conductors .... Part C
Service-Entrance Conductors .... Part D
Service Equipment — General .... Part E
Service Equipment — Disconnecting Means .... Part F
Service Equipment — Overcurrent Protection .... Part G
Services over 600 Volts, Nominal .... Part H

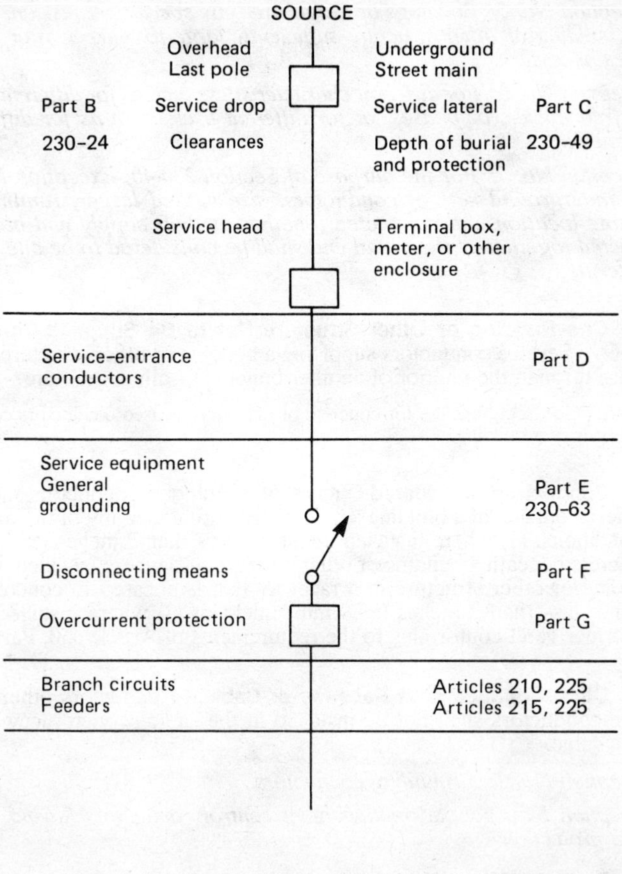
SOURCE
Overhead
Last pole
Underground
Street main
Part B
Service drop
Service lateral
Part C
230-24
Clearances
Depth of burial
and protection
230-49
Service head
Terminal box,
meter, or other
enclosure
Service-entrance
conductors
Part D
Service equipment
General
grounding
Part E
230-63
Disconnecting means
Part F
Overcurrent protection
Part G
Branch circuits
Feeders
Articles 210, 225
Articles 215, 225

Diagram 230-1. Services.

Exception No. 3: Multiple-Occupancy Buildings. By special permission, in multiple-occupancy buildings where there is no available space for service equipment accessible to all the occupants.

Exception No. 4: Capacity Requirements. Two or more services shall be permitted:

a. Where the capacity requirements are in excess of 2000 amperes at a supply voltage of 600 volts or less; or

b. Where the load requirements of a single-phase installation are greater than the serving agency normally supplies through one service; or

c. By special permission.

Exception No. 5: Buildings of Large Area. By special permission, for a single building or other structure sufficiently large to make two or more services necessary.

Exception No. 6: For different characteristics, such as for different voltages, frequencies, or phases, or for different uses, such as for different rate schedules.

Exception No. 7: For the purpose of Section 230-40, Exception No. 2 only, underground sets of conductors, size $^1/_0$ and larger, running to the same location and connected together at their supply end but not connected together at their load end shall be considered to be one service lateral.

230-3. One Building or Other Structure Not to Be Supplied Through Another.   Service conductors supplying a building or other structure shall not pass through the interior of another building or other structure.

(FPN):   See Section 230-6 for concrete or masonry-encased conductors considered outside of a building.

230-6. Conductors Considered Outside of Building.   Conductors shall be considered outside of a building or other structure under any of the following conditions: (1) where installed under not less than 2 inches (50.8 mm) of concrete beneath a building or other structure; (2) where installed within a building or other structure in a raceway that is encased in concrete or brick not less than 2 inches (50.8 mm) thick; or (3) where installed in a transformer vault conforming to the requirements of Article 450, Part C.

230-7. Other Conductors in Raceway or Cable.   Conductors other than service conductors shall not be installed in the same service raceway or service cable.

Exception No. 1: Grounding conductors.

Exception No. 2: Load management control conductors having overcurrent protection.

230-8. Raceway Seal.   Where a service raceway enters from an underground distribution system, it shall be sealed in accordance with Section 300-5. Spare or unused raceways shall also be sealed. Sealants shall be identified for use with the cable insulation, shield, or other components.

230-9. Clearance from Building Openings. Service conductors installed as open conductors or multiconductor cable without an overall outer jacket shall have a clearance of not less than 3 feet (914 mm) from windows that are designed to be opened, doors, porches, fire escapes, or similar locations.

Exception: Conductors run above the top level of a window shall be permitted to be less than the 3 feet (914 mm) requirement above.

Overhead service conductors shall not be installed beneath openings through which materials may be moved, such as openings in farm and commercial buildings, and shall not be installed where they will obstruct entrance to these building openings.

230-10. Service Conductors. The conductors from the service point to the service disconnecting means shall be considered service conductors.

B. Overhead Service-Drop Conductors

230-21. Overhead Supply. Overhead conductors to a building or other structure (such as a pole) on which a meter or disconnecting means is installed shall be considered as a service drop and installed accordingly.

(FPN): Example: Farm loads in Part D of Article 220.

230-22. Insulation or Covering. Service conductors shall normally withstand exposure to atmospheric and other conditions of use without detrimental leakage of current. Individual conductors shall be insulated or covered with an extruded thermoplastic or thermosetting insulating material.

Exception: The grounded conductor of a multiconductor cable shall be permitted to be bare.

230-23. Size and Rating.

(a) **General.** Conductors shall have sufficient ampacity to carry the load without a temperature rise detrimental to the covering or insulation of the conductors and shall have adequate mechanical strength.

(b) **Minimum Size.** The conductors shall not be smaller than No. 8 copper or No. 6 aluminum or copper-clad aluminum.

Exception: For installations to supply only limited loads of a single branch circuit such as small polyphase power, controlled water heaters and the like, they shall not be smaller than No. 12 hard-drawn copper or equivalent.

(c) **Grounded Conductors.** The grounded conductor shall not be less than the minimum size as required by Section 250-23(b).

230-24. Clearances. The vertical clearances of all service-drop conductors shall be based on conductor temperature of 60°F (15°C), no wind, with final unloaded sag in the wire, conductor, or cable.

Service-drop conductors shall not be readily accessible and shall comply with (a) through (d) below for services not over 600 volts, nominal.

(a) Above Roofs. Conductors shall have a vertical clearance of not less than 8 feet (2.44 m) above the roof surface. The vertical clearance above the roof level shall be maintained for a distance not less than 3 feet (914 mm) in all directions from the edge of the roof.

Exception No. 1: The area above a roof surface subject to pedestrian or vehicular traffic shall have a vertical clearance from the roof surface in accordance with the clearance requirements of Section 230-24(b).

Exception No. 2: Where the voltage between conductors does not exceed 300 and the roof has a slope of not less than 4 inches (102 mm) in 12 inches (305 mm), a reduction in clearance to 3 feet (914 mm) shall be permitted.

Exception No. 3: Where the voltage between conductors does not exceed 300, a reduction in clearance above only the overhanging portion of the roof to not less than 18 inches (457 mm) shall be permitted if (1) not more than 6 feet (1.83 m) of service-drop conductors, 4 feet (1.22 m) horizontally, pass above the roof overhang, and (2) they are terminated at a through-the-roof raceway or approved support.

(FPN):   See Section 230-28 for mast supports.

Exception No. 4: The requirement for maintaining the vertical clearance 3 feet (914 mm) from the edge of the roof shall not apply to the final conductor span where the service drop is attached to the side of a building.

(b) Vertical Clearance from Ground. Service-drop conductors where not in excess of 600 volts, nominal, shall have the following minimum clearance from final grade.

10 feet (3.05 m) — at the electric service entrance to buildings, or at the drip loop of the building electric entrance, or above areas or sidewalks accessible only to pedestrians, measured from final grade or other accessible surface only for service-drop cables supported on and cabled together with a grounded bare messenger and limited to 150 volts to ground.

12 feet (3.66 m) — over residential property and driveways, and those commercial areas not subject to truck traffic where the voltage is limited to 300 volts to ground.

15 feet (4.57 m) — for those areas listed in the 12-foot (3.66-m) classification where the voltage exceeds 300 volts to ground.

18 feet (5.49 m) — over public streets, alleys, roads, parking areas subject to truck traffic, driveways on other than residential property, and other land traversed by vehicles such as cultivated, grazing, forest, and orchard.

(c) Clearance from Building Openings. See Section 230-9.

(d) Clearance from Swimming Pools. See Section 680-8.

230-26. Point of Attachment. The point of attachment of the service-drop conductors to a building or other structure shall provide the minimum clearances as specified in Section 230-24. In no case shall this point of attachment be less than 10 feet (3.05 m) above finished grade.

230-27. Means of Attachment. Multiconductor cables used for service drops shall be attached to buildings or other structures by fittings identified for use with service conductors. Open conductors shall be attached to fittings identified for use with service conductors or to noncombustible, nonabsorbent insulators securely attached to the building or other structure.

230-28. Service Masts as Supports. Where a service mast is used for the support of service-drop conductors, it shall be of adequate strength or be supported by braces or guys to withstand safely the strain imposed by the service drop. Where raceway-type service masts are used, all raceway fittings shall be identified for use with service masts.

(FPN): It is the intent of this section to allow only power service-drop conductors to be attached to a service mast.

230-29. Supports Over Buildings. Service-drop conductors passing over a roof shall be securely supported by substantial structures. Where practicable, such supports shall be independent of the building.

C. Underground Service-Lateral Conductors

230-30. Insulation. Service-lateral conductors shall withstand exposure to atmospheric and other conditions of use without detrimental leakage of current. Service-lateral conductors shall be insulated for the applied voltage.

Exception: A grounded conductor shall be permitted to be uninsulated as follows:

a. Bare copper used in a raceway.

b. Bare copper for direct burial where bare copper is judged to be suitable for the soil conditions.

c. Bare copper for direct burial without regard to soil conditions where part of a cable assembly identified for underground use.

d. Aluminum or copper-clad aluminum without individual insulation or covering where part of a cable assembly identified for underground use in a raceway or for direct burial.

230-31. Size and Rating.

(a) General. Service-lateral conductors shall have sufficient ampacity to carry the current for the load as computed in accordance with Article 220 and shall have adequate mechanical strength.

(b) Minimum Size. The conductors shall not be smaller than No. 8 copper or No. 6 aluminum or copper-clad aluminum.

Exception: For installations to supply only limited loads of a single branch circuit such as small polyphase power, controlled water heaters and the like, they shall not be smaller than No. 12 copper or No. 10 aluminum or copper-clad aluminum.

(c) Grounded Conductors. The grounded conductor shall not be less than the minimum size required by Section 250-23(b).

(FPN): Reasonable efficiency of operation can be provided when voltage drop is taken into consideration in sizing the service-lateral conductors.

230-32. Protection Against Damage. Underground service-lateral conductors shall be protected against damage in accordance with Section 300-5. Service-lateral conductors entering a building shall be installed in accordance with Section 230-6 or protected by a raceway wiring method identified in Section 230-43.

D. Service-Entrance Conductors

230-40. Number of Service-Entrance Conductor Sets.   Each service drop or lateral shall supply only one set of service-entrance conductors.

Exception No. 1: Buildings with more than one occupancy shall be permitted to have one set of service-entrance conductors run to each occupancy or to a group of occupancies.

Exception No. 2: Where two to six service disconnecting means in separate enclosures are grouped at one location and supply separate loads from one service drop or lateral, one set of service-entrance conductors shall be permitted to supply each or several such service equipment enclosures.

230-41. Insulation of Service-Entrance Conductors.   Service-entrance conductors shall normally withstand exposure to atmospheric and other conditions of use without detrimental leakage of current. Service-entrance conductors entering or on the exterior of buildings or other structures shall be insulated.

Exception: A grounded conductor shall be permitted to be uninsulated as follows:

a. Bare copper used in a raceway or part of a service cable assembly.

b. Bare copper for direct burial where bare copper is judged to be suitable for the soil conditions.

c. Bare copper for direct burial without regard to soil conditions where part of a cable assembly identified for underground use.

d. Aluminum or copper-clad aluminum without individual insulation or covering where part of a cable assembly or identified for underground use in a raceway or for direct burial.

230-42. Size and Rating.

(a) General.   Service-entrance conductors shall be of sufficient size to carry the loads as computed in Article 220. Ampacity shall be determined from Section 310-15.

Exception: The maximum allowable current of approved busways shall be that value for which the busway has been listed or labeled.

(b) Ungrounded Conductors.   Ungrounded conductors shall have an ampacity of not less than:

(1) 100 amperes, 3-wire for a service to a one-family dwelling with six or more 2-wire branch circuits.

(2) 100 amperes, 3-wire for a service to a one-family dwelling with an initial net computed load of 10 kVA or more.

(3) 60 amperes for other loads.

Exception No. 1: For loads consisting of not more than two 2-wire branch circuits, No. 8 copper or No. 6 aluminum or copper-clad aluminum.

Exception No. 2: By special permission, for loads limited by demand or by the source of supply, No. 8 copper or No. 6 aluminum or copper-clad aluminum.

Exception No. 3: For limited loads of a single branch circuit, No. 12 copper or No. 10 aluminum or copper-clad aluminum, but in no case smaller than the branch-circuit conductors.

(c) Grounded Conductors. The grounded conductor shall not be less | than the minimum size as required by Section 250-23(b).

230-43. Wiring Methods for 600 Volts, Nominal, or Less. Service-entrance conductors shall be installed in accordance with the applicable requirements of this Code covering the type of wiring method used and limited to the following methods: (1) open wiring on insulators; (2) rigid metal conduit; (3) intermediate metal conduit; (4) electrical metallic tubing; (5) service-entrance cables; (6) wireways; (7) busways; (8) auxiliary gutters; (9) rigid nonmetallic conduit; (10) cablebus; (11) Type MC cable; (12) mineral-insulated, metal-sheathed cable; (13) flexible metal conduit or liquidtight flexible metal conduit not over 6 feet (1.83 m) long between raceways, or between raceway and service equipment, with equipment bonding jumper routed with the flexible metal conduit according to provisions of Section 250-79(a), (c), (d), and (f); or (14) liquidtight flexible nonmetallic conduit. |

Approved cable tray systems shall be permitted to support cables approved for use as service-entrance conductors. |

230-46. Unspliced Conductors. Service-entrance conductors shall not be spliced.

Exception No. 1: Clamped or bolted connections in metering equipment enclosures shall be permitted.

Exception No. 2: Where service-entrance conductors are tapped to supply two to six disconnecting means grouped at a common location.

Exception No. 3: At a properly enclosed junction point where an underground wiring method is changed to another type of wiring method.

Exception No. 4: A connection shall be permitted where service conductors are extended from a service drop to an outside meter location and returned to connect to the service-entrance conductors of an existing installation.

Exception No. 5: Where the service-entrance conductors consist of busway, connections shall be permitted as required to assemble the various sections and fittings.

230-49. Protection Against Physical Damage — Underground. Underground service-entrance conductors shall be protected against physical damage in accordance with Section 300-5.

230-50. Protection of Open Conductors and Cables Against Damage — Aboveground. Service-entrance conductors installed aboveground shall be protected against physical damage as specified in (a) or (b) below.

(a) Service-Entrance Cables. Service-entrance cables, where subject to physical damage, such as where installed in exposed places near sidewalks, walkways, driveways, or coal chutes, or where subject to contact with awnings, shutters, swinging signs, or similar objects, shall be protected in any of the following ways: (1) by rigid metal conduit; (2) by intermediate metal conduit; (3) by rigid nonmetallic conduit suitable for the location; (4) by electrical metallic tubing; or (5) by other approved means.

(b) Other than Service-Entrance Cable. Individual open conductors and cables other than service-entrance cables shall not be installed within 10 feet (3.05 m) of grade level or where exposed to physical damage.

Exception: Type MI and Type MC cable shall be permitted within 10 feet (3.05 m) of grade level where not exposed to physical damage or where protected in accordance with Section 300-5(d).

230-51. Mounting Supports. Cables or individual open service conductors shall be supported as specified in (a), (b), or (c) below.

(a) Service-Entrance Cables. Service-entrance cables shall be supported by straps or other approved means within 12 inches (305 mm) of every service head, gooseneck, or connection to a raceway or enclosure and at intervals not exceeding 30 inches (762 mm).

(b) Other Cables. Cables that are not approved for mounting in contact with a building or other structure shall be mounted on insulating supports installed at intervals not exceeding 15 feet (4.57 m) and in a manner that will maintain a clearance of not less than 2 inches (50.8 mm) from the surface over which they pass.

(c) Individual Open Conductors. Individual open conductors shall be installed in accordance with Table 230-51(c). Where exposed to the weather, the conductors shall be mounted on insulators or on insulating supports attached to racks, brackets, or other approved means. Where not exposed to the weather, the conductors shall be mounted on glass or porcelain knobs.

230-52. Individual Conductors Entering Buildings or Other Structures. Where individual open conductors enter a building or other structure, they shall enter through roof bushings or through the wall in an upward slant through individual, noncombustible, nonabsorbent insulating tubes. Drip loops shall be formed on the conductors before they enter the tubes.

230-53. Raceways to Drain. Where exposed to the weather, raceways enclosing service-entrance conductors shall be raintight and arranged to drain. Where embedded in masonry, raceways shall be arranged to drain.

Exception: As permitted in Section 350-2.

Table 230-51(c).　Supports and Clearances for Individual Open Service Conductors

Maximum Volts	Maximum Distance in Feet Between Supports	Minimum Clearances In Inches	
		Between Conductors	From Surface
600	9	6	2
600	15	12	2
300	4½	3	2
600*	4½*	2½*	1*

For SI units: one inch = 25.4 millimeters; one foot = 0.3048 meter.
*Where not exposed to weather.

230-54. Connections at Service Head.

(a) Raintight Service Head. Service raceways shall be equipped with a raintight service head at the point of connection to service-drop conductors.

(b) Service Cable Equipped with Raintight Service Head or Gooseneck. Service cables, either (1) unless continuous from pole to service equipment or meter, shall be equipped with a raintight service head, or (2) shall be formed in a gooseneck and taped and painted or taped with a self-sealing, weather-resistant thermoplastic.

(c) Service Heads Above Service-Drop Attachment. Service heads, and goosenecks in service-entrance cables, shall be located above the point of attachment of the service-drop conductors to the building or other structure.

Exception: Where it is impracticable to locate the service head above the point of attachment, the service head location shall be permitted not farther than 24 inches (610 mm) from the point of attachment.

(d) Secured. Service cables shall be held securely in place.

(e) Separately Bushed Openings. Service heads shall have conductors of different potential brought out through separately bushed openings.

(f) Drip Loops. Drip loops shall be formed on individual conductors. To prevent the entrance of moisture, service-entrance conductors shall be connected to the service-drop conductors either (1) below the level of the service head, or (2) below the level of the termination of the service-entrance cable sheath.

(g) Arranged that Water Will Not Enter Service Raceway or Equipment. Service-drop conductors and service-entrance conductors shall be arranged so that water will not enter service raceway or equipment.

230-55. Termination at Service Equipment. Any service raceway or cable shall terminate at the inner end in a box, cabinet, or equivalent fitting that effectively encloses all energized metal parts.

Exception: Where the service disconnecting means is mounted on a switchboard having exposed busbars on the back, a raceway shall be permitted to terminate at a bushing.

230-56. Service-Entrance Conductor with the Higher Voltage-to-Ground.
On a 4-wire delta-connected service where the midpoint of one phase is grounded, the service-entrance conductor having the higher phase voltage to ground shall be durably and permanently marked by an outer finish that is orange in color or by other effective means.

E. Service Equipment — General

230-62. Service Equipment — Enclosed or Guarded. Energized parts of service equipment shall be enclosed as specified in (a) below, or guarded as specified in (b) below.

(a) Enclosed. Energized parts shall be enclosed so that they will not be exposed to accidental contact or guarded as in (b) below.

(b) Guarded. Energized parts that are not enclosed shall be installed on a switchboard, panelboard, or control board and guarded in accordance with Sections 110-17 and 110-18. Such an enclosure shall be provided with means for locking or sealing doors providing access to energized parts.

230-63. Grounding and Bonding. Service equipment, raceways, cable armor, cable sheaths, etc., and any service conductor that is to be grounded shall be grounded in accordance with the following parts of Article 250.

Part B. Circuit and System Grounding.
Part C. Location of System Grounding Connections.
Part D. Enclosure Grounding.
Part F. Methods of Grounding.
Part G. Bonding.
Part H. Grounding Electrode System.
Part J. Grounding Conductors.

230-64. Working Space. Sufficient working space shall be provided in the vicinity of the service equipment to permit safe operation, inspection, and repairs. In no case shall this be less than that specified by Section 110-16.

230-65. Available Short-Circuit Current. Service equipment shall be suitable for the short-circuit current available at its supply terminals.

230-66. Marking. Service equipment shall be marked to identify it as being suitable for use as service equipment.

F. Service Equipment — Disconnecting Means

230-70. General. Means shall be provided to disconnect all conductors in a building or other structure from the service-entrance conductors.

(a) Location. The service disconnecting means shall be installed at a readily accessible location either outside of a building or structure, or inside nearest the point of entrance of the service conductors.

(b) Marking. Each service disconnecting means shall be permanently marked to identify it as a service disconnecting means.

(c) **Suitable For Use.** Each service disconnecting means shall be suitable for the prevailing conditions. Service equipment installed in hazardous (classified) locations shall comply with the requirements of Articles 500 through 517.

230-71. Maximum Number of Disconnects.

(a) **General.** The service disconnecting means for each service permitted by Section 230-2, or for each set of service-entrance conductors permitted by Section 230-40, Exception No. 1, shall consist of not more than six switches or six circuit breakers mounted in a single enclosure, in a group of separate enclosures, or in or on a switchboard. There shall be no more than six disconnects per service grouped in any one location.

Exception: For the purpose of this section, disconnecting means used solely for the control circuit of the ground-fault protection system, installed as part of the listed equipment, shall not be considered a service disconnecting means.

(b) **Single-Pole Units.** Two or three single-pole switches or breakers, capable of individual operation, shall be permitted on multiwire circuits, one pole for each ungrounded conductor, as one multipole disconnect, provided they are equipped with "handle ties" or a "master handle" to disconnect all conductors of the service with no more than six operations of the hand.

(FPN):   See Section 384-16(a) for service equipment in panelboards.

230-72. Grouping of Disconnects.

(a) **General.** The two to six disconnects as permitted in Section 230-71 shall be grouped. Each disconnect shall be marked to indicate the load served.

Exception: One of the two to six service disconnecting means permitted in Section 230-71, where used only for a water pump also intended to provide fire protection, shall be permitted to be located remote from the other disconnecting means.

(b) **Additional Service Disconnecting Means.** The one or more additional service disconnecting means for fire pumps or for emergency, legally required standby, or optional standby services permitted by Section 230-2 shall be installed sufficiently remote from the one to six service disconnecting means for normal service to minimize the possibility of simultaneous interruption of supply.

(FPN):   See Sections 700-12(d) and (e) for emergency system services.

(c) **Access to Occupants.** In a multiple-occupancy building, each occupant shall have access to the occupant's service disconnecting means.

Exception: In a multiple-occupancy building where electric service and electrical maintenance are provided by the building management and where these are under continuous building management supervision, the service disconnecting means supplying more than one occupancy shall be permitted to be accessible to authorized management personnel only.

230-74. Simultaneous Opening of Poles. Each service disconnecting means shall simultaneously disconnect all ungrounded service conductors from the premises wiring system.

230-75. Disconnection of Grounded Conductor. Where the service disconnecting means does not disconnect the grounded conductor from the premises wiring, other means shall be provided for this purpose in the service equipment. A terminal or bus to which all grounded conductors can be attached by means of pressure connectors shall be permitted for this purpose.

In a multisection switchboard, disconnects for the grounded conductor shall be permitted to be in any section of the switchboard, provided any such switchboard section is marked.

230-76. Manually or Power Operable. The service disconnecting means for ungrounded service conductors shall consist of either (1) a manually operable switch or circuit breaker equipped with a handle or other suitable operating means, or (2) a power-operated switch or circuit breaker provided the switch or circuit breaker can be opened by hand in the event of a power supply failure.

230-77. Indicating. The service disconnecting means shall plainly indicate whether it is in the open or closed position.

230-78. Externally Operable. An enclosed service disconnecting means shall be externally operable without exposing the operator to contact with energized parts.

Exception: A power-operated switch or circuit breaker shall not be required to be externally operable by hand to a closed position.

230-79. Rating of Disconnect. The service disconnecting means shall have a rating not less than the load to be carried, determined in accordance with Article 220. In no case shall the rating be lower than specified in (a), (b), (c), or (d) below.

(a) One-Circuit Installation. For installations to supply only limited loads of a single branch circuit, the service disconnecting means shall have a rating of not less than 15 amperes.

(b) Two-Circuit Installations. For installations consisting of not more than two 2-wire branch circuits, the service disconnecting means shall have a rating of not less than 30 amperes.

(c) One-Family Dwelling. For a one-family dwelling, the service disconnecting means shall have a rating of not less than 100 amperes, 3-wire under either of the following conditions: (1) where the initial computed load is 10 kVA or more, or (2) where the initial installation consists of six or more 2-wire branch circuits.

(d) All Others. For all other installations, the service disconnecting means shall have a rating of not less than 60 amperes.

230-80. Combined Rating of Disconnects. Where the service disconnecting means consists of more than one switch or circuit breaker, as permitted by Section 230-71, the combined ratings of all the switches or circuit breakers used shall not be less than the rating required for a single switch or circuit breaker.

230-81. Connection to Terminals. The service conductors shall be connected to the service disconnecting means by pressure connectors, clamps, or other approved means. Connections that depend upon solder shall not be used.

230-82. Equipment Connected to the Supply Side of Service Disconnect. Equipment shall not be connected to the supply side of the service disconnecting means.

(FPN): It is not the intent of this section to require individual meter socket enclosures to be suitable for use as service equipment.

Exception No. 1: Cable limiters or other current-limiting devices.

Exception No. 2: Fuses and disconnecting means or circuit breakers suitable for use as service equipment, in meter pedestals or otherwise provided and connected in series with the ungrounded service conductors and located away from the building supplied.

Exception No. 3: Meters nominally rated not in excess of 600 volts, provided all metal housings and service enclosures are grounded in accordance with Article 250.

Exception No. 4: Instrument transformers (current and voltage), high-impedance shunts, surge-protective devices identified for use on the supply side of the service disconnect, load management devices, and surge arresters.

Exception No. 5: Taps used only to supply load management devices, circuits for emergency systems, stand-by power systems, fire pump equipment, and fire and sprinkler alarms if provided with service equipment and installed in accordance with requirements for service-entrance conductors.

Exception No. 6: Solar photovoltaic systems or interconnected electric power production sources. See Articles 690 or 705 as applicable.

Exception No. 7: Where the service disconnecting means is power operable, the control circuit shall be permitted to be connected ahead of the service disconnecting means if suitable overcurrent protection and disconnecting means are provided.

Exception No. 8: Ground-fault protection systems where installed as part of listed equipment, if suitable overcurrent protection and disconnecting means are provided.

230-83. Transfer Equipment. Transfer equipment, including transfer switches, shall operate such that all ungrounded conductors of one source of supply are disconnected before any ungrounded conductors of the second source are connected.

Exception No. 1: Where manual equipment identified for the purpose or suitable automatic equipment is utilized, two or more sources shall be permitted to be connected in parallel through transfer equipment.

Exception No. 2: Where parallel operation is used and suitable automatic or manual control equipment is provided.

G. Service Equipment — Overcurrent Protection

230-90. Where Required. Each ungrounded service conductor shall have overload protection.

(a) Ungrounded Conductor. Such protection shall be provided by an overcurrent device in series with each ungrounded service conductor having a rating or setting not higher than the allowable ampacity of the conductor.

Exception No. 1: For motor-starting currents, ratings in conformity with Sections 430-52, 430-62, and 430-63 shall be permitted.

Exception No. 2: Fuses and circuit breakers with a rating or setting in conformity with Section 240-3(b) or (c) and Section 240-6.

Exception No. 3: Not more than six circuit breakers or six sets of fuses shall be permitted as the overcurrent device to provide the overload protection. The sum of the ratings of the circuit breakers or fuses shall be permitted to exceed the ampacity of the service conductors, provided the calculated load in accordance with Article 220 does not exceed the ampacity of the service conductors.

Exception No. 4: Fire Pumps. Where the service to the fire pump room is judged to be outside of buildings, these provisions shall not apply. Overcurrent protection for fire pump services shall be selected or set to carry locked-rotor current of the motor(s) indefinitely.

(FPN): See Centrifugal Fire Pumps, NFPA 20-1990 (ANSI).

A set of fuses shall be considered all the fuses required to protect all the ungrounded conductors of a circuit. Single-pole circuit breakers, grouped in accordance with Section 230-71(b), shall be considered as one protective device.

(b) Not in Grounded Conductor. No overcurrent device shall be inserted in a grounded service conductor except a circuit breaker that simultaneously opens all conductors of the circuit.

230-91. Location of Overcurrent Protection.

(a) General. The service overcurrent device shall be an integral part of the service disconnecting means or shall be located immediately adjacent thereto.

(b) Access to Occupants. In a multiple-occupancy building, each occupant shall have access to the overcurrent protective devices.

Exception: As permitted in Section 240-24(b), Exception.

230-92. Locked Service Overcurrent Devices. Where the service overcurrent devices are locked or sealed, or otherwise not readily accessible, branch-circuit overcurrent devices shall be installed on the load side, shall be mounted in a readily accessible location, and shall be of lower ampere | rating than the service overcurrent device.

230-93. Protection of Specific Circuits. Where necessary to prevent tampering, an automatic overcurrent device protecting service conductors supplying only a specific load, such as a water heater, shall be permitted to be locked or sealed where located so as to be accessible.

230-94. Relative Location of Overcurrent Device and Other Service Equipment. The overcurrent device shall protect all circuits and devices.

Exception No. 1: The service switch shall be permitted on the supply side.

Exception No. 2: High-impedance shunt circuits, lightning arresters, surge protective capacitors, and instrument transformers (current and voltage) shall be permitted to be connected and installed on the supply side of the service disconnecting means as permitted in Section 230-82.

Exception No. 3: Circuits for emergency supply and load management devices shall be permitted to be connected on the supply side of the service overcurrent device where separately provided with overcurrent protection.

Exception No. 4: Circuits used only for the operation of fire alarm, other protective signaling systems, or the supply to fire pump equipment shall be permitted to be connected on the supply side of the service overcurrent device where separately provided with overcurrent protection.

Exception No. 5: Meters nominally rated not in excess of 600 volts, provided all metal housings and service enclosures are grounded in accordance with Article 250.

Exception No. 6: Where service equipment is power operable, the control circuit shall be permitted to be connected ahead of the service equipment if suitable overcurrent protection and disconnecting means are provided.

230-95. Ground-Fault Protection of Equipment. Ground-fault protection of equipment shall be provided for solidly grounded wye electrical services of more than 150 volts to ground, but not exceeding 600 volts phase-to-phase for each service disconnecting means rated 1000 amperes or more.

Exception No. 1: The provisions of this section shall not apply to a service disconnecting means for a continuous industrial process where a nonorderly shutdown will introduce additional or increased hazards.

Exception No. 2: The provisions of this section shall not apply to fire pumps.

(a) Setting. The ground-fault protection system shall operate to cause the service disconnecting means to open all ungrounded conductors of the faulted circuit. The maximum setting of the ground-fault protection shall be 1200 amperes, and the maximum time delay shall be one second for ground-fault currents equal to or greater than 3000 amperes.

(b) Fuses. If a switch and fuse combination is used, the fuses employed shall be capable of interrupting any current higher than the interrupting capacity of the switch during a time when the ground-fault protective system will not cause the switch to open.

(FPN No. 1): As used in this section, the rating of the service disconnecting means is considered to be the rating of the largest fuse that can be installed or the highest continuous current trip setting for which the actual overcurrent device installed in a circuit breaker is rated or can be adjusted.

(FPN No. 2): It is recognized that ground-fault protection may be desirable for service disconnecting means rated less than 1000 amperes on solidly grounded wye systems having more than 150 volts to ground, not exceeding 600 volts phase-to-phase.

(FPN No. 3): As used in this section, solidly grounded means that the grounded conductor (neutral) is grounded without inserting any resistor or impedance device.

(FPN No. 4): Ground-fault protection that functions to open the service disconnecting means will afford no protection from faults on the line side of the protective element. It serves only to limit damage to conductors and equipment on the load side in the event of an arcing ground fault on the load side of the protective element.

(FPN No. 5): This added protective equipment at the service equipment will make it necessary to review the overall wiring system for proper selective overcurrent protection coordination. Additional installations of ground-fault protective equipment will be needed on feeders and branch circuits where maximum continuity of electrical service is necessary.

(FPN No. 6): Where ground-fault protection is provided for the service disconnecting means and interconnection is made with another supply system by a transfer device, means or devices may be needed to assure proper ground-fault sensing by the ground-fault protection equipment.

(c) Performance Testing. The ground-fault protection system shall be performance tested when first installed on site. The test shall be conducted in accordance with instructions which shall be provided with the equipment. A written record of this test shall be made and shall be available to the authority having jurisdiction.

H. Services Exceeding 600 Volts, Nominal

230-200. General. Service conductors and equipment used on circuits exceeding 600 volts, nominal, shall comply with all applicable provisions of the preceding sections of this article and with the following sections, which supplement or modify the preceding sections. In no case shall the provisions of this article apply to equipment on the supply side of the service point.

(FPN): For clearances of conductors of over 600 volts, nominal, see National Electrical Safety Code, ANSI C2-1990.

230-202. Service-Entrance Conductors. Service-entrance conductors to buildings or enclosures shall be installed to conform to the following:

(a) **Conductor Size.** Service-entrance conductors shall not be smaller than No. 6 unless in cable. Conductors in cable shall not be smaller than No. 8.

(b) **Wiring Methods.** Service-entrance conductors shall be installed by means of one of the following wiring methods: (1) in rigid metal conduit; (2) in intermediate metal conduit; (3) in rigid nonmetallic conduit; (4) as multiconductor cable identified as service cable; (5) as open conductors where supported on insulators and where either accessible only to qualified persons or where effectively guarded against accidental contact; (6) in cablebus; or (7) in busways.

Underground service-entrance conductors shall conform to Section 710-4(b).

Cable tray systems shall be permitted to support cables identified as service-entrance conductors. See Article 318.

(FPN): See Section 310-6 for shielding of solid dielectric insulated conductors.

(c) **Open Work.** Open wire services shall be installed in accordance with the provisions of Article 710, Part D.

(d) **Supports.** Service-entrance conductors and their supports, including insulators, shall have strength and stability sufficient to ensure maintenance of adequate clearance with abnormal currents in case of short circuits.

(e) **Guarding.** Open wires shall be guarded to make them accessible only to qualified persons.

(f) **Service Cable.** Where cable conductors emerge from a metal sheath or raceway, the insulation of the conductors shall be protected from moisture and physical damage by a pothead or other approved means.

(g) **Draining Raceways.** Unless conductors identified for use in wet locations are used, raceways embedded in masonry or exposed to the weather shall be arranged to drain.

(h) **Over 15,000 Volts.** Where the voltage exceeds 15,000 volts between conductors, they shall enter either metal-enclosed switchgear or a transformer vault conforming to the requirements of Sections 450-41 through 450-48.

230-203. Warning Signs. Signs with the words "Danger High Voltage—Keep Out" shall be posted in plain view where unauthorized persons might come in contact with energized parts.

230-204. Isolating Switches.

(a) **Where Required.** Where oil switches or air, oil, vacuum, or sulfur hexafluoride circuit breakers constitute the service disconnecting means, an air-break isolating switch shall be installed on the supply side of the disconnecting means and all associated service equipment.

Exception: Where such equipment is mounted on removable truck panels or metal-enclosed switchgear units, which cannot be opened unless the circuit is disconnected, and which, when removed from the normal operating position, automatically disconnect the circuit breaker or switch from all energized parts.

(b) Fuses as Isolating Switch. Where fuses are of the type that can be operated as a disconnecting switch, a set of such fuses shall be permitted as the isolating switch where: (1) the oil disconnecting means is a nonautomatic switch, and (2) the set of fuses disconnect the oil switch and all associated service equipment from the service conductors.

(c) Accessible to Qualified Persons Only. The isolating switch shall be accessible to qualified persons only.

(d) Grounding Connection. Isolating switches shall be provided with a means for readily connecting the load side conductors to ground when disconnected from the source of supply.

A means for grounding the load side conductors shall not be required for any duplicate isolating switch installed and maintained by the electric supply company.

230-205. Disconnecting Means.

(a) Location. The service disconnecting means shall be located in accordance with Section 230-70 or Section 230-208(b).

Exception: Where under single management, the service disconnecting means shall be permitted to be located in a separate building or structure on the same premises. In such case, the service disconnecting means shall be capable of being electrically opened by a readily accessible control device located as near as practicable to where the feeder conductors enter the building served. The control device shall be permanently marked to identify its function and shall provide visual indication of the "on" or "off" status of the remote service disconnect.

(b) Type. The service disconnecting means shall simultaneously disconnect all ungrounded conductors and shall have a fault-closing rating not less than the maximum short-circuit current available at its supply terminals.

Where fused switches or separately mounted fuses are installed, the fuse characteristics shall be permitted to contribute to the fault-closing rating of the disconnecting means.

230-206. Overcurrent Devices as Disconnecting Means.
Where the circuit breaker or alternative for it specified in Section 230-208 for service overcurrent devices meets the requirements specified in Section 230-205, they shall constitute the service disconnecting means.

230-207. Equipment in Secondaries.
Where the primary service equipment supplies one or more transformers whose secondary windings connect to a common bus of bars or wires, and the primary load-interrupter switch or circuit breaker is capable of being opened and closed from a point outside the transformer vault, the disconnecting means and overcurrent protection shall not be required in the secondary circuit if the primary fuse or circuit breaker is rated or set to protect the secondary circuit.

230-208. Overcurrent Protection Requirements. Service-entrance conductors shall have a short-circuit protective device in each ungrounded conductor on the load side of, or as an integral part of, the service-entrance switch. The protective device shall be capable of detecting and interrupting all values of current in excess of its trip setting or melting point, that can occur at its location. A fuse rated in continuous amperes not to exceed three times the ampacity of the conductor, or a circuit breaker with a trip setting of not more than six times the ampacity of the conductors, shall be considered as providing the required short-circuit protection.

(FPN): See Tables 310-69 through 310-84 for ampacities of conductors rated 2001 volts and above.

Overcurrent devices shall conform to the following:

(a) In Vault or Consisting of Metal-Enclosed Switchgear. Where the service equipment is installed in a transformer vault meeting the provisions of Sections 450-41 through 450-48, or consists of metal-enclosed switchgear, the overcurrent protection and disconnecting means shall be one of the following:

(1) A nonautomatic oil switch, oil fuse cutout, or air load-interrupter switch shall be permitted with fuses. The interrupting rating of this switch shall equal or exceed the continuous current rating of the fuse.

(2) An automatic trip circuit breaker of suitable current-carrying and interrupting capacity.

(3) A switch capable of interrupting the no-load current of the transformer supplied through the switch and suitable fuses shall be permitted provided the switch is interlocked with a single switch or circuit breaker on the secondary circuit of the transformer so that the primary switch cannot be opened when the secondary circuit is closed.

(b) Not in Vault or Not Consisting of Metal-Enclosed Switchgear. Where the service equipment is not in a vault or metal-enclosed switchgear, the overcurrent protection and disconnecting means shall be either of the following:

(1) An air load-interrupter switch or other switch capable of interrupting the rated circuit load shall be permitted with fuses on a pole or elevated structure outside the building, provided the switch is operable by persons using the building.

(2) An automatic-trip circuit breaker of suitable ampacity and interrupting capacity. The circuit breaker shall be located outside the building as near as practicable to where the feeder conductors enter the building. The location shall be permitted on a pole, roof, foundation, or other structure.

(c) Fuses. Fuses shall have an interrupting rating no less than the maximum available short-circuit current in the circuit at their supply terminals.

(d) Circuit Breakers. Circuit breakers shall be free to open in case the circuit is closed on an overload. This can be accomplished by means such as trip-free circuit breakers. A service circuit breaker shall indicate clearly whether it is open or closed, and shall have an interrupting rating no less than the maximum available short-circuit current at its supply terminals.

Overcurrent relays shall be furnished in connection with current transformers in one of the following combinations:

(1) Three overcurrent relays operated from current transformers in each phase.

(2) Two overcurrent relays operated by current from current transformers in any two phases and one overcurrent relay sensitive to ground-fault current that is operated by the sum of the currents from current transformers in each phase.

(3) Two overcurrent relays operated by current from current transformers in any two phases and one overcurrent relay sensitive to ground-fault current that is operated from a current transformer that links all three phase conductors and the grounded circuit conductor (neutral), if provided.

(e) Enclosed Overcurrent Devices. The restriction to 80 percent of rating for an enclosed overcurrent device on continuous loads shall not apply to overcurrent devices installed in services operating at over 600 volts.

230-209. Surge Arresters (Lightning Arresters).   Surge arresters installed in accordance with the requirements of Article 280 shall be permitted on each ungrounded overhead service conductor.

230-210. Service Equipment—General Provisions.   Service equipment including instrument transformers shall conform to Article 710, Part B.

230-211. Metal-Enclosed Switchgear.   Metal-enclosed switchgear shall consist of a substantial metal structure and a sheet metal enclosure. Where installed over a combustible floor, suitable protection thereto shall be provided.

ARTICLE 240 — OVERCURRENT PROTECTION

240-1. Scope.   Parts A through G of this article provide the general requirements for overcurrent protection and overcurrent protective devices not more than 600 volts, nominal. Part H covers overcurrent protection over 600 volts, nominal.

(FPN): Overcurrent protection for conductors and equipment is provided to open the circuit if the current reaches a value that will cause an excessive or dangerous temperature in conductors or conductor insulation. See also Sections 110-9 and 110-10 for requirements for interrupting capacity and protection against fault currents.

A. General

240-2. Protection of Equipment.   Equipment shall be protected against overcurrent in accordance with the article in this Code covering the type of equipment as specified in the following list.

240-3. Protection of Conductors. Conductors, other than flexible cords and fixture wires, shall be protected against overcurrent in accordance with their ampacities as specified in Section 310-15, unless otherwise permitted in (a) through (m) below.

(a) Power Loss Hazard. Conductor overload protection shall not be required where the interruption of the circuit would create a hazard, such as in a material handling magnet circuit or fire pump circuit. Short-circuit protection shall be provided.

(FPN):   See Installation of Centrifugal Fire Pumps, NFPA 20 - 1990 (ANSI).

(b) Devices Rated 800 Amperes or Less. Conductors not part of a multioutlet branch circuit supplying receptacles for cord- and plug-connected portable loads and where the ampacity of the conductors does not correspond with the standard ampere rating of a fuse or a circuit breaker without overload trip adjustments above its rating (but that may have other trip or rating adjustments), the next higher standard device rating shall be permitted only if this rating does not exceed 800 amperes.

(c) Devices Rated Over 800 Amperes. Where the overcurrent device is rated over 800 amperes, the ampacity of the conductors it protects shall be equal to or greater than the rating of the overcurrent device as defined in Section 240-6.

(d) Tap Conductors. Tap conductors shall be permitted to be protected against overcurrent in accordance with Sections 210-19(c), 240-21, 364-11, 364-12, and 430-53(d).

(e) Motor-Operated Appliance Circuit Conductors. Motor-operated appliance circuit conductors shall be permitted to be protected against overcurrent in accordance with Parts B and D of Article 422.

(f) Motor and Motor-Control Circuit Conductors. Motor and motor-control circuit conductors shall be permitted to be protected against overcurrent in accordance with Parts C, D, E, and F of Article 430.

(g) Phase Converter Supply Conductors. Phase converter supply conductors for motor loads and nonmotor loads shall be permitted to be protected against overcurrent in accordance with Section 455-7.

(h) Air-Conditioning and Refrigeration Equipment Circuit Conductors. Air-conditioning and refrigeration equipment circuit conductors shall be permitted to be protected against overcurrent in accordance with Parts C and F of Article 440.

(i) Transformer Secondary Conductors. Single-phase (other than 2-wire) and multiphase transformer secondary conductors are not considered to be protected by the primary overcurrent protection. Conductors supplied by the secondary side of a single-phase transformer having a 2-wire (single-voltage) secondary shall be permitted to be protected by overcurrent protection provided on the primary (supply) side of the transformer, provided this protection is in accordance with Section 450-3 and does not exceed the value determined by multiplying the secondary conductor ampacity by the secondary to primary transformer voltage ratio.

(j) Capacitor Circuit Conductors. Capacitor circuit conductors shall be permitted to be protected against overcurrent in accordance with Sections 460-8(b) and 460-25(a) through (d).

(k) Electric Welder Circuit Conductors. Welder circuit conductors shall be permitted to be protected against overcurrent in accordance with Sections 630-12, 630-22, and 630-32.

(l) Remote-Control, Signaling, and Power-Limited Circuit Conductors. Remote-control, signaling, and power-limited circuit conductors shall be permitted to be protected against overcurrent in accordance with Sections 725-12, 725-13, 725-31, 725-35, and 725-36.

(m) Fire Protective Signaling System Circuit Conductors. Fire protective signaling system circuit conductors shall be permitted to be protected against overcurrent in accordance with Sections 760-12, 760-13, 760-21, 760-23, and 760-24.

240-4. Protection of Flexible Cords and Fixture Wires. Flexible cord, including tinsel cord and extension cords, shall be protected against overcurrent in accordance with their ampacities as specified in Tables 400-5(A) and 400-5(B). Fixture wire shall be protected against overcurrent in accordance with its ampacity as specified in Table 402-5. Supplementary overcurrent protection as in Section 240-10 shall be permitted to be an acceptable means for providing this protection.

Exception No. 1: Where a flexible cord or a tinsel cord approved for and used with a specific listed appliance or portable lamps is connected to a branch circuit of Article 210 in accordance with the following:

20-ampere circuits, tinsel cord or No. 18 cord and larger.

30-ampere circuits, No. 16 cord and larger.

40-ampere circuits, cord of 20-ampere capacity and over.

50-ampere circuits, cord of 20-ampere capacity and over.

Exception No. 2: Where fixture wire is connected to 120-volt or higher branch circuit of Article 210 in accordance with the following:

20-ampere circuits, No. 18 up to 50 feet (15.2 m) of run length.

20-ampere circuits, No. 16 up to 100 feet (30.5 m) of run length.

20-ampere circuits, No. 14 and larger.

30-ampere circuits, No. 14 and larger.

40-ampere circuits, No. 12 and larger.

50-ampere circuits, No. 12 and larger.

Exception No. 3: Flexible cord used in listed extension cord sets having No. 16 or larger conductors shall be considered to be protected by 20-ampere branch-circuit overcurrent protection.

240-6. Standard Ampere Ratings.

(a) Fuses and Fixed Trip Circuit Breakers. The standard ampere ratings for fuses and inverse time circuit breakers shall be considered 15, 20, 25, 30, 35, 40, 45, 50, 60, 70, 80, 90, 100, 110, 125, 150, 175, 200, 225, 250, 300, 350, 400, 450, 500, 600, 700, 800, 1000, 1200, 1600, 2000, 2500, 3000, 4000, 5000, and 6000 amperes.

Exception: Additional standard ratings for fuses shall be considered 1, 3, 6, 10, and 601.

(b) Adjustable Trip Circuit Breakers. The rating of an adjustable trip circuit breaker having external means for adjusting the long-time pickup (ampere rating or overload) setting shall be the maximum setting possible.

Exception: Circuit breakers that have removable and sealable covers over the adjusting means, or are located behind bolted equipment enclosure doors, or are located behind locked doors accessible only to qualified personnel, shall be permitted to have ampere ratings equal to the adjusted (set) long-time pickup settings.

(FPN): It is not the intent to prohibit the use of nonstandard ampere ratings for fuses and inverse time circuit breakers.

240-8. Fuses or Circuit Breakers in Parallel. Fuses, circuit breakers, or combinations thereof shall not be connected in parallel.

Exception: Circuit breakers or fuses, factory-assembled in parallel, and listed as a unit.

240-9. Thermal Devices. Thermal relays and other devices not designed to open short circuits shall not be used for the protection of conductors against overcurrent due to short circuits or grounds, but the use of such devices shall be permitted to protect motor-branch-circuit conductors from overload if protected in accordance with Section 430-40.

240-10. Supplementary Overcurrent Protection. Where supplementary overcurrent protection is used for lighting fixtures, appliances, and other equipment or for internal circuits and components of equipment, it shall not be used as a substitute for branch-circuit overcurrent devices or in place of the branch-circuit protection specified in Article 210. Supplementary overcurrent devices shall not be required to be readily accessible.

240-11. Definition of Current-Limiting Overcurrent Protective Device. A current-limiting overcurrent protective device is a device that, when interrupting currents in its current-limiting range, will reduce the current flowing in the faulted circuit to a magnitude substantially less than that obtainable in the same circuit if the device were replaced with a solid conductor having comparable impedance.

240-12. Electrical System Coordination. Where an orderly shutdown is required to minimize hazard(s) to personnel and equipment, a system of coordination based on the following two conditions shall be permitted:

(1) Coordinated short-circuit protection.

(2) Overload indication based on monitoring systems or devices.

(FPN): Coordination is defined as properly localizing a fault condition to restrict outages to the equipment affected, accomplished by choice of selective fault-protective devices. The monitoring system may cause the condition to go to alarm, allowing corrective action or an orderly shutdown, thereby minimizing personnel hazard and equipment damage.

240-13. Ground-Fault Protection of Equipment. Ground-fault protection of equipment shall be provided in accordance with the provisions of Section 230-95 for solidly grounded wye electrical systems of more than 150 volts to ground, but not exceeding 600 volts phase-to-phase for each building or structure main disconnecting means rated 1000 amperes or more.

Exception No. 1: The provisions of this section shall not apply to a disconnecting means for a continuous industrial process where a nonorderly shutdown will introduce additional or increased hazards.

Exception No. 2: The provisions of this section shall not apply to fire pumps.

Exception No. 3: The provisions of this section shall not apply if the disconnect is protected by the service ground-fault protection, and this protection is not nullified by subsequent grounded circuit conductor connections to additional grounding electrodes, as may be permitted in Section 250-24.

B. Location

240-20. Ungrounded Conductors.

(a) Overcurrent Device Required. A fuse or an overcurrent trip unit of a circuit breaker shall be connected in series with each ungrounded conductor. A combination of a current transformer and overcurrent relay shall be considered equivalent to an overcurrent trip unit.

(FPN): For motor circuits, see Parts C, D, F, and J of Article 430.

(b) Circuit Breaker as Overcurrent Device. Circuit breakers shall open all ungrounded conductors of the circuit.

Exception: In grounded systems, individual single-pole circuit breakers with handle ties shall be permitted as the protection for each ungrounded conductor for line-to-line connected loads of the following:

a. Single-phase circuits,

b. 3- wire direct-current circuits, or

c. Lighting or appliance branch circuits connected to 4-wire, 3-phase systems or 5-wire, 2-phase systems, provided such lighting or appliance circuits are supplied from a system having a grounded neutral and no conductor operates at a voltage greater than permitted in Section 210-6.

(c) Closed-Loop Power Distribution Systems. Listed devices providing equivalent overcurrent protection in closed-loop power distribution systems shall be permitted as a substitute for fuses or circuit breakers.

240-21. Location in Circuit.
An overcurrent device shall be connected in each ungrounded circuit conductor as follows:

(a) Feeder and Branch-Circuit Conductors. Feeder and branch-circuit conductors shall be protected by overcurrent-protective devices connected at the point the conductors receive their supply, unless otherwise permitted in (b) through (m) below.

(b) Feeder Taps Not Over 10 Feet (3.05 m) Long. Conductors shall be permitted to be tapped, without overcurrent protection at the tap, to a feeder or transformer secondary where all the following conditions are met:

(1) The length of the tap conductors does not exceed 10 feet (3.05 m).

(2) The ampacity of the tap conductors is:

a. Not less than the combined computed loads on the circuits supplied by the tap conductors, and

b. Not less than the rating of the device supplied by the tap conductors, or not less than the rating of the overcurrent-protective device at the termination of the tap conductors.

(3) The tap conductors do not extend beyond the switchboard, panelboard, or control devices they supply.

(4) Except at the point of connection to the feeder, the tap conductors are enclosed in a raceway, which shall extend from the tap to the enclosure of an enclosed switchboard, panelboard, or control devices, or to the back of an open switchboard.

(5) For field installations where the tap conductors leave the enclosure or vault in which the tap is made, the rating of the overcurrent device on the line side of the tap conductors shall not exceed 1000 percent of the tap conductor's ampacity.

(FPN): See Sections 384-16(a) and (d) for lighting and appliance branch-circuit panelboards.

(c) Feeder Taps Not Over 25 Feet (7.62 m) Long. Conductors shall be permitted to be tapped, without overcurrent protection at the tap, to a feeder where all the following conditions are met:

(1) The length of the tap conductors does not exceed 25 feet (7.62 m).

(2) The ampacity of the tap conductors is not less than $1/3$ of the rating of the overcurrent device protecting the feeder conductors.

(3) The tap conductors terminate in a single circuit breaker or a single set of fuses that will limit the load to the ampacity of the tap conductors. This device shall be permitted to supply any number of additional overcurrent devices on its load side.

(4) The tap conductors are suitably protected from physical damage or are enclosed in a raceway.

(d) Transformer Feeder Taps With Primary Plus Secondary Not Over 25 Feet (7.62 m) Long. Conductors supplying a transformer shall be permitted to be tapped, without overcurrent protection at the tap, from a feeder where all the following conditions are met:

(1) The conductors supplying the primary of a transformer have an ampacity at least $1/3$ of the rating of the overcurrent device protecting the feeder conductors.

(2) The conductors supplied by the secondary of the transformer shall have an ampacity that, when multiplied by the ratio of the secondary-to-primary voltage, is at least $1/3$ of the rating of the overcurrent device protecting the feeder conductors.

(3) The total length of one primary plus one secondary conductor, excluding any portion of the primary conductor that is protected at its ampacity, is not over 25 feet (7.62 m).

(4) The primary and secondary conductors are suitably protected from physical damage.

(5) The secondary conductors terminate in a single circuit breaker or set of fuses that will limit the load to that allowed in Section 310-15.

(e) Feeder Taps Over 25 Feet (7.62 m) Long. Conductors over 25 feet (7.62 m) long shall be permitted to be tapped from feeders in high bay manufacturing buildings over 35 feet (10.67 m) high at walls, where conditions of maintenance and supervision ensure that only qualified persons will service the systems. Conductors tapped, without overcurrent protection at the tap, to a feeder shall be permitted to be not over 25 feet (7.62 m) long horizontally and not over 100 feet (30.5 m) total length where all the following conditions are met:

(1) The ampacity of the tap conductors is not less than ⅓ of the rating of the overcurrent device protecting the feeder conductors.

(2) The tap conductors terminate at a single circuit breaker or a single set of fuses that will limit the load to the ampacity of the tap conductors. This single overcurrent device shall be permitted to supply any number of additional overcurrent devices on its load side.

(3) The tap conductors are suitably protected from physical damage or are enclosed in a raceway.

(4) The tap conductors are continuous from end-to-end and contain no splices.

(5) The tap conductors are sized No. 6 copper or No. 4 aluminum or larger.

(6) The tap conductors do not penetrate walls, floors, or ceilings.

(7) The tap is made no less than 30 feet (9.14 m) from the floor.

(f) Branch-Circuit Taps. Taps to individual outlets and circuit conductors supplying a single electric household range shall be permitted to be protected by the branch-circuit overcurrent devices where in accordance with the requirements of Sections 210-19, 210-20, and 210-24.

(g) Busway Taps. Busways and busway taps shall be permitted to be protected against overcurrent in accordance with Sections 364-10 through 364-14.

(h) Motor Circuit Taps. Motor-branch-circuit conductors shall be permitted to be protected against overcurrent in accordance with Sections 430-28 and 430-53.

(i) Conductors From Generator Terminals. Conductors from generator terminals shall be permitted to be protected against overcurrent in accordance with Section 445-5.

(j) Transformer Secondary Conductors of Separately Derived Systems for Industrial Installations. Conductors connected to a transformer secondary of a separately derived system for industrial installations shall be considered to be protected against overcurrent where all the following conditions are met:

(1) The length of the secondary conductors does not exceed 25 feet (7.62 m).

(2) The ampacity of the secondary conductors is not less than the secondary current rating of the transformer, and the sum of the ratings of the overcurrent devices does not exceed the ampacity of the secondary conductors.

(3) All overcurrent devices are grouped.

(4) The tap conductors are suitably protected from physical damage.

(m) Outside Feeder Taps. Outside conductors tapped to a feeder or connected to a transformer secondary shall be permitted to be protected by complying with all of the following conditions:

(1) The conductors are suitably protected from physical damage.

(2) The conductors terminate at a single circuit breaker or a single set of fuses that will limit the load to the ampacity of the conductors. This single overcurrent device shall be permitted to supply any number of additional overcurrent devices on its load side.

(3) The overcurrent device for the conductors is an integral part of a disconnecting means or shall be located immediately adjacent thereto.

(4) The disconnecting means for the conductors are installed at a readily accessible location either outside of a building or structure, or inside nearest the point of entrance of the conductors.

(n) Service Conductors. Service-entrance conductors shall be permitted to be protected by overcurrent devices in accordance with Section 230-91.

240-22. Grounded Conductors.   No overcurrent device shall be connected in series with any conductor that is intentionally grounded.

Exception No. 1: Where the overcurrent device opens all conductors of the circuit, including the grounded conductor, and is so designed that no pole can operate independently.

Exception No. 2: Where required by Sections 430-36 and 430-37 for motor overload protection.

240-23. Change in Size of Grounded Conductor.   Where a change occurs in the size of the ungrounded conductor, a similar change shall be permitted to be made in the size of the grounded conductor.

240-24. Location in or on Premises.

(a) Readily Accessible. Overcurrent devices shall be readily accessible.

Exception No. 1: For busways as provided in Section 364-12.

Exception No. 2: For supplementary overcurrent protection as described in Section 240-10.

Exception No. 3: For service overcurrent devices as described in Section 230-92.

(b) Occupant to Have Ready Access. Each occupant shall have ready access to all overcurrent devices protecting the conductors supplying that occupancy.

Exception: In a multiple-occupancy building where electric service and electrical maintenance are provided by the building management and where these are under continuous building management supervision, the service overcurrent devices and feeder overcurrent devices supplying more than one occupancy shall be permitted to be accessible to authorized management personnel only.

(c) Not Exposed to Physical Damage. Overcurrent devices shall be located where they will not be exposed to physical damage.

(FPN):   See Section 110-11, Deteriorating Agents.

(d) Not in Vicinity of Easily Ignitible Material. Overcurrent devices shall not be located in the vicinity of easily ignitible material such as in clothes closets.

(e) Not Located in Bathrooms. In dwelling units and guest rooms of hotels and motels, branch-circuit overcurrent devices shall not be located in bathrooms as defined in Section 210-8.

C. Enclosures

240-30. General.   Overcurrent devices shall be enclosed in cabinets or cutout boxes.

Exception No. 1: Where a part of an assembly that provides equivalent protection.

Exception No. 2: Where mounted on open-type switchboards, panelboards, or control boards that are in rooms or enclosures free from dampness and easily ignitible material and accessible only to qualified personnel.

Exception No. 3: The operating handle of a circuit breaker shall be permitted to be accessible without opening a door or cover.

240-32. Damp or Wet Locations.   Enclosures for overcurrent devices in damp or wet locations shall comply with Section 373-2(a).

240-33. Vertical Position.   Enclosures for overcurrent devices shall be mounted in a vertical position.

Exception: Where this is shown to be impracticable and complies with Section 240-81.

D. Disconnecting and Guarding

240-40. Disconnecting Means for Fuses.   Disconnecting means shall be provided on the supply side of all fuses in circuits of over 150 volts to ground and cartridge fuses in circuits of any voltage, where accessible to other than qualified persons, so that each individual circuit containing fuses can be independently disconnected from the source of electric energy.

Exception No. 1: A device provided for current-limiting on the supply side of the service disconnecting means as permitted by Section 230-82.

Exception No. 2: A single disconnecting means shall be permitted on the supply side of more than one set of fuses as provided by Section 430-112 for group operation of motors and in Section 424-22 for fixed electric space heating equipment.

240-41. Arcing or Suddenly Moving Parts. Arcing or suddenly moving parts shall comply with (a) and (b) below.

(a) Location. Fuses and circuit breakers shall be so located or shielded that persons will not be burned or otherwise injured by their operation.

(b) Suddenly Moving Parts. Handles or levers of circuit breakers, and similar parts that may move suddenly in such a way that persons in the vicinity are likely to be injured by being struck by them, shall be guarded or isolated.

E. Plug Fuses, Fuseholders, and Adapters

240-50. General.

(a) Maximum Voltage. Plug fuses and fuseholders shall not be used in circuits exceeding 125 volts between conductors.

Exception: In circuits supplied by a system having a grounded neutral and having no conductor at over 150 volts to ground.

(b) Marking. Each fuse, fuseholder, and adapter shall be marked with its ampere rating.

(c) Hexagonal Configuration. Plug fuses of 15-ampere and lower rating shall be identified by a hexagonal configuration of the window, cap, or other prominent part to distinguish them from fuses of higher ampere ratings.

(d) No Energized Parts. Plug fuses, fuseholders, and adapters shall have no exposed energized parts after fuses or fuses and adapters have been installed.

(e) Screw Shell. The screw shell of a plug-type fuseholder shall be connected to the load side of the circuit.

240-51. Edison-Base Fuses.

(a) Classification. Plug fuses of the Edison-base type shall be classified at not over 125 volts and 30 amperes and below.

(b) Replacement Only. Plug fuses of the Edison-base type shall be used only for replacements in existing installations where there is no evidence of overfusing or tampering.

240-52. Edison-Base Fuseholders. Fuseholders of the Edison-base type shall be installed only where they are made to accept Type S fuses by the use of adapters.

240-53. Type S Fuses. Type S fuses shall be of the plug type and shall comply with (a) and (b) below.

(a) Classification. Type S fuses shall be classified at not over 125 volts and 0 to 15 amperes, 16 to 20 amperes, and 21 to 30 amperes.

(b) Noninterchangeable. Type S fuses of an ampere classification as specified in (a) above shall not be interchangeable with a lower ampere classification. They shall be so designed that they cannot be used in any fuseholder other than a Type S fuseholder or a fuseholder with a Type S adapter inserted.

240-54. Type S Fuses, Adapters, and Fuseholders.

(a) To Fit Edison-Base Fuseholders. Type S adapters shall fit Edison-base fuseholders.

(b) To Fit Type S Fuses Only. Type S fuseholders and adapters shall be so designed that either the fuseholder itself or the fuseholder with a Type S adapter inserted cannot be used for any fuse other than a Type S fuse.

(c) Nonremovable. Type S adapters shall be so designed that once inserted in a fuseholder, they cannot be removed.

(d) Nontamperable. Type S fuses, fuseholders, and adapters shall be so designed that tampering or shunting (bridging) would be difficult.

(e) Interchangeability. Dimensions of Type S fuses, fuseholders, and adapters shall be standardized to permit interchangeability regardless of the manufacturer.

F. Cartridge Fuses and Fuseholders

240-60. General.

(a) Maximum Voltage—300-Volt Type. Cartridge fuses and fuseholders of the 300-volt type shall not be used in circuits of over 300 volts between conductors.

Exception: In single-phase line-to-neutral circuits supplied from 3-phase, 4-wire, solidly grounded neutral systems where the line-to-neutral voltage does not exceed 300V.

(b) Noninterchangeable—0-6000 Ampere Cartridge Fuseholders. Fuseholders shall be so designed that it will be difficult to put a fuse of any given class into a fuseholder that is designed for a current lower, or voltage higher, than that of the class to which the fuse belongs. Fuseholders for current-limiting fuses shall not permit insertion of fuses that are not current-limiting.

(c) Marking. Fuses shall be plainly marked, either by printing on the fuse barrel or by a label attached to the barrel, showing the following: (1) ampere rating; (2) voltage rating; (3) interrupting rating where other than 10,000 amperes; (4) "current-limiting" where applicable; and (5) the name or trademark of the manufacturer.

Exception: Interrupting rating marking shall not be required on fuses used for supplementary protection.

240-61. Classification. Cartridge fuses and fuseholders shall be classified according to voltage and amperage ranges. Fuses rated 600 volts, nominal, or less, shall be permitted to be used for voltages at or below their ratings.

G. Circuit Breakers

240-80. Method of Operation. Circuit breakers shall be trip free and capable of being closed and opened by manual operation. Their normal method of operation by other than manual means, such as electrical or pneumatic, shall be permitted if means for manual operation is also provided.

Exception: As provided in Section 230-76(2) for circuit breakers used as service disconnecting means.

240-81. Indicating. Circuit breakers shall clearly indicate whether they are in the open "off" or closed "on" position.

Where circuit breaker handles are operated vertically rather than rotationally or horizontally, the "up" position of the handle shall be the "on" position.

240-82. Nontamperable. A circuit breaker shall be of such design that any alteration of its trip point (calibration) or the time required for its operation will require dismantling of the device or breaking of a seal for other than intended adjustments.

240-83. Marking.

(a) Durable and Visible. Circuit breakers shall be marked with their ampere rating in a manner that will be durable and visible after installation. Such marking shall be permitted to be made visible by removal of a trim or cover.

(b) Location. Circuit breakers rated at 100 amperes or less and 600 volts or less shall have the ampere rating molded, stamped, etched, or similarly marked into their handles or escutcheon areas.

(c) Interrupting Rating. Every circuit breaker having an interrupting rating other than 5000 amperes shall have its interrupting rating shown on the circuit breaker.

Exception: Interrupting rating marking shall not be required on circuit breakers used for supplementary protection.

If a circuit breaker is used on a circuit having an available fault current higher than its marked interrupting rating by being connected on the load side of an acceptable overcurrent-protective device having the higher rating, this additional series combination interrupting rating shall be marked on the end use equipment, such as switchboards and panelboards.

(d) Circuit Breakers Used as Switches. Where used as switches in 120-volt and 277-volt fluorescent lighting circuits, circuit breakers shall be listed and shall be marked "SWD."

(e) Voltage Marking. Circuit breakers shall be marked with a voltage rating no less than the nominal system voltage that is indicative of their capability to interrupt fault currents between phases or phase to ground.

(FPN): A circuit breaker with a straight voltage rating, e.g., 240V or 480V, may be applied in a circuit in which the nominal voltage between any two conductors does not exceed the circuit breaker's voltage rating; except that a two-pole circuit breaker is not suitable for protecting a 3-phase corner-grounded delta circuit unless it is marked 1-phase/3-phase to indicate such suitability.

A circuit breaker with a slash rating, e.g., 120/240V or 480Y/277V, may only be applied in a circuit in which the nominal voltage to ground from any conductor does not exceed the lower of the two values of the circuit breaker's voltage rating and the nominal voltage between any two conductors does not exceed the higher value of the circuit breaker's voltage rating.

H. Overcurrent Protection Over 600 Volts, Nominal

240-100. Feeders.   Feeders shall have a short-circuit protective device in each ungrounded conductor or comply with Section 230-208(d)(2) or (d)(3). The protective device(s) shall be capable of detecting and interrupting all values of current that can occur at their location in excess of their trip setting or melting point. In no case shall the fuse rating in continuous amperes exceed three times, or the long-time trip element setting of a breaker, six times, the ampacity of the conductor.

Conductors tapped to a feeder shall be permitted to be protected by the feeder overcurrent device where that overcurrent device also protects the tap conductors.

(FPN):   The operating time of the protective device, the available short-circuit current, and the conductor used will need to be coordinated to prevent damaging or dangerous temperatures in conductors or conductor insulation under short-circuit conditions.

240-101. Branch Circuits.   Branch circuits shall have a short-circuit protective device in each ungrounded conductor or comply with Section 230-208(d)(2) or (d)(3). The protective device(s) shall be capable of detecting and interrupting all values of current that can occur at their location in excess of their trip setting or melting point.

ARTICLE 250 — GROUNDING

> **NOTICE**
> Following the issuance of this edition an appeal was filed with respect to Section 250-45 of this article — refer to page 70-iii.

A. General

250-1. Scope.   This article covers general requirements for grounding and bonding of electrical installations, and specific requirements in (a) through (f) below.

(a) Systems, circuits, and equipment required, permitted, or not permitted to be grounded.

(b) Circuit conductor to be grounded on grounded systems.

(c) Location of grounding connections.

(d) Types and sizes of grounding and bonding conductors and electrodes.

(e) Methods of grounding and bonding.

(f) Conditions under which guards, isolation, or insulation may be substituted for grounding.

(FPN No. 1):   Systems and circuit conductors are grounded to limit voltages due to lightning, line surges, or unintentional contact with higher voltage lines, and to stabilize the voltage to ground during normal operation. Equipment grounding conductors are bonded to the system grounded conductor to provide a low impedance path for fault current that will facilitate the operation of overcurrent devices under ground-fault conditions.

(FPN No. 2):   Conductive materials enclosing electrical conductors or equipment, or forming part of such equipment, are grounded to limit the voltage to ground on these materials and bonded to facilitate the operation of overcurrent devices under ground-fault conditions. See Section 110-10.

250-2. Application of Other Articles. In other articles applying to particular cases of installation of conductors and equipment, there are requirements that are in addition to those of this article or are modifications of them:

	Article	Section
Agricultural Buildings		547-8
Appliances		422-16
Branch Circuits		210-5
		210-6
		210-7
Cablebus		365-9
Circuits and Equipment Operating at Less Than 50 Volts	720	
Class 1, Class 2, and Class 3 Remote-Control, Signaling, and Power-Limited Circuits		725-20
		725-43
Closed-Loop and Programmed Power Distribution		780-3
Communications Circuits	800	
Community Antenna Television and Radio Distribution Systems		820-33
		820-40
		820-41
Conductors for General Wiring	310	
Cranes and Hoists		610
Electronic Computer/Data Processing Equipment		645-15
Electrically Driven or Controlled Irrigation Machines		675-11(c)
		675-12
		675-13
		675-14
		675-15
Electric Signs and Outline Lighting	600	
Electrolytic Cells	668	
Elevators, Dumbwaiters, Escalators, Moving Walks, Wheelchair Lifts, and Stairway Chair Lifts	620	
Fire Protective Signaling Systems		760-6
Fixed Electric Heating Equipment for Pipelines and Vessels		427-21
		427-29
		427-48
Fixed Electric Space Heating Equipment		424-14
Fixed Outdoor Electric Deicing and Snow-Melting Equipment		426-27
Fixtures and Lighting Equipment		410-17
		410-18
		410-19
		410-21
		410-105(b)
Flexible Cords and Cables		400-22
		400-23

B. Circuit and System Grounding

250-3. Direct-Current Systems.

(a) Two-Wire Direct-Current Systems. Two-wire dc systems supplying premises wiring shall be grounded.

Exception No. 1: A system equipped with a ground detector and supplying only industrial equipment in limited areas.

Exception No. 2: A system operating at 50 volts or less between conductors.

Exception No. 3: A system operating at over 300 volts between conductors.

Exception No. 4: A rectifier-derived dc system supplied from an ac system complying with Section 250-5.

Exception No. 5: DC fire protective signaling circuits having a maximum current of 0.030 amperes as specified in Article 760, Part C.

(b) Three-Wire Direct-Current Systems. The neutral conductor of all 3-wire dc systems supplying premises wiring shall be grounded.

250-5. Alternating-Current Circuits and Systems to Be Grounded. AC circuits and systems shall be grounded as provided for in (a), (b), (c), or (d) below. Other circuits and systems shall be permitted to be grounded.

(FPN): An example of a system permitted to be grounded is a corner-grounded delta transformer connection. See Section 250-25(4) for conductor to be grounded.

(a) Alternating-Current Circuits of Less than 50 Volts. AC circuits of less than 50 volts shall be grounded under any of the following conditions:

(1) Where supplied by transformers if the transformer supply system exceeds 150 volts to ground.

(2) Where supplied by transformers if the transformer supply system is ungrounded.

(3) Where installed as overhead conductors outside of buildings.

(b) Alternating-Current Systems of 50 Volts to 1000 Volts. AC systems of 50 volts to 1000 volts supplying premises wiring and premises wiring systems shall be grounded under any of the following conditions:

(1) Where the system can be so grounded that the maximum voltage to ground on the ungrounded conductors does not exceed 150 volts.

(2) Where the system is 3-phase, 4-wire wye connected in which the neutral is used as a circuit conductor.

(3) Where the system is 3-phase, 4-wire delta connected in which the midpoint of one phase winding is used as a circuit conductor.

(4) Where a grounded service conductor is uninsulated in accordance with the exceptions to Sections 230-22, 230-30, and 230-41.

Exception No. 1: Electric systems used exclusively to supply industrial electric furnaces for melting, refining, tempering, and the like.

Exception No. 2: Separately derived systems used exclusively for rectifiers supplying only adjustable speed industrial drives.

Exception No. 3: Separately derived systems supplied by transformers that have a primary voltage rating less than 1000 volts, provided that all of the following conditions are met:

a. The system is used exclusively for control circuits.

b. The conditions of maintenance and supervision assure that only qualified persons will service the installation.

c. *Continuity of control power is required.*

d. *Ground detectors are installed on the control system.*

Exception No. 4: Isolated systems as permitted or required in Articles 517 and 668.

(FPN): The proper use of suitable ground detectors on ungrounded systems can provide additional protection.

Exception No. 5: High-impedance grounded neutral systems in which a grounding impedance, usually a resistor, limits the ground-fault current to a low value. High-impedance grounded neutral systems shall be permitted for 3-phase ac systems of 480 volts to 1000 volts where all of the following conditions are met:

a. *The conditions of maintenance and supervision assure that only qualified persons will service the installation.*

b. *Continuity of power is required.*

c. *Ground detectors are installed on the system.*

d. *Line-to-neutral loads are not served.*

(c) Alternating-Current Systems of 1 kV and Over. AC systems supplying mobile or portable equipment shall be grounded as specified in Section 250-154. Where supplying other than portable equipment, such systems shall be permitted to be grounded. Where such systems are grounded, they shall comply with the applicable provisions of this article.

(d) Separately Derived Systems. A premises wiring system whose power is derived from generator, transformer, or converter windings and has no direct electrical connection, including a solidly connected grounded circuit conductor, to supply conductors originating in another system, if required to be grounded as in (a) or (b) above, shall be grounded as specified in Section 250-26.

(FPN No. 1): An alternate alternating-current power source such as an on-site generator is not a separately derived system if the neutral is solidly interconnected to a service-supplied system neutral.

(FPN No. 2): For systems that are not separately derived and are not required to be grounded as specified in Section 250-26, see Section 445-5 for minimum size of conductors that must carry fault current.

250-6. Portable and Vehicle-Mounted Generators.

(a) Portable Generators. Under the following conditions, the frame of a portable generator shall not be required to be grounded and shall be permitted to serve as the grounding electrode for a system supplied by the generator:

(1) The generator supplies only equipment mounted on the generator or cord- and plug-connected equipment through receptacles mounted on the generator, or both, and

(2) The noncurrent-carrying metal parts of equipment and the equipment grounding conductor terminals of the receptacles are bonded to the generator frame.

(b) Vehicle-Mounted Generators. Under the following conditions, the frame of a vehicle shall be permitted to serve as the grounding electrode for a system supplied by a generator located on the vehicle:

(1) The frame of the generator is bonded to the vehicle frame, and

(2) The generator supplies only equipment located on the vehicle or cord- and plug-connected equipment through receptacles mounted on the vehicle or both equipment located on the vehicle and cord- and plug-connected equipment through receptacles mounted on the vehicle or on the generator, and

(3) The noncurrent-carrying metal parts of equipment and the equipment grounding conductor terminals of the receptacles are bonded to the generator frame, and

(4) The system complies with all other provisions of this article.

(c) Neutral Conductor Bonding. A neutral conductor shall be bonded to the generator frame where the generator is a component of a separately derived system. The bonding of any conductor other than a neutral within the generator to its frame shall not be required.

(FPN):   For grounding of portable generators supplying fixed wiring systems, see Section 250-5(d).

250-7. Circuits Not to Be Grounded.   The following circuits shall not be grounded:

(a) Cranes. Circuits for electric cranes operating over combustible fibers in Class III locations, as provided in Section 503-13.

(b) Health Care Facilities. Circuits as provided in Article 517.

(c) Electrolytic Cells. Circuits as provided in Article 668.

C. Location of System Grounding Connections

250-21. Objectionable Current over Grounding Conductors.

(a) Arrangement to Prevent Objectionable Current. The grounding of electric systems, circuit conductors, surge arresters, and conductive noncurrent-carrying materials and equipment shall be installed and arranged in a manner that will prevent an objectionable flow of current over the grounding conductors or grounding paths.

(b) Alterations to Stop Objectionable Current. If the use of multiple grounding connections results in an objectionable flow of current, one or more of the following alterations shall be permitted to be made, provided that the requirements of Section 250-51 are met:

(1) Discontinue one or more but not all of such grounding connections.

(2) Change the locations of the grounding connections.

(3) Interrupt the continuity of the conductor or conductive path interconnecting the grounding connections.

(4) Take other suitable remedial action satisfactory to the authority having jurisdiction.

(c) Temporary Currents Not Classified as Objectionable Currents. Temporary currents resulting from accidental conditions, such as ground-fault currents, that occur only while the grounding conductors are performing their intended protective functions shall not be classified as objectionable current for the purposes specified in (a) and (b) above.

(d) Limitations to Permissible Alterations. The provisions of this section shall not be considered as permitting electronic equipment being operated on ac systems or branch circuits that are not grounded as required by this article. Currents that introduce noise or data errors in electronic equipment shall not be considered the objectionable currents addressed in this section.

250-22. Point of Connection for Direct-Current Systems. DC systems to be grounded shall have the grounding connection made at one or more supply stations. A grounding connection shall not be made at individual services nor at any point on premises wiring.

Exception: Where the dc system source is located on the premises, a grounding connection shall be made either (1) at the source or the first system disconnecting means or overcurrent device, or (2) by another means that accomplishes equivalent system protection and that utilizes equipment listed and identified for the use.

250-23. Grounding Service-Supplied Alternating-Current Systems.

(a) System Grounding Connections. A premises wiring system that is supplied by an ac service and is required to be grounded by Section 250-5 shall have at each service a grounding electrode conductor connected to a grounding electrode that complies with Part H of Article 250. The grounding electrode conductor shall be connected to the grounded service conductor at any accessible point from the load end of the service drop or service lateral to and including the terminal or bus to which the grounded service conductor is connected at the service disconnecting means. Where the transformer supplying the service is located outside the building, at least one additional grounding connection shall be made from the grounded service conductor to a grounding electrode, either at the transformer or elsewhere outside the building. A grounding connection shall not be made to any grounded circuit conductor on the load side of the service disconnecting means.

(FPN): See definition of "Service Drop" and "Service Lateral"; also Section 230-21.

Exception No. 1: A grounding electrode conductor shall be connected to the grounded conductor of a separately derived system in accordance with the provisions of Section 250-26(b).

Exception No. 2: A grounding conductor connection shall be made at each separate building where required by Section 250-24.

Exception No. 3: For ranges, counter-mounted cooking units, wall-mounted ovens, clothes dryers, and meter enclosures as permitted by Section 250-61.

Exception No. 4: For services that are dual fed (double ended) in a common enclosure or grouped together in separate enclosures and employing a secondary tie, a single grounding electrode connection to the tie point of the grounded circuit conductors from each power source shall be permitted.

Exception No. 5: Where the main bonding jumper specified in Sections 250-53(b) and 250-79 is a wire or busbar, and is installed from the neutral bar or bus to the equipment grounding terminal bar or bus in the service equipment, the grounding electrode conductor shall be permitted to be connected to the equipment grounding terminal bar or bus to which the main bonding jumper is connected.

Exception No. 6: As covered in Section 250-27 for high-impedance grounded neutral systems grounding connection requirements.

(b) Grounded Conductor Brought to Service Equipment. Where an ac system operating at less than 1000 volts is grounded at any point, the grounded conductor shall be run to each service disconnecting means and shall be bonded to each disconnecting means enclosure. This conductor shall be routed with the phase conductors and shall not be smaller than the required grounding electrode conductor specified in Table 250-94, and, in addition, for service-entrance phase conductors larger than 1100 kcmil copper or 1750 kcmil aluminum, the grounded conductor shall not be smaller than $12\frac{1}{2}$ percent of the area of the largest service-entrance phase conductor. Where the service-entrance phase conductors are paralleled, the size of the grounded conductor shall be based on the equivalent area for parallel conductors as indicated in this section.

(FPN):   See Section 310-4 for grounded conductors connected in parallel.

Exception No. 1: The grounded conductor shall not be required to be larger than the largest ungrounded service-entrance phase conductor.

Exception No. 2: As covered in Section 250-27 for high-impedance grounded neutral systems grounding connection requirements.

Exception No. 3: Where more than one service disconnecting means are located in an assembly listed for use as service equipment, one grounded conductor shall be required to be run to the assembly, and it shall be bonded to the assembly enclosure.

250-24. Two or More Buildings or Structures Supplied from a Common Service.

(a) Grounded Systems. Where two or more buildings or structures are supplied from a common service, the grounded system in each building or structure shall have a grounding electrode as described in Part H connected to the metal enclosure of the building disconnecting means and to the ac system grounded circuit conductor on the supply side of the building or structure disconnecting means.

Exception No. 1: A grounding electrode at separate buildings or structures shall not be required where only one branch circuit is supplied and there is no equipment in the building or structure that requires grounding.

Exception No. 2: A grounded circuit conductor connection to the grounding electrode shall not be required at a separate building or structure if an equipment grounding conductor is run with the circuit conductors for grounding any noncurrent-carrying equipment, interior metal piping systems, and building or structural metal frames, and the equipment grounding conductor is bonded at a separate building or structure disconnecting means to existing grounding electrodes described in Part H. Where there are no existing electrodes, a grounding electrode meeting the requirements of Part H shall be installed where the building or structure supplies more than one branch circuit. Where livestock is housed, that portion of the equipment grounding conductor run underground to the disconnecting means shall be insulated or covered copper.

(FPN): See Section 547-8(a), Exception for special grounding requirements for agricultural buildings.

(b) Ungrounded Systems. Where two or more buildings or structures are supplied by a common service from an ungrounded system, each building or structure shall have a grounding electrode as described in Part H connected to the metal enclosure of the building or structure disconnecting means.

Exception No. 1: A grounding electrode at separate buildings or structures shall not be required where only one branch circuit is supplied and there is no equipment in the building or structure that requires grounding.

Exception No. 2: A grounding electrode and grounding electrode conductor connection to the metal enclosure of the building or structure disconnecting means shall not be required, provided all of the following conditions are met:

a. An equipment grounding conductor is run with the circuit conductors to the building or structure disconnecting means for grounding any noncurrent-carrying equipment, interior metal piping systems, and building or structural metal frames.

b. There are no existing grounding electrodes as described in Part H.

c. The building or structure supplies only one branch circuit.

d. Where livestock is housed, that portion of the equipment grounding conductor run underground to the disconnecting means shall be insulated or covered copper.

(FPN): See Section 547-8(a), Exception for special grounding requirements for agricultural buildings.

(c) Disconnecting Means Located in Separate Building or Structure on the Same Premises. Where one or more disconnecting means supply one or more additional buildings or structures under single management, and where these disconnecting means are located remote from those buildings or structures in accordance with the provisions of Section 225-8(b), Exception Nos. 1 and 2 or Section 230-205(a), Exception, all of the following conditions shall be met:

(1) The connection of the grounded circuit conductor to the grounding electrode at a separate building or structure shall not be made;

(2) An equipment grounding conductor for grounding any noncurrent-carrying equipment, interior metal piping systems, and building or structural metal frames is run with the circuit conductors to a separate building or structure and bonded to existing grounding electrodes described in Part H, or, where there are no existing electrodes, a grounding electrode meeting the requirements of Part H shall be installed where a separate building or structure supplies more than one branch circuit;

(3) Bonding the equipment grounding conductor to the grounding electrode at a separate building or structure shall be made in a junction box, panelboard, or similar enclosure located immediately inside or outside the separate building or structure.

Exception No. 1: A grounding electrode at a separate building or structure shall not be required where only one branch circuit is supplied and there is no equipment in the building or structure that requires grounding.

Exception No. 2: Where livestock is housed, that portion of the equipment grounding conductor run underground to the disconnecting means shall be insulated or covered copper.

(d) Grounding Conductor. The size of the grounding conductor to the grounding electrode(s) shall not be less than given in Table 250-95, and the installation shall be in accordance with Sections 250-92(a) and (b).

Exception No. 1: The grounding conductor shall not be required to be larger than the largest ungrounded supply conductor.

Exception No. 2: Where connected to made electrodes as in Section 250-83(c) or (d), that portion of the grounding conductor that is the sole connection between the electrode(s) and the grounded or grounding conductor or the metal enclosure of the building disconnecting means shall not be required to be larger than No. 6 copper or No. 4 aluminum.

250-25. Conductor to Be Grounded — Alternating-Current Systems.   For ac premises wiring systems, the conductor to be grounded shall be as specified in (1) through (5) below.

(1) Single-phase, 2-wire: one conductor.

(2) Single-phase, 3-wire: the neutral conductor.

(3) Multiphase systems having one wire common to all phases: the common conductor.

(4) Multiphase systems requiring one grounded phase: one phase conductor.

(5) Multiphase systems in which one phase is used as in (2) above: the neutral conductor.

Grounded conductors shall be identified by the means specified in Article 200.

250-26. Grounding Separately Derived Alternating-Current Systems. A separately derived ac system that is required to be grounded by Section 250-5 shall be grounded as specified in (a) through (d) below.

(a) Bonding Jumper. A bonding jumper, sized in accordance with Section 250-79(d) for the derived phase conductors, shall be used to connect the equipment grounding conductors of the derived system to the grounded conductor. Except as permitted by Exception No. 4 or No. 5 of Section 250-23(a), this connection shall be made at any point on the separately derived system from the source to the first system disconnecting means or overcurrent device; or it shall be made at the source of a separately derived system that has no disconnecting means or overcurrent devices.

Exception No. 1: The size of the bonding jumper for a system that supplies a Class 1, Class 2, or Class 3 circuit, and is derived from a transformer rated not more than 1000 volt-amperes, shall not be smaller than the derived phase conductors and shall not be smaller than No. 14 copper or No. 12 aluminum.

Exception No. 2: As covered in Section 250-27 and Exception No. 5 for Section 250-5(b) for high-impedance grounded neutral systems grounding connection requirements.

(b) Grounding Electrode Conductor. A grounding electrode conductor, sized in accordance with Section 250-94 for the derived phase conductors, shall be used to connect the grounded conductor of the derived system to the grounding electrode as specified in (c) below. Except as permitted by Exception No. 4 of Section 250-23(a), this connection shall be made at any point on the separately derived system from the source to the first system disconnecting means or overcurrent device; or it shall be made at the source of a separately derived system that has no disconnecting means or overcurrent devices.

Exception No. 1: A grounding electrode conductor shall not be required for a system that supplies a Class 1, Class 2, or Class 3 circuit, and is derived from a transformer rated not more than 1000 volt-amperes, provided the system grounded conductor is bonded to the transformer frame or enclosure by a jumper sized in accordance with Section 250-26, Exception No. 1 for (a) above, and the transformer frame or enclosure is grounded by one of the means specified in Section 250-57.

Exception No. 2: As covered in Section 250-27 and Exception No. 5 for Section 250-5(b) for high-impedance grounded neutral systems grounding connection requirements.

(c) Grounding Electrode. The grounding electrode shall be as near as practicable to and preferably in the same area as the grounding conductor connection to the system. The grounding electrode shall be (1) the nearest available effectively grounded structural metal member of the structure; or (2) the nearest available effectively grounded metal water pipe; or (3) other electrodes as specified in Sections 250-81 and 250-83 where electrodes specified by (1) or (2) above are not available.

(d) Grounding Methods. In all other respects, grounding methods shall comply with requirements prescribed in other parts of this Code.

250-27. High-Impedance Grounded Neutral System Connections. High-impedance grounded neutral systems as permitted in Exception No. 5 of Section 250-5(b) shall comply with the provisions of (a) through (f) below.

(a) Grounding Impedance Location. The grounding impedance shall be installed between the grounding electrode conductor and the system neutral. Where a neutral is not available, the grounding impedance shall be installed between the grounding electrode conductor and the neutral derived from a grounding transformer.

(b) Neutral Conductor. The neutral conductor from the neutral point of the transformer or generator to its connection point to the grounding impedance shall be fully insulated.

The neutral conductor shall have an ampacity of not less than the maximum current rating of the grounding impedance. In no case shall the neutral conductor be smaller than No. 8 copper or No. 6 aluminum, or copper-clad aluminum.

(c) System Neutral Connection. The system neutral conductor shall not be connected to ground except through the grounding impedance.

(FPN): The impedance is normally selected to limit the ground-fault current to a value slightly greater than or equal to the capacitive charging current of the system. This value of impedance will also limit transient overvoltages to safe values. For guidance, refer to criteria for limiting transient overvoltages in Recommended Practice for Grounding of Industrial and Commercial Power Systems, ANSI/IEEE 142-1982.

(d) Neutral Conductor Routing. The conductor connecting the neutral point of the transformer or generator to the grounding impedance shall be permitted to be installed in a separate raceway. It shall not be required to run this conductor with the phase conductors to the first system disconnecting means or overcurrent device.

(e) Equipment Bonding Jumper. The equipment bonding jumper (the connection between the equipment grounding conductors and the grounding impedance) shall be an unspliced conductor run from the first system disconnecting means or overcurrent device to the grounded side of the grounding impedance.

(f) Grounding Electrode Conductor Location. The grounding electrode conductor shall be attached at any point from the grounded side of the grounding impedance to the equipment grounding connection at the service equipment or first system disconnecting means.

D. Enclosure Grounding

250-32. Service Raceways and Enclosures. Metal enclosures for service conductors and equipment shall be grounded.

Exception: A metal elbow that is installed in an underground installation of rigid nonmetallic conduit and is isolated from possible contact by a minimum cover of 18 inches (457 mm) to any part of the elbow.

250-33. Other Conductor Enclosures. Metal enclosures for other than service conductors shall be grounded.

Exception No. 1: Metal enclosures for conductors added to existing installations of open wire, knob-and-tube wiring, and nonmetallic-sheathed cable, that do not provide an equipment ground, if in runs of less than 25 feet (7.62 m), if free from probable contact with ground, grounded metal, metal lath, or other conductive material, and if guarded against contact by persons, shall not be required to be grounded.

Exception No. 2: Short sections of metal enclosures used to provide support or protection of cable assemblies from physical damage shall not be required to be grounded.

Exception No. 3: Enclosures not required to be grounded by Section 250-43(i) shall not be required to be grounded.

Exception No. 4: A metal elbow that is installed in an underground installation of rigid nonmetallic conduit and is isolated from possible contact by a minimum cover of 18 inches (457 mm) to any part of the elbow.

E. Equipment Grounding

250-42. Equipment Fastened in Place or Connected by Permanent Wiring Methods (Fixed). Exposed noncurrent-carrying metal parts of fixed equipment likely to become energized shall be grounded under any of the conditions in (a) through (f) below.

(a) Vertical and Horizontal Distances. Where within 8 feet (2.44 m) vertically or 5 feet (1.52 m) horizontally of ground or grounded metal objects and subject to contact by persons.

(b) Wet or Damp Locations. Where located in a wet or damp location and not isolated.

(c) Electrical Contact. Where in electrical contact with metal.

(d) Hazardous (Classified) Locations. Where in a hazardous (classified) location as covered by Articles 500 through 517.

(e) Wiring Methods. Where supplied by a metal-clad, metal-sheathed, metal-raceway, or other wiring method that provides an equipment ground, except as permitted by Section 250-33 for short sections of metal enclosures.

(f) Over 150 Volts to Ground. Where equipment operates with any terminal at over 150 volts to ground.

Exception No. 1: Enclosures for switches or circuit breakers used for other than service equipment and accessible to qualified persons only.

Exception No. 2: Metal frames of electrically heated appliances, exempted by special permission, in which case the frames shall be permanently and effectively insulated from ground.

Exception No. 3: Distribution apparatus, such as transformer and capacitor cases, mounted on wooden poles, at a height exceeding 8 feet (2.44 m) above ground or grade level.

Exception No. 4: Listed equipment protected by a system of double insulation, or its equivalent, shall not be required to be grounded. Where such a system is employed, the equipment shall be distinctively marked.

250-43. Fastened in Place or Connected by Permanent Wiring Methods (Fixed) — Specific. Exposed, noncurrent-carrying metal parts of the kinds of equipment described in (a) through (j) below, and noncurrent-carrying metal parts of equipment and enclosures described in (k) and (l) below, shall be grounded regardless of voltage.

(a) Switchboard Frames and Structures. Switchboard frames and structures supporting switching equipment.

Exception: Frames of 2-wire dc switchboards where effectively insulated from ground.

(b) Pipe Organs. Generator and motor frames in an electrically operated pipe organ.

Exception: Where the generator is effectively insulated from ground and from the motor driving it.

(c) Motor Frames. Motor frames, as provided by Section 430-142.

(d) Enclosures for Motor Controllers. Enclosures for motor controllers.

Exception No. 1: Enclosures attached to ungrounded portable equipment.

Exception No. 2: Lined covers of snap switches.

(e) Elevators and Cranes. Electric equipment for elevators and cranes.

(f) Garages, Theaters, and Motion Picture Studios. Electric equipment in commercial garages, theaters, and motion picture studios.

Exception: Pendant lampholders supplied by circuits not over 150 volts to ground.

(g) Electric Signs. Electric signs and associated equipment.

(h) Motion Picture Projection Equipment. Motion picture projection equipment.

(i) Remote-Control, Signaling, and Fire Protective Signaling Circuits. Equipment supplied by Class 1, Class 2, and Class 3 remote-control and signaling circuits, and by fire protective signaling circuits, shall be grounded where system grounding is required by Part B of this article.

(j) Lighting Fixtures. Lighting fixtures as provided in Part E of Article 410.

(k) Motor-Operated Water Pumps. Motor-operated water pumps including the submersible type.

(l) Metal Well Casings. Where a submersible pump is used in a metal well casing, the well casing shall be bonded to the pump circuit equipment grounding conductor.

250-44. Nonelectric Equipment. The metal parts of nonelectric equipment described in (a) through (e) below shall be grounded.

(a) Cranes. Frames and tracks of electrically operated cranes.

(b) Elevator Cars. Frames of nonelectrically driven elevator cars to which electric conductors are attached.

(c) Electric Elevators. Hand-operated metal shifting ropes or cables of electric elevators.

(d) Metal Partitions. Metal partitions, grill work, and similar metal enclosures around equipment of 1 kV and over between conductors except substations or vaults under the sole control of the supply company.

(e) Mobile Homes and Recreational Vehicles. Mobile homes and recreational vehicles as required in Articles 550 and 551.

(FPN): Where extensive metal in or on buildings may become energized and is subject to personal contact, adequate bonding and grounding will provide additional safety.

250-45. Equipment Connected by Cord and Plug. Under any of the conditions described in (a) through (d) below, exposed noncurrent-carrying metal parts of cord- and plug-connected equipment likely to become energized shall be grounded.

(a) In Hazardous (Classified) Locations. In hazardous (classified) locations (see Articles 500 through 517).

(b) Over 150 Volts to Ground. Where operated at over 150 volts to ground.

Exception No. 1: Motors, where guarded.

Exception No. 2: Metal frames of electrically heated appliances, exempted by special permission, in which case the frames shall be permanently and effectively insulated from ground.

Exception No. 3: Listed equipment protected by a system of double insulation, or its equivalent, shall not be required to be grounded. Where such a system is employed, the equipment shall be distinctively marked.

(c) In Residential Occupancies. In residential occupancies: (1) refrigerators, freezers, and air conditioners; (2) clothes-washing, clothes-drying, dish-washing machines, kitchen waste disposers, sump pumps, electrical aquarium equipment; (3) hand-held motor-operated tools, stationary and fixed motor-operated tools, light industrial motor-operated tools; (4) motor-operated appliances of the following types: hedge clippers, lawn mowers, snow blowers, and wet scrubbers; (5) portable handlamps.

Exception: Listed tools and listed appliances protected by a system of double insulation, or its equivalent, shall not be required to be grounded. Where such a system is employed, the equipment shall be distinctively marked.

(d) In Other than Residential Occupancies. In other than residential occupancies: (1) refrigerators, freezers, and air conditioners; (2) clothes-washing, clothes-drying, dish-washing machines, electronic computer/data processing equipment, sump pumps, electrical aquarium equipment; (3) hand-held motor-operated tools, stationary and fixed motor-operated tools, light industrial motor-operated tools; (4) motor-operated appliances of the following types: hedge clippers, lawn mowers, snow blowers, and wet scrubbers; (5) cord- and plug-connected appliances used in damp or wet locations or by persons standing on the ground or on metal floors or working inside of metal tanks or boilers; (6) tools likely to be used in wet or conductive locations; and (7) portable handlamps.

Exception No. 1: Tools and portable handlamps likely to be used in wet or conductive locations shall not be required to be grounded where supplied through an isolating transformer with an ungrounded secondary of not over 50 volts.

Exception No. 2: Listed, hand-held, motor-operated tools, listed stationary and fixed motor-operated tools, listed, light industrial, motor-operated tools, and listed appliances protected by an approved system of double insulation or its equivalent shall not be required to be grounded. Where such a system is employed, the equipment shall be distinctively marked.

(FPN): With reference to (c) and (d), portable tools or appliances are not intended to be used in damp, wet, or conductive locations unless they are grounded, double insulated, or supplied through an isolating transformer.

250-46. Spacing from Lightning Rods. Metal raceways, enclosures, frames, and other noncurrent-carrying metal parts of electric equipment shall be kept at least 6 feet (1.83 m) away from lightning rod conductors, or they shall be bonded to the lightning rod conductors.

(FPN): See Sections 250-86 and 800-40(b)(3)(3). For further information, see Lightning Protection Code, NFPA 780-1992 (ANSI), which contains detailed information on grounding lightning protection systems.

F. Methods of Grounding

250-50. Equipment Grounding Conductor Connections. Equipment grounding conductor connections at the source of separately derived systems shall be made in accordance with Section 250-26(a). Equipment grounding conductor connections at service equipment shall be made as indicated in (a) or (b) below.

(a) For Grounded System. The connection shall be made by bonding the equipment grounding conductor to the grounded service conductor and the grounding electrode conductor.

(b) For Ungrounded System. The connection shall be made by bonding the equipment grounding conductor to the grounding electrode conductor.

Exception for (a) and (b) above: For replacement of nongrounding-type receptacles with grounding-type receptacles and for branch-circuit extensions only in existing installations that do not have an equipment grounding conductor in the branch circuit, the grounding conductor of a grounding-type receptacle outlet shall be permitted to be grounded to any accessible point on the grounding electrode system as described in Section 250-81.

(FPN): See Section 210-7(d), Exception for the use of a ground-fault circuit-interrupter-type of receptacle.

250-51. Effective Grounding Path. The path to ground from circuits, equipment, and metal enclosures for conductors shall (1) be permanent and continuous; (2) have capacity to conduct safely any fault current likely to be imposed on it; and (3) have sufficiently low impedance to limit the voltage to ground and to facilitate the operation of the circuit protective devices.

The earth shall not be used as the sole equipment grounding conductor.

250-53. Grounding Path to Grounding Electrode at Services.

(a) Grounding Electrode Conductor. A grounding electrode conductor shall be used to connect the equipment grounding conductors, the service-equipment enclosures, and, where the system is grounded, the grounded service conductor to the grounding electrode.

Exception: As covered in Section 250-27 for high-impedance grounded neutral system connections.

(FPN):   See Section 250-23(a) for alternating-current system grounding connections. |

(b) Main Bonding Jumper. For a grounded system, an unspliced main bonding jumper shall be used to connect the equipment grounding conductor and the service-disconnect enclosure to the grounded conductor of the system at each service disconnect.

Exception No. 1: Where more than one service disconnecting means are located in an assembly listed for use as service equipment, one grounded conductor shall be required to be run to the assembly, and it shall be bonded to the assembly enclosure.

Exception No. 2: As covered in Section 250-27 for high-impedance grounded neutral system connections.

250-54. Common Grounding Electrode.
Where an ac system is connected to a grounding electrode in or at a building as specified in Sections 250-23 and 250-24, the same electrode shall be used to ground conductor enclosures and equipment in or on that building. Where separate services supply a building and are required to be connected to a grounding electrode, the same grounding electrode shall be used.

Two or more grounding electrodes that are effectively bonded together shall be considered as a single grounding electrode system in this sense.

250-55. Underground Service Cable.
Where served from a continuous underground metal-sheathed cable system, the sheath or armor of underground service cable metallically connected to the underground system, or underground service conduit containing a metal-sheathed cable bonded to the underground system, shall not be required to be grounded at the building and shall be permitted to be insulated from the interior conduit or piping.

250-56. Short Sections of Raceway.
Isolated sections of metal raceway or cable armor, where required to be grounded, shall be grounded in accordance with Section 250-57.

250-57. Equipment Fastened in Place or Connected by Permanent Wiring Methods (Fixed)—Grounding.
Noncurrent-carrying metal parts of equipment, raceways, and other enclosures, where required to be grounded, shall be grounded by one of the methods indicated in (a) or (b) below.

Exception: Where equipment, raceways, and enclosures are grounded by connection to the grounded circuit conductor as permitted by Sections 250-24, 250-60, and 250-61.

(a) **Equipment Grounding Conductor Types.** By any of the equipment grounding conductors permitted by Section 250-91(b).

(b) **With Circuit Conductors.** By an equipment grounding conductor contained within the same raceway, cable, or cord or otherwise run with the circuit conductors. Bare, covered, or insulated equipment grounding conductors shall be permitted. Individually covered or insulated equipment grounding conductors shall have a continuous outer finish that is either green or green with one or more yellow stripes.

Exception No. 1: An insulated or covered conductor larger than No. 6 copper or aluminum shall, at the time of installation, be permitted to be permanently identified as an equipment grounding conductor at each end and at every point where the conductor is accessible. Identification shall be accomplished by one of the following:

a. Stripping the insulation or covering from the entire exposed length,

b. Coloring the exposed insulation or covering green, or

c. Marking the exposed insulation or covering with green colored tape or green colored adhesive labels.

Exception No. 2: For direct-current circuits only, the equipment grounding conductor shall be permitted to be run separately from the circuit conductors.

Exception No. 3: As provided in the exception to Sections 250-50(a) and (b), the equipment grounding conductor shall be permitted to be run separately from the circuit conductors.

Exception No. 4: Where the conditions of maintenance and supervision assure that only qualified persons will service the installation, one or more insulated conductors in a multiconductor cable shall, at the time of installation, be permitted to be permanently identified as equipment grounding conductors at each end and at every point where the conductors are accessible by one of the following means:

a. Stripping the insulation from the entire exposed length,

b. Coloring the exposed insulation green, or

c. Marking the exposed insulation with green tape or green colored adhesive labels.

(FPN No. 1):   See Section 250-79 for equipment bonding jumper requirements.

(FPN No. 2):   See Section 400-7 for use of cords for fixed equipment.

250-58. Equipment Considered Effectively Grounded.   Under the conditions specified in (a) and (b) below, the noncurrent-carrying metal parts of the equipment shall be considered effectively grounded.

(a) **Equipment Secured to Grounded Metal Supports.** Electric equipment secured to and in electrical contact with a metal rack or structure provided for its support and grounded by one of the means indicated in Section 250-57. The structural metal frame of a building shall not be used as the required equipment grounding conductor for ac equipment.

(b) Metal Car Frames. Metal car frames supported by metal hoisting cables attached to or running over metal sheaves or drums of elevator machines that are grounded by one of the methods indicated in Section 250-57.

250-59. Cord- and Plug-Connected Equipment. Noncurrent-carrying metal parts of cord- and plug-connected equipment, where required to be grounded, shall be grounded by one of the methods indicated in (a), (b), or (c) below.

(a) By Means of the Metal Enclosure. By means of the metal enclosure of the conductors supplying such equipment if a grounding-type attachment plug with one fixed grounding contact is used for grounding the metal enclosure, and if the metal enclosure of the conductors is secured to the attachment plug and to equipment by approved connectors.

Exception: A self-restoring grounding contact shall be permitted on grounding-type attachment plugs used on the power supply cord of portable hand-held, hand-guided, or hand-supported tools or appliances.

(b) By Means of a Grounding Conductor. By means of an equipment grounding conductor run with the power supply conductors in a cable assembly or flexible cord properly terminated in grounding-type attachment plug with one fixed grounding contact. An uninsulated equipment grounding conductor shall be permitted, but, if individually covered, the covering shall have a continuous outer finish that is either green or green with one or more yellow stripes.

Exception: A self-restoring grounding contact shall be permitted on grounding-type attachment plugs used on the power supply cord of portable hand-held, hand-guided, or hand-supported tools or appliances.

(c) Separate Flexible Wire or Strap. By means of a separate flexible wire or strap, insulated or bare, protected as well as practicable against physical damage, where part of equipment.

250-60. Frames of Ranges and Clothes Dryers. Frames of electric ranges, wall-mounted ovens, counter-mounted cooking units, clothes dryers, and outlet or junction boxes that are part of the circuit for these appliances shall be grounded in the manner specified by Section 250-57 or 250-59; or, except for mobile homes and recreational vehicles, shall be permitted to be grounded to the grounded circuit conductor if all of the conditions indicated in (a) through (d) below are met.

(a) The supply circuit is 120/240-volt, single-phase, 3-wire; or 208Y/120-volt derived from a 3-phase, 4-wire wye-connected system.

(b) The grounded conductor is not smaller than No. 10 copper or No. 8 aluminum.

(c) The grounded conductor is insulated; or the grounded conductor is uninsulated and part of a Type SE service-entrance cable and the branch circuit originates at the service equipment.

(d) Grounding contacts of receptacles furnished as part of the equipment are bonded to the equipment.

250-61. Use of Grounded Circuit Conductor for Grounding Equipment.

(a) **Supply-Side Equipment.** A grounded circuit conductor shall be permitted to ground noncurrent-carrying metal parts of equipment, raceways, and other enclosures at any of the following locations:

(1) On the supply side of the service disconnecting means.

(2) On the supply side of the main disconnecting means for separate buildings as provided in Section 250-24.

(3) On the supply side of the disconnecting means or overcurrent devices of a separately derived system.

(b) **Load-Side Equipment.** A grounded circuit conductor shall not be used for grounding noncurrent-carrying metal parts of equipment on the load side of the service disconnecting means or on the load side of a separately derived system disconnecting means or the overcurrent devices for a separately derived system not having a main disconnecting means.

Exception No. 1: The frames of ranges, wall-mounted ovens, counter-mounted cooking units, and clothes dryers under the conditions specified by Section 250-60.

Exception No. 2: As permitted in Section 250-24 for separate buildings.

Exception No. 3: It shall be permissible to ground meter enclosures by connection to the grounded circuit conductor on the load-side of the service disconnect if:

a. No service ground-fault protection is installed; and

b. All meter enclosures are located near the service disconnecting means.

Exception No. 4: As required in Sections 710-72(e)(1) and 710-74.

Exception No. 5: DC systems shall be permitted to be grounded on the load side of the disconnecting means or overcurrent device in accordance with Section 250-22, Exception.

250-62. Multiple Circuit Connections. Where equipment is required to be grounded, and is supplied by separate connection to more than one circuit or grounded premises wiring system, a means for grounding shall be provided for each such connection as specified in Sections 250-57 and 250-59.

G. Bonding

250-70. General. Bonding shall be provided where necessary to assure electrical continuity and the capacity to conduct safely any fault current likely to be imposed.

250-71. Service Equipment.

(a) **Bonding of Service Equipment.** The noncurrent-carrying metal parts of equipment indicated in (1), (2), and (3) below shall be effectively bonded together.

(1) Except as permitted in Section 250-55, the service raceways, cable trays, cablebus framework, or service cable armor or sheath.

(2) All service equipment enclosures containing service conductors, including meter fittings, boxes, or the like, interposed in the service raceway or armor.

(3) Any metallic raceway or armor enclosing a grounding electrode conductor as permitted in Section 250-92(a). Bonding shall apply at each end and to all intervening raceways, boxes, and enclosures between the service equipment and the grounding electrode.

(b) Bonding to Other Systems. An accessible means external to enclosures for connecting intersystem bonding and grounding conductors shall be provided at the service by at least one of the following means:

(1) Exposed metallic service raceways.

(2) Exposed grounding electrode conductor.

(3) Approved means for the external connection of a copper or other corrosion-resistant bonding or grounding conductor to the service raceway or equipment.

For the purposes of providing an accessible means for intersystem bonding, the disconnecting means at a separate building or structure as permitted in Section 250-24 and the disconnecting means at a mobile home as permitted in Section 550-23(a), Exception No. 1 shall be considered the service equipment.

(FPN No. 1): A No. 6 copper conductor with one end bonded to the service raceway or equipment and with 6 inches (152 mm) or more of the other end made accessible on the outside wall is an example of the approved means covered in (b)(3).

(FPN No. 2): See Sections 800-40 and 820-40 for bonding and grounding requirements for communications and CATV circuits.

250-72. Method of Bonding Service Equipment. Electrical continuity at service equipment shall be assured by one of the methods specified in (a) through (e) below.

(a) Grounded Service Conductor. Bonding equipment to the grounded service conductor in a manner provided in Section 250-113.

(b) Threaded Connections. Connections utilizing threaded couplings or threaded bosses on enclosures shall be made up wrenchtight where rigid metal conduit or intermediate metal conduit is involved.

(c) Threadless Couplings and Connectors. Threadless couplings and connectors made up tight for rigid metal conduit, intermediate metal conduit and electrical metallic tubing. Standard locknuts or bushings shall not be used for the bonding required by this section.

(d) Bonding Jumpers. Bonding jumpers meeting the other requirements of this article shall be used around concentric or eccentric knockouts that are punched or otherwise formed so as to impair the electrical connection to ground.

(e) Other Devices. Other approved devices, such as bonding-type locknuts and bushings.

250-73. Metal Armor or Tape of Service Cable. The metal covering of service cable having an uninsulated grounded service conductor in continuous electrical contact with its metallic armor or tape shall be considered to be grounded.

250-74. Connecting Receptacle Grounding Terminal to Box. An equipment bonding jumper shall be used to connect the grounding terminal of a grounding-type receptacle to a grounded box.

Exception No. 1: Where the box is surface mounted, direct metal-to-metal contact between the device yoke and the box shall be permitted to ground the receptacle to the box. This exception shall not apply to cover-mounted receptacles unless the box and cover combination are listed as providing satisfactory ground continuity between the box and the receptacle.

Exception No. 2: Contact devices or yokes designed and listed for the purpose shall be permitted in conjunction with the supporting screws to establish the grounding circuit between the device yoke and flush-type boxes.

Exception No. 3: Floor boxes designed for and listed as providing satisfactory ground continuity between the box and the device.

Exception No. 4: Where required for the reduction of electrical noise (electromagnetic interference) on the grounding circuit, a receptacle in which the grounding terminal is purposely insulated from the receptacle mounting means shall be permitted. The receptacle grounding terminal shall be grounded by an insulated equipment grounding conductor run with the circuit conductors. This grounding conductor shall be permitted to pass through one or more panelboards without connection to the panelboard grounding terminal as permitted in Section 384-20, Exception so as to terminate within the same building or structure directly at an equipment grounding conductor terminal of the applicable derived system or service.

(FPN): Use of an isolated equipment grounding conductor does not relieve the requirement for grounding the raceway system and outlet box.

250-75. Bonding Other Enclosures. Metal raceways, cable trays, cable armor, cable sheath, enclosures, frames, fittings, and other metal noncurrent-carrying parts that are to serve as grounding conductors with or without the use of supplementary equipment grounding conductors shall be effectively bonded where necessary to assure electrical continuity and the capacity to conduct safely any fault current likely to be imposed on them. Any nonconductive paint, enamel, or similar coating shall be removed at threads, contact points, and contact surfaces or be connected by means of fittings so designed as to make such removal unnecessary.

Exception: Where required for the reduction of electrical noise (electromagnetic interference) on the grounding circuit, an equipment enclosure supplied by a branch circuit shall be permitted to be isolated from a raceway containing circuits supplying only that equipment by one or more listed nonmetallic raceway fittings located at the point of attachment of the raceway to the equipment enclosure. The metal raceway shall comply with provisions of this article and shall be supplemented by an internal insu-

lated equipment grounding conductor installed in accordance with Section 250-74, Exception No. 4 to ground the equipment enclosure.

(FPN): Use of an isolated equipment grounding conductor does not relieve the requirement for grounding the raceway system.

250-76. Bonding for Over 250 Volts. For circuits of over 250 volts to ground, the electrical continuity of metal raceways and cables with metal sheaths that contain any conductor other than service conductors shall be assured by one or more of the methods specified for services in Sections 250-72(b) through (e).

Exception: Where oversized, concentric, or eccentric knockouts are not encountered, the following methods shall be permitted:

a. Threadless couplings and connectors for cables with metal sheaths.

b. Two locknuts, on rigid metal conduit, or intermediate metal conduit, one inside and one outside of boxes and cabinets.

c. Fittings with shoulders that seat firmly against the box or cabinet, such as electrical metallic tubing connectors, flexible metal conduit connectors, and cable connectors, with one locknut on the inside of boxes and cabinets.

250-77. Bonding Loosely Jointed Metal Raceways. Expansion joints and telescoping sections of raceways shall be made electrically continuous by equipment bonding jumpers or other means.

250-78. Bonding in Hazardous (Classified) Locations. Regardless of the voltage of the electrical system, the electrical continuity of noncurrent-carrying metal parts of equipment, raceways, and other enclosures in any hazardous (classified) location as defined in Article 500 shall be assured by any of the methods specified for services in Sections 250-72(b) through (e) that are approved for the wiring method used.

250-79. Main and Equipment Bonding Jumpers.

(a) Material. Main and equipment bonding jumpers shall be of copper or other corrosion-resistant material. A main bonding jumper shall be a wire, bus, screw, or similar suitable conductor.

(b) Construction. Where a main bonding jumper is a screw only, the | screw shall be identified with a green finish that shall be visible with the screw installed.

(c) Attachment. Main and equipment bonding jumpers shall be attached in the manner specified by the applicable provisions of Section 250-113 for circuits and equipment and by Section 250-115 for grounding electrodes.

(d) Size — Equipment Bonding Jumper on Supply Side of Service and Main Bonding Jumper. The bonding jumper shall not be smaller than the sizes shown in Table 250-94 for grounding electrode conductors. Where the service-entrance phase conductors are larger than 1100 kcmil copper or 1750 kcmil aluminum, the bonding jumper shall have an area not less than 12½ percent of the area of the largest phase conductor except that, where the phase conductors and the bonding jumper are of different materials (copper or aluminum), the minimum size of the bonding jumper shall be based on the assumed use of phase conductors of the same material as the bonding jumper and with an ampacity equivalent to that of the installed phase conductors. Where the service-entrance conductors are paralleled in two or more raceways or cables, the equipment bonding jumper, where routed with the raceways or cables, shall be run in parallel. The size of the bonding jumper for each raceway or cable shall be based on the size of the service-entrance conductors in each raceway or cable.

The bonding jumper for a grounding electrode conductor raceway or cable armor as covered in Section 250-92(b) shall be the same size or larger than the required enclosed grounding electrode conductor.

(e) Size — Equipment Bonding Jumper on Load Side of Service. The equipment bonding jumper on the load side of the service overcurrent devices shall not be smaller than the sizes listed in Table 250-95. A single common continuous equipment bonding jumper shall be permitted to bond two or more raceways or cables where the bonding jumper is sized in accordance with Table 250-95 for the largest overcurrent device supplying circuits therein.

Exception: The equipment bonding jumper shall not be required to be larger than the circuit conductors supplying the equipment, but shall not be smaller than No. 14.

(f) Installation — Equipment Bonding Jumper. The equipment bonding jumper shall be permitted to be installed inside or outside of a raceway or enclosure. Where installed on the outside, the length of the equipment bonding jumper shall not exceed 6 feet (1.83 m) and shall be routed with the raceway or enclosure. Where installed inside of a raceway, the equipment bonding jumper shall comply with the requirements of Sections 250-114 and 310-12(b).

250-80. Bonding of Piping Systems.

(a) Metal Water Piping. The interior metal water piping system shall be bonded to the service equipment enclosure, the grounded conductor at the service, the grounding electrode conductor where of sufficient size, or to the one or more grounding electrodes used. The bonding jumper shall be sized in accordance with Table 250-94 and installed in accordance with Sections 250-92(a) and (b). The points of attachment of the bonding jumper shall be accessible.

Exception: In buildings of multiple occupancy, where the interior metal water piping system for the individual occupancies is metallically isolated from all other occupancies by use of nonmetallic water piping, the interior metal water piping system for each occupancy shall be permitted to be bonded to the panelboard or switchboard enclosure (other than service equipment) supplying that occupancy. The bonding jumper shall be sized in accordance with Table 250-95.

(b) Other Metal Piping. Interior metal piping that may become energized shall be bonded to the service equipment enclosure, the grounded conductor at the service, the grounding electrode conductor where of sufficient size, or to the one or more grounding electrodes used. The bonding jumper shall be sized in accordance with Table 250-95 using the rating of the circuit that may energize the piping.

The equipment grounding conductor for the circuit that may energize the piping shall be permitted to serve as the bonding means.

(FPN): Bonding all piping and metal air ducts within the premises will provide additional safety.

H. Grounding Electrode System

250-81. Grounding Electrode System. If available on the premises at each building or structure served, each item (a) through (d) below, and any made electrodes in accordance with Sections 250-83(c) and (d), shall be bonded together to form the grounding electrode system. Interior metal water piping located more than 5 feet (152 cm) from the point of entrance to the building shall not be used as a conductor to interconnect the electrodes and the grounding electrode conductor. The bonding jumper shall be installed in accordance with Sections 250-92(a) and (b), shall be sized in accordance with Section 250-94, and shall be connected in the manner specified in Section 250-115. The unspliced grounding electrode conductor shall be permitted to run to any convenient grounding electrode available in the grounding electrode system. It shall be sized for the largest grounding electrode conductor required among all the available electrodes.

Exception No. 1: It shall be permitted to splice the grounding electrode conductor by means of irreversible compression-type connectors listed for the purpose or the exothermic welding process.

Exception No. 2: In industrial and commercial buildings where conditions of maintenance and supervision ensure that only qualified persons will service the installation and the entire length of the interior metal water pipe that is being used for the conductor is exposed.

(FPN): See Section 547-8 for special grounding and bonding requirements for agricultural buildings.

(a) Metal Underground Water Pipe. A metal underground water pipe in direct contact with the earth for 10 feet (3.05 m) or more (including any metal well casing effectively bonded to the pipe) and electrically continuous (or made electrically continuous by bonding around insulating joints or sections or insulating pipe) to the points of connection of the grounding electrode conductor and the bonding conductors. Continuity of the grounding path or the bonding connection to interior piping shall not rely on water meters. A metal underground water pipe shall be supplemented by an additional electrode of a type specified in Section 250-81 or in Section 250-83. The supplemental electrode shall be permitted to be bonded to the grounding electrode conductor, the grounded service-entrance conductor, the grounded service raceway, or any grounded service enclosure.

Where the supplemental electrode is a made electrode as in Section 250-83(c) or (d), that portion of the bonding jumper that is the sole connection to the supplemental grounding electrode shall not be required to be larger than No. 6 copper wire or No. 4 aluminum wire.

Exception: The supplemental electrode shall be permitted to be bonded to the interior metal water piping at any convenient point as covered in Section 250-81, Exception No. 2.

(b) Metal Frame of the Building. The metal frame of the building, where effectively grounded.

(FPN): Effectively grounded means intentionally connected to earth through a ground connection or connections of sufficiently low impedance and having sufficient current-carrying capacity to prevent the buildup of voltages that may result in undue hazard to connected equipment or to persons.

(c) Concrete-Encased Electrode. An electrode encased by at least 2 inches (50.8 mm) of concrete, located within and near the bottom of a concrete foundation or footing that is in direct contact with the earth, consisting of at least 20 feet (6.1 m) of one or more bare or zinc galvanized or other electrically conductive coated steel reinforcing bars or rods of not less than $\frac{1}{2}$ inch (12.7 mm) diameter, or consisting of at least 20 feet (6.1 m) of bare copper conductor not smaller than No. 4.

(d) Ground Ring. A ground ring encircling the building or structure, in direct contact with the earth at a depth below earth surface not less than $2\frac{1}{2}$ feet (762 mm), consisting of at least 20 feet (6.1 m) of bare copper conductor not smaller than No. 2.

250-83. Made and Other Electrodes. Where none of the electrodes specified in Section 250-81 is available, one or more of the electrodes specified in (b) through (d) below shall be used. Where practicable, made electrodes shall be embedded below permanent moisture level. Made electrodes shall be free from nonconductive coatings, such as paint or enamel. Where more than one electrode is used, each electrode of one grounding system (including that used for lightning rods) shall not be less than 6 feet (1.83 m) from any other electrode of another grounding system.

(FPN): Two or more electrodes that are effectively bonded together are to be treated as a single electrode system in this sense.

(a) Metal Underground Gas Piping System. A metal underground gas piping system shall not be used as a grounding electrode.

(b) Other Local Metal Underground Systems or Structures. Other local metal underground systems or structures, such as piping systems and underground tanks.

(c) Rod and Pipe Electrodes. Rod and pipe electrodes shall not be less than 8 feet (2.44 m) in length and shall consist of the following materials, and shall be installed in the following manner:

(1) Electrodes of pipe or conduit shall not be smaller than $\frac{3}{4}$-inch trade size and, where of iron or steel, shall have the outer surface galvanized or otherwise metal-coated for corrosion protection.

(2) Electrodes of rods of iron or steel shall be at least $\frac{5}{8}$ inch (15.87 mm) in diameter. Stainless steel rods less than $\frac{5}{8}$ inch (15.87 mm) in diameter, nonferrous rods, or their equivalent shall be listed and shall not be less than $\frac{1}{2}$ inch (12.7 mm) in diameter.

(3) The electrode shall be installed such that at least 8 feet (2.44 m) of length is in contact with the soil. It shall be driven to a depth of not less than 8 feet (2.44 m) except that, where rock bottom is encountered, the electrode shall be driven at an oblique angle not to exceed 45 degrees from the vertical or shall be buried in a trench that is at least $2\frac{1}{2}$ feet (762 mm) deep. The upper end of the electrode shall be flush with or below ground level unless the aboveground end and the grounding electrode conductor attachment are protected against physical damage as specified in Section 250-117.

(d) Plate Electrodes. Each plate electrode shall expose not less than 2 square feet (0.186 sq m) of surface to exterior soil. Electrodes of iron or steel plates shall be at least ¼ inch (6.35 mm) in thickness. Electrodes of nonferrous metal shall be at least 0.06 inch (1.52 mm) in thickness.

(e) Aluminum Electrodes. Aluminum electrodes shall not be permitted.

250-84. Resistance of Made Electrodes. A single electrode consisting of a rod, pipe, or plate that does not have a resistance to ground of 25 ohms or less shall be augmented by one additional electrode of any of the types specified in Section 250-81 or 250-83. Where multiple rod, pipe, or plate electrodes are installed to meet the requirements of this section, they shall be not less than 6 feet (1.83 m) apart.

(FPN): The paralleling efficiency of rods longer than 8 feet (2.44 m) is improved by spacing greater than 6 feet (1.83 m).

250-86. Use of Lightning Rods. Lightning rod conductors and driven pipes, rods, or other made electrodes used for grounding lightning rods shall not be used in lieu of the made grounding electrodes required by Section 250-83 for grounding wiring systems and equipment. This provision shall not prohibit the required bonding together of grounding electrodes of different systems.

(FPN No. 1): See Sections 250-46, 800-40(d), 810-21(j), and 820-40(d) for bonding of electrodes.

(FPN No. 2): Bonding together of all separate grounding electrodes will limit potential differences between them and between their associated wiring systems.

J. Grounding Conductors

250-91. Material. The material for grounding conductors shall be as specified in (a), (b), and (c) below.

(a) Grounding Electrode Conductor. The grounding electrode conductor shall be of copper, aluminum, or copper-clad aluminum. The material selected shall be resistant to any corrosive condition existing at the installation or shall be suitably protected against corrosion. The conductor shall be solid or stranded, insulated, covered, or bare and shall be installed in one continuous length without a splice or joint.

Exception No. 1: Splices in busbars shall be permitted.

Exception No. 2: Where a service consists of more than a single enclosure as permitted in Section 230-40, Exception No. 2, it shall be permissible to connect taps to the grounding electrode conductor. Each such tap conductor shall extend to the inside of each such enclosure. The grounding electrode conductor shall be sized in accordance with Section 250-94, but the tap conductors shall be permitted to be sized in accordance with the grounding electrode conductors specified in Section 250-94 for the largest conductor serving the respective enclosures. The tap conductors shall be connected to the grounding electrode conductor in such a manner that the grounding electrode conductor remains without a splice or joint.

Exception No. 3: It shall be permitted to splice the grounding electrode conductor by means of irreversible compression-type connectors listed for the purpose or the exothermic welding process.

(b) Types of Equipment Grounding Conductors. The equipment grounding conductor run with or enclosing the circuit conductors shall be one or more or a combination of the following: (1) a copper or other corrosion-resistant conductor. This conductor shall be solid or stranded; insulated, covered, or bare; and in the form of a wire or a busbar of any shape; (2) rigid metal conduit; (3) intermediate metal conduit; (4) electrical metallic tubing; (5) flexible metal conduit where both the conduit and fittings are listed for grounding; (6) armor of Type AC cable; (7) the copper sheath of mineral-insulated, metal-sheathed cable; (8) the metallic sheath or the combined metallic sheath and grounding conductors of Type MC cable; (9) cable trays as permitted in Sections 318-3(c) and 318-7; (10) cablebus framework as permitted in Section 365-2(a); (11) other electrically continuous metal raceways listed for grounding.

Exception No. 1: Listed flexible metal conduit and listed flexible metallic tubing shall be permitted for grounding if all the following conditions are met:

a. The total combined length of flexible metal conduit and flexible metallic tubing and liquidtight flexible metal conduit in the same ground return path does not exceed 6 ft (1.83 m).

b. The circuit conductors contained therein are protected by overcurrent devices rated at 20 amperes or less.

c. The conduit or tubing is terminated in fittings listed for grounding.

Exception No. 2: Listed liquidtight flexible metal conduit shall be permitted as a grounding means in the $1^1/_4$-inch and smaller trade sizes if the total combined length of liquidtight flexible metal conduit and flexible metal conduit and flexible metallic tubing in the same ground return path does not exceed 6 ft (1.83 m), the conduit is terminated in fittings listed for grounding, and the circuit conductors contained therein are protected by overcurrent devices rated at 20 amperes or less for $3/_8$-inch and $1/_2$-inch trade sizes and 60 amperes or less for $3/_4$-inch through $1^1/_4$-inch trade sizes.

Exception No. 3: For direct-current circuits only, the equipment grounding conductor shall be permitted to be run separately from the circuit conductors.

(c) Supplementary Grounding. Supplementary grounding electrodes shall be permitted to augment the equipment grounding conductors specified in Section 250-91(b), but the earth shall not be used as the sole equipment grounding conductor.

250-92. Installation. Grounding conductors shall be installed as specified in (a), (b), and (c) below.

(a) Grounding Electrode Conductor. A grounding electrode conductor or its enclosure shall be securely fastened to the surface on which it is carried. A No. 4, copper or aluminum, or larger conductor shall be protected if exposed to severe physical damage. A No. 6 grounding conductor that is free from exposure to physical damage shall be permitted to be run along the surface of the building construction without metal covering or protection where it is securely fastened to the construction; otherwise, it shall be in rigid metal conduit, intermediate metal conduit, rigid nonmetallic conduit, electrical metallic tubing, or cable armor. Grounding conductors smaller than No. 6 shall be in rigid metal conduit, intermediate metal conduit, rigid nonmetallic conduit, electrical metallic tubing, or cable armor.

Insulated or bare aluminum or copper-clad aluminum grounding conductors shall not be used where in direct contact with masonry or the earth or where subject to corrosive conditions. Where used outside, aluminum or copper-clad aluminum grounding conductors shall not be installed within 18 inches (457 mm) of the earth.

(b) Enclosures for Grounding Electrode Conductors. Metal enclosures for grounding electrode conductors shall be electrically continuous from the point of attachment to cabinets or equipment to the grounding electrode, and shall be securely fastened to the ground clamp or fitting. Metal enclosures that are not physically continuous from cabinet or equipment to the grounding electrode shall be made electrically continuous by bonding each end to the grounding conductor. Where a raceway is used as protection for a grounding conductor, the installation shall comply with the requirements of the appropriate raceway article.

(c) Equipment Grounding Conductor. An equipment grounding conductor shall be installed as follows.

(1) Where it consists of a raceway, cable tray, cable armor, or cable sheath or where it is a wire within a raceway or cable, it shall be installed in accordance with the applicable provisions in this Code using fittings for joints and terminations approved for use with the type raceway or cable used. All connections, joints, and fittings shall be made tight using suitable tools.

(2) Where it is a separate equipment grounding conductor as provided in the exception for Sections 250-50(a) and (b), it shall be installed in accordance with (a) above in regard to restrictions for aluminum and also in regard to protection from physical damage.

Exception: Sizes smaller than No. 6 shall not be required to be enclosed in a raceway or armor where run in the hollow spaces of a wall or partition or where otherwise installed so as not to be subject to physical damage.

250-93. Size of Direct-Current System Grounding Conductor. The size of the grounding conductor for a dc system shall be as specified in (a) through (c) below.

(a) Not Be Smaller than the Neutral Conductor. Where the dc system consists of a 3-wire balancer set or a balancer winding with overcurrent protection as provided in Section 445-4(d), the grounding conductor shall not be smaller than the neutral conductor.

(b) Not Be Smaller than the Largest Conductor. Where the dc system is other than as in (a) above, the grounding conductor shall not be smaller than the largest conductor supplied by the system.

(c) Not Be Smaller than No. 8. In no case shall the grounding conductor be smaller than No. 8 copper or No. 6 aluminum.

250-94. Size of Alternating-Current Grounding Electrode Conductor. The size of the grounding electrode conductor of a grounded or ungrounded ac system shall not be less than given in Table 250-94.

Table 250-94.
Grounding Electrode Conductor for AC Systems

Size of Largest Service-Entrance Conductor or Equivalent Area for Parallel Conductors		Size of Grounding Electrode Conductor	
Copper	Aluminum or Copper-Clad Aluminum	Copper	Aluminum or Copper-Clad Aluminum*
2 or smaller	1/0 or smaller	8	6
1 or 1/0	2/0 or 3/0	6	4
2/0 or 3/0	4/0 or 250 kcmil	4	2
Over 3/0 thru 350 kcmil	Over 250 kcmil thru 500 kcmil	2	1/0
Over 350 kcmil thru 600 kcmil	Over 500 kcmil thru 900 kcmil	1/0	3/0
Over 600 kcmil thru 1100 kcmil	Over 900 kcmil thru 1750 kcmil	2/0	4/0
Over 1100 kcmil	Over 1750 kcmil	3/0	250 kcmil

Where multiple sets of service-entrance conductors are used as permitted in Section 230-40, Exception No. 2, the equivalent size of the largest service-entrance conductor shall be determined by the largest sum of the areas of the corresponding conductors of each set.

Where there are no service-entrance conductors, the grounding electrode conductor size shall be determined by the equivalent size of the largest service-entrance conductor required for the load to be served.

*See installation restrictions in Section 250-92(a).

(FPN): See Section 250-23(b) for size of alternating-current system grounded conductor brought to service equipment.

Exception No. 1: Grounded Systems.

a. Where connected to made electrodes as in Section 250-83 (c) or (d), that portion of the grounding electrode conductor that is the sole connection to the grounding electrode shall not be required to be larger than No. 6 copper wire or No. 4 aluminum wire.

b. Where connected to a concrete-encased electrode as in Section 250-81(c), that portion of the grounding electrode conductor that is the sole connection to the grounding electrode shall not be required to be larger than No. 4 copper wire.

c. Where connected to a ground ring as in Section 250-81(d), that portion of the grounding electrode conductor that is the sole connection to the grounding electrode shall not be required to be larger than the conductor used for the ground ring.

Exception No. 2: Ungrounded Systems.

a. Where connected to made electrodes as in Section 250-83 (c) or (d), that portion of the grounding electrode conductor that is the sole connection to the grounding electrode shall not be required to be larger than No. 6 copper wire or No. 4 aluminum wire.

b. Where connected to a concrete-encased electrode as in Section 250-81(c), that portion of the grounding electrode conductor that is the sole connection to the grounding electrode shall not be required to be larger than No. 4 copper wire.

c. Where connected to a ground ring as in Section 250-81(d), that portion of the grounding electrode conductor that is the sole connection to the grounding electrode shall not be required to be larger than the conductor used for the ground ring.

250-95. Size of Equipment Grounding Conductors. Copper, aluminum, or copper-clad aluminum equipment grounding conductors shall not be less than shown in Table 250-95.

Where conductors are run in parallel in multiple raceways or cables, as permitted in Section 310-4, the equipment grounding conductor, where used, shall be run in parallel. Each parallel equipment grounding conductor shall be sized on the basis of the ampere rating of the overcurrent device protecting the circuit conductors in the raceway in accordance with Table 250-95.

When conductors are adjusted in size to compensate for voltage drop, equipment grounding conductors, where required, shall be adjusted proportionately according to circular mil area.

Where a single equipment grounding conductor is run with multiple circuits in the same raceway, it shall be sized for the largest overcurrent device protecting conductors in the raceway or cable.

Where the overcurrent device consists of an instantaneous trip circuit breaker or a motor short-circuit protector, as allowed in Section 430-52, the equipment grounding conductor size shall be permitted to be based on the rating of the motor overload protective device but not less than the size shown in Table 250-95.

Exception No. 1: An equipment grounding conductor not smaller than No. 18 copper and not smaller than the circuit conductors and part of fixture wires or cords in accordance with Section 240-4.

Exception No. 2: The equipment grounding conductor shall not be required to be larger than the circuit conductors supplying the equipment.

Exception No. 3: Where a raceway or a cable armor or sheath is used as the equipment grounding conductor, as provided in Sections 250-51, 250-57(a), 250-73, and 250-91(b).

250-97. Outline Lighting. Isolated noncurrent-carrying metal parts of outline lighting systems shall be permitted to be bonded together by a No. 14 copper or No. 12 aluminum conductor protected from physical damage, where a conductor complying with Section 250-95 is used to ground the group.

Table 250-95. Minimum Size Equipment Grounding Conductors for Grounding Raceway and Equipment

Rating or Setting of Automatic Overcurrent Device in Circuit Ahead of Equipment, Conduit, etc., Not Exceeding (Amperes)	Size	
	Copper Wire No.	Aluminum or Copper-Clad Aluminum Wire No.*
15	14	12
20	12	10
30	10	8
40	10	8
60	10	8
100	8	6
200	6	4
300	4	2
400	3	1
500	2	1/0
600	1	2/0
800	1/0	3/0
1000	2/0	4/0
1200	3/0	250 kcmil
1600	4/0	350 ”
2000	250 kcmil	400 ”
2500	350 ”	600 ”
3000	400 ”	600 ”
4000	500 ”	800 ”
5000	700 ”	1200 ”
6000	800 ”	1200 ”

* See installation restrictions in Section 250-92(a).

250-99. Equipment Grounding Conductor Continuity.

(a) Separable Connections. Separable connections such as those provided in draw-out equipment or attachment plugs and mating connectors and receptacles shall provide for first-make, last-break of the equipment grounding conductor.

Exception: Interlocked equipment, plugs, receptacles, and connectors that preclude energization without grounding continuity.

(b) Switches. No automatic cutout or switch shall be placed in the equipment grounding conductor of a premises wiring system.

Exception: Where the opening of the cutout or switch disconnects all sources of energy.

K. Grounding Conductor Connections

250-112. To Grounding Electrode. The connection of a grounding electrode conductor to a grounding electrode shall be accessible and made in a manner that will assure a permanent and effective ground. Where necessary to assure this for a metal piping system used as a grounding electrode, effective bonding shall be provided around insulated joints and sections and around any equipment that is likely to be disconnected for repairs or replacement. Bonding conductors shall be of sufficient length to permit removal of such equipment while retaining the integrity of the bond.

Exception: An encased or buried connection to a concrete-encased, driven, or buried grounding electrode shall not be required to be accessible.

250-113. To Conductors and Equipment. Grounding conductors and bonding jumpers shall be connected by exothermic welding, listed pressure connectors, listed clamps, or other listed means. Connection devices or fittings that depend solely on solder shall not be used.

250-114. Continuity and Attachment of Branch-Circuit Equipment Grounding Conductors to Boxes. Where more than one equipment grounding conductor enters a box, all such conductors shall be spliced or joined within the box or to the box with devices suitable for the use. Connections depending solely on solder shall not be used. Splices shall be made in accordance with Section 110-14(b) except that insulation shall not be required. The arrangement of grounding connections shall be such that the disconnection or the removal of a receptacle, fixture, or other device fed from the box will not interfere with or interrupt the grounding continuity.

Exception: The equipment grounding conductor permitted in Section 250-74, Exception No. 4 shall not be required to be connected to the other equipment grounding conductors or to the box.

(a) Metal Boxes. A connection shall be made between the one or more equipment grounding conductors and a metal box by means of a grounding screw that shall be used for no other purpose, or a listed grounding device.

(b) Nonmetallic Boxes. One or more equipment grounding conductors brought into a nonmetallic outlet box shall be so arranged that a connection can be made to any fitting or device in that box requiring grounding.

250-115. Connection to Electrodes. The grounding conductor shall be connected to the grounding electrode by exothermic welding, listed lugs, listed pressure connectors, listed clamps, or other listed means. Connections depending on solder shall not be used. Ground clamps shall be listed for the materials of the grounding electrode and the grounding electrode conductor and, where used on pipe, rod, or other buried electrodes, shall also be listed for direct soil burial. Not more than one conductor shall be connected to the grounding electrode by a single clamp or fitting unless the clamp or fitting is listed for multiple conductors. One of the methods indicated in (a), (b), (c), or (d) below shall be used.

(a) Bolted Clamp. A listed bolted clamp of cast bronze or brass, or plain or malleable iron.

(b) Pipe Fitting, Pipe Plug, etc. A pipe fitting, pipe plug, or other approved device screwed into a pipe or pipe fitting.

(c) Sheet-Metal-Strap Type Ground Clamp. A listed sheet-metal-strap type ground clamp having a rigid metal base that seats on the electrode and having a strap of such material and dimensions that it is not likely to stretch during or after installation.

(d) Other Means. An equally substantial approved means.

250-117. Protection of Attachment. Ground clamps or other fittings shall be approved for general use without protection or shall be protected from ordinary physical damage as indicated in (a) or (b) below.

(a) Not Likely to Be Damaged. Installations where they are not likely to be damaged.

(b) Protective Covering. Enclosing in metal, wood, or equivalent protective covering.

250-118. Clean Surfaces. Nonconductive coatings (such as paint, lacquer, and enamel) on equipment to be grounded shall be removed from threads and other contact surfaces to assure good electrical continuity or be connected by means of fittings so designed as to make such removal unnecessary.

250-119. Identification of Wiring Device Terminals. The terminal for the connection of the equipment grounding conductor shall be identified by (1) a green-colored, not readily removable terminal screw with a hexagonal head; (2) a green-colored, hexagonal, not readily removable terminal nut; or (3) a green-colored pressure wire connector. If the terminal for the grounding conductor is not visible, the conductor entrance hole shall be marked with the word "green" or otherwise identified by a distinctive green color.

L. Instrument Transformers, Relays, Etc.

250-121. Instrument Transformer Circuits. Secondary circuits of current and potential instrument transformers shall be grounded where the primary windings are connected to circuits of 300 volts or more to ground and, where on switchboards, shall be grounded irrespective of voltage.

Exception: Circuits where the primary windings are connected to circuits of less than 1000 volts with no live parts or wiring exposed or accessible to other than qualified persons.

250-122. Instrument Transformer Cases. Cases or frames of instrument transformers shall be grounded where accessible to other than qualified persons.

Exception: Cases or frames of current transformers, the primaries of which are not over 150 volts to ground and that are used exclusively to supply current to meters.

250-123. Cases of Instruments, Meters, and Relays — Operating at Less than 1000 Volts. Instruments, meters, and relays operating with windings or working parts at less than 1000 volts shall be grounded as specified in (a), (b), or (c) below.

(a) Not on Switchboards. Instruments, meters, and relays not located on switchboards, operating with windings or working parts at 300 volts or more to ground, and accessible to other than qualified persons, shall have the cases and other exposed metal parts grounded.

(b) On Dead-Front Switchboards. Instruments, meters, and relays (whether operated from current and potential transformers or connected directly in the circuit) on switchboards having no live parts on the front of the panels shall have the cases grounded.

(c) On Live-Front Switchboards. Instruments, meters, and relays (whether operated from current and potential transformers or connected directly in the circuit) on switchboards having exposed live parts on the front of panels shall not have their cases grounded. Mats of insulating rubber or other suitable floor insulation shall be provided for the operator where the voltage to ground exceeds 150.

250-124. Cases of Instruments, Meters, and Relays — Operating Voltage 1 kV and Over. Where instruments, meters, and relays have current-carrying parts of 1 kV and over to ground, they shall be isolated by elevation or protected by suitable barriers, grounded metal, or insulating covers or guards. Their cases shall not be grounded.

Exception: Cases of electrostatic ground detectors where the internal ground segments of the instrument are connected to the instrument case and grounded and the ground detector is isolated by elevation.

250-125. Instrument Grounding Conductor. The grounding conductor for secondary circuits of instrument transformers and for instrument cases shall not be smaller than No. 12 copper or No. 10 aluminum. Cases of instrument transformers, instruments, meters, and relays that are mounted directly on grounded metal surfaces of enclosures or grounded metal switchboard panels shall be considered to be grounded, and no additional grounding conductor shall be required.

M. Grounding of Systems and Circuits of 1 kV and Over (High Voltage)

250-150. General. Where high-voltage systems are grounded, they shall comply with all applicable provisions of the preceding sections of this article and with the following sections, which supplement and modify the preceding sections.

250-151. Derived Neutral Systems. A system neutral derived from a grounding transformer shall be permitted to be used for grounding high-voltage systems.

250-152. Solidly Grounded Neutral Systems.

(a) Neutral Conductor. The minimum insulation level for neutral conductors of solidly grounded systems shall be 600 volts.

Exception No. 1: Bare copper conductors shall be permitted to be used for the neutral of service entrances and the neutral of direct buried portions of feeders.

Exception No. 2: Bare conductors shall be permitted for the neutral of overhead portions installed outdoors.

(FPN): See Section 225-4 for conductor covering where within 10 feet (3.05 m) of any building or other structure.

(b) Multiple Grounding. The neutral of a solidly grounded neutral system shall be permitted to be grounded at more than one point for:

(1) Services.

(2) Direct buried portions of feeders employing a bare copper neutral.

(3) Overhead portion installed outdoors.

(c) Neutral Grounding Conductor. The neutral grounding conductor shall be permitted to be a bare conductor if isolated from phase conductors and protected from physical damage.

250-153. Impedance Grounded Neutral Systems. Impedance grounded neutral systems shall comply with the provisions of (a) through (d) below.

(a) Location. The grounding impedance shall be inserted in the grounding conductor between the grounding electrode of the supply system and the neutral point of the supply transformer or generator.

(b) Identified and Insulated. Where the neutral conductor of an impedance grounded neutral system is used, it shall be identified, as well as fully insulated with the same insulation as the phase conductors.

(c) System Neutral Connection. The system neutral shall not be connected to ground, except through the neutral grounding impedance.

(d) Equipment Grounding Conductors. Equipment grounding conductors shall be permitted to be bare and shall be connected to the ground bus and grounding electrode conductor at the service-entrance equipment and extended to the system ground.

250-154. Grounding of Systems Supplying Portable or Mobile Equipment. Systems supplying portable or mobile high-voltage equipment, other than substations installed on a temporary basis, shall comply with (a) through (f) below.

(a) Portable or Mobile Equipment. Portable or mobile high-voltage equipment shall be supplied from a system having its neutral grounded through an impedance. Where a delta-connected high-voltage system is used to supply portable or mobile equipment, a system neutral shall be derived.

(b) Exposed Noncurrent-Carrying Metal Parts. Exposed noncurrent-carrying metal parts of portable or mobile equipment shall be connected by an equipment grounding conductor to the point at which the system neutral impedance is grounded.

(c) Ground-Fault Current. The voltage developed between the portable or mobile equipment frame and ground by the flow of maximum ground-fault current shall not exceed 100 volts.

(d) Ground-Fault Detection and Relaying. Ground-fault detection and relaying shall be provided to automatically deenergize any high-voltage system component that has developed a ground fault. The continuity of the equipment grounding conductor shall be continuously monitored so as to deenergize automatically the high-voltage feeder to the portable or mobile equipment upon loss of continuity of the equipment grounding conductor.

(e) Isolation. The grounding electrode to which the portable or mobile equipment system neutral impedance is connected shall be isolated from and separated in the ground by at least 20 feet (6.1 m) from any other system or equipment grounding electrode, and there shall be no direct connection between the grounding electrodes, such as buried pipe, fence, etc.

(f) Trailing Cable and Couplers. High-voltage trailing cable and couplers for interconnection of portable or mobile equipment shall meet the requirements of Part C of Article 400 for cables and Section 710-45 for couplers.

250-155. Grounding of Equipment. All noncurrent-carrying metal parts of fixed, portable, and mobile equipment and associated fences, housings, enclosures, and supporting structures shall be grounded.

Exception No. 1: Where isolated from ground and located so as to prevent any person who can make contact with ground from contacting such metal parts when the equipment is energized.

Exception No. 2: Pole-mounted distribution apparatus as provided in Section 250-42, Exception No. 3.

Grounding conductors not an integral part of a cable assembly shall not be smaller than No. 6 copper or No. 4 aluminum.

ARTICLE 280 — SURGE ARRESTERS

A. General

280-1. Scope. This article covers general requirements, installation requirements, and connection requirements for surge arresters installed on premises wiring systems.

280-2. Definition. A surge arrester is a protective device for limiting surge voltages by discharging or bypassing surge current, and it also prevents continued flow of follow current while remaining capable of repeating these functions.

280-3. Number Required. Where used at a point on a circuit, a surge arrester shall be connected to each ungrounded conductor. A single installation of such surge arresters shall be permitted to protect a number of interconnected circuits provided that no circuit is exposed to surges while disconnected from the surge arresters.

280-4. Surge Arrester Selection.

(a) On Circuits of Less than 1000 Volts. The rating of the surge arrester shall be equal to or greater than the maximum continuous phase-to-ground power frequency voltage available at the point of application.

(b) On Circuits of 1 kV and Over—Silicon Carbide Types. The rating of a silicon carbide-type surge arrester shall be not less than 125 percent of the maximum continuous phase-to-ground voltage available at the point of application.

(FPN No. 1): For further information on surge arresters, see Guide for the Application of Gapped Silicon—Carbide Surge Arresters for Alternating-Current Systems, C62.2-1988, and Standard for Metal Oxide Surge Arresters for Alternating-Current Power Circuits, ANSI/IEEE C62.11-1986.

(FPN No. 2): The selection of a properly rated metal oxide arrester is based on considerations of maximum continuous operating voltage and the magnitude and duration of overvoltages at the arrester location as affected by phase-to-ground faults, system grounding techniques, switching surges, and other causes. See manufacturer's application rules for selection of the specific arrester to be used at a particular location.

B. Installation

280-11. Location. Surge arresters shall be permitted to be located indoors or outdoors and shall be made inaccessible to unqualified persons.

Exception: Surge arresters listed for installation in accessible locations.

280-12. Routing of Surge Arrester Connections. The conductor used to connect the surge arrester to line or bus and to ground shall not be any longer than necessary and shall avoid unnecessary bends.

C. Connecting Surge Arresters

280-21. Installed at Services of Less than 1000 Volts. Line and ground connecting conductors shall not be smaller than No. 14 copper or No. 12 aluminum. The arrester grounding conductor shall be connected to one of the following: (1) the grounded service conductor; (2) the grounding electrode conductor; (3) the grounding electrode for the service; or (4) the equipment grounding terminal in the service equipment.

280-22. Installed on the Load Side Services of Less than 1000 Volts. Line and ground connecting conductors shall not be smaller than No. 14 copper or No. 12 aluminum. A surge arrester shall be permitted to be connected between any two conductors (ungrounded conductor(s), grounded conductor, grounding conductor). The grounded conductor and the grounding conductor shall be interconnected only by the normal operation of the surge arrester during a surge.

280-23. Circuits of 1 kV and Over—Surge-Arrester Conductors. The conductor between the surge arrester and the line and surge arrester and the grounding connection shall not be smaller than No. 6 copper or aluminum.

280-24. Circuits of 1 kV and Over—Interconnections. The grounding conductor of a surge arrester protecting a transformer that supplies a secondary distribution system shall be interconnected as specified in (a), (b), or (c) below.

(a) Metallic Interconnections. A metallic interconnection shall be made to the secondary grounded circuit conductor or the secondary circuit grounding conductor provided that, in addition to the direct grounding connection at the surge arrester:

(1) The grounded conductor of the secondary has elsewhere a grounding connection to a continuous metal underground water piping system. However, in urban water-pipe areas where there are at least four water-pipe connections on the neutral and not fewer than four such connections in each mile of neutral, the metallic interconnection shall be permitted to be made to the secondary neutral with omission of the direct grounding connection at the surge arrester.

(2) The grounded conductor of the secondary system is a part of a multiground neutral system of which the primary neutral has at least four ground connections in each mile of line in addition to a ground at each service.

(b) Through Spark Gap or Device. Where the surge arrester grounding conductor is not connected as in (a) above or where the secondary is not grounded as in (a) above but is otherwise grounded as in Sections 250-81 and 250-83, an interconnection shall be made through a spark gap or listed device as follows:

(1) For ungrounded or unigrounded primary systems, the spark gap or listed device shall have a 60-hertz breakdown voltage of at least twice the primary circuit voltage but not necessarily more than 10 kV, and there shall be at least one other ground on the grounded conductor of the secondary not less than 20 feet (6.1 m) distant from the surge arrester grounding electrode.

(2) For multigrounded neutral primary systems, the spark gap or listed device shall have a 60-hertz breakdown or not more than 3 kV, and there shall be at least one other ground on the grounded conductor of the secondary not less than 20 feet (6.1 m) distant from the surge arrester grounding electrode.

(c) By Special Permission. An interconnection of the surge arrester ground and the secondary neutral, other than as provided in (a) or (b) above, shall be permitted to be made only by special permission.

280-25. Grounding. Except as indicated in this article, surge arrester grounding connections shall be made as specified in Article 250. Grounding conductors shall not be run in metal enclosures unless bonded to both ends of such enclosure.

Chapter 3.   Wiring Methods and Materials

ARTICLE 300 — WIRING METHODS

A. General Requirements

300-1. Scope.

(a) All Wiring Installations. This article covers wiring methods for all wiring installations.

Exception No. 1: As permitted in Article 504, Intrinsically Safe Systems. |

Exception No. 2: Only those sections referenced in Article 725 shall apply to Class 1, Class 2, and Class 3 circuits.

Exception No. 3: Only those sections referenced in Article 760 shall apply to fire protective signaling circuits.

Exception No. 4: Only those sections referenced in Article 770 shall apply to optical fiber cables.

Exception No. 5: Only those sections referenced in Article 800 shall apply to communications systems.

Exception No. 6: Only those sections referenced in Article 810 shall apply to radio and television equipment.

Exception No. 7: Only those sections referenced in Article 820 shall apply to community antenna television and radio distribution systems.

(b) Integral Parts of Equipment. The provisions of this article are not intended to apply to the conductors that form an integral part of equipment, such as motors, controllers, motor control centers, or factory-assembled control equipment.

300-2. Limitations.

(a) Voltage. Wiring methods specified in Chapter 3 shall be used for voltages 600 volts, nominal, or less where not specifically limited in some section of Chapter 3. They shall be permitted for voltages over 600 volts, nominal, where specifically permitted elsewhere in this Code.

(b) Temperature. Temperature limitation of conductors shall be in accordance with Section 310-10.

300-3. Conductors.

(a) Single Conductors. Single conductors specified in Table 310-13 shall only be permitted to be installed where part of a recognized wiring method of Chapter 3.

(b) Conductors of the Same Circuit. All conductors of the same circuit and, where used, the neutral and all equipment grounding conductors shall be contained within the same raceway, cable tray, trench, cable, or cord.

Exception No. 1: The conductors of single-conductor Type MI cable with a nonmagnetic sheath and installed in accordance with Section 330-16 shall be permitted to be run in separate cables.

Exception No. 2 to (b): Column type panelboards using auxiliary gutters and pull boxes with neutral terminations.

Exception No. 3 to (a) and (b): As permitted in Sections 250-57(b), 250-79(f), 300-5(i), 300-20(a), 318-8(d), and 339-3(a)(2).

(c) Conductors of Different Systems.

(1) 600 Volts, Nominal, or Less. Conductors of 600 volts, nominal, or less, alternating-current circuits, and direct-current circuits shall be permitted to occupy the same equipment wiring enclosure, cable, or raceway. All conductors shall have an insulation rating equal to at least the maximum circuit voltage applied to any conductor within the enclosure, cable, or raceway.

Exception: For solar photovoltaic systems in accordance with Section 690-4(b).

(FPN):   See Section 725-52(a)(2) for Class 2 and Class 3 circuit conductors.

(2) Over 600 Volts, Nominal. Conductors of circuits rated over 600 volts, nominal, shall not occupy the same equipment wiring enclosure, cable, or raceway with conductors of circuits rated 600 volts, nominal, or less.

(FPN):   See Section 300-32, Conductors of Different Systems — over 600 volts, nominal.

Exception No. 1: Secondary wiring to electric-discharge lamps of 1000 volts or less, if insulated for the secondary voltage involved, shall be permitted to occupy the same fixture enclosure as the branch-circuit conductors.

Exception No. 2: Primary leads of electric-discharge lamp ballasts, insulated for the primary voltage of the ballast, where contained within the individual wiring enclosure, shall be permitted to occupy the same fixture enclosure as the branch-circuit conductors.

Exception No. 3: Excitation, control, relay, and ammeter conductors used in connection with any individual motor or starter shall be permitted to occupy the same enclosure as the motor circuit conductors.

300-4. Protection Against Physical Damage.   Where subject to physical damage, conductors shall be adequately protected.

(a) Cables and Raceways Through Wood Members.

(1) Bored Holes. In both exposed and concealed locations, where a cable or raceway-type wiring method is installed through bored holes in joists, rafters, or wood members, holes shall be bored so that the edge of the hole is not less than $1\frac{1}{4}$ inches (31.8 mm) from the nearest edge of the wood member. Where this distance cannot be maintained, the cable or raceway shall be protected from penetration by screws or nails by a steel plate or bushing, at least $\frac{1}{16}$ inch (1.59 mm) thick, and of appropriate length and width installed to cover the area of the wiring.

Exception: Raceways as covered in Articles 345, 346, 347, and 348.

(2) Notches in Wood. Where there is no objection because of weakening the building structure, in both exposed and concealed locations, cables or raceways shall be permitted to be laid in notches in wood studs, joists, rafters, or other wood members where the cable or raceway at those points is protected against nails or screws by a steel plate at least $\frac{1}{16}$ inch (1.59 mm) thick installed before the building finish is applied.

Exception: Raceways as covered in Articles 345, 346, 347, and 348.

(b) Nonmetallic-Sheathed Cables and Electrical Nonmetallic Tubing through Metal Framing Members.

(1) Nonmetallic-Sheathed Cable. In both exposed and concealed locations where nonmetallic-sheathed cables pass through either factory or field punched, cut or drilled slots or holes in metal members, the cable shall be protected by bushings or grommets securely fastened in the opening prior to installation of the cable.

(2) Nonmetallic-Sheathed Cable and Electrical Nonmetallic Tubing. Where nails or screws are likely to penetrate nonmetallic-sheathed cable, or electrical nonmetallic tubing, a steel sleeve, steel plate, or steel clip not less than $\frac{1}{16}$ inch (1.59 mm) in thickness shall be used to protect the cable or tubing.

(c) Cables through Spaces Behind Panels Designed to Allow Access. Cables, or raceway-type wiring methods, installed behind panels designed to allow access shall be supported according to their applicable articles.

(d) Cables and Raceways Parallel to Framing Members. In both exposed and concealed locations, where a cable- or raceway-type wiring method is installed parallel to framing members, such as joists, rafters, or studs, the cable or raceway shall be installed and supported so that the nearest outside surface of the cable or raceway is not less than $1\frac{1}{4}$ inches (31.8 mm) from the nearest edge of the framing member where nails or screws are likely to penetrate. Where this distance cannot be maintained, the cable or raceway shall be protected from penetration by nails or screws by a steel plate, sleeve, or equivalent at least $\frac{1}{16}$ inch (1.59 mm) thick.

Exception No. 1: Raceways as covered in Articles 345, 346, 347, and 348.

Exception No. 2: For concealed work in finished buildings, or finished panels for prefabricated buildings where such supporting is impracticable, it shall be permissible to fish the cables between access points.

Exception No. 3: For mobile homes and recreational vehicles.

300-5. Underground Installations.

(a) Minimum Cover Requirements. Direct buried cable or conduit or other raceways shall be installed to meet the minimum cover requirements of Table 300-5.

(b) Grounding. All underground installations shall be grounded and bonded in accordance with Article 250 of this Code.

(c) Underground Cables Under Buildings. Underground cable installed under a building shall be in a raceway that is extended beyond the outside walls of the building.

Table 300-5. Minimum Cover Requirements, 0 to 600 Volts, Nominal, Burial in Inches
(Cover is defined as the shortest distance measured between a point on the top surface of any direct buried conductor, cable, conduit, or other raceway and the top surface of finished grade, concrete, or similar cover.)

Location of Wiring Method or Circuit	Type of Wiring Method or Circuit				
	1 Direct Burial Cables or Conductors	2 Rigid Metal Conduit or Intermediate Metal Conduit	3 Rigid Nonmetallic Conduit Approved for Direct Burial without Concrete Encasement or Other Approved Raceways	4 Residential Branch Circuits Rated 120 Volts or less with GFCI Protection and Maximum Overcurrent Protection of 20 Amperes	5 Circuits for Control of Irrigation and Landscape Lighting Limited to Not More than 30 Volts and Installed with Type UF or in Other Identified Cable or Raceway
All Locations Not Specified Below	24	6	18	12	6
In Trench Below 2-Inch Thick Concrete or Equivalent	18	6	12	6	6
Under a Building	0 (In Raceway Only)	0	0	0 (In Raceway Only)	0 (In Raceway Only)
Under Minimum of 4-Inch Thick Concrete Exterior Slab with No Vehicular Traffic and the Slab Extending Not Less than 6 Inches beyond the Underground Installation	18	4	4	6 (Direct Burial) 4 (In Raceway)	6 (Direct Burial) 4 (In Raceway)

Table 300-5. (Continued)

Location of Wiring Method or Circuit					
Under Streets, Highways, Roads, Alleys, Driveways, and Parking Lots	24	24	24	24	24
One- and Two-Family Dwelling Driveways and Outdoor Parking Areas, and Used Only for Dwelling-Related Purposes	18	18	18	12	18
In or under Airport Runways, Including Adjacent Areas Where Trespassing Prohibited	18	18	18	18	18
In Solid Rock Where Covered by Minimum of 2 Inches Concrete Extending Down to Rock	2 (In Raceway Only)	2 (In Raceway Only)	2	2 (In Raceway Only)	2 (In Raceway Only)

Note 1. For SI units: one inch = 25.4 millimeters

Note 2. Raceways approved for burial only where concrete encased shall require concrete envelope not less than 2 inches thick.

Note 3. Lesser depths shall be permitted where cables and conductors rise for terminations or splices or where access is otherwise required.

Note 4. Where one of the wiring method types listed in columns 1-3 is combined with one of the circuit types in columns 4 and 5, the shallower depth of burial shall be permitted.

(d) Protection from Damage. Direct buried conductors and cables emerging from the ground shall be protected by enclosures or raceways extending from the minimum cover distance required by Section 300-5(a) below grade to a point at least 8 feet (2.44 m) above finished grade. In no case shall the protection be required to exceed 18 inches (457 mm) below finished grade.

Conductors entering a building shall be protected to the point of entrance.

Where the enclosure or raceway is subject to physical damage, the conductors shall be installed in rigid metal conduit, intermediate metal conduit, Schedule 80 rigid nonmetallic conduit, or equivalent.

(e) Splices and Taps. Direct buried conductors or cables shall be permitted to be spliced or tapped without the use of splice boxes. The splices or taps shall be made by approved methods and with identified materials.

(f) Backfill. Backfill containing large rock, paving materials, cinders, large or sharply angular substance, or corrosive material shall not be placed in an excavation where materials may damage raceways, cables, or other substructures or prevent adequate compaction of fill or contribute to corrosion of raceways, cables, or other substructures.

Where necessary to prevent physical damage to the raceway or cable, protection shall be provided in the form of granular or selected material, suitable running boards, suitable sleeves, or other approved means.

(g) Raceway Seals. Conduits or raceways through which moisture may contact energized live parts shall be sealed or plugged at either or both ends.

(FPN):   Presence of hazardous gases or vapors may also necessitate sealing of underground conduits or raceways entering buildings.

(h) Bushing. A bushing, or terminal fitting, with an integral bushed opening shall be used at the end of a conduit or other raceway that terminates underground where the conductors or cables emerge as a direct burial wiring method. A seal incorporating the physical protection characteristics of a bushing shall be permitted to be used in lieu of a bushing.

(i) Single Conductors. All conductors of the same circuit and, where used, the neutral and all equipment grounding conductors shall be installed in the same raceway or shall be installed in close proximity in the same trench.

Exception No. 1: Conductors in parallel in raceways shall be permitted, but each raceway shall contain all conductors of the same circuit including grounding conductors.

Exception No. 2: Isolated phase installations shall be permitted in nonmetallic raceways in close proximity where conductors are paralleled as permitted in Section 310-4 and the conditions of Section 300-20 are met.

300-6. Protection Against Corrosion.   Metal raceways, cable armor, boxes, cable sheathing, cabinets, elbows, couplings, fittings, supports, and support hardware shall be of materials suitable for the environment in which they are to be installed.

(a) General. Ferrous raceways, cable armor, boxes, cable sheathing, cabinets, metal elbows, couplings, fittings, supports, and support hardware shall be suitably protected against corrosion inside and outside (except threads at joints) by a coating of approved corrosion-resistant material such as zinc, cadmium, or enamel. Where protected from corrosion solely by enamel, they shall not be used out-of-doors or in wet locations as described in (c) below. Where boxes or cabinets have an approved system of organic coatings and are marked "Raintight," "Rainproof" or "Outdoor Type," they shall be permitted out-of-doors.

Exception: Threads at joints shall be permitted to be coated with an identified electrically conductive compound.

(b) In Concrete or in Direct Contact with the Earth. Ferrous or nonferrous metal raceways, cable armor, boxes, cable sheathing, cabinets, elbows, couplings, fittings, supports, and support hardware shall be permitted to be installed in concrete or in direct contact with the earth, or in areas subject to severe corrosive influences where made of material judged suitable for the condition, or where provided with corrosion protection approved for the condition.

(c) Indoor Wet Locations. In portions of dairies, laundries, canneries, and other indoor wet locations, and in locations where walls are frequently washed or where there are surfaces of absorbent materials, such as damp paper or wood, the entire wiring system, where installed exposed, including all boxes, fittings, conduits, and cable used therewith, shall be mounted so that there is at least ¼-inch (6.35-mm) air space between it and the wall or supporting surface.

Exception: Nonmetallic raceways, boxes, and fittings shall be permitted to be installed without the airspace on a concrete, masonry, tile, or similar surface.

(FPN): In general, areas where acids and alkali chemicals are handled and stored may present such corrosive conditions, particularly when wet or damp. Severe corrosive conditions may also be present in portions of meat-packing plants, tanneries, glue houses, and some stables; installations immediately adjacent to a seashore and swimming pool areas; areas where chemical de-icers are used; and storage cellars or rooms for hides, casings, fertilizer, salt, and bulk chemicals.

300-7. Raceways Exposed to Different Temperatures.

(a) Sealing. Where portions of an interior raceway system are exposed to widely different temperatures, as in refrigerating or cold-storage plants, circulation of air from a warmer to a colder section through the raceway shall be prevented.

(b) Expansion Joints. Raceways shall be provided with expansion joints where necessary to compensate for thermal expansion and contraction.

300-8. Installation of Conductors with Other Systems. Raceways or cable trays containing electric conductors shall not contain any pipe, tube, or equal for steam, water, air, gas, drainage, or any service other than electrical.

300-9. Grounding Metal Enclosures. Metal raceways, boxes, cabinets, cable armor, and fittings shall be grounded as required in Article 250.

300-10. Electrical Continuity of Metal Raceways and Enclosures. Metal raceways, cable armor, and other metal enclosures for conductors shall be metallically joined together into a continuous electric conductor and shall be so connected to all boxes, fittings, and cabinets as to provide effective electrical continuity. Raceways and cable assemblies shall be mechanically secured to boxes, fittings, cabinets, and other enclosures.

Exception No. 1: As provided in Section 370-17(c) for nonmetallic boxes.

Exception No. 2: As provided in Section 250-33, Exception No. 2 for metal enclosures.

Exception No. 3: As provided in Section 250-75, Exception where permitted for reduction of electrical noise.

300-11. Securing and Supporting.

(a) Secured in Place. Raceways, cable assemblies, boxes, cabinets, and fittings shall be securely fastened in place. Support wires that do not provide secure support shall not be permitted as the sole support.

Branch-circuit wiring associated with equipment that is located within, supported by, or secured to a fire-rated floor or roof/ceiling assembly shall not be secured to the ceiling support wires.

Branch-circuit wiring associated with equipment that is located within, supported by, or secured to a non-fire-rated floor or roof/ceiling assembly shall be permitted to be supported by the ceiling support wires.

Exception: As permitted elsewhere in this Code.

(b) Raceways Used as Means of Support. Raceways shall not be used as a means of support for other raceways, cables, or nonelectric equipment.

Exception No. 1: Where the raceways or means of support are identified for the purpose.

(FPN): See Article 318 for cable trays.

Exception No. 2: Raceways containing power supply conductors for electrically controlled equipment shall be permitted to support Class 2 circuit conductors or cables that are solely for the purpose of connection to the equipment control circuits.

Exception No. 3: As permitted in Section 370-23 for boxes or conduit bodies or Section 410-16(f) for fixtures.

300-12. Mechanical Continuity—Raceways and Cables. Metal or nonmetallic raceways, cable armors, and cable sheaths shall be continuous between cabinets, boxes, fittings, or other enclosures or outlets.

Exception: Short sections of raceways used to provide support or protection of cable assemblies from physical damage.

300-13. Mechanical and Electrical Continuity—Conductors.

(a) General. Conductors in raceways shall be continuous between outlets, devices, etc., and there shall be no splice or tap within a raceway itself.

Exception No. 1: As provided in Section 374-8 for auxiliary gutters.

Exception No. 2: As provided in Section 362-7 for wireways.

Exception No. 3: As provided in Section 300-15(a), Exception No. 1 for boxes or fittings.

Exception No. 4: As provided in Section 352-7 for surface metal raceways.

Exception No. 5: Busways as covered in Article 364.

(b) Device Removal. In multiwire branch circuits, the continuity of a grounded conductor shall not be dependent upon device connections, such as lampholders, receptacles, etc., where the removal of such devices would interrupt the continuity.

300-14. Length of Free Conductors at Outlets, Junctions, and Switch Points. At least 6 inches (152 mm) of free conductor shall be left at each outlet, junction, and switch point for splices or the connection of fixtures or devices.

Exception: Conductors that are not spliced or terminated at the outlet, junction, or switch point.

300-15. Boxes or Fittings—Where Required.

(a) Box or Fitting. A box or fitting shall be installed at each conductor splice connection point, outlet, switch point, junction point, or pull point for the connection of conduit, electrical metallic tubing, surface raceway, or other raceways.

Exception No. 1: A box or fitting shall not be required for a conductor splice connection in surface raceways, wireways, header-ducts, multioutlet assemblies, auxiliary gutters, cable trays, and conduit bodies complying with Sections 370-16(c) and 370-28 and having removable covers that are accessible after installation.

Exception No. 2: As permitted in Section 410-31 where a fixture is used as a raceway.

(b) Box Only. A box shall be installed at each conductor splice connection point, outlet, switch point, junction point, or pull point for the connection of Type AC cable, Type MC cable, mineral-insulated, metal-sheathed cable, nonmetallic-sheathed cable, or other cables, at the connection point between any such cable system and a raceway system and at each outlet and switch point for concealed knob-and-tube wiring.

Exception No. 1: Where cables enter or exit from conduit or tubing that is used to provide cable support or protection against physical damage. A fitting shall be provided on the end(s) of the conduit or tubing to protect the wires or cables from abrasion.

Exception No. 2: As permitted by Section 336-16 for insulated outlet devices supplied by nonmetallic-sheathed cable.

Exception No. 3: Where accessible fittings are used for straight-through splices in mineral-insulated, metal-sheathed cable.

Exception No. 4: A wiring device with integral enclosure identified for the use having brackets that securely fasten the device to walls or ceilings of conventional on-site frame construction for use with nonmetallic-sheathed cable shall be permitted without a separate box.

(FPN):   See Sections 336-15, Exception No. 2, 545-10, 550-10(j), and 551-47(e), Exception No. 1.

Exception No. 5: Where metallic manufactured wiring systems are used.

Exception No. 6: A conduit body shall be permitted in lieu of a box where installed to comply with Section 370-16(c) and Section 370-28.

Exception No. 7: Where a device identified and listed as suitable for installation without a box is used with a closed-loop power distribution system.

Exception No. 8: A fitting identified for the use shall be permitted in lieu of a box where accessible after installation and where the conductors are not spliced or terminated.

(c) Fittings and Connectors. Fittings and connectors shall be used only with the specific wiring methods for which they are designed and listed.

(d) Equipment. An integral junction box or wiring compartment as part of listed equipment shall be permitted at an outlet in lieu of a box.

300-16. Raceway or Cable to Open or Concealed Wiring.

(a) Box or Fitting. A box or terminal fitting having a separately bushed hole for each conductor shall be used wherever a change is made from conduit, electrical metallic tubing, nonmetallic-sheathed cable, Type AC cable, Type MC cable, or mineral-insulated, metal-sheathed cable and surface raceway wiring to open wiring or to concealed knob-and-tube wiring. A fitting used for this purpose shall contain no taps or splices and shall not be used at fixture outlets.

(b) Bushing. A bushing shall be permitted in lieu of a box or terminal fitting at the end of a conduit or electrical metallic tubing where the raceway terminates behind an open (unenclosed) switchboard or at an unenclosed control and similar equipment. The bushing shall be of the insulating type for other than lead-sheathed conductors.

300-17. Number and Size of Conductors in Raceway.   The number and size of conductors in any raceway shall not be more than will permit dissipation of the heat and ready installation or withdrawal of the conductors without damage to the conductors or to their insulation.

(FPN):   See the following sections of this Code: electrical nonmetallic tubing, 331-6; conduit, 345-7 and 346-6; electrical metallic tubing, 348-6; rigid nonmetallic conduit, 347-11; flexible metallic tubing, 349-12; flexible metal conduit, 350-3; liquidtight flexible metal conduit, 351-6; liquidtight nonmetallic flexible conduit, 351-25; surface raceways, 352-4 and 352-25; underfloor raceways, 354-5; cellular metal floor raceways, 356-5; cellular concrete floor raceways, 358-11; wireways, 362-5; auxiliary gutters, 374-5; fixture wire, 402-7; theaters, 520-5; signs, 600-21(d); elevators, 620-33; sound recording, 640-3 and 640-4; Class 1, Class 2, and Class 3 circuits, Article 725; fire protective signaling circuits, Article 760, and optical fiber cables, Article 770.

300-18. Raceway Installations. Raceways shall be installed complete between outlet, junction, or splicing points prior to the installation of conductors.

Exception No. 1: Exposed raceways having a removable cover.

Exception No. 2: Where required to facilitate the installation of utilization equipment.

Exception No. 3: Prewired assemblies in accordance with Articles 349 and 350.

300-19. Supporting Conductors in Vertical Raceways.

(a) Spacing Intervals—Maximum. Conductors in vertical raceways shall be supported if the vertical rise exceeds the values in Table 300-19(a). One cable support shall be provided at the top of the vertical raceway or as close to the top as practical. Intermediate supports shall be provided as necessary to limit supported conductor lengths to not greater than those values specified in Table 300-19(a).

Exception: Steel wire armor cable shall be supported at the top of the riser with a cable support that clamps the steel wire armor. A safety device shall be permitted at the lower end of the riser to hold the cable in the event there is slippage of the cable in the wire armored cable support. Additional wedge-type supports shall be permitted to relieve the strain on the equipment terminals caused by expansion of the cable under load.

Table 300-19(a). Spacings for Conductor Supports

	Conductors		
AWG or Circular-Mil Size of Wire	Support of Conductors in Vertical Raceways	Aluminum or Copper-Clad Aluminum	Copper
18 AWG through 8 AWG	Not greater than	.... 100 feet	... 100 feet
6 AWG through 1/0 AWG	" " "	.... 200 feet	... 100 feet
2/0 AWG through 4/0 AWG	" " "	.... 180 feet	... 80 feet
Over 4/0 AWG through 350 kcmil ..	" " "	.... 135 feet	... 60 feet
Over 350 kcmil through 500 kcmil ..	" " "	.... 120 feet	... 50 feet
Over 500 kcmil through 750 kcmil ..	" " "	.... 95 feet	... 40 feet
Over 750 kcmil......................	" " "	.... 85 feet	... 35 feet

For SI units: one foot = 0.3048 meter.

(b) Support Methods. One of the following methods of support shall be used:

(1) By clamping devices constructed of or employing insulating wedges inserted in the ends of the conduits. Where clamping of insulation does not adequately support the cable, the conductor also shall be clamped.

(2) By inserting boxes at the required intervals in which insulating supports are installed and secured in a satisfactory manner to withstand the weight of the conductors attached thereto, the boxes being provided with covers.

(3) In junction boxes, by deflecting the cables not less than 90 degrees and carrying them horizontally to a distance not less than twice the diameter of the cable, the cables being carried on two or more insulating supports, and additionally secured thereto by tie wires if desired. Where this method is used, cables shall be supported at intervals not greater than 20 percent of those mentioned in the preceding tabulation.

(4) By a method of equal effectiveness.

300-20. Induced Currents in Metal Enclosures or Metal Raceways.

(a) Conductors Grouped Together. Where conductors carrying alternating current are installed in metal enclosures or metal raceways, they shall be so arranged as to avoid heating the surrounding metal by induction. To accomplish this, all phase conductors and, where used, the neutral and all equipment grounding conductors shall be grouped together.

Exception No. 1: As permitted in Section 250-50, Exception for equipment grounding connections.

Exception No. 2: As permitted in Section 426-42 and Section 427-47 for skin effect heating.

(b) Individual Conductors. Where a single conductor carrying alternating current passes through metal with magnetic properties, the inductive effect shall be minimized by (1) cutting slots in the metal between the individual holes through which the individual conductors pass, or (2) passing all the conductors in the circuit through an insulating wall sufficiently large for all of the conductors of the circuit.

Exception: In the case of circuits supplying vacuum or electric-discharge lighting systems or signs, or X-ray apparatus, the currents carried by the conductors are so small that the inductive heating effect can be ignored where these conductors are placed in metal enclosures or pass through metal.

(FPN): Because aluminum is not a magnetic metal, there will be no heating due to hysteresis; however, induced currents will be present. They will not be of sufficient magnitude to require grouping of conductors or special treatment in passing conductors through aluminum wall sections.

300-21. Spread of Fire or Products of Combustion. Electrical installations in hollow spaces, vertical shafts, and ventilation or air-handling ducts shall be so made that the possible spread of fire or products of combustion will not be substantially increased. Openings around electrical penetrations through fire-resistant-rated walls, partitions, floors, or ceilings shall be firestopped using approved methods to maintain the fire resistance rating.

(FPN): Directories of electrical construction materials published by qualified testing laboratories contain many listing installation restrictions necessary to maintain the fire resistive rating of assemblies where penetrations or openings are made. An example is the 24-inch (610-mm) minimum horizontal separation between boxes on opposite sides of the wall. Assistance in complying with Section 300-21 can be found in these directories and product listings.

300-22. Wiring in Ducts, Plenums, and Other Air-Handling Spaces. The provisions of this section apply to the installation and uses of electric wiring and equipment in ducts, plenums, and other air-handling spaces.

(FPN): See Article 424, Part F, Duct Heaters.

(a) Ducts for Dust, Loose Stock, or Vapor Removal. No wiring systems of any type shall be installed in ducts used to transport dust, loose stock, or flammable vapors. No wiring system of any type shall be installed in any duct, or shaft containing only such ducts, used for vapor removal or for ventilation of commercial-type cooking equipment.

(b) Ducts or Plenums Used for Environmental Air. Only wiring methods consisting of Type MI cable, Type MC cable employing a smooth or corrugated impervious metal sheath without an overall nonmetallic covering, electrical metallic tubing, flexible metallic tubing, intermediate metal conduit, or rigid metal conduit shall be installed in ducts or plenums specifically fabricated to transport environmental air. Flexible metal conduit and liquidtight flexible metal conduit shall be permitted, in lengths not to exceed 4 feet (1.22 m), to connect physically adjustable equipment and devices permitted to be in these ducts and plenum chambers. The connectors used with flexible metal conduit shall effectively close any openings in the connection. Equipment and devices shall be permitted within such ducts or plenum chambers only if necessary for their direct action upon, or sensing of, the contained air. Where equipment or devices are installed and illumination is necessary to facilitate maintenance and repair, enclosed gasketed-type fixtures shall be permitted.

(c) Other Space Used for Environmental Air. Section 300-22(c) applies to space used for environmental air-handling purposes other than ducts and plenums as specified in Sections 300-22(a) and 300-22(b). Only totally enclosed nonventilated insulated busway having no provisions for plug-in connections and wiring methods consisting of Type MI cable, Type MC cable without an overall nonmetallic covering, Type AC cable, or other factory-assembled multiconductor control or power cable that is specifically listed for the use shall be installed in such other space.

Other type cables and conductors shall be installed in electrical metallic tubing, flexible metallic tubing, intermediate metal conduit, rigid metal conduit, flexible metal conduit, or, where accessible, surface metal raceway or wireway with metal covers or solid bottom metal cable tray with solid metal covers.

Electric equipment with a metal enclosure or with a nonmetallic enclosure listed for the use and having adequate fire-resistant and low-smoke-producing characteristics, and associated wiring material suitable for the ambient temperature, shall be permitted to be installed in such other space unless prohibited elsewhere in this Code.

(FPN): The space over a hung ceiling used for environmental air-handling purposes is an example of the type of other space to which Section 300-22(c) applies.

Exception No. 1: Liquidtight flexible metal conduit in single lengths not exceeding 6 feet (1.83 m).

Exception No. 2: Integral fan systems specifically identified for such use.

Exception No. 3: This section does not include habitable rooms or areas of buildings, the prime purpose of which is not air handling.

Exception No. 4: Listed prefabricated cable assemblies of metallic manufactured wiring systems without nonmetallic sheath shall be permitted where listed for this use.

Exception No. 5: This section does not include the joist or stud spaces in dwelling units where wiring passes through such spaces perpendicular to the long dimension of such spaces.

(d) Data Processing Systems. Electric wiring in air-handling areas beneath raised floors for data processing systems shall comply with Article 645.

300-23. Panels Designed to Allow Access.   Cables, raceways, and equipment installed behind panels designed to allow access including suspended ceiling panels shall be so arranged and secured as to allow the removal of panels and access to the equipment.

B. Requirements for Over 600
Volts, Nominal

300-31. Covers Required.   Suitable covers shall be installed on all boxes, fittings, and similar enclosures to prevent accidental contact with energized parts or physical damage to parts or insulation.

300-32. Conductors of Different Systems.   Conductors of high-voltage and low-voltage systems shall not occupy the same wiring enclosure or pull and junction boxes.

Exception No. 1: In motors, switchgear and control assemblies, and similar equipment.

Exception No. 2: In manholes, if extra-low or low-voltage conductors are separated from high-voltage conductors.

300-34. Conductor Bending Radius.   The conductor shall not be bent to a radius less than eight times the overall diameter for nonshielded conductors or twelve times the diameter for shielded or lead-covered conductors during or after installation.

300-35. Protection Against Induction Heating.   Metallic raceways and associated conductors shall be so arranged as to avoid heating of the raceway in accordance with the applicable provisions of Section 300-20.

300-36. Grounding.   Wiring and equipment installations shall be grounded in accordance with the applicable provisions of Article 250.

300-37. Underground Installations.   The minimum cover requirements shall be in accordance with Section 710-4(b).

ARTICLE 305 — TEMPORARY WIRING

305-1. Scope. The provisions of this article apply to temporary electrical power and lighting wiring methods that may be of a class less than would be required for a permanent installation.

305-2. Other Articles. Except as specifically modified in this article, all other requirements of this Code for permanent wiring shall apply to temporary wiring installations.

305-3. Time Constraints.

(a) During the Period of Construction. Temporary electrical power and lighting installations shall be permitted during the period of construction, remodeling, maintenance, repair, or demolition of buildings, structures, equipment, or similar activities.

(b) 90 Days. Temporary electrical power and lighting installations shall be permitted for a period not to exceed 90 days for Christmas decorative lighting, carnivals, and similar purposes.

(c) Emergencies and Tests. Temporary electrical power and lighting installations shall be permitted during emergencies and for tests, experiments, and developmental work.

(d) Removal. Temporary wiring shall be removed immediately upon completion of construction or purpose for which the wiring was installed.

305-4. General.

(a) Services. Services shall be installed in conformance with Article 230.

(b) Feeders. Feeders shall be protected as provided in Article 240. They shall originate in an approved distribution center. Conductors shall be permitted within cable assemblies; or, cords or cables of a type identified in | Table 400-4 for hard usage or extra-hard usage. Where the voltage does not exceed 150 volts to ground and where not subject to physical damage, feeders shall be permitted to be run as open conductors if supported on insulators at intervals of not more than 10 feet (3.05 m).

Exception: Where installed for the purposes specified in Section 305-3(c).

(c) Branch Circuits. All branch circuits shall originate in an approved power outlet or panelboard. Conductors shall be permitted within cable assemblies; or, multiconductor cord or cable of a type identified in Table 400-4 for hard usage or extra-hard usage. All conductors shall be protected as provided in Article 240. Where the voltage does not exceed 150 volts to ground and where not subject to physical damage, branch circuits shall be permitted to be run as open conductors if supported on insulators at intervals of not more than 10 feet (3.05 m). No open wiring branch circuit conductors shall be laid on the floor or ground.

Exception: Where installed for the purposes specified in Section 305-3(c).

(d) Receptacles. All receptacles shall be of the grounding type. Unless installed in a continuous grounded metal raceway or metal-covered cable, all branch circuits shall contain a separate equipment grounding conductor, and all receptacles shall be electrically connected to the equipment grounding conductors. Receptacles on construction sites shall not be installed on branch circuits that supply temporary lighting. Receptacles shall not be connected to the same ungrounded conductor of multiwire circuits that supply temporary lighting.

(e) Disconnecting Means. Suitable disconnecting switches or plug connectors shall be installed to permit the disconnection of all ungrounded conductors of each temporary circuit. Multiwire branch circuits shall be provided with a means to disconnect simultaneously all ungrounded conductors at the power outlet or panelboard where the branch circuit originated. Approved handle ties shall be permitted.

(f) Lamp Protection. All lamps for general illumination shall be protected from accidental contact or breakage by a suitable fixture or lampholder with a guard.

Brass shell, paper-lined sockets, or other metal-cased sockets shall not be used unless the shell is grounded.

(g) Splices. On construction sites, a box shall not be required for splices or junction connections where the circuit conductors are multiconductor cord or cable assemblies or open conductors. See Sections 110-14(b) and 400-9. A box, conduit body, or terminal fitting having a separately bushed hole for each conductor shall be used wherever a change is made to a conduit or tubing system or a metal-sheathed cable system.

(h) Protection from Accidental Damage. Flexible cords and cables shall be protected from accidental damage. Sharp corners and projections shall be avoided. Where passing through doorways or other pinch points, protection shall be provided to avoid damage.

(i) Termination(s) at Devices. Cables entering enclosures containing devices requiring termination shall be secured to the box with fittings designed for the purpose.

305-5. Grounding. All grounding shall conform with Article 250.

305-6. Ground-Fault Protection for Personnel. Ground-fault protection for personnel on construction sites shall be provided to comply with (a) or (b) below.

(a) Ground-Fault Circuit-Interrupters. All 125-volt, single-phase, 15- and 20-ampere receptacle outlets that are not a part of the permanent wiring of the building or structure and that are in use by personnel shall have ground-fault circuit-interrupter protection for personnel. If a receptacle or receptacles are installed as part of the permanent wiring of the building or structure and used for temporary electric power, GFCI protection for personnel shall be provided.

Exception: Receptacles on a 2-wire, single-phase portable or vehicle-mounted generator rated not more than 5 kW, where the circuit conductors of the generator are insulated from the generator frame and all other grounded surfaces.

(b) Assured Equipment Grounding Conductor Program. A written procedure shall be continuously enforced at the construction site by one or more designated persons to assure that equipment grounding conductors for all cord sets, receptacles that are not a part of the permanent wiring of the building or structure, and equipment connected by cord and plug are installed and maintained in accordance with the applicable requirements of Sections 210-7(c), 250-45, 250-59, and 305-4(d).

(1) The following tests shall be performed on all cord sets, receptacles that are not part of the permanent wiring of the building or structure, and cord- and plug-connected equipment required to be grounded.

a. All equipment grounding conductors shall be tested for continuity and shall be electrically continuous.

b. Each receptacle and attachment plug shall be tested for correct attachment of the equipment grounding conductor. The equipment grounding conductor shall be connected to its proper terminal.

c. All required tests shall be performed:

1. Before first use on the construction site.
2. When there is evidence of damage.
3. Before equipment is returned to service following any repairs.
4. At intervals not exceeding 3 months.

(2) The tests required in (1) above shall be recorded and made available to the authority having jurisdiction.

305-7. Guarding.   For wiring over 600 volts, nominal, suitable fencing, barriers, or other effective means shall be provided to prevent access of other than authorized and qualified personnel.

ARTICLE 310 — CONDUCTORS FOR GENERAL WIRING

310-1. Scope.   This article covers general requirements for conductors and their type designations, insulations, markings, mechanical strengths, ampacity ratings, and uses. These requirements do not apply to conductors that form an integral part of equipment, such as motors, motor controllers, and similar equipment, or to conductors specifically provided for elsewhere in this Code.

(FPN):   For flexible cords and cables, see Article 400. For fixture wires, see Article 402.

310-2. Conductors.

(a) Insulated.   Conductors shall be insulated.

Exception: Where covered or bare conductors are specifically permitted elsewhere in this Code.

(FPN):   See Section 250-152 for insulation of neutral conductors of a solidly grounded high-voltage system.

(b) Conductor Material.   Conductors in this article shall be of aluminum, copper-clad aluminum, or copper unless otherwise specified.

310-3. Stranded Conductors.   Where installed in raceways, conductors of size No. 8 and larger shall be stranded.

Exception: As permitted or required elsewhere in this Code.

310-4. Conductors in Parallel. Aluminum, copper-clad aluminum, or copper conductors of size No. 1/0 and larger, comprising each phase, neutral or grounded circuit conductor, shall be permitted to be connected in parallel (electrically joined at both ends to form a single conductor).

Exception No. 1: As permitted in Section 620-12(a)(1), Exception.

Exception No. 2: Conductors in sizes smaller than No. 1/0 shall be permitted to be run in parallel to supply control power to indicating instruments, contactors, relays, solenoids, and similar control devices provided (a) they are contained within the same raceway or cable; (b) the ampacity of each individual conductor is sufficient to carry the entire load current shared by the parallel conductors; and (c) the overcurrent protection is such that the ampacity of each individual conductor will not be exceeded if one or more of the parallel conductors become inadvertently disconnected.

Exception No. 3: Conductors in sizes smaller than No. 1/0 shall be permitted to be run in parallel for frequencies of 360 hertz and higher where conditions (a), (b), and (c) of Exception No. 2 are met.

The paralleled conductors in each phase, neutral or grounded circuit conductor shall:

(1) Be the same length;

(2) Have the same conductor material;

(3) Be the same size in circular mil area;

(4) Have the same insulation type;

(5) Be terminated in the same manner.

Where run in separate raceways or cables, the raceways or cables shall have the same physical characteristics.

(FPN): Differences in inductive reactance and unequal division of current can be minimized by choice of materials, methods of construction, and orientation of conductors. It is not the intent to require that conductors of one phase, neutral or grounded circuit conductor be the same as those of another phase, neutral or grounded circuit conductor to achieve balance.

Where equipment grounding conductors are used with conductors in parallel, they shall comply with the requirements of this section except that they shall be sized in accordance with Section 250-95.

Where conductors are used in parallel, space in enclosures shall be given consideration (see Articles 370 and 373).

Conductors installed in parallel shall comply with the provisions of Article 310, Note 8(a), Notes to Ampacity Tables of 0 to 2000 Volts.

310-5. Minimum Size of Conductors.   The minimum size of conductors shall be as shown in Table 310-5.

Exception No. 1: For flexible cords as permitted by Section 400-12.

Exception No. 2: For fixture wire as permitted by Section 410-24.

Exception No. 3: For fractional horsepower motors as permitted by Section 430-22.

Exception No. 4: For cranes and hoists as permitted by Section 610-14.

Exception No. 5: For elevator control and signaling circuits as permitted by Section 620-12.

Table 310-5.

Voltage Rating of Conductor—Volts	Minimum Conductor Size—AWG
0 through 2000	14 Copper 12 Aluminum or Copper-Clad Aluminum
2001 through 5000	8
5001 through 8000	6
8001 through 15000	2
15001 through 28000	1
28001 through 35000	1/0

Exception No. 6: For Class 1, Class 2, and Class 3 circuits as permitted by Sections 725-16 and 725-37.

Exception No. 7: For fire protective signaling circuits as permitted by Sections 760-16, 760-25, and 760-51.

Exception No. 8: For Type V cables, the minimum conductor sizes are: No. 12 for 2000-volt rating, No. 10 for 3000-volt rating, and No. 8 for 4000-volt rating.

Exception No. 9: For motor control circuits as permitted by Section 430-72.

310-6. Shielding. Solid dielectric insulated conductors operated above 2000 volts in permanent installations shall have ozone-resistant insulation and shall be shielded. All metallic insulation shields shall be grounded through an effective grounding path meeting the requirements of Section 250-51. Shielding shall be for the purpose of confining the voltage stresses to the insulation.

Exception: Nonshielded insulated conductors listed by a qualified testing laboratory shall be permitted for use up to 8000 volts under the following conditions:

a. Conductors shall have insulation resistant to electric discharge and surface tracking, or the insulated conductor(s) shall be covered with a material resistant to ozone, electric discharge, and surface tracking.

b. Where used in wet locations, the insulated conductor(s) shall have an overall nonmetallic jacket or a continuous metallic sheath.

c. Where operated at 5001 to 8000 volts, the insulated conductor(s) shall have a nonmetallic jacket over the insulation. The insulation shall have a specific inductive capacity no greater than 3.6, and the jacket shall have a specific inductive capacity no greater than 10 and no less than 6.

d. Insulation and jacket thicknesses shall be in accordance with Table 310-63.

310-7. Direct Burial Conductors. Conductors used for direct burial applications shall be of a type identified for such use.

Cables rated above 2000 volts shall be shielded.

Exception: Nonshielded multiconductor cables rated 2001-5000 volts shall be permitted if the cable has an overall metallic sheath or armor.

The metallic shield, sheath, or armor shall be grounded through an effective grounding path meeting the requirements of Section 250-51.

(FPN No. 1): See Section 300-5 for installation requirements for conductors rated 600 volts or less.

(FPN No. 2): See Section 710-4(b) for installation requirements for conductors rated over 600 volts.

310-8. Wet Locations.

(a) Insulated Conductors. Insulated conductors used in wet locations shall be (1) lead-covered; (2) Types RHW, TW, THW, THHW, THWN, XHHW; or (3) of a type listed for use in wet locations.

(b) Cables. Cables of one or more conductors used in wet locations shall be of a type listed for use in wet locations.

Conductors used for direct burial applications shall be of a type listed for such use.

310-9. Corrosive Conditions.
Conductors exposed to oils, greases, vapors, gases, fumes, liquids, or other substances having a deleterious effect upon the conductor or insulation shall be of a type suitable for the application.

310-10. Temperature Limitation of Conductors.
No conductor shall be used in such a manner that its operating temperature will exceed that designated for the type of insulated conductor involved. In no case shall conductors be associated together in such a way with respect to type of circuit, the wiring method employed, or the number of conductors that the limiting temperature of any conductor is exceeded.

(FPN): The temperature rating of a conductor (see Tables 310-13 and 310-61) is the maximum temperature, at any location along its length, that the conductor can withstand over a prolonged time period without serious degradation. The allowable ampacity tables, the ampacity tables of Article 310 and the ampacity tables of Appendix B, the correction factors at the bottom of these tables, and the notes to the tables provide guidance for coordinating conductor sizes, types, allowable ampacities, ampacities, ambient temperatures, and number of associated conductors.

The principal determinants of operating temperature are:

1. Ambient temperature. Ambient temperature may vary along the conductor length as well as from time to time.

2. Heat generated internally in the conductor as the result of load current flow.

3. The rate at which generated heat dissipates into the ambient medium. Thermal insulation that covers or surrounds conductors will affect the rate of heat dissipation.

4. Adjacent load-carrying conductors. Adjacent conductors have the dual effect of raising the ambient temperature and impeding heat dissipation.

310-11. Marking.

(a) Required Information. All conductors and cables shall be marked to indicate the following information, using the applicable method described in (b) below.

(1) The maximum rated voltage for which the conductor was listed.

(2) The proper type letter or letters for the type of wire or cable as specified elsewhere in this Code.

(3) The manufacturer's name, trademark, or other distinctive marking by which the organization responsible for the product can be readily identified.

(4) The AWG size or circular-mil area.

(b) Method of Marking.

(1) Surface Marking. The following conductors and cables shall be durably marked on the surface. The AWG size or circular mil area shall be repeated at intervals not exceeding 24 inches (610 mm). All other markings shall be repeated at intervals not exceeding 40 inches (1.02 m).

   a. Single- and multiconductor rubber- and thermoplastic- insulated wire and cable.

   b. Nonmetallic-sheathed cable.

   c. Service-entrance cable.

   d. Underground feeder and branch-circuit cable.

   e. Tray cable.

   f. Irrigation cable.

   g. Power-limited tray cable.

(2) Marker Tape. Metal-covered multiconductor cables shall employ a marker tape located within the cable and running for its complete length.

Exception No. 1: Mineral-insulated, metal-sheathed cable.

Exception No. 2: Type AC cable.

Exception No. 3: The information required in Section 310-11(a) shall be permitted to be durably marked on the outer nonmetallic covering of Type MC or Type PLTC cables at intervals not exceeding 40 inches (1.02 m).

Exception No. 4: The information required in Section 310-11(a) shall be permitted to be durably marked on a nonmetallic covering under the metallic sheath of Type PLTC cable at intervals not exceeding 40 inches (1.02 m).

(FPN):   Included in the group of metal-covered cables are Type AC cable (Article 333), Type MC cable (Article 334), and lead-sheathed cable.

(3) Tag Marking. The following conductors and cables shall be marked by means of a printed tag attached to the coil, reel, or carton:

   a. Mineral-insulated, metal-sheathed cable.

   b. Switchboard wires.

   c. Metal-covered, single-conductor cables.

   d. Conductors having outer surface of asbestos.

   e. Type AC cable.

(4) Optional Marking of Wire Size. For the following multiconductor cables, the information required in (a) (4) above shall be permitted to be marked on the surface of the individual insulated conductors:

    a. Type MC cable.
    b. Tray cable.
    c. Irrigation cable.
    d. Power-limited tray cable.
    e. Power-limited fire protective signaling cable.

(c) Suffixes to Designate Number of Conductors. A type letter or letters used alone shall indicate a single insulated conductor. The following letter suffixes shall indicate the following:

D—For two insulated conductors laid parallel within an outer nonmetallic covering.

M—For an assembly of two or more insulated conductors twisted spirally within an outer nonmetallic covering.

(d) Optional Markings. Conductor types listed in Table 310-13 shall be permitted to be surface marked to indicate special characteristics of the cable materials.

(FPN): Examples of these markings include but are not limited to "LS" for limited-smoke and markings such as "sunlight-resistant."

310-12. Conductor Identification.

(a) Grounded Conductors. Insulated conductors of No. 6 or smaller, intended for use as grounded conductors of circuits, shall have an outer identification of a white or natural gray color. Multiconductor flat cable No. 4 or larger shall be permitted to employ an external ridge on the grounded conductor.

Exception No. 1: Fixture wires as outlined in Article 402.

Exception No. 2: Mineral-insulated, metal-sheathed cable.

Exception No. 3: A conductor identified as required by Section 210-5(a) for branch circuits.

Exception No. 4: Where the conditions of maintenance and supervision assure that only qualified persons will service the installation, grounded conductors in multiconductor cables shall be permitted to be permanently identified at their terminations at the time of installation by a distinctive white marking or other equally effective means.

For aerial cable, the identification shall be as above, or by means of a ridge so located on the exterior of the cable as to identify it.

Wires having their outer covering finished to show a white or natural gray color but having colored tracer threads in the braid, identifying the source of manufacture, shall be considered as meeting the provisions of this section.

(FPN): For identification requirements for conductors larger than No. 6, see Section 200-6.

(b) Equipment Grounding Conductors. Bare, covered, or insulated grounding conductors shall be permitted. Individually covered or insulated grounding conductors shall have a continuous outer finish that is either green, or green with one or more yellow stripes.

Exception No. 1: An insulated or covered conductor larger than No. 6 shall, at the time of installation, be permitted to be permanently identified as a grounding conductor at each end and at every point where the conductor is accessible. Identification shall be accomplished by one of the following means:

a. Stripping the insulation or covering from the entire exposed length;

b. Coloring the exposed insulation or covering green; or

c. Marking the exposed insulation or covering with green colored tape or green colored adhesive labels.

Exception No. 2: Where the conditions of maintenance and supervision assure that only qualified persons will service the installation, an insulated conductor in a multiconductor cable shall, at the time of installation, be permitted to be permanently identified as a grounding conductor at each end and at every point where the conductor is accessible by one of the following means:

a. Stripping the insulation from the entire exposed length;

b. Coloring the exposed insulation green; or

c. Marking the exposed insulation with green colored tape or green colored adhesive labels.

(c) Ungrounded Conductors. Conductors that are intended for use as ungrounded conductors, whether used as single conductors or in multiconductor cables, shall be finished to be clearly distinguishable from grounded and grounding conductors. Ungrounded conductors shall be distinguished by colors other than white, natural gray, or green; or by a combination of color plus distinguishing marking. Distinguishing markings shall also be in a color other than white, natural gray, or green, and shall consist of a stripe or stripes or a regularly spaced series of identical marks. Distinguishing markings shall not conflict in any manner with the surface markings required by Section 310-11(b)(1).

Exception: As permitted by Section 200-7.

310-13. Conductor Constructions and Applications. Insulated conductors shall comply with the applicable provisions of one or more of the following: Tables 310-13, 310-61, 310-62, 310-63, and 310-64.

These conductors shall be permitted for use in any of the wiring methods recognized in Chapter 3 and as specified in their respective tables.

(FPN):   Thermoplastic insulation may stiffen at temperatures colder than minus 10°C (plus 14°F), requiring that care be exercised during installation at such temperatures. Thermoplastic insulation may also be deformed at normal temperatures where subjected to pressure, requiring care be exercised during installation and at points of support. Thermoplastic insulation, where used on dc circuits in wet locations, may result in electroendosmosis between conductor and insulation.

310-14. Aluminum Conductor Material. Solid aluminum conductors No. 8, 10, and 12 shall be made of an AA-8000 series electrical grade aluminum alloy conductor material. Stranded aluminum conductors No. 8 through 1000 kcmil marked as Type XHHW, THW, THHW, THWN, THHN, service-entrance Type SE Style U and SE Style R shall be made of an AA-8000 series electrical grade aluminum alloy conductor material.

Table 310-13.　Conductor Application and Insulations

Trade Name	Type Letter	Max. Operating Temp.	Application Provisions	Insulation	AWG or kcmil	Thickness of Insulation Mils	Outer Covering****
Fluorinated Ethylene Propylene	FEP or FEPB	90° C 194° F 200° C 392° F	Dry and damp locations. Dry locations—special applications.†	Fluorinated Ethylene Propylene Fluorinated Ethylene Propylene	14-10 8-2 14-8 6-2	20 30 14 14	None Glass braid Asbestos or other suitable braid material
Mineral Insulation (Metal Sheathed)	MI	90° F 194° F 250° C 482° F	Dry and wet locations. For special application.†	Magnesium Oxide	18-16****** 16-10 9-4 3-500	23 36 50 55	Copper or Alloy Steel
Moisture-, Heat-, and Oil-Resistant Thermoplastic	MTW††	60° C 140° F 90° C 194° F	Machine tool wiring in wet locations as permitted in NFPA 79. (See Article 670.) Machine tool wiring in dry locations as permitted in NFPA 79. (See Article 670.)	Flame-Retardant, Moisture-, Heat-, and Oil-Resistant Thermoplastic	22-12 10 8 6 4-2 1-4/0 213-500 501-1000	(A) 30 30 45 60 60 80 95 110 ／ (B) 15 20 30 30 40 50 60 70	(A) None (B) Nylon jacket or equivalent
Paper		85° C 185° F	For underground service conductors, or by special permission.	Paper			Lead sheath

*****Some insulations do not require an outer covering.
******For signaling circuits permitting 300-volt insulation.
†Where environmental conditions require maximum conductor operating temperatures above 90°C.
††Insulation and outer coverings that meet the requirements of flame-retardant, limited-smoke and are so listed shall be permitted to be designated limited-smoke with the suffix /LS after the Code type designation.

Table 310-13. (Continued)

Trade Name	Type Letter	Max Operating Temperature	Application Provisions	Insulation	AWG/MCM — Thickness (mils)	Outer Covering
Perfluoro-alkoxy	PFA	90° C / 194° F, 200° C / 392° F	Dry and damp locations. Dry locations—special applications.†	Perfluoro-alkoxy	14-10 .... 20, 8-2 .... 30, 1-4/0 45	None
Perfluoro-alkoxy	PFAH	250° C / 482° F	Dry locations only. Only for leads within apparatus or within raceways connected to apparatus. (Nickel or nickel-coated copper only.)	Perfluoro-alkoxy	14-10 .... 20, 8-2 .... 30, 1-4/0 45	None
Heat-Resistant or Cross-Linked Synthetic Polymer	RH	75° C / 167° F	Dry and damp locations.		14-12** .... 30, 10 .... 45, 8-2 .... 60, 1-4/0 .... 80, 213-500 .... 95, 501-1000 .... 110, 1001-2000 .... 125, For 601-2000 volts, see Table 310-62.	Moisture-resistant, flame-retardant, non-metallic covering*
Heat-Resistant or Cross-Linked Synthetic Polymer	RHH††	90° C / 194° F	Dry and damp locations.	Heat-Resistant or Cross-Linked Synthetic Polymer		
Moisture- and Heat-Resistant or Cross-Linked Synthetic Polymer	RHW††, †††	75° C / 167° F	Dry and wet locations. Where over 2000 volts insulation, shall be ozone-resistant	Moisture- and Heat-Resistant or Cross-Linked Synthetic Polymer	14-10 .... 45, 8-2 .... 60, 1-4/0 .... 80, 213-500 .... 95, 501-1000 .... 110, 1001-2000 .... 125, For 601-2000 volts, see Table 310-62.	Moisture-resistant, flame-retardant, non-metallic covering*

*Some rubber insulations do not require an outer covering.

**For size Nos. 14-12, RHH shall be 45 mils thickness insulation.

†Where environmental conditions require maximum conductor operating temperatures above 90°C.

††Insulation and outer coverings that meet the requirements of flame-retardant, limited-smoke and are so listed shall be permitted to be designated limited-smoke with the suffix /LS after the Code type designation.

†††Listed wire types designated with the suffix "-2," such as RHW-2, shall be permitted to be used at a continuous 90°C-operating temperature, wet or dry.

Table 310-13. (Continued)

Trade Name	Type Letter	Max. Operating Temp.	Application Provisions	Insulation	AWG or kcmil	Thickness of Insulation Mils		Outer Covering****
Moisture- and Heat-Resistant or Cross-Linked Synthetic Polymer	RHW-2	90° C 194° F	Dry and wet locations.	Moisture- and Heat-Resistant or Cross-Linked Synthetic Polymer	14-10 8-2 1-4/0 213-500 501-1000 1001-2000 For 601-2000 volts, see Table 310-62.	45 60 80 95 110 125		Moisture-resistant, flame-Resistant, nonmetallic covering*
Silicone-Asbestos	SA	90° C 194° F 125° C 257° F	Dry and damp locations. For special application.†	Silicone Rubber	14-10 8-2 1-4/0 213-500 501-1000 1001-2000	45 60 80 95 110 125		Asbestos, glass, or other suitable braid material
Synthetic Heat-Resistant	SIS††	90° C 194° F	Switchboard wiring only.	Heat-Resistant Cross-Linked Synthetic Polymer	14-10 8 6-2 1-4/0	30 45 60 80		None
Thermoplastic and Asbestos	TA	90° C 194° F	Switchboard wiring only.	Thermoplastic and Asbestos	14-8 6-2 1-4/0	Th'pl'. 20 30 40	Asb. 20 25 30	Flame-retardant, nonmetallic covering
Thermoplastic and Fibrous Outer Braid	TBS	90° C 194° F	Switchboard wiring only.	Thermoplastic	14-10 8 6-2 1-4/0	30 45 60 80		Flame-retardant, nonmetallic covering

*Some rubber insulations do not require an outer covering.

****Some insulations do not require an outer covering.

†Where environmental conditions require maximum conductor operating temperatures above 90°C.

††Insulation and outer coverings that meet the requirements of flame-retardant, limited-smoke and are so listed shall be permitted to be designated limited-smoke with the suffix /LS after the Code type designation.

Table 310-13. (Continued)

Trade Name	Type Letter	Max. Operating Temp.	Application Provisions	Insulation	Size / Thickness (mils)	Outer Covering
Extended Polytetra-fluoroethylene	TFE	250° C 482° F	Dry locations only. Only for leads within apparatus or within raceways connected to apparatus, or as open wiring. (Nickel or nickel-coated copper only.)	Extruded Polytetra-fluoro-ethylene	14-10 .... 20 8-2 .... 30 1-4/0 45	None
Heat-Resistant Thermoplastic	THHN††	90° C 194° F	Dry and damp locations.	Flame-Retardant, Heat-Resistant Thermoplastic	14-12 .... 15 10 .... 20 8-6 .... 30 4-2 .... 40 1-4/0 .... 50 250-500 .... 60 501-1000 70	Nylon jacket or equivalent
Moisture- and Heat-Resistant Thermoplastic	THHW	75° C 167° F 90° C 194° F	Wet location. Dry location.	Flame-Retardant, Moisture- and Heat-Resistant Thermoplastic	14-10 .... 45 8-2 .... 60 1-4/0 .... 80 213-500 .... 95 501-1000 110	None
Moisture- and Heat-Resistant Thermoplastic	THW††, †††	75° C 167° F 90° C 194° F	Dry and wet locations. Special applications within electric discharge lighting equipment. Limited to 1000 open-circuit volts or less. (Size 14-8 only as permitted in Section 410-31.)	Flame-Retardant, Moisture- and Heat-Resistant Thermoplastic	14-10 .... 45 8-2 .... 60 1-4/0 .... 80 213-500 .... 95 501-1000 ...... 110 1001-2000 125	None

††Insulation and outer coverings that meet the requirements of flame-retardant, limited-smoke and are so listed shall be permitted to be designated limited-smoke with the suffix /LS after the Code type designation.

†††Listed wire types designated with the suffix "-2," such as RHW-2, shall be permitted to be used at a continuous 90°C-operating temperature, wet or dry.

Table 310-13. (Continued)

Trade Name	Type Letter	Max. Operating Temp.	Application Provisions	Insulation	AWG or kcmil	Thickness of Insulation Mils	Outer Covering****
Moisture- and Heat-Resistant Thermoplastic	THWN ††, †††	75° C 167° F	Dry and wet locations.	Flame-Retardant, Moisture- and Heat-Resistant Thermoplastic	14-12 10 8-6 4-2 1-4/0 250-500 501-1000	15 20 30 40 50 60 70	Nylon jacket or equivalent
Moisture-Resistant Thermoplastic	TW††	60° C 140° F	Dry and wet locations.	Flame-Retardant, Moisture-Resistant Thermoplastic	14-10 8 6-2 1-4/0 213-500 501-1000 1001-2000	30 45 60 80 95 110 125	None
Underground Feeder and Branch-Circuit Cable—Single Conductor. (For Type UF cable employing more than one conductor, see Article 339.)	UF	60° C 140° F / 75° C 167° F**	See Article 339.	Moisture-Resistant / Moisture- and Heat-Resistant	14-10 8-2 1-4/0	60* 80* 95*	Integral with insulation

*Includes integral jacket.

**For ampacity limitation, see Section 339-5.

****Some insulations do not require an outer covering.

††Insulation and outer coverings that meet the requirements of flame-retardant, limited-smoke and are so listed shall be permitted to be designated limited-smoke with the suffix /LS after the Code type designation.

†††Listed wire types designated with the suffix "-2", such as RHW-2, shall be permitted to be used at a continuous 90°C-operating temperature, wet or dry.

Table 310-13. (Continued)

Trade Name	Type Letter	Max. Operating Temperature	Application Provisions	Insulation	AWG or MCM	Thickness of Insulation (Mils)	Outer Covering
Underground Service-Entrance Cable—Single Conductor. (For Type USE cable employing more than one conductor, see Article 338.)	USE†††	75° C 167° F	See Article 338.	Heat- and Moisture-Resistant	12-10 8-2 1-4/0 213-500 501-1000 1001-2000	45 60 80 95*** 110 125	Moisture-resistant non-metallic covering [See Section 338-1(b).]
Heat-Resistant Cross-Linked Synthetic Polymer	XHH ††	90° C 194° F	Dry and damp locations.	Flame-Retardant Cross-Linked Synthetic Polymer	14-10 8-2 1-4/0 213-500 501-1000 1001-2000	30 45 55 65 80 95	None
Moisture- and Heat-Resistant Cross-Linked Synthetic Polymer	XHHW ††, †††	90° C 194° F 75° C 167° F	Dry and damp locations. Wet locations.	Flame-Retardant Cross-Linked Synthetic Polymer	14-10 8-2 1-4/0 213-500 501-1000 1001-2000	30 45 55 65 80 95	None

***Insulation thickness shall be permitted to be 80 mils for listed Type USE conductors that have been subjected to special investigations.

The nonmetallic covering over individual rubber-covered conductors of aluminum-sheathed cable and of lead-sheathed or multiconductor cable shall not be required to be flame-retardant. For Type MC cable, see Section 334-20. For nonmetallic-sheathed cable, see Section 336-25. For Type UF cable, see Section 339-1.

††Insulation and outer coverings that meet the requirements of flame-retardant, limited-smoke and are so listed shall be permitted to be designated limited-smoke with the suffix /LS after the Code type designation.

†Listed wire types designated with the suffix "-2," such as RHW-2 , shall be permitted to be used at a continuous 90°C-operating temperature, wet or dry.

Table 310-13. (Continued)

Trade Name	Type Letter	Max. Operating Temp.	Application Provisions	Insulation	AWG or kcmil	Thickness of Insulation Mils	Outer Covering****
Moisture- and Heat-Resistant Cross-Linked Synthetic Polymer	XHHW-2	90° C 194° F	Dry and wet locations.	Flame-Retardant Cross-Linked Synthetic Polymer	14-10 8-2 1-4/0 213-500 501-1000 1001-2000	30 45 55 65 80 95	None
Modified Ethylene Tetrafluoroethylene	Z	90° C 194° F 150° C 302° F	Dry and damp locations. Dry locations—special applications.†	Modified Ethylene Tetrafluoroethylene	14-12 10 8-4 3-1 1/0-4/0	15 20 25 35 45	None
Modified Ethylene Tetrafluoroethylene	ZW†††	75° C 167° F 90° C 194° F 150° C 302° F	Wet locations. Dry and damp locations. Dry locations—special applications.†	Modified Ethylene Tetrafluoroethylene	14-10 8-2	30 45	None

****Some insulations do not require an outer covering.

†Where environmental conditions require maximum conductor operating temperatures above 90°C.

†††Listed wire types designated with the suffix "-2," such as RHW-2, shall be permitted to be used at a continuous 90°C-operating temperature, wet or dry.

310-15. Ampacities. Ampacities for conductors shall be permitted to be determined by (a) or (b) below.

(FPN): Ampacities provided by this section do not take voltage drop into consideration. Conductors of circuits as defined in Article 100, sized to prevent a voltage drop exceeding 5 percent, will provide reasonable efficiency of operation.

(a) General. Ampacities for conductors rated 0 to 2000 volts shall be as specified in the Allowable Ampacity Tables 310-16 through 310-19 and their accompanying notes. Ampacities for solid dielectric insulated conductors rated 2001 through 35,000 volts shall be as specified in Tables 310-67 through 310-84 and their accompanying notes.

(FPN): Tables 310-16 through 310-19 are application tables that are for use in determining conductor sizes on loads calculated in accordance with Article 220. Allowable ampacities result from consideration of one or more of the following:

1. Temperature compatibility with connected equipment, especially at the connection points.

2. Coordination with circuit and system overcurrent protection.

3. Compliance with the requirements of product listings or certifications. See Section 110-3(b).

4. Preservation of the safety benefits of established industry practices and standardized procedures.

(b) Engineering Supervision. Under engineering supervision, conductor ampacities shall be permitted to be calculated by means of the following general formula:

$$I = \sqrt{\frac{TC - (TA + DELTA\ TD)}{RDC\ (1 + YC)RCA}}$$

Where:

TC = Conductor temperature in degrees C.

TA = Ambient temperature in degrees C.

DELTA TD = Dielectric loss temperature rise.

RDC = DC resistance of conductor at temperature TC.

YC = Component ac resistance resulting from skin effect and proximity effect.

RCA = Effective thermal resistance between conductor and surrounding ambient.

(FPN): See Appendix B for examples of formula applications.

(c) Selection of Ampacity. Where more than one calculated or tabulated ampacity could apply for a given circuit length, the lowest value shall be used.

Exception: Where two different ampacities apply to adjacent portions of a circuit, the higher ampacity shall be permitted to be used beyond the point of transition, a distance equal to 10 feet (3.05 m) or 10 percent of the circuit length figured at the higher ampacity, whichever is less.

(FPN): See Section 110-14(c) for conductor temperature limitations due to termination provisions.

(d) Electrical Ducts. Electrical ducts as used in Article 310 shall include any of the electrical conduits recognized in Chapter 3 as suitable for use underground; and other raceways round in cross section, listed for underground use; embedded in earth or concrete.

Table 310-16. Allowable Ampacities of Insulated Conductors Rated 0-2000 Volts, 60° to 90°C (140° to 194°F) Not More Than Three Conductors in Raceway or Cable or Earth (Directly Buried), Based on Ambient Temperature of 30°C (86°F)

Size	Temperature Rating of Conductor. See Table 310-13.						Size
	60°C (140°F)	75°C (167°F)	90°C (194°F)	60°C (140°F)	75°C (167°F)	90°C (194°F)	
AWG kcmil	TYPES TW†, UF†	TYPES FEPW†, RH†, RHW†, THHW†, THW†, THWN†, XHHW† USE†, ZW†	TYPES TA, TBS, SA SIS, FEP†, FEPB†, MI RHH†, RHW-2, THHN†, THHW†, THW-2, THWN-2, USE-2, XHH, XHHW† XHHW-2, ZW-2	TYPES TW†, UF†	TYPES RH†, RHW†, THHW†, THW†, THWN†, XHHW†, USE†	TYPES TA, TBS, SA, SIS, THHN†, THHW†, THW-2, THWN-2, RHH†, RHW-2 USE-2 XHH, XHHW XHHW-2, ZW-2	AWG kcmil
	COPPER			ALUMINUM OR COPPER-CLAD ALUMINUM			
18	....	....	14	....	....	....	....
16	....	....	18	....	....	....	....
14	20†	20†	25†	....	....	....	....
12	25†	25†	30†	20†	20†	25†	12
10	30	35†	40†	25	30†	35†	10
8	40	50	55	30	40	45	8
6	55	65	75	40	50	60	6
4	70	85	95	55	65	75	4
3	85	100	110	65	75	85	3
2	95	115	130	75	90	100	2
1	110	130	150	85	100	115	1
1/0	125	150	170	100	120	135	1/0
2/0	145	175	195	115	135	150	2/0
3/0	165	200	225	130	155	175	3/0
4/0	195	230	260	150	180	205	4/0
250	215	255	290	170	205	230	250
300	240	285	320	190	230	255	300
350	260	310	350	210	250	280	350
400	280	335	380	225	270	305	400
500	320	380	430	260	310	350	500
600	355	420	475	285	340	385	600
700	385	460	520	310	375	420	700
750	400	475	535	320	385	435	750
800	410	490	555	330	395	450	800
900	435	520	585	355	425	480	900
1000	455	545	615	375	445	500	1000
1250	495	590	665	405	485	545	1250
1500	520	625	705	435	520	585	1500
1750	545	650	735	455	545	615	1750
2000	560	665	750	470	560	630	2000

CORRECTION FACTORS

Ambient Temp. °C	For ambient temperatures other than 30°C (86°F), multiply the allowable ampacities shown above by the appropriate factor shown below.						Ambient Temp. °F
21-25	1.08	1.05	1.04	1.08	1.05	1.04	70-77
26-30	1.00	1.00	1.00	1.00	1.00	1.00	78-86
31-35	.91	.94	.96	.91	.94	.96	87-95
36-40	.82	.88	.91	.82	.88	.91	96-104
41-45	.71	.82	.87	.71	.82	.87	105-113
46-50	.58	.75	.82	.58	.75	.82	114-122
51-55	.41	.67	.76	.41	.67	.76	123-131
56-60	....	.58	.71	....	.58	.71	132-140
61-70	....	.33	.58	....	.33	.58	141-158
71-80	....	....	.41	....	....	.41	159-176

†Unless otherwise specifically permitted elsewhere in this Code, the overcurrent protection for conductor types marked with an obelisk (†) shall not exceed 15 amperes for No. 14, 20 amperes for No. 12, and 30 amperes for No. 10 copper; or 15 amperes for No. 12 and 25 amperes for No. 10 aluminum and copper-clad aluminum after any correction factors for ambient temperature and number of conductors have been applied.

Table 310-17. Allowable Ampacities of Single Insulated Conductors, Rated 0 through 2000 Volts, In Free Air Based on Ambient Air Temperature of 30°C (86°F)

Size	Temperature Rating of Conductor. See Table 310-13.						Size
	60°C (140°F)	75°C (167°F)	90°C (194°F)	60°C (140°F)	75°C (167°F)	90°C (194°F)	
AWG kcmil	TYPES TW†, UF†	TYPES FEPW†, RH†, RHW†, THHW†, THW†, THWN†, XHHW†, ZW†	TYPES TA, TBS, SA SIS, FEP†, FEPB†, MI, RHH†, RHW-2, THHN†, THHW†, THW-2, THWN-2, USE-2, XHH, XHHW†, XHHW-2, ZW-2	TYPES TW†, UF†	TYPES RH†, RHW†, THHW†, THW†, THWN†, XHHW†	TYPES TA, TBS, SA, SIS, THHN†, THHW†, THW-2, THWN-2, RHH†, RHW-2, USE-2, XHH, XHHW†, XHHW-2, ZW-2	AWG kcmil
	COPPER			ALUMINUM OR COPPER-CLAD ALUMINUM			
18	. . . .	. . . .	18	. . . .	. . . .	. . . .	. . . .
16	. . . .	. . . .	24	. . . .	. . . .	. . . .	. . . .
14	25†	30†	35†	. . . .	. . . .	. . . .	. . . .
12	30†	35†	40†	25†	30†	35†	12
10	40†	50†	55†	35†	40†	40†	10
8	60	70	80	45	55	60	8
6	80	95	105	60	75	80	6
4	105	125	140	80	100	110	4
3	120	145	165	95	115	130	3
2	140	170	190	110	135	150	2
1	165	195	220	130	155	175	1
1/0	195	230	260	150	180	205	1/0
2/0	225	265	300	175	210	235	2/0
3/0	260	310	350	200	240	275	3/0
4/0	300	360	405	235	280	315	4/0
250	340	405	455	265	315	355	250
300	375	445	505	290	350	395	300
350	420	505	570	330	395	445	350
400	455	545	615	355	425	480	400
500	515	620	700	405	485	545	500
600	575	690	780	455	540	615	600
700	630	755	855	500	595	675	700
750	655	785	885	515	620	700	750
800	680	815	920	535	645	725	800
900	730	870	985	580	700	785	900
1000	780	935	1055	625	750	845	1000
1250	890	1065	1200	710	855	960	1250
1500	980	1175	1325	795	950	1075	1500
1750	1070	1280	1445	875	1050	1185	1750
2000	1155	1385	1560	960	1150	1335	2000
CORRECTION FACTORS							
Ambient Temp. °C	For ambient temperatures other than 30°C (86°F), multiply the allowable ampacities shown above by the appropriate factor shown below.						Ambient Temp. °F
21-25	1.08	1.05	1.04	1.08	1.05	1.04	70-77
26-30	1.00	1.00	1.00	1.00	1.00	1.00	78-86
31-35	.91	.94	.96	.91	.94	.96	87-95
36-40	.82	.88	.91	.82	.88	.91	96-104
41-45	.71	.82	.87	.71	.82	.87	105-113
46-50	.58	.75	.82	.58	.75	.82	114-122
51-55	.41	.67	.76	.41	.67	.76	123-131
56-60	. . . .	.58	.71	. . . .	.58	.71	132-140
61-70	. . . .	.33	.58	. . . .	.33	.58	141-158
71-80	. . . .	. . . .	.41	. . . .	. . . .	.41	159-176

†Unless otherwise specifically permitted elsewhere in this Code, the overcurrent protection for conductor types marked with an obelisk (†) shall not exceed 15 amperes for No. 14, 20 amperes for No. 12, and 30 amperes for No. 10 copper; or 15 amperes for No. 12 and 25 amperes for No. 10 aluminum and copper-clad aluminum.

Table 310-18.   Allowable Ampacities of Three Single Insulated Conductors Rated 0 through 2000 Volts, 150° to 250°C (302° to 482°F), in Raceway or Cable Based on Ambient Air Temperature of 40°C (104°F)

Size	Temperature Rating of Conductor. See Table 310-13.				Size
	150°C (302°F)	200°C (392°F)	250°C (482°F)	150°C (302°F)	
	TYPE Z	TYPES FEP, FEPB, PFA	TYPES PFAH, TFE	TYPE Z	
AWG kcmil			NICKEL OR NICKEL-COATED COPPER	ALUMINUM OR COPPER-CLAD ALUMINUM	AWG kcmil
	COPPER				
14	34	36	39	. . . .	14
12	43	45	54	30	12
10	55	60	73	44	10
8	76	83	93	57	8
6	96	110	117	75	6
4	120	125	148	94	4
3	143	152	166	109	3
2	160	171	191	124	2
1	186	197	215	145	1
1/0	215	229	244	169	1/0
2/0	251	260	273	198	2/0
3/0	288	297	308	227	3/0
4/0	332	346	361	260	4/0
250	. . . .	. . . .	. . . .	. . . .	250
300	. . . .	. . . .	. . . .	. . . .	300
350	. . . .	. . . .	. . . .	. . . .	350
400	. . . .	. . . .	. . . .	. . . .	400
500	. . . .	. . . .	. . . .	. . . .	500
600	. . . .	. . . .	. . . .	. . . .	600
700	. . . .	. . . .	. . . .	. . . .	700
750	. . . .	. . . .	. . . .	. . . .	750
800	. . . .	. . . .	. . . .	. . . .	800
1000	. . . .	. . . .	. . . .	. . . .	1000
1500	. . . .	. . . .	. . . .	. . . .	1500
2000	. . . .	. . . .	. . . .	. . . .	2000

CORRECTION FACTORS

Ambient Temp. °C	For ambient temperatures other than 40°C (104°F), multiply the allowable ampacities shown above by the appropriate factor shown below.				Ambient Temp. °F
41-50	.95	.97	.98	.95	105-122
51-60	.90	.94	.95	.90	123-140
61-70	.85	.90	.93	.85	141-158
71-80	.80	.87	.90	.80	159-176
81-90	.74	.83	.87	.74	177-194
91-100	.67	.79	.85	.67	195-212
101-120	.52	.71	.79	.52	213-248
121-140	.30	.61	.72	.30	249-284
141-160	. . . .	.50	.65	. . . .	285-320
161-180	. . . .	.35	.58	. . . .	321-356
181-200	. . . .	. . . .	.49	. . . .	357-392
201-225	. . . .	. . . .	.35	. . . .	393-437

Table 310-19.   Allowable Ampacities for Single Insulated Conductors Rated 0 through 2000 Volts, 150° to 250°C (302° to 482°F), in Free Air Based on Ambient Air Temperature of 40°C (104°F)

Size	Temperature Rating of Conductor. See Table 310-13.						Size
	150°C (302°F)	200°C (392°F)	Bare or covered conductors	250°C (482°F)	150°C (302°F)	Bare or covered conductors	
AWG kcmil	TYPE Z	TYPES FEP, FEPB, PFA		TYPES PFAH, TFE	TYPE Z		AWG kcmil
	COPPER			NICKEL OR NICKEL-COATED COPPER	ALUMINUM OR COPPER-CLAD ALUMINUM		
14	46	54	30	59	. . . .	25	14
12	60	68	35	78	47	30	12
10	80	90	50	107	63	35	10
8	106	124	70	142	83	55	8
6	155	165	95	205	112	75	6
4	190	220	125	278	148	100	4
3	214	252	150	327	170	120	3
2	255	293	175	381	198	135	2
1	293	344	200	440	228	160	1
1/0	339	399	235	532	263	185	1/0
2/0	390	467	275	591	305	215	2/0
3/0	451	546	320	708	351	250	3/0
4/0	529	629	370	830	411	285	4/0
250	. . . .	. . . .	415	. . . .	. . . .	325	250
300	. . . .	. . . .	460	. . . .	. . . .	360	300
350	. . . .	. . . .	520	. . . .	. . . .	405	350
400	. . . .	. . . .	560	. . . .	. . . .	435	400
500	. . . .	. . . .	635	. . . .	. . . .	495	500
600	. . . .	. . . .	710	. . . .	. . . .	560	600
700	. . . .	. . . .	780	. . . .	. . . .	615	700
750	. . . .	. . . .	805	. . . .	. . . .	635	750
800	. . . .	. . . .	835	. . . .	. . . .	660	800
900	. . . .	. . . .	865	. . . .	. . . .	715	900
1000	. . . .	. . . .	895	. . . .	. . . .	770	1000
1500	. . . .	. . . .	1205	. . . .	. . . .	980	1500
2000	. . . .	. . . .	1420	. . . .	. . . .	1215	2000

CORRECTION FACTORS

Ambient Temp. °C	For ambient temperatures other than 40°C (104°F), multiply the allowable ampacities shown above by the appropriate factor shown below.						Ambient Temp. °F
41-50	.95	.97	. . . .	.98	.95	. . . .	105-122
51-60	.90	.94	. . . .	.95	.90	. . . .	123-140
61-70	.85	.90	. . . .	.93	.85	. . . .	141-158
71-80	.80	.87	. . . .	.90	.80	. . . .	159-176
81-90	.74	.83	. . . .	.87	.74	. . . .	177-194
91-100	.67	.79	. . . .	.85	.67	. . . .	195-212
101-120	.52	.71	. . . .	.79	.52	. . . .	213-248
121-140	.30	.61	. . . .	.72	.30	. . . .	249-284
141-160	. . . .	.50	. . . .	.65	. . . .	. . . .	285-320
161-180	. . . .	.35	. . . .	.58	. . . .	. . . .	321-356
181-200	. . . .	. . . .	. . . .	.49	. . . .	. . . .	357-392
201-225	. . . .	. . . .	. . . .	.35	. . . .	. . . .	393-437

Notes to Ampacity Tables of 0 to 2000 Volts

1. Explanation of Tables. For explanation of type letters and for recognized size of conductors for the various conductor insulations, see Section 310-13. For installation requirements, see Sections 310-1 through 310-10 and the various articles of this Code. For flexible cords, see Tables 400-4, 400-5(A), and 400-5(B).

3. 120/240 Volts, 3-Wire, Single-Phase Dwelling Services and Feeders. For dwelling units, conductors, as listed below, shall be permitted to be utilized as 120/240-volt, 3-wire, single-phase service-entrance conductors, service lateral conductors, and feeder conductors that supply the total load to a dwelling unit and installed in raceway or cable with or without an equipment grounding conductor. The grounded conductor shall be permitted to be not more than two AWG sizes smaller than the ungrounded conductors for application of this note, provided the requirements of Sections 215-2, 220-22, and 230-42 are met.

Conductor Types and Sizes
RH-RHH-RHW-THHW-THW-THWN-THHN-XHHW-USE

Copper	Aluminum or Copper-Clad AL	Rating in Amps
AWG	AWG	
4	2	100
3	1	110
2	1/0	125
1	2/0	150
1/0	3/0	175
2/0	4/0	200
3/0	250 kcmil	225
4/0	300 kcmil	250
250 kcmil	350 kcmil	300
350 kcmil	500 kcmil	350
400 kcmil	600 kcmil	400

5. Bare Conductors. Where bare conductors are used with insulated conductors, their allowable ampacities shall be limited to those permitted for the adjacent insulated conductors.

6. Mineral-Insulated, Metal-Sheathed Cable. The temperature limitation on which the ampacities of mineral-insulated, metal-sheathed cable are based is determined by the insulating materials used in the end seal. Termination fittings incorporating unimpregnated, organic, insulating materials are limited to 90°C (194°F) operation.

7. Type MTW Machine Tool Wire.

(FPN): For the allowable ampacities of Type MTW wire, see Table 13-5(a) in the Electrical Standard for Industrial Machinery, NFPA 79-1991.

8. Adjustment Factors.

(a) More than Three Current-Carrying Conductors in a Raceway or Cable. Where the number of current-carrying conductors in a raceway or cable exceeds three, the allowable ampacities shall be reduced as shown in the following table:

Number of Current-Carrying Conductors	Percent of Values in Tables as Adjusted for Ambient Temperature if Necessary
4 through 6	80
7 through 9	70
10 through 20	50
21 through 30	45
31 through 40	40
41 and above	35

Where single conductors or multiconductor cables are stacked or bundled longer than 24 inches (610 mm) without maintaining spacing and are not installed in raceways, the allowable ampacity of each conductor shall be reduced as shown in the above table.

Exception No. 1: Where conductors of different systems, as provided in Section 300-3, are installed in a common raceway or cable, the derating factors shown above shall apply to the number of power and lighting (Articles 210, 215, 220, and 230) conductors only.

Exception No. 2: For conductors installed in cable trays, the provisions of Section 318-11 shall apply.

Exception No. 3: Derating factors shall not apply to conductors in nipples having a length not exceeding 24 inches (610 mm).

Exception No. 4: Derating factors shall not apply to underground conductors entering or leaving an outdoor trench if those conductors have physical protection in the form of rigid metal conduit, intermediate metal conduit, or rigid nonmetallic conduit having a length not exceeding 10 feet (3.05 m) above grade and the number of conductors does not exceed four.

Exception No. 5: For other loading conditions, adjustment factors and ampacities shall be permitted to be calculated under Section 310-15(b).

(FPN): See Appendix B, Table B-310-11 for adjustment factors for more than three current-carrying conductors in a raceway or cable with load diversity.

(b) More than One Conduit, Tube, or Raceway. Spacing between conduits, tubing, or raceways shall be maintained.

9. Overcurrent Protection.

Where the standard ratings and settings of overcurrent devices do not correspond with the ratings and settings allowed for conductors, the next higher standard rating and setting shall be permitted.

Exception: As limited in Section 240-3.

10. Neutral Conductor.

(a)   A neutral conductor that carries only the unbalanced current from other conductors of the same circuit need not be counted when applying the provisions of Note 8.

(b)   In a 3-wire circuit consisting of 2-phase wires and the neutral of a 4-wire, 3-phase wye-connected system, a common conductor carries approximately the same current as the other conductors and shall be counted when applying the provisions of Note 8.

(c)   On a 4-wire, 3-phase wye circuit where the major portion of the load consists of nonlinear loads such as electric-discharge lighting, electronic computer/data processing, or similar equipment, there are harmonic currents present in the neutral conductor, and the neutral shall be considered to be a current-carrying conductor.

11. Grounding or Bonding Conductor.   A grounding or bonding conductor shall not be counted when applying the provisions of Note 8.

Table 310-61.   Conductor Application and Insulation

Trade Name	Type Letter	Maximum Operating Temperature	Application Provision	Insulation	Outer Covering
Medium voltage solid dielectric	MV-75* MV-85* MV-90*	75° C 85° C 90° C	Dry or wet locations rated 2001 volts and higher	Thermoplastic or Thermosetting	Jacket, Sheath, or Armor

*Insulation and outer coverings that meet the requirements as flame-retardant limited-smoke and are so listed shall be permitted to be designated limited-smoke with the suffix LS after the Code type designation.

Table 310-62.   Thickness of Insulation for 601-2000-Volt Nonshielded Types RHH and RHW, in Mils

Conductor Size AWG-kcmil	A	B
14-10	80	60
8	80	70
6-2	95	70
1-2/0	110	90
3/0-4/0	110	90
213-500	125	105
501-1000	140	120

Note: Column A insulations are limited to natural, SBR, and butyl rubbers.
Note: Column B insulations are materials such as cross-linked polyethylene, ethylene propylene rubber, and composites thereof.

Table 310-63. Thickness of Insulation and Jacket for Nonshielded Solid Dielectric Insulated Conductors Rated 2001 to 8000 Volts, in Mils

Conductor-Size AWG-kcmil	2001-5000 Volts						5001-8000 Volts 100 Percent Insulation Level Wet or Dry Locations		
	Dry Locations Single Conductor			Wet or Dry Locations			Single Conductor		Multi-Conductor*
	Without Jacket	With Jacket		Single Conductor		Multi-Conductor*			
	Insulation	Insulation	Jacket	Insulation	Jacket	Insulation	Insulation	Jacket	Insulation
8	110	90	30	125	80	90	180	80	180
6	110	90	30	125	80	90	180	80	180
4-2	110	90	45	125	80	90	180	95	180
1-2/0	110	90	45	125	95	90	180	95	180
3/0-4/0	110	90	65	125	95	90	180	110	180
213-500	120	90	65	140	110	90	210	110	210
501-750	130	90	65	155	125	90	235	125	235
751-1000	130	90	65	155	125	90	250	140	250

*Note: Under a common overall covering such as a jacket, sheath, or armor.

Table 310-64. Thickness of Insulation for Shielded Solid Dielectric Insulated Conductors Rated 2001 to 35,000 Volts, in Mils

Conductor Size AWG-kcmil	2001-5000 Volts	5001-8000		8001-15,000		15,001-25,000		25,001-28,000	28,001-35,000
		100 Per-cent insu-lation level	133 Per-cent insu-lation level	100 Per-cent insu-lation level	133 Per-cent insu-lation level	100 Per-cent insu-lation level	133 Per-cent insu-lation level	100 Per-cent insu-lation level	100 Per-cent insu-lation level
8	90	—	—	—	—	—	—	—	—
6-4	90	115	140	—	—	—	—	—	—
2	90	115	140	175	215	—	—	—	—
1	90	115	140	175	215	260	345	280	—
1/0-1000	90	115	140	175	215	260	345	280	345

*Definitions:

100 Percent Insulation Level. Cables in this category shall be permitted to be applied where the system is provided with relay protection such that ground faults will be cleared as rapidly as possible but, in any case, within 1 minute. While these cables are applicable to the great majority of cable installations that are on grounded systems, they shall be permitted to be used also on other systems for which the application of cables is acceptable, provided the above clearing requirements are met in completely deenergizing the faulted section.

133 Percent Insulation Level. This insulation level corresponds to that formerly designated for ungrounded systems. Cables in this category shall be permitted to be applied in situations where the clearing time requirements of the 100 percent level category cannot be met, and yet there is adequate assurance that the faulted section will be deenergized in a time not exceeding 1 hour. Also, they shall be permitted to be used where additional insulation strength over the 100 percent level category is desirable.

Table 310-67.   Ampacities of Insulated Single Copper Conductor Cables ●
Triplexed in Air Based on Conductor Temperature of 90°C (194°F)
and Ambient Air Temperature of 40°C (104°F)

Conductor Size AWG-kcmil	2001-5000 Volts Ampacity	5001-35,000 Volts Ampacity
8	65	—
6	90	100
4	120	130
2	160	170
1	185	195
1/0	215	225
2/0	250	260
3/0	290	300
4/0	335	345
250	375	380
350	465	470
500	580	580
750	750	730
1000	880	850

Table 310-68.   Ampacities of Insulated Single Aluminum Conductor Cables
Triplexed in Air Based on Conductor Temperature of 90°C (194°F)
and Ambient Air Temperature of 40°C (104°F)

Conductor Size AWG-kcmil	2001-5000 Volts Ampacity	5001-35,000 Volts Ampacity
8	50	—
6	70	75
4	90	100
2	125	130
1	145	150
1/0	170	175
2/0	195	200
3/0	225	230
4/0	265	270
250	295	300
350	365	370
500	460	460
750	600	590
1000	715	700

Table 310-69.   Ampacities of Insulated Single Copper Conductor Isolated in Air Based on Conductor Temperature of 90°C (194°F) and Ambient Air Temperature of 40°C (104°F)

Conductor Size AWG-kcmil	2001-5000 Volts Ampacity	5001-15,000 Volts Ampacity	15,001-35,000 Volts Ampacity
8	83	—	—
6	110	110	—
4	145	150	—
2	190	195	—
1	225	225	225
1/0	260	260	260
2/0	300	300	300
3/0	345	345	345
4/0	400	400	395
250	445	445	440
350	550	550	545
500	695	685	680
750	900	885	870
1000	1075	1060	1040
1250	1230	1210	1185
1500	1365	1345	1315
1750	1495	1470	1430
2000	1605	1575	1535

Table 310-70.   Ampacities of Insulated Single Aluminum Conductor Isolated in Air Based on Conductor Temperature of 90°C (194°F) and Ambient Air Temperature of 40°C (104°F)

Conductor Size AWG-kcmil	2001-5000 Volts Ampacity	5001-15,000 Volts Ampacity	15,001-35,000 Volts Ampacity
8	64	—	—
6	85	87	—
4	115	115	—
2	150	150	—
1	175	175	175
1/0	200	200	200
2/0	230	235	230
3/0	270	270	270
4/0	310	310	310
250	345	345	345
350	430	430	430
500	545	535	530
750	710	700	685
1000	855	840	825
1250	980	970	950
1500	1105	1085	1060
1750	1215	1195	1165
2000	1320	1295	1265

Table 310-71. Ampacities of an Insulated Three-Conductor Copper Cable Isolated in Air Based on Conductor Temperature of 90°C (194°F) and Ambient Air Temperature of 40°C (104°F)

Conductor Size AWG-kcmil	2001-5000 Volts Ampacity	5001-35,000 Volts Ampacity
8	59	—
6	79	93
4	105	120
2	140	165
1	160	185
1/0	185	215
2/0	215	245
3/0	250	285
4/0	285	325
250	320	360
350	395	435
500	485	535
750	615	670
1000	705	770

Table 310-72. Ampacities of an Insulated Three-Conductor Aluminum Cable Isolated in Air Based on Conductor Temperature of 90°C (194°F) and Ambient Air Temperature of 40°C (104°F)

Conductor Size AWG-kcmil	2001-5000 Volts Ampacity	5001-35,000 Volts Ampacity
8	46	—
6	61	72
4	81	95
2	110	125
1	125	145
1/0	145	170
2/0	170	190
3/0	195	220
4/0	225	255
250	250	280
350	310	345
500	385	425
750	495	540
1000	585	635

**Table 310-73.   Ampacities of an Insulated Triplexed or Three Single
Conductor Copper Cables in Isolated Conduit in Air
Based on Conductor Temperature of 90°C (194°F)
and Ambient Air Temperature of 40°C (104°F)**

Conductor Size AWG-kcmil	2001-5000 Volts Ampacity	5001-35,000 Volts Ampacity
8	55	—
6	75	83
4	97	110
2	130	150
1	155	170
1/0	180	195
2/0	205	225
3/0	240	260
4/0	280	295
250	315	330
350	385	395
500	475	480
750	600	585
1000	690	675

**Table 310-74.   Ampacities of Insulated Triplexed or Three Single
Conductor Aluminum Cables in Isolated Conduit in Air
Based on Conductor Temperature of 90°C (194°F)
and Ambient Air Temperature of 40°C (104°F)**

Conductor Size AWG-kcmil	2001-5000 Volts Ampacity	5001-35,000 Volts Ampacity
8	43	—
6	58	65
4	76	84
2	100	115
1	120	130
1/0	140	150
2/0	160	175
3/0	190	200
4/0	215	230
250	250	255
350	305	310
500	380	385
750	490	485
1000	580	565

**Table 310-75.   Ampacities of an Insulated Three-Conductor
Copper Cable in Isolated Conduit in Air
Based on Conductor Temperature of 90°C (194°F)
and Ambient Air Temperature of 40°C (104°F)**

Conductor Size AWG-kcmil	2001-5000 Volts Ampacity	5001-35,000 Volts Ampacity
8	52	—
6	69	83
4	91	105
2	125	145
1	140	165
1/0	165	195
2/0	190	220
3/0	220	250
4/0	255	290
250	280	315
350	350	385
500	425	470
750	525	570
1000	590	650

**Table 310-76.   Ampacities of an Insulated Three-Conductor
Aluminum Cable in Isolated Conduit in Air
Based on Conductor Temperature of 90°C (194°F)
and Ambient Air Temperature of 40°C (104°F)**

Conductor Size AWG-kcmil	2001-5000 Volts Ampacity	5001-35,000 Volts Ampacity
8	41	—
6	53	64
4	71	84
2	96	115
1	110	130
1/0	130	150
2/0	150	170
3/0	170	195
4/0	200	225
250	220	250
350	275	305
500	340	380
750	430	470
1000	505	550

Table 310-77. Ampacities of Three Single Insulated Copper Conductors in Underground Electrical Ducts (Three Conductors per Electrical Duct) Based on Ambient Earth Temperature of 20°C (68°F), Electrical Duct Arrangement per Figure 310-1, 100 percent Load Factor, Thermal Resistance (RHO) of 90, Conductor Temperature 90°C (194°F)

One Circuit (See Figure 310-1, Detail 1) Size AWG-kcmil	2001-5000 Volts Ampacity	5001-35,000 Volts Ampacity
8	64	—
6	85	90
4	110	115
2	145	155
1	170	175
1/0	195	200
2/0	220	230
3/0	250	260
4/0	290	295
250	320	325
350	385	390
500	470	465
750	585	565
1000	670	640
Three Circuits (See Figure 310-1, Detail 2) Size		
8	56	—
6	73	77
4	95	99
2	125	130
1	140	145
1/0	160	165
2/0	185	185
3/0	210	210
4/0	235	240
250	260	260
350	315	310
500	375	370
750	460	440
1000	525	495
Six Circuits (See Figure 310-1, Detail 3) Size		
8	48	—
6	62	64
4	80	82
2	105	105
1	115	120
1/0	135	135
2/0	150	150
3/0	170	170
4/0	195	190
250	210	210
350	250	245
500	300	290
750	365	350
1000	410	390

Table 310-78.   Ampacities of Three Single Insulated Aluminum Conductors in Underground Electrical Ducts (Three Conductors per Electrical Duct) Based on Ambient Earth Temperature of 20°C (68°F), Electrical Duct Arrangement per Figure 310-1, 100 percent Load Factor, Thermal Resistance (RHO) of 90, Conductor Temperature of 90°C (194°F)

One Circuit (See Figure 310-1, Detail 1) Size AWG-kcmil	2001-5000 Volts Ampacity	5001-35,000 Volts Ampacity
8	50	—
6	66	70
4	86	91
2	115	120
1	130	135
1/0	150	155
2/0	170	175
3/0	195	200
4/0	225	230
250	250	250
350	305	305
500	370	370
750	470	455
1000	545	525
Three Circuits (See Figure 310-1, Detail 2) Size		
8	44	—
6	57	60
4	74	77
2	96	100
1	110	110
1/0	125	125
2/0	145	145
3/0	160	165
4/0	185	185
250	205	200
350	245	245
500	295	290
750	370	355
1000	425	405
Six Circuits (See Figure 310-1, Detail 3) Size		
8	38	—
6	48	50
4	62	64
2	80	80
1	91	90
1/0	105	105
2/0	115	115
3/0	135	130
4/0	150	150
250	165	165
350	195	195
500	240	230
750	290	280
1000	335	320

Table 310-79. Ampacities of Three Insulated Copper Conductors Cabled within an Overall Covering (Three-Conductor Cable) in Underground Electrical Ducts (One Cable per Electrical Duct) Based on Ambient Earth Temperature of 20°C (68°F), Electrical Duct Arrangement per Figure 310-1, 100 percent Load Factor, Thermal Resistance (RHO) of 90, Conductor Temperature 90°C (194°F)

One Circuit (See Figure 310-1, Detail 1) Size AWG-kcmil	2001-5000 Volts Ampacity	5001-35,000 Volts Ampacity
8	59	—
6	78	88
4	100	115
2	135	150
1	155	170
1/0	175	195
2/0	200	220
3/0	230	250
4/0	265	285
250	290	310
350	355	375
500	430	450
750	530	545
1000	600	615
Three Circuits (See Figure 310-1, Detail 2) Size		
8	53	—
6	69	75
4	89	97
2	115	125
1	135	140
1/0	150	160
2/0	170	185
3/0	195	205
4/0	225	230
250	245	255
350	295	305
500	355	360
750	430	430
1000	485	485
Six Circuits (See Figure 310-1, Detail 3) Size		
8	46	—
6	60	63
4	77	81
2	98	105
1	110	115
1/0	125	130
2/0	145	150
3/0	165	170
4/0	185	190
250	200	205
350	240	245
500	290	290
750	350	340
1000	390	380

Table 310-80. Ampacities of Three Insulated Aluminum Conductors Cabled within an Overall Covering (Three-Conductor Cable) in Underground Electrical Ducts (One Cable per Electrical Duct) Based on Ambient Earth Temperature of 20°C (68°F), Electrical Duct Arrangement per Figure 310-1, 100 percent Load Factor, Thermal Resistance (RHO) of 90, Conductor Temperature 90°C (194°F)

One Circuit (See Figure 310-1, Detail 1) Size AWG-kcmil	2001-5000 Volts Ampacity	5001-35,000 Volts Ampacity
8	46	—
6	61	69
4	80	89
2	105	115
1	120	135
1/0	140	150
2/0	160	170
3/0	180	195
4/0	205	220
250	230	245
350	280	295
500	340	355
750	425	440
1000	495	510
Three Circuits (See Figure 310-1, Detail 2) Size		
8	41	—
6	54	59
4	70	75
2	90	100
1	105	110
1/0	120	125
2/0	135	140
3/0	155	160
4/0	175	180
250	190	200
350	230	240
500	280	285
750	345	350
1000	400	400
Six Circuits (See Figure 310-1, Detail 3) Size		
8	36	—
6	46	49
4	60	63
2	77	80
1	87	90
1/0	99	105
2/0	110	115
3/0	130	130
4/0	145	150
250	160	160
350	190	190
500	230	230
750	280	275
1000	320	315

**Table 310-81.   Ampacities of Single Insulated Copper Conductors
Directly Buried in Earth
Based on Ambient Earth Temperature of 20°C (68°F),
Arrangement per Figure 310-1, 100 percent Load Factor,
Thermal Resistance (RHO) of 90, Conductor Temperature 90°C (194°F)**

Conductor Size AWG-kcmil	2001-5000 Volts Ampacity	5001-35,000 Volts Ampacity
One Circuit— 3 Conductors (See Figure 310-1, Detail 9)		
8	110	—
6	140	130
4	180	170
2	230	210
1	260	240
1/0	295	275
2/0	335	310
3/0	385	355
4/0	435	405
250	470	440
350	570	535
500	690	650
750	845	805
1000	980	930
Two Circuits— 6 Conductors (See Figure 310-1, Detail 10)		
8	100	—
6	130	120
4	165	160
2	215	195
1	240	225
1/0	275	255
2/0	310	290
3/0	355	330
4/0	400	375
250	435	410
350	520	495
500	630	600
750	775	740
1000	890	855

For SI units: one inch = 25.4 millimeters.

Table 310-82.   Ampacities of Single Insulated Aluminum Conductors Directly Buried in Earth
Based on Ambient Earth Temperature of 20°C (68°F),
Arrangement per Figure 310-1, 100 percent Load Factor,
Thermal Resistance (RHO) of 90, Conductor Temperature 90°C (194°F)

Conductor Size AWG-kcmil	2001-5000 Volts Ampacity	5001-35,000 Volts Ampacity
One Circuit— 3 Conductors (See Figure 310-1, Detail 9)		
8	85	—
6	110	100
4	140	130
2	180	165
1	205	185
1/0	230	215
2/0	265	245
3/0	300	275
4/0	340	315
250	370	345
350	445	415
500	540	510
750	665	635
1000	780	740
Two Circuits— 6 Conductors (See Figure 310-1, Detail 10)		
8	80	—
6	100	95
4	130	125
2	165	155
1	190	175
1/0	215	200
2/0	245	225
3/0	275	255
4/0	310	290
250	340	320
350	410	385
500	495	470
750	610	580
1000	710	680

For SI units: one inch = 25.4 millimeters.

Table 310-83. Ampacities of Three Insulated Copper Conductors Cabled within an Overall Covering (Three-Conductor Cable), Directly Buried in Earth
Based on Ambient Earth Temperature of 20°C (68°F),
Arrangement per Figure 310-1, 100 percent Load Factor,
Thermal Resistance (RHO) of 90, Conductor Temperature 90°C (194°F)

Conductor Size AWG-kcmil	2001-5000 Volts Ampacity	5001-35,000 Volts Ampacity
One Circuit (See Figure 310-1, Detail 5)		
8	85	—
6	105	115
4	135	145
2	180	185
1	200	210
1/0	230	240
2/0	260	270
3/0	295	305
4/0	335	350
250	365	380
350	440	460
500	530	550
750	650	665
1000	730	750
Two Circuits (See Figure 310-1, Detail 10)		
8	80	—
6	100	105
4	130	135
2	165	170
1	185	195
1/0	215	220
2/0	240	250
3/0	275	280
4/0	310	320
250	340	350
350	410	420
500	490	500
750	595	605
1000	665	675

For SI units: one inch = 25.4 millimeters.

**Table 310-84.   Ampacities of Three Insulated Aluminum Conductors
Cabled within an Overall Covering (Three-Conductor Cable),
Directly Buried in Earth
Based on Ambient Earth Temperature of 20°C (68°F),
Arrangement per Figure 310-1, 100 percent Load Factor,
Thermal Resistance (RHO) of 90, Conductor Temperature 90°C (194°F)**

Conductor Size AWG-kcmil	2001-5000 Volts Ampacity	5001-35,000 Volts Ampacity
One Circuit (See Figure 310-1, Detail 5)		
8	65	—
6	80	90
4	105	115
2	140	145
1	155	165
1/0	180	185
2/0	205	210
3/0	230	240
4/0	260	270
250	285	300
350	345	360
500	420	435
750	520	540
1000	600	620
Two Circuits (See Figure 310-1, Detail 6)		
8	60	—
6	75	80
4	100	105
2	130	135
1	145	150
1/0	165	170
2/0	190	195
3/0	215	220
4/0	245	250
250	265	275
350	320	330
500	385	395
750	480	485
1000	550	560

For SI units: one inch = 25.4 millimeters.

Notes to Tables 310-69 through 310-84

(FPN): For ampacities calculated in accordance with the following Notes 1 and 2, reference IEEE/ICEA "Power Cable Ampacities," Vol. I and II (IPCEA Pub. No. P-46-426) and the "References" therein for availability of all factors and constants.

1. Ambients Not in Tables. Ampacities at ambient temperatures other than those shown in the tables shall be determined by means of the following formula:

$$I_2 = I_1 \sqrt{\frac{TC - TA_2 - DELTA\ TD}{TC - TA_1 - DELTA\ TC}}$$

Where:

I_1 = Ampacity from tables at ambient TA_1.

I_2 = Ampacity at desired ambient TA_2.

TC = Conductor temperature in degrees C.

TA_1 = Surrounding ambient from tables in degrees C.

TA_2 = Desired ambient in degrees C.

DELTA TD = Dielectric loss temperature rise.

2. Grounded Shields. Ampacities shown in Tables 310-69, 310-70, 310-81, and 310-82 are for cable with shields grounded at one point only. Where shields are grounded at more than one point, ampacities shall be adjusted to take into consideration the heating due to shield currents.

3. Electrical Duct Bank Configuration. Ampacities shown in Tables 310-77, 310-78, 310-79, and 310-80 shall apply only where the cables are located in the outer electrical ducts of the electrical duct bank. Ampacities for cables located in the inner electrical ducts of the electrical duct bank shall be determined by special calculations.

4. Burial Depth of Underground Circuits. For applications where burial depths must be deeper than shown in the underground ampacity tables, the following ampacity derating factor shall be permitted to be used: 6 percent per increased foot of depth for all values of Rho. No rating change is required where the burial depth is decreased.

5. Thermal Resistivity. Thermal resistivity, as used in this Code, refers to the heat transfer capability through a substance by conduction. It is the reciprocal of thermal conductivity and is designated Rho and expressed in the units °C-cm/watt.

6. Electrical Ducts Utilized in Figure 310-1. Spacings between electrical ducts (raceways) as defined in Figure 310-1 shall be permitted to be less than specified in Figure 310-1, where ducts (raceways) enter equipment enclosures from underground, without reducing the ampacity of conductors contained within such ducts (raceways).

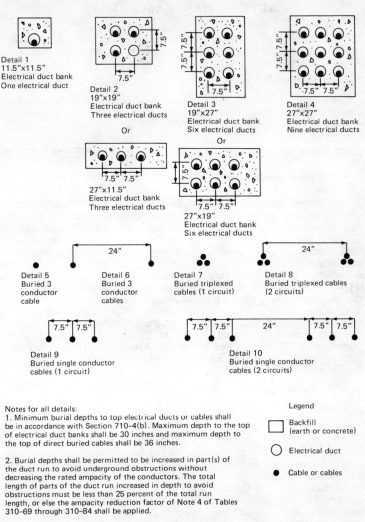

Detail 1
11.5"x11.5"
Electrical duct bank
One electrical duct

Detail 2
19"x19"
Electrical duct bank
Three electrical ducts

Or

Detail 3
19"x27"
Electrical duct bank
Six electrical ducts

Detail 4
27"x27"
Electrical duct bank
Nine electrical ducts

27"x11.5"
Electrical duct bank
Three electrical ducts

Or

27"x19"
Electrical duct bank
Six electrical ducts

Detail 5
Buried 3
conductor
cable

Detail 6
Buried 3
conductor
cables

Detail 7
Buried triplexed
cables (1 circuit)

Detail 8
Buried triplexed cables
(2 circuits)

Detail 9
Buried single conductor
cables (1 circuit)

Detail 10
Buried single conductor
cables (2 circuits)

Notes for all details:
1. Minimum burial depths to top electrical ducts or cables shall be in accordance with Section 710-4(b). Maximum depth to the top of electrical duct banks shall be 30 inches and maximum depth to the top of direct buried cables shall be 36 inches.

2. Burial depths shall be permitted to be increased in part(s) of the duct run to avoid underground obstructions without decreasing the rated ampacity of the conductors. The total length of parts of the duct run increased in depth to avoid obstructions must be less than 25 percent of the total run length, or else the ampacity reduction factor of Note 4 of Tables 310-69 through 310-84 shall be applied.

3. For SI units: one inch = 25.4 millimeters; one foot = 305 millimeters.

Legend

☐ Backfill
(earth or concrete)

○ Electrical duct

● Cable or cables

Figure 310-1.   Cable Installation Dimensions for Use with Tables 310-77 through 310-84.

ARTICLE 318 — CABLE TRAYS

318-1. Scope. This article covers cable tray systems including ladders, troughs, channels, solid bottom trays, and other similar structures.

318-2. Definition.

Cable Tray System. A unit or assembly of units or sections, and associated fittings, forming a rigid structural system used to support cables and raceways.

318-3. Uses Permitted.

(a) Wiring Methods. The following shall be permitted to be installed in cable tray systems under the conditions described in their respective articles:

(1) Mineral-insulated, metal-sheathed cable (Article 330); (2) electrical nonmetallic tubing (Article 331); (3) armored cable (Article 333); (4) metal-clad cable (Article 334); (5) nonmetallic-sheathed cable (Article 336); (6) shielded, nonmetallic-sheathed cable (Article 337); (7) multiconductor service-entrance cable (Article 338); (8) multiconductor underground feeder and branch-circuit cable (Article 339); (9) power and control tray cable (Article 340); (10) power-limited tray cable (Sections 725-50, 725-51, and 725-53); (11) other factory-assembled, multiconductor control, signal, or power cables that are specifically approved for installation in cable trays; (12) intermediate metal conduit (Article 345); (13) rigid metal conduit (Article 346); (14) rigid nonmetallic conduit (Article 347); (15) electrical metallic tubing (Article 346): (14) rigid nonmetallic conduit (Article 347); (15) electrical metallic tubing (Article 348); (16) flexible metallic tubing (Article 349); (17) flexible metal conduit (Article 350); or (18) liquidtight flexible metal conduit and liquidtight flexible nonmetallic conduit (Article 351).

(b) In Industrial Establishments. In industrial establishments only, where conditions of maintenance and supervision assure that only qualified persons will service the installed cable tray system, any of the cables in (1) and (2) below shall be permitted to be installed in ladder, ventilated trough, 4-inch (102-mm) ventilated channel cable trays, or 6-inch (152-mm) ventilated channel cable trays.

(1) Single Conductor. Single conductor cable shall be No. 1/0 or larger and shall be of a type listed and marked on the surface for use in cable trays. Where Nos. 1/0 through 4/0 single conductor cables are installed in ladder cable tray, the maximum allowable rung spacing for the ladder cable tray shall be 9 inches (229 mm). Where exposed to direct rays of the sun, cables shall be identified as being sunlight-resistant.

Exception No. 1: Welding cables as permitted in Article 630, Part E.

Exception No. 2: Single conductors used as equipment grounding conductors shall be permitted to be No. 4 or larger.

(2) Multiconductor. Multiconductor cables Type MV (Article 326) where exposed to direct rays of the sun shall be identified as being sunlight-resistant.

(c) **Equipment Grounding Conductors.** Metal in cable trays, as defined in Table 318-7(b) (2), shall be permitted to be used as equipment grounding conductors in commercial and industrial establishments, only where continuous maintenance and supervision assure that qualified persons will service the installed cable tray system.

(d) **Hazardous (Classified) Locations.** Cable trays in hazardous (classified) locations shall contain only the cable types permitted in Sections 501-4, 502-4, 503-3, and 504-20.

(e) **Nonmetallic Cable Tray.** Nonmetallic cable tray shall be permitted in corrosive areas and in areas requiring voltage isolation.

(FPN): It is not the intent of this article to limit the use of cable trays to industrial establishments only.

318-4. Uses Not Permitted. Cable tray systems shall not be used in hoistways or where subject to severe physical damage. Cable tray systems shall not be used in environmental air-handling spaces except to support wiring methods recognized in Section 300-22(c).

318-5. Construction Specifications.

(a) **Strength and Rigidity.** Cable trays shall have suitable strength and rigidity to provide adequate support for all contained wiring.

(b) **Smooth Edges.** Cable trays shall not have sharp edges, burrs, or projections that may damage the insulation or jackets of the wiring.

(c) **Corrosion Protection.** Cable trays shall be made of corrosion-resistant material or, if made of metal, shall be adequately protected against corrosion.

(d) **Side Rails.** Cable trays shall have side rails or equivalent structural members.

(e) **Fittings.** Cable trays shall include fittings or other suitable means for changes in direction and elevation of runs.

(f) **Nonmetallic Cable Tray.** Nonmetallic cable trays shall be made of flame-retardant material.

318-6. Installation.

(a) **Complete System.** Cable trays shall be installed as a complete system. Field bends or modifications shall be so made that the electrical continuity of the cable tray system and support for the cables shall be maintained.

(FPN): It is the intent to permit discontinuous segments and termination of cable tray installations where the system provides for the support of cables in accordance with their corresponding articles and where adequate bonding is provided in the cable tray system design.

(b) **Completed Before Installation.** Each run of cable tray shall be completed before the installation of cables.

(c) **Supports.** Supports shall be provided to prevent stress on cables where they enter raceways or other enclosures from cable tray systems.

(d) Covers. In portions of runs where additional protection is required, covers or enclosures providing the required protection shall be of a material compatible with the cable tray.

(e) Multiconductor Cables Rated 600 Volts or Less. Multiconductor cables rated 600 volts or less shall be permitted to be installed in the same cable tray.

(f) Cables Rated Over 600 Volts. Cables rated over 600 volts shall not be installed in the same cable tray with cables rated 600 volts or less.

Exception No. 1: Where separated by a solid fixed barrier of a material compatible with the cable tray.

Exception No. 2: Where cables over 600 volts are Type MC.

(g) Through Partitions and Walls. Cable trays shall be permitted to extend transversely through partitions and walls or vertically through platforms and floors in wet or dry locations where the installations, complete with installed cables, are made in accordance with the requirements of Section 300-21.

(h) Exposed and Accessible. Cable trays shall be exposed and accessible except as permitted by Section 318-6(g).

(i) Adequate Access. Sufficient space shall be provided and maintained about cable trays to permit adequate access for installing and maintaining the cables.

(FPN):   See Section 310-10 for temperature limitation of conductors.

(j) Incidental Support from Cable Tray. Where approved and designed to support the load, cable trays shall be permitted as a means of incidental support for raceways that are supported in accordance with Articles 345, 346, 347, and 348.

(FPN):   See Section 300-11(b) for restrictions concerning raceways used as a means of support.

318-7. Grounding.

(a) Metallic Cable Trays. Metallic cable trays that support electrical conductors shall be grounded as required for conductor enclosures in Article 250.

(b) Steel or Aluminum Cable Tray Systems. Steel or aluminum cable tray systems shall be permitted to be used as equipment grounding conductors provided that all the following requirements are met:

(1) The cable tray sections and fittings shall be identified for grounding purposes.

(2) The minimum cross-sectional area of cable trays shall conform to the requirements in Table 318-7(b)(2).

(3) All cable tray sections and fittings shall be legibly and durably marked to show the cross-sectional area of metal in channel cable trays, or cable trays of one-piece construction, and the total cross-sectional area of both side rails for ladder or trough cable trays.

(4) Cable tray sections, fittings, and connected raceways shall be bonded in accordance with Section 250-75 using bolted mechanical connectors or bonding jumpers sized and installed in accordance with Section 250-79.

Table 318-7(b)(2).
Metal Area Requirements for Cable Trays
Used as Equipment Grounding Conductors

Maximum Fuse Ampere Rating, Circuit Breaker Ampere Trip Setting, or Circuit Breaker Protective Relay Ampere Trip Setting for Ground-Fault Protection of Any Cable Circuit in the Cable Tray System	Minimum Cross-Sectional Area of Metal* in Square Inches	
	Steel Cable Trays	Aluminum Cable Trays
60	0.20	0.20
100	0.40	0.20
200	0.70	0.20
400	1.00	0.40
600	1.50**	0.40
1000	—	0.60
1200	—	1.00
1600	—	1.50
2000	—	2.00**

For SI units: one square inch = 645 square millimeters.

*Total cross-sectional area of both side rails for ladder or trough cable trays; or the minimum cross-sectional area of metal in channel cable trays or cable trays of one-piece construction.

**Steel cable trays shall not be used as equipment grounding conductors for circuits with ground-fault protection above 600 amperes. Aluminum cable trays shall not be used as equipment grounding conductors for circuits with ground-fault protection above 2000 amperes.

318-8. Cable Installation.

(a) Cable Splices. Cable splices made and insulated by approved methods shall be permitted to be located within a cable tray provided they are accessible and do not project above the side rails.

(b) Fastened Securely. In other than horizontal runs, the cables shall be fastened securely to transverse members of the cable trays.

(c) Bushed Conduit and Tubing. A box shall not be required where cables or conductors are installed in bushed conduit and tubing used for support or for protection against physical damage.

(d) Connected in Parallel. Where single conductor cables comprising each phase or neutral of a circuit are connected in parallel as permitted in Section 310-4, the conductors shall be installed in groups consisting of not more than one conductor per phase or neutral to prevent current unbalance in the paralleled conductors due to inductive reactance.

Single conductors shall be securely bound in circuit groups to prevent excessive movement due to fault-current magnetic forces.

Exception: Where single conductors are cabled together, such as triplexed assemblies.

318-9. Number of Multiconductor Cables, Rated 2000 Volts or Less, in Cable Trays. The number of multiconductor cables, rated 2000 volts or less, permitted in a single cable tray shall not exceed the requirements of this section. The conductor sizes herein apply to both aluminum and copper conductors.

(a) Any Mixture of Cables. Where ladder or ventilated trough cable trays contain multiconductor power or lighting cables, or any mixture of multiconductor power, lighting, control, and signal cables, the maximum number of cables shall conform to the following:

(1) Where all of the cables are No. 4/0 or larger, the sum of the diameters of all cables shall not exceed the cable tray width, and the cables shall be installed in a single layer.

(2) Where all of the cables are smaller than No. 4/0, the sum of the cross-sectional areas of all cables shall not exceed the maximum allowable cable fill area in Column 1 of Table 318-9, for the appropriate cable tray width.

(3) Where No. 4/0 or larger cables are installed in the same cable tray with cables smaller than No. 4/0, the sum of the cross-sectional areas of all cables smaller than No. 4/0 shall not exceed the maximum allowable fill area resulting from the computation in Column 2 of Table 318-9, for the appropriate cable tray width. The No. 4/0 and larger cables shall be installed in a single layer, and no other cables shall be placed on them.

(b) Multiconductor Control and/or Signal Cables Only. Where a ladder or ventilated trough cable tray, having a usable inside depth of 6 inches (152 mm) or less, contains multiconductor control and/or signal cables only, the sum of the cross-sectional areas of all cables at any cross section shall not exceed 50 percent of the interior cross-sectional area of the cable tray. A depth of 6 inches (152 mm) shall be used to compute the allowable interior cross-sectional area of any cable tray that has a usable inside depth of more than 6 inches (152 mm).

(c) Solid Bottom Cable Trays Containing Any Mixture. Where solid bottom cable trays contain multiconductor power or lighting cables, or any mixture of multiconductor power, lighting, control, and signal cables, the maximum number of cables shall conform to the following:

(1) Where all of the cables are No. 4/0 or larger, the sum of the diameters of all cables shall not exceed 90 percent of the cable tray width, and the cables shall be installed in a single layer.

(2) Where all of the cables are smaller than No. 4/0, the sum of the cross-sectional areas of all cables shall not exceed the maximum allowable cable fill area in Column 3 of Table 318-9 for the appropriate cable tray width.

(3) Where No. 4/0 or larger cables are installed in the same cable tray with cables smaller than No. 4/0, the sum of the cross-sectional areas of all cables smaller than No. 4/0 shall not exceed the maximum allowable fill area resulting from the computation in Column 4 of Table 318-9 for the appropriate cable tray width. The No. 4/0 and larger cables shall be installed in a single layer, and no other cables shall be placed on them.

(d) Solid Bottom Cable Tray Multiconductor Control and Signal Cables Only. Where a solid bottom cable tray, having a usable inside depth of 6 inches (152 mm) or less, contains multiconductor control and/or signal cables only, the sum of the cross-sectional areas of all cables at any cross section shall not exceed 40 percent of the interior cross-sectional area of the cable tray. A depth of 6 inches (152 mm) shall be used to compute the allowable interior cross-sectional area of any cable tray that has a usable inside depth of more than 6 inches (152 mm).

Table 318-9. Allowable Cable Fill Area for Multiconductor Cables in Ladder, Ventilated Trough, or Solid Bottom Cable Trays for Cables Rated 2000 Volts or Less

	Maximum Allowable Fill Area in Square Inches for Multiconductor Cables			
	Ladder or Ventilated Trough Cable Trays, Section 318-9(a)		Solid Bottom Cable Trays, Section 318-9(c)	
Inside Width of Cable Tray (Inches)	Column 1 Applicable for Section 318-9(a)(2) Only (Square Inches)	Column 2* Applicable for Section 318-9(a)(3) Only (Square Inches)	Column 3 Applicable for Section 318-9(c)(2) Only (Square Inches)	Column 4* Applicable for Section 318-9(c)(3) Only (Square Inches)
6	7	$7-(1.2\ Sd)$**	5.5	$5.5-Sd$**
12	14	$14-(1.2\ Sd)$	11.0	$11.0-Sd$
18	21	$21-(1.2\ Sd)$	16.5	$16.5-Sd$
24	28	$28-(1.2\ Sd)$	22.0	$22.0-Sd$
30	35	$35-(1.2\ Sd)$	27.5	$27.5-Sd$
36	42	$42-(1.2\ Sd)$	33.0	$33.0-Sd$

For SI units: one square inch = 645 square millimeters.

* The maximum allowable fill areas in Columns 2 and 4 shall be computed. For example, the maximum allowable fill, in square inches, for a 6-inch (152-mm) wide cable tray in Column 2 shall be: 7 minus (1.2 multiplied by Sd).

** The term Sd in Columns 2 and 4 is equal to the sum of the diameters, in inches, of all Nos. 4/0 AWG and larger multiconductor cables in the same cable tray with smaller cables.

(e) Ventilated Channel Cable Trays. Where ventilated channel cable trays contain multiconductor cables of any type, the combined cross-sectional area of all cables shall not exceed 1.3 square inches (839 sq mm) in 3-inch (76-mm) wide channel trays, 2.5 square inches (1613 sq mm) in 4-inch (102-mm) wide channel trays, or 3.8 square inches (2452 sq mm) in 6-inch (152-mm) wide channel trays.

Exception: Where only one multiconductor cable is installed in a ventilated channel tray, the cross-sectional area of the cable shall not exceed 2.3 square inches (1484 sq mm) in a 3-inch (76-mm) wide channel tray, 4.5 square inches (2903 sq mm) in a 4-inch (102-mm) wide channel tray, or 7.0 square inches (4516 sq mm) in a 6-inch (152-mm) wide channel tray.

318-10. Number of Single Conductor Cables, Rated 2000 Volts or Less, in Cable Trays. The number of single conductor cables, rated 2000 volts or less, permitted in a single cable tray section shall not exceed the requirements of this section. The single conductors, or conductor assemblies, shall be evenly distributed across the cable tray. The conductor sizes herein apply to both aluminum and copper conductors.

(a) Ladder or Ventilated Trough Cable Trays. Where ladder or ventilated trough cable trays contain single conductor cables, the maximum number of single conductors shall conform to the following:

(1) Where all of the cables are 1000 kcmil or larger, the sum of the diameters of all single conductor cables shall not exceed the cable tray width.

(2) Where all of the cables are from 250 kcmil up to 1000 kcmil, the sum of the cross-sectional areas of all single conductor cables shall not exceed the maximum allowable cable fill area in Column 1 of Table 318-10, for the appropriate cable tray width.

(3) Where 1000 kcmil or larger single conductor cables are installed in the same cable tray with single conductor cables smaller than 1000 kcmil, the sum of the cross-sectional areas of all cables smaller than 1000 kcmil shall not exceed the maximum allowable fill area resulting from the computation in Column 2 of Table 318-10, for the appropriate cable tray width.

(4) Where any of the conductor cables are Nos. 1/0 through 4/0, the sum of the diameters of all single conductor cables shall not exceed the cable tray width, and they shall be installed in a single layer.

(b) Ventilated Channel Cable Trays. Where 3-inch (76-mm), 4-inch (102-mm), or 6-inch (152-mm) wide ventilated channel cable trays contain single conductor cables, the sum of the diameters of all single conductors shall not exceed the inside width of the channel.

318-11. Ampacity of Cables, Rated 2000 Volts or Less, in Cable Trays.

(a) Multiconductor Cables. The ampacity of multiconductor cables, nominally rated 2000 volts or less, installed according to the requirements of Section 318-9, shall comply with the allowable ampacities of Tables 310-16 and 310-18. The derating factors of Article 310, Note 8(a) of the Notes to Ampacity Tables of 0 to 2000 Volts shall apply only to multiconductor cables with more than three current-carrying conductors. Derating shall be limited to the number of current-carrying conductors in the cable and not to the number of conductors in the cable tray.

Exception No. 1: Where cable trays are continuously covered for more than 6 feet (1.83 m) with solid unventilated covers, not over 95 percent of the allowable ampacities of Tables 310-16 and 310-18 shall be permitted for multiconductor cables.

Exception No. 2: Where multiconductor cables are installed in a single layer in uncovered trays, with a maintained spacing of not less than one cable diameter between cables, the ampacity shall not exceed the allowable ambient temperature corrected ampacities of multiconductor cables, with not more than three insulated conductors rated 0-2000 volts in free air, in accordance with Section 310-15(b).

(FPN):   See Table B-310-3 in Appendix B.

Table 318-10. Allowable Cable Fill Area for Single Conductor Cables in Ladder or Ventilated Trough Cable Trays For Cables Rated 2000 Volts or Less

Inside Width of Cable Tray (Inches)	Maximum Allowable Fill Area in Square Inches for Single Conductor Cables in Ladder or Ventilated Trough Cable Trays	
	Column 1 Applicable for Section 318-10(a)(2) Only (Square Inches)	Column 2* Applicable for Section 318-10(a)(3) Only (Square Inches)
6	6.50	6.50−(1.1 Sd)**
12	13.0	13.0 −(1.1 Sd)
18	19.5	19.5 −(1.1 Sd)
24	26.0	26.0 −(1.1 Sd)
30	32.5	32.5 −(1.1 Sd)
36	39.0	39.0 −(1.1 Sd)

For SI units: one square inch = 645 square millimeters.

*The maximum allowable fill areas in Column 2 shall be computed. For example, the maximum allowable fill, in square inches, for a 6-inch (152-mm) wide cable tray shall be: 6.5 minus (1.1 multiplied by Sd).

**The term Sd in Column 2 is equal to the sum of the diameters, in inches, of all 1000 kcmil and larger single conductor cables in the same ladder or ventilated trough cable tray with small cables.

(b) Single Conductor Cables. The derating factors of Article 310, Note 8 (a) of the Notes to Ampacity Tables of 0 to 2000 Volts shall not apply to the ampacity of cables in cable trays. The ampacity of single conductor cables, or single conductors cabled together (triplexed, quadruplexed, etc.), nominally rated 2000 volts or less, shall comply with the following:

(1) Where installed according to the requirements of Section 318-10, the ampacities for 600 kcmil and larger single conductor cables in uncovered cable trays shall not exceed 75 percent of the allowable ampacities in Tables 310-17 and 310-19. Where cable trays are continuously covered for more than 6 feet (1.83 m) with solid unventilated covers, the ampacities for 600 kcmil and larger cables shall not exceed 70 percent of the allowable ampacities in Tables 310-17 and 310-19.

(2) Where installed according to the requirements of Section 318-10, the ampacities for No. 1/0 through 500 kcmil single conductor cables in uncovered cable trays shall not exceed 65 percent of the allowable ampacities in Tables 310-17 and 310-19. Where cable trays are continuously covered for more than 6 feet (1.83 m) with solid unventilated covers, the ampacities for No. 1/0 through 500 kcmil cables shall not exceed 60 percent of the allowable ampacities in Tables 310-17 and 310-19.

(3) Where single conductors are installed in a single layer in uncovered cable trays, with a maintained space of not less than one cable diameter between individual conductors, the ampacity of Nos. 1/0 and larger cables shall not exceed the allowable ampacities in Tables 310-17 and 310-19.

(4) Where single conductors are installed in a triangular configuration in uncovered cable trays, with a maintained space of not less than 2.15 times one conductor diameter (2.15 × O.D.) between circuits, the ampacity of Nos. 1/0 and larger cables shall not exceed the allowable ampacities of two or three single insulated conductors rated 0 through 2000 volts supported on a messenger in accordance with Section 310-15(b).

(FPN):　See Table B-310-2 in Appendix B.

318-12. Number of Type MV and Type MC Cables (2001 Volts or Over) in Cable Trays.　The number of cables, rated 2001 volts or over, permitted in a single cable tray shall not exceed the requirements of this section.

The sum of the diameters of single conductor and multiconductor cables shall not exceed the cable tray width, and the cables shall be installed in a single layer. Where single conductor cables are triplexed, quadruplexed, or bound together in circuit groups, the sum of the diameters of the single conductors shall not exceed the cable tray width, and these groups shall be installed in single layer arrangement.

318-13. Ampacity of Type MV and Type MC Cables (2001 Volts or Over) in Cable Trays.　The ampacity of cables, rated 2001 volts, nominal, or over, installed according to Section 318-12 shall not exceed the requirements of this section.

(a) Multiconductor Cables (2001 Volts or Over).　The ampacity of multiconductor cables shall comply with the allowable ampacities of Tables 310-75 and 310-76.

Exception No. 1: Where cable trays are continuously covered for more than 6 feet (1.83 m) with solid unventilated covers, not more than 95 percent of the allowable ampacities of Tables 310-75 and 310-76 shall be permitted for multiconductor cables.

Exception No. 2: Where multiconductor cables are installed in a single layer in uncovered cable trays, with a maintained spacing of not less than one cable diameter between cables, the ampacity shall not exceed the allowable ampacities of Tables 310-71 and 310-72.

(b) Single Conductor Cables (2001 Volts or Over).　The ampacity of single conductor cables, or single conductors cabled together (triplexed, quadruplexed, etc.), shall comply with the following:

(1) The ampacities for Nos. 1/0 and larger single conductor cables in uncovered cable trays shall not exceed 75 percent of the allowable ampacities in Tables 310-69 and 310-70. Where the cable trays are covered for more than 6 feet (1.83 m) with solid unventilated covers, the ampacities for Nos. 1/0 and larger single conductor cables shall not exceed 70 percent of the allowable ampacities in Tables 310-69 and 310-70.

(2) Where single conductor cables are installed in a single layer in uncovered cable trays, with a maintained space of not less than one cable diameter between individual conductors, the ampacity of Nos. 1/0 and larger cables shall not exceed the allowable ampacities in Tables 310-69 and 310-70.

(3) Where single conductors are installed in a triangular configuration in uncovered cable trays, with a maintained space of not less than 2.15 times one conductor diameter (2.15 × O.D.) between circuits, the ampacity of Nos. 1/0 and larger cables shall not exceed the allowable ampacities in Tables 310-67 and 310-68.

ARTICLE 320 — OPEN WIRING ON INSULATORS

320-1. Definition. Open wiring on insulators is an exposed wiring method using cleats, knobs, tubes, and flexible tubing for the protection and support of single insulated conductors run in or on buildings, and not concealed by the building structure.

320-2. Other Articles. Open wiring on insulators shall comply with this article and also with the applicable provisions of other articles in this Code, especially Articles 225 and 300.

320-3. Uses Permitted. Open wiring on insulators shall be permitted on systems of 600 volts, nominal, or less, only for industrial or agricultural establishments, indoors or outdoors, in wet or dry locations, where subject to corrosive vapors, and for services.

320-5. Conductors.

(a) Type. Conductors shall be of a type specified by Article 310.

(b) Ampacity. The ampacity shall comply with Section 310-15.

320-6. Conductor Supports. Conductors shall be rigidly supported on noncombustible, nonabsorbent insulating materials and shall not contact any other objects. Supports shall be installed as follows: (1) within 6 inches (152 mm) from a tap or splice; (2) within 12 inches (305 mm) of a dead-end connection to a lampholder or receptacle; (3) at intervals not exceeding 4½ feet (1.37 m) and at closer intervals sufficient to provide adequate support where likely to be disturbed.

Exception No. 1: Supports for conductors No. 8 or larger installed across open spaces shall be permitted up to 15 feet (4.57 m) apart if noncombustible, nonabsorbent insulating spacers are used at least every 4½ feet (1.37 m) to maintain at least 2½ inches (64 mm) between conductors.

Exception No. 2: Where not likely to be disturbed in buildings of mill construction, No. 8 and larger conductors shall be permitted to be run across open spaces if supported from each wood cross member on approved insulators maintaining 6 inches (152 mm) between conductors.

Exception No. 3: In industrial establishments only, where conditions of maintenance and supervision assure that only qualified persons will service the system, conductors of sizes 250 kcmil and larger shall be permitted to be run across open spaces where supported on intervals up to 30 feet (9.1 m) apart.

320-7. Mounting of Conductor Supports. Where nails are used to mount knobs, they shall not be smaller than 10 penny. Where screws are used to mount knobs, or where nails or screws are used to mount cleats, they shall be of a length sufficient to penetrate the wood to a depth equal to at least one-half the height of the knob and the full thickness of the cleat. Cushion washers shall be used with nails.

320-8. Tie Wires. No. 8 or larger conductors supported on solid knobs shall be securely tied thereto by tie wires having an insulation equivalent to that of the conductor.

320-10. Flexible Nonmetallic Tubing. In dry locations where not exposed to severe physical damage, conductors shall be permitted to be separately enclosed in flexible nonmetallic tubing. The tubing shall be in continuous lengths not exceeding 15 feet (4.57 m) and secured to the surface by straps at intervals not exceeding 4½ feet (1.37 m).

320-11. Through Walls, Floors, Wood Cross Members, etc. Open conductors shall be separated from contact with walls, floors, wood cross members, or partitions through which they pass by tubes or bushings of noncombustible, nonabsorbent insulating material. Where the bushing is shorter than the hole, a waterproof sleeve of noninductive material shall be inserted in the hole and an insulating bushing slipped into the sleeve at each end in such a manner as to keep the conductors absolutely out of contact with the sleeve. Each conductor shall be carried through a separate tube or sleeve.

(FPN):   See Section 310-10 for temperature limitation of conductors.

320-12. Clearance from Piping, Exposed Conductors, etc. Open conductors shall be separated at least 2 inches (50.8 mm) from metal raceways, piping, or other conducting material, and from any exposed lighting, power, or signaling conductor, or shall be separated therefrom by a continuous and firmly fixed nonconductor in addition to the insulation of the conductor. Where any insulating tube is used, it shall be secured at the ends. Where practicable, conductors shall pass over rather than under any piping subject to leakage or accumulations of moisture.

320-13. Entering Spaces Subject to Dampness, Wetness, or Corrosive Vapors. Conductors entering or leaving locations subject to dampness, wetness, or corrosive vapors shall have drip loops formed on them and shall then pass upward and inward from the outside of the buildings, or from the damp, wet, or corrosive location, through noncombustible, nonabsorbent insulating tubes.

(FPN):   See Section 230-52 for individual conductors entering buildings or other structures.

320-14. Protection from Physical Damage. Conductors within 7 feet (2.13 m) from the floor shall be considered exposed to physical damage. Where open conductors cross ceiling joists and wall studs and are exposed to physical damage, they shall be protected by one of the following methods: (1) by guard strips not less than 1 inch (25.4 mm) nominal in thickness and at least as high as the insulating supports, placed on each side of and close to the wiring; (2) by a substantial running board at least ½ inch (12.7 mm) thick back of the conductors with side protections. Running boards shall extend at least 1 inch (25.4 mm) outside the conductors, but not more than 2 inches (50.8 mm), and the protecting sides shall be at least 2 inches (50.8 mm) high and at least 1 inch (25.4 mm) nominal in thickness; (3) by boxing made as above and furnished with a cover kept at least 1 inch (25.4 mm) away from the conductors within. Where protecting vertical conductors on side walls, the boxing shall be closed at the top and the holes through which the conductors pass shall be bushed; (4) by rigid metal conduit, intermediate metal conduit, rigid nonmetallic conduit, or electrical metallic tubing, in which case the rules of Article 345, 346, 347, or 348 shall apply; or by metal piping, in which case the conductors shall be encased in continuous lengths of approved flexible tubing. The conductors passing through metal enclosures shall be so grouped that current in both directions is approximately equal.

320-15. Unfinished Attics and Roof Spaces. Conductors in unfinished attics and roof spaces shall comply with (a) or (b) below.

(a) Accessible by Stairway or Permanent Ladder. Conductors shall be installed along the side of or through bored holes in floor joists, studs, or rafters. Where run through bored holes, conductors in the joists and in studs or rafters to a height of not less than 7 feet (2.13 m) above the floor or floor joists shall be protected by substantial running boards extending not less than 1 inch (25.4 mm) on each side of the conductors. Running boards shall be securely fastened in place. Running boards and guard strips shall not be required for conductors installed along the sides of joists, studs, or rafters.

(b) Not Accessible by Stairway or Permanent Ladder. Conductors shall be installed along the sides of or through bored holes in floor joists, studs, or rafters.

Exception: In buildings completed before wiring is installed and having head room at all points of less than 3 feet (914 mm).

320-16. Switches. Surface-type snap switches shall be mounted in accordance with Section 380-10(a), and boxes shall not be required. Other type switches shall be installed in accordance with Section 380-4.

ARTICLE 321 — MESSENGER SUPPORTED WIRING

321-1. Definition. Messenger supported wiring is an exposed wiring support system using a messenger wire to support insulated conductors by any one of the following: (1) a messenger with rings and saddles for conductor support; (2) a messenger with a field-installed lashing material for conductor support; (3) factory-assembled aerial cable; (4) multiplex cables utilizing a bare conductor, factory assembled and twisted with one or more insulated conductors, such as duplex, triplex, or quadruplex type of construction.

321-2. Other Articles. Messenger supported wiring shall comply with this article and also with the applicable provisions of other articles in this Code, especially Articles 225 and 300.

321-3. Uses Permitted.

(a) **Cable Types.** The following shall be permitted to be installed in messenger supported wiring under the conditions described in the article referenced for each: (1) mineral-insulated, metal-sheathed cable (Article 330); (2) metal-clad cable (Article 334); (3) multiconductor service-entrance cable (Article 338); (4) multiconductor underground feeder and branch-circuit cable (Article 339); (5) power and control tray cable (Article 340); (6) power-limited tray cable (Sections 725-51 and 725-53); (7) other factory-assembled, multiconductor control, signal, or power cables that are identified for the use.

(b) **In Industrial Establishments.** In industrial establishments only, where conditions of maintenance and supervision assure that only competent individuals will service the installed messenger supported wiring, the following shall be permitted:

(1) Any of the conductor types shown in Table 310-13 or Table 310-62.

(2) MV Cable.

Where exposed to weather, conductors shall be listed for use in wet locations.

Where exposed to direct rays of the sun, conductors or cables shall be sunlight-resistant.

(c) **Hazardous (Classified) Locations.** Messenger supported wiring shall be permitted to be used in hazardous (classified) locations where the contained cables are permitted for such use in Sections 501-4, 502-4, 503-3, and 504-20.

321-4. Uses Not Permitted. Messenger supported wiring shall not be used in hoistways or where subject to severe physical damage.

321-5. Ampacity. The ampacity shall be determined by Section 310-15.

321-6. Messenger Support. The messenger shall be supported at dead ends and at intermediate locations so as to eliminate tension on the conductors. The conductors shall not be permitted to come into contact with the messenger supports or any structural members, walls, or pipes.

321-7. Grounding. The messenger shall be grounded as required by Sections 250-32 and 250-33 for enclosure grounding.

321-8. Conductor Splices and Taps. Conductor splices and taps made and insulated by approved methods shall be permitted in messenger supported wiring.

ARTICLE 324 — CONCEALED KNOB-AND-TUBE WIRING

324-1. Definition. Concealed knob-and-tube wiring is a wiring method using knobs, tubes, and flexible nonmetallic tubing for the protection and support of single insulated conductors concealed in hollow spaces of walls and ceilings of buildings.

324-2. Other Articles. Concealed knob-and-tube wiring shall comply with this article and also with the applicable provisions of other articles in this Code, especially Article 300.

324-3. Uses Permitted. Concealed knob-and-tube wiring shall be permitted to be used only for extensions of existing installations and elsewhere only by special permission under the following conditions:

(1) In the hollow spaces of walls and ceilings.

(2) In unfinished attic and roof spaces as provided in Section 324-11.

324-4. Uses Not Permitted. Concealed knob-and-tube wiring shall not be used in commercial garages, theaters and similar locations, motion picture studios, hazardous (classified) locations, or in the hollow spaces of walls, ceilings, and attics where such spaces are insulated by loose, rolled, or foamed-in-place insulating material that envelops the conductors.

324-5. Conductors.

(a) Type. Conductors shall be of a type specified by Article 310.

(b) Ampacity. The ampacity shall be determined by Section 310-15.

324-6. Conductor Supports. Conductors shall be rigidly supported on noncombustible, nonabsorbent insulating materials and shall not contact any other objects. Supports shall be installed as follows: (1) within 6 inches (152 mm) of each side of each tap or splice, and (2) at intervals not exceeding 4½ feet (1.37 m).

Exception: If it is not practicable to provide supports in dry locations, it shall be permissible to fish conductors through hollow spaces if each conductor is individually enclosed in flexible nonmetallic tubing. The tubing shall be in continuous lengths between supports, between boxes, or between a support and a box.

324-7. Tie Wires. Where solid knobs are used, conductors shall be securely tied thereto by tie wires having insulation equivalent to that of the conductor.

324-8. Conductor Clearances. A clearance of not less than 3 inches (76 mm) shall be maintained between conductors and of not less than 1 inch (25.4 mm) between the conductor and the surface over which it passes.

Exception: Where space is too limited to provide the above minimum clearances, such as at meters, panelboards, outlets, and switch points, the conductors shall be individually enclosed in flexible nonmetallic tubing, which shall be in continuous lengths between the last support or box and the terminal point.

324-9. Through Walls, Floors, Wood Cross Members, etc. Conductors shall comply with Section 320-11 where passing through holes in structural members. Where passing through wood cross members in plastered partitions, conductors shall be protected by noncombustible, nonabsorbent, insulating tubes extending not less than 3 inches (76 mm) beyond the wood member.

324-10. Clearance from Piping, Exposed Conductors, etc. Conductors shall comply with Section 320-12 for clearances from other exposed conductors, piping, etc.

324-11. Unfinished Attics and Roof Spaces. Conductors in unfinished attics and roof spaces shall comply with (a) or (b) below.

(FPN): See Section 310-10 for temperature limitation of conductors.

(a) Accessible by Stairway or Permanent Ladder. Conductors shall be installed along the side of or through bored holes in floor joists, studs, or rafters. Where run through bored holes, conductors in the joists and in studs or rafters to a height of not less than 7 feet (2.13 m) above the floor or floor joists shall be protected by substantial running boards extending not less than 1 inch (25.4 mm) on each side of the conductors. Running boards shall be securely fastened in place. Running boards and guard strips shall not be required where conductors are installed along the sides of joists, studs, or rafters.

(b) Not Accessible by Stairway or Permanent Ladder. Conductors shall be installed along the sides of or through bored holes in floor joists, studs, or rafters.

Exception: In buildings completed before wiring is installed and having head room at all points of less than 3 feet (914 mm).

324-12. Splices. Splices shall be soldered unless approved splicing devices are used. In-line or strain splices shall not be used.

324-13. Boxes. Outlet boxes shall comply with Article 370.

324-14. Switches. Switches shall comply with Sections 380-4 and 380-10(b).

ARTICLE 325 — INTEGRATED GAS SPACER CABLE

Type IGS

A. General

325-1. Definition. Type IGS cable is a factory assembly of one or more conductors, each individually insulated and enclosed in a loose fit nonmetallic flexible conduit as an integrated gas spacer cable rated 0-600 volts.

325-2. Other Articles. The Type IGS cable shall comply with this article and also with the applicable provisions of other articles in this Code.

325-3. Uses Permitted. The Type IGS cable shall be permitted for use underground, including direct burial in the earth, as service-entrance conductors, or as feeder or branch-circuit conductors.

325-4. Uses Not Permitted. The Type IGS cable shall not be used as interior wiring or exposed in contact with buildings.

B. Installation

325-11. Bending Radius. Where the coilable nonmetallic conduit and cable is bent for installation purposes or is flexed or bent during shipment or installation, the radii of bends measured to the inside of the bend shall not be less than specified in Table 325-11.

Table 325-11.   Minimum Radii of Bends

Conduit Trade Size	Minimum Radii
2 inch	24 inches (610 mm)
3 inch	35 inches (889 mm)
4 inch	45 inches (1.14 m)

325-12. Bends. A run of Type IGS cable between pull boxes or terminations shall not contain more than the equivalent of four quarter bends (360 degrees, total), including those bends located immediately at the pull box or terminations.

325-13. Fittings. Terminations and splices for the Type IGS cable shall be identified as a type that is suitable for maintaining the gas pressure within the conduit. A valve and cap shall be provided for each length of the cable and conduit to check the gas pressure or to inject gas into the conduit.

325-14. Ampacity. The ampacity of the Type IGS cable and conduit shall not exceed values shown in Table 325-14 for single conductor or multiconductor cable.

C. Construction Specifications

325-20. Conductors. The conductors shall be solid aluminum rods, laid parallel, consisting of one to nineteen $1/2$-inch (12.7-mm) diameter rods.

The minimum conductor size shall be 250 kcmil and the maximum size shall be 4750 kcmil.

Table 325-14. Ampacity Type IGS

Size kcmil	Amperes
250	119
500	168
750	206
1000	238
1250	266
1500	292
1750	315
2000	336
2250	357
2500	376
3000	412
3250	429
3500	445
3750	461
4000	476
4250	491
4500	505
4750	519

325-21. Insulation. The insulation shall be dry kraft paper tapes and a pressurized sulfur hexafluoride gas (SF_6), both approved for electrical use. The nominal gas pressure shall be 20 pounds per square inch gauge (psig) (138 kPa gauge).

The thickness of the paper spacer shall be as specified in Table 325-21.

Table 325-21. Paper Spacer Thickness

Size kcmil	Thickness Inches (mm)
250-1000	.040 (1.02)
1250-4750	.060 (1.52)

325-22. Conduit. The conduit shall be an approved medium density polyethylene identified as suitable for use with natural gas rated pipe in 2 inches, 3 inches, or 4 inches trade size. The percent fill dimensions for the conduit are shown in Table 325-22.

The size of the conduit permitted for each conductor size shall be calculated for a percent fill not to exceed Table 1, Chapter 9.

Table 325-22.   Conduit Dimensions

Conduit Trade Size, Inches	Outside Diameter, Inches (mm)	Inside Diameter, Inches (mm)
2	2.375 (60)	1.947 (49.46)
3	3.500 (89)	2.886 (73.30)
4	4.500 (114)	3.710 (94.23)

325-23. Grounding.   The Type IGS cable shall comply with Article 250.

325-24. Marking.   The provisions of Section 310-11 shall apply for the Type IGS cable.

ARTICLE 326 — MEDIUM VOLTAGE CABLE

Type MV

326-1. Definition.   Type MV is a single or multiconductor solid dielectric insulated cable rated 2001 volts or higher.

326-2. Other Articles.   In addition to the provisions of this article, Type MV cable shall comply with the applicable provisions of this Code, especially those of Articles 300, 305, 310, 318, 501, and 710.

326-3. Uses Permitted.   Type MV cables shall be permitted for use on power systems rated up to 35,000 volts, nominal, in wet or dry locations, in raceways, cable trays as specified in Section 318-3(b)(1), or directly buried in accordance with Section 710-4(b) and in messenger supported wiring.

326-4. Uses Not Permitted.   Type MV cable shall not be used unless identified for the use (1) where exposed to direct sunlight, and (2) in cable trays.

326-5. Construction.   Type MV cables shall have copper, aluminum, or copper-clad aluminum conductors and shall be constructed in accordance with Article 310.

326-6. Ampacity.   The ampacity of Type MV cable shall be in accordance with Section 310-15.

Exception: The ampacity of Type MV cable installed in cable tray shall be in accordance with Section 318-13.

326-7. Marking.   Medium voltage cable shall be marked as required in Section 310-11.

ARTICLE 328 — FLAT CONDUCTOR CABLE

Type FCC

A. General

328-1. Scope. This article covers a field-installed wiring system for branch circuits incorporating Type FCC cable and associated accessories as defined by the article. The wiring system is designed for installation under carpet squares.

328-2. Definitions.

Type FCC Cable: Type FCC cable consists of three or more flat copper conductors placed edge-to-edge and separated and enclosed within an insulating assembly.

FCC System: A complete wiring system for branch circuits that is designed for installation under carpet squares. The FCC system includes Type FCC cable and associated shielding, connectors, terminators, adapters, boxes, and receptacles.

Cable Connector: A connector designed to join Type FCC cables without using a junction box.

Insulating End: An insulator designed to electrically insulate the end of a Type FCC cable.

Top Shield: A grounded metal shield covering undercarpet components of the FCC system for the purposes of providing protection against physical damage.

Bottom Shield: A protective layer that is installed between the floor and Type FCC flat conductor cable to protect the cable from physical damage and may or may not be incorporated as an integral part of the cable.

Transition Assembly: An assembly to facilitate connection of the FCC system to other approved wiring systems, incorporating (1) a means of electrical interconnection, and (2) a suitable box or covering for providing electrical safety and protection against physical damage.

Metal Shield Connections: Means of connection designed to electrically and mechanically connect a metal shield to another metal shield, to a receptacle housing or self-contained device, or to a transition assembly.

328-3. Other Articles.   The FCC systems shall conform with applicable provisions of Articles 210, 220, 240, 250, and 300.

328-4. Uses Permitted.

(a) Branch Circuits. Use of FCC systems shall be permitted both for general-purpose and appliance branch circuits and for individual branch circuits.

(b) Floors. Use of FCC systems shall be permitted on hard, sound, smooth, continuous floor surfaces made of concrete, ceramic, or composition flooring, wood, and similar materials.

(c) Walls. Use of FCC systems shall be permitted on wall surfaces in surface metal raceways.

(d) Damp Locations. Use of FCC systems in damp locations shall be permitted.

(e) Heated Floors. Materials used for floors heated in excess of 30°C (86°F) shall be identified as suitable for use at these temperatures.

328-5. Uses Not Permitted. FCC systems shall not be used (1) outdoors or in wet locations; (2) where subject to corrosive vapors; (3) in any hazardous (classified) location; or (4) in residential, school, and hospital buildings.

328-6. Branch-Circuit Ratings.

(a) Voltage. Voltage between ungrounded conductors shall not exceed 300 volts. Voltage between ungrounded conductors and the grounded conductor shall not exceed 150 volts.

(b) Current. General-purpose and appliance branch circuits shall have ratings not exceeding 20 amperes. Individual branch circuits shall have ratings not exceeding 30 amperes.

B. Installation

328-10. Coverings. Floor-mounted Type FCC cable, cable connectors, and insulating ends shall be covered with carpet squares no larger than 36 inches (914 mm) square. Those carpet squares that are adhered to the floor shall be attached with release-type adhesives.

328-11. Cable Connections and Insulating Ends. All Type FCC cable connections shall use connectors identified for their use, installed such that electrical continuity, insulation, and sealing against dampness and liquid spillage are provided. All bare cable ends shall be insulated and sealed against dampness and liquid spillage using listed insulating ends.

328-12. Shields.

(a) Top Shield. A metal top shield shall be installed over all floor-mounted Type FCC cable, connectors, and insulating ends. The top shield shall completely cover all cable runs, corners, connectors, and ends.

(b) Bottom Shield. A bottom shield shall be installed beneath all Type FCC cable, connectors, and insulating ends.

328-13. Enclosure and Shield Connections. All metal shields, boxes, receptacle housings, and self-contained devices shall be electrically continuous to the equipment grounding conductor of the supplying branch circuit. All such electrical connections shall be made with connectors identified for this use. The electrical resistivity of such shield system shall not be more than that of one conductor of the Type FCC cable used in the installation.

328-14. Receptacles. All receptacles, receptacle housings, and self-contained devices used with the FCC system shall be identified for this use and shall be connected to the Type FCC cable and metal shields. Connection from any grounding conductor of the Type FCC cable shall be made to the shield system at each receptacle.

328-15. Connection to Other Systems. Power feed, grounding connection, and shield system connection between the FCC system and other wiring systems shall be accomplished in a transition assembly identified for this use.

328-16. Anchoring. All FCC system components shall be firmly anchored to the floor or wall using an adhesive or mechanical anchoring system identified for this use. Floors shall be prepared to assure adherence of the FCC system to the floor until the carpet squares are placed.

328-17. Crossings. Crossings of more than two Type FCC cable runs shall not be permitted at any one point. Crossings of a Type FCC cable over or under a flat communications or signal cable shall be permitted. In each case, a grounded layer of metal shielding shall separate the two cables, and crossings of more than two flat cables shall not be permitted at any one point.

328-18. System Height. Any portion of an FCC system with a height above floor level exceeding 0.090 inches (2.29 mm) shall be tapered or feathered at the edges to floor level.

328-19. FCC Systems Alterations. Alterations to FCC systems shall be permitted. New cable connectors shall be used at new connection points to make alterations. It shall be permitted to leave unused cable runs and associated cable connectors in place and energized. All cable ends shall be covered with insulating ends.

328-20. Polarization of Connections. All receptacles and connections shall be constructed and installed so as to maintain proper polarization of the system.

C. Construction

328-30. Type FCC Cable. Type FCC cable shall be approved for use with the FCC system and shall consist of three, four, or five flat copper conductors, one of which shall be an equipment grounding conductor. The insulating material of the cable shall be moisture-resistant and flame-retardant.

328-31. Markings. Type FCC cable shall be clearly and durably marked on both sides at intervals of not more than 24 inches (610 mm) with the information required by Section 310-11(a) and with the following additional information: (1) material of conductors; (2) maximum temperature rating; and (3) ampacity.

328-32. Conductor Identification.

(a) Colors. Conductors shall be clearly and durably marked on both sides throughout their length as specified in Section 310-12.

(b) Order. For a 2-wire FCC system with grounding, the grounding conductor shall be central.

328-33. Corrosion Resistance.
Metal components of the system shall be either (1) corrosion-resistant; (2) coated with corrosion-resistant materials; or (3) insulated from contact with corrosive substances.

328-34. Insulation.
All insulating materials in the FCC systems shall be identified for their use.

328-35. Shields.

(a) Materials and Dimensions. All top and bottom shields shall be of designs and materials identified for their use. Top shields shall be metal. Both metallic and nonmetallic materials shall be permitted for bottom shields.

(b) Resistivity. Metal shields shall have cross-sectional areas that provide for electrical resistivity of not more than that of one conductor of the Type FCC cable used in the installation.

(c) Metal-Shield Connectors. Metal shields shall be connected to each other and to boxes, receptacle housings, self-contained devices, and transition assemblies using metal-shield connectors.

328-36. Receptacles and Housings.
Receptacle housings and self-contained devices designed either for floor mounting or for in- or on-wall mounting shall be permitted for use with the FCC system. Receptacle housings and self-contained devices shall incorporate means for facilitating entry and termination of Type FCC cable and for electrically connecting the housing or device with the metal shield. Receptacles and self-contained devices shall comply with Section 210-7. Power and communications outlets installed together in common housing shall be permitted in accordance with Section 800-52(c)(2), Exception No. 2.

328-37. Transition Assemblies.
All transition assemblies shall be identified for their use. Each assembly shall incorporate means for facilitating entry of the Type FCC cable into the assembly, for connecting the Type FCC cable to grounded conductors, and for electrically connecting the assembly to the metal cable shields and to equipment grounding conductors.

ARTICLE 330 — MINERAL-INSULATED, METAL-SHEATHED CABLE

Type MI

A. General

330-1. Definition. Type MI mineral-insulated, metal-sheathed cable is a factory assembly of one or more conductors insulated with a highly compressed refractory mineral insulation and enclosed in a liquidtight and gastight continuous copper or alloy steel sheath.

330-2. Other Articles. Type MI cable shall comply with this article and also with the applicable provisions of other articles in this Code, especially Article 300.

330-3. Uses Permitted. Type MI cable shall be permitted as follows: (1) for services, feeders, and branch circuits; (2) for power, lighting, control, and signal circuits; (3) in dry, wet, or continuously moist locations; (4) indoors or outdoors; (5) where exposed or concealed; (6) embedded in plaster, concrete, fill, or other masonry, whether above or below grade; (7) in any hazardous (classified) location; (8) where exposed to oil and gasoline; (9) where exposed to corrosive conditions not deteriorating to its sheath; (10) in underground runs where suitably protected against physical damage and corrosive conditions.

330-4. Uses Not Permitted. Type MI cable shall not be used where exposed to destructive corrosive conditions.

Exception: Where protected by materials suitable for the conditions.

B. Installation

330-10. Wet Locations. Where installed in wet locations, Type MI cable shall comply with Section 300-6(c).

330-11. Through Joists, Studs, or Rafters. Type MI cable shall comply with Section 300-4 where installed through studs, joists, rafters, or similar wood members.

330-12. Supports. Type MI cable shall be securely supported at intervals not exceeding 6 feet (1.83 m) by straps, staples, hangers, or similar fittings so designed and installed as not to damage the cable.

Exception: Where cable is fished in.

330-13. Bends. Bends in Type MI cable shall be so made as not to damage the cable. The radius of the inner edge of any bend shall not be less than shown below.

(1) Five times the external diameter of the metallic sheath for cable not more than ¾ inch (19 mm) in external diameter.

(2) Ten times the external diameter of the metallic sheath for cable greater than ³/₄ inch (19 mm) but not more than 1 inch (25.4 mm) in external diameter.

330-14. Fittings. Fittings used for connecting Type MI cable to boxes, cabinets, or other equipment shall be identified for such use. Where single-conductor cables enter ferrous metal boxes or cabinets, the installation shall comply with Section 300-20 to prevent inductive heating.

330-15. Terminal Seals. Where Type MI cable terminates, an approved seal shall be provided immediately after stripping to prevent the entrance of moisture into the insulation. The conductors extending beyond the sheath shall be individually provided with an approved insulating material.

330-16. Single Conductors. Where single-conductor cables are used, all phase conductors and, where used, the neutral conductor shall be grouped together to minimize induced voltage on the sheath. Where single-conductor cables enter ferrous enclosures, the installation shall comply with Section 300-20 to prevent heating from induction.

C. Construction Specifications

330-20. Conductors. Type MI cable conductors shall be of solid copper or nickel-clad copper with a resistance corresponding to standard AWG sizes.

330-21. Insulation. The conductor insulation in Type MI cable shall be a highly compressed refractory mineral that provides proper spacing for all conductors.

330-22. Outer Sheath. The outer sheath shall be of a continuous construction to provide mechanical protection and moisture seal. Where made of copper, it shall provide an adequate path for equipment grounding purposes. Where made of steel, an equipment grounding conductor in accordance with Article 250 shall be provided.

ARTICLE 331 — ELECTRICAL NONMETALLIC TUBING

A. General

331-1. Definition. Electrical nonmetallic tubing is a pliable corrugated raceway of circular cross-section with integral or associated couplings, connectors and fittings listed for the installation of electric conductors. It is composed of a material that is resistant to moisture, chemical atmospheres, and is flame-retardant.

A pliable raceway is a raceway that can be bent by hand with a reasonable force, but without other assistance.

(FPN): It is intended that the material used has ignitibility, flammability, smoke generation, and toxicity characteristics that do not exceed those of rigid (nonplasticized) polyvinyl chloride.

331-2. Other Articles. Installations for electrical nonmetallic tubing shall comply with the provisions of the applicable sections of Article 300. Where equipment grounding is required by Article 250, a separate equipment grounding conductor shall be installed in the raceway.

331-3. Uses Permitted. The use of electrical nonmetallic tubing and fittings shall be permitted:

(1) In any building not exceeding three floors above grade.

  a. For exposed work, where not subject to physical damage.

  b. Concealed within walls, floors, and ceilings.

(FPN): See Section 336-4(a) for definition of first floor.

(2) In any building exceeding three floors above grade, electrical nonmetallic tubing shall be concealed within walls, floors, and ceilings where the walls, floors, and ceilings provide a thermal barrier of material that has at least a 15-minute finish rating as identified in listings of fire-rated assemblies.

(3) In locations subject to severe corrosive influences as covered in Section 300-6 and where subject to chemicals for which the materials are specifically approved.

(4) In concealed, dry, and damp locations not prohibited by Section 331-4.

(5) Above suspended ceilings where the suspended ceilings provide a thermal barrier of material that has at least a 15-minute finish rating as identified in listings of fire-rated assemblies, except as permitted in Section 331-3(1)a.

(6) Embedded in poured concrete, provided fittings identified for this purpose are used for connections.

(7) For wet locations indoors or in a concrete slab on or below grade, with fittings listed for the purpose.

(FPN): Extreme cold may cause some types of nonmetallic conduits to become brittle and, therefore, more susceptible to damage from physical contact.

331-4. Uses Not Permitted. Electrical nonmetallic tubing shall not be used:

(1) In hazardous (classified) locations.

Exception: Except as permitted by Section 504-20.

(2) For the support of fixtures and other equipment.

(3) Where subject to ambient temperatures exceeding those for which the tubing is listed.

(4) For conductors whose insulation temperature limitations would exceed those for which the tubing is listed.

(5) For direct earth burial.

(6) Where the voltage is over 600 volts.

(7) In exposed locations, except as permitted by Sections 331-3(1), 331-3(5), and 331-3(7). |

(8) In theaters and similar locations, except as provided in Articles 518 and 520.

B. Installation

331-5. Size.

(a) Minimum. Tubing smaller than $1/2$-inch electrical trade size shall not be used.

(b) Maximum. Tubing larger than 2-inch electrical trade size shall not be | used.

331-6. Number of Conductors in Tubing. The number of conductors in a single tubing shall not exceed that permitted by the percentage fill in Table 1, Chapter 9.

331-7. Trimming. All cut ends of tubing shall be trimmed inside and outside to remove rough edges.

331-8. Joints. All joints between lengths of tubing and between tubing and couplings, fittings and boxes shall be by an approved method.

331-9. Bends — How Made. Bends of electrical nonmetallic tubing shall be so made that the tubing will not be damaged and that the internal diameter of the tubing will not be effectively reduced. Bends shall be permitted to be made manually without auxiliary equipment, and the radius of the curve of the inner edge of such bends shall not be less than shown in Table 346-10.

331-10. Bends — Number in One Run. There shall not be more than the equivalent of four quarter bends (360 degrees total) between pull points, e.g., conduit bodies and boxes.

331-11. Supports. Electrical nonmetallic tubing shall be installed as a complete system as provided in Article 300 and shall be securely fastened in place within 3 feet (914 mm) of each outlet box, device box, junction box, cabinet, or fitting. |

Tubing shall be secured at least every 3 feet (914 mm).

Exception No. 1: Where support is being provided at least every 3 feet (914 mm) by openings through framing members.

Exception No. 2: Lengths not exceeding a distance of 6 feet (1.83 m) from a fixture terminal connection for tap connections to lighting fixtures. |

331-12. Boxes and Fittings. Boxes and fittings shall comply with the applicable provisions of Article 370.

331-13. Splices and Taps. Splices and taps shall be made only in junction boxes, outlet boxes, device boxes, or conduit bodies. See Article 370. |

331-14. Bushings. Where a tubing enters a box, fitting, or other enclosure, a bushing or adapter shall be provided to protect the wire from abrasion unless the box, fitting, or enclosure design provides equivalent protection.

(FPN): See Section 373-6(c) for the protection of conductors size No. 4 or larger.

C. Construction Specifications

331-15. General. Electrical nonmetallic tubing shall be clearly and durably marked at least every 10 feet (3.05 m) as required in the first sentence of Section 110-21. The type of material shall also be included in the marking. Tubing that has limited-smoke-producing characteristics shall be permitted to be identified with the suffix LS.

ARTICLE 333 — ARMORED CABLE

Type AC

A. General

333-1. Definition. Type AC cable is a fabricated assembly of insulated conductors in a flexible metallic enclosure. See Section 333-19.

333-2. Other Articles. Type AC cable shall comply with this article and also with the applicable provisions of other articles in this Code, especially Article 300.

333-3. Uses Permitted. Except where otherwise specified in this Code and where not subject to physical damage, Type AC cable shall be permitted for branch circuits and feeders in both exposed and concealed work and in cable trays where identified for such use.

Type AC cable shall be permitted in dry locations and embedded in plaster finish on brick or other masonry, except in damp or wet locations. It shall be permissible to run or fish this cable in the air voids of masonry block or tile walls where such walls are not exposed or subject to excessive moisture or dampness or are below grade line.

333-4. Uses Not Permitted. Type AC cable shall not be used where prohibited elsewhere in this Code, including (1) in theaters and similar locations, except as provided in Article 518, Places of Assembly; (2) in motion picture studios; (3) in any hazardous (classified) location except as permitted by Sections 501-4(b) and 504-20; (4) where exposed to corrosive fumes or vapors; (5) on cranes or hoists, except as provided in Section 610-11, Exception No. 3; (6) in storage battery rooms; (7) in hoistways or on elevators, except as provided in Section 620-21; or (8) in commercial garages where prohibited in Article 511.

B. Installation

333-7. Support. Type AC cable shall be secured by approved staples, straps, hangers, or similar fittings so designed and installed as not to damage the cable at intervals not exceeding 4½ feet (1.37 m) and within 12 inches (305 mm) of every outlet box, junction box, cabinet or fitting.

Exception No. 1: Where cable is fished.

Exception No. 2: Lengths of not more than 2 feet (610 mm) at terminals where flexibility is necessary.

Exception No. 3: Lengths of not more than 6 feet (1.83 m) from an outlet for connections within an accessible ceiling to lighting fixtures or equipment.

Exception No. 4: Where installed in cable trays, Type AC cable shall comply with Section 318-8(b).

333-8. Bending Radius. All bends shall be made so that the cable will not be damaged, and the radius of the curve of the inner edge of any bend shall not be less than five times the diameter of the Type AC cable.

333-9. Boxes and Fittings. At all points where the armor of AC cable terminates, a fitting shall be provided to protect wires from abrasion, unless the design of the outlet boxes or fittings is such as to afford equivalent protection, and, in addition, an approved insulating bushing or its equivalent approved protection shall be provided between the conductors and the armor. The connector or clamp by which the Type AC cable is fastened to boxes or cabinets shall be of such design that the insulating bushing or its equivalent will be visible for inspection. This bushing shall not be required with lead-covered cables where so installed that the lead sheath will be visible for inspection. Where change is made from Type AC cable to other cable or raceway wiring methods, a box, fitting, or conduit body shall be installed at junction points as required in Section 300-15.

333-10. Through or Parallel to Framing Members. Type AC cable shall comply with Section 300-4 where installed through or parallel to studs, joists, rafters, or similar wood or metal members.

333-11. Exposed Work. Exposed runs of cable shall closely follow the surface of the building finish or of running boards.

Exception No. 1: Lengths of not more than 24 inches (610 mm) at terminals where flexibility is necessary.

Exception No. 2: On the underside of floor joists in basements where supported at each joist and so located as not to be subject to physical damage.

Exception No. 3: Lengths of not more than 6 feet (1.83 m) from an outlet for connection within an accessible ceiling to lighting fixtures or other equipment.

333-12. In Accessible Attics. Type AC cables in accessible attics or roof spaces shall be installed as specified in (a) and (b) below.

(a) Where Run Across the Top of Floor Joists. Where run across the top of floor joists, or within 7 feet (2.13 m) of floor or floor joists across the face of rafters or studding, in attics and roof spaces that are accessible, the cable shall be protected by substantial guard strips that are at least as high as the cable. Where this space is not accessible by permanent stairs or ladders, protection shall only be required within 6 feet (1.83 m) of the nearest edge of the scuttle hole or attic entrance.

(b) Cable Installed Parallel to Framing Members. Where the cable is installed parallel to the sides of rafters, studs, or floor joists, neither guard strips nor running boards shall be required, and the installation shall also comply with Section 300-4(d).

C. Construction Specifications

333-19. Construction. Type AC cable shall have an armor of flexible metal tape. The insulated conductors shall be in accordance with Section 333-20. Cables of the AC type shall have an internal bonding strip of copper or aluminum in intimate contact with the armor for its entire length.

333-20. Conductors. Insulated conductors shall be of a type listed in Table 310-13 or those identified for use in this cable. In addition, the conductors shall have an overall moisture-resistant and fire retardant fibrous covering. For Type ACT, a moisture-resistant fibrous covering shall be required only on the individual conductors. The ampacity shall be determined by Section 310-15.

Exception: Armored cable installed in thermal insulation shall have conductors rated at 90°C (194°F). The ampacity of cable installed in these applications shall be that of 60°C (140°F) conductors.

333-21. Grounding. Type AC cable shall provide an adequate path for equipment grounding as required by Section 250-51.

333-22. Marking. The provisions of Section 310-11 shall apply, except that Type AC cable shall have ready identification of the manufacturer by distinctive external markings on the cable sheath throughout its entire length. Cables that are flame-retardant and have limited-smoke characteristics shall be permitted to be identified with the suffix LS.

ARTICLE 334 — METAL-CLAD CABLE

Type MC

A. General

334-1. Definition. Type MC cable is a factory assembly of one or more conductors, each individually insulated and enclosed in a metallic sheath of interlocking tape, or a smooth or corrugated tube.

334-2. Other Articles. Metal-clad cable shall comply with this article and also with the applicable provisions of other articles in this Code, especially Article 300.

Type MC cable shall be permitted for systems in excess of 600 volts, nominal. See Section 300-2(a).

334-3. Uses Permitted. Except where otherwise specified in this Code and where not subject to physical damage, Type MC cables shall be permitted as follows: (1) for services, feeders, and branch circuits; (2) for power, lighting, control, and signal circuits; (3) indoors or outdoors; (4) where exposed or concealed; (5) direct buried where identified for such use; (6) in cable tray; (7) in any approved raceway; (8) as open runs of cable; (9) as aerial cable on a messenger; (10) in hazardous (classified) locations as permitted in Articles 501, 502, 503, and 504; (11) in dry locations; and (12) in wet locations where any of the following conditions are met:

(1) The metallic covering is impervious to moisture.

(2) A lead sheath or moisture-impervious jacket is provided under the metal covering.

(3) The insulated conductors under the metallic covering are approved for use in wet locations.

Exception: See Section 501-4(b), Exception.

(FPN):   See Section 300-6 for protection against corrosion.

334-4. Uses Not Permitted. Type MC cable shall not be used where exposed to destructive corrosive conditions, such as direct burial in the earth, in concrete, or where exposed to cinder fills, strong chlorides, caustic alkalis, or vapors of chlorine or of hydrochloric acids.

Exception: Where the metallic sheath is suitable for the conditions or is protected by material suitable for the conditions.

B. Installation

334-10. Installation. Type MC cable shall be installed in compliance with Articles 300, 710, and 725 as applicable.

(a) Support. Type MC cable shall be supported and secured at intervals not exceeding 6 feet (1.83 m).

Exception No. 1: Where installed as branch circuits in dwelling units, Type MC cable shall be secured within 12 inches (305 mm) from every outlet box, junction box, cabinet, or fitting.

Exception No. 2: Where Type MC cable is fished.

(b) Cable Tray. Type MC cable installed in cable tray shall comply with Article 318.

(c) Direct Buried. Direct buried cable shall comply with Section 300-5 or 710-4, as appropriate.

(d) Installed as Service-Entrance Cable. Type MC cable installed as service-entrance cable shall comply with Article 230.

(e) Installed Outside of Buildings or as Aerial Cable. Type MC cable installed outside of buildings or as aerial cable shall comply with Article 225 and Article 321.

(f) Through or Parallel to Joists, Studs, and Rafters. Type MC cable shall comply with Section 300-4 where installed through or parallel to joists, studs, rafters, or similar wood or metal members.

(g) In Accessible Attics. The installation of Type MC cable in accessible attics or roof spaces shall also comply with Section 333-12.

334-11. Bending Radius. All bends shall be so made that the cable will not be damaged, and the radius of the curve of the inner edge of any bend shall not be less than shown below.

(a) Smooth Sheath.

(1) Ten times the external diameter of the metallic sheath for cable not more than $3/4$ inch (19 mm) in external diameter;

(2) Twelve times the external diameter of the metallic sheath for cable more than $3/4$ inch (19 mm) but not more than $1^1/2$ inches (38 mm) in external diameter; and

(3) Fifteen times the external diameter of the metallic sheath for cable more than $1^1/2$ inches (38 mm) in external diameter.

(b) Interlocked-type Armor or Corrugated Sheath. Seven times the external diameter of the metallic sheath.

(c) Shielded Conductors. Twelve times the overall diameter of one of the individual conductors or seven times the overall diameter of the multiconductor cable, whichever is greater.

334-12. Fittings. Fittings used for connecting Type MC cable to boxes, cabinets, or other equipment shall be identified for such use. Where single-conductor cables enter ferrous metal boxes or cabinets, the installation shall comply with Section 300-20 to prevent inductive heating.

334-13. Ampacity. The ampacity of Type MC cable shall be in accordance with Section 310-15.

Exception No. 1: The ampacities for Type MC cable installed in cable tray shall be determined in accordance with Sections 318-11 and 318-13.

Exception No. 2: The ampacities of No. 18 and No. 16 conductors shall be in accordance with Table 402-5.

(FPN): See Section 310-10 for temperature limitation of conductors.

C. Construction Specifications

334-20. Conductors. The conductors shall be of copper, aluminum, or copper-clad aluminum, solid or stranded.

The minimum conductor size shall be No. 18 copper and No. 12 aluminum or copper-clad aluminum.

334-21. Insulation. The insulated conductors shall comply with (a) or (b) below.

(a) 600 Volts. Insulated conductors in sizes No. 18 and 16 shall be of a type listed in Table 402-3, with a maximum operating temperature not less than 90°C (194°F), and as permitted by Section 725-16. Conductors larger than No. 16 shall be of a type listed in Table 310-13 or of a type identified for use in MC cable.

(b) Over 600 Volts. Insulated conductors shall be of a type listed in Tables 310-61 through 310-67.

334-22. Metallic Sheath. The metallic covering shall be one of the following types: smooth metallic sheath, welded and corrugated metallic sheath, interlocking metal tape armor. The metallic sheath shall be continuous and close fitting.

Supplemental protection of an outer covering of corrosion-resistant material shall be permitted and shall be required where such protection is needed. The sheath shall not be used as a current-carrying conductor.

(FPN):   See Section 300-6 for protection against corrosion.

334-23. Grounding. Type MC cable shall provide an adequate path for equipment grounding as required by Article 250.

334-24. Marking. The provisions of Section 310-11 shall apply. Cables that are flame-retardant and have limited-smoke characteristics shall be permitted to be identified with the suffix LS.

ARTICLE 336 — NONMETALLIC-SHEATHED CABLE

Types NM and NMC

A. General

336-1. Definition. Nonmetallic-sheathed cable is a factory assembly of two or more insulated conductors having an outer sheath of moisture-resistant, flame-retardant, nonmetallic material.

336-2. Other Articles. In addition to the provisions of this article, installations of nonmetallic-sheathed cable shall comply with the other applicable provisions of this Code, especially Articles 300 and 310.

336-3. Uses Permitted. Type NM and Type NMC cables shall be permitted to be used in one- and two-family dwellings, multifamily dwellings, and other structures, except as prohibited in Section 336-4. Where installed in cable trays, cables shall be identified for this use.

(FPN):   See Section 310-10 for temperature limitation of conductors.

(a) Type NM. Type NM cable shall be permitted for both exposed and concealed work in normally dry locations. It shall be permissible to install or fish Type NM cable in air voids in masonry block or tile walls where such walls are not exposed or subject to excessive moisture or dampness.

(b) Type NMC. Type NMC cable shall be permitted (1) for both exposed and concealed work in dry, moist, damp, or corrosive locations; (2) in outside and inside walls of masonry block or tile; (3) in a shallow chase in masonry, concrete, or adobe protected against nails or screws by a steel plate at least $\frac{1}{16}$ inch (1.59 mm) thick and covered with plaster, adobe, or similar finish.

336-4. Uses Not Permitted.

(a) Type NM or NMC. Types NM and NMC cables shall not be used (1) in any dwelling or structure exceeding three floors above grade; (2) as service-entrance cable; (3) in commercial garages having hazardous (classified) locations as provided in Section 511-3; (4) in theaters and similar locations, except as provided in Article 518, Places of Assembly; (5) in motion picture studios; (6) in storage battery rooms; (7) in hoistways; (8) embedded in poured cement, concrete, or aggregate; or (9) in any hazardous (classified) location except as permitted by Sections 501-4(b), Exception and 504-20. For the purpose of this article, the first floor of a building shall be that floor that has 50 percent or more of the exterior wall surface area level with or above finished grade. One additional level that is the first level and not designed for human habitation and used only for vehicle parking, storage, or similar use shall be permitted.

(b) Type NM. Type NM cable shall not be installed (1) where exposed to corrosive fumes or vapors; (2) where embedded in masonry, concrete, adobe, fill, or plaster; (3) in a shallow chase in masonry, concrete, or adobe and covered with plaster, adobe, or similar finish.

B. Installation

336-10. Exposed Work — General.   In exposed work, except as provided in Sections 300-11(a), 336-12, and 336-13, the cable shall be installed as specified in (a) and (b) below.

(a) To Follow Surface. The cable shall closely follow the surface of the building finish or of running boards.

(b) Protection from Physical Damage. The cable shall be protected from physical damage where necessary by conduit, electrical metallic tubing, Schedule 80 PVC rigid nonmetallic conduit, pipe, guard strips, or other means. Where passing through a floor, the cable shall be enclosed in rigid metal conduit, intermediate metal conduit, electrical metallic tubing, Schedule 80 PVC rigid nonmetallic conduit, or other metal pipe extending at least 6 inches (152 mm) above the floor.

336-11. Through or Parallel to Joists, Studs, and Rafters.   Type NM or NMC cable shall comply with Section 300-4 where installed through or parallel to joists, studs, rafters, or similar wood or metal members.

336-12. In Unfinished Basements.   Where the cable is run at angles with joists in unfinished basements, it shall be permissible to secure cables not smaller than two No. 6 or three No. 8 conductors directly to the lower edges of the joists. Smaller cables shall either be run through bored holes in joists or on running boards. Where run parallel to the joists, cable of any size shall be secured to the sides of joists in accordance with Section 300-4(d).

336-13. In Accessible Attics.   The installation of cable in accessible attics or roof spaces shall also comply with Section 333-12.

336-14. Bends.   Bends in cable shall be so made, and other handling shall be such, that the protective coverings of the cable will not be damaged, and no bend shall have a radius less than five times the diameter of the cable.

336-15. Supports.   Nonmetallic-sheathed cable shall be secured by staples, cable ties, straps, or similar fittings so designed and installed as not to damage the cable. Cable shall be secured in place at intervals not exceeding 4½ feet (1.37 m) and within 12 inches (305 mm) from every cabinet, box, or fitting. Two conductor cables shall not be stapled on edge. Cables run through holes in wooden joists, rafters, or studs shall be considered to be supported.

(FPN):   See Section 370-17(c) for support where nonmetallic boxes are used.

Exception No. 1: For concealed work in finished buildings or finished panels for prefabricated buildings where such supporting is impracticable, it shall be permissible to fish the cable between access points.

Exception No. 2: A wiring device identified for the use, without a separate outlet box, incorporating an integral cable clamp shall be permitted where the cable is secured in place at intervals not exceeding 4½ feet (1.37 m) and within 12 inches (305 mm) from the wiring device wall opening, and there shall be at least a 12-inch (305-mm) loop of unbroken cable or 6 inches (152 mm) of a cable end available on the interior side of the finished wall to permit replacement.

336-16. Devices of Insulating Material.   Switch, outlet, and tap devices of insulating material shall be permitted to be used without boxes in exposed cable wiring and for rewiring in existing buildings where the cable is concealed and fished. Openings in such devices shall form a close fit around the outer covering of the cable, and the device shall fully enclose that part of the cable from which any part of the covering has been removed.

Where connections to conductors are by binding-screw terminals, there shall be available as many terminals as conductors.

Exception: Where cables are clamped within the structure and terminals are of a type identified for use with multiconductors.

336-17. Boxes of Insulating Material.   Nonmetallic outlet boxes shall be permitted as provided in Section 370-3.

336-18. Devices with Integral Enclosures.   Wiring devices with integral enclosures identified for such use shall be permitted as provided in Section 300-15(b), Exception No. 4.

C. Construction Specifications

336-25. Construction.   The insulated conductors of Types NM and NMC cable shall be in conformity with Section 336-26. The conductors shall be sizes No. 14 through 2 with copper conductors or sizes No. 12 through 2 with aluminum or copper-clad aluminum conductors. In addition to the insulated conductors, the cable shall be permitted to have an insulated or bare conductor for equipment grounding purposes only. Where provided, the grounding conductor shall be sized in accordance with Article 250.

(a) Type NM.   The overall covering shall be flame-retardant and moisture-resistant.

(b) Type NMC.   The overall covering shall be flame-retardant, moisture-resistant, fungus-resistant, and corrosion-resistant.

336-26. Conductors.   The insulated conductors shall be one of the types listed in Table 310-13 that is suitable for branch-circuit wiring or one that is identified for use in these cables.

Conductors shall be rated at 90°C (194°F). The ampacity of Types NM and NMC cable shall be that of 60°C- (140°F-) conductors and shall comply with Section 310-15.

336-27. Marking.   In addition to the provisions of Section 310-11, the cable shall have a distinctive marking on the exterior for its entire length specifying the cable type. Cables that are flame-retardant and have limited-smoke characteristics shall be permitted to be identified with the suffix LS.

ARTICLE 337 — SHIELDED
NONMETALLIC-SHEATHED CABLE

Type SNM

337-1. Definition.   Type SNM shielded nonmetallic-sheathed cable is a factory assembly of two or more insulated conductors in an extruded core of moisture-resistant, flame-resistant nonmetallic material, covered with an overlapping spiral metal tape and wire shield and jacketed with an extruded moisture-, flame-, oil-, corrosion-, fungus-, and sunlight-resistant nonmetallic material.

337-2. Other Articles.   In addition to the provisions of this article, installation of Type SNM cable shall conform to other applicable provisions, such as Articles 300, 318, 501, and 502.

337-3. Uses Permitted.   Type SNM cable shall be used only as follows: (1) where operating temperatures do not exceed the rating marked on the cable; (2) in cable trays or in raceways; (3) in hazardous (classified) locations where permitted in Articles 500 through 516; or (4) as permitted in Article 547.

(FPN):   See Section 310-10 for temperature limitation of conductors.

337-4. Bends. Bends in Type SNM cable shall be so made as not to damage the cable or its covering. The radius of the inner edge shall not be less than five times the cable diameter.

337-5. Handling. Type SNM cable shall be handled in such a manner as not to damage the cable or its covering.

337-6. Fittings. Fittings for connecting Type SNM cable to enclosures or equipment shall be identified for this use.

337-7. Bonding. The wire shield shall be bonded to the frame or enclosure of the utilization equipment and to the ground bus or connection at the power supply point. This bonding shall be accomplished using fittings (Section 337-6) or by other approved bonding methods [Section 501-16(a)].

337-8. Construction. The conductors of Type SNM cable shall be Type TFN, TFFN, THHN, or THWN in sizes No. 18 through No. 2 copper and No. 12 through No. 2 in aluminum or copper-clad aluminum. Conductor sizes shall be permitted to be mixed in individual cables. The flat overlapping metal tapes shall be spiraled with a long lay. The shield wires shall have a total cross-sectional area as required by Article 250 and not less than the largest circuit conductor in the cable.

The outer jacket shall be water-, oil-, flame-, corrosion-, fungus-, and sunlight-resistant, and suitable for installation in cable trays.

337-9. Marking. Type SNM cable shall have a distinctive marking on its exterior surface for its entire length indicating its type and maximum operating temperature. It shall comply with the general marking requirements of Section 310-11. Cables that are flame-retardant and have limited-smoke characteristics shall be permitted to be identified with the suffix LS.

The conductors shall each be numbered for identification from each other by durable marking on two sides 180 degrees apart every 6 inches (152 mm) of length, with alternate legends inverted to facilitate reading from both sides.

ARTICLE 338 — SERVICE-ENTRANCE CABLE

Types SE and USE

338-1. Definition. Service-entrance cable is a single conductor or multi-conductor assembly provided with or without an overall covering, primarily used for services, and is of the following types:

(a) **Type SE.** Type SE, having a flame-retardant, moisture-resistant covering.

(b) **Type USE.** Type USE, identified for underground use, having a moisture-resistant covering, but not required to have a flame-retardant covering.

Cabled single-conductor Type USE constructions recognized for underground use may have a bare copper conductor cabled with the assembly. Type USE single, parallel, or cabled conductor assemblies recognized for underground use may have a bare copper concentric conductor applied. These constructions do not require an outer overall covering.

Exception: Type USE shall be permitted for service laterals above ground outside of buildings.

(FPN No. 1):   See Section 230-41, Exception, item b for directly buried, uninsulated service-entrance conductors.

(FPN No. 2):   See Section 300-5(d) for protection from physical damage.

(c) One Uninsulated Conductor. If Type SE or USE cable consists of two or more conductors, one shall be permitted to be uninsulated.

338-2. Uses Permitted as Service-Entrance Conductors. Service-entrance cable used as service-entrance conductors shall be installed as required by Article 230.

338-3. Uses Permitted as Branch Circuits or Feeders.

(a) Grounded Conductor Insulated. Type SE service-entrance cables shall be permitted in interior wiring systems where all of the circuit conductors of the cable are of the rubber-covered or thermoplastic type.

(b) Grounded Conductor Not Insulated. Type SE service-entrance cables without individual insulation on the grounded circuit conductor shall not be used as a branch circuit or as a feeder within a building, except a cable that has a final nonmetallic outer covering and is supplied by alternating current at not over 150 volts to ground shall be permitted (1) as a branch circuit to supply only a range, wall-mounted oven, counter-mounted cooking unit, or clothes dryer as covered in Section 250-60 or (2) as a feeder to supply only other buildings on the same premises.

Type SE service-entrance cable shall be permitted for use where the fully insulated conductors are used for circuit wiring and the uninsulated conductor is used for equipment grounding purposes.

(c) Temperature Limitations. Type SE service-entrance cable used to supply appliances shall not be subject to conductor temperatures in excess of the temperature specified for the type of insulation involved.

338-4. Interior Installation Methods.   In addition to the provisions of this article, Type SE service-entrance cable used for interior wiring shall be installed in accordance with the provisions of Article 336 and shall comply with the applicable provisions of Article 300.

(FPN):   See Section 310-10 for temperature limitation of conductors.

338-5. Marking. Service-entrance cable shall be marked as required in Section 310-11. Cable with the neutral conductor smaller than the ungrounded conductors shall be so marked.

ARTICLE 339 — UNDERGROUND FEEDER AND BRANCH-CIRCUIT CABLE

Type UF

339-1. Description and Marking.

(a) Description. Underground feeder and branch-circuit cable shall be an approved Type UF cable in sizes No. 14 copper or No. 12 aluminum or copper-clad aluminum through No. 4/0. The conductors of Type UF shall be one of the moisture-resistant types listed in Table 310-13 that is suitable for branch-circuit wiring or one that is identified for such use. In addition to the insulated conductors, the cable shall be permitted to have an approved size of insulated or bare conductor for equipment grounding purposes only. The overall covering shall be flame-retardant, moisture-, fungus-, and corrosion-resistant, and suitable for direct burial in the earth.

(b) Marking. In addition to the provisions of Section 310-11, the cable shall have a distinctive marking on the exterior for its entire length specifying the cable type.

339-2. Other Articles.

In addition to the provisions of this article, installations of underground feeder and branch-circuit cable (Type UF) shall comply with other applicable provisions of this Code, especially Article 300 and Section 310-13.

339-3. Use.

(a) Uses Permitted.

(1) Type UF cable shall be permitted for use underground, including direct burial in the earth, as feeder or branch-circuit cable where provided with overcurrent protection of the rated ampacity as required in Section 339-4.

(2) Where single-conductor cables are installed, all cables of the feeder circuit, subfeeder circuit, or branch circuit, including the neutral and equipment grounding conductor, if any, shall be run together in the same trench or raceway.

Exception: For solar photovoltaic systems in accordance with Section 690-31.

(3) For underground requirements, see Section 300-5.

(4) Type UF cable shall be permitted for interior wiring in wet, dry, or corrosive locations under the recognized wiring methods of this Code, and, where installed as nonmetallic-sheathed cable, the installation and conductor requirements shall comply with the provisions of Article 336 and shall be of the multiconductor type.

Exception: Single-conductor cables shall be permitted as the nonheating leads for heating cables as provided in Section 424-43 and in solar photovoltaic systems in accordance with Section 690-31.

Type UF cable supported by cable trays shall be of the multiconductor type.

(FPN): See Section 310-10 for temperature limitation of conductors.

(b) Uses Not Permitted. Type UF cable shall not be used (1) as service-entrance cables; (2) in commercial garages; (3) in theaters; (4) in motion picture studios; (5) in storage battery rooms; (6) in hoistways; (7) in any hazardous (classified) location; (8) embedded in poured cement, concrete, or aggregate, except where embedded in plaster as nonheating leads as provided in Article 424; (9) where exposed to direct rays of the sun, unless identified as sunlight-resistant.

Exception: See Section 501-4(b), Exception.

339-4. Overcurrent Protection.   Overcurrent protection shall be provided in accordance with provisions of Section 240-3.

339-5. Ampacity.   The ampacity of Type UF cable shall be that of 60°C (140°F) conductors in accordance with Section 310-15.

ARTICLE 340 — POWER AND CONTROL TRAY CABLE

Type TC

340-1. Definition.   Type TC power and control tray cable is a factory assembly of two or more insulated conductors, with or without associated bare or covered grounding conductors under a nonmetallic sheath, approved for installation in cable trays, in raceways, or where supported by a messenger wire.

340-2. Other Articles.   In addition to the provisions of this article, installations of Type TC tray cable shall comply with other applicable articles of this Code, especially Articles 300 and 318.

340-3. Construction.   The insulated conductors of Type TC tray cable shall be in sizes No. 18 through 1000 kcmil copper and sizes No. 12 through 1000 kcmil aluminum or copper-clad aluminum. Insulated conductors of sizes No. 14 and larger copper and sizes No. 12 and larger aluminum or copper-clad aluminum shall be one of the types listed in Table 310-13 or 310-62 that is suitable for branch circuit and feeder circuits or one that is identified for such use. Insulated conductors of sizes No. 18 and No. 16 copper shall be in accordance with Section 725-16. The outer sheath shall be a flame-retardant, nonmetallic material. A metallic sheath shall not be permitted either under or over the nonmetallic sheath. Where installed in wet locations, Type TC cable shall be resistant to moisture and corrosive agents.

Exception: Where used for fire protective signaling systems, conductors shall be in accordance with Section 760-16.

340-4. Use Permitted.   Type TC tray cable shall be permitted to be used (1) for power, lighting, control, and signal circuits; (2) in cable trays, or in raceways, or where supported in outdoor locations by a messenger wire; (3) in cable trays in hazardous (classified) locations as permitted in Articles

318, 501, 502, and 504 in industrial establishments where the conditions of maintenance and supervision assure that only qualified persons will service the installation; (4) for Class 1 circuits as permitted in Article 725; (5) for nonpower-limited fire protective signaling circuits if conductors comply with the requirements of Section 760-16.

(FPN):   See Section 310-10 for temperature limitation of conductors.

340-5. Uses Not Permitted.   Type TC tray cable shall not be (1) installed where they will be exposed to physical damage; (2) installed as open cable on brackets or cleats; (3) used where exposed to direct rays of the sun, unless identified as sunlight-resistant; (4) direct buried, unless identified for such use.

340-6. Marking.   The cable shall be marked in accordance with Section 310-11. Cables that are flame-retardant and have limited-smoke characteristics shall be permitted to be identified with the suffix LS.

340-7. Ampacity.   The ampacities of the conductors of Type TC tray cable shall be determined from Section 402-5 for conductors smaller than No. 14 and from Section 318-11.

340-8. Bends.   Bends in Type TC cable shall be made so as not to damage the cable.

ARTICLE 342 — NONMETALLIC EXTENSIONS

342-1. Definition.   Nonmetallic extensions are an assembly of two insulated conductors within a nonmetallic jacket or an extruded thermoplastic covering. The classification includes both surface extensions, intended for mounting directly on the surface of walls or ceilings, and aerial cable, containing a supporting messenger cable as an integral part of the cable assembly.

342-2. Other Articles.   In addition to the provisions of this article, nonmetallic extensions shall be installed in accordance with the applicable provisions of this Code.

342-3. Uses Permitted.   Nonmetallic extensions shall be permitted only where all of the following conditions are met:

(a) **From an Existing Outlet.**   The extension is from an existing outlet on a 15- or 20-ampere branch circuit in conformity with the requirements of Article 210.

(b) **Exposed and in a Dry Location.**   The extension is run exposed and in a dry location.

(c) **Nonmetallic Surface Extensions.**   For nonmetallic surface extensions, the building is occupied for residential or office purposes and does not exceed the height limitations specified in Section 336-4(a).

(c1) (Alternate to (c)). For aerial cable, the building is occupied for industrial purposes, and the nature of the occupancy requires a highly flexible means for connecting equipment.

(FPN): See Section 310-10 for temperature limitation of conductors.

342-4. Uses Not Permitted. Nonmetallic extensions shall not be used:

(a) Aerial Cable. As aerial cable to substitute for one of the general wiring methods specified by this Code.

(b) Unfinished Areas. In unfinished basements, attics, or roof spaces.

(c) Voltage Between Conductors. Where the voltage between conductors exceeds 150 volts for nonmetallic surface extension and 300 volts for aerial cable.

(d) Corrosive Vapors. Where subject to corrosive vapors.

(e) Through a Floor or Partition. Where run through a floor or partition, or outside the room in which it originates.

342-5. Splices and Taps. Extensions shall consist of a continuous unbroken length of the assembly, without splices, and without exposed conductors between fittings. Taps shall be permitted where approved fittings completely covering the tap connections are used. Aerial cable and its tap connectors shall be provided with an approved means for polarization. Receptacle-type tap connectors shall be of the locking-type.

342-6. Fittings. Each run shall terminate in a fitting that covers the end of the assembly. All fittings and devices shall be of a type identified for the use.

342-7. Installation. Nonmetallic extensions shall be installed as specified in (a) and (b) below.

(a) Nonmetallic Surface Extensions.

(1) One or more extensions shall be permitted to be run in any direction from an existing outlet, but not on the floor or within 2 inches (50.8 mm) from the floor.

(2) Nonmetallic surface extensions shall be secured in place by approved means at intervals not exceeding 8 inches (203 mm).

Exception: Where connection to the supplying outlet is made by means of an attachment plug, the first fastening shall be permitted 12 inches (305 mm) or less from the plug.

There shall be at least one fastening between each two adjacent outlets supplied. An extension shall be attached only to woodwork or plaster finish, and shall not be in contact with any metal work or other conductive material other than with metal plates on receptacles.

(3) A bend that reduces the normal spacing between the conductors shall be covered with a cap to protect the assembly from physical damage.

(b) Aerial Cable.

(1) Aerial cable shall be supported by its messenger cable and securely attached at each end with approved clamps and turnbuckles. Intermediate supports shall be provided at not more than 20-foot (6.1-m) intervals. Cable tension shall be adjusted to eliminate excessive sag. The cable shall have a clearance of not less than 2 inches (50.8 mm) from steel structural members or other conductive material.

(2) Aerial cable shall have a clearance of not less than 10 feet (3.05 m) above floor areas accessible to pedestrian traffic, and not less than 14 feet (4.27 m) above floor areas accessible to vehicular traffic.

(3) Cable suspended over work benches, not accessible to pedestrian traffic, shall have a clearance of not less than 8 feet (2.44 m) above the floor.

(4) Aerial cables shall be permitted as a means to support lighting fixtures where the total load on the supporting messenger cable does not exceed that for which the assembly is intended.

(5) The supporting messenger cable, where installed in conformity with the applicable provisions of Article 250 and if properly identified as an equipment grounding conductor, shall be permitted to ground equipment. The messenger cable shall not be used as a branch-circuit conductor.

342-8. Marking.   Nonmetallic extensions shall be marked in accordance with Section 110-21.

ARTICLE 343 — PREASSEMBLED CABLE
IN NONMETALLIC CONDUIT

A. General

343-1. Description.   Preassembled cable in nonmetallic conduit is a listed factory assembly of conductors or cables inside a nonmetallic, smooth wall conduit with a circular cross section.

The nonmetallic conduit shall be composed of a material that is resistant to moisture and corrosive agents. It shall also be capable of being supplied on reels without damage or distortion and shall be of sufficient strength to withstand abuse, such as impact or crushing, in handling and during installation without damage to conduit or conductors.

343-2. Other Articles.   Installations for preassembled cable in nonmetallic conduit shall comply with the provisions of the applicable sections of Article 300. Where equipment grounding is required by Article 250, an assembly containing a separate equipment grounding conductor shall be used.

343-3. Uses Permitted.   The use of preassembled cable in nonmetallic conduit and fittings shall be permitted:

(1) For direct burial underground installation. For minimum cover requirements, see Tables 300-5 and 710-4(b) under rigid nonmetallic conduit.

(2) Encased or embedded in concrete.

(3) In cinder fill.

(4) In underground locations subject to severe corrosive influences as covered in Section 300-6 and where subject to chemicals for which the assembly is specifically approved.

343-4. Uses Not Permitted. Preassembled cable in nonmetallic conduit shall not be used:

(1) In exposed locations.

(2) Inside buildings.

Exception: The conductor portion of the assembly shall be permitted to extend within the building for termination purposes.

(3) In hazardous (classified) locations.

Exception: As covered in Sections 503-3(a), 504-20, 514-8, and 515-5, and in Class I, Division 2 locations as permitted in the exception to Section 501-4(b).

B. Installation

343-5. Size.

(a) Minimum. Preassembled cable in nonmetallic conduit smaller than $\frac{1}{2}$-inch (13-mm) electrical trade size shall not be used.

(b) Maximum. Preassembled cable in nonmetallic conduit larger than 2-inch (50-mm) electrical trade size shall not be used.

343-6. Trimming. For termination, the conduit shall be trimmed away from the conductors or cables using an approved method that will not damage the conductor or cable insulation or jacket. All ends shall be trimmed inside and out to remove rough edges.

343-7. Joints. All joints between conduit, fittings, and boxes shall be made by an approved method.

343-8. Conductor Terminations. All terminations between the conductors or cables and equipment shall be made by an approved method for that type of conductor or cable.

343-9. Bushings. Where the preassembled cable in nonmetallic conduit enters a box, fitting, or other enclosure, a bushing or adapter shall be provided to protect the conductor or cable from abrasion unless the design of the box, fitting, or enclosure provides equivalent protection.

(FPN): See Section 373-6(c) for the protection of conductors size No. 4 or larger.

343-10. Bends—How Made. Bends of preassembled cable in nonmetallic conduit shall be manually made so that the conduit will not be damaged and the internal diameter of the conduit will not be effectively reduced.

343-11. Bends—Number in One Run. There shall not be more than the equivalent of four quarter bends (360 degrees total) between termination points.

343-12. Splices and Taps. Splices and taps shall be made in junction boxes or other enclosures. See Article 370.

C. Construction

343-13. General. Preassembled cable in nonmetallic conduit is an assembly that is provided in continuous lengths shipped in a coil, reel, or carton.

343-14. Conductors and Cables. Conductors and cables used in preassembled cable in nonmetallic conduit shall be as follows:

(a) Conductors or cables of 600 volts or less shall be suitable for use in wet locations. Alternating-current and direct-current circuits shall be permitted. All conductors shall have an insulation rating equal to at least the maximum nominal circuit voltage of any conductor or cable within the conduit.

(b) Over 600 Volts. Conductors or cables rated over 600 volts shall not occupy the same conduit with conductors or cables of circuits rated 600 volts or less. All conductors or cables shall be suitable for use in wet locations.

(FPN):   See Section 300-3.

343-15. Conductor Fill. The maximum number of conductors or cables in preassembled cable in nonmetallic conduit shall not exceed the number indicated in Table 1 of Chapter 9 for the sizes involved.

343-16. Marking. Preassembled cable in nonmetallic conduit shall be clearly and durably marked at least every 10 feet (3.05 m) as required by Section 110-21. The type of conduit material shall also be included in the marking.

Identification of conductors or cables used in the assembly shall be provided on a tag attached to each end of the assembly or to the side of a reel.

ARTICLE 345 — INTERMEDIATE METAL CONDUIT

A. General

345-1. Definition. Intermediate metal conduit is a metal raceway of circular cross section with integral or associated couplings, connectors, and fittings approved for the installation of electrical conductors.

345-2. Other Articles. Installations for intermediate metal conduit shall comply with the provisions of the applicable sections of Article 300.

345-3. Uses Permitted.

(a) All Atmospheric Conditions and Occupancies. Use of intermediate metal conduit shall be permitted under all atmospheric conditions and occupancies. Where practicable, dissimilar metals in contact anywhere in the system shall be avoided to eliminate the possibility of galvanic action. Intermediate metal conduit shall be permitted as an equipment grounding conductor.

(FPN): See Section 250-91(b) for types of equipment grounding conductors.

Exception: Aluminum fittings and enclosures shall be permitted to be used with steel intermediate metal conduit.

(b) Corrosion Protection. Intermediate metal conduit, elbows, couplings, and fittings shall be permitted to be installed in concrete, in direct contact with the earth, or in areas subject to severe corrosive influences where protected by corrosion protection and judged suitable for the condition.

(FPN): See Section 300-6 for protection against corrosion.

(c) Cinder Fill. Intermediate metal conduit shall be permitted to be installed in or under cinder fill where subject to permanent moisture where protected on all sides by a layer of noncinder concrete not less than 2 inches (50.8 mm) thick; where the conduit is not less than 18 inches (457 mm) under the fill; or where protected by corrosion protection and judged suitable for the condition.

(FPN): See Section 300-6 for protection against corrosion.

B. Installation

345-5. Wet Locations. All supports, bolts, straps, screws, etc., shall be of corrosion-resistant materials or protected against corrosion by corrosion-resistant materials.

(FPN): See Section 300-6 for protection against corrosion.

345-6. Size.

(a) Minimum. Conduit smaller than ½-inch electrical trade size shall not be used.

(b) Maximum. Conduit larger than 4-inch electrical trade size shall not be used.

345-7. Number of Conductors in Conduit. The number of conductors in a single conduit shall not exceed that permitted by the percentage fill specified in Table 1, Chapter 9, using the conduit dimensions of Table 4, Chapter 9.

345-8. Reaming and Threading. All cut ends of conduits shall be reamed or otherwise finished to remove rough edges. Where conduit is threaded in the field, a standard cutting die with a ¾-inch taper per foot (1 in 16) shall be used.

(FPN): See Standards for Pipe Threads, General Purpose (Inch), ANSI/ASME B.1.20.1-1983.

345-9. Couplings and Connectors.

(a) Threadless. Threadless couplings and connectors used with conduit shall be made tight. Where buried in masonry or concrete, they shall be the concrete-tight type. Where installed in wet locations, they shall be the raintight type.

(b) Running Threads. Running threads shall not be used on conduit for connection at couplings.

345-10. Bends — How Made. Bends of intermediate metal conduit shall be so made that the conduit will not be damaged and that the internal diameter of the conduit will not be effectively reduced. The radius of the curve of the inner edge of any field bend shall not be less than indicated in Table 346-10.

Exception: For field bends for conductors without lead sheath and made with a single operation (one shot) bending machine designed for the purpose, the minimum radius shall not be less than that indicated in Table 346-10, Exception.

345-11. Bends — Number in One Run. There shall not be more than the equivalent of four quarter bends (360 degrees total) between pull points, e.g., conduit bodies and boxes.

345-12. Supports. Intermediate metal conduit shall be installed as a complete system as provided in Article 300 and shall be securely fastened in place. Conduit shall be supported at least every 10 feet (3.05 m). In addition, conduit shall be securely fastened within 3 feet (914 mm) of each outlet box, junction box, device box, cabinet, conduit body, or other conduit termination. Fastening shall be permitted to be increased to a distance of 5 feet (1.52 m) where structural members do not readily permit fastening within 3 feet (914 mm).

(FPN): Support provided by a bored or punched hole in a framing member is intended to meet the support requirements of this section.

Exception No. 1: If made up with threaded couplings, it shall be permissible to support straight runs of intermediate metal conduit in accordance with Table 346-12, provided such supports prevent transmission of stresses to termination where conduit is deflected between supports.

Exception No. 2: The distance between supports shall be permitted to be increased to 20 feet (6.1 m) for exposed vertical risers from industrial machinery, provided that the conduit is made up with threaded couplings, is firmly supported at the top and bottom of the riser, and no other means of intermediate support is readily available.

345-13. Boxes and Fittings. See Article 370.

345-14. Splices and Taps. Splices and taps shall be made only in junction boxes, outlet boxes, device boxes, or conduit bodies. See Article 370.

345-15. Bushings. Where a conduit enters a box, fitting, or other enclosure, a bushing shall be provided to protect the wire from abrasion unless the design of the box, fitting, or enclosure is such as to afford equivalent protection.

(FPN): See Section 373-6(c) for the protection of conductors Nos. 4 and larger at bushings.

C. Construction Specifications

345-16. General. Intermediate metal conduit shall comply with (a) through (c) below.

(a) Standard Lengths. Intermediate metal conduit as shipped shall be in standard lengths of 10 feet (3.05 m), including coupling, one coupling to be furnished with each length. For specific applications or use, it shall be permissible to ship lengths shorter or longer than 10 feet (3.05 m), with or without couplings.

(b) Corrosion-Resistant Material. Nonferrous conduit of corrosion-resistant material shall have suitable markings.

(c) Marking. Each length shall be clearly and durably identified at 2½-foot (762-mm) intervals with the letters IMC. Each length shall be marked as required in the first sentence of Section 110-21.

ARTICLE 346 — RIGID METAL CONDUIT

346-1. Use. The use of rigid metal conduit shall be permitted under all atmospheric conditions and occupancies subject to the following:

(a) Protected by Enamel. Ferrous raceways and fittings protected from corrosion solely by enamel shall be permitted only indoors and in occupancies not subject to severe corrosive influences.

(b) Dissimilar Metals. Where practicable, dissimilar metals in contact anywhere in the system shall be avoided to eliminate the possibility of galvanic action.

Exception: Aluminum fittings and enclosures shall be permitted to be used with steel rigid metal conduit, and steel fittings and enclosures shall be permitted to be used with aluminum rigid metal conduit.

(c) Corrosion Protection. Ferrous or nonferrous metal conduit, elbows, couplings, and fittings shall be permitted to be installed in concrete, in direct contact with the earth, or in areas subject to severe corrosive influences where protected by corrosion protection and judged suitable for the condition.

(FPN): See Section 300-6 for protection against corrosion.

346-2. Other Articles. Installations of rigid metal conduit shall comply with the applicable provisions of Article 300.

A. Installation

346-3. Cinder Fill. Conduit shall not be used in or under cinder fill where subject to permanent moisture.

Exception No. 1: Where of corrosion-resistant material suitable for the purpose.

Exception No. 2: Where protected on all sides by a layer of noncinder concrete at least 2 inches (50.8 mm) thick.

Exception No. 3: Where the conduit is at least 18 inches (457 mm) under the fill.

346-4. Wet Locations. All supports, bolts, straps, screws, etc., shall be of corrosion-resistant materials or protected against corrosion by corrosion-resistant materials.

(FPN): See Section 300-6 for protection against corrosion.

346-5. Minimum Size. Conduit smaller than 1/2-inch electrical trade size shall not be used.

Exception: For enclosing the leads of motors as permitted in Section 430-145(b).

346-6. Number of Conductors in Conduit. The number of conductors permitted in a single conduit shall not exceed the percentage fill specified in Table 1, Chapter 9.

(FPN): For conductor cross-sectional area, see Tables 5, 5A, 6, and 8 and the applicable Notes to Tables at the beginning of Chapter 9.

346-7. Reaming and Threading.

(a) Reamed. All cut ends of conduits shall be reamed or otherwise finished to remove rough edges.

(b) Threaded. Where conduit is threaded in the field, a standard conduit cutting die with a 3/4-inch taper per foot (1 in 16) shall be used.

(FPN): See Standards for Pipe Threads, General Purpose (Inch), ANSI/ASME, B.1.20.1-1983.

346-8. Bushings. Where a conduit enters a box, fitting, or other enclosure, a bushing shall be provided to protect the wire from abrasion unless the box, fitting, or enclosure design provides equivalent protection.

(FPN): See Section 373-6(c) for the protection of conductors at bushings.

346-9. Couplings and Connectors.

(a) Threadless. Threadless couplings and connectors used with conduit shall be made tight. Where buried in masonry or concrete, they shall be of the concrete-tight type. Where installed in wet locations, they shall be of the raintight type.

(b) Running Threads. Running threads shall not be used on conduit for connection at couplings.

346-10. Bends — How Made. Bends of rigid metal conduit shall be so made that the conduit will not be damaged, and that the internal diameter of the conduit will not be effectively reduced. The radius of the curve of the inner edge of any field bend shall not be less than shown in Table 346-10.

Exception: For field bends for conductors without lead sheath and made with a single operation (one shot) bending machine designed for the purpose, the minimum radius shall not be less than indicated in Table 346-10, Exception.

346-11. Bends — Number in One Run. There shall not be more than the equivalent of four quarter bends (360 degrees total) between pull points, e.g., conduit bodies and boxes.

Table 346-10. Radius of Conduit Bends (Inches)

Size of Conduit (In.)	Conductors without Lead Sheath (In.)	Conductors with Lead Sheath (In.)
½	4	6
¾	5	8
1	6	11
1¼	8	14
1½	10	16
2	12	21
2½	15	25
3	18	31
3½	21	36
4	24	40
5	30	50
6	36	61

For SI units: (Radius) one inch = 25.4 millimeters.

Table 346-10, Exception.
Radius of Conduit Bends (Inches)

Size of Conduit (In.)	Radius to Center of Conduit (In.)
½	4
¾	4½
1	5¾
1¼	7¼
1½	8¼
2	9½
2½	10½
3	13
3½	15
4	16
5	24
6	30

For SI units: (Radius) one inch = 25.4 millimeters.

346-12. Supports. Rigid metal conduit shall be installed as a complete system as provided in Article 300 and shall be securely fastened in place. Conduit shall be supported at least every 10 feet (3.05 m). In addition, conduit shall be securely fastened within 3 feet (914 mm) of each outlet box, junction box, device box, cabinet, conduit body, or other conduit termination. Fastening shall be permitted to be increased to a distance of 5 feet (1.52 m) where structural members do not readily permit fastening within 3 feet (914 mm).

(FPN): Support provided by a bored or punched hole in a framing member is intended to meet the support requirements of this section.

Exception No. 1: If made up with threaded couplings, it shall be permissible to support straight runs of rigid metal conduit in accordance with Table 346-12, provided such supports prevent transmission of stresses to termination where conduit is deflected between supports.

Exception No. 2: The distance between supports shall be permitted to be increased to 20 feet (6.1 m) for exposed vertical risers from industrial machinery, provided that the conduit is made up with threaded couplings, is firmly supported at the top and bottom of the riser, and no other means of intermediate support is readily available.

Table 346-12.   Supports for Rigid Metal Conduit

Conduit Size (Inches)	Maximum Distance Between Rigid Metal Conduit Supports (Feet)
1/2-3/4	10
1	12
1 1/4-1 1/2	14
2-2 1/2	16
3 and larger	20

For SI units: (Supports) one foot = 0.3048 meter.

346-13. Boxes and Fittings.   Boxes and fittings shall comply with the applicable provisions of Article 370.

346-14. Splices and Taps.   Splices and taps shall be made only in junction boxes, outlet boxes, device boxes, or conduit bodies. See Article 370. |

B. Construction Specifications

346-15. General.   Rigid metal conduit shall comply with (a) through (c) below.

(a) Standard Lengths.   Rigid metal conduit as shipped shall be in standard lengths of 10 feet (3.05 m), including coupling, one coupling to be furnished with each length. Each length shall be reamed and threaded on each end. For specific applications or uses, it shall be permissible to ship standard lengths or lengths shorter or longer than 10 feet (3.05 m) with or without couplings and with or without threads.

(b) Corrosion-Resistant Material.   Nonferrous conduit of corrosion-resistant material shall have suitable markings.

(c) Durably Identified.   Each length shall be clearly and durably identified in every 10 feet (3.05 m) as required in the first sentence of Section 110-21.

ARTICLE 347 — RIGID NONMETALLIC CONDUIT

347-1. Description. This article shall apply to a type of conduit and fittings of suitable nonmetallic material that is resistant to moisture and chemical atmospheres. For use aboveground, it shall also be flame-retardant, resistant to impact and crushing, resistant to distortion from heat under conditions likely to be encountered in service, and resistant to low temperature and sunlight effects. For use underground, the material shall be acceptably resistant to moisture and corrosive agents and shall be of sufficient strength to withstand abuse, such as by impact and crushing, in handling and during installation. Conduits listed for the purpose shall be permitted to be installed underground in continuous lengths from a reel. Where intended for direct burial, without encasement in concrete, the material shall also be capable of withstanding continued loading that is likely to be encountered after installation.

347-2. Uses Permitted. The use of listed rigid nonmetallic conduit and fittings shall be permitted under the following conditions:

(FPN): Extreme cold may cause some nonmetallic conduits to become brittle and therefore more suceptible to damage from physical contact.

(a) Concealed. In walls, floors, and ceilings.

(b) Corrosive Influences. In locations subject to severe corrosive influences as covered in Section 300-6 and where subject to chemicals for which the materials are specifically approved.

(c) Cinders. In cinder fill.

(d) Wet Locations. In portions of dairies, laundries, canneries, or other wet locations and in locations where walls are frequently washed, the entire conduit system including boxes and fittings used therewith shall be so installed and equipped as to prevent water from entering the conduit. All supports, bolts, straps, screws, etc., shall be of corrosion-resistant materials or be protected against corrosion by approved corrosion-resistant materials.

(e) Dry and Damp Locations. In dry and damp locations not prohibited by Section 347-3.

(f) Exposed. For exposed work where not subject to physical damage if identified for such use.

(g) Underground Installations. For underground installations, see Sections 300-5 and 710-4(b).

347-3. Uses Not Permitted. Rigid nonmetallic conduit shall not be used:

(a) Hazardous (Classified) Locations. In hazardous (classified) locations, except as covered in Sections 503-3(a), 504-20, 514-8, and 515-5; and in Class I, Division 2 locations as permitted in the exception to Section 501-4(b).

(b) Support of Fixtures. For the support of fixtures or other equipment.

Exception: Rigid nonmetallic conduit shall be permitted to support nonmetallic conduit bodies no larger than the largest trade size of an entering raceway. The conduit bodies shall not contain devices or support fixtures.

(c) Physical Damage. Where subject to physical damage unless identified for such use.

(d) Ambient Temperatures. Where subject to ambient temperatures exceeding those for which the conduit is approved.

(e) Insulation Temperature Limitations. For conductors whose insulation temperature limitations would exceed those for which the conduit is approved.

(f) Theaters and Similar Locations. In theaters and similar locations, except as provided in Articles 518 and 520.

347-4. Other Articles. Installation of rigid nonmetallic conduit shall comply with the applicable provisions of Article 300. Where equipment grounding is required by Article 250, a separate equipment grounding conductor shall be installed in the conduit.

Exception: As permitted in Section 250-91(b), Exception No. 3 for direct-current circuits.

A. Installations

347-5. Trimming. All cut ends shall be trimmed inside and outside to remove rough edges.

347-6. Joints. All joints between lengths of conduit, and between conduit and couplings, fittings, and boxes, shall be made by an approved method.

347-8. Supports. Rigid nonmetallic conduit shall be installed as a complete system as provided in Section 300-18 and shall be supported as required in Table 347-8. In addition, conduit shall be securely fastened within 3 feet (914 mm) of each outlet box, junction box, device box, conduit body, or other conduit termination. Rigid nonmetallic conduit shall be fastened so that movement from thermal expansion or contraction will be permitted.

Table 347-8.   Support of Rigid Nonmetallic Conduit

Conduit Size (Inches)	Maximum Spacing Between Supports (Feet)
½-1	3
1¼-2	5
2½-3	6
3½-5	7
6	8

For SI units: (Supports) one foot = 0.3048 meter.

347-9. Expansion Joints. Expansion joints for rigid nonmetallic conduit shall be provided to compensate for thermal expansion and contraction.

Exception: Where the computed length change due to thermal expansion or contraction, according to Table 10, Chapter 9, is less than .25 inches

(6.36 mm) in a straight run between securely mounted items such as boxes, cabinets, elbows, or other conduit terminations, an expansion joint shall not be required.

(FPN): See Table 10 in Chapter 9 for expansion characteristics of PVC rigid nonmetallic conduit.

347-10. Minimum Size. No conduit smaller than 1/2-inch electrical trade size shall be used.

347-11. Number of Conductors. The number of conductors permitted in a single conduit shall not exceed the percentage fill specified in Table 1, Chapter 9.

(FPN): For conductor cross-sectional area, see Tables 5, 5A, 6, and 8 and the applicable Notes to Tables at the beginning of Chapter 9.

347-12. Bushings. Where a conduit enters a box or other fitting, a bushing or adapter shall be provided to protect the wire from abrasion unless the design of the box or fitting is such as to provide equivalent protection.

(FPN): See Section 373-6(c) for the protection of conductors No. 4 and larger at bushings.

347-13. Bends — How Made. Bends of rigid nonmetallic conduit shall be so made that the conduit will not be damaged and that the internal diameter of the conduit will not be effectively reduced. Field bends shall be made only with bending equipment identified for the purpose, and the radius of the curve of the inner edge of such bends shall not be less than shown in Table 346-10.

347-14. Bends — Number in One Run. There shall not be more than the equivalent of four quarter bends (360 degrees total) between pull points, e.g., conduit bodies and boxes.

347-15. Boxes and Fittings. Boxes and fittings shall comply with the applicable provisions of Article 370.

347-16. Splices and Taps. Splices and taps shall be made only in junction boxes, outlet boxes, device boxes, or conduit bodies. See Article 370.

B. Construction Specifications

347-17. General. Rigid nonmetallic conduit shall comply with the following:

Marking. Each length of nonmetallic conduit shall be clearly and durably marked at least every 10 feet (3.05 m) as required in the first sentence of Section 110-21. The type of material shall also be included in the marking unless it is visually identifiable. For conduit recognized for use aboveground, these markings shall be permanent. For conduit limited to underground use only, these markings shall be sufficiently durable to remain legible until the material is installed. Conduit shall be permitted to be surface marked to indicate special characteristics of the material.

(FPN): Examples of these optional markings include but are not limited to "LS" for limited-smoke and markings such as "sunlight-resistant."

ARTICLE 348 — ELECTRICAL METALLIC TUBING

348-1. Use. The use of electrical metallic tubing shall be permitted for both exposed and concealed work. Electrical metallic tubing shall not be used (1) where, during installation or afterward, it will be subject to severe physical damage; (2) where protected from corrosion solely by enamel; (3) in cinder concrete or cinder fill where subject to permanent moisture unless protected on all sides by a layer of noncinder concrete at least 2 inches (50.8 mm) thick or unless the tubing is at least 18 inches (457 mm) under the fill; (4) in any hazardous (classified) location except as permitted by Sections 502-4, 503-3, and 504-20; or (5) for the support of fixtures or other equipment. | Where practicable, dissimilar metals in contact anywhere in the system shall be avoided to eliminate the possibility of galvanic action.

Exception: Aluminum fittings and enclosures shall be permitted to be used with steel electrical metallic tubing.

Ferrous or nonferrous electrical metallic tubing, elbows, couplings, and fittings shall be permitted to be installed in concrete, in direct contact with the earth, or in areas subject to severe corrosive influences where protected by corrosion protection and judged suitable for the condition.

(FPN): See Section 300-6 for protection against corrosion.

348-2. Other Articles. Installations of electrical metallic tubing shall comply with the applicable provisions of Article 300.

A. Installation

348-4. Wet Locations. All supports, bolts, straps, screws, etc., shall be of corrosion-resistant materials or protected against corrosion by corrosion-resistant materials.

(FPN): See Section 300-6 for protection against corrosion.

348-5. Size.

(a) Minimum. Tubing smaller than $1/2$-inch electrical trade size shall not be used.

Exception: For enclosing the leads of motors as permitted in Section | 430-145(b).

(b) Maximum. The maximum size of tubing shall be the 4-inch electrical trade size.

348-6. Number of Conductors in Tubing. The number of conductors permitted in a single tubing shall not exceed the percentage fill specified in Table 1, Chapter 9.

(FPN): For conductor cross-sectional area, see Tables 5, 5A, 6, and 8 and the | applicable Notes to Tables at the beginning of Chapter 9.

348-7. Threads. Electrical metallic tubing shall not be threaded. Where integral couplings are utilized, such couplings shall be permitted to be factory threaded.

348-8. Couplings and Connectors. Couplings and connectors used with tubing shall be made up tight. Where buried in masonry or concrete, they shall be concrete-tight type. Where installed in wet locations, they shall be of the raintight type.

348-9. Bends — How Made. Bends in the tubing shall be so made that the tubing will not be damaged and the internal diameter of the tubing will not be effectively reduced. The radius of the curve of the inner edge of any field bend shall not be less than shown in Table 346-10.

Exception: For field bends made with a bending machine designed for the purpose, the minimum radius shall not be less than indicated in Table 346-10, Exception.

348-10. Bends — Number in One Run. There shall not be more than the equivalent of four quarter bends (360 degrees total) between pull points, e.g., conduit bodies and boxes.

348-11. Reaming. All cut ends of electrical metallic tubing shall be reamed or otherwise finished to remove rough edges.

348-12. Supports. Electrical metallic tubing shall be installed as a complete system as provided in Article 300 and shall be securely fastened in place at least every 10 feet (3.05 m) and within 3 feet (914 mm) of each outlet box, junction box, device box, cabinet, conduit body, or other tubing terminations.

Exception: Fastening of unbroken lengths shall be permitted to be increased to a distance of 5 feet (1.52 m) where structural members do not readily permit fastening within 3 feet (914 mm).

348-13. Boxes and Fittings. Boxes and fittings shall comply with the applicable provisions of Article 370.

348-14. Splices and Taps. Splices and taps shall be made only in junction boxes, outlet boxes, device boxes, or conduit bodies. See Article 370.

B. Construction Specifications

348-15. General. Electrical metallic tubing shall comply with (a) through (d) below.

(a) Cross Section. The tubing, and elbows and bends for use with the tubing, shall have a circular cross section.

(b) Finish. Tubing shall have such a finish or treatment of outer surfaces as will provide an approved durable means of readily distinguishing it, after installation, from rigid metal conduit.

(c) Connectors. Where the tubing is coupled together by threads, the connector shall be so designed as to prevent bending of the tubing at any part of the thread.

(d) Marking. Electrical metallic tubing shall be clearly and durably marked at least every 10 feet (3.05 m) as required in the first sentence of Section 110-21.

ARTICLE 349 — FLEXIBLE METALLIC TUBING

A. General

349-1. Scope. The provisions of this article apply to a raceway for electric conductors that is circular in cross section, flexible, metallic, and liquidtight without a nonmetallic jacket.

349-2. Other Articles. Installations of flexible metallic tubing shall comply with the provisions of the applicable sections of Article 300 and Section 110-21.

349-3. Uses Permitted. Flexible metallic tubing shall be permitted to be used (1) in dry locations; (2) in accessible locations where protected from physical damage or concealed, such as above suspended ceilings; (3) for 1000 volts maximum; and (4) in branch circuits.

349-4. Uses Not Permitted. Flexible metallic tubing shall not be used (1) in hoistways; (2) in storage battery rooms; (3) in hazardous (classified) locations unless otherwise permitted under other articles in this Code; (4) underground for direct earth burial, or embedded in poured concrete or aggregate; (5) where subject to physical damage; and (6) in lengths over 6 feet (1.83 m).

B. Construction and Installation

349-10. Size.

(a) Minimum. Flexible metallic tubing smaller than $\frac{1}{2}$-inch electrical trade size shall not be used.

Exception No. 1: $\frac{3}{8}$-inch trade size shall be permitted to be installed in accordance with Sections 300-22(b) and (c).

Exception No. 2: $\frac{3}{8}$-inch trade size shall be permitted in lengths not in excess of 6 feet (1.83 m) as part of an approved assembly or for lighting fixtures. See Section 410-67(c).

(b) Maximum. The maximum size of flexible metallic tubing shall be the $\frac{3}{4}$-inch trade size.

349-12. Number of Conductors.

(a) $\frac{1}{2}$-Inch and $\frac{3}{4}$-Inch Flexible Metallic Tubing. The number of conductors permitted in $\frac{1}{2}$-inch and $\frac{3}{4}$-inch trade sizes of flexible metallic tubing shall not exceed the percentage of fill specified in Table 1, Chapter 9.

(b) $\frac{3}{8}$-Inch Flexible Metallic Tubing. The number of conductors permitted in $\frac{3}{8}$-inch trade size flexible metallic tubing shall not exceed that permitted in Table 350-3.

(FPN): For conductor cross-sectional area, see Tables 5, 5A, 6, and 8 and the applicable Notes to Tables at the beginning of Chapter 9.

349-16. Grounding. See Section 250-91(b), Exception No. 1.

349-17. Splices and Taps. Splices and taps shall be made only in junction boxes, outlet boxes, device boxes, or conduit bodies. See Article 370.

349-18. Fittings. Flexible metallic tubing shall be used only with approved terminal fittings. Fittings shall effectively close any openings in the connection.

(FPN): See Sections 300-22(b) and (c) for use in ducts, plenums, and other spaces used for environmental air.

349-20. Bends.

(a) Infrequent Flexing Use. Where the flexible metallic tubing shall be infrequently flexed in service after installation, the radii of bends measured to the inside of the bend shall not be less than specified in Table 349-20(a).

Table 349-20(a).   Minimum Radii for Flexing Use

Trade Size	Minimum Radii
³⁄₈ inch	10 inches
¹⁄₂ inch	12¹⁄₂ inches
³⁄₄ inch	17¹⁄₂ inches

For SI units: (Radii) one inch = 25.4 millimeters.

(b) Fixed Bends. Where the flexible metallic tubing is bent for installation purposes and is not flexed or bent as required by use after installation, the radii of bends measured to the inside of the bend shall not be less than specified in Table 349-20(b).

Table 349-20(b).   Minimum Radii for Fixed Bends

Trade Size	Minimum Radii
³⁄₈ inch	3¹⁄₂ inches
¹⁄₂ inch	4 inches
³⁄₄ inch	5 inches

For SI units: (Radii) one inch = 25.4 millimeters.

ARTICLE 350 — FLEXIBLE METAL CONDUIT

350-1. Other Articles. Installations of flexible metal conduit shall comply with the applicable provisions of Article 300.

350-2. Uses Not Permitted. Flexible metal conduit shall not be used (1) in wet locations unless conductors are of lead-covered type or of other types approved for the specific conditions and the installation is such that water is not likely to enter other raceways or enclosures to which the conduit is connected; (2) in hoistways, other than provided in Section 620-21; (3) in storage-battery rooms; (4) in any hazardous (classified) location other than permitted in Sections 501-4(b) and 504-20; (5) where rubber-covered conductors are exposed to oil, gasoline, or other materials having a deteriorating effect on rubber; (6) underground or embedded in poured concrete or aggregate; or (7) where subject to physical damage.

350-3. Minimum Size. Flexible metal conduit less than ½-inch electrical trade size shall not be used.

Exception No. 1: For enclosing the leads of motors as permitted in Section 430-145(b).

Exception No. 2: Flexible metal conduit of 3/8-inch nominal trade size shall be permitted in lengths not in excess of 6 feet (1.83 m) as a part of a listed assembly, or for tap connections to lighting fixtures as required in Section 410-67(c), or for utilization equipment.

Exception No. 3: Flexible metal conduit of 3/8-inch nominal trade size shall be permitted for manufactured wiring systems as permitted in Section 604-6(a).

Exception No. 4: As permitted in Section 620-21, Exception No. 5.

Exception No. 5: Flexible metal conduit of 3/8-inch nominal trade size shall be permitted as part of a listed assembly to connect wired fixture sections as permitted in Section 410-77(c).

Table 350-3. Maximum Number of Insulated Conductors in 3/8-Inch Flexible Metal Conduit*

Col. A = With fitting inside conduit.
Col. B = With fitting outside conduit.

Size AWG	Types RFH-2, SF-2		Types TF, XHHW, AF, TW		Types TFN, THHN, THWN		Types FEP, FEPB, PF, PGF	
	A	B	A	B	A	B	A	B
18	..	3	3	7	4	8	5	8
16	..	2	2	4	3	7	4	8
14	..	..	..	4	3	7	3	7
12	..	..	..	3	..	4	..	4
10	..	..	..	..	..	2	..	3

* In addition, one uninsulated equipment grounding conductor of the same size shall be permitted.

350-4. Supports. Flexible metal conduit shall be secured by an approved means at intervals not exceeding 4½ feet (1.37 m) and within 12 inches (305 mm) on each side of every outlet box, junction box, cabinet, or fitting.

Exception No. 1: Where flexible metal conduit is fished.

Exception No. 2: Lengths not exceeding 3 feet (914 mm) at terminals where flexibility is necessary.

Exception No. 3: Lengths not exceeding 6 feet (1.83 m) from a fixture terminal connection for tap connections to lighting fixtures as required in Section 410-67(c).

350-5. Grounding. Flexible metal conduit shall be permitted as a grounding means as covered in Section 250-91(b). Where an equipment bonding jumper is required around flexible metal conduit, it shall be installed in accordance with Section 250-79.

Exception No. 1: Where used to connect equipment where flexibility is required, an equipment grounding conductor shall be installed.

Exception No. 2: Listed flexible metal conduit shall be permitted as a grounding means if the total length in any ground return path is 6 feet (1.83 m) or less, the conduit is terminated in fittings listed for grounding, and the circuit conductors contained therein are protected by overcurrent devices rated at 20 amperes or less.

350-6. Bends — Number in One Run. There shall not be more than the equivalent of four quarter bends (360 degrees total) between pull points, e.g., conduit bodies and boxes.

350-7. Number of Conductors. The number of conductors permitted in a single ½-inch through 4-inch trade size shall not exceed the percentage of fill specified in Table 1, Chapter 9. See Table 350-3 for ⅜-inch flexible metal conduit.

350-8. Fittings. Fittings listed for flexible metal conduit shall be used. Angle connectors shall not be used for concealed raceway installations.

350-9. Trimming. All cut ends of conduit shall be trimmed or otherwise finished to remove rough edges, except where fittings that thread into the convolutions are used.

350-10. Splices and Taps. Splices and taps shall be made only in junction boxes, outlet boxes, device boxes, or conduit bodies.

ARTICLE 351 — LIQUIDTIGHT FLEXIBLE METAL CONDUIT AND LIQUIDTIGHT FLEXIBLE NONMETALLIC CONDUIT

351-1. Scope. This article covers liquidtight flexible metal conduit and liquidtight flexible nonmetallic conduit.

A. Liquidtight Flexible Metal Conduit

351-2. Definition. Liquidtight flexible metal conduit is a raceway of circular cross section having an outer liquidtight, nonmetallic, sunlight-resistant jacket over an inner flexible metal core with associated couplings, connectors, and fittings and approved for the installation of electric conductors.

351-3. Other Articles. Installations of liquidtight flexible metal conduit shall comply with the applicable provisions of Article 300 and with the specific sections of Articles 350, 501, 502, 503, and 553 referenced below.

(FPN): For marking requirements, see Section 110-21.

351-4. Use.

(a) Permitted. Liquidtight flexible metal conduit shall be permitted to be used in exposed or concealed locations:

(1) Where conditions of installation, operation, or maintenance require flexibility or protection from liquids, vapors, or solids.

(2) As permitted by Sections 501-4(b), 502-4, 503-3, and 504-20 and in other hazardous (classified) locations where specifically approved, and by Section 553-7(b).

(3) For direct burial where listed and marked for the purpose.

(b) Not Permitted. Liquidtight flexible metal conduit shall not be used:

(1) Where subject to physical damage.

(2) Where any combination of ambient and conductor temperature will produce an operating temperature in excess of that for which the material is approved.

351-5. Size.

(a) Minimum. Liquidtight flexible metal conduit smaller than $\frac{1}{2}$-inch electrical trade size shall not be used.

Exception: $\frac{3}{8}$-inch size shall be permitted as covered in Section 350-3.

(b) Maximum. The maximum size of liquidtight flexible metal conduit shall be the 4-inch trade size.

351-6. Number of Conductors.

(a) Single Conduit. The number of conductors permitted in a single conduit, $\frac{1}{2}$- through 4-inch trade sizes, shall not exceed the percentage of fill specified in Table 1, Chapter 9.

(b) $\frac{3}{8}$-Inch Liquidtight Flexible Metal Conduit. The number of conductors permitted in $\frac{3}{8}$-inch liquidtight flexible metal conduit shall not exceed that permitted in Table 350-3.

351-7. Fittings. Liquidtight flexible metal conduit shall be used only with approved terminal fittings. Angle connectors shall not be used for concealed raceway installations.

351-8. Supports. Where liquidtight flexible metal conduit is installed as a fixed raceway, it shall be secured at intervals not exceeding $4\frac{1}{2}$ feet (1.37 m) and within 12 inches (305 mm) on each side of every outlet box, junction box, cabinet, or fitting.

Exception No. 1: Where liquidtight flexible metal conduit is fished.

Exception No. 2: Lengths not exceeding 3 feet (914 mm) at terminals where flexibility is necessary.

Exception No. 3: Lengths not exceeding 6 feet (1.83 m) from a fixture terminal connection for tap conductors to lighting fixtures as required in Section 410-67(c).

351-9. Grounding. Liquidtight flexible metal conduit shall be permitted as a grounding conductor where both the conduit and the fittings are approved for grounding. Where an equipment bonding jumper is required around liquidtight flexible metal conduit, it shall be installed in accordance with Section 250-79.

Exception No. 1: Where used to connect equipment and flexibility is required, an equipment grounding conductor shall be installed.

Exception No. 2: Listed liquidtight flexible metal conduit shall be permitted as a grounding means in the 1¹/₄-inch and smaller trade sizes if the total length of all liquidtight flexible metal conduit in any ground return path is 6 feet (1.83 m) or less, the conduit is terminated in fittings listed for grounding, and the circuit conductors contained therein are protected by overcurrent devices rated at 20 amperes or less for ³/₈-inch and ¹/₂-inch trade sizes and 60 amperes or less for ³/₄-inch through 1¹/₄-inch trade sizes.

(FPN): See Sections 501-16(b), 502-16(b), and 503-16(b) for types of equipment grounding conductors.

351-10. Bends — Number in One Run. There shall not be more than the equivalent of four quarter bends (360 degrees total) between pull points, e.g., conduit bodies and boxes.

351-11. Splices and Taps. Splices and taps shall be made only in junction boxes, outlet boxes, device boxes, or conduit bodies. See Article 370.

B. Liquidtight Flexible Nonmetallic Conduit

351-22. Definition. Liquidtight flexible nonmetallic conduit is a raceway of circular cross section of various types:

(1) A smooth seamless inner core and cover bonded together and having one or more reinforcement layers between the core and cover.

(2) A smooth inner surface with integral reinforcement within the conduit wall.

(3) A corrugated internal and external surface without integral reinforcement within the conduit wall.

This conduit is flame-resistant and, with fittings, is approved for the installation of electrical conductors.

351-23. Use.

(a) Permitted. Liquidtight flexible nonmetallic conduit shall be permitted to be used in exposed or concealed locations:

(FPN): Extreme cold may cause some types of nonmetallic conduits to become brittle and therefore more susceptible to damage from physical contact.

(1) Where flexibility is required for installation, operation, or maintenance;

(2) Where protection of the contained conductors is required from vapors, liquids, or solids;

(3) For outdoor locations where listed and marked as suitable for the purpose.

(FPN): For marking requirements, see Section 110-21.

(4) For direct burial where listed and marked for the purpose.

(b) Not Permitted. Liquidtight flexible nonmetallic conduit shall not be used:

(1) Where subject to physical damage;

(2) Where any combination of ambient and conductor temperatures is in excess of that for which the liquidtight flexible nonmetallic conduit is approved;

(3) In lengths longer than 6 feet (1.83 m).

Exception: Where longer length is essential for a required degree of flexibility.

(4) Where voltage of the contained conductors is in excess of 600 volts, nominal.

Exception: As permitted in Section 600-31(a), Exception for electric signs over 600 volts.

351-24. Size. The sizes of liquidtight flexible nonmetallic conduit shall be electrical trade sizes ½ inch to 4 inch inclusive.

Exception No. 1: ³/₈-inch size for enclosing the leads of motors as permitted in Section 430-145(b).

Exception No. 2: ³/₈-inch trade size shall be permitted in lengths not in excess of 6 feet (1.83 m) as a part of a listed assembly, for tap connections to lighting fixtures as required in Section 410-67(c), or for utilization equipment. See Table 350-3 for conduit fill.

Exception No. 3: ³/₈-inch trade size for electric sign conductors on insulators in accordance with Section 600-31(a).

351-25. Number of Conductors. The number of conductors permitted in a single conduit shall be in accordance with the percentage fill specified in Table 1, Chapter 9.

351-26. Fittings. Liquidtight flexible nonmetallic conduit shall be used only with terminal fittings identified for such use. Angle connectors shall not be used for concealed raceway installations.

351-28. Equipment Grounding. Where an equipment grounding conductor is required for the circuits installed in liquidtight flexible nonmetallic conduit, it shall be permitted to be installed on the inside or outside of the conduit. Where installed on the outside, the length of the equipment grounding conductor shall not exceed 6 feet (1.83 m) and shall be routed with the raceway or enclosure. Fittings and boxes shall be bonded or grounded in accordance with Article 250.

351-29. Splices and Taps.   Splices and taps shall be made only in junction boxes, outlet boxes, device boxes, or conduit bodies. See Article 370.

351-30. Bends — Number in One Run.   There shall not be more than the equivalent of four quarter bends (360 degrees total) between pull points, e.g., conduit bodies and boxes.

ARTICLE 352 — SURFACE METAL RACEWAYS AND SURFACE NONMETALLIC RACEWAYS

A. Surface Metal Raceways

352-1. Use.   The use of surface metal raceways shall be permitted in dry locations. They shall not be used (1) where subject to severe physical damage unless otherwise approved; (2) where the voltage is 300 volts or more between conductors unless the metal has a thickness of not less than .040 inch (1.02 mm); (3) where subject to corrosive vapors; (4) in hoistways; (5) in any hazardous (classified) location except Class I, Division 2 locations as permitted in the exception to Section 501-4(b); nor (6) concealed except as follows:

Exception No. 1: Surface metal raceways shall be permitted for under-plaster extensions where identified for such use.

Exception No. 2: As permitted in Section 645-5(d)(2).

(FPN):   See definition of "Exposed — (As applied to wiring methods)" in Article 100.

352-2. Other Articles.   Surface metal raceways shall comply with the applicable provisions of Article 300.

352-3. Size of Conductors.   No conductor larger than that for which the raceway is designed shall be installed in surface metal raceway.

352-4. Number of Conductors in Raceways.   The number of conductors installed in any raceway shall be no greater than the number for which the raceway is designed.

The derating factors of Article 310, Note 8 (a) of Notes to Ampacity Tables of 0 to 2000 Volts shall not apply to conductors installed in surface metal raceways where all of the following conditions are met: (1) the cross-sectional area of the raceway exceeds 4 square inches (2580 sq mm); (2) the current-carrying conductors do not exceed thirty in number; (3) the sum of the cross-sectional areas of all contained conductors does not exceed 20 percent of the interior cross-sectional area of the surface metal raceway.

(FPN):   For conductor cross-sectional area, see Tables 5, 5A, 6, and 8 and the applicable Notes to Tables at the beginning of Chapter 9.

352-5. Extension Through Walls and Floors.   It shall be permissible to extend unbroken lengths of surface metal raceways through dry walls, dry partitions, and dry floors.

(FPN):   See Section 353-3 for multioutlet assemblies.

352-6. Combination Raceways.   Where combination surface metal raceways are used both for signaling and for lighting and power circuits, the different systems shall be run in separate compartments identified by sharply contrasting colors of the interior finish, and the same relative position of compartments shall be maintained throughout the premises.

352-7. Splices and Taps.   Splices and taps shall be permitted in surface metal raceway having a removable cover that is accessible after installation. The conductors, including splices and taps, shall not fill the raceway to more than 75 percent of its area at that point. Splices and taps in surface metal raceways without removable covers shall be made only in junction boxes. All splices and taps shall be made by approved methods.

352-8. General.   Surface metal raceways shall be of such construction as will distinguish them from other raceways. Surface metal raceways and their elbows, couplings, and similar fittings shall be so designed that the sections can be electrically and mechanically coupled together and installed without subjecting the wires to abrasion.

Where covers and accessories of nonmetallic materials are used on surface metal raceways, they shall be identified for such use.

352-9. Grounding.   Surface metal raceway enclosures providing a transition from other wiring methods shall have a means for connecting an equipment grounding conductor.

B. Surface Nonmetallic Raceways

352-21. Description.   Part B of this article shall apply to a type of surface nonmetallic raceway and fittings of suitable nonmetallic material that is resistant to moisture and chemical atmospheres. It shall also be flame-retardant, resistant to impact and crushing, resistant to distortion from heat under conditions likely to be encountered in service, and resistant to low-temperature effects. Surface nonmetallic raceways that have limited-smoke-producing characteristics shall be permitted to be identified with the suffix LS.

352-22. Use.   The use of surface nonmetallic raceways shall be permitted in dry locations. They shall not be used (1) where concealed; (2) where subject to severe physical damage; (3) where the voltage is 300 volts or more between conductors, unless listed for higher voltage; (4) in hoistways; (5) in any hazardous (classified) location except Class I, Division 2 locations as permitted in the exception to Section 501-4(b); (6) where subject to ambient temperatures exceeding those for which the nonmetallic raceway is listed; nor (7) for conductors whose insulation temperature limitations would exceed those for which the nonmetallic raceway is listed.

352-23. Other Articles.   Surface nonmetallic raceways shall comply with the applicable provisions of Article 300.

352-24. Size of Conductors. No conductor larger than that for which the raceway is designed shall be installed in surface nonmetallic raceway.

352-25. Number of Conductors in Raceways. The number of conductors installed in any raceway shall be no greater than the number for which the raceway is designed.

352-26. Combination Raceways. Where combination surface nonmetallic raceways are used both for signaling and for lighting and power circuits, the different systems shall be run in separate compartments, identified by printed legend or by sharply contrasting colors of the interior finish, and the same relative position of compartments shall be maintained throughout the premises.

352-27. General. Surface nonmetallic raceways shall be of such construction as will distinguish them from other raceways. Surface nonmetallic raceways and their elbows, couplings, and similar fittings shall be so designed that the sections can be mechanically coupled together and installed without subjecting the wires to abrasion.

352-28. Extension Through Walls and Floors. It shall be permissible to extend unbroken lengths of surface nonmetallic raceways through dry walls, dry partitions, and dry floors.

(FPN): See Section 353-3 for multioutlet assemblies.

352-29. Splices and Taps. Splices and taps shall be permitted in surface nonmetallic raceways having a removable cover that is accessible after installation. The conductors, including splices and taps, shall not fill the raceway to more than 75 percent of its area at that point. Splices and taps in surface nometallic raceways without removable covers shall be made only in junction boxes. All splices and taps shall be made by approved methods.

ARTICLE 353 — MULTIOUTLET ASSEMBLY

353-1. Other Articles. A multioutlet assembly shall comply with applicable provisions of Article 300.

(FPN): See definition in Article 100.

353-2. Use. The use of a multioutlet assembly shall be permitted in dry locations. It shall not be installed (1) where concealed, except that it shall be permissible to surround the back and sides of a metal multioutlet assembly by the building finish or recess a nonmetallic multioutlet assembly in a baseboard; (2) where subject to severe physical damage; (3) where the voltage is 300 volts or more between conductors unless the assembly is of metal having a thickness of not less than .040 inch (1.02 mm); (4) where subject to corrosive vapors; (5) in hoistways; nor (6) in any hazardous (classified) locations except Class I, Division 2 locations as permitted in the exception to Section 501-4(b).

353-3. Metal Multioutlet Assembly Through Dry Partitions. It shall be permissible to extend a metal multioutlet assembly through (not run within) dry partitions, if arrangements are made for removing the cap or cover on all exposed portions and no outlet is located within the partitions.

ARTICLE 354 — UNDERFLOOR RACEWAYS

354-1. Other Articles. Underfloor raceways shall comply with the applicable provisions of Article 300.

354-2. Use. The installation of underfloor raceways shall be permitted beneath the surface of concrete or other flooring material or in office occupancies, where laid flush with the concrete floor and covered with linoleum or equivalent floor covering. Underfloor raceways shall not be installed (1) where subject to corrosive vapors, nor (2) in any hazardous (classified) locations, except as permitted by Section 504-20 and in Class I, Division 2 locations as permitted in the exception to Section 501-4(b). Unless made of a material judged suitable for the condition or unless corrosion protection approved for the condition is provided, ferrous or nonferrous metal underfloor raceways, junction boxes, and fittings shall not be installed in concrete or in areas subject to severe corrosive influences.

354-3. Covering. Raceway coverings shall comply with (a) through (d) below.

(a) **Raceways Not Over 4 Inches (102 mm) Wide.** Half-round and flat-top raceways not over 4 inches (102 mm) in width shall have not less than $3/4$ inch (19 mm) of concrete or wood above the raceway.

Exception: As permitted in (c) and (d) below for flat-top raceways.

(b) **Raceways Over 4 Inches (102 mm) Wide but Not Over 8 Inches (203 mm) Wide.** Flat-top raceways over 4 inches (102 mm) but not over 8 inches (203 mm) wide with a minimum of 1 inch (25.4 mm) spacing between raceways shall be covered with concrete to a depth of not less than 1 inch (25.4 mm). Raceways spaced less than 1 inch (25.4 mm) apart shall be covered with concrete to a depth of $1\frac{1}{2}$ inches (38 mm).

(c) **Trench-type Raceways Flush with Concrete.** Trench-type flush raceways with removable covers shall be permitted to be laid flush with the floor surface. Such approved raceways shall be so designed that the cover plates will provide adequate mechanical protection and rigidity equivalent to junction box covers.

(d) **Other Raceways Flush with Concrete.** In office occupancies, approved metal flat-top raceways, if not over 4 inches (102 mm) in width, shall be permitted to be laid flush with the concrete floor surface, provided they are covered with substantial linoleum not less than $1/16$ inch (1.59 mm) in thickness or with equivalent floor covering. Where more than one and not more than three single raceways are each installed flush with the concrete, they shall be contiguous with each other and joined to form a rigid assembly.

354-4. Size of Conductors.   No conductor larger than that for which the raceway is designed shall be installed in underfloor raceways.

354-5. Maximum Number of Conductors in Raceway.   The combined cross-sectional area of all conductors or cables shall not exceed 40 percent of the interior cross-sectional area of the raceway.

(FPN):   For conductor cross-sectional area, see Tables 5, 5A, 6, and 8 and the applicable Notes to Tables at the beginning of Chapter 9.

354-6. Splices and Taps.   Splices and taps shall be made only in junction boxes.

For the purposes of this section, so-called loop wiring (continuous, unbroken conductor connecting the individual outlets) shall not be considered to be a splice or tap.

Exception: Splices and taps shall be permitted in trench-type flush raceway having a removable cover that is accessible after installation. The conductors, including splices and taps, shall not fill the raceway more than 75 percent of its area at that point.

354-7. Discontinued Outlets.   When an outlet is abandoned, discontinued, or removed, the sections of circuit conductors supplying the outlet shall be removed from the raceway. No splices or reinsulated conductors, such as would be the case with abandoned outlets on loop wiring, shall be allowed in raceways.

354-8. Laid in Straight Lines.   Underfloor raceways shall be laid so that a straight line from the center of one junction box to the center of the next junction box will coincide with the centerline of the raceway system. Raceways shall be firmly held in place to prevent disturbing this alignment during construction.

354-9. Markers at Ends.   A suitable marker shall be installed at or near each end of each straight run of raceways to locate the last insert.

354-10. Dead Ends.   Dead ends of raceways shall be closed.

354-13. Junction Boxes.   Junction boxes shall be leveled to the floor grade and sealed to prevent the free entrance of water or concrete. Junction boxes used with metal raceways shall be metal and shall be electrically continuous with the raceways.

354-14. Inserts.   Inserts shall be leveled and sealed to prevent the entrance of concrete. Inserts used with metal raceways shall be metal and shall be electrically continuous with the raceway. Inserts set in or on fiber raceways before the floor is laid shall be mechanically secured to the raceway. Inserts set in fiber raceways after the floor is laid shall be screwed into the raceway. When cutting through the raceway wall and setting inserts, chips and other dirt shall not be allowed to remain in the raceway, and tools shall be used that are so designed as to prevent the tool from entering the raceway and damaging conductors that may be in place.

354-15. Connections to Cabinets and Wall Outlets. Connections between raceways and distribution centers and wall outlets shall be made by means of flexible metal conduit where not installed in concrete, rigid metal conduit, intermediate metal conduit, electrical metallic tubing, or approved fittings. Where a metallic underfloor raceway system provides for the termination of an equipment grounding conductor, rigid nonmetallic conduit, electrical nonmetallic tubing, or liquidtight flexible nonmetallic conduit where not installed in concrete, shall be permitted.

ARTICLE 356 — CELLULAR METAL FLOOR RACEWAYS

356-1. Definitions. For the purposes of this article, a "cellular metal floor raceway" shall be defined as the hollow spaces of cellular metal floors, together with suitable fittings, that may be approved as enclosures for electric conductors. A "cell" shall be defined as a single, enclosed tubular space in a cellular metal floor member, the axis of the cell being parallel to the axis of the metal floor member. A "header" shall be defined as a transverse raceway for electric conductors, providing access to predetermined cells of a cellular metal floor, thereby permitting the installation of electric conductors from a distribution center to the cells.

356-2. Uses Not Permitted. Conductors shall not be installed in cellular metal floor raceways (1) where subject to corrosive vapor; (2) in any hazardous (classified) location except as permitted by Section 504-20, and in Class I, Division 2 locations as permitted in the exception to Section 501-4(b); nor (3) in commercial garages, other than for supplying ceiling outlets or extensions to the area below the floor but not above.

(FPN):   See Section 300-8 for installation of conductors with other systems.

356-3. Other Articles. Cellular metal floor raceways shall comply with the applicable provisions of Article 300.

A. Installation

356-4. Size of Conductors. No conductor larger than No. 1/0 shall be installed, except by special permission.

356-5. Maximum Number of Conductors in Raceway. The combined cross-sectional area of all conductors or cables shall not exceed 40 percent of the interior cross-sectional area of the cell or header.

(FPN):   For conductor cross-sectional area, see Tables 5, 5A, 6, and 8 and the | applicable Notes to Tables at the beginning of Chapter 9.

356-6. Splices and Taps. Splices and taps shall be made only in header access units or junction boxes.

For the purposes of this section, so-called loop wiring (continuous unbroken conductor connecting the individual outlets) shall not be considered to be a splice or tap.

356-7. Discontinued Outlets. When an outlet is abandoned, discontinued, or removed, the sections of circuit conductors supplying the outlet shall be removed from the raceway. No splices or reinsulated conductors, such as would be the case with abandoned outlets on loop wiring, shall be allowed in raceways.

356-8. Markers. A suitable number of markers shall be installed for the future locating of cells.

356-9. Junction Boxes. Junction boxes shall be leveled to the floor grade and sealed against the free entrance of water or concrete. Junction boxes used with these raceways shall be of metal and shall be electrically continuous with the raceway.

356-10. Inserts. Inserts shall be leveled to the floor grade and sealed against the entrance of concrete. Inserts shall be of metal and shall be electrically continuous with the raceway. In cutting through the cell wall and setting inserts, chips and other dirt shall not be allowed to remain in the raceway, and tools shall be used that are designed to prevent the tool from entering the cell and damaging the conductors.

356-11. Connection to Cabinets and Extensions from Cells. Connections between raceways and distribution centers and wall outlets shall be made by means of flexible metal conduit where not installed in concrete, rigid metal conduit, intermediate metal conduit, electrical metallic tubing, or approved fittings. Where there are provisions for the termination of an equipment grounding conductor, nonmetallic conduit, electrical nonmetallic tubing, or liquidtight flexible nonmetallic conduit where not installed in concrete, shall be permitted.

B. Construction Specifications

356-12. General. Cellular metal floor raceways shall be so constructed that adequate electrical and mechanical continuity of the complete system will be secured. They shall provide a complete enclosure for the conductors. The interior surfaces shall be free from burrs and sharp edges, and surfaces over which conductors are drawn shall be smooth. Suitable bushings or fittings having smooth rounded edges shall be provided where conductors pass.

ARTICLE 358 — CELLULAR CONCRETE FLOOR RACEWAYS

358-1. Scope. This article covers cellular concrete floor raceways, the hollow spaces in floors constructed of precast cellular concrete slabs, together with suitable metal fittings designed to provide access to the floor cells.

358-2. Definitions. A "cell" shall be defined as a single, enclosed tubular space in a floor made of precast cellular concrete slabs, the direction of the cell being parallel to the direction of the floor member. A "header" shall be defined as transverse metal raceways for electric conductors, providing access to predetermined cells of a precast cellular concrete floor, thereby permitting the installation of electric conductors from a distribution center to the floor cells.

358-3. Other Articles. Cellular concrete floor raceways shall comply with the applicable provisions of Article 300.

358-4. Uses Not Permitted. Conductors shall not be installed in precast cellular concrete floor raceways (1) where subject to corrosive vapor; (2) in any hazardous (classified) location except as permitted by Section 504-20, and in Class I, Division 2 locations as permitted in the exception to Section 501-4(b); nor (3) in commercial garages, other than for supplying ceiling outlets or extensions to the area below the floor but not above.

(FPN): See Section 300-8 for installation of conductors with other systems.

358-5. Header. The header shall be installed in a straight line at right angles to the cells. The header shall be mechanically secured to the top of the precast cellular concrete floor. The end joints shall be closed by a metal closure fitting and sealed against the entrance of concrete. The header shall be electrically continuous throughout its entire length and shall be electrically bonded to the enclosure of the distribution center.

358-6. Connection to Cabinets and Other Enclosures. Connections from headers to cabinets and other enclosures shall be made by means of metal raceways and approved fittings.

358-7. Junction Boxes. Junction boxes shall be leveled to the floor grade and sealed against the free entrance of water or concrete. Junction boxes shall be of metal and shall be mechanically and electrically continuous with the header.

358-8. Markers. A suitable number of markers shall be installed for the future location of cells.

358-9. Inserts. Inserts shall be leveled and sealed against the entrance of concrete. Inserts shall be of metal and shall be fitted with receptacles of the grounded type. A grounding conductor shall connect the insert receptacles to a positive ground connection provided on the header. When cutting through the cell wall for setting inserts or other purposes (such as providing access openings between header and cells), chips and other dirt shall not be allowed to remain in the raceway, and the tool used shall be so designed as to prevent the tool from entering the cell and damaging the conductors.

358-10. Size of Conductors. No conductor larger than No. 1/0 shall be installed, except by special permission.

358-11. Maximum Number of Conductors. The combined cross-sectional area of all conductors or cables shall not exceed 40 percent of the cross-sectional area of the cell or header.

358-12. Splices and Taps. Splices and taps shall be made only in header access units or junction boxes.

For the purposes of this section, so-called loop wiring (continuous unbroken conductor connecting the individual outlets) shall not be considered to be a splice or tap.

358-13. Discontinued Outlets. When an outlet is abandoned, discontinued, or removed, the sections of circuit conductors supplying the outlet shall be removed from the raceway. No splices or reinsulated conductors, such as would be the case of abandoned outlets on loop wiring, shall be allowed in raceways.

ARTICLE 362 — METAL WIREWAYS AND NONMETALLIC WIREWAYS

A. Metal Wireways

362-1. Definition. Wireways are sheet-metal troughs with hinged or removable covers for housing and protecting electric wires and cable and in which conductors are laid in place after the wireway has been installed as a complete system.

362-2. Use. Wireways shall be permitted only for exposed work. Wireways installed in wet locations shall be of raintight construction. Wireways shall not be installed (1) where subject to severe physical damage or corrosive vapor, nor (2) in any hazardous (classified) location except as permitted by Sections 501-4(b), 502-4(b), and 504-20.

Exception: Wireways shall be permitted in concealed spaces in accordance with Section 640-4, Exception, item c.

362-3. Other Articles. Installations of wireways shall comply with the applicable provisions of Article 300.

362-4. Size of Conductors. No conductor larger than that for which the wireway is designed shall be installed in any wireway.

362-5. Number of Conductors. Wireways shall not contain more than thirty current-carrying conductors at any cross section. Conductors for signaling circuits or controller conductors between a motor and its starter and used only for starting duty shall not be considered as current-carrying conductors.

The sum of cross-sectional areas of all contained conductors at any cross section of the wireway shall not exceed 20 percent of the interior cross-sectional area of the wireway.

The derating factors specified in Article 310, Note 8 (a) of Notes to Ampacity Tables of 0 to 2000 Volts shall not be applicable to the thirty current-carrying conductors at 20 percent fill specified above.

Exception No. 1: Where the derating factors specified in Article 310, Note 8 (a) of Notes to Ampacity Tables of 0 to 2000 Volts are applied, the number of current-carrying conductors shall not be limited, but the sum of the cross-sectional areas of all contained conductors at any cross section of the wireway shall not exceed 20 percent of the interior cross-sectional area of the wireway.

Exception No. 2: As provided in Section 520-5, the thirty-conductor limitation shall not apply to theaters and similar locations.

Exception No. 3: As provided in Section 620-32, the 20-percent fill limitation shall not apply to elevators and dumbwaiters.

(FPN): For conductor cross-sectional area, see Tables 5, 5A, 6, and 8 and the applicable Notes to Tables at the beginning of Chapter 9.

362-6. Deflected Insulated Conductors. Where insulated conductors are deflected within a wireway, either at the ends or where conduits, fittings, or other raceways or cables enter or leave the wireway, or where the direction of the wireway is deflected greater than 30 degrees, dimensions corresponding to Section 373-6 shall apply.

362-7. Splices and Taps. Splices and taps shall be permitted within a wireway provided they are accessible. The conductors, including splices and taps, shall not fill the wireway to more than 75 percent of its area at that point.

362-8. Supports. Wireways shall be supported at intervals not to exceed 5 feet (1.52 m), or for individual lengths longer than 5 feet (1.52 m) at each end or joint, unless listed for other support intervals. In no case shall the distance between supports exceed 10 feet (3.05 m).

Exception: Vertical runs of wireways shall be securely supported at intervals not exceeding 15 feet (4.57 m) and shall have not more than one joint between supports. Adjoining wireway sections shall be securely fastened together to provide a rigid joint.

362-9. Extension Through Walls. Unbroken lengths of wireway shall be permitted to pass transversely through walls if in unbroken lengths where passing through.

362-10. Dead Ends. Dead ends of wireways shall be closed.

362-11. Extensions from Wireways. Extensions from wireways shall be made with cord pendants or any wiring method in Chapter 3 that includes a means for equipment grounding. Where a separate equipment grounding conductor is employed, connection of the equipment grounding conductors in the wiring method to the wireway shall comply with Sections 250-113 and 250-118. Where rigid nonmetallic conduit, electrical nonmetallic tubing, or liquidtight flexible nonmetallic conduit is used, connection of the equipment grounding conductor in the nonmetallic raceway to a metal wireway shall comply with Sections 250-113 and 250-118.

362-12. Marking. Wireways shall be marked so that their manufacturer's name or trademark will be visible after installation.

362-13. Grounding. Grounding shall be according to the provisions of Article 250.

B. Nonmetallic Wireways

362-14. Definition. Nonmetallic wireways are flame-retardant, nonmetallic troughs with removable covers for housing and protecting electric wires and cables in which conductors are laid in place after the wireway has been installed as a complete system.

362-15. Uses Permitted. The use of nonmetallic wireways shall be permitted:

(1) Only for exposed work.

Exception: Nonmetallic wireways shall be permitted in concealed spaces in accordance with Section 640-4, Exception, item c.

(2) Where subject to corrosive vapors.

(3) In wet locations where listed for the purpose.

(FPN): Extreme cold may cause nonmetallic wireways to become brittle and, therefore, more susceptible to damage from physical contact.

362-16. Uses Not Permitted. Nonmetallic wireways shall not be used:

(1) Where subject to physical damage.

(2) In any hazardous (classified) location.

Exception: As permitted in Section 504-20.

(3) Where exposed to sunlight unless listed and marked as suitable for the purpose.

(4) Where subject to ambient temperatures other than those for which nonmetallic wireway is listed.

362-17. Other Articles. Installations of nonmetallic wireways shall comply with the applicable provisions of Article 300. Where equipment grounding is required by Article 250, a separate equipment grounding conductor shall be installed in the nonmetallic wireway.

362-18. Size of Conductors. No conductor larger than that for which the nonmetallic wireway is designed shall be installed in any nonmetallic wireway.

362-19. Number of Conductors. The sum of cross-sectional areas of all contained conductors at any cross section of the nonmetallic wireway shall not exceed 20 percent of the interior cross-sectional area of the nonmetallic wireway. Conductors for signaling circuits or controller conductors between a motor and its starter and used only for starting duty shall not be considered as current-carrying conductors.

The derating factors specified in Article 310, Note 8(a) of Notes to Ampacity Tables of 0 to 2000 Volts shall be applicable to the current-carrying conductors at the 20 percent fill specified above.

(FPN): For conductor cross-sectional area, see Tables 5, 5A, 6, and 8 and the applicable Notes to Tables at the beginning of Chapter 9.

362-20. Deflected Insulated Conductors. Where insulated conductors are deflected within a nonmetallic wireway, either at the ends or where conduits, fittings, or other raceways or cables enter or leave the nonmetallic wireway, or where the direction of the nonmetallic wireway is deflected greater than 30 degrees, dimensions corresponding to Section 373-6 shall apply.

362-21. Splices and Taps. Splices and taps shall be permitted within a nonmetallic wireway provided they are accessible. The conductors, including splices and taps, shall not fill the nonmetallic wireway to more than 75 percent of its area at that point.

362-22. Supports. Nonmetallic wireways shall be supported at intervals not to exceed 3 feet (914 mm), and at each end or joint, unless listed for other support intervals. In no case shall the distance between supports exceed 10 feet (3.05 m).

Exception: Vertical runs of nonmetallic wireways shall be securely supported at intervals not exceeding 4 feet (1.22 m), unless listed for other support intervals, and shall have not more than one joint between supports. Adjoining nonmetallic wireway sections shall be securely fastened together to provide a rigid joint.

362-23. Expansion Joints. Expansion joints for nonmetallic wireway shall be provided for thermal expansion and contraction.

Exception: Where the length change due to thermal expansion or contraction is less than 0.25 inch (6.35 mm) in a straight run, an expansion joint shall not be required.

362-24. Extension Through Walls. Unbroken lengths of nonmetallic wireway shall be permitted to pass transversely through walls if in unbroken lengths where passing through.

362-25. Dead Ends. Dead ends of nonmetallic wireways shall be closed.

362-26. Extensions from Nonmetallic Wireways. Extensions from nonmetallic wireways shall be made using any wiring method in Chapter 3. A separate equipment grounding conductor shall be installed in any of the wiring methods used for the extension.

362-27. Marking. Nonmetallic wireways shall be marked so that the manufacturer's name or trademark and interior cross-sectional area in square inches shall be visible after installation. Nonmetallic wireways that have limited-smoke-producing characteristics shall be permitted to be identified with the suffix LS.

ARTICLE 363 — FLAT CABLE ASSEMBLIES

Type FC

363-1. Definition. Type FC, a flat cable assembly, is an assembly of parallel conductors formed integrally with an insulating material web specifically designed for field installation in surface metal raceway.

363-2. Other Articles. In addition to the provisions of this article, installation of Type FC cable shall conform with the applicable provisions of Articles 210, 220, 250, 300, 310, and 352.

363-3. Uses Permitted. Flat cable assemblies shall be permitted only as branch circuits to supply suitable tap devices for lighting, small appliances, or small power loads. Flat cable assemblies shall be installed for exposed work only. Flat cable assemblies shall be installed in locations where they will not be subjected to severe physical damage.

363-4. Uses Not Permitted. Flat cable assemblies shall not be installed (1) where subject to corrosive vapors unless suitable for the application; (2) in hoistways; (3) in any hazardous (classified) location; or (4) outdoors or in wet or damp locations unless identified for use in wet locations.

363-5. Installation. Flat cable assemblies shall be installed in the field only in surface metal raceways identified for the use. The channel portion of the surface metal raceway systems shall be installed as complete systems before the flat cable assemblies are pulled into the raceways.

363-6. Number of Conductors. The flat cable assemblies shall consist of either two, three, or four conductors.

363-7. Size of Conductors. Flat cable assemblies shall have conductors of No. 10 special stranded copper wires.

363-8. Conductor Insulation. The entire flat cable assembly shall be formed to provide a suitable insulation covering all of the conductors and using one of the materials recognized in Table 310-13 for general branch-circuit wiring.

363-9. Splices. Splices shall be made in approved junction boxes using approved wiring methods.

363-10. Taps. Taps shall be made between any phase conductor and the neutral or any other phase conductor by means of devices and fittings identified for the use. Tap devices shall be rated at not less than 15 amperes, or more than 300 volts to ground, and they shall be color-coded in accordance with the requirements of Section 363-20.

363-11. Dead Ends. Each flat cable assembly dead end shall be terminated in an end-cap device identified for the use.

The dead-end fitting for the enclosing surface metal raceway shall be identified for the use.

363-12. Fixture Hangers. Fixture hangers installed with the flat cable assemblies shall be identified for the use.

363-13. Fittings. Fittings to be installed with flat cable assemblies shall be designed and installed to prevent physical damage to the cable assemblies.

363-14. Extensions. All extensions from flat cable assemblies shall be made by approved wiring methods, within the junction boxes, installed at either end of the flat cable assembly runs.

363-15. Supports. The flat cable assemblies shall be supported by means of their special design features, within the surface metal raceways.

The surface metal raceways shall be supported as required for the specific raceway to be installed.

363-16. Rating. The rating of the branch circuit shall not exceed 30 amperes.

363-17. Marking. In addition to the provisions of Section 310-11, Type FC cable shall have the temperature rating durably marked on the surface at intervals not exceeding 24 inches (610 mm).

363-18. Protective Covers. Where a flat cable assembly is installed less than 8 feet (2.44 m) from the floor, it shall be protected by a metal cover identified for the use.

363-19. Identification. The grounded conductor shall be identified throughout its length by means of a distinctive and durable white or natural gray marking.

363-20. Terminal Block Identification. Terminal blocks identified for the use shall have distinctive and durable markings for color or word coding. The grounded conductor section shall have a white marking or other suitable designation. The next adjacent section of the terminal block shall have a black marking or other suitable designation. The next section shall have a red marking or other suitable designation. The final or outer section, opposite the grounded conductor section of the terminal block, shall have a blue marking or other suitable designation.

ARTICLE 364 — BUSWAYS

A. General Requirements

364-1. Scope. This article covers service-entrance, feeder, and branch-circuit busways and associated fittings.

364-2. Definition. For the purpose of this article, a busway is considered to be a grounded metal enclosure containing factory-mounted, bare or insulated conductors, which are usually copper or aluminum bars, rods, or tubes.

(FPN): For cablebus, refer to Article 365.

364-3. Other Articles. Installations of busways shall comply with the applicable provisions of Article 300.

364-4. Use.

(a) Use Permitted. Busways shall be installed only where they are located in the open and are visible.

Exception No. 1: Busways shall be permitted to be installed behind panels if means of access are provided and if all the following conditions are met.

a. No overcurrent devices are installed on the busway other than for individual fixtures or other loads.

b. The space behind the access panels is not used for air-handling purposes.

c. The busway is totally enclosed, nonventilating type.

d. Busway is so installed that the joints between sections and fittings are accessible for maintenance purposes.

Exception No. 2: Busways shall be permitted to be installed behind access panels in accordance with Section 300-22(c).

(b) Use Prohibited. Busways shall not be installed (1) where subject to severe physical damage or corrosive vapors; (2) in hoistways; (3) in any hazardous (classified) location, unless specifically approved for such use [see Section 501-4(b)]; nor (4) outdoors or in wet or damp locations unless identified for such use.

364-5. Support.   Busways shall be securely supported at intervals not exceeding 5 feet (1.52 m) unless otherwise designed and marked.

364-6. Through Walls and Floors.   It shall be permissible to extend unbroken lengths of busway through dry walls. It shall be permissible to extend busways vertically through dry floors if totally enclosed (unventilated) where passing through and for a minimum distance of 6 feet (1.83 m) above the floor to provide adequate protection from physical damage.

364-7. Dead Ends.   A dead end of a busway shall be closed.

364-8. Branches from Busways.   Branches from busways shall be made with busways, rigid metal conduit, intermediate metal conduit, rigid nonmetallic conduit, electrical nonmetallic tubing, flexible metal conduit, liquidtight flexible metal conduit, electrical metallic tubing, metal surface raceway, metal-clad cable, or listed bus drop cable; or with suitable cord assemblies approved for hard usage for the connection of portable equipment or for the connection of stationary equipment to facilitate their interchange. Flexible cord assembly connections shall be permitted to be made directly to the load end terminals of a busway plug-in device, provided a suitable tension take-up device is employed on the cord and the length of the cord from the busway plug-in device to the tension take-up device does not exceed 6 feet (1.83 m). Where a nonmetallic raceway is used, connection of equipment grounding conductors in the nonmetallic raceway to the busway shall comply with Sections 250-113 and 250-118.

364-9. Overcurrent Protection.   Overcurrent protection shall be provided in accordance with Sections 364-10 through 364-14.

364-10. Rating of Overcurrent Protection — Feeders. Where the allowable current rating of the busway does not correspond to a standard rating of the overcurrent device, the next higher rating shall be permitted.

364-11. Reduction in Ampacity Size of Busway. Overcurrent protection shall be required where busways are reduced in ampacity.

Exception: For industrial establishments only, omission of overcurrent protection shall be permitted at points where busways are reduced in ampacity, provided that the length of the busway having the smaller ampacity does not exceed 50 feet (15.2 m) and has an ampacity at least equal to 1/3 the rating or setting of the overcurrent device next back on the line, and provided further that such busway is free from contact with combustible material.

364-12. Feeder or Branch Circuits. Where a busway is used as a feeder, devices or plug-in connections for tapping off feeder or branch circuits from the busway shall contain the overcurrent devices required for the protection of the feeder or branch circuits. The plug-in device shall consist of an externally operable circuit breaker or an externally operable fusible switch. Where such devices are mounted out of reach and contain disconnecting means, suitable means such as ropes, chains, or sticks shall be provided for operating the disconnecting means from the floor.

Exception No. 1: As permitted in Section 240-21 for taps.

Exception No. 2: For fixed or semi-fixed lighting fixtures, where the branch-circuit overcurrent device is part of the fixture cord plug on cord-connected fixtures.

Exception No. 3: Where fixtures without cords are plugged directly into the busway and the overcurrent device is mounted on the fixture.

364-13. Rating of Overcurrent Protection — Branch Circuits. A busway shall be permitted as a branch circuit of any one of the types described in Article 210. When so used, the rating or setting of the overcurrent device protecting the busway shall determine the ampere rating of the branch circuit, and the circuit shall in all respects conform with the requirements of Article 210 that apply to branch circuits of that rating.

364-14. Length of Busways Used as Branch Circuits. Busways used as branch circuits and designed so that loads can be connected at any point shall be limited to such lengths as will provide that in normal use the circuits will not be overloaded.

364-15. Marking. Busways shall be marked with the voltage and current rating for which they are designed, and with the manufacturer's name or trademark in such manner as to be visible after installation.

B. Requirements for Over 600 Volts, Nominal

364-21. Identification. Each bus run shall be provided with a permanent nameplate on which the following information shall be provided: (1) rated voltage; (2) rated continuous current; if bus is forced-cooled, both the normal forced-cooled rating and the self-cooled (not forced-cooled) rating for the same temperature rise shall be given; (3) rated frequency; (4) rated impulse withstand voltage; (5) rated 60-Hz withstand voltage (dry); (6) rated momentary current; and (7) manufacturer's name or trademark.

(FPN): See Switchgear Assemblies, ANSI C37.20-1969 (R1982), for construction and testing requirements for metal-enclosed buses.

364-22. Grounding. Metal-enclosed bus shall be grounded in accordance with Article 250.

364-23. Adjacent and Supporting Structures. Metal-enclosed busways shall be installed so that temperature rise from induced circulating currents in any adjacent metallic parts will not be hazardous to personnel or constitute a fire hazard.

364-24. Neutral. Neutral bus, where required, shall be sized to carry all neutral load current, including harmonic currents, and shall have adequate momentary and short-circuit rating consistent with system requirements.

364-25. Barriers and Seals. Bus runs having sections located both inside and outside of buildings shall have a vapor seal at the building wall to prevent interchange of air between indoor and outdoor sections.

Exception: Vapor seals shall not be required in forced-cooled bus.

Fire barriers shall be provided where fire walls, floors, or ceilings are penetrated.

(FPN): See Section 300-21 for information concerning the spread of fire or products of combustion.

364-26. Drain Facilities. Drain plugs, filter drains, or similar methods shall be provided to remove condensed moisture from low points in bus run.

364-27. Ventilated Bus Enclosures. Ventilated bus enclosures shall be installed in accordance with Article 710, Part D, unless designed so that foreign objects inserted through any opening will be deflected from energized parts.

364-28. Terminations and Connections. Where bus enclosures terminate at machines cooled by flammable gas, seal-off bushings, baffles, or other means shall be provided to prevent accumulation of flammable gas in the bus enclosures.

Flexible or expansion connections shall be provided in long, straight runs of bus to allow for temperature expansion or contraction, or where the bus run crosses building vibration insulation joints.

All conductor termination and connection hardware shall be accessible for installation, connection, and maintenance.

364-29. Switches. Switching devices or disconnecting links provided in the bus run shall have the same momentary rating as the bus. Disconnecting links shall be plainly marked to be removable only when bus is deenergized. Switching devices that are not load break shall be interlocked to prevent operation under load, and disconnecting link enclosures shall be interlocked to prevent access to energized parts.

364-30. Wiring 600 Volts or Less, Nominal. Secondary control devices and wiring that are provided as part of the metal-enclosed bus run shall be insulated by fire-retardant barriers from all primary circuit elements with the exception of short lengths of wire, such as at instrument transformer terminals.

ARTICLE 365 — CABLEBUS

365-1. Definition. Cablebus is an approved assembly of insulated conductors with fittings and conductor terminations in a completely enclosed, ventilated protective metal housing. The assembly is designed to carry fault current and to withstand the magnetic forces of such current.

(FPN): Cablebus is ordinarily assembled at the point of installation from components furnished or specified by the manufacturer in accordance with instructions for the specific job.

365-2. Use.

(a) 600 Volts or Less. Cablebus shall be permitted at any voltage or current for which spaced conductors are rated and shall be installed for exposed work only. Cablebus installed outdoors or in corrosive, wet, or damp locations shall be identified for such use. Cablebus shall not be installed in hoistways or hazardous (classified) locations unless specifically approved for such use. Cablebus shall be permitted to be used for branch circuits, feeders, and services.

Cablebus framework, where adequately bonded, shall be permitted as the equipment grounding conductor for branch circuits and feeders.

(b) Over 600 Volts. Cablebus shall be permitted for systems in excess of 600 volts, nominal. See Section 710-4(a).

365-3. Conductors.

(a) Types of Conductors. The current-carrying conductors in cablebus shall have an insulation rating of 75°C or higher of an approved type and suitable for the application in accordance with Articles 310 and 710.

(b) Ampacity of Conductors. The ampacity of conductors in cablebus shall be in accordance with Tables 310-17 and 310-19.

(c) Size and Number of Conductors. The size and number of conductors shall be that for which the cablebus is designed, and in no case smaller than No. 1/0.

(d) Conductor Supports. The insulated conductors shall be supported on blocks or other mounting means designed for the purpose.

The individual conductors in a cablebus shall be supported at intervals not greater than 3 feet (914 mm) for horizontal runs and 1½ feet (457 mm) for vertical runs. Vertical and horizontal spacing between supported conductors shall not be less than one conductor diameter at the points of support.

365-5. Overcurrent Protection. Where the allowable ampacity of cablebus conductors does not correspond to a standard rating of an overcurrent device, the next higher ampere-rated overcurrent device shall be permitted.

365-6. Support and Extension Through Walls and Floors.

(a) Support. Cablebus shall be securely supported at intervals not exceeding 12 feet (3.66 m).

Exception: Where spans longer than 12 feet (3.66 m) are required, the structure shall be specifically designed for the required span length.

(b) Transversely Routed. It shall be permissible to extend cablebus transversely through partitions or walls, other than fire walls, provided the section within the wall is continuous, protected against physical damage, and unventilated.

(c) Through Dry Floors and Platforms. Except where fire stops are required, it shall be permissible to extend cablebus vertically through dry floors and platforms, provided the cablebus is totally enclosed at the point where it passes through the floor or platform and for a distance of 6 feet (1.83 m) above the floor or platform.

(d) Through Floors and Platforms in Wet Locations. Except where fire stops are required, it shall be permissible to extend cablebus vertically through floors and platforms in wet locations where (1) there are curbs or other suitable means to prevent water flow through the floor or platform opening, and (2) where the cablebus is totally enclosed at the point where it passes through the floor or platform and for a distance of 6 feet (1.83 m) above the floor or platform.

365-7. Fittings. A cablebus system shall include approved fittings for (1) changes in horizontal or vertical direction of the run; (2) dead ends; (3) terminations in or on connected apparatus or equipment or the enclosures for such equipment; and (4) additional physical protection where required, such as guards for severe mechanical exposure.

365-8. Conductor Terminations. Approved terminating means shall be used for connections to cablebus conductors.

365-9. Grounding. Sections of cablebus shall be electrically bonded either by inherent design of the mechanical joints or by applied bonding means.

(FPN): See Section 250-75 for bonding of metal noncurrent-carrying parts.

A cablebus installation shall be grounded in accordance with Sections 250-32 and 250-33.

365-10. Marking. Each section of cablebus shall be marked with the manufacturer's name or trade designation and the maximum diameter, number, voltage rating, and ampacity of the conductors to be installed. Markings shall be so located as to be visible after installation.

ARTICLE 370 — OUTLET, DEVICE, PULL AND JUNCTION BOXES, CONDUIT BODIES AND FITTINGS

A. Scope and General

370-1. Scope. This article covers the installation and use of all boxes, conduit bodies, and fittings as required by Section 300-15, and boxes, conduit bodies, and fittings referred to in Section 300-15 used as outlet, junction, or pull boxes, depending on their use. Cast, sheet metal, nonmetallic, and other boxes such as FS, FD, and larger boxes are not classified as conduit bodies.

(FPN):   For systems over 600 volts, nominal, see Part D of this article.

370-2. Round Boxes.   Round boxes shall not be used where conduits or connectors requiring the use of locknuts or bushings are to be connected to the side of the box.

370-3. Nonmetallic Boxes.   Nonmetallic boxes shall be permitted only with open wiring on insulators, concealed knob-and-tube wiring, nonmetallic-sheathed cable, and nonmetallic raceways.

Exception No. 1: Where internal bonding means are provided between all entries, nonmetallic boxes shall be permitted to be used with metal raceways or metal-jacketed cables.

Exception No. 2: Where integral bonding means with a provision for attaching a grounding jumper inside the box are provided between all threaded entries in nonmetallic boxes listed for the purpose, nonmetallic boxes shall be permitted to be used with metal raceways or metal-jacketed cables.

370-4. Metal Boxes.   All metal boxes shall be grounded in accordance with the provisions of Article 250.

370-5. Short Radius Conduit Bodies.   Conduit bodies such as capped elbows and service-entrance elbows enclosing conductors No. 6 or smaller, and that are only intended to enable the installation of the raceway and the contained conductors, shall not contain splices, taps, or devices and shall be of sufficient size to provide free space for all conductors enclosed in the fitting.

B. Installation

370-15. Damp, Wet, or Hazardous (Classified) Locations.

(a) Damp or Wet Locations.   In damp or wet locations, boxes, conduit bodies, and fittings shall be so placed or equipped as to prevent moisture from entering or accumulating within the box, conduit body, or fitting. Boxes, conduit bodies, and fittings installed in wet locations shall be listed for use in wet locations.

(b) Hazardous (Classified) Locations.   Installations in hazardous (classified) locations shall conform to Articles 500 through 517.

(FPN No. 1):   For boxes in floors, see Section 370-27(b).

(FPN No. 2):   For protection against corrosion, see Section 300-6.

370-16. Number of Conductors in Outlet, Device, and Junction Boxes, and Conduit Bodies.   Boxes shall be of sufficient size to provide free space for all conductors enclosed in the box.

The provisions of this section shall not apply to terminal housings supplied with motors. See Section 430-12.

Boxes and conduit bodies containing conductors, size No. 4 or larger, shall also comply with the provisions of Section 370-28.

(a) Standard Boxes. The maximum number of conductors permitted in standard boxes shall be as is listed in Table 370-16(a). See Section 370-28 where boxes or conduit bodies are used as junction or pull boxes.

(1) Table 370-16(a) shall apply where no fittings or devices, such as fixture studs, cable clamps, hickeys, switches, or receptacles, are contained in the box and where no equipment grounding conductors are part of the wiring within the box. Where one or more of these types of fittings, such as fixture studs, cable clamps, or hickeys are contained in the box, the number of conductors shown in the table shall be reduced by one for each type of fitting; an additional deduction of two conductors shall be made for each mounting yoke or strap containing one or more devices or equipment; and a further deduction of one conductor shall be made for one or more equipment grounding conductors entering the box. Where a second set of equipment grounding conductors, as permitted by Section 250-74, Exception No. 4, is present in the box, then an additional deduction of one conductor shall be made. A conductor running through the box shall be counted as one conductor, and each conductor originating outside of the box and terminating inside the box is counted as one conductor. Conductors, no part of which leaves the box, shall not be counted. The volume of a wiring enclosure (box) shall be the total volume of the assembled sections, and, where used, the space provided by plaster rings, domed covers, extension rings, etc., that are marked with their volume in cubic inches, or are made from boxes the dimensions of which are listed in Table 370-16(a).

Exception: Where an equipment grounding conductor or not over four fixture wires smaller than No. 14, or both, enter a box from a fixture canopy and terminate within that box, it shall be permitted to omit these conductors from the calculations. Where branch-circuit conductors enter a fixture canopy from a box and terminate within that canopy, it shall be permitted to count these conductors as not leaving the box.

(2) For combinations of conductor sizes shown in Table 370-16(a), the maximum number of conductors permitted shall be computed using the volume per conductor listed in Table 370-16(b), with the deductions provided for in Section 370-16(a)(1). For fixture studs, cable clamps, or hickeys, these deductions shall be based on the largest conductor entering the box; for each yoke or strap containing one or more devices or equipment, the deduction shall be based on the largest conductor connected to device(s) or equipment supported by that yoke or strap; and for equipment grounding conductors, the deductions shall be based on the largest equipment grounding conductor entering the box. The maximum number and size of conductors listed in Table 370-16(a) shall not be exceeded.

(b) Other Boxes. Boxes 100 cubic inches (1640 cm^3) or less other than those described in Table 370-16(a) and nonmetallic boxes shall be durably and legibly marked by the manufacturer with their cubic inch capacity. The maximum number of conductors permitted shall be computed using the volume per conductor listed in Table 370-16(b), with the deductions provided for in Section 370-16(a)(1). For fixture studs, cable clamps, or hickeys, these deductions shall be based on the largest conductor entering the box; for each yoke or strap containing one or more devices or equipment, the deduction shall be based on the largest conductor connected to device(s) or equipment supported by that yoke or strap; and for equipment grounding conductors, the deductions shall be based on the largest equipment grounding conductor entering the box. Boxes described in Table 370-16(a) that have a larger cubic inch capacity than is designated in the table shall be permitted to have their cubic inch capacity marked as required by this section, and the maximum number of conductors permitted shall be computed using the volume per conductor listed in Table 370-16(b).

Table 370-16(a).   Metal Boxes

Box Dimension, Inches Trade Size or Type	Min. Cu. In. Cap.	Maximum Number of Conductors						
		No. 18	No. 16	No. 14	No. 12	No. 10	No. 8	No. 6
4 x 1¼ Round or Octagonal	12.5	8	7	6	5	5	4	2
4 x 1½ Round or Octagonal	15.5	10	8	7	6	6	5	3
4 x 2⅛ Round or Octagonal	21.5	14	12	10	9	8	7	4
4 x 1¼ Square	18.0	12	10	9	8	7	6	3
4 x 1½ Square	21.0	14	12	10	9	8	7	4
4 x 2⅛ Square	30.3	20	17	15	13	12	10	6
4¹¹⁄₁₆ x 1¼ Square	25.5	17	14	12	11	10	8	5
4¹¹⁄₁₆ x 1½ Square	29.5	19	16	14	13	11	9	5
4¹¹⁄₁₆ x 2⅛ Square	42.0	28	24	21	18	16	14	8
3 x 2 x 1½ Device	7.5	5	4	3	3	3	2	1
3 x 2 x 2 Device	10.0	6	5	5	4	4	3	2
3 x 2 x 2¼ Device	10.5	7	6	5	4	4	3	2
3 x 2 x 2½ Device	12.5	8	7	6	5	5	4	2
3 x 2 x 2¾ Device	14.0	9	8	7	6	5	4	2
3 x 2 x 3½ Device	18.0	12	10	9	8	7	6	3
4 x 2⅛ x 1½ Device	10.3	6	5	5	4	4	3	2
4 x 2⅛ x 1⅞ Device	13.0	8	7	6	5	5	4	2
4 x 2⅛ x 2 ⅛ Device	14.5	9	8	7	6	5	4	2
3¾ x 2 x 2½ Masonry Box/Gang	14.0	9	8	7	6	5	4	2
3¾ x 2 x 3½ Masonry Box/Gang	21.0	14	12	10	9	8	7	4
FS—Minimum Internal Depth 1¾ Single Cover/Gang	13.5	9	7	6	6	5	4	2
FD—Minimum Internal Depth 2⅜ Single Cover/Gang	18.0	12	10	9	8	7	6	3
FS—Minimum Internal Depth 1¾ Multiple Cover/Gang	18.0	12	10	9	8	7	6	3
FD—Minimum Internal Depth 2⅜ Multiple Cover/Gang	24.0	16	13	12	10	9	8	4

For SI units: one cubic inch = 16.4 cm.³

Table 370-16(b).   Volume Required per Conductor

Size of Conductor	Free Space within Box for Each Conductor
No. 18	1.5 cubic inches
No. 16	1.75 cubic inches
No. 14	2. cubic inches
No. 12	2.25 cubic inches
No. 10	2.5 cubic inches
No. 8	3. cubic inches
No. 6	5. cubic inches

For SI units: one cubic inch = 16.4 cm.[3]

(c) Conduit Bodies. Conduit bodies enclosing No. 6 conductors or smaller other than short radius conduit bodies as described in Section 370-5 shall have a cross-sectional area not less than twice the cross-sectional area of the largest conduit or tubing to which it is attached. The maximum number of conductors permitted shall be the maximum number permitted by Table 1 of Chapter 9 for the conduit to which it is attached.

Conduit bodies shall not contain splices, taps, or devices unless they are durably and legibly marked by the manufacturer with their cubic inch capacity. The maximum number of conductors permitted shall be computed using the provisions of Section 370-16(b) for other than standard boxes. Conduit bodies shall be supported in a rigid and secure manner.

370-17. Conductors Entering Boxes, Conduit Bodies, or Fittings. Conductors entering boxes, conduit bodies, or fittings shall be protected from abrasion and shall comply with (a) through (d) below.

(a) Openings to Be Closed. Openings through which conductors enter shall be adequately closed.

(b) Metal Boxes, Conduit Bodies, and Fittings. Where metal outlet boxes, conduit bodies, or fittings are installed with open wiring or concealed knob-and-tube wiring, conductors shall enter through insulating bushings or, in dry locations, through flexible tubing extending from the last insulating support and firmly secured to the box, conduit body, or fitting. Where raceway or cable is installed with metal outlet boxes, conduit bodies, or fittings, the raceway or cable shall be secured to such boxes, conduit bodies, and fittings.

(c) Nonmetallic Boxes. Nonmetallic boxes shall be suitable for the lowest temperature rated conductor entering the box. Where nonmetallic boxes are used with open wiring or concealed knob-and-tube wiring, the conductors shall enter the box through individual holes. Where flexible tubing is used to encase the conductors, the tubing shall extend from the last insulating support to no less than 1/4 inch (6.35 mm) inside the box. Where nonmetallic-sheathed cable is used, the cable assembly, including the sheath, shall extend into the box no less than 1/4 inch (6.35 mm) through a nonmetallic-sheathed cable knockout opening. In all instances, all permitted wiring methods shall be secured to the boxes.

Exception: Where nonmetallic-sheathed cable is used with boxes no larger than a nominal size 2¹/₄ inch by 4 inch mounted in walls and where the cable is fastened within 8 inches (203 mm) of the box measured along the sheath and where the sheath extends into the box no less than ¹/₄ inch (6.35 mm), securing the cable to the box shall not be required. Multiple cable entries shall be permitted in a single knockout opening.

(d) Conductors No. 4 or Larger. Installation shall comply with Section 373-6(c).

370-18. Unused Openings. Unused openings in boxes, conduit bodies, and fittings shall be effectively closed to afford protection substantially equivalent to that of the wall of the box, conduit body, or fitting. Metal plugs or plates used with nonmetallic boxes, conduit bodies, or fittings shall be recessed at least ¹/₄ inch (6.35 mm) from the outer surface.

370-19. Boxes Enclosing Flush Devices. Boxes used to enclose flush devices shall be of such design that the devices will be completely enclosed on back and sides, and that substantial support for the devices will be provided. Screws for supporting the box shall not be used in attachment of the device contained therein.

370-20. In Wall or Ceiling. In walls or ceilings of concrete, tile, or other noncombustible material, boxes and fittings shall be so installed that the front edge of the box or fitting will not set back of the finished surface more than ¹/₄ inch (6.35 mm). In walls and ceilings constructed of wood or other combustible material, outlet boxes and fittings shall be flush with the finished surface or project therefrom.

370-21. Repairing Plaster and Dry Wall or Plasterboard. Plaster, dry wall, or plasterboard surfaces that are broken or incomplete shall be repaired so there will be no gaps or open spaces greater than ¹/₈ inch (3.18 mm) at the edge of the box or fitting.

370-22. Exposed Surface Extensions. Surface extensions from an outlet box of a concealed wiring system shall be made by mounting and mechanically securing a box or extension ring over the concealed box. Where required, equipment grounding shall be in accordance with Article 250.

Exception: A surface extension shall be permitted to be made from the cover of a concealed box where the cover is designed so that it cannot fall off, or be removed if its security means becomes loose, and if the wiring method is flexible and so arranged that any required grounding continuity is independent of the connection between the box and cover.

370-23. Supports. Enclosures within the scope of Article 370 shall be rigidly and securely fastened in place in accordance with (a) through (g) below.

(a) Surface Mounting. They shall be fastened to the surface upon which they are mounted unless such surface does not provide adequate support, in which case they shall be supported in accordance with (b).

(b) Structural Mounting. They shall be rigidly supported from a structural member of the building either directly or by using a metal or wood brace. Support wires that do not provide rigid support shall not be permitted as the sole support.

(1) Nails, where used as a fastening means, shall be permitted to pass through the interior of the enclosure if located within $\frac{1}{4}$ inch (6.35 mm) of the back or ends of the enclosure.

(2) Metal braces shall be protected against corrosion and formed from metal not less than .020 inch (508 micrometers) thick uncoated. Wood braces shall have a cross section not less than nominal 1 inch (25.4 mm) by 2 inches (50.8 mm).

(c) Nonstructural Mounting. It shall be permissible to make a flush installation in existing covered surfaces where adequate support is provided by clamps, anchors, or fittings. Framing members of suspended ceiling systems shall be permitted as the support if the framing members are adequately supported and securely fastened to each other and to the building structure. Enclosures so supported shall be fastened to the framing member by mechanical means such as bolts, screws, or rivets. Clips identified for use with the type of ceiling framing member(s) and enclosure(s) shall also be permitted.

(d) Raceway Supported Enclosure(s), Without Devices or Fixtures. Enclosures that are not over 100 cubic inches (1640 cm^3) in size and that have threaded entries or have hubs identified for the purpose, and that do not contain devices or support fixtures, shall be considered to be adequately supported where two or more conduits are threaded wrenchtight into the enclosure or hubs, and where each conduit is supported within 3 feet (914 mm) of the enclosure on two or more sides so as to provide the rigid and secure installation intended by this section of the Code.

Exception: Conduit or electrical metallic tubing shall be permitted to support conduit bodies provided the conduit bodies are not larger than the largest trade size of the conduit or electrical metallic tubing.

Such enclosures shall also be considered to be adequately supported if they comply with subsection (e) following.

(e) Raceway Supported Enclosures, With Devices or Fixtures. Enclosures that are not over 100 cubic inches (1640 cm^3) in size and that have threaded entries or have hubs identified for the purpose, and that support fixtures or contain devices, or both, shall be considered adequately supported where two or more conduits are threaded wrenchtight into the enclosure or hubs and where each conduit is supported within 18 inches (457 mm) of the enclosure so as to provide the rigid and secure installation intended by this section of the Code.

Exception: Conduit shall be permitted to support conduit bodies provided the conduit bodies are not larger than the largest trade size of the conduit.

(f) Enclosure(s) in Concrete or Masonry. Enclosure(s) shall be permitted to be supported by being embedded.

(g) Pendant Boxes. Boxes shall be supported from a multiconductor cord or cable in an approved manner that protects the conductors against strain, such as a strain relief connector threaded into a box with a hub.

370-24. Depth of Outlet Boxes. No box shall have an internal depth of less than $\frac{1}{2}$ inch (12.7 mm). Boxes intended to enclose flush devices shall have an internal depth of not less than $\frac{15}{16}$ inch (23.8 mm).

370-25. Covers and Canopies. In completed installations, each outlet box shall have a cover, faceplate, or fixture canopy.

(a) Nonmetallic or Metal Covers and Plates. Nonmetallic or metal covers and plates shall be permitted with nonmetallic outlet boxes. Where metal covers or plates are used, they shall comply with the grounding requirements of Section 250-42.

(FPN): See Sections 410-18(a) and 410-56(c) for metal faceplates.

(b) Exposed Combustible Wall or Ceiling Finish. Where a fixture canopy or pan is used, any combustible wall or ceiling finish exposed between the edge of the canopy or pan and the outlet box shall be covered with noncombustible material.

(c) Flexible Cord Pendants. Covers of outlet boxes and conduit bodies having holes through which flexible cord pendants pass shall be provided with bushings designed for the purpose or shall have smooth, well-rounded surfaces on which the cords may bear. So-called hard-rubber or composition bushings shall not be used.

370-27. Outlet Boxes.

(a) Boxes at Lighting Fixture Outlets. Boxes used at lighting fixture outlets shall be designed for the purpose. At every outlet used exclusively for lighting, the box shall be so designed or installed that a lighting fixture may be attached.

(b) Floor Boxes. Boxes listed specifically for this application shall be used for receptacles located in the floor.

Exception: Boxes located in elevated floors of show windows and similar locations where the authority having jurisdiction judges them to be free from physical damage, moisture, and dirt.

(c) Boxes at Fan Outlets. Outlet boxes shall not be used as the sole support for ceiling (paddle) fans.

Exception: Boxes listed for the application shall be permitted as the sole means of support.

370-28. Pull and Junction Boxes. Boxes and conduit bodies used as pull or junction boxes shall comply with (a) through (d) of this section.

(a) Minimum Size. For raceways containing conductors of No. 4 or larger, and for cables containing conductors of No. 4 or larger, the minimum dimensions of pull or junction boxes installed in a raceway or cable run shall comply with the following:

(1) Straight Pulls. In straight pulls, the length of the box shall not be less than eight times the trade diameter of the largest raceway.

(2) Angle or U Pulls. Where angle or U pulls are made, the distance between each raceway entry inside the box and the opposite wall of the box shall not be less than six times the trade diameter of the largest raceway in a row. This distance shall be increased for additional entries by the amount of the sum of the diameters of all other raceway entries in the same row on the same wall of the box. Each row shall be calculated individually, and the single row that provides the maximum distance shall be used.

Exception: Where a raceway or cable entry is in the wall of a box or conduit body opposite to a removable cover and where the distance from that wall to the cover is in conformance with the column for one wire per terminal in Table 373-6(a).

The distance between raceway entries enclosing the same conductor shall not be less than six times the trade diameter of the larger raceway.

When transposing cable size into raceway size in (a)(1) and (a)(2) above, the minimum trade size raceway required for the number and size of conductors in the cable shall be used.

(3) Boxes or conduit bodies of dimensions less than those required in (a)(1) and (a)(2) above shall be permitted for installations of combinations of conductors that are less than the maximum conduit or tubing fill (of conduits or tubing being used) permitted by Table 1 of Chapter 9, provided the box or conduit body has been approved for and is permanently marked with the maximum number and maximum size of conductors permitted.

Exception: Terminal housings supplied with motors, which shall comply with the provisions of Section 430-12.

(b) Conductors in Pull or Junction Boxes. In pull boxes or junction boxes having any dimension over 6 feet (1.83 m), all conductors shall be cabled or racked up in an approved manner.

(FPN): See Section 373-6(c) for insulation of conductors at bushings.

(c) Covers. All pull boxes, junction boxes, conduit bodies, and fittings shall be provided with covers compatible with the box, conduit body, or fitting construction and suitable for the conditions of use. Where metal covers are used, they shall comply with the grounding requirements of Section 250-42.

(d) Permanent Barriers. Where permanent barriers are installed in a box, each section shall be considered as a separate box.

370-29. Conduit Bodies, Junction, Pull, and Outlet Boxes to Be Accessible. Conduit bodies, junction, pull, and outlet boxes shall be so installed that the wiring contained in them can be rendered accessible without removing any part of the building or, in underground circuits, without excavating sidewalks, paving, earth, or other substance that is to be used to establish the finished grade.

Exception: Listed boxes shall be permitted where covered by gravel, light aggregate, or noncohesive granulated soil if their location is effectively identified and accessible for excavation.

C. Construction Specifications

370-40. Metal Boxes, Conduit Bodies, and Fittings.

(a) Corrosion-Resistant. Metal boxes, conduit bodies, and fittings shall be corrosion-resistant or shall be well-galvanized, enameled, or otherwise properly coated inside and out to prevent corrosion.

(FPN): See Section 300-6 for limitation in the use of boxes and fittings protected from corrosion solely by enamel.

(b) Thickness of Metal. Sheet steel boxes not over 100 cubic inches (1640 cm^3) in size shall be made from steel not less than 0.0625 inch (1.59 mm) thick. The wall of a malleable iron box or conduit body and a die-cast or permanent-mold cast aluminum, brass, bronze, or zinc box or conduit body shall not be less than $3/32$ inch (2.38 mm) thick. Other cast metal boxes or conduit bodies shall have a wall thickness not less than $1/8$ inch (3.17 mm).

Exception No. 1: Listed boxes and conduit bodies shown to have equivalent strength and characteristics shall be permitted to be made of thinner or other metals.

Exception No. 2: The walls of listed short radius conduit bodies, as covered in Section 370-5, shall be permitted to be made of thinner metal.

(c) Metal Boxes Over 100 Cubic Inches. Metal boxes over 100 cubic inches (1640 cm^3) in size shall be constructed so as to be of ample strength and rigidity. If of sheet steel, the metal thickness shall not be less than 0.053 inch (1.35 mm) uncoated.

(d) Grounding Provisions. A means shall be provided in each metal box, designed for use with nonmetallic raceways and nonmetallic cable systems, for the connection of an equipment grounding conductor.

370-41. Covers.

Metal covers shall be of the same material as the box or fitting with which they are used, or they shall be lined with firmly attached insulating material not less than $1/32$ inch (0.79 mm) thick, or they shall be listed for the purpose. Metal covers shall be the same thickness as the boxes or fittings for which they are used, or they shall be listed for the purpose. Covers of porcelain or other approved insulating materials shall be permitted if of such form and thickness as to afford the required protection and strength.

370-42. Bushings.

Covers of outlet boxes, conduit bodies, and outlet fittings having holes through which flexible cord pendants may pass shall be provided with approved bushings or shall have smooth, well-rounded surfaces upon which the cord may bear. Where conductors other than flexible cord may pass through a metal cover, a separate hole equipped with a bushing of suitable insulating material shall be provided for each conductor. Such separate holes shall be connected by a slot as required by Section 300-20.

(FPN): For alternating-current conductors, see Section 300-20(b).

370-43. Nonmetallic Boxes. Provisions for supports or other mounting means for nonmetallic boxes shall be outside of the box, or the box shall be so constructed as to prevent contact between the conductors in the box and the supporting screws.

370-44. Marking. All boxes and conduit bodies, covers, extension rings, plaster rings, and the like shall be durably and legibly marked with the manufacturer's name or trademark.

D. Pull and Junction Boxes for Use on Systems Over 600 Volts, Nominal

370-70. General. In addition to the generally applicable provisions of Article 370, the rules in Sections 370-71 and 370-72 shall apply.

370-71. Size of Pull and Junction Boxes. Pull and junction boxes shall provide adequate space and dimensions for the installation of conductors in accordance with the following:

(a) For Straight Pulls. The length of the box shall not be less than forty-eight times the outside diameter, over sheath, of the largest shielded or lead-covered conductor or cable entering the box. The length shall not be less than thirty-two times the outside diameter of the largest nonshielded conductor or cable.

(b) For Angle or U Pulls. The distance between each cable or conductor entry inside the box and the opposite wall of the box shall not be less than thirty-six times the outside diameter, over sheath, of the largest cable or conductor. This distance shall be increased for additional entries by the amount of the sum of the outside diameters, over sheath, of all other cables or conductor entries through the same wall of the box.

The distance between a cable or conductor entry and its exit from the box shall be not less than thirty-six times the outside diameter, over sheath, of that cable or conductor.

Exception No. 1: Terminal housings supplied with motors, which shall comply with the provisions of Section 430-12.

Exception No. 2: Where cables are nonshielded and not lead-covered, the distance of thirty-six times the outside diameter shall be permitted to be reduced to twenty-four times the outside diameter.

Exception No. 3: Where a conductor or cable entry is in the wall of a box opposite to a removable cover and where the distance from that wall to the cover is in conformance with the provisions of Section 300-34.

(c) Removable Sides. One or more sides of any pull box shall be removable.

370-72. Construction and Installation Requirements.

(a) Corrosion Protection. Boxes shall be made of material inherently resistant to corrosion or shall be suitably protected, both internally and externally, by enameling, galvanizing, plating, or other means.

(b) Passing Through Partitions. Suitable bushings, shields, or fittings having smooth, rounded edges shall be provided where conductors or cables pass through partitions and at other locations where necessary.

(c) Complete Enclosure. Boxes shall provide a complete enclosure for the contained conductors or cables.

(d) Wiring Is Accessible. Boxes shall be so installed that the wiring is accessible without removing any part of the building. Working space shall be provided in accordance with Section 110-34.

(e) Suitable Covers. Boxes shall be closed by suitable covers securely fastened in place. Underground box covers that weigh over 100 pounds (45.4 kg) shall be considered as meeting this requirement. Covers for boxes shall be permanently marked "DANGER HIGH VOLTAGE — KEEP OUT." The marking shall be on the outside of the box cover and shall be readily visible. Letters shall be block type at least $1/2$ inch (12.7 mm) in height.

(f) Suitable for Expected Handling. Boxes and their covers shall be capable of withstanding the handling to which they may likely be subjected.

ARTICLE 373 — CABINETS, CUTOUT BOXES, AND METER SOCKET ENCLOSURES

373-1. Scope. This article covers the installation and construction specifications of cabinets, cutout boxes, and meter socket enclosures.

A. Installation

373-2. Damp, Wet, or Hazardous (Classified) Locations.

(a) Damp and Wet Locations. In damp or wet locations, cabinets and cutout boxes of the surface type shall be so placed or equipped as to prevent moisture or water from entering and accumulating within the cabinet or cutout box, and shall be mounted so there is at least $1/4$-inch (6.35-mm) air space between the enclosure and the wall or other supporting surface. Cabinets or cutout boxes installed in wet locations shall be weatherproof.

(FPN): For protection against corrosion, see Section 300-6.

(b) Hazardous (Classified) Locations. Installations in hazardous (classified) locations shall conform to Articles 500 through 517.

373-3. Position in Wall. In walls of concrete, tile, or other noncombustible material, cabinets shall be so installed that the front edge of the cabinet will not set back of the finished surface more than $1/4$ inch (6.35 mm). In walls constructed of wood or other combustible material, cabinets shall be flush with the finished surface or project therefrom.

373-4. Unused Openings. Unused openings in cabinet or cutout boxes shall be effectively closed to afford protection substantially equivalent to that of the wall of the cabinet or cutout box. Where metal plugs or plates are used with nonmetallic cabinets or cutout boxes, they shall be recessed at least $1/4$ inch (6.35 mm) from the outer surface.

373-5. Conductors Entering Cabinets or Cutout Boxes. Conductors entering cabinets or cutout boxes shall be protected from abrasion and shall comply with (a) through (c) below.

(a) Openings to Be Closed. Openings through which conductors enter shall be adequately closed.

(b) Metal Cabinets and Cutout Boxes. Where metal cabinets or cutout boxes are installed with open wiring or concealed knob-and-tube wiring, conductors shall enter through insulating bushings or, in dry locations, through flexible tubing extending from the last insulating support and firmly secured to the cabinet or cutout box.

(c) Cables. Where cable is used, each cable shall be secured to the cabinet or cutout box.

373-6. Deflection of Conductors. Conductors at terminals or conductors entering or leaving cabinets or cutout boxes and the like shall comply with (a) through (c) below.

(a) Width of Wiring Gutters. Conductors shall not be deflected within a cabinet or cutout box unless a gutter having a width in accordance with Table 373-6(a) is provided. Conductors in parallel in accordance with Section 310-4 shall be judged on the basis of the number of conductors in parallel.

Table 373-6(a). Minimum Wire-Bending Space at Terminals and Minimum Width of Wiring Gutters in Inches

| AWG or Circular-Mil Size of Wire | Wires per Terminal | | | | |
	1	2	3	4	5
14-10	Not Specified	—	—	—	—
8-6	1½	—	—	—	—
4-3	2	—	—	—	—
2	2½	—	—	—	—
1	3	—	—	—	—
1/0-2/0	3½	5	7	—	—
3/0-4/0	4	6	8	—	—
250 kcmil	4½	6	8	10	—
300-350 kcmil	5	8	10	12	—
400-500 kcmil	6	8	10	12	14
600-700 kcmil	8	10	12	14	16
750-900 kcmil	8	12	14	16	18
1000-1250 kcmil	10	—	—	—	—
1500-2000 kcmil	12	—	—	—	—

For SI units: one inch = 25.4 millimeters.

Bending space at terminals shall be measured in a straight line from the end of the lug or wire connector (in the direction that the wire leaves the terminal) to the wall, barrier, or obstruction.

(b) Wire Bending Space at Terminals.   Wire bending space at each terminal shall be provided in accordance with (1) or (2) below.

(1) Table 373-6(a) shall apply where the conductor does not enter or leave the enclosure through the wall opposite its terminal.

Exception No. 1: A conductor shall be permitted to enter or leave an enclosure through the wall opposite its terminal provided the conductor enters or leaves the enclosure where the gutter joins an adjacent gutter that has a width that conforms to Table 373-6(b) for that conductor.

Exception No. 2: A conductor not larger than 350 kcmil shall be permitted to enter or leave an enclosure containing only a meter socket(s) through the wall opposite its terminal, provided the terminal is a lay-in type where either:

a. The terminal is directly facing the enclosure wall and offset is not greater than 50 percent of the bending space specified in Table 373-6(a), or

b. The terminal is directed toward the opening in the enclosure and is within a 45-degree angle of directly facing the enclosure wall.

(FPN):   Offset is the distance measured along the enclosure wall from the axis of the centerline of the terminal to a line passing through the center of the opening in the enclosure.

(2) Table 373-6(b) shall apply where the conductor enters or leaves the enclosure through the wall opposite its terminal.

(c) Insulated Fittings.   Where ungrounded conductors of No. 4 or larger enter a raceway in a cabinet, pull box, junction box, or auxiliary gutter, the conductors shall be protected by a substantial fitting providing a smoothly rounded insulating surface, unless the conductors are separated from the raceway fitting by substantial insulating material securely fastened in place.

Exception: Where threaded hubs or bosses that are an integral part of an enclosure provide a smoothly rounded or flared entry for conductors.

Conduit bushings constructed wholly of insulating material shall not be used to secure a raceway. The insulating fitting or insulating material shall have a temperature rating not less than the insulation temperature rating of the installed conductors.

373-7. Space in Enclosures.   Cabinets and cutout boxes shall have sufficient space to accommodate all conductors installed in them without crowding.

373-8. Enclosures for Switches or Overcurrent Devices.   Enclosures for switches or overcurrent devices shall not be used as junction boxes, auxiliary gutters, or raceways for conductors feeding through or tapping off to other switches or overcurrent devices.

Exception: Where adequate space is provided so that the conductors do not fill the wiring space at any cross section to more than 40 percent of the cross-sectional area of the space, and so that the conductors, splices, and taps do not fill the wiring space at any cross section to more than 75 percent of the cross-sectional area of the space.

Table 373-6(b).　Minimum Wire-Bending Space at Terminals for Section 373-6(b)(2) in Inches

Wire Size (AWG or kcmil)	Wires per Terminal							
	1		2		3		4 or More	
14-10	Not Specified		—		—		—	
8		1½	—		—		—	
6	2		—		—		—	
4	3		—		—		—	
3	3		—		—		—	
2	3½		—		—		—	
1	4½		—		—		—	
1/0	5½		5½		7		—	
2/0	6		6		7½		—	
3/0	6½	(½)	6½	(½)	8			
4/0	7	(1)	7½	(1½)	8½	(½)	—	
250	8½	(2)	8½	(2)	9	(1)	10	
300	10	(3)	10	(2)	11	(1)	12	
350	12	(3)	12	(3)	13	(3)	14	(2)
400	13	(3)	13	(3)	14	(3)	15	(3)
500	14	(3)	14	(3)	15	(3)	16	(3)
600	15	(3)	16	(3)	18	(3)	19	(3)
700	16	(3)	18	(3)	20	(3)	22	(3)
750	17	(3)	19	(3)	22	(3)	24	(3)
800	18		20		22		24	
900	19		22		24		24	
1000	20		—		—		—	
1250	22		—		—		—	
1500	24		—		—		—	
1750	24		—		—		—	
2000	24		—		—		—	

For SI units: one inch = 25.4 millimeters.

Bending space at terminals shall be measured in a straight line from the end of the lug or wire connector in a direction perpendicular to the enclosure wall.

For removable and lay-in wire terminals intended for only one wire, bending space shall be permitted to be reduced by the number of inches shown in parentheses.

373-9. Side or Back Wiring Spaces or Gutters.　Cabinets and cutout boxes shall be provided with back wiring spaces, gutters, or wiring compartments as required by Section 373-11(c) and (d).

B. Construction Specifications

373-10. Material.　Cabinets and cutout boxes shall comply with (a) through (c) below.

(a) Metal Cabinets and Cutout Boxes.　Metal cabinets and cutout boxes shall be protected both inside and outside against corrosion.

(FPN):   For protection against corrosion, see Section 300-6.

(b) Strength. The design and construction of cabinets and cutout boxes shall be such as to secure ample strength and rigidity. If constructed of sheet steel, the metal thickness shall not be less than 0.053 inch (1.35 mm) uncoated.

(c) Nonmetallic Cabinets. Nonmetallic cabinets shall be listed or they shall be submitted for approval prior to installation.

373-11. Spacing.   The spacing within cabinets and cutout boxes shall comply with (a) through (d) below.

(a) General. Spacing within cabinets and cutout boxes shall be sufficient to provide ample room for the distribution of wires and cables placed in them, and for a separation between metal parts of devices and apparatus mounted within them as follows:

(1) Base. Other than at points of support, there shall be an air space of at least $\frac{1}{16}$ inch (1.59 mm) between the base of the device and the wall of any metal cabinet or cutout box in which the device is mounted.

(2) Doors. There shall be an air space of at least 1 inch (25.4 mm) between any live metal part, including live metal parts of enclosed fuses, and the door.

Exception: Where the door is lined with an approved insulating material or is of a thickness of metal not less than 0.093 inches (2.36 mm) uncoated, the air space shall not be less than $\frac{1}{2}$ inch (12.7 mm).

(3) Live Parts. There shall be an air space of at least $\frac{1}{2}$ inch (12.7 mm) between the walls, back, gutter partition, if of metal, or door of any cabinet or cutout box and the nearest exposed current-carrying part of devices mounted within the cabinet where the voltage does not exceed 250. This spacing shall be increased to at least 1 inch (25.4 mm) for voltages 251 to 600, nominal.

Exception: As permitted in (2) above.

(b) Switch Clearance. Cabinets and cutout boxes shall be deep enough to allow the closing of the doors when 30-ampere branch-circuit panelboard switches are in any position; when combination cutout switches are in any position; or when other single-throw switches are opened as far as their construction will permit.

(c) Wiring Space. Cabinets and cutout boxes that contain devices or apparatus connected within the cabinet or box to more than eight conductors, including those of branch circuits, meter loops, feeder circuits, power circuits, and similar circuits, but not including the supply circuit or a continuation thereof, shall have back-wiring spaces or one or more side-wiring spaces, side gutters, or wiring compartments.

(d) Wiring Space — Enclosure. Side-wiring spaces, side gutters, or side-wiring compartments of cabinets and cutout boxes shall be made tight enclosures by means of covers, barriers, or partitions extending from the bases of the devices, contained in the cabinet to the door, frame, or sides of the cabinet.

Exception: Where the enclosure contains only those conductors that are led from the cabinet at points directly opposite their terminal connections to devices within the cabinet.

Partially enclosed back-wiring spaces shall be provided with covers to complete the enclosure. Wiring spaces that are required by (c) above, and that are exposed when doors are open, shall be provided with covers to complete the enclosure. Where adequate space is provided for feed-through conductors and for splices as required in Section 373-8, Exception, additional barriers shall not be required.

ARTICLE 374 — AUXILIARY GUTTERS

374-1. Use. Auxiliary gutters shall be permitted to supplement wiring spaces at meter centers, distribution centers, switchboards, and similar points of wiring systems and may enclose conductors or busbars but shall not be used to enclose switches, overcurrent devices, appliances, or other similar equipment.

374-2. Extension Beyond Equipment. An auxiliary gutter shall not extend a greater distance than 30 feet (9.14 m) beyond the equipment that it supplements.

Exception: As provided in Section 620-35 for elevators.

(FPN): For wireways, see Article 362. For busways, see Article 364.

374-3. Supports. Gutters shall be supported throughout their entire length at intervals not exceeding 5 feet (1.52 m).

374-4. Covers. Covers shall be securely fastened to the gutter.

374-5. Number of Conductors. Auxiliary gutters shall not contain more than thirty current-carrying conductors at any cross section. The sum of the cross-sectional areas of all contained conductors at any cross section of an auxiliary gutter shall not exceed 20 percent of the interior cross-sectional area of the auxiliary gutter.

Exception No. 1: As provided in Section 620-35 for elevators.

Exception No. 2: Conductors for signaling circuits or controller conductors between a motor and its starter and used only for starting duty shall not be considered as current-carrying conductors.

Exception No. 3: Where the correction factors specified in Article 310, Note 8(a) of Notes to Ampacity Tables of 0 to 2000 Volts are applied, there shall be no limit on the number of current-carrying conductors, but the sum of the cross-sectional area of all contained conductors at any cross section of the auxiliary gutter shall not exceed 20 percent of the interior cross-sectional area of the auxiliary gutter.

(FPN): For conductor cross-sectional area, see Tables 5, 5A, 6, and 8 and the applicable Notes to Tables at the beginning of Chapter 9.

374-6. Ampacity of Conductors. Where the number of current-carrying conductors contained in the auxiliary gutter is thirty or less, the correction factors specified in Article 310, Note 8(a) of Notes to Ampacity Tables of 0 to 2000 Volts shall not apply. The current carried continuously in bare copper bars in auxiliary gutters shall not exceed 1000 amperes per square inch (645 sq mm) of cross section of the conductor. For aluminum bars, the current carried continuously shall not exceed 700 amperes per square inch (645 sq mm) of cross section of the conductor.

374-7. Clearance of Bare Live Parts. Bare conductors shall be securely and rigidly supported so that the minimum clearance between bare current-carrying metal parts of opposite polarities mounted on the same surface will not be less than 2 inches (50.8 mm), nor less than 1 inch (25.4 mm) for parts that are held free in the air. A clearance not less than 1 inch (25.4 mm) shall be secured between bare current-carrying metal parts and any metal surface. Adequate provisions shall be made for the expansion and contraction of busbars.

374-8. Splices and Taps. Splices and taps shall comply with (a) through (d) below.

(a) Within Gutters. Splices or taps shall be permitted within gutters where they are accessible by means of removable covers or doors. The conductors, including splices and taps, shall not fill the gutter to more than 75 percent of its area.

(b) Bare Conductors. Taps from bare conductors shall leave the gutter opposite their terminal connections, and conductors shall not be brought in contact with uninsulated current-carrying parts of opposite polarity.

(c) Suitably Identified. All taps shall be suitably identified at the gutter as to the circuit or equipment that they supply.

(d) Overcurrent Protection. Tap connections from conductors in auxiliary gutters shall be provided with overcurrent protection as required in Section 240-21.

374-9. Construction and Installation. Auxiliary gutters shall comply with (a) through (f) below.

(a) Electrical and Mechanical Continuity. Gutters shall be so constructed and installed that adequate electrical and mechanical continuity of the complete system will be secured.

(b) Substantial Construction. Gutters shall be of substantial construction and shall provide a complete enclosure for the contained conductors. All surfaces, both interior and exterior, shall be suitably protected from corrosion. Corner joints shall be made tight and, where the assembly is held together by rivets or bolts, shall be spaced not more than 12 inches (305 mm) apart.

(c) Smooth Rounded Edges. Suitable bushings, shields, or fittings having smooth, rounded edges shall be provided where conductors pass between gutters, through partitions, around bends, between gutters and cabinets or junction boxes, and at other locations where necessary to prevent abrasion of the insulation of the conductors.

(d) Deflected Insulated Conductors. Where insulated conductors are deflected within an auxiliary gutter, either at the ends or where conduits, fittings, or other raceways or cables enter or leave the gutter, or where the direction of the gutter is deflected greater than 30 degrees, dimensions corresponding to Section 373-6 shall apply.

(e) Outdoor Use. Auxiliary gutters installed in wet locations shall be of raintight construction.

(f) Grounding. Grounding shall be according to the provisions of Article 250.

ARTICLE 380 — SWITCHES

A. Installation

380-1. Scope. The provisions of this article shall apply to all switches, switching devices, and circuit breakers where used as switches.

380-2. Switch Connections.

(a) Three-Way and Four-Way Switches. Three-way and four-way switches shall be so wired that all switching is done only in the ungrounded circuit conductor. Where in metal raceways or metal-jacketed cables, wiring between switches and outlets shall be in accordance with Section 300-20(a).

Exception: Switch loops shall not require a grounded conductor.

(b) Grounded Conductors. Switches or circuit breakers shall not disconnect the grounded conductor of a circuit.

Exception No. 1: Where the switch or circuit breaker simultaneously disconnects all conductors of the circuit.

Exception No. 2: Where the switch or circuit breaker is so arranged that the grounded conductor cannot be disconnected until all the ungrounded conductors of the circuit have been disconnected.

380-3. Enclosure. Switches and circuit breakers shall be of the externally operable type mounted in an enclosure listed for the intended use. The minimum wire bending space at terminals and minimum gutter space provided in switch enclosures shall be as required in Section 373-6.

Exception: Pendant- and surface-type snap switches and knife switches mounted on an open-face switchboard or panelboard.

380-4. Wet Locations. A switch or circuit breaker in a wet location or outside of a building shall be enclosed in a weatherproof enclosure or cabinet that shall comply with Section 373-2(a): Switches shall not be installed within wet locations in tub or shower spaces unless installed as part of a listed assembly.

380-5. Time Switches, Flashers, and Similar Devices. Time switches, flashers, and similar devices shall be of the enclosed type or shall be mounted in cabinets or boxes or equipment enclosures. Energized parts shall be barriered to prevent operator exposure when making manual adjustments or switching.

Exception: Where mounted so they are accessible only to qualified persons and so located in an enclosure that any energized parts within 6 inches (152 mm) of the manual adjustment or switch are covered by suitable barriers.

380-6. Position of Knife Switches.

(a) Single-Throw Knife Switches. Single-throw knife switches shall be so placed that gravity will not tend to close them. Single-throw knife switches, approved for use in the inverted position, shall be provided with a locking device that will ensure that the blades remain in the open position when so set.

(b) Double-Throw Knife Switches. Double-throw knife switches shall be permitted to be mounted so that the throw will be either vertical or horizontal. Where the throw is vertical, a locking device shall be provided to hold the blades in the open position when so set.

(c) Connection of Knife Switches. Single-throw knife switches shall be so connected that the blades are deenergized when the switch is in the open position.

Exception: Where the load side of the switch is connected to circuits or equipment, the inherent nature of which may provide a backfeed source of power. For such installations, a permanent sign shall be installed on the switch enclosure or immediately adjacent to open switches that reads: | *"WARNING—LOAD SIDE OF SWITCH MAY BE ENERGIZED BY BACKFEED."*

380-7. Indicating. General-use and motor-circuit switches and circuit breakers, where mounted in an enclosure as described in Section 380-3, shall clearly indicate whether they are in the open "off" or closed "on" position.

Where these switch or circuit breaker handles are operated vertically rather than rotationally or horizontally, the up position of the handle shall be the "on" position.

Exception: Double-throw switches.

380-8. Accessibility and Grouping.

(a) Location. All switches and circuit breakers used as switches shall be so located that they may be operated from a readily accessible place. They shall be so installed that the center of the grip of the operating handle of the switch or circuit breaker, when in its highest position, will not be more than $6\frac{1}{2}$ feet (1.98 m) above the floor or working platform.

Exception No. 1: On busway installations, fused switches and circuit breakers shall be permitted to be located at the same level as the busway. Suitable means shall be provided to operate the handle of the device from the floor.

Exception No. 2: Switches installed adjacent to motors, appliances, or other equipment that they supply shall be permitted to be located higher than specified in the foregoing and to be accessible by portable means.

Exception No. 3: Hookstick operable isolating switches shall be permitted at greater heights. |

(b) Voltage Between Adjacent Switches. Snap switches shall not be grouped or ganged in enclosures unless they can be so arranged that the voltage between adjacent switches does not exceed 300, or unless they are installed in enclosures equipped with permanently installed barriers between adjacent switches.

380-9. Faceplates for Flush-Mounted Snap Switches. Flush snap switches, that are mounted in ungrounded metal boxes and located within reach of conducting floors or other conducting surfaces, shall be provided with faceplates of nonconducting, noncombustible material. Metal faceplates shall be of ferrous metal not less than 0.030 inch (0.762 mm) in thickness or of nonferrous metal not less than 0.040 inch (1.016 mm) in thickness. Faceplates of insulating material shall be noncombustible and not less than 0.10 inch (2.54 mm) in thickness, but they shall be permitted to be less than 0.10 inch (2.54 mm) in thickness if formed or reinforced to provide adequate mechanical strength. Faceplates shall be installed so as to completely cover the wall opening and seat against the wall surface.

380-10. Mounting of Snap Switches.

(a) Surface-type. Snap switches used with open wiring on insulators shall be mounted on insulating material that will separate the conductors at least ½ inch (12.7 mm) from the surface wired over.

(b) Box Mounted. Flush-type snap switches mounted in boxes that are set back of the wall surface as permitted in Section 370-20 shall be installed so that the extension plaster ears are seated against the surface of the wall. Flush-type snap switches mounted in boxes that are flush with the wall surface or project therefrom shall be so installed that the mounting yoke or strap of the switch is seated against the box.

380-11. Circuit Breakers as Switches. A hand-operable circuit breaker equipped with a lever or handle, or a power-operated circuit breaker capable of being opened by hand in the event of a power failure, shall be permitted to serve as a switch if it has the required number of poles. Note: See provisions contained in Section 240-81 and Section 240-83.

380-12. Grounding of Enclosures. Metal enclosures for switches or circuit breakers shall be grounded as specified in Article 250. Where nonmetallic enclosures are used with metal-sheathed cables or metallic conduits, provision shall be made for grounding continuity. Metal face plates for snap switches shall be effectively grounded where used with a wiring method that includes or provides an equipment ground.

380-13. Knife Switches.

(a) Isolating Switches. Knife switches rated at over 1200 amperes at 250 volts or less, and at over 600 amperes at 251 to 600 volts, shall be used only as isolating switches and shall not be opened under load.

(b) To Interrupt Currents. To interrupt currents over 1200 amperes at 250 volts, nominal, or less, or over 600 amperes at 251 to 600 volts, nominal, a circuit breaker or a switch of special design listed for such purpose shall be used.

(c) General-Use Switches. Knife switches of ratings less than specified in (a) and (b) above shall be considered general-use switches.

(FPN):   See definition of general-use switch in Article 100.

(d) Motor-Circuit Switches. Motor-circuit switches shall be permitted to be of the knife-switch type.

(FPN):   See definition of a motor-circuit switch in Article 100.

380-14. Rating and Use of Snap Switches. Snap switches shall be used within their ratings and as indicated in (a) through (c) below:

(a) AC General-Use Snap Switch. A form of general-use snap switch suitable only for use on alternating-current circuits for controlling the following:

(1) Resistive and inductive loads, including electric-discharge lamps, not exceeding the ampere rating of the switch at the voltage involved.

(2) Tungsten-filament lamp loads not exceeding the ampere rating of the switch at 120 volts.

(3) Motor loads not exceeding 80 percent of the ampere rating of the switch at its rated voltage.

(b) AC-DC General-Use Snap Switch. A form of general-use snap switch suitable for use on either ac or dc circuits for controlling the following:

(1) Resistive loads not exceeding the ampere rating of the switch at the voltage applied.

(2) Inductive loads not exceeding 50 percent of the ampere rating of the switch at the applied voltage. Switches rated in horsepower are suitable for controlling motor loads within their rating at the voltage applied.

(3) Tungsten-filament lamp loads not exceeding the ampere rating of the switch at the applied voltage if "T" rated.

(FPN No. 1):   For switches on signs and outline lighting, see Section 600-2.

(FPN No. 2):   For switches controlling motors, see Sections 430-83, 430-109, and 430-110.

(c) CO/ALR Snap Switches. Snap switches rated 20 amperes or less directly connected to aluminum conductors shall be listed and marked CO/ALR.

B. Construction Specifications

380-15. Marking. Switches shall be marked with the current and voltage and, if horsepower rated, the maximum rating for which they are designed.

380-16. 600-Volt Knife Switches. Auxiliary contacts of a renewable or quick-break type or the equivalent shall be provided on all knife switches rated 600 volts designed for use in breaking current over 200 amperes.

380-17. Fused Switches. A fused switch shall not have fuses in parallel except as permitted in Section 240-8, Exception.

380-18. Wire Bending Space. The wire bending space required by Section 380-3 shall meet Table 373-6(b) spacings to the enclosure wall opposite the line and load terminals.

ARTICLE 384 — SWITCHBOARDS AND PANELBOARDS

384-1. Scope. This article covers (1) all switchboards, panelboards, and distribution boards installed for the control of light and power circuits, and (2) battery-charging panels supplied from light or power circuits.

Exception: Switchboards or portions thereof used exclusively to control signaling circuits operated by batteries.

384-2. Other Articles. Switches, circuit breakers, and overcurrent devices used on switchboards, panelboards, and distribution boards, and their enclosures, shall comply with this article and also with the requirements of Articles 240, 250, 370, 373, 380, and other articles that apply. Switchboards and panelboards in hazardous (classified) locations shall comply with the requirements of Articles 500 through 517.

384-3. Support and Arrangement of Busbars and Conductors.

(a) Conductors and Busbars on a Switchboard, Panelboard, or Control Board. Conductors and busbars on a switchboard, panelboard, or control board shall be so located as to be free from physical damage and shall be held firmly in place. Other than the required interconnections and control wiring, only those conductors that are intended for termination in a vertical section of a switchboard shall be located in that section. Barriers shall be placed in all service switchboards that will isolate the service busbars and terminals from the remainder of the switchboard.

Exception: Conductors shall be permitted to travel horizontally through vertical sections of switchboards where such conductors are isolated from busbars by a barrier.

(b) Overheating and Inductive Effects. The arrangement of busbars and conductors shall be such as to avoid overheating due to inductive effects.

(c) Used As Service Equipment. Each switchboard, or panelboard, if used as service equipment shall be provided with a main bonding jumper sized in accordance with Section 250-79(d) or the equivalent placed within the panelboard or one of the sections of the switchboard for connecting the grounded service conductor on its supply side to the switchboard or panelboard frame. All sections of a switchboard shall be bonded together using an equipment grounding conductor sized in accordance with Table 250-95.

Exception: As covered in Section 250-27 for high-impedance grounded neutral system connections.

(d) Terminals. Terminals in switchboards and panelboards shall be so located that it will not be necessary to reach across or beyond an ungrounded line bus in order to make connections.

(e) High-Leg Marking. On a switchboard or panelboard supplied from a 4-wire delta-connected system, where the midpoint of one phase is grounded, that phase busbar or conductor having the higher voltage to ground shall be durably and permanently marked by an outer finish that is orange in color, or by other effective means.

(f) Phase Arrangement. The phase arrangement on 3-phase buses shall be A, B, C from front to back, top to bottom, or left to right, as viewed from the front of the switchboard or panelboard. The B phase shall be that phase having the higher voltage to ground on 3-phase, 4-wire delta-connected systems. Other busbar arrangements shall be permitted for additions to existing installations and shall be marked.

Exception: Equipment within the same single section or multisection switchboard or panelboard as the meter on 3-phase, 4-wire delta-connected systems shall be permitted to have the same phase configuration as the metering equipment.

(g) Minimum Wire Bending Space. The minimum wire bending space at terminals and minimum gutter space provided in panelboards and switchboards shall be as required in Section 373-6.

384-4. Installation. Equipment within the scope of Article 384 shall be located in rooms or spaces dedicated to such equipment. Such space shall include that space described in Section 110-16 and, in addition, shall include an exclusively dedicated space extending 25 feet (7.62 m) from the floor or to the structural ceiling with a width and depth that of the equipment. No piping, ducts, or equipment foreign to the electrical equipment or architectural appurtenances shall be permitted to be installed in, enter, or pass through such spaces or rooms.

(FPN No. 1):   It is not the intent to mandate a dedicated room.

(FPN No. 2):   This section is not intended to prohibit sprinkler protection for the electrical installation.

(FPN No. 3):   For the purpose of this section, dropped, suspended, and similar ceilings not intended to add strength to the building structure are not structural ceilings.

(FPN No. 4):   It is not the intent that any of the provisions of this rule or the exceptions thereto allow any equipment to be located in the working space described in Section 110-16.

Exception No. 1: Control equipment that by its very nature or because of other rules of this Code must be adjacent to or within sight of its operating machinery.

Exception No. 2: Ventilating, heating, or cooling equipment that serves the electrical rooms or spaces.

Exception No. 3: Equipment located throughout industrial plants that is isolated from foreign equipment by height or physical enclosures or covers that will afford adequate mechanical protection from vehicular traffic, accidental contact by unauthorized personnel, or accidental spillage or leakage from piping systems.

Exception No. 4: Outdoor electrical equipment located in weatherproof enclosures protected from accidental contact by unauthorized personnel or vehicular traffic or accidental spillage or leakage from piping systems.

A. Switchboards

384-5. Location of Switchboards. Switchboards that have any exposed live parts shall be located in permanently dry locations and then only where under competent supervision and accessible only to qualified persons. Switchboards shall be so located that the probability of damage from equipment or processes is reduced to a minimum.

384-6. Switchboards in Damp or Wet Locations. Switchboards in damp or wet locations shall be installed to comply with Section 373-2(a).

384-7. Location Relative to Easily Ignitible Material. Switchboards shall be so placed as to reduce to a minimum the probability of communicating fire to adjacent combustible materials. Where installed over a combustible floor, suitable protection thereto shall be provided.

384-8. Clearances.

(a) From Ceiling. A space of 3 feet (914 mm) or more shall be provided between the top of any switchboard and any combustible ceiling.

Exception No. 1: Where a noncombustible shield is provided between the switchboard and the ceiling.

Exception No. 2: Totally enclosed switchboards.

(b) Around Switchboards. Clearances around switchboards shall comply with the provisions of Section 110-16.

384-9. Conductor Insulation. An insulated conductor used within a switchboard shall be listed, flame-retardant, and shall be rated not less than the voltage applied to it and not less than the voltage applied to other conductors or busbars with which it may come in contact.

384-10. Clearance for Conductors Entering Bus Enclosures. Where conduits or other raceways enter a switchboard, floor standing panelboard, or similar enclosure at the bottom, sufficient space shall be provided to permit installation of conductors in the enclosure. The wiring space shall not be less than shown in the following table where the conduit or raceways enter or leave the enclosure below the busbars, their supports, or other obstructions. The conduit or raceways, including their end fittings, shall not rise more than 3 inches (76 mm) above the bottom of the enclosure.

Conductor	Minimum Spacing Between Bottom of Enclosure and Busbars, Their Supports, or Other Obstructions (Inches)
Insulated busbars, their supports, or other obstructions.....	8 (203 mm)
Noninsulated busbars..	10 (254 mm)

384-11. Grounding Switchboard Frames. Switchboard frames and structures supporting switching equipment shall be grounded.

Exception: Frames of 2-wire, direct-current switchboards shall not be required to be grounded if effectively insulated from ground.

384-12. Grounding of Instruments, Relays, Meters, and Instrument Transformers on Switchboards. Instruments, relays, meters, and instrument transformers located on switchboards shall be grounded as specified in Sections 250-121 through 250-125.

B. Panelboards

384-13. General. All panelboards shall have a rating not less than the minimum feeder capacity required for the load computed in accordance with Article 220. Panelboards shall be durably marked by the manufacturer with the voltage and the current rating and the number of phases for which they are designed and with the manufacturer's name or trademark in such a manner as to be visible after installation, without disturbing the interior parts or wiring. All panelboard circuits and circuit modifications shall be legibly identified as to purpose or use on a circuit directory located on the face or inside of the panel doors.

(FPN): See Section 110-22 for additional requirements.

384-14. Lighting and Appliance Branch-Circuit Panelboard. For the purposes of this article, a lighting and appliance branch-circuit panelboard is one having more than 10 percent of its overcurrent devices rated 30 amperes or less, for which neutral connections are provided.

384-15. Number of Overcurrent Devices on One Panelboard. Not more than forty-two overcurrent devices (other than those provided for in the mains) of a lighting and appliance branch-circuit panelboard shall be installed in any one cabinet or cutout box.

A lighting and appliance branch-circuit panelboard shall be provided with physical means to prevent the installation of more overcurrent devices than that number for which the panelboard was designed, rated, and approved.

For the purposes of this article, a 2-pole circuit breaker shall be considered two overcurrent devices; a 3-pole breaker shall be considered three overcurrent devices.

384-16. Overcurrent Protection.

(a) Lighting and Appliance Branch-Circuit Panelboard Individually Protected. Each lighting and appliance branch-circuit panelboard shall be individually protected on the supply side by not more than two main circuit breakers or two sets of fuses having a combined rating not greater than that of the panelboard.

Exception No. 1: Individual protection for a lighting and appliance panelboard shall not be required if the panelboard feeder has overcurrent protection not greater than the rating of the panelboard.

Exception No. 2: For existing installations, individual protection for lighting and appliance branch-circuit panelboards shall not be required where such panelboards are used as service equipment in supplying an individual residential occupancy.

(b) Snap Switches Rated at 30 Amperes or Less. Panelboards equipped with snap switches rated at 30 amperes or less shall have overcurrent protection not in excess of 200 amperes.

(c) Continuous Load. The total load on any overcurrent device located in a panelboard shall not exceed 80 percent of its rating where, in normal operation, the load will continue for three hours or more.

Exception: Where the assembly, including the overcurrent device, is listed for continuous operation at 100 percent of its rating.

(d) Supplied through a Transformer. Where a panelboard is supplied through a transformer, the overcurrent protection required in (a) and (b) above shall be located on the secondary side of the transformer.

Exception: A panelboard supplied by the secondary side of a single-phase transformer having a 2-wire (single-voltage) secondary shall be considered as protected by overcurrent protection provided on the primary (supply) side of the transformer, provided this protection is in accordance with Section 450-3(b)(1) and does not exceed the value determined by multiplying the panelboard rating by the secondary-to-primary voltage ratio.

(e) Delta Breakers. A 3-phase disconnect or overcurrent device shall not be connected to the bus of any panelboard that has less than 3-phase buses.

(FPN):   This is intended to prohibit the use of "delta breakers" in panelboards.

(f) Back-Fed Devices. Plug-in-type overcurrent protection devices or plug-in-type main lug assemblies that are back fed shall be secured in place by an additional fastener that requires other than a pull to release the device from the mounting means on the panel.

384-17. Panelboards in Damp or Wet Locations.   Panelboards in damp or wet locations shall be installed to comply with Section 373-2(a).

384-18. Enclosure.   Panelboards shall be mounted in cabinets, cutout boxes, or enclosures designed for the purpose and shall be dead front.

Exception: Panelboards other than of the dead-front externally operable type shall be permitted where accessible only to qualified persons.

384-19. Relative Arrangement of Switches and Fuses.   In panelboards, fuses of any type shall be installed on the load side of any switches.

Exception: As provided in Section 230-94 for use as service equipment.

384-20. Grounding of Panelboards.   Panelboard cabinets and panelboard frames, if of metal, shall be in physical contact with each other and shall be grounded in accordance with Article 250 or Section 384-3(c). Where the panelboard is used with nonmetallic raceway or cable or where separate grounding conductors are provided, a terminal bar for the grounding conductors shall be secured inside the cabinet. The terminal bar shall be bonded to the cabinet and panelboard frame, if of metal, otherwise it shall be connected to the grounding conductor that is run with the conductors feeding the panelboard.

Grounding conductors shall not be connected to a terminal bar provided for grounded conductors (may be a neutral) unless the bar is identified for the purpose and is located where connection is made from the grounded conductor to a grounding electrode as permitted or required by Article 250.

Exception: Where an isolated equipment grounding conductor is provided as permitted by Section 250-74, Exception No. 4, this insulated equipment grounding conductor that is run with the circuit conductors shall be permitted to pass through the panelboard without being connected to the panelboard's equipment grounding terminal bar.

C. Construction Specifications

384-30. Panels.   The panels of switchboards shall be made of moisture-resistant, noncombustible material.

384-31. Busbars.   Insulated or bare busbars shall be rigidly mounted.

384-32. Protection of Instrument Circuits.   Instruments, pilot lights, potential transformers, and other switchboard devices with potential coils shall be supplied by a circuit that is protected by standard overcurrent devices rated 15 amperes or less.

Exception No. 1: Where the operation of the overcurrent device might introduce a hazard in the operation of devices.

Exception No. 2: For ratings of 2 amperes or less, special types of enclosed fuses shall be permitted.

384-33. Component Parts.   Switches, fuses, and fuseholders used on panelboards shall comply with the applicable requirements of Articles 240 and 380.

384-34. Knife Switches.   Exposed blades of knife switches shall be deenergized when open.

384-35. Wire Bending Space in Panelboards.   The enclosure for a panelboard shall have the top and bottom wire bending space sized in accordance with Table 373-6(b) for the largest conductor entering or leaving the enclosure. Side wire bending space shall be in accordance with Table 373-6(a) for the largest conductor to be terminated in that space.

Exception No. 1: Either the top or bottom wire bending space shall be permitted to be sized in accordance with Table 373-6(a) for a lighting and appliance branch-circuit panelboard rated 225 amperes or less.

Exception No. 2: Either the top or bottom wire bending space for any panelboard shall be permitted to be sized in accordance with Table 373-6(a) where at least one side wire bending space is sized in accordance with Table 373-6(b) for the largest conductor to be terminated in any side wire bending space.

Exception No. 3: The top and bottom wire bending space shall be permitted to be sized in accordance with Table 373-6(a) spacings if the panelboard is designed and constructed for wiring using only one single 90-degree bend for each conductor, including the neutral, and the wiring diagram shows and specifies the method of wiring that shall be used.

Exception No. 4: Either the top or the bottom wire bending space, but not both, shall be permitted to be sized in accordance with Table 373-6(a) where there are no conductors terminated in that space.

384-36. Minimum Spacings.   The distance between bare metal parts, busbars, etc., shall not be less than specified in Table 384-36.

Exception No. 1: At switches or circuit breakers.

Exception No. 2: Inherent spacings in listed components.

Where close proximity does not cause excessive heating, parts of the same polarity at switches, enclosed fuses, etc., shall be permitted to be placed as close together as convenience in handling will allow.

Table 384-36.   Minimum Spacings Between Bare Metal Parts

	Opposite Polarity Where Mounted on the Same Surface	Opposite Polarity Where Held Free in Air	Live Parts to Ground*
Not over 125 volts, nominal..................	$3/4$ inch	$1/2$ inch	$1/2$ inch
Not over 250 volts, nominal..................	$1 1/4$ inch	$3/4$ inch	$1/2$ inch
Not over 600 volts, nominal..................	2 inches	1 inch	1 inch

For SI units: one inch = 25.4 millimeters.

*For spacing between live parts and doors of cabinets, see Sections 373-11(a) (1), (2), and (3).

Chapter 4. Equipment for General Use

ARTICLE 400 — FLEXIBLE CORDS AND CABLES

A. General

400-1. Scope. This article covers general requirements, applications, and construction specifications for flexible cords and flexible cables.

400-2. Other Articles. Flexible cords and flexible cables shall comply with this article and with the applicable provisions of other articles of this Code.

400-3. Suitability. Flexible cords and cables and their associated fittings shall be suitable for the conditions of use and location.

400-4. Types. Flexible cords and flexible cables shall conform to the description in Table 400-4. Types of flexible cords and flexible cables other than those listed in the table shall be the subject of special investigation.

Notes to Table 400-4

1. Except for Types HPN, SP-1, SP-2, SP-3, SPE-1, SPE-2, SPE-3, SPT-1, SPT-2, SPT-3, TPT, and 3-conductor parallel versions of SRD, SRDE, SRDT, individual conductors are twisted together.

2. Types TPT, TS, and TST shall be permitted in lengths not exceeding 8 feet (2.44 m) where attached directly, or by means of a special type of plug, to a portable appliance rated at 50 watts or less and of such nature that extreme flexibility of the cord is essential.

3. Rubber-filled or varnished cambric tapes shall be permitted as a substitute for the inner braids.

4. Types G, S, SC, SCE, SCT, SE, SEO, SEOO, SO, SOO, ST, STO, STOO, and W shall be permitted for use on theater stages, in garages, and elsewhere where flexible cords are permitted by this Code.

5. Elevator traveling cables for operating control and signal circuits shall contain nonmetallic fillers as necessary to maintain concentricity. Cables shall have steel supporting members as required for suspension by Section 620-41. In locations subject to excessive moisture or corrosive vapors or gases, supporting members of other materials shall be permitted. Where steel supporting members are used, they shall run straight through the center of the cable assembly and shall not be cabled with the copper strands of any conductor.

In addition to conductors used for control and signaling circuits, Types E, EO, ET, ETLB, ETP, and ETT elevator cables shall be permitted to incorporate in the construction one or more No. 20 telephone conductor pairs, one or more coaxial cables, and/or one or more optical fibers. The No. 20 conductor pairs shall be permitted to be covered with suitable shielding for telephone, audio, or higher frequency communication circuits; the coaxial cables consist of a center conductor, insulation, and

Table 400-4.　Flexible Cords and Cables
(See Section 400-4)

Trade Name	Type Letter	Size AWG	No. of Conductors	Insulation	Nominal Insulation Thickness* AWG	Mils	Braid on Each Conductor	Outer Covering	Use — Pendant or Portable	Use — Dry Locations	Use — Not Hard Usage
Lamp Cord	C	18-10	2 or more	Thermoset or Thermoplastic	18-16	30	Cotton	None	Pendant or Portable	Dry Locations	Not Hard Usage
					14-10	45					
Elevator Cable	E See Note 5.	20-14	2 or More	Thermoset	20-16	20	Cotton	Three Cotton, Outer one Flame-Retardant & Moisture-Resistant, See Note 3.	Elevator Lighting and Control	Nonhazardous Locations	
	See Note 9.				20-16	20	Flexible Nylon Jacket				
Elevator Cable	EO See Note 5.	20-14	2 or More	Thermoset	20-16	20	Cotton	Three Cotton, Outer one Flame-Retardant & Moisture-Resistant, See Note 3.		Nonhazardous Locations	
					14	30		One Cotton and a Neoprene Jacket See Note 3.	Elevator Lighting and Control	Hazardous (Classified) Locations	

See Notes 1 through 9.
*See Note 8.

Table 400-4. (Continued)

Trade Name	Type Letter	Size AWG or kcmil	Number of Conductors	Insulation	AWG	Mils	Braid on Each Conductor	Outer Covering	Use
Elevator Cable	ET See Note 5.	20-14	2 or More	Thermoplastic	20-16	20	Rayon	Three Cotton or equivalent, Outer one Flame-Retardant & Moisture-Resist. See Note 3.	Nonhazardous Locations
	ETLB See Note 5.				14	30	None		
	ETP See Note 5.			Thermoplastic			Rayon	Thermoplastic	Hazardous (Classified) Locations
	ETT See Note 5.			Thermoplastic			None	One Cotton or equivalent and a Thermoplastic Jacket	Hazardous (Classified) Locations
Portable Power Cable	G	8-500 kcmil	2-6 plus Grounding Conductor(s)	Thermoset	8-2 1-4/0 250 kcmil 500 kcmil	60 80 95		Oil-Resistant Thermoset	Portable, Extra Hard Usage, and as Permitted in Sections 520-68(a) and 530-12.
Heater Cord	HPD	18-12	2, 3, or 4	Thermoset with Asbestos or All Thermoset	18-16 14-12	15 30	None	Cotton or Rayon	Portable Heaters / Dry Locations / Not Hard Usage
Parallel Heater Cord	HPN See Note 6.	18-12	2 or 3	Thermosetting	18-16 14 12	45 80 95	None	Thermosetting	Portable / Damp Locations / Not Hard Usage

See Notes 1 through 9.
* See Note 8.

Table 400-4. (Continued)

Trade Name	Type Letter	Size AWG	No. of Conductors	Insulation	Nominal Insulation Thickness* AWG	Nominal Insulation Thickness* Mils	Braid on Each Conductor	Outer Covering	Use	
Thermoset Jacketed Heater Cord	HS	14-12	2, 3, or 4	Thermoset with Asbestos or All Thermoset	18-16 (Thermoset/Asbestos)	15	None	Cotton and Thermoset	Portable or Portable Heaters	Extra Hard Usage
	HSJ See Note 7.	18-12			18-16 (All Thermoset)	30			Damp Locations	Hard Usage
	HSJO See Note 7.				14-12 (Thermoset/Asbestos)	30		Cotton and Oil-Resistant		
	HSO	14-12			14-12 (All Thermoset)	45				Extra Hard Usage
Twisted Portable Cord	PD	18-10	2 or more	Thermoset or Thermoplastic	18-16 / 14-10	30 / 45	Cotton	Cotton or Rayon	Pendant or Portable / Dry Locations	Not Hard Usage
Hard Service Cord	S See Note 4.	18-2	2 or more	Thermoset	18-16 / 14-10 / 8-2	30 / 45 / 60	None	Thermoset	Pendant or Portable / Damp Locations	Extra Hard Usage
Flexible Stage and Lighting Power Cable	SC	8-250 kcmil	1 or more	Thermoset	8-2 / 1-4/0 / 250 kcmil	60 / 80 / 95		Thermoset**	Portable, Extra Hard Usage, and as permitted in Articles 520 and 530	
Flexible Stage and Lighting Power Cable	SCE	8-250 kcmil	1 or more	Thermoplastic Elastomer	8-2 / 1-4/0 / 250 kcmil	60 / 80 / 95		Thermoplastic Elastomer**	Portable, Extra Hard Usage, and as permitted in Articles 520 and 530	
Flexible Stage and Lighting Power Cable	SCT	8-250 kcmil	1 or more	Thermoplastic	8-2 / 1-4/0 / 250 kcmil	60 / 80 / 95		Thermoplastic**	Portable, Extra Hard Usage, and as permitted in Articles 520 and 530	

See Notes 1 through 9.

* See Note 8.

**The required outer covering on some single-conductor cables may be integral with the insulation.

Table 400-4. (Continued)

Trade Name	Type Letter	Size AWG	No. of Conductors	Insulation	AWG	Braid on Each Conductor	Outer Covering	Use
Hard Service Cord	SE See Note 4.	18-2	2 or more	Thermoplastic Elastomer	18-16 14-10 8-2 / 30 45 60	None	Thermoplastic Elastomer	Pendant or Portable / Damp Locations / Extra Hard Usage
	SEO See Note 4.							
	SEOO See Note 4.			Oil-Resistant Thermoplastic Elastomer			Oil-Resistant Thermoplastic Elastomer	
Junior Hard Service Cord	SJ	18-10	2,3,4, or 5	Thermoset	8-12 / 30	None	Thermoset	Pendant or Portable / Damp Locations / Hard Usage
	SJE			Thermoplastic Elastomer			Thermoplastic Elastomer	
	SJEO						Oil-Resistant Thermoplastic Elastomer	
	SJEOO			Oil-Resistant Thermoplastic Elastomer				
	SJO			Thermoset			Oil-Resistant Thermoset	

See Notes 1 through 9.
*See Note 8.

Table 400-4. (Continued)

Trade Name	Type Letter	Size AWG	No. of Conductors	Insulation	Nominal Insulation Thickness* AWG	Nominal Insulation Thickness* Mils	Braid on Each Conductor	Outer Covering	Use Pendant or Portable	Use Damp Locations	Use Extra Hard Usage
Junior Hard Service Cord	SJOO	18-10	2, 3, 4, or 5	Oil-Resistant Thermoset	8-12	30	None	Oil-Resistant Thermoset	Pendant or Portable	Damp Locations	Extra Hard Usage
	SJT			Thermoplastic or Thermoset	10	45		Thermoplastic			
	SJTO			Thermoset or Thermoplastic				Oil-Resistant Thermoplastic			
	SJTOO			Oil-Resistant Thermoplastic or Thermoset				Oil-Resistant Thermoplastic			
Hard Service Cord	SO See Note 4.	18-2	2 or more	Thermoset	18-16	30		Oil-Resistant Thermoset	Pendant or Portable	Damp Locations	Extra Hard Usage
	SOO See Note 4.			Oil-Resistant Thermoset	14-10	45		Oil-Resistant Thermoset			
					8-2	60					

See Notes 1 through 9.
* See Note 8.

Table 400-4. (Continued)

Trade Name	Type Letter	Size AWG	Number of Conductors	Insulation	AWG	Thickness	Braid	Outer Covering	Use		
All Thermoset Parallel Cord	SP-1 See Note 6.	18	2 or 3	Thermoset	18	30	None	Thermoset	Pendant or Portable	Damp Locations	Not Hard Usage
	SP-2 See Note 6.	18-16			18-16	45					
	SP-3 See Note 6.	18-10		Thermoset	18-16 14 12 10	60 80 95 110	None	Thermoset	Refrigerators, Room Air Conditioners, and as permitted in Section 422-8(d)	Damp Locations	Not Hard Usage
All Elastomer (Thermoplastic) Parallel Cord	SPE-1 See Note 6.	18	2 or 3	Thermoplastic Elastomer	18	30	None	Thermoplastic Elastomer	Pendant or Portable	Damp Locations	Not Hard Usage
	SPE-2 See Note 6.	18-16			18-16	45					
	SPE-3 See Note 6.	18-10		Thermoplastic Elastomer	18-16 14 12 10	60 80 95 110	None	Thermoplastic Elastomer	Refrigerators, Room Air Conditioners, and as permitted in Section 422-8(d)	Damp Locations	Not Hard Usage
All Plastic Parallel Cord	SPT-1 See Note 6.	18		Thermoplastic	18	30	None	Thermoplastic	Pendant or Portable	Damp Locations	Not Hard Usage

See Notes 1 through 9.
* See Note 8.

Table 400-4. (Continued)

Trade Name	Type Letter	Size AWG	No. of Conductors	Insulation	Nominal Insulation Thickness*		Braid on Each Conductor	Outer Covering	Use		
					AWG	Mils					
All Plastic Parallel Cord	SPT-2 See Note 6.	18-16	2 or 3	Thermoplastic	18-16	45	None	Thermoplastic	Pendant or Portable	Damp Locations	Not Hard Usage
	SPT-3 See Note 6.	18-10		Thermoplastic	18-16 14 12 10	60 80 95 110	None	Thermoplastic	Refrigerators, Room Air Conditioners, and as permitted in Section 422-8(d)	Damp Locations	Not Hard Usage
Range, Dryer Cable	SRD	10-4	3 or 4	Thermoset	10-4	45	None	Thermoset	Portable	Damp Locations	Ranges, Dryers
	SRDE	10-4	3 or 4	Thermoplastic Elastomer			None	Thermoplastic Elastomer	Portable	Damp Locations	Ranges, Dryers
	SRDT	10-4	3 or 4	Thermoplastic			None	Thermoplastic	Portable	Damp Locations	Ranges, Dryers
Hard Service Cord	ST See Note 4.	18-2	2 or more	Thermoplastic or Thermoset	18-16 14-10 8-2	30 45 60	None	Thermoplastic	Pendant or Portable	Damp Locations	Extra Hard Usage
	STO See Note 4.							Oil-Resistant Thermoplastic			
	STOO See Note 4.			Oil-Resistant Thermoplastic or Thermoset							

See Notes 1 through 9.
* See Note 8.

Table 400-4. (Continued)

Trade Name	Type Letter	Size AWG or kcmil	Number of Conductors	Insulation	Nominal Insulation Thickness AWG	Mils	Braid	Outer Covering	Use
Vacuum Cleaner Cord	SV See Note 6.	18-17	2 or 3	Thermoset	18-17	15	None	Thermoset	Pendant or Portable; Damp Locations; Not Hard Usage
	SVE See Note 6.			Thermoplastic Elastomer				Thermoplastic Elastomer	
	SVEO See Note 6.			Oil-Resistant Thermoplastic Elastomer				Oil-Resistant Thermoplastic Elastomer	
	SVEOO See Note 6.			Oil-Resistant Thermoplastic Elastomer				Oil-Resistant Thermoplastic Elastomer	
	SVO			Thermoset				Oil-Resistant Thermoset	
	SVOO			Oil-Resistant Thermoset				Oil-Resistant Thermoset	
	SVT See Note 6.			Thermoset or Thermoplastic				Thermoplastic	
	SVTO See Note 6.			Thermoset or Thermoplastic				Oil-Resistant Thermoplastic	
	SVTOO			Oil-Resistant Thermoplastic or Thermoset				Oil-Resistant Thermoplastic	
Parallel Tinsel Cord	TPT See Note 2.	27	2	Thermoplastic	27	30	None	Thermoplastic	Attached to an Appliance; Damp Locations; Not Hard Usage

See Notes 1 through 9.

* See Note 8.

Table 400-4. (Continued)

Trade Name	Type Letter	Size AWG	No. of Conductors	Insulation	Nominal Insulation Thickness*		Braid on Each Conductor	Outer Covering	Use		
					AWG	Mils					
Jacketed Tinsel Cord	TS See Note 2.	27	2	Thermoset	27	15	None	Thermoset	Attached to an Appliance	Damp Locations	Not Hard Usage
	TST See Note 2.	27	2	Thermoplastic			None	Thermoplastic	Attached to an Appliance	Damp Locations	Not Hard Usage
Portable Power Cable	W	8-500 kcmil	1-6	Thermoset	8-2 = 1-4/0 = 250 kcmil 500 kcmil	60 80 95		Oil-Resistant Thermoset	Portable, Extra Hard Usage, and as Permitted in Sections 520-68 and 530-12.		

See Notes 1 through 9.
* See Note 8.

shield for use in video or other radio frequency communication circuits. The optical fiber shall be suitably covered with flame-retardant thermoplastic. The insulation of the conductors shall be rubber or thermoplastic of thickness not less than specified for the other conductors of the particular type of cable. Metallic shields shall have their own protective covering. Where used, these components shall be permitted to be incorporated in any layer of the cable assembly but shall not run straight through the center.

6. The third conductor in these cables shall be used for equipment grounding purposes only.

7. The individual conductors of all cords, except those of heat-resistant cords, shall have a thermoset or thermoplastic insulation, except that the grounding conductor where used shall be in accordance with Section 400-23(b). Unvulcanized rubber compounds shall be permitted to be used for all sizes of heater cord Types HSJ and HSJO, and for sizes No. 18 and 16 Type HPD.

8. Where the voltage between any two conductors exceeds 300, but does not exceed 600, flexible cord of Nos. 10 and smaller shall have thermoset or thermoplastic insulation on the individual conductors at least 45 mils in thickness, unless Type S, SE, SEO, SEOO, SO, SOO, ST, STO, or STOO cord is used.

9. Insulations and outer coverings that meet the requirements as flame-retardant, limited-smoke, and are so listed, shall be permitted to be designated limited-smoke with the suffix /LS after the Code type designation.

400-5. Ampacities for Flexible Cords and Cables. Table 400-5(A) provides the allowable ampacities and Table 400-5(B) provides the ampacities for flexible cords and cables with not more than three current-carrying conductors. If the number of current-carrying conductors exceeds three, the allowable ampacity or the ampacity of each conductor shall be reduced from the 3-conductor rating as shown in the following table:

Number of Conductors	Percent of Value in Tables 400-5(A) and 400-5(B)
4 through 6	80
7 through 9	70
10 through 20	50
21 through 30	45
31 through 40	40
41 and above	35

Ultimate Insulation Temperature. In no case shall conductors be associated together in uch a way with respect to the kind of circuit, the wiring method used, or the number of conductors such that the limiting temperature of the conductors is exceeded.

A neutral conductor that carries only the unbalanced current from other conductors of the same circuit need not be considered as a current-carrying conductor.

In a 3-wire circuit consisting of two phase wires and the neutral of a 4-wire, 3-phase wye-connected system, a common conductor carries approximately the same current as the other conductors and shall be considered a current-carrying conductor.

On a 4-wire, 3-phase wye circuit where the major portion of the load consists of nonlinear loads such as electric-discharge lighting, electronic computer/data processing, or similar equipment, there are harmonic currents present in the neutral conductor, and the neutral shall be considered to be a current-carrying conductor.

An equipment grounding conductor shall not be considered a current-carrying conductor.

Where a single conductor is used for both equipment grounding and to carry unbalanced current from other conductors, as provided for in Section 250-60 for electric ranges and electric clothes dryers, it shall not be considered as a current-carrying conductor.

Exception: For other loading conditions, adjustment factors shall be permitted to be calculated under Section 310-15(b).

(FPN): See Appendix B, Table B-310-11 for adjustment factors for more than three current-carrying conductors in a raceway or cable with load diversity.

400-6. Markings.

(a) Standard Markings. Flexible cords and cables shall be marked by means of a printed tag attached to the coil reel or carton. The tag shall contain the information required in Section 310-11(a).

Types S, SC, SCE, SCT, SE, SEO, SEOO, SJ, SJE, SJEO, SJEOO, SJO, SJT, SJTO, SJTOO, SO, SOO, ST, STO, and STOO flexible cords and G and W flexible cables shall be durably marked on the surface at intervals not exceeding 24 inches (610 mm) with the type designation, size, and number of conductors.

(b) Optional Markings. Flexible cords and cable types listed in Table 400-4 shall be permitted to be surface marked to indicate special characteristics of the cable materials.

(FPN): Examples of these markings include, but are not limited to, "LS" for limited-smoke and markings such as "sunlight-resistant."

400-7. Uses Permitted.

(a) Uses. Flexible cords and cables shall be used only for (1) pendants; (2) wiring of fixtures; (3) connection of portable lamps or appliances; (4) elevator cables; (5) wiring of cranes and hoists; (6) connection of stationary equipment to facilitate their frequent interchange; (7) prevention of the transmission of noise or vibration; (8) appliances where the fastening means and mechanical connections are specifically designed to permit ready removal for maintenance and repair, and the appliance is intended or identified for flexible cord connection; (9) data processing cables as permitted by Section 645-5; (10) connection of moving parts; or (11) temporary wiring as permitted in Sections 305-4(b) and 305-4(c).

Table 400-5(A). Allowable Ampacity for Flexible Cords and Cables
[Based on Ambient Temperature of 30°C (86°F). See Section 400-13 and Table 400-4.]

Size AWG	Thermoset Type TS / Thermoplastic Types TPT, TST	Thermoset Types C, E, EO, PD, S, SJ, SJO, SJOO, SO, SOO, SP-1, SP-2, SP-3, SRD, SV, SVO, SVOO / Thermoplastic Types ET, ETLB, ETT, SE, SEO, SJE, SJEO, SJT, SJTO, SJTOO, SPE-1, SPE-2, SPE-3, SPT-1, SPT-2, SPT-3, ST, SRDE, SRDT, STO. STOO, SVE, SVEO, SVT, SVTO, SVTOO		Types AFS, AFSJ, HPD, HPN, HS, HSJ, HSJO, HSO
		A†	B†	
27*	0.5	..	..	..
20	..	5**	7**	..
18	..	7	10	10
17	..	..	12	..
16	..	10	13	15
15	..	..	..	17
14	..	15	18	20
12	..	20	25	30
10	..	25	30	35
8	..	35	40	..
6	..	45	55	..
4	..	60	70	..
2	..	80	95	..

* Tinsel cord.
** Elevator cables only.

† The allowable currents under subheading A apply to 3-conductor cords and other multiconductor cords connected to utilization equipment so that only 3 conductors are current-carrying. The allowable currents under subheading B apply to 2-conductor cords and other multiconductor cords connected to utilization equpment so that only 2 conductors are current-carrying.

(FPN): It is intended that this table be used in conjunction with applicable end-use product standards to ensure selection of the proper size and type.

(b) Attachment Plugs. Where used as permitted in subsections (a)(3), (a)(6), and (a)(8) of this section, each flexible cord shall be equipped with an attachment plug and shall be energized from a receptacle outlet.

Exception: As permitted in Section 364-8.

400-8. Uses Not Permitted. Unless specifically permitted in Section 400-7, flexible cords and cables shall not be used (1) as a substitute for the fixed wiring of a structure; (2) where run through holes in walls, ceilings, or floors; (3) where run through doorways, windows, or similar openings; (4) where attached to building surfaces; (5) where concealed behind building walls, ceilings, or floors; or (6) where installed in raceways, except as otherwise permitted in this Code.

Exception: Flexible cord and cable shall be permitted to have one connection to the building surface for a suitable tension take-up device. Length of the cord or cable from the supply termination to the take-up device shall be limited to 6 feet (1.83 m).

Table 400-5(B).   Ampacity of Cable Types SC, SCE, SCT, G, and W.
[Based on Ambient Temperature of 30°C (86°F). See Table 400-4.]
Temperature Rating of Cable

Size AWG/ kcmil	60°C (140°F)			75°C (167°F)			90°C (194°F)		
	D	E	F	D	E	F	D	E	F
8	60	55	48	70	65	57	80	74	65
6	80	72	63	95	88	77	105	99	87
4	105	96	84	125	115	101	140	130	114
3	120	113	99	145	135	118	165	152	133
2	140	128	112	170	152	133	190	174	152
1	165	150	131	195	178	156	220	202	177
1/0	195	173	151	230	207	181	260	234	205
2/0	225	199	174	265	238	208	300	271	237
3/0	260	230	201	310	275	241	350	313	274
4/0	300	265	232	360	317	277	405	361	316
250	340	296	259	405	354	310	455	402	352
300	375	330	289	445	395	346	505	449	393
350	420	363	318	505	435	381	570	495	433
400	455	392	343	545	469	410	615	535	468
500	515	448	392	620	537	470	700	613	536

The ampacities under subheading D are permitted for single-conductor Types SC, SCE, SCT, and W cable only where the individual conductors are not installed in raceways and are not in physical contact with each other except in lengths not to exceed 24 inches (610 mm) where passing through the wall of an enclosure.

The ampacities under subheading E apply to 2-conductor cables and other multiconductor cables connected to utilization equipment so that only 2 conductors are current-carrying. The ampacities under subheading F apply to 3-conductor cables and other multiconductor cables connected to utilization equipment so that only 3 conductors are current-carrying.

400-9. Splices.   Flexible cord shall be used only in continuous lengths without splice or tap when initially installed in applications permitted by Section 400-7(a). The repair of hard-service cord (see Column 1, Table 400-4) Nos. 14 and larger shall be permitted if conductors are spliced in accordance with Section 110-14(b) and the completed splice retains the insulation, outer sheath properties, and usage characteristics of the cord being spliced.

400-10. Pull at Joints and Terminals.   Flexible cords shall be so connected to devices and to fittings that tension will not be transmitted to joints or terminals.

(FPN):   Some methods of preventing pull on a cord from being transmitted to joints or terminals are (1) knotting the cord, (2) winding with tape, and (3) fittings designed for the purpose.

400-11. In Show Windows and Show Cases.   Flexible cords used in show windows and show cases shall be Type AFS, S, SE, SEO, SEOO, SJ, SJE, SJEO, SJEOO, SJO, SJOO, SJT, SJTO, SJTOO, SO, SOO, ST, STO, or STOO.

Exception No. 1: For the wiring of chain-supported lighting fixtures.

Exception No. 2: As supply cords for portable lamps and other merchandise being displayed or exhibited.

400-12. Minimum Size.   The individual conductors of a flexible cord or cable shall not be smaller than the sizes in Table 400-4.

400-13. Overcurrent Protection.   Flexible cords not smaller than No. 18, and tinsel cords or cords having equivalent characteristics of smaller size approved for use with specific appliances, shall be considered as protected against overcurrent by the overcurrent devices described in Section 240-4.

400-14. Protection from Damage.   Flexible cords and cables shall be protected by bushings or fittings where passing through holes in covers, outlet boxes, or similar enclosures.

B. Construction Specifications

400-20. Labels.   Flexible cords shall be examined and tested at the factory and labeled before shipment.

400-21. Nominal Insulation Thickness.   The nominal thickness of insulation for conductors of flexible cords and cables shall not be less than specified in Table 400-4.

400-22. Grounded-Conductor Identification.   One conductor of flexible cords that is intended to be used as a grounded circuit conductor shall have a continuous marker readily distinguishing it from the other conductor or conductors. The identification shall consist of one of the methods indicated in (a) through (f) below.

(a) Colored Braid.   A braid finished to show a white or natural gray color and the braid on the other conductor or conductors finished to show a readily distinguishable solid color or colors.

(b) Tracer in Braid.   A tracer in a braid of any color contrasting with that of the braid and no tracer in the braid of the other conductor or conductors. No tracer shall be used in the braid of any conductor of a flexible cord that contains a conductor having a braid finished to show white or natural gray.

Exception: In the case of Types C and PD, and cords having the braids on the individual conductors finished to show white or natural gray. In such cords, the identifying marker shall be permitted to consist of the solid white or natural gray finish on one conductor, provided there is a colored tracer in the braid of each other conductor.

(c) Colored Insulation. A white or natural gray insulation on one conductor and insulation of a readily distinguishable color or colors on the other conductor or conductors for cords having no braids on the individual conductors.

For jacketed cords furnished with appliances, one conductor having its insulation colored light blue, with the other conductors having their insulation of a readily distinguishable color other than white or natural gray.

Exception: Cords that have insulation on the individual conductors integral with the jacket.

It shall be permissible to cover the insulation with an outer finish to provide the desired color.

(d) Colored Separator. A white or natural gray separator on one conductor and a separator of a readily distinguishable solid color on the other conductor or conductors of cords having insulation on the individual conductors integral with the jacket.

(e) Tinned Conductors. One conductor having the individual strands tinned and the other conductor or conductors having the individual strands untinned for cords having insulation on the individual conductors integral with the jacket.

(f) Surface Marking. One or more stripes, ridges, or grooves so located on the exterior of the cord as to identify one conductor for cords having insulation on the individual conductors integral with the jacket.

400-23. Equipment Grounding-Conductor Identification. A conductor intended to be used as an equipment grounding conductor shall have a continuous identifying marker readily distinguishing it from the other conductor or conductors. Conductors having a continuous green color or a continuous green color with one or more yellow stripes shall not be used for other than equipment grounding purposes. The identifying marker shall consist of one of the methods in (a) or (b) below.

(a) Colored Braid. A braid finished to show a continuous green color or a continuous green color with one or more yellow stripes.

(b) Colored Insulation or Covering. For cords having no braids on the individual conductors, an insulation of a continuous green color or a continuous green color with one or more yellow stripes.

400-24. Attachment Plugs. Where a flexible cord is provided with an equipment grounding conductor and equipped with an attachment plug, the attachment plug shall comply with Sections 250-59(a) and (b).

C. Portable Cables Over 600 Volts, Nominal

400-30. Scope. This part applies to multiconductor portable cables used to connect mobile equipment and machinery.

400-31. Construction.

(a) Conductors. The conductors shall be No. 8 copper or larger and shall employ flexible stranding.

(b) Shields. Cables operated at over 2000 volts shall be shielded. Shielding shall be for the purpose of confining the voltage stresses to the insulation.

(c) Equipment Grounding Conductor(s). Equipment grounding conductor(s) shall be provided. The total area shall not be less than that of the size of the equipment grounding conductor required in Section 250-95.

400-32. Shielding. All shields shall be grounded.

400-33. Grounding. Grounding conductors shall be connected in accordance with Part K of Article 250.

400-34. Minimum Bending Radii. The minimum bending radii for portable cables during installation and handling in service shall be adequate to prevent damage to the cable.

400-35. Fittings. Connectors used to connect lengths of cable in a run shall be of a type that lock firmly together. Provisions shall be made to prevent opening or closing these connectors while energized. Suitable means shall be used to eliminate tension at connectors and terminations.

400-36. Splices and Terminations. Portable cables shall not contain splices unless the splices are of the permanent molded, vulcanized types in accordance with Section 110-14(b). Terminations on portable cables rated over 600 volts, nominal, shall be accessible only to authorized and qualified personnel.

ARTICLE 402 — FIXTURE WIRES

402-1. Scope. This article covers general requirements and construction specifications for fixture wires.

402-2. Other Articles. Fixture wires shall comply with this article and also with the applicable provisions of other articles of this Code.

(FPN): For application in lighting fixtures, see Article 410.

402-3. Types. Fixture wires shall be of a type listed in Table 402-3, and they shall comply with all requirements of that table. The fixture wires listed in Table 402-3 are all suitable for service at 600 volts, nominal, unless otherwise specified.

(FPN): Thermoplastic insulation may stiffen at temperatures colder than minus 10°C (plus 14°F), requiring care be exercised during installation at such temperatures. Thermoplastic insulation may also be deformed at normal temperatures where subjected to pressure, requiring care be exercised during installation and at points of support.

402-5. Allowable Ampacities for Fixture Wires. The ampacity of fixture wire shall be as specified in Table 402-5.

No conductor shall be used under such conditions that its operating temperature will exceed the temperature specified in Table 402-3 for the type of insulation involved.

(FPN): See Section 310-10 for temperature limitation of conductors.

Table 402-3. Fixture Wire

Trade Name	Type Letter	Insulation	AWG	Thickness of Moisture-Resistant Insulation Mils	Thickness of Asbestos Mils	Outer Covering	Max. Operating Temp.	Application Provisions
Asbestos Covered Heat-ered Heat-Resistant Fixture Wire	AF	Impregnated Asbestos or Moisture-Resistant Insulation and Impregnated Asbestos	18-14 — 12-10	— / 20 — / 25	30 / 10 45 / 20	None	150°C 302°F	Fixture wiring. Limited to 300 volts and indoor dry locations.
Heat-Resistant Rubber-Covered Fixture Wire Flexible Stranding	FFH-2	Heat-Resistant Rubber Heat-Resistant Latex Rubber	18-16 18-16		30 18	Nonmetallic Covering	75°C 167°F	Fixture wiring, and as permitted in Section 725-16.
ECTFE Solid or 7-Strand	HF	Ethylene Chloro-Trifluoro-Ethylene	18-14		15	None	150°C 302°F	Fixture wiring, and as permitted in Section 725-16
ECTFE Flexible Stranding	HFF	Ethylene Chloro-Trifluoro-Ethylene	18-14		15	None	150°C 302°F	Fixture wiring, and as permitted in Section 725-16.

Table 402-3 (Continued)

Trade Name	Type Letter	Insulation	AWG	Thickness of Insulation	Mils	Outer Covering	Max. Operating Temp.	Application Provisions
Tape Insulated Fixture Wire—Solid or 7-Strand	KF-1	Aromatic Polyimide Tape	18-10	....	5.5	None	200°C 392°F	Fixture Wiring. Limited to 300 volts.
	KF-2	Aromatic Polyimide Tape	18-10	....	8.4	None	200°C 392°F	Fixture wiring, and as permitted in Sections 725-16 and 760-16.
Tape Insulated Fixture Wire—Flexible Stranding	KFF-1	Aromatic Polyimide Tape	18-10	....	5.5	None	200°C 392°F	Fixture Wiring. Limited to 300 volts.
	KFF-2	Aromatic Polyimide Tape	18-10	....	8.4	None	200°C 392°F	Fixture wiring, and as permitted in Section 725-16.
Perfluoro-alkoxy—Solid or 7-Strand (Nickel or Nickel-Coated Copper)	PAF	Perfluoro-alkoxy	18-14	....	20	None	250°C 482°F	Fixture wiring, and as permitted in Sections 725-16 and 760-16. (Nickel or nickel-coated copper)
Perfluoro-alkoxy—Flexible Stranding	PAFF	Perfluoro-alkoxy	18-14	....	20	None	150°C 302°F	Fixture wiring, and as permitted in Section 725-16.
Fluorinated Ethylene Propylene Fixture Wire—Solid or 7-Strand	PF	Fluorinated Ethylene Propylene	18-14	....	20	None	200°C 392°F	Fixture wiring, and as permitted in Sections 725-16 and 760-16.

Table 402-3 (Continued)

Trade Name	Type Letter	Insulation	AWG	Thickness of Insulation	Mils	Outer Covering	Max. Operating Temp.	Application Provisions
Fluorinated Ethylene Propylene Fixture Wire—Flexible Stranding	PFF	Fluorinated Ethylene Propylene	18-14	....	20	None	150°C 302°F	Fixture wiring, and as permitted in Section 725-16.
Fluorinated Ethylene Propylene Fixture Wire—Solid or 7-Strand	PGF	Fluorinated Ethylene Propylene	18-14	....	14	Glass Braid	200°C 392°F	Fixture wiring, and as permitted in Sections 725-16 and 760-16.
Fluorinated Ethylene Propylene Fixture Wire—Flexible Stranding	PGFF	Fluorinated Ethylene Propylene	18-14	....	14	Glass Braid	150°C 302°F	Fixture wiring, and as permitted in Section 725-16.
Extruded Polytetra-Fluoroethylene Solid or 7-Strand (Nickel or Nickel-Coated Copper)	PTF	Extruded Polytetra-Fluoro-ethylene	18-14	....	20	None	250°C 482°F	Fixture wiring, and as permitted in Sections 725-16 and 760-16. (Nickel or nickel-coated copper)

Table 402-3 (Continued)

Trade Name	Type Letter	Insulation	AWG	Thickness of Insulation	Mils	Outer Covering	Max. Operating Temp.	Application Provisions
Extruded Polytetra-Fluoroethylene Flexible Stranding 26-36 AWG Silver or Nickel-Coated Copper	PTFF	Extruded Polytetra-Fluoroethylene	18-14	....	20	None	150°C 302°F	Fixture wiring, and as permitted in Section 725-16. (Silver or nickel-coated copper)
Heat-Resistant Rubber-Covered Fixture Wire—Solid or 7-Strand	RFH-1	Heat-Resistant Rubber	18	....	15	Nonmetallic Covering	75°C 167°F	Fixture wiring. Limited to 300 volts.
	RFH-2	Heat-Resistant Rubber	18-16	....	30	Nonmetallic Covering	75°C 167°F	Fixture wiring, and as permitted in Sections 725-16 and 760-16.
		Heat-Resistant Latex Rubber	18-16	....	18			
Heat-Resistant Cross-Linked Synthetic Polymer-Insulated Fixture Wire—Solid or Stranded	RFHH-2*	Cross-Linked Synthetic Polymer	18-16	....	30			
	—		18-16	....	45			Fixture wiring, and as permitted in Sections 725-16 and 760-16. Multiconductor cable as permitted in Sections 725-16 and 760-16.
	RFHH-3*					None or Nonmetallic Covering	90°C 194°F	

* Insulations and outer coverings that meet the requirements of flame-retardant, limited-smoke and are so listed shall be permitted to be designated limited-smoke with the suffix /LS after the Code type designation.

Table 402-3 (Continued)

Trade Name	Type Letter	Insulation	AWG	Thickness of Insulation	Mils	Outer Covering	Max. Operating Temp.	Application Provisions
Silicone Insulated Fixture Wire—Solid or 7-Strand	SF-1	Silicone Rubber	18	....	15	Nonmetallic Covering	200°C 392°F	Fixture wiring. Limited to 300 volts.
	SF-2	Silicone Rubber	18-14	....	30	Nonmetallic Covering	200°C 392°F	Fixture wiring, and as permitted in Sections 725-16 and 760-16.
Silicone Insulated Fixture Wire—Flexible Stranding	SFF-1	Silicone Rubber	18	....	15	Nonmetallic Covering	150°C 302°F	Fixture wiring. Limited to 300 volts.
	SFF-2	Silicone Rubber	18-14	....	30	Nonmetallic Covering	150°C 302°F	Fixture wiring, and as permitted in Section 725-16.
Thermoplastic Covered Fixture Wire—Solid or 7-Strand	TF*	Thermoplastic	18-16	....	30	None	60°C 140°F	Fixture wiring, and as permitted in Sections 725-16 and 760-16.
Thermoplastic Covered Fixture Wire—Flexible Stranding	TFF*	Thermoplastic	18-16	....	30	None	60°C 140°F	Fixture wiring, and as permitted in Section 725-16.
Heat-Resistant Thermoplastic-Covered Fixture Wire—Solid or 7-Strand	TFN*	Thermoplastic	18-16	....	15	Nylon-Jacketed or equivalent	90°C 194°F	Fixture wiring, and as permitted in Sections 725-16 and 760-16.

*Insulations and outer coverings that meet the requirements of flame-retardant limited-smoke and are so listed shall be permitted to be designated limited-smoke with the suffix /LS after the Code type designation.

Table 402-3 (Continued)

Trade Name	Type Letter	Insulation	AWG	Thickness of Insulation	Mils	Outer Covering	Max. Operating Temp.	Application Provisions
Heat-Resistant Thermoplastic Covered Fixture Wire—Flexible Stranded	TFFN*	Thermoplastic	18-16	....	15	Nylon-Jacketed or equivalent	90°C 194°F	Fixture wiring, and as permitted in Section 725-16.
Cross-Linked Polyolefin Insulated Fixture Wire—Solid or 7-Strand	XF*	Cross-Linked Polyolefin	18-14 12-10	....	30 45	None	150°C 302°F	Fixture wiring. Limited to 300 volts.
Cross-Linked Polyolefin Insulated Fixture Wire—Flexible Stranded	XFF*	Cross-Linked Polyolefin	18-14 12-10	....	30 45	None	150°C 302°F	Fixture wiring. Limited to 300 volts.
Modified ETFE Solid or 7-Strand	ZF	Modified Ethylene Tetrafluoro-Ethylene	18-14	....	15	None	150°C 302°F	Fixture wiring, and as permitted in Sections 725-16 and 760-16.
Flexible Stranding	ZFF	Modified Ethylene Tetrafluoro-Ethylene	18-14	....	15	None	150°C 302°F	Fixture wiring, and as permitted in Section 725-16.
High Temp. Modified ETFE—Solid or 7-Strand	ZHF	Modified Ethylene Tetrafluoro-ethylene	18-14	....	15	None	200°C 392°F	Fixture wiring, and as permitted in Sections 725-16 and 760-16

* Insulations and outer coverings that meet the requirements of flame-retardant, limited-smoke and are so listed shall be permitted to be designated limited-smoke with the suffix /LS after the Code type designation.

Table 402-5.

Size (AWG)	Allowable Ampacity
18	6
16	8
14	17
12	23
10	28

402-6. Minimum Size. Fixture wires shall not be smaller than No. 18.

402-7. Number of Conductors in Conduit or Tubing. The number of fixture wires permitted in a single conduit or tubing shall be as shown in Table 2 of Chapter 9.

402-8. Grounded-Conductor Identification. One conductor of fixture wires that is intended to be used as a grounded conductor shall be identified by means of stripes or by the means described in Sections 400-22(a) through (e).

402-9. Marking.

(a) **Required Information.** All fixture wires shall be marked to indicate the information required in Section 310-11(a).

(b) **Method of Marking.** Thermoplastic-insulated fixture wire shall be durably marked on the surface at intervals not exceeding 24 inches (610 mm). All other fixture wire shall be marked by means of a printed tag attached to the coil, reel, or carton.

(c) **Optional Marking.** Fixture wire types listed in Table 402-3 shall be permitted to be surface marked to indicate special characteristics of the cable materials.

(FPN): Examples of these markings include, but are not limited to, "LS" for limited-smoke or markings such as "sunlight-resistant."

402-10. Uses Permitted. Fixture wires shall be permitted (1) for installation in lighting fixtures and in similar equipment where enclosed or protected and not subject to bending or twisting in use, or (2) for connecting lighting fixtures to the branch-circuit conductors supplying the fixtures.

402-11. Uses Not Permitted. Fixture wires shall not be used as branch-circuit conductors.

Exception: As permitted by Section 725-16 for Class 1 circuits and Section 760-16 for fire protective signaling circuits.

402-12. Overcurrent Protection. Overcurrent protection for fixture wires shall be as specified in Section 240-4.

ARTICLE 410 — LIGHTING FIXTURES, LAMPHOLDERS, LAMPS, AND RECEPTACLES

A. General

410-1. Scope. This article covers lighting fixtures, lampholders, pendants, receptacles, incandescent filament lamps, arc lamps, electric-discharge lamps, the wiring and equipment forming part of such lamps, fixtures, and lighting installations.

410-2. Application to Other Articles. Equipment for use in hazardous (classified) locations shall conform to Articles 500 through 517.

410-3. Live Parts. Fixtures, lampholders, lamps, and receptacles shall have no live parts normally exposed to contact. Exposed accessible terminals in lampholders, receptacles, and switches shall not be installed in metal fixture canopies or in open bases of portable table or floor lamps.

Exception: Cleat-type lampholders and receptacles located at least 8 feet (2.44 m) above the floor shall be permitted to have exposed contacts.

B. Fixture Locations

410-4. Fixtures in Specific Locations.

(a) Wet and Damp Locations. Fixtures installed in wet or damp locations shall be so installed that water cannot enter or accumulate in wiring compartments, lampholders, or other electrical parts. All fixtures installed in wet locations shall be marked, "Suitable for Wet Locations." All fixtures installed in damp locations shall be marked, "Suitable for Wet Locations" or "Suitable for Damp Locations."

Installations underground or in concrete slabs or masonry in direct contact with the earth, and locations subject to saturation with water or other liquids, such as locations exposed to weather and unprotected, vehicle washing areas, and like locations, shall be considered to be wet locations with respect to the above requirement.

Interior locations protected from weather but subject to moderate degrees of moisture, such as some basements, some barns, some cold-storage warehouses and the like, the partially protected locations under canopies, marquees, roofed open porches, and the like, shall be considered to be damp locations with respect to the above requirement.

(FPN): See Article 680 for lighting fixtures in swimming pools, fountains, and similar installations.

(b) Corrosive Locations. Fixtures installed in corrosive locations shall be of a type suitable for such locations.

(FPN): See Section 210-7 for receptacles in fixtures.

(c) In Ducts or Hoods. Fixtures shall be permitted to be installed in cooking hoods of nonresidential occupancies where all of the following conditions are met:

(1) The fixture shall be identified for use within commercial cooking hoods and installed so that the temperature limits of the materials used are not exceeded.

(2) The fixture shall be so constructed that all exhaust vapors, grease, oil or cooking vapors are excluded from the lamp and wiring compartment. Diffusers shall be resistant to thermal shock.

(3) Parts of the fixture exposed within the hood shall be corrosion-resistant or protected against corrosion, and the surface shall be smooth so as not to collect deposits and facilitate cleaning.

(4) Wiring methods and materials supplying the fixture(s) shall not be exposed within the cooking hood.

(FPN): See Section 110-11 for conductors and equipment exposed to deteriorating agents.

(d) Pendants. No parts of cord-connected fixtures, hanging fixtures, or pendants shall be located within a zone measured 3 feet (914 mm) horizontally and 8 feet (2.44 m) vertically from the top of the bathtub rim. This zone is all encompassing and includes the zone directly over the tub.

410-5. Fixtures Near Combustible Material. Fixtures shall be so constructed, or installed, or equipped with shades or guards that combustible material will not be subjected to temperatures in excess of 90°C (194°F).

410-6. Fixtures Over Combustible Material. Lampholders installed over highly combustible material shall be of the unswitched type. Unless an individual switch is provided for each fixture, lampholders shall be located at least 8 feet (2.44 m) above the floor, or shall be so located or guarded that the lamps cannot be readily removed or damaged.

410-7. Fixtures in Show Windows. Externally wired fixtures shall not be used in a show window.

Exception: Fixtures of the chain-supported type shall be permitted to be externally wired.

410-8. Fixtures in Clothes Closets.

(a) Definition.

Storage Space: Storage space shall be defined as a volume bounded by the sides and back closet walls and planes extending from the closet floor vertically to a height of 6 feet (1.83 m) or the highest clothes-hanging rod and parallel to the walls at a horizontal distance of 24 inches (610 mm) from the sides and back of the closet walls respectively, and continuing vertically to the closet ceiling parallel to the walls at a horizontal distance of 12 inches (305 mm) or the width of the shelf, whichever is greater.

(FPN): See Figure 410-8.

For a closet that permits access to both sides of a hanging rod, the storage space shall include the volume below the highest rod extending 12 inches (305 mm) on either side of the rod on a plane horizontal to the floor extending the entire length of the rod.

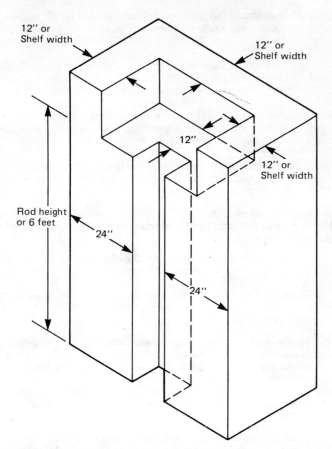

For SI units: one inch = 25.4 millimeters; one foot = 0.3048 meter.

Figure 410-8. Closet Storage Space.

(b) **Fixture Types Permitted.** Listed fixtures of the following types shall be permitted to be installed in a closet:

(1) A surface-mounted or recessed incandescent fixture with a completely enclosed lamp.

(2) A surface-mounted or recessed fluorescent fixture.

(c) **Fixture Types Not Permitted.** Incandescent fixtures with open or partially enclosed lamps and pendant fixtures or lampholders shall not be permitted.

(d) **Location.** Fixtures in clothes closets shall be permitted to be installed as follows:

(1) Surface-mounted incandescent fixtures installed on the wall above the door or on the ceiling, provided there is a minimum clearance of 12 inches (305 mm) between the fixture and the nearest point of a storage space.

(2) Surface-mounted fluorescent fixtures installed on the wall above the door or on the ceiling, provided there is a minimum clearance of 6 inches (152 mm) between the fixture and the nearest point of a storage space.

(3) Recessed incandescent fixtures with a completely enclosed lamp installed in the wall or the ceiling, provided there is a minimum clearance of 6 inches (152 mm) between the fixture and the nearest point of a storage space.

(4) Recessed fluorescent fixtures installed in the wall or on the ceiling, provided there is a minimum clearance of 6 inches (152 mm) between the fixture and the nearest point of a storage space.

410-9. Space for Cove Lighting. Coves shall have adequate space and shall be so located that lamps and equipment can be properly installed and maintained.

C. Provisions at Fixture Outlet Boxes, Canopies, and Pans

410-10. Space for Conductors. Canopies and outlet boxes taken together shall provide adequate space so that fixture conductors and their connecting devices can be properly installed.

410-11. Temperature Limit of Conductors in Outlet Boxes. Fixtures shall be of such construction or so installed that the conductors in outlet boxes shall not be subjected to temperatures greater than that for which the conductors are rated.

Branch-circuit wiring shall not be passed through an outlet box that is an integral part of an incandescent fixture unless the fixture is identified for through-wiring.

410-12. Outlet Boxes to Be Covered. In a completed installation, each outlet box shall be provided with a cover unless covered by means of a fixture canopy, lampholder, receptacle, or similar device.

410-13. Covering of Combustible Material at Outlet Boxes. Any combustible wall or ceiling finish exposed between the edge of a fixture canopy or pan and an outlet box shall be covered with noncombustible material.

410-14. Connection of Electric-Discharge Lighting Fixtures.

(a) Independently of the Outlet Box. Where electric-discharge lighting fixtures are supported independently of the outlet box, they shall be connected through metal raceways, nonmetallic raceways, Type MC cable, Type AC cable, Type MI cable, or nonmetallic-sheathed cables.

Exception: Cord-connected fixtures shall be permitted as provided in Sections 410-30(b) and (c).

(b) Access to Boxes. Electric-discharge lighting fixtures surface-mounted over concealed outlet, pull, or junction boxes shall be installed with suitable openings in back of the fixture to provide access to the boxes.

D. Fixture Supports

410-15. Supports.

(a) General. Fixtures, lampholders, and receptacles shall be securely supported. A fixture that weighs more than 6 pounds (2.72 kg) or exceeds 16 inches (406 mm) in any dimension shall not be supported by the screw shell of a lampholder.

(b) Metal Poles Supporting Lighting Fixtures. Metal poles shall be permitted to be used to support lighting fixtures and enclose supply conductors, provided that the following conditions are met:

(1) An accessible handhole, not less than 2 inches (50.8 mm) × 4 inches (102 mm), having a raintight cover, shall provide access to the supply raceway or cable termination within the pole or pole base. Where raceway risers or cable is not installed within the pole, a threaded fitting or nipple shall be brazed or welded to the pole opposite the handhole for the supply connection. Other poles shall be permitted to be field-welded, brazed, or tapped. Such poles shall be capped or covered.

Exception: It shall be permitted to omit the handhole required in (b)(1) above on a metal pole 20 feet (6.10 m) or less in height above grade if the pole is provided with a hinged base. The grounding terminal shall be accessible and be within the hinged base. Both parts of the hinged pole shall be bonded in accordance with Section 250-75.

(2) A terminal for grounding the pole shall be provided; it shall be accessible from the handhole.

Exception: It shall be permitted to omit the handhole and grounding terminal required in (b)(1) and (b)(2) above where the supply wiring method continues without splice or pull point to a fixture mounted on a metal pole 8 feet (2.44 m) or less in height above grade, and where the interior of the pole and any splices are accessible by the removal of the fixture.

(3) Metal raceways or other equipment grounding conductors shall be bonded to the pole with an equipment grounding conductor recognized by Section 250-91(b) and sized in accordance with Section 250-95.

(4) Conductors in vertical metal poles used as raceways shall be supported as provided in Section 300-19.

410-16. Means of Support.

(a) Outlet Boxes. Where the outlet box or fitting will provide adequate support, a fixture shall be attached thereto or be supported as required by Section 370-23 for boxes. A fixture that weighs more than 50 pounds (22.7 kg) shall be supported independently of the outlet box.

(b) Inspection. Fixtures shall be so installed that the connections between the fixture conductors and the circuit conductors can be inspected without requiring the disconnection of any part of the wiring.

Exception: Fixtures connected by attachment plugs and receptacles.

(c) Suspended Ceilings. Framing members of suspended ceiling systems used to support fixtures shall be securely fastened to each other and shall be securely attached to the building structure at appropriate intervals. Fixtures shall be securely fastened to the ceiling framing member by mechanical means, such as bolts, screws, or rivets. Clips identified for use with the type of ceiling framing member(s) and fixture(s) shall also be permitted.

(d) Fixture Studs. Fixture studs that are not a part of outlet boxes, hickeys, tripods, and crowfeet shall be made of steel, malleable iron, or other material suitable for the application.

(e) Insulating Joints. Insulating joints that are not designed to be mounted with screws or bolts shall have an exterior metal casing, insulated from both screw connections.

(f) Raceway Fittings. Raceway fittings used to support lighting fixture(s) shall be capable of supporting the weight of the complete fixture assembly and lamp(s).

(g) Busways. Fixtures shall be permitted to be connected to busways in accordance with Section 364-12.

(h) Trees. Outdoor lighting fixtures and associated equipment shall be permitted to be supported by trees.

(FPN No. 1): See Section 225-26.

(FPN No. 2): See Section 300-5(d).

E. Grounding

410-17. General. Fixtures and lighting equipment shall be grounded as provided in Part E of this article.

410-18. Exposed Fixture Parts.

(a) With Exposed Conductive Parts. The exposed conductive parts of lighting fixtures and equipment directly wired or attached to outlets supplied by a wiring method that provides an equipment ground shall be grounded.

(b) Made of Insulating Material. Fixtures directly wired or attached to outlets supplied by a wiring method that does not provide a ready means for grounding shall be made of insulating material and shall have no exposed conductive parts.

410-19. Equipment Over 150 Volts to Ground.

(a) Metal Fixtures, Transformers, and Transformer Enclosures. Metal fixtures, transformers, and transformer enclosures on circuits operating at over 150 volts to ground shall be grounded.

(b) Other Exposed Metal Parts. Other exposed metal parts shall be grounded or insulated from ground and other conducting surfaces and inaccessible to unqualified persons.

Exception: Lamp tie wires, mounting screws, clips, and decorative bands on glass lamps spaced not less than $1^{1}/_{2}$ inches (38 mm) from lamp terminals shall not be required to be grounded.

410-20. Equipment Grounding Conductor Attachment. Fixtures with exposed metal parts shall be provided with a means for connecting an equipment grounding conductor for such fixtures.

410-21. Methods of Grounding. Fixtures and equipment shall be considered grounded where mechanically connected to an equipment grounding conductor as specified in Section 250-91(b) and sized in accordance with Section 250-95.

F. Wiring of Fixtures

410-22. Fixture Wiring — General. Wiring on or within fixtures shall be neatly arranged and shall not be exposed to physical damage. Excess wiring shall be avoided. Conductors shall be so arranged that they shall not be subjected to temperatures above those for which they are rated.

410-23. Polarization of Fixtures. Fixtures shall be so wired that the screw shells of lampholders will be connected to the same fixture or circuit conductor or terminal. The grounded conductor, where connected to a screw-shell lampholder, shall be connected to the screw shell.

410-24. Conductors.

(a) Insulation. Fixtures shall be wired with conductors having insulation suitable for the environmental conditions, current, voltage, and temperature to which the conductors will be subjected.

(b) Conductor Size. Fixture conductors shall not be smaller than No. 18.

(FPN No. 1):   For ampacity of fixture wire, see Section 402-5.

(FPN No. 2):   For maximum operating temperature and voltage limitation of fixture wires, see Section 402-3.

410-25. Conductors for Certain Conditions.

(a) Mogul-Base Lampholders. Fixtures provided with mogul-base, screw-shell lampholders and operating at not over 300 volts between conductors shall be wired with Type AF, SF-1, SF-2, SFF-1, SFF-2, PF, PGF, PFF, PGFF, PTF, PTFF, PAF, PAFF, XF, XFF, ZF, or ZFF fixture wire.

(b) Other than Mogul-Base, Screw-Shell Lampholders. Fixtures provided with other than mogul-base, screw-shell lampholders and operating at not over 300 volts between conductors shall be wired with Type AF, SF-1, SF-2, PF, PGF, PFF, PGFF, PTF, PTFF, PAF, PAFF, XF, XFF, ZF, or ZFF fixture wire or Type AFC or AFPD flexible cord.

Exception No. 1: Where temperatures do not exceed 90°C (194°F), Types TFN and TFFN fixture wire shall be permitted.

Exception No. 2: Where temperatures exceed 60°C (140°F) but are not higher than 75°C (167°F), Types RH and RHW rubber-covered wire and Types RFH-1, RFH-2, and FFH-2 fixture wire shall be permitted.

Exception No. 3: Where temperatures do not exceed 60°C (140°F), Type TW thermoplastic wire, Types TF and TFF fixture wire shall be permitted, including fixtures of decorative types on which lamps of not over 60-watt rating are used in connection with imitation candles.

(FPN): See Table 402-3 and Section 402-3 for fixture wires and conductors; and Table 400-5(A) for flexible cords.

410-27. Pendant Conductors for Incandescent Filament Lamps.

(a) **Support.** Pendant lampholders with permanently attached leads, where used for other than festoon wiring, shall be hung from separate stranded rubber-covered conductors that are soldered directly to the circuit conductors but supported independently thereof.

(b) **Size.** Such pendant conductors shall not be smaller than No. 14 for mogul-base or medium-base screw-shell lampholders, nor smaller than No. 18 for intermediate or candelabra-base lampholders.

Exception: Listed Christmas tree and decorative lighting outfits shall be permitted to be smaller than No. 18.

(c) **Twisted or Cabled.** Pendant conductors longer than 3 feet (914 mm) shall be twisted together where not cabled in a listed assembly.

410-28. Protection of Conductors and Insulation.

(a) **Properly Secured.** Conductors shall be secured in a manner that will not tend to cut or abrade the insulation.

(b) **Protection Through Metal.** Conductor insulation shall be protected from abrasion where it passes through metal.

(c) **Fixture Stems.** Splices and taps shall not be located within fixture arms or stems.

(d) **Splices and Taps.** No unnecessary splices or taps shall be made within or on a fixture.

(FPN): For approved means of making connections, see Section 110-14.

(e) **Stranding.** Stranded conductors shall be used for wiring on fixture chains and on other movable or flexible parts.

(f) **Tension.** Conductors shall be so arranged that the weight of the fixture or movable parts will not put a tension on the conductors.

410-29. Cord-Connected Showcases.

Individual showcases, other than fixed, shall be permitted to be connected by flexible cord to permanently installed receptacles, and groups of not more than six such showcases shall be permitted to be coupled together by flexible cord and separable locking-type connectors with one of the group connected by flexible cord to a permanently installed receptacle.

The installation shall comply with (a) through (e) of this section.

(a) **Cord Requirements.** Flexible cord shall be hard-service type, having conductors not smaller than the branch-circuit conductors, having ampacity at least equal to the branch-circuit overcurrent device, and having an equipment grounding conductor.

(FPN): See Table 250-95 for size of equipment grounding conductor.

(b) Receptacles, Connectors, and Attachment Plugs. Receptacles, connectors, and attachment plugs shall be of a listed grounding type rated 15 or 20 amperes.

(c) Support. Flexible cords shall be secured to the undersides of showcases so that (1) wiring will not be exposed to mechanical damage; (2) a separation between cases not in excess of 2 inches (50.8 mm), nor more than 12 inches (305 mm) between the first case and the supply receptacle, will be assured; and (3) the free lead at the end of a group of showcases will have a female fitting not extending beyond the case.

(d) No Other Equipment. Equipment other than showcases shall not be electrically connected to showcases.

(e) Secondary Circuit(s). Where showcases are cord-connected, the secondary circuit(s) of each electric-discharge lighting ballast shall be limited to one showcase.

410-30. Cord-Connected Lampholders and Fixtures.

(a) Lampholders. Where a metal lampholder is attached to a flexible cord, the inlet shall be equipped with an insulating bushing that, if threaded, shall not be smaller than nominal $\frac{3}{8}$-inch pipe size. The cord hole shall be of a size appropriate for the cord, and all burrs and fins shall be removed in order to provide a smooth bearing surface for the cord.

Bushing having holes $\frac{9}{32}$ inch (7.14 mm) in diameter shall be permitted for use with plain pendant cord and holes $\frac{13}{32}$ inch (10.3 mm) in diameter with reinforced cord.

(b) Adjustable Fixtures. Fixtures that require adjusting or aiming after installation shall not be required to be equipped with an attachment plug or cord connector provided the exposed cord is of the hard usage or extra-hard usage type and is not longer than that required for maximum adjustment. The cord shall not be subject to strain or physical damage.

(c) Electric-Discharge Fixtures.

(1) A listed fixture or a listed fixture assembly shall be permitted to be cord-connected if located directly below the outlet box and the cord is continuously visible for its entire length outside the fixture and is not subject to strain or physical damage. Such cord-equipped fixtures shall terminate at the outer end of the cord in a grounding-type attachment plug (cap) or busway plug.

Exception: A listed fixture or a listed fixture assembly incorporating cord and canopy shall not be required to terminate at the outer end in an attachment plug or busway plug.

(2) Electric-discharge lighting fixtures provided with mogul-base, screw-shell lampholders shall be permitted to be connected to branch circuits of 50 amperes or less by cords complying with Section 240-4. Receptacles and attachment plugs shall be permitted to be of lower ampere rating than the branch circuit but not less than 125 percent of the fixture full-load current.

(3) Electric-discharge lighting fixtures equipped with a flanged surface inlet shall be permitted to be supplied by cord pendants equipped with cord connectors. Inlets and connectors shall be permitted to be of lower ampere rating than the branch circuit but not less than 125 percent of the fixture load current.

410-31. Fixtures as Raceways. Fixtures shall not be used as a raceway for circuit conductors.

Exception No. 1: Fixtures listed for use as a raceway.

Exception No. 2: Fixtures designed for end-to-end assembly to form a continuous raceway or fixtures connected together by recognized wiring methods shall be permitted to carry through conductors of a 2-wire or multiwire branch circuit supplying the fixtures.

Exception No. 3: One additional 2-wire branch circuit separately supplying one or more of the connected fixtures described in Exception No. 2 shall be permitted to be carried through the fixtures.

(FPN): See Article 100 for definition of "Multiwire Branch Circuit."

Branch-circuit conductors within 3 inches (76 mm) of a ballast within the ballast compartment shall have an insulation temperature rating not lower than 90°C (194°F), such as Types RHH, THW, THHN, THHW, FEP, FEPB, SA, and XHHW.

G. Construction of Fixtures

410-34. Combustible Shades and Enclosures. Adequate air space shall be provided between lamps and shades or other enclosures of combustible material.

410-35. Fixture Rating.

(a) Marking. All fixtures requiring ballasts or transformers shall be plainly marked with their electrical rating and the manufacturer's name, trademark, or other suitable means of identification. A fixture requiring supply wire rated higher than 90°C (194°F) shall be so marked, in letters 1/4 inch (6.35 mm) high prominently displayed on the fixture and shipping carton or equivalent.

(b) Electrical Rating. The electrical rating shall include the voltage and frequency and shall indicate the current rating of the unit, including the ballast, transformer, or autotransformer.

410-36. Design and Material. Fixtures shall be constructed of metal, wood, or other material suitable for the application and shall be so designed and assembled as to secure requisite mechanical strength and rigidity. Wiring compartments, including their entrances, shall be such that conductors may be drawn in and withdrawn without physical damage.

410-37. Nonmetallic Fixtures. In all fixtures not made entirely of metal or noncombustible material, wiring compartments shall be lined with metal.

Exception: Where armored or lead-covered conductors with suitable fittings are used.

410-38. Mechanical Strength.

(a) Tubing for Arms. Tubing used for arms and stems where provided with cut threads shall not be less than 0.040 inch (0.1 mm) in thickness and where provided with rolled (pressed) threads shall not be less than 0.025 inch (0.635 mm) in thickness. Arms and other parts shall be fastened to prevent turning.

(b) Metal Canopies. Metal canopies supporting lampholders, shades, etc., exceeding 8 pounds (3.63 kg), or incorporating attachment-plug receptacles, shall not be less than 0.020 inch (508 micrometers) in thickness. Other canopies shall not be less than 0.016 inch (406 micrometers) if made of steel and not less than 0.020 inch (508 micrometers) if of other metals.

(c) Canopy Switches. Pull-type canopy switches shall not be inserted in the rims of metal canopies that are less than 0.025 inch (635 micrometers) in thickness unless the rims are reinforced by the turning of a bead or the equivalent. Pull-type canopy switches, whether mounted in the rims or elsewhere in sheet metal canopies, shall not be located more than 3½ inches (89 mm) from the center of the canopy. Double set-screws, double canopy rings, a screw ring, or equal method shall be used where the canopy supports a pull-type switch or pendant receptacle.

The above thickness requirements shall apply to measurements made on finished (formed) canopies.

410-39. Wiring Space.
Bodies of fixtures, including portable lamps, shall provide ample space for splices and taps and for the installation of devices, if any. Splice compartments shall be of nonabsorbent, noncombustible material.

410-42. Portable Lamps.

(a) General. Portable lamps shall be wired with flexible cord, recognized by Section 400-4, and an attachment plug of the polarized or grounding type. Where used with Edison-base lampholders, the grounded conductor shall be identified and attached to the screw shell and the identified blade of the attachment plug.

(b) Portable Handlamps. In addition to the provisions of Section 410-42(a), portable handlamps shall comply with the following: (1) metal shell, paper-lined lampholders shall not be used; (2) handlamps shall be equipped with a handle of molded composition or other insulating material; (3) handlamps shall be equipped with a substantial guard attached to the lampholder or handle; (4) metallic guards shall be grounded by the means of an equipment grounding conductor run with circuit conductors within the power-supply cord.

410-44. Cord Bushings.
A bushing or the equivalent shall be provided where flexible cord enters the base or stem of a portable lamp. The bushing shall be of insulating material unless a jacketed type of cord is used.

410-45. Tests.
All wiring shall be free from short circuits and grounds and shall be tested for these defects prior to being connected to the circuit.

410-46. Live Parts. Exposed live parts within porcelain fixtures shall be suitably recessed and so located as to make it improbable that wires will come in contact with them. There shall be a spacing of at least ½ inch (12.7 mm) between live parts and the mounting plane of the fixture.

H. Installation of Lampholders

410-47. Screw-Shell Type. Lampholders of the screw-shell type shall be installed for use as lampholders only. Where supplied by a circuit having a grounded conductor, the grounded conductor shall be connected to the screw shell.

410-48. Double-Pole Switched Lampholders. Where supplied by the ungrounded conductors of a circuit, the switching device of lampholders of the switched type shall simultaneously disconnect both conductors of the circuit.

410-49. Lampholders in Wet or Damp Locations. Lampholders installed in wet or damp locations shall be of the weatherproof type.

J. Construction of Lampholders

410-50. Insulation. The outer metal shell and the cap shall be lined with insulating material that shall prevent the shell and cap from becoming a part of the circuit. The lining shall not extend beyond the metal shell more than ⅛ inch (3.17 mm), but shall prevent any current-carrying part of the lamp base from being exposed when a lamp is in the lampholding device.

410-51. Lead Wires. Lead wires, furnished as a part of weatherproof lampholders and intended to be exposed after installation, shall be of approved stranded, rubber-covered conductors not smaller than No. 14 and shall be sealed in place or otherwise made raintight.

Exception: No. 18 rubber-covered conductors shall be permitted for candelabra sockets.

410-52. Switched Lampholders. Switched lampholders shall be of such construction that the switching mechanism interrupts the electrical connection to the center contact. The switching mechanism shall also be permitted to interrupt the electrical connection to the screw shell if the connection to the center contact is simultaneously interrupted.

K. Lamps and Auxiliary Equipment

410-53. Bases, Incandescent Lamps. An incandescent lamp for general use on lighting branch circuits shall not be equipped with a medium base if rated over 300 watts, nor with a mogul base if rated over 1500 watts. Special bases or other devices shall be used for over 1500 watts.

410-54. Electric-Discharge Lamp Auxiliary Equipment.

   (a) Enclosures. Auxiliary equipment for electric-discharge lamps shall be enclosed in noncombustible cases and treated as sources of heat.

(b) Switching. Where supplied by the ungrounded conductors of a circuit, the switching device of auxiliary equipment shall simultaneously disconnect all conductors.

410-55. Arc Lamps. Arc lamps used in theaters shall comply with Section 520-61, and arc lamps used in projection machines shall comply with Section 540-20. Arc lamps used on constant-current systems shall comply with the general requirements of Article 710.

L. Receptacles, Cord Connectors, and Attachment Plugs (Caps)

410-56. Rating and Type.

(a) Receptacles. Receptacles installed for the attachment of portable cords shall be rated at not less than 15 amperes, 125 volts, or 15 amperes, 250 volts, and shall be of a type not suitable for use as lampholders.

Exception: The use of receptacles of 10-ampere, 250-volt rating used in nonresidential occupancies for the supply of equipment other than portable hand tools, portable handlamps, and extension cords shall be permitted.

(b) CO/ALR Receptacles. Receptacles rated 20 amperes or less and directly connected to aluminum conductors shall be marked CO/ALR.

(c) Isolated Ground Receptacles. Receptacles intended for the reduction of electrical noise (electromagnetic interference) as permitted in Section 250-74, Exception No. 4 shall be identified by an orange color or an orange triangle located on the face of the receptacle.

(d) Faceplates. Metal faceplates shall be of ferrous metal not less than 0.030 inch (762 micrometers) in thickness or of nonferrous metal not less than 0.040 inch (1 mm) in thickness. Metal faceplates shall be grounded. Faceplates of insulating material shall be noncombustible and not less than 0.10 inch (2.54 mm) in thickness but shall be permitted to be less than 0.10 inch (2.54 mm) in thickness if formed or reinforced to provide adequate mechanical strength.

(e) Position of Receptacle Faces. After installation, receptacle faces shall be flush with or project from faceplates of insulating material and shall project a minimum of 0.015 inch (381 micrometers) from metal faceplates. Faceplates shall be installed so as to completely cover the opening and seat against the mounting surface. Receptacles mounted in boxes that are set back of the wall surface, as permitted in Section 370-20, shall be installed so that the mounting yoke or strap of the receptacle is held rigidly at the surface of the wall. Receptacles mounted in boxes that are flush with the wall surface or project therefrom shall be so installed that the mounting yoke or strap of the receptacle is seated against the box or raised box cover.

(f) Attachment Plugs. All 15- and 20-ampere attachment plugs and connectors shall be so constructed that there are no exposed current-carrying parts except the prongs, blades, or pins. The cover for wire terminations shall be a part, which is essential for the operation of an attachment plug or connector (dead-front construction).

(g) Attachment Plug Ejector Mechanisms. Attachment plug ejector mechanisms shall not adversely affect engagement of the blades of the attachment plug with the contacts of the receptacle.

(h) Noninterchangeability. Receptacles, cord connectors, and attachment plugs shall be constructed so that the receptacle or cord connectors will not accept an attachment plug with a different voltage or current rating than that for which the device is intended. Nongrounding-type receptacles and connectors shall not accept grounding-type attachment plugs.

Exception: A 20-ampere T-slot receptacle or cord connector shall be permitted to accept a 15-ampere attachment plug of the same voltage rating.

(i) Receptacles in Raised Covers. Receptacles installed in raised covers shall not be secured solely by a single screw.

Exception: Devices or assemblies listed and identified for such use.

410-57. Receptacles in Damp or Wet Locations.

(a) Damp Locations. A receptacle installed outdoors in a location protected from the weather or in other damp locations shall have an enclosure for the receptacle that is weatherproof when the receptacle is covered (attachment plug cap not inserted and receptacle covers closed).

An installation suitable for wet locations shall also be considered suitable for damp locations.

A receptacle shall be considered to be in a location protected from the weather where located under roofed open porches, canopies, marquees, and the like, and will not be subjected to a beating rain or water run-off.

(b) Wet Locations. A receptacle installed outdoors where exposed to weather or in other wet locations shall be in a weatherproof enclosure, the integrity of which is not affected when the receptacle is in use (attachment plug cap inserted).

Exception: An enclosure that is weatherproof only when a self-closing receptacle cover is closed shall be permitted to be used for a receptacle installed outdoors where the receptacle is provided for use with portable tools or other portable equipment normally connected to the outlet only when attended.

(c) Protection for Floor Receptacles. Standpipes of floor receptacles shall allow floor-cleaning equipment to be operated without damage to receptacles.

(d) Flush Mounting with Faceplate. The enclosure for a receptacle installed in an outlet box flush-mounted on a wall surface shall be made weatherproof by means of a weatherproof faceplate assembly that provides a watertight connection between the plate and the wall surface.

(e) Installation. A receptacle outlet installed outdoors shall be located so that water accumulation is not likely to touch the outlet cover or plate.

410-58. Grounding-type Receptacles, Adapters, Cord Connectors, and Attachment Plugs.

(a) Grounding Poles. Grounding-type receptacles, cord connectors, and attachment plugs shall be provided with one fixed grounding pole in addition to the circuit poles.

Exception: The grounding contacting pole of grounding-type attachment plugs on the power-supply cords of portable hand-held, hand-guided, or hand-supported tools or appliances shall be permitted to be of the movable, self-restoring type on circuits operating at not over 150 volts between any two conductors nor over 150 volts between any conductor and ground.

(b) Grounding-Pole Identification. Grounding-type receptacles, adapters, cord connections, and attachment plugs shall have a means for connection of a grounding conductor to the grounding pole. A terminal for connection to the grounding pole shall be designated by:

(1) A green-colored hexagonal-headed or shaped terminal screw or nut, not readily removable; or

(2) A green-colored pressure wire connector body (a wire barrel); or

(3) A similar green-colored connection device, in the case of adapters. The grounding terminal of a grounding adapter shall be a green-colored rigid ear, lug, or similar device. The grounding connection shall be so designed that it cannot make contact with current-carrying parts of the receptacle, adapter, or attachment plug. The adapter shall be polarized.

(4) If the terminal for the equipment grounding conductor is not visible, the conductor entrance hole shall be marked with the word "green" or otherwise identified by a distinctive green color.

(c) Grounding Terminal Use. A grounding terminal or grounding-type device shall not be used for purposes other than grounding.

(d) Grounding-Pole Requirements. Grounding-type attachment plugs and mating cord connectors and receptacles shall be so designed that the grounding connection is made before the current-carrying connections. Grounding-type devices shall be designed so grounding poles of attachment plugs cannot be brought into contact with current-carrying parts of receptacles or cord connectors.

(e) Use. Grounding-type attachment plugs shall be used only where an equipment ground is to be provided.

M. Special Provisions for Flush and Recessed Fixtures

410-64. General. Fixtures installed in recessed cavities in walls or ceilings shall comply with Sections 410-65 through 410-72.

410-65. Temperature.

(a) Combustible Material. Fixtures shall be so installed that adjacent combustible material will not be subjected to temperatures in excess of 90°C (194°F).

(b) Fire-Resistant Construction. Where a fixture is recessed in fire-resistant material in a building of fire-resistant construction, a temperature higher than 90°C (194°F), but not higher than 150°C (302°F), shall be considered acceptable if the fixture is plainly marked that it is listed for that service.

(c) Recessed Incandescent Fixtures. Incandescent fixtures shall have thermal protection and shall so be identified as thermally protected.

Exception No. 1: Recessed incandescent fixtures identified for use and installed in poured concrete.

Exception No. 2: Listed recessed incandescent fixtures that provide, by construction design, the equivalent temperature performance characteristics of thermally protected fixtures and are so identified.

410-66. Clearance and Installation.

(a) Clearance. Recessed portions of lighting fixture enclosures, other than at the points of support, shall be spaced at least ½ inch (12.7 mm) from combustible materials.

Exception: Recessed fixtures identified as suitable for insulation to be in direct contact with the fixture.

(b) Installation. Thermal insulation shall not be installed within 3 inches (76 mm) of the recessed fixture enclosure, wiring compartment, or ballast, and shall not be so installed above the fixture so as to entrap heat and prevent the free circulation of air.

Exception: Recessed fixtures identified as suitable for insulation to be in direct contact with the fixture.

410-67. Wiring.

(a) General. Conductors having insulation suitable for the temperature encountered shall be used.

(b) Circuit Conductors. Branch-circuit conductors having an insulation suitable for the temperature encountered shall be permitted to terminate in the fixture.

(c) Tap Conductors. Tap conductors of a type suitable for the temperature encountered shall be permitted to run from the fixture terminal connection to an outlet box placed at least 1 foot (305 mm) from the fixture. Such tap conductors shall be in suitable raceway or Type AC or MC cable of at least 4 feet (1.22 m) but not more than 6 feet (1.83 m) in length.

N. Construction of Flush and Recessed Fixtures

410-68. Temperature. Fixtures shall be so constructed that adjacent combustible material will not be subject to temperatures in excess of 90°C (194°F).

410-69. Housing. Sheet metal of flush and recessed fixture housings shall be protected against corrosion and shall not be less than No. 22 MSG.

Exception: A wiring compartment cover shall be permitted to be of a thinner material provided it is within the No. 22 MSG housing and does not support any current-carrying components.

410-70. Lamp Wattage Marking. Incandescent lamp fixtures shall be marked to indicate the maximum allowable wattage of lamps. The markings shall be permanently installed, in letters at least $1/4$ inch (6.35 mm) high, and shall be located where visible during relamping.

410-71. Solder Prohibited. No solder shall be used in the construction of a fixture box.

410-72. Lampholders. Lampholders of the screw-shell type shall be of porcelain or other suitable insulating materials. Where used, cements shall be of the high-heat type.

P. Special Provisions for Electric-Discharge Lighting Systems of 1000 Volts or Less

410-73. General.

(a) Open-Circuit Voltage of 1000 Volts or Less. Equipment for use with electric-discharge lighting systems and designed for an open-circuit voltage of 1000 volts or less shall be of a type intended for such service.

(b) Considered as Energized. The terminals of an electric-discharge lamp shall be considered as energized where any lamp terminal is connected to a circuit of over 300 volts.

(c) Transformers of the Oil-Filled Type. Transformers of the oil-filled type shall not be used.

(d) Additional Requirements. In addition to complying with the general requirements for lighting fixtures, such equipment shall comply with Part P of this article.

(e) Thermal Protection. Where fluorescent fixtures are installed indoors, the ballasts shall have thermal protection integral within the ballast. Replacement ballasts for all fluorescent fixtures installed indoors shall also have thermal protection integral within the ballast.

Exception No. 1: Fluorescent fixtures using straight tubular lamps with simple reactance ballasts.

Exception No. 2: Ballasts for use in exit fixtures and so identified.

Exception No. 3: Egress lighting energized only during an emergency.

(f) High-Intensity Discharge Fixtures. Recessed high-intensity discharge fixtures shall be thermally protected and shall be so identified. Where fixtures, whether recessed or otherwise, are operated by remote ballasts, the ballasts shall also be thermally protected.

Exception: Recessed high-intensity discharge fixtures identified for use and installed in poured concrete.

(FPN): The thermal protection required by Section 410-73 may be accomplished by means other than a thermal protector.

410-74. Direct-Current Equipment. Fixtures installed on direct-current circuits shall be equipped with auxiliary equipment and resistors especially designed and for direct-current operation, and the fixtures shall be so marked.

410-75. Open-Circuit Voltage Exceeding 300 Volts. Equipment having an open-circuit voltage exceeding 300 volts shall not be installed in dwelling occupancies unless such equipment is so designed that there will be no exposed live parts when lamps are being inserted, are in place, or are being removed.

410-76. Fixture Mounting.

(a) **Exposed Ballasts.** Fixtures having exposed ballasts or transformers shall be so installed that such ballasts or transformers will not be in contact with combustible material.

(b) **Combustible Low-Density Cellulose Fiberboard.** Where a surface-mounted fixture containing a ballast is to be installed on combustible low-density cellulose fiberboard, it shall be listed for this condition or shall be spaced not less than $1\frac{1}{2}$ inches (38 mm) from the surface of the fiberboard. Where such fixtures are partially or wholly recessed, the provisions of Sections 410-64 through 410-72 shall apply.

(FPN): Combustible low-density cellulose fiberboard includes sheets, panels, and tiles that have a density of 20 pounds per cubic foot (320.36 kg/cu m) or less, and that are formed of bonded plant fiber material but does not include solid or laminated wood, nor fiberboard that has a density in excess of 20 pounds per cubic foot (320.36 kg/cu m) or is a material that has been integrally treated with fire-retarding chemicals to the degree that the flame spread in any plane of the material will not exceed 25, determined in accordance with tests for surface burning characteristics of building materials. See Test Method for Surface Burning Characteristics of Building Materials, ANSI/ASTM E84-1984.

410-77. Equipment Not Integral with Fixture.

(a) **Metal Cabinets.** Auxiliary equipment, including reactors, capacitors, resistors, and similar equipment, where not installed as part of a lighting fixture assembly, shall be enclosed in accessible, permanently installed metal cabinets.

(b) **Separate Mounting.** Separately mounted ballasts that are intended for direct connection to a wiring system shall not be required to be separately enclosed.

(c) **Wired Fixture Sections.** Wired fixture sections are paired, with a ballast(s) supplying a lamp or lamps in both. For interconnection between paired units, it shall be permissible to use $\frac{3}{8}$-inch flexible metal conduit in lengths not exceeding 25 feet (7.62 m) in conformance with Article 350. Fixture wire operating at line voltage, supplying only the ballast(s) of one of the paired fixtures, shall be permitted in the same raceway as the lamp supply wires of the paired fixtures.

410-78. Autotransformers. An autotransformer that is used to raise the voltage to more than 300 volts, as part of a ballast for supplying lighting units, shall be supplied only by a grounded system.

410-79. Switches. Snap switches shall comply with Section 380-14.

Q. Special Provisions for Electric-Discharge Lighting Systems of More than 1000 Volts

410-80. General.

(a) Open-Circuit Voltage Exceeding 1000 Volts. Equipment for use with electric-discharge lighting systems and designed for an open-circuit voltage exceeding 1000 volts shall be of a type intended for such service.

(b) Dwelling Occupancies. Equipment having an open-circuit voltage exceeding 1000 volts shall not be installed in or on dwelling occupancies.

(c) Live Parts. The terminal of an electric-discharge lamp shall be considered as a live part where any lamp terminal is connected to a circuit of over 300 volts.

(d) Additional Requirements. In addition to complying with the general requirements for lighting fixtures, such equipment shall comply with Part Q of this article.

(FPN):   For signs and outline lighting, see Article 600.

410-81. Control.

(a) Disconnection. Fixtures or lamp installations shall be controlled either singly or in groups by an externally operable switch or circuit breaker that opens all ungrounded primary conductors.

(b) Within Sight or Locked Type. The switch or circuit breaker shall be located within sight from the fixtures or lamps, or it shall be permitted elsewhere if it is provided with a means for locking in the open position.

410-82. Lamp Terminals and Lampholders.   Parts that must be removed for lamp replacement shall be hinged or held captive. Lamps or lampholders will be so designed that there shall be no exposed live parts when lamps are being inserted or are being removed.

410-83. Transformer Ratings.   Transformers and ballasts shall have a secondary open-circuit voltage of not over 15,000 volts with an allowance on test of 1000 volts additional. The secondary-current rating shall not be more than 120 milliamperes if the open-circuit voltage is over 7500 volts, and not more than 240 milliamperes if the open-circuit voltage is 7500 volts or less.

410-84. Transformer Type.   Transformers shall be enclosed and listed.

410-85. Transformer Secondary Connections.   The high-voltage windings of transformers shall not be connected in series or in parallel.

Exception: Two transformers, each having one end of its high-voltage winding grounded and connected to the enclosure, shall be permitted to have their high-voltage windings connected in series to form the equivalent of a midpoint grounded transformer. The grounded ends shall be connected by insulated conductors not smaller than No. 14.

410-86. Transformer Locations.

(a) **Accessible.** Transformers shall be accessible after installation.

(b) **Secondary Conductors.** Transformers shall be installed as near to the lamps as practicable to keep the secondary conductors as short as possible.

(c) **Adjacent to Combustible Materials.** Transformers shall be so located that adjacent combustible materials will not be subjected to temperatures in excess of 90°C (194°F).

410-87. Transformer Loading. The lamps connected to any transformer shall be of such length and characteristics as not to cause a condition of continuous overvoltage on the transformer.

410-88. Wiring Method — Secondary Conductors. Conductors shall be installed in accordance with Section 600-31.

410-89. Lamp Supports. Lamps shall be adequately supported as required in Section 600-33.

410-90. Exposure to Damage. Lamps shall not be located where normally exposed to physical damage.

410-91. Marking. Each fixture or each secondary circuit of tubing having an open-circuit voltage of over 1000 volts shall have a clearly legible marking in letters not less than ¼ inch (6.35 mm) high reading "Caution. . . .volts." The voltage indicated shall be the rated open-circuit voltage.

410-92. Switches. Snap switches shall comply with Section 380-14.

R. Lighting Track

410-100. Definition. Lighting track is a manufactured assembly designed to support and energize lighting fixtures that are capable of being readily repositioned on the track. Its length may be altered by the addition or subtraction of sections of track.

410-101. Installation.

(a) **Lighting Track.** Lighting track shall be permanently installed and permanently connected to a branch circuit. Only lighting track fittings shall be installed on lighting track. Lighting track fittings shall not be equipped with general-purpose receptacles.

(b) **Connected Load.** The connected load on lighting track shall not exceed the rating of the track. Lighting track shall be supplied by a branch circuit having a rating not more than that of the track.

(c) **Locations Not Permitted.** Lighting track shall not be installed (1) where subject to physical damage; (2) in wet or damp locations; (3) where subject to corrosive vapors; (4) in storage battery rooms; (5) in hazardous (classified) locations; (6) where concealed; (7) where extended through walls or partitions; (8) less than 5 feet (1.52 m) above the finished floor except where protected from physical damage or track operating at less than 30 volts RMS open-circuit voltage.

(d) Support. Fittings identified for use on lighting track shall be designed specifically for the track on which they are to be installed. They shall be securely fastened to the track, maintain polarization and grounding, and shall be designed to be suspended directly from the track.

410-102. Track Load. For branch-circuit calculations, a maximum of 2 feet (609.6 mm) of lighting track or fraction thereof shall be considered 180 VA. Where multicircuit track is installed, the load requirement of this section shall be considered to be divided equally between the circuits.

Exception: Where installed in dwelling unit(s) or the guest rooms of hotels or motels.

410-103. Heavy-Duty Track. Heavy-duty lighting track is lighting track identified for use exceeding 20 amperes. Each fitting attached to a heavy-duty lighting track shall have individual overcurrent protection.

410-104. Fastening. Lighting track shall be securely mounted so that each fastening will be suitable for supporting the maximum weight of fixtures that can be installed. Unless identified for supports at greater intervals, a single section 4 feet (1.22 m) or shorter in length shall have two supports, and, where installed in a continuous row, each individual section of not more than 4 feet (1.22 m) in length shall have one additional support.

410-105. Construction Requirements.

(a) Construction. The housing for the lighting track system shall be of substantial construction to maintain rigidity. The conductors shall be installed within the track housing permitting insertion of a fixture, and designed to prevent tampering and accidental contact with live parts. Components of lighting track systems of different voltages shall not be interchangeable. The track conductors shall be a minimum No. 12 or equal, and shall be copper. The track system ends shall be insulated and capped.

Exception: Fittings that incorporate an integral device to reduce the line voltage for a lower voltage lamp.

(b) Grounding. Lighting track shall be grounded in accordance with Article 250, and the track sections shall be securely coupled to maintain continuity of the circuitry, polarization, and grounding throughout.

ARTICLE 422 — APPLIANCES

A. General

422-1. Scope. This article covers electric appliances used in any occupancy.

422-2. Live Parts. Appliances shall have no live parts normally exposed to contact.

Exception: Toasters, grills, or other appliances in which the current-carrying parts at high temperatures are necessarily exposed.

422-3. Other Articles. All requirements of this Code shall apply where applicable. Appliances for use in hazardous (classified) locations shall comply with Articles 500 through 517.

The requirements of Article 430 shall apply to the installation of motor-operated appliances and the requirements of Article 440 shall apply to the installation of appliances containing hermetic refrigerant motor-compressor(s), except as specifically amended in this article.

B. Branch-Circuit Requirements

422-4. Branch-Circuit Sizing. This section specifies sizes of conductors capable of carrying appliance current without overheating under the conditions specified. This section shall not apply to conductors that form an integral part of an appliance.

(a) Individual Circuits. The rating of an individual branch circuit shall not be less than the marked rating of the appliance or the marked rating of an appliance having combined loads as provided in Section 422-32.

Exception No. 1: For motor-operated appliances not having a marked rating, the branch-circuit size shall be in accordance with Part B of Article 430.

Exception No. 2: For an appliance, other than a motor-operated appliance, that is continuously loaded, the branch-circuit rating shall not be less than 125 percent of the marked rating; or not less than 100 percent if the branch-circuit device and its assembly are listed for continuous loading at 100 percent of its rating.

Exception No. 3: Branch circuits for household cooking appliances shall be permitted to be in accordance with Table 220-19.

(b) Circuits Supplying Two or More Loads. For branch circuits supplying appliance and other loads, the rating shall be determined in accordance with Section 210-23.

422-5. Branch-Circuit Overcurrent Protection. Branch circuits shall be protected in accordance with Section 240-3.

If a protective device rating is marked on an appliance, the branch-circuit overcurrent device rating shall not exceed the protective device rating marked on the appliance.

C. Installation of Appliances

422-6. General. All appliances shall be installed in an approved manner.

422-7. Central Heating Equipment. Central heating equipment other than fixed electric space heating equipment shall be supplied by an individual branch circuit.

Exception: Auxiliary equipment such as a pump, valve, humidifier, or electrostatic air cleaner directly associated with the heating equipment shall be permitted to be connected to the same branch circuit.

422-8. Flexible Cords.

(a) **Heater Cords.** All cord- and plug-connected smoothing irons and electrically heated appliances that are rated at more than 50 watts and produce temperatures in excess of 121°C (250°F) on surfaces with which the cord is likely to be in contact shall be provided with one of the types of approved heater cords listed in Table 400-4.

(b) **Other Heating Appliances.** All other cord- and plug-connected electrically heated appliances shall be connected with one of the approved types of cord listed in Table 400-4, selected in accordance with the usage specified in that table.

(c) **Other Appliances.** Flexible cord shall be permitted (1) for connection of appliances to facilitate their frequent interchange or to prevent the transmission of noise or vibration or (2) to facilitate the removal or disconnection of appliances that are fastened in place, where the fastening means and mechanical connections are specifically designed to permit ready removal for maintenance or repair, and the appliance is intended or identified for flexible cord connection.

(d) Specific Appliances.

(1) Electrically operated kitchen waste disposers intended for dwelling unit use shall be permitted to be cord-and-plug connected with a flexible cord identified for the purpose, terminated with a grounding-type attachment plug where all of the following conditions are met:

a. The length of the cord shall not be less than 18 inches (457 mm) and not over 36 inches (914 mm).

b. Receptacles shall be located to avoid physical damage to the flexible cord.

c. The receptacle shall be accessible.

(2) Built-in dishwashers and trash compactors intended for dwelling use shall be permitted to be cord-and-plug connected with a flexible cord identified for the purpose, terminated with a grounding-type attachment plug where all of the following conditions are met:

a. The length of the cord shall be 3 to 4 feet (0.914 to1.22 m).

b. Receptacles shall be located to avoid physical damage to the flexible cord.

c. The receptacle shall be located in the space occupied by the appliance or adjacent thereto.

d. The receptacle shall be accessible.

Exception: Listed kitchen waste disposers, dishwashers, and trash compactors protected by a system of double insulation, or its equivalent, shall not be required to be grounded. Where such a system is employed, the equipment shall be distinctively marked.

(3) Cord- and plug-connected high-pressure spray washing machines shall be provided with factory-installed ground-fault circuit-interrupter protection for personnel. The ground-fault circuit-interrupter shall be an integral part of the attachment plug or shall be located in the supply cord within 12 inches (305 mm) of the attachment plug.

Exception No. 1: A high-pressure spray washer protected by a system of double insulation if such a unit is provided with a permanent warning indicating that it shall be connected to a receptacle protected by a ground-fault circuit-interrupter.

Exception No. 2: A high-pressure spray washer rated for a 3-phase supply system.

Exception No. 3: A high-pressure spray washer rated over 250 volts.

422-9. Cord- and Plug-Connected Immersion Heaters. Electric heaters of the cord- and plug-connected immersion type shall be so constructed and installed that current-carrying parts are effectively insulated from electrical contact with the substance in which they are immersed.

422-10. Protection of Combustible Material. Each electrically heated appliance that is intended by size, weight, and service to be located in a fixed position shall be so placed as to provide ample protection between the appliance and adjacent combustible material.

422-11. Stands for Cord- and Plug-Connected Appliances. Each smoothing iron and other cord- and plug-connected electrically heated appliance intended to be applied to combustible material shall be equipped with an approved stand, which shall be permitted to be a separate piece of equipment or a part of the appliance.

422-12. Signals for Heated Appliances. In other than dwelling-type occupancies, each electrically heated appliance or group of appliances intended to be applied to combustible material shall be provided with a signal.

Exception: If an appliance is provided with an integral temperature-limiting device.

422-13. Flatirons. Electrically heated smoothing irons shall be equipped with an identified temperature-limiting means.

422-14. Water Heaters.

(a) Storage- and Instantaneous-type Water Heaters. Each storage- or instantaneous-type water heater shall be equipped with a temperature-limiting means in addition to its control thermostat to disconnect all ungrounded conductors, and such means shall be (1) installed to sense maximum water temperature and (2) either a trip-free, manually reset type or a type having a replacement element. Such water heaters shall be marked to require the installation of a temperature and pressure relief valve.

(FPN): See Relief Valves and Automatic Gas Shutoff Devices for Hot Water Supply Systems, ANSI Z21.22-1986.

Exception: Water heaters with supply water temperature of 82°C (180°F) or above and a capacity of 60 kW or above and identified as being suitable for this use; and water heaters with a capacity of 1 gallon (3.785 L) or less and identified as being suitable for such use.

(b) Storage-type Water Heaters. A branch circuit supplying a fixed storage-type water heater having a capacity of 120 gallons (454.2 L) or less shall have a rating not less than 125 percent of the nameplate rating of the water heater.

(FPN): For branch-circuit sizing, see Section 422-4(a), Exception No. 2.

422-15. Infrared Lamp Industrial Heating Appliances.

(a) 300 Watts or Less. Infrared heating lamps rated at 300 watts or less shall be permitted with lampholders of the medium-base, unswitched porcelain type or other types identified as suitable for use with infrared heating lamps rated 300 watts or less.

(b) Over 300 Watts. Screw-shell lampholders shall not be used with infrared lamps over 300 watts rating.

Exception: Lampholders identified as suitable for use with infrared heating lamps rated more than 300 watts.

(c) Lampholders. Lampholders shall be permitted to be connected to any of the branch circuits of Article 210 and, in industrial occupancies, shall be permitted to be operated in series on circuits of over 150 volts to ground provided the voltage rating of the lampholders is not less than the circuit voltage.

Each section, panel, or strip carrying a number of infrared lampholders (including the internal wiring of such section, panel, or strip) shall be considered an appliance. The terminal connection block of each such assembly shall be considered an individual outlet.

422-16. Grounding. Appliances required by Article 250 to be grounded shall have exposed noncurrent-carrying metal parts grounded in the manner specified in Article 250.

(FPN): See Sections 250-42, 250-43, and 250-45 for equipment grounding of refrigerators and freezers and Sections 250-57 and 250-60 for equipment grounding of electric ranges, wall-mounted ovens, counter-mounted cooking units, and clothes dryers.

422-17. Wall-Mounted Ovens and Counter-Mounted Cooking Units.

(a) Permitted to Be Cord- and Plug-Connected or Permanently Connected. Wall-mounted ovens and counter-mounted cooking units complete with provisions for mounting and for making electrical connections shall be permitted to be permanently connected or, only for ease in servicing or for installation, cord-and plug-connected.

(b) Separable Connector or a Plug and Receptacle Combination. A separable connector or a plug and receptacle combination in the supply line to an oven or cooking unit shall:

(1) Not be installed as the disconnecting means required by Section 422-20.

(2) Be approved for the temperature of the space in which it is located.

422-18. Support of Ceiling Fans. Listed ceiling fans that do not exceed 35 pounds (15.88 kg) in weight, with or without accessories, shall be permitted to be supported by outlet boxes identified for such use and supported in accordance with Sections 370-23 and 370-27.

422-19. Other Installation Methods.   Appliances employing methods of installation other than covered by this article shall be permitted to be used only by special permission.

D. Control and Protection of Appliances

422-20. Disconnecting Means.   A means shall be provided to disconnect each appliance from all ungrounded conductors in accordance with the following sections of Part D. If an appliance is supplied by more than one source, the disconnecting means shall be grouped and identified.

422-21. Disconnection of Permanently Connected Appliances.

(a) Rated at Not Over 300 Volt Amperes or ¹/₈ Horsepower.   For permanently connected appliances rated at not over 300 volt amperes or ¹/₈ horsepower, the branch-circuit overcurrent device shall be permitted to serve as the disconnecting means.

(b) Permanently Connected Appliances of Greater Rating.   For permanently connected appliances of greater rating, the branch-circuit switch or circuit breaker shall be permitted to serve as the disconnecting means where the switch or circuit breaker is within sight from the appliance or is capable of being locked in the open position.

(FPN No. 1):   For motor-driven appliances of more than ¹/₈ horsepower, see Section 422-27.

(FPN No. 2):   For appliances employing unit switches, see Section 422-25.

422-22. Disconnection of Cord- and Plug-Connected Appliances.

(a) Separable Connector or an Attachment Plug and Receptacle.   For cord- and plug-connected appliances, an accessible separable connector or an accessible plug and receptacle shall be permitted to serve as the disconnecting means. Other cord- and plug-connected appliances shall be provided with disconnecting means in accordance with Section 422-21.

(b) Connection at the Rear Base of a Range.   For cord- and plug-connected household electric ranges, an attachment plug and receptacle connection at the rear base of a range, if it is accessible from the front by removal of a drawer, shall be considered as meeting the intent of Section 422-22(a).

(c) Rating.   The rating of a receptacle or of a separable connector shall not be less than the rating of any appliance connected thereto.

Exception: Demand factors authorized elsewhere in this Code shall be permitted to be applied.

(d) Requirements for Attachment Plugs and Connectors.   Attachment plugs and connectors shall conform to the following:

(1) Live Parts.   They shall be so constructed and installed as to guard against inadvertent contact with live parts.

(2) Interrupting Capacity.   They shall be capable of interrupting their rated current without hazard to the operator.

(3) Interchangeability. They shall be so designed that they will not fit into receptacles of lesser rating.

422-23. Polarity in Cord- and Plug-Connected Appliances. If the appliance is provided with a manually operated, line-connected, single-pole switch for appliance on-off operation, an Edison-base lampholder, or a 15- or 20-ampere receptacle, the attachment plug shall be of the polarized or grounding type.

(FPN): For polarity of Edison-base lampholders, see Section 410-42(a). |

422-24. Cord- and Plug-Connected Appliances Subject to Immersion. Cord- and plug-connected portable freestanding hydromassage units and hand-held hair dryers shall be constructed to provide protection for personnel against electrocution when immersed while in the "on" or "off" position.

422-25. Unit Switch(es) as Disconnecting Means. A unit switch(es) with a marked "off" position that is a part of an appliance and disconnects all ungrounded conductors shall be permitted as the disconnecting means required by this article where other means for disconnection are provided in the following types of occupancies:

(a) Multifamily Dwellings. In multifamily dwellings, the other disconnecting means shall be within the dwelling unit, or on the same floor as the dwelling unit in which the appliance is installed, and shall be permitted to control lamps and other appliances.

(b) Two-Family Dwellings. In two-family dwellings, the other disconnecting means shall be permitted either inside or outside of the dwelling unit in which the appliance is installed. In this case, an individual switch or circuit breaker for the dwelling unit shall be permitted and shall also be permitted to control lamps and other appliances.

(c) One-Family Dwellings. In one-family dwellings, the service disconnecting means shall be permitted to be the other disconnecting means.

(d) Other Occupancies. In other occupancies, the branch-circuit switch or circuit breaker, where readily accessible for servicing of the appliance, shall be permitted as the other disconnecting means.

422-26. Switch and Circuit Breaker to Be Indicating. Switches and circuit breakers used as disconnecting means shall be of the indicating type.

422-27. Disconnecting Means for Motor-Driven Appliances. If a switch or circuit breaker serves as the disconnecting means for a permanently connected motor-driven appliance of more than 1/8 horsepower, it shall be located within sight from the motor controller and shall comply with Part I of Article 430.

Exception: A switch or circuit breaker that serves as the other disconnecting means as required in Section 422-25(a), (b), (c), or (d) shall be permitted to be out of sight from the motor controller of an appliance provided with a unit switch(es) with a marked "off" position and that disconnects all ungrounded conductors.

422-28. Overcurrent Protection.

(a) Appliances. Appliances shall be protected against overcurrent in accordance with (b) through (f) below and Sections 422-4 and 422-5.

Exception: Motors of motor-operated appliances shall be provided with overload protection in accordance with Part C of Article 430. Hermetic refrigerant motor-compressors in air-conditioning or refrigerating equipment shall be provided with overload protection in accordance with Part F of Article 440. Where appliance overcurrent protective devices that are separate from the appliance are required, data for selection of these devices shall be marked on the appliance. The minimum marking shall be that specified in Sections 430-7 and 440-4.

(b) Household-type Appliance with Surface Heating Elements. A household-type appliance with surface heating elements having a maximum demand of more than 60 amperes computed in accordance with Table 220-19 shall have its power supply subdivided into two or more circuits, each of which is provided with overcurrent protection rated at not over 50 amperes.

(c) Infrared Lamp Commercial and Industrial Heating Appliances. Infrared lamp commercial and industrial heating appliances shall have overcurrent protection not exceeding 50 amperes.

(d) Open-Coil or Exposed Sheathed-Coil Types of Surface Heating Elements in Commercial-type Heating Appliances. Open-coil or exposed sheathed-coil types of surface heating elements in commercial-type heating appliances shall be protected by overcurrent protective devices rated at not over 50 amperes.

(e) Single Nonmotor-Operated Appliance. If the branch circuit supplies a single nonmotor-operated appliance, the rating of overcurrent protection shall (1) not exceed that marked on the appliance; (2) if the overcurrent protection rating is not marked and the appliance is rated over 13.3 amperes, not exceed 150 percent of the appliance rated current; or (3) if the overcurrent protection rating is not marked and the appliance is rated 13.3 amperes or less, not exceed 20 amperes.

Exception: Where 150 percent of appliance rating does not correspond to a standard overcurrent device ampere rating, the next higher standard rating shall be permitted.

(f) Electric Heating Appliances Employing Resistance-type Heating Elements Rated More than 48 Amperes. Electric heating appliances employing resistance-type heating elements rated more than 48 amperes shall have the heating elements subdivided. Each subdivided load shall not exceed 48 amperes and shall be protected at not more than 60 amperes.

These supplementary overcurrent protective devices shall be (1) factory-installed within or on the heater enclosure or provided as a separate assembly by the heater manufacturer; (2) accessible, but need not be readily accessible; and (3) suitable for branch-circuit protection.

The main conductors supplying these overcurrent protective devices shall be considered branch-circuit conductors.

Exception No. 1: Household-type appliances with surface heating elements as covered in Section 422-28(b) and commercial-type heating appliances as covered in Section 422-28(d).

Exception No. 2: Commercial kitchen and cooking appliances using sheathed-type heating elements not covered in Section 422-28(d) shall be permitted to be subdivided into circuits not exceeding 120 amperes and protected at not more than 150 amperes where one of the following is met:

a. Elements are integral with and enclosed within a cooking surface;

b. Elements are completely contained within an enclosure identified as suitable for this use; or

c. Elements are contained within an ASME rated and stamped vessel.

Exception No. 3: Water heaters and steam boilers employing resistance-type immersion electric heating elements contained in an ASME rated and stamped vessel shall be permitted to be subdivided into circuits not exceeding 120 amperes and protected at not more than 150 amperes.

E. Marking of Appliances

422-29. Cord- and Plug-Connected Pipe Heating Assemblies. Cord- and plug-connected pipe heating assemblies intended to prevent freezing of piping shall be listed.

422-30. Nameplate.

(a) Nameplate Marking. Each electric appliance shall be provided with a nameplate giving the identifying name and the rating in volts and amperes, or in volts and watts. If the appliance is to be used on a specific frequency or frequencies, it shall be so marked.

Where motor overload protection external to the appliance is required, the appliance shall be so marked.

(FPN): See Section 422-28(a), Exception for overcurrent protection requirements.

(b) To Be Visible. Marking shall be located so as to be visible or easily accessible after installation.

422-31. Marking of Heating Elements. All heating elements that are rated over one ampere, replaceable in the field, and a part of an appliance shall be legibly marked with the ratings in volts and amperes, or in volts and watts, or with the manufacturer's part number.

422-32. Appliances Consisting of Motors and Other Loads. Appliances shall be marked in accordance with (a) or (b) below.

(a) Marking. In addition to the marking required in Section 422-30, the marking on an appliance consisting of a motor with other load(s) or motors with or without other load(s) shall specify the minimum supply circuit conductor ampacity and the maximum rating of the circuit overcurrent protective device.

Exception No. 1: Appliances factory-equipped with cords and attachment plugs, complying with Section 422-30.

Exception No. 2: An appliance where both the minimum supply circuit conductor ampacity and maximum rating of the circuit overcurrent protective device are not more than 15 amperes and complies with Section 422-30.

(b) Alternate Marking Method. An alternate marking method shall be permitted to specify the rating of the largest motor in volts and amperes, and the additional load(s) in volts and amperes, or volts and watts in addition to the marking required in Section 422-30.

Exception No. 1: Appliances factory-equipped with cords and attachment plugs, complying with Section 422-30.

Exception No. 2: The ampere rating of a motor $1/8$ horsepower or less or a nonmotor load 1 ampere or less shall be permitted to be omitted unless such loads constitute the principal load.

ARTICLE 424 — FIXED ELECTRIC SPACE HEATING EQUIPMENT

A. General

424-1. Scope. This article covers fixed electric equipment used for space heating. For the purpose of this article, heating equipment shall include heating cable, unit heaters, boilers, central systems, or other approved fixed electric space heating equipment. This article shall not apply to process heating and room air conditioning.

424-2. Other Articles. All requirements of this Code shall apply where applicable. Fixed electric space heating equipment for use in hazardous (classified) locations shall comply with Articles 500 through 517. Fixed electric space heating equipment incorporating a hermetic refrigerant motor-compressor shall also comply with Article 440.

424-3. Branch Circuits.

(a) Branch-Circuit Requirements. Individual branch circuits shall be permitted to supply any size fixed electric space heating equipment.

Branch circuits supplying two or more outlets for fixed electric space heating equipment shall be rated 15, 20, or 30 amperes.

Exception: In other than residential occupancies, fixed infrared heating equipment shall be permitted to be supplied from branch circuits rated not over 50 amperes.

(b) Branch-Circuit Sizing. The ampacity of the branch-circuit conductors and the rating or setting of overcurrent protective devices supplying fixed electric space heating equipment consisting of resistance elements with or without a motor shall not be less than 125 percent of the total load of the motors and the heaters. The rating or setting of overcurrent protective devices shall be permitted in accordance with Section 240-3(b). A contactor, thermostat, relay, or similar device, listed for continuous operation at 100 percent of its rating, shall be permitted to supply its full-rated load as provided in Section 210-22(c), Exception.

The size of the branch-circuit conductors and overcurrent protective devices supplying fixed electric space heating equipment including a hermetic refrigerant motor-compressor with or without resistance units shall be computed in accordance with Sections 440-34 and 440-35.

The provisions of this section shall not apply to conductors that form an integral part of approved fixed electric space heating equipment.

B. Installation

424-9. General. All fixed electric space heating equipment shall be installed in an approved manner.

Permanently installed electric baseboard heaters equipped with factory-installed receptacle outlets, or outlets provided as a separate listed assembly, shall be permitted in lieu of a receptacle outlet(s) that is required by Section 210-50(b). Such receptacle outlets shall not be connected to the heater circuits.

(FPN): Listed baseboard heaters include instructions that may not permit their installation below receptacle outlets.

424-10. Special Permission. Fixed electric space heating equipment and systems installed by methods other than covered by this article shall be permitted only by special permission.

424-11. Supply Conductors. Fixed electric space heating equipment requiring supply conductors with over 60°C insulation shall be clearly and permanently marked. This marking shall be plainly visible after installation and shall be permitted to be adjacent to the field-connection box.

424-12. Locations.

(a) Exposed to Severe Physical Damage. Fixed electric space heating equipment shall not be used where exposed to severe physical damage unless adequately protected.

(b) Damp or Wet Locations. Heaters and related equipment installed in damp or wet locations shall be approved for such locations and shall be constructed and installed so that water cannot enter or accumulate in or on wired sections, electrical components, or duct work.

(FPN No. 1): See Section 110-11 for equipment exposed to deteriorating agents.

(FPN No. 2): See Section 680-27 for pool deck areas.

424-13. Spacing from Combustible Materials. Fixed electric space heating equipment shall be installed to provide the required spacing between the equipment and adjacent combustible material, unless it has been found to be acceptable where installed in direct contact with combustible material.

424-14. Grounding. All exposed noncurrent-carrying metal parts of fixed electric space heating equipment likely to become energized shall be grounded as required in Article 250.

C. Control and Protection of Fixed Electric Space Heating Equipment

424-19. Disconnecting Means. Means shall be provided to disconnect the heater, motor controller(s), and supplementary overcurrent protective device(s) of all fixed electric space heating equipment from all ungrounded conductors. Where heating equipment is supplied by more than one source, the disconnecting means shall be grouped and identified.

(a) Heating Equipment with Supplementary Overcurrent Protection. The disconnecting means for fixed electric space heating equipment with supplementary overcurrent protection shall be within sight from the supplementary overcurrent protective device(s), on the supply side of these devices, if fuses, and, in addition, shall comply with either (1) or (2) below.

(1) Heater Containing No Motor Rated Over $\frac{1}{8}$ Horsepower. The above disconnecting means or unit switches complying with Section 424-19(c) shall be permitted to serve as the required disconnecting means for both the motor controller(s) and heater under either item a or b below.

a. The disconnecting means provided is also within sight from the motor controller(s) and the heater; or

b. The disconnecting means provided shall be capable of being locked in the open position.

(2) Heater Containing a Motor(s) Rated Over $\frac{1}{8}$ Horsepower. The above disconnecting means shall be permitted to serve as the required disconnecting means for both the motor controller(s) and heater by one of the means specified in items a through d below.

a. Where the disconnecting means is also in sight from the motor controller(s) and the heater.

b. Where the disconnecting means is not within sight from the heater, a separate disconnecting means shall be installed, or the disconnecting means shall be capable of being locked in the open position, or unit switches complying with Section 424-19(c) shall be permitted.

c. Where the disconnecting means is not within sight from the motor controller location, a disconnecting means complying with Section 430-102 shall be provided.

d. Where the motor is not in sight from the motor controller location, Section 430-102(b) shall apply.

(b) Heating Equipment Without Supplementary Overcurrent Protection.

(1) Without Motor or with Motor Not Over $\frac{1}{8}$ Horsepower. For fixed electric space heating equipment without a motor rated over $\frac{1}{8}$ horsepower, the branch-circuit switch or circuit breaker shall be permitted to serve as the disconnecting means, where the switch or circuit breaker is within sight from the heater or is capable of being locked in the open position.

(2) Over $\frac{1}{8}$ Horsepower. For motor-driven electric space heating equipment with a motor rated over $\frac{1}{8}$ horsepower, a disconnecting means shall be located within sight from the motor controller.

Exception: As permitted by Section 424-19(a)(2).

(c) Unit Switch(es) as Disconnecting Means. A unit switch(es) with a marked "off" position that is part of a fixed heater and disconnects all ungrounded conductors shall be permitted as the disconnecting means required by this article where other means for disconnection are provided in the following types of occupancies.

(1) Multifamily Dwellings. In multifamily dwellings, the other disconnecting means shall be within the dwelling unit, or on the same floor as the dwelling unit in which the fixed heater is installed, and shall also be permitted to control lamps and appliances.

(2) Two-Family Dwellings. In two-family dwellings, the other disconnecting means shall be permitted either inside or outside of the dwelling unit in which the fixed heater is installed. In this case, an individual switch or circuit breaker for the dwelling unit shall be permitted and shall also be permitted to control lamps and appliances.

(3) One-Family Dwellings. In one-family dwellings, the service disconnecting means shall be permitted to be the other disconnecting means.

(4) Other Occupancies. In other occupancies, the branch-circuit switch or circuit breaker, where readily accessible for servicing of the fixed heater, shall be permitted as the other disconnecting means.

424-20. Thermostatically Controlled Switching Devices.

(a) Serving as Both Controllers and Disconnecting Means. Thermostatically controlled switching devices and combination thermostats and manually controlled switches shall be permitted to serve as both controllers and disconnecting means, provided all of the following conditions are met:

(1) Provided with a marked "off" position.

(2) Directly open all ungrounded conductors when manually placed in the "off" position.

(3) Designed so that the circuit cannot be energized automatically after the device has been manually placed in the "off" position.

(4) Located as specified in Section 424-19.

(b) Thermostats that Do Not Directly Interrupt All Ungrounded Conductors. Thermostats that do not directly interrupt all ungrounded conductors and thermostats that operate remote-control circuits shall not be required to meet the requirements of (a). These devices shall not be permitted as the disconnecting means.

424-21. Switch and Circuit Breaker to Be Indicating.   Switches and circuit breakers used as disconnecting means shall be of the indicating type.

424-22. Overcurrent Protection.

(a) Branch-Circuit Devices. Electric space heating equipment, other than such motor-operated equipment as required by Articles 430 and 440 to have additional overcurrent protection, shall be permitted to be protected against overcurrent where supplied by one of the branch circuits in Article 210.

(b) Resistance Elements. Resistance-type heating elements in electric space heating equipment shall be protected at not more than 60 amperes. Equipment rated more than 48 amperes and employing such elements shall have the heating elements subdivided, and each subdivided load shall not exceed 48 amperes. Where a subdivided load is less than 48 amperes, the rating of the supplementary overcurrent protective device shall comply with Section 424-3(b).

Exception: As provided in Section 424-72(a).

(c) Overcurrent Protective Devices. The supplementary overcurrent protective devices for the subdivided loads specified in (b) above shall be (1) factory-installed within or on the heater enclosure or supplied for use with the heater as a separate assembly by the heater manufacturer; (2) accessible, but shall not be required to be readily accessible; and (3) suitable for branch-circuit protection.

(FPN):   See Section 240-10.

Where cartridge fuses are used to provide this overcurrent protection, a single disconnecting means shall be permitted to be used for the several subdivided loads.

(FPN No. 1):   For supplementary overcurrent protection, see Section 240-10.

(FPN No. 2):   For disconnecting means for cartridge fuses in circuits of any voltage, see Section 240-40.

(d) Branch-Circuit Conductors. The conductors supplying the supplementary overcurrent protective devices shall be considered branch-circuit conductors.

Exception: For heaters rated 50 kW or more, the conductors supplying the supplementary overcurrent protective devices specified in (c) above shall be permitted to be sized at not less than 100 percent of the nameplate rating of the heater, provided all of the following conditions are met:

a. The heater is marked with a minimum conductor size; and

b. The conductors are not smaller than the marked minimum size; and

c. A temperature-actuated device controls the cyclic operation of the equipment.

(e) Conductors for Subdivided Loads. Field-wired conductors between the heater and the supplementary overcurrent protective devices shall be sized at not less than 125 percent of the load served. The supplementary overcurrent protective devices specified in (c) shall protect these conductors in accordance with Section 240-3.

Exception: For heaters rated 50 kW or more, the ampacity of field-wired conductors between the heater and the supplementary overcurrent protective devices shall be permitted to be not less than 100 percent of the load of their respective subdivided circuits, provided all of the following conditions are met:

a. The heater is marked with a minimum conductor size; and

b. The conductors are not smaller than the marked minimum size; and

c. A temperature-activated device controls the cyclic operation of the equipment.

D. Marking of Heating Equipment

424-28. Nameplate.

(a) Marking Required. Each unit of fixed electric space heating equipment shall be provided with a nameplate giving the identifying name and the normal rating in volts and watts, or in volts and amperes.

Electric space heating equipment intended for use on alternating current only or direct current only shall be marked to so indicate. The marking of equipment consisting of motors over ⅛ horsepower and other loads shall specify the rating of the motor in volts, amperes, and frequency, and the heating load in volts and watts, or in volts and amperes.

(b) Location. This nameplate shall be located so as to be visible or easily accessible after installation.

424-29. Marking of Heating Elements.

All heating elements that are replaceable in the field and are a part of an electric heater shall be legibly marked with the ratings in volts and watts, or in volts and amperes.

E. Electric Space Heating Cables

424-34. Heating Cable Construction.

Heating cables shall be furnished complete with factory-assembled nonheating leads at least 7 feet (2.13 m) in length.

424-35. Marking of Heating Cables.

Each unit shall be marked with the identifying name or identification symbol, catalog number, ratings in volts and watts, or in volts and amperes.

Each unit length of heating cable shall have a permanent legible marking on each nonheating lead located within 3 inches (76 mm) of the terminal end. The lead wire shall have the following color identification to indicate the circuit voltage on which it is to be used: 120-volt nominal, yellow; 208-volt nominal, blue; 240-volt nominal, red; and 277-volt nominal, brown.

424-36. Clearances of Wiring in Ceilings.

Wiring located above heated ceilings shall be spaced not less than 2 inches (50.8 mm) above the heated ceiling and shall be considered as operating at an ambient of 50°C (122°F). The ampacity of conductors shall be computed on the basis of the correction factors shown in the 0-2000 volt ampacity tables of Article 310.

Exception: Wiring above heated ceilings and located above thermal insulation having a minimum thickness of 2 inches (50.8 mm) shall not require correction for temperature.

424-37. Location of Branch-Circuit and Feeder Wiring in Exterior Walls.

Wiring methods shall comply with Article 300 and Section 310-10.

424-38. Area Restrictions.

(a) Shall Not Extend Beyond the Room or Area. Heating cables shall not extend beyond the room or area in which they originate.

(b) Uses Prohibited. Cables shall not be installed in closets, over walls or partitions that extend to the ceiling, or over cabinets whose clearance from the ceiling is less than the minimum horizontal dimension of the cabinet to the nearest cabinet edge that is open to the room or area.

Exception: Isolated single runs of cable shall be permitted to pass over partitions where they are embedded.

(c) In Closet Ceilings as Low Temperature Heat Sources to Control Relative Humidity. This provision shall not prevent the use of cable in closet ceilings as low temperature heat sources to control relative humidity, provided they are used only in those portions of the ceiling that are unobstructed to the floor by shelves or other permanent fixtures.

424-39. Clearance from Other Objects and Openings. Heating elements of cables shall be separated at least 8 inches (203 mm) from the edge of outlet boxes and junction boxes that are to be used for mounting surface lighting fixtures. A clearance of not less than 2 inches (50.8 mm) shall be provided from recessed fixtures and their trims, ventilating openings, and other such openings in room surfaces. Sufficient area shall be provided to assure that no heating cable will be covered by any surface-mounted units.

424-40. Splices. Embedded cables shall be spliced only where necessary and only by approved means, and in no case shall the length of the heating cable be altered.

424-41. Installation of Heating Cables on Dry Board, in Plaster, and on Concrete Ceilings.

(a) Shall Not Be Installed in Walls. Cables shall not be installed in walls.

Exception: Isolated single runs of cable shall be permitted to run down a vertical surface to reach a dropped ceiling.

(b) Adjacent Runs. Adjacent runs of cable not exceeding 2¾ watts per foot (305 mm) shall be installed not less than 1½ inches (38 mm) on centers.

(c) Surfaces to Be Applied. Heating cables shall be applied only to gypsum board, plaster lath, or other fire-resistant material. With metal lath or other electrically conductive surfaces, a coat of plaster shall be applied to completely separate the metal lath or conductive surface from the cable.

(FPN): See also (f) below.

(d) Splices. All heating cables, the splice between the heating cable and nonheating leads, and 3-inch (76-mm) minimum of the nonheating lead at the splice shall be embedded in plaster or dry board in the same manner as the heating cable.

(e) Ceiling Surface. The entire ceiling surface shall have a finish of thermally noninsulating sand plaster having a nominal thickness of ½ inch (12.7 mm), or other noninsulating material identified as suitable for this use and applied according to specified thickness and directions.

(f) Secured. Cables shall be secured at intervals not exceeding 16 inches (406 mm) by means of approved stapling, tape, plaster, nonmetallic spreaders, or other approved means. Staples or metal fasteners that straddle the cable shall not be used with metal lath or other electrically conductive surfaces.

Exception: Cables identified to be secured at intervals not to exceed 6 feet (1.83 m).

(g) Dry Board Installations. In dry board installations, the entire ceiling below the heating cable shall be covered with gypsum board not exceeding $\frac{1}{2}$ inch (12.7 mm) thickness. The void between the upper layer of gypsum board, plaster lath, or other fire-resistant material and the surface layer of gypsum board shall be completely filled with thermally conductive, nonshrinking plaster or other approved material or equivalent thermal conductivity.

(h) Free from Contact with Conductive Surfaces. Cables shall be kept free from contact with metal or other electrically conductive surfaces.

(i) Joists. In dry board applications, cable shall be installed parallel to the joist, leaving a clear space centered under the joist of $2\frac{1}{2}$ inches (64 mm) (width) between centers of adjacent runs of cable. A surface layer of gypsum board shall be mounted so that the nails or other fasteners do not pierce the heating cable.

(j) Crossing Joists. Cables shall cross joists only at the ends of the room.

Exception: Where the cable is required to cross joists elsewhere in order to satisfy the manufacturer's instructions that the installer avoid placing the cable too close to ceiling penetrations and light fixtures.

424-42. Finished Ceilings. Finished ceilings shall not be covered with decorative panels or beams constructed of materials that have thermal insulating properties, such as wood, fiber, or plastic. Finished ceilings shall be permitted to be covered with paint, wallpaper, or other approved surface finishes.

424-43. Installation of Nonheating Leads of Cables.

(a) Free Nonheating Leads. Free nonheating leads of cables shall be installed in accordance with approved wiring methods from the junction box to a location within the ceiling. Such installations shall be permitted to be single conductors in approved raceways, single or multiconductor Type UF, Type NMC, Type MI, or other approved conductors.

(b) Leads in Junction Box. Not less than 6 inches (152 mm) of free nonheating lead shall be within the junction box. The marking of the leads shall be visible in the junction box.

(c) Excess Leads. Excess leads of heating cables shall not be cut but shall be secured to the underside of the ceiling and embedded in plaster or other approved material, leaving only a length sufficient to reach the junction box with not less than 6 inches (152 mm) of free lead within the box.

424-44. Installation of Cables in Concrete or Poured Masonry Floors.

(a) Watts per Linear Foot. Heating cables shall not exceed $16\frac{1}{2}$ watts per linear foot (305 mm) of cable.

(b) Spacing Between Adjacent Runs. The spacing between adjacent runs of cable shall be not less than 1 inch (25.4 mm) on centers.

(c) Secured in Place. Cables shall be secured in place by nonmetallic frames or spreaders or other approved means while the concrete or other finish is applied.

Cables shall not be installed where they bridge expansion joints unless protected from expansion and contraction.

(d) Spacings Between Heating Cable and Metal Embedded in the Floor. Spacings shall be maintained between the heating cable and metal embedded in the floor.

Exception: Grounded metal-clad cable shall be permitted to be in contact with metal embedded in the floor.

(e) Leads Protected. Leads shall be protected where they leave the floor by rigid metal conduit, intermediate metal conduit, rigid nonmetallic conduit, electrical metallic tubing, or by other approved means.

(f) Bushings or Approved Fittings. Bushings or approved fittings shall be used where the leads emerge within the floor slab.

424-45. Inspection and Tests. Cable installations shall be made with due care to prevent damage to the cable assembly and shall be inspected and approved before cables are covered or concealed.

F. Duct Heaters

424-57. General. Part F shall apply to any heater mounted in the air stream of a forced-air system where the air-moving unit is not provided as an integral part of the equipment.

424-58. Identification. Heaters installed in an air duct shall be identified as suitable for the installation.

424-59. Airflow. Means shall be provided to assure uniform and adequate airflow over the face of the heater in accordance with the manufacturer's instructions.

(FPN): Heaters installed within 4 feet (1.22 m) of the outlet of an air-moving device, heat pump, air conditioner, elbows, baffle plates, or other obstructions in duct work may require turning vanes, pressure plates, or other devices on the inlet side of the duct heater to assure an even distribution of air over the face of the heater.

424-60. Elevated Inlet Temperature. Duct heaters intended for use with elevated inlet air temperature shall be identified as suitable for use at the elevated temperatures.

424-61. Installation of Duct Heaters with Heat Pumps and Air Conditioners. Heat pumps and air conditioners having duct heaters closer than 4 feet (1.22 m) to the heat pump or air conditioner shall have both the duct heater and heat pump or air conditioner identified as suitable for such installation and so marked.

424-62. Condensation. Duct heaters used with air conditioners or other air-cooling equipment that may result in condensation of moisture shall be identified as suitable for use with air conditioners.

424-63. Fan Circuit Interlock. Means shall be provided to ensure that the fan circuit is energized when any heater circuit is energized. However, time- or temperature-controlled delay in energizing the fan motor shall be permitted.

424-64. Limit Controls. Each duct heater shall be provided with an approved, integral, automatic-reset temperature-limiting control or controllers to deenergize the circuit or circuits.

In addition, an integral independent supplementary control or controllers shall be provided in each duct heater that will disconnect a sufficient number of conductors to interrupt current flow. This device shall be manually resettable or replaceable.

424-65. Location of Disconnecting Means. Duct heater controller equipment shall be accessible with the disconnecting means installed at or within sight from the controller.

Exception: As permitted by Section 424-19(a).

424-66. Installation. Duct heaters shall be installed in accordance with the manufacturer's instructions in a manner so that operation will not create a hazard to persons or property. Furthermore, duct heaters shall be located with respect to building construction and other equipment so as to permit access to the heater. Sufficient clearance shall be maintained to permit replacement of controls and heating elements and for adjusting and cleaning of controls and other parts requiring such attention. See Section 110-16.

(FPN): For additional installation information, see Installation of Air Conditioning and Ventilating Systems, NFPA 90A-1989 (ANSI) and Installation of Warm Air Heating and Air Conditioning Systems, NFPA 90B-1989 (ANSI).

G. Resistance-type Boilers

424-70. Scope. The provisions in Part G of this article shall apply to boilers employing resistance-type heating elements. Electrode-type boilers shall not be considered as employing resistance-type heating elements. See Part H of this article.

424-71. Identification. Resistance-type boilers shall be identified as suitable for the installation.

424-72. Overcurrent Protection.

(a) Boiler Employing Resistance-type Immersion Heating Elements in an ASME Rated and Stamped Vessel. A boiler employing resistance-type immersion heating elements contained in an ASME rated and stamped vessel shall have the heating elements protected at not more than 150 amperes. Such a boiler rated more than 120 amperes shall have the heating elements subdivided into loads not exceeding 120 amperes.

Where a subdivided load is less than 120 amperes, the rating of the overcurrent protective device shall comply with Section 424-3(b).

(b) Boiler Employing Resistance-type Heating Elements Rated More than 48 Amperes and Not Contained in an ASME Rated and Stamped Vessel. A boiler employing resistance-type heating elements not contained in an ASME rated and stamped vessel shall have the heating elements protected at not more than 60 amperes. Such a boiler rated more than 48 amperes shall have the heating elements subdivided into loads not exceeding 48 amperes.

Where a subdivided load is less than 48 amperes, the rating of the overcurrent protective device shall comply with Section 424-3(b).

(c) Supplementary Overcurrent Protective Devices. The supplementary overcurrent protective devices for the subdivided loads as required by Sections 424-72(a) and (b) shall be (1) factory-installed within or on the boiler enclosure or provided as a separate assembly by the boiler manufacturer; and (2) accessible, but need not be readily accessible; and (3) suitable for branch-circuit protection.

Where cartridge fuses are used to provide this overcurrent protection, a single disconnecting means shall be permitted for the several subdivided circuits. See Section 240-40.

(d) Conductors Supplying Supplementary Overcurrent Protective Devices. The conductors supplying these supplementary overcurrent protective devices shall be considered branch-circuit conductors.

Exception: Where the heaters are rated 50 kW or more, the conductors supplying the overcurrent protective device specified in (c) above shall be permitted to be sized at not less than 100 percent of the nameplate rating of the heater, provided all of the following conditions are met:

a. The heater is marked with a minimum conductor size; and

b. The conductors are not smaller than the marked minimum size; and

c. A temperature- or pressure-actuated device controls the cyclic operation of the equipment.

(e) Conductors for Subdivided Loads. Field-wired conductors between the heater and the supplementary overcurrent protective devices shall be sized at not less than 125 percent of the load served. The supplementary overcurrent protective devices specified in (c) shall protect these conductors in accordance with Section 240-3.

Exception: For heaters rated 50 kW or more, the ampacity of field-wired conductors between the heater and the supplementary overcurrent protective devices shall be permitted to be not less than 100 percent of the load of their respective subdivided circuits, provided all of the following conditions are met:

a. The heater is marked with a minimum conductor size; and

b. The conductors are not smaller than the marked minimum size; and

c. A temperature-activated device controls the cyclic operation of the equipment.

424-73. Over-Temperature Limit Control. Each boiler designed, so that in normal operation there is no change in state of the heat transfer medium, shall be equipped with a temperature sensitive limiting means. It shall be installed to limit maximum liquid temperature and shall directly or indirectly disconnect all ungrounded conductors to the heating elements. Such means shall be in addition to a temperature regulating system and other devices protecting the tank against excessive pressure.

424-74. Over-Pressure Limit Control. Each boiler designed, so that in normal operation there is a change in state of the heat transfer medium from liquid to vapor, shall be equipped with a pressure sensitive limiting means. It shall be installed to limit maximum pressure and shall directly or indirectly disconnect all ungrounded conductors to the heating elements. Such means shall be in addition to a pressure regulating system and other devices protecting the tank against excessive pressure.

424-75. Grounding. All noncurrent-carrying metal parts shall be grounded in accordance with Article 250. Means for connection of equipment grounding conductor(s) sized in accordance with Table 250-95 shall be provided.

H. Electrode-type Boilers

424-80. Scope. The provisions in Part H of this article shall apply to boilers for operation at 600 volts, nominal, or less, in which heat is generated by the passage of current between electrodes through the liquid being heated.

(FPN): For over 600 volts see Part G of Article 710.

424-81. Identification. Electrode-type boilers shall be identified as suitable for the installation.

424-82. Branch-Circuit Requirements. The size of branch-circuit conductors and overcurrent protective devices shall be calculated on the basis of 125 percent of the total load (motors not included). A contactor, relay, or other device, approved for continuous operation at 100 percent of its rating, shall be permitted to supply its full-rated load. See Section 210-22(c), Exception. The provisions of this section shall not apply to conductors that form an integral part of an approved boiler.

Exception: For an electrode boiler rated 50 kW or more, the conductors supplying the boiler electrode(s) shall be permitted to be sized at not less than 100 percent of the nameplate rating of the electrode boiler, provided all the following conditions are met:

a. The electrode boiler is marked with a minimum conductor size; and

b. The conductors are not smaller than the marked minimum size; and

c. A temperature- or pressure-actuated device controls the cyclic operation of the equipment.

424-83. Over-Temperature Limit Control. Each boiler designed, so that in normal operation there is no change in state of the heat transfer medium, shall be equipped with a temperature sensitive limiting means. It shall be installed to limit maximum liquid temperature and shall directly or indirectly interrupt all current flow through the electrodes. Such means shall be in addition to the temperature regulating system and other devices protecting the tank against excessive pressure.

424-84. Over-Pressure Limit Control. Each boiler designed, so that in normal operation there is a change in state of the heat transfer medium from liquid to vapor, shall be equipped with a pressure sensitive limiting means. It shall be installed to limit maximum pressure and shall directly or indirectly interrupt all current flow through the electrodes. Such means shall be in addition to a pressure regulating system and other devices protecting the tank against excessive pressure.

424-85. Grounding. For those boilers designed such that fault currents do not pass through the pressure vessel, and the pressure vessel is electrically isolated from the electrodes, all exposed noncurrent-carrying metal parts, including the pressure vessel, supply, and return connecting piping, shall be grounded in accordance with Article 250.

For all other designs, the pressure vessel containing the electrodes shall be isolated and electrically insulated from ground.

424-86. Markings. All electrode-type boilers shall be marked to show (1) the manufacturer's name; (2) the normal rating in volts, amperes, and kilowatts; (3) the electrical supply required specifying frequency, number of phases, and number of wires; (4) the marking: "Electrode-type Boiler"; (5) a warning marking — "ALL POWER SUPPLIES SHALL BE DISCONNECTED BEFORE SERVICING, INCLUDING SERVICING THE PRESSURE VESSEL."

The nameplate shall be located so as to be visible after installation.

J. Electric Radiant Heating Panels and Heating Panel Sets

424-90. Scope. The provisions of Part J of this article shall apply to radiant heating panels and heating panel sets.

424-91. Definitions.

(a) Heating Panel. A heating panel is a complete assembly provided with a junction box or a length of flexible conduit for connection to a branch circuit.

(b) Heating Panel Set. A heating panel set is a rigid or nonrigid assembly provided with nonheating leads or a terminal junction assembly identified as being suitable for connection to a wiring system.

424-92. Markings.

(1) Markings shall be permanent and in a location that is visible prior to application of panel finish.

(2) Each unit shall be identified as suitable for the installation.

(3) Each unit shall be marked with the identifying name or identification symbol, catalog number, and rating in volts and watts, or in volts and amperes.

(4) The manufacturers of heating panels or heating panel sets shall provide marking labels that indicate that the space heating installation incorporates heating panels or heating panel sets and instructions that the labels shall be affixed to the panelboards to identify which branch circuits supply the circuits to those space heating installations.

Exception: Heating panels and heating panel set installations that are visible and distinguishable after installation shall not be required to have labels provided.

424-93. Installation.

(a) General.

(1) Heating panels and heating panel sets shall be installed in accordance with the manufacturer's instructions.

(2) The heating portion shall not:

a. Be installed in or behind surfaces where subject to physical damage.

b. Be run through or above walls, partitions, cupboards, or similar portions of structures that extend to the ceiling.

c. Be run in or through thermal insulation, but shall be permitted to be in contact with the surface of thermal insulation.

(3) Edges of panels and panel sets shall be separated by not less than 8 inches (203 mm) from the edges of any outlet boxes and junction boxes that are to be used for mounting surface lighting fixtures. A clearance of not less than 2 inches (50.8 mm) shall be provided from recessed fixtures and their trims, ventilating openings, and other such openings in room surfaces. Sufficient area shall be provided to assure that no heating panel or heating panel set is to be covered by any surface-mounted units.

Exception: Heating panels and panel sets listed and marked for lesser clearances shall be permitted to be installed at the marked clearances.

(4) After the heating panels or heating panel sets are installed and inspected, it shall be permitted to install a surface that has been identified by the manufacturer's instructions as being suitable for the installation. The surface shall be secured so that the nails or other fastenings do not pierce the heating panels or heating panel sets.

(5) Surfaces permitted by Section 424-93(a)(4) shall be permitted to be covered with paint, wallpaper, or other approved surfaces identified in the manufacturer's instructions as being suitable.

(b) Heating Panel Sets.

(1) Heating panel sets shall be permitted to be secured to the lower face of joists or mounted in between joists, headers, or nailing strips.

(2) Heating panel sets shall be installed parallel to joists or nailing strips.

(3) Nailing or stapling of heating panel sets shall be done only through the unheated portions provided for this purpose. Heating panel sets shall not be cut through or nailed through any point closer than 1/4 inch (6.35 mm) to the element. Nails, staples, or other fasteners shall not be used where they penetrate current-carrying parts.

(4) Heating panel sets shall be installed as complete units unless identified as suitable for field cutting in an approved manner.

424-94. Clearances of Wiring in Ceilings.
Wiring located above heated ceilings shall be spaced not less than 2 inches (50.8 mm) above the heated ceiling and shall be considered as operating at an ambient of 50°C (122°F). The ampacity shall be computed on the basis of the correction factors given in the 0-2000 volt ampacity tables of Article 310.

Exception: Wiring above heated ceilings and located above thermal insulations having a minimum thickness of 2 inches (50.8 mm) shall not require correction for temperature.

424-95. Location of Branch-Circuit and Feeder Wiring in Walls.

(a) Exterior Walls. Wiring methods shall comply with Article 300 and Section 310-10.

(b) Interior Walls. Any wiring behind heating panels or heating panel sets located in interior walls or partitions shall be considered as operating at an ambient of 40°C (104°F), and the ampacity shall be computed on the basis of the correction factors given in the 0-2000 volt ampacity tables of Article 310.

424-96. Connection to Branch-Circuit Conductors.

(a) General. Heating panels or heating panel sets assembled together in the field to form a heating installation in one room or area shall be connected in accordance with the manufacturer's instructions.

(b) Heating Panels. Heating panels shall be connected to branch-circuit wiring by an approved wiring method.

(c) Heating Panel Sets.

(1) Heating panel sets shall be connected to branch-circuit wiring by a method identified as being suitable for the purpose.

(2) A heating panel set provided with terminal junction assembly shall be permitted to have the nonheating leads attached at the time of installation in accordance with the manufacturer's instructions.

424-97. Nonheating Leads.
Excess nonheating leads of heating panels or heating panel sets shall be permitted to be cut to the required length. They shall meet the installation requirements of the wiring method employed in accordance with Section 424-96. Nonheating leads shall be an integral part of a heating panel and a heating panel set and shall not be subjected to the ampacity requirements of Section 424-3(b) for branch circuits.

424-98. Installation in Concrete or Poured Masonry.

(a) Maximum Heated Area. Heating panels or heating panel sets shall not exceed 33 watts per square foot (0.093 sq m) of heated area.

(b) Secured in Place and Identified as Suitable. Heating panels or heating panel sets shall be secured in place by means specified in the manufacturer's instructions and identified as suitable for the installation.

(c) **Expansion Joints.** Heating panels or heating panel sets shall not be installed where they bridge expansion joints unless provision is made for expansion and contraction.

(d) **Spacings.** Spacings shall be maintained between heating panels or heating panel sets and metal embedded in the floor.

Exception: Grounded metal-clad heating panels shall be permitted to be in contact with metal embedded in the floor.

(e) **Protection of Leads.** Leads shall be protected where they leave the floor by rigid metal conduit, intermediate metal conduit, rigid nonmetallic conduit, electrical metallic tubing, or by other approved means.

(f) **Bushings or Fittings Required.** Bushings or approved fittings shall be used where the leads emerge within the floor slabs.

424-99. Installation Under Floor Covering.

(a) **Identification.** Heating panels or heating panel sets for installation under floor covering shall be identified as suitable for installation under floor covering.

(b) **Maximum Heated Area.** Heating panels or panel sets, installed under floor covering, shall not exceed 15 watts per square foot (0.093 m^2) of heated area.

(c) **Installation.** Listed heating panels or panel sets, if installed under floor covering, shall be installed on floor surfaces that are smooth and flat in accordance with the manufacturer's instructions and shall also comply with the following:

(1) **Expansion Joints.** Heating panels or heating panel sets shall not be installed where they bridge expansion joints unless protected from expansion and contraction.

(2) **Connection to Conductors.** Heating panels and heating panel sets shall be connected to branch-circuit and supply wiring by wiring methods recognized in Chapter 3.

(3) **Anchoring.** Heating panels and heating panel sets shall be firmly anchored to the floor using an adhesive or anchoring system identified for this use.

(4) **Coverings.** After heating panels or heating panel sets are installed and inspected, they shall be permitted to be covered by a floor covering that has been identified by the manufacturer as being suitable for the installation. The covering shall be secured to the heating panel or heating panel sets with release-type adhesives or by means identified for this use.

(5) **Fault Protection.** A device to open all ungrounded conductors supplying the heating panels or heating panel sets, provided by the manufacturer, shall function when a low or high resistance line-to-line, line to grounded conductor, or line to ground fault occurs, such as the result of a penetration of the element/element assembly.

(FPN): An integral grounding shield may be required to provide this protection.

ARTICLE 426 — FIXED OUTDOOR ELECTRIC DEICING AND SNOW-MELTING EQUIPMENT

A. General

426-1. Scope. The requirements of this article shall apply to electrically energized heating systems and the installation of these systems.

(a) Embedded. Embedded in driveways, walks, steps, and other areas.

(b) Exposed. Exposed on drainage systems, bridge structures, roofs, and other structures.

426-2. Definitions. For the purpose of this article:

Heating System. A complete system consisting of components such as heating elements, fastening devices, nonheating circuit wiring, leads, temperature controllers, safety signs, junction boxes, raceways, and fittings.

Resistance Heating Element. A specific separate element to generate heat that is embedded in or fastened to the surface to be heated.

(FPN): Tubular heaters, strip heaters, heating cable, heating tape, and heating panels are examples of resistance heaters.

Impedance Heating System. A system in which heat is generated in a pipe or rod, or combination of pipes and rods, by causing current to flow through the pipe or rod by direct connection to an ac voltage source from a dual-winding transformer. The pipe or rod shall be permitted to be embedded in the surface to be heated, or constitute the exposed components to be heated.

Skin Effect Heating System. A system in which heat is generated on the inner surface of a ferromagnetic envelope embedded in or fastened to the surface to be heated.

(FPN): Typically, an electrically insulated conductor is routed through and connected to the envelope at the other end. The envelope and the electrically insulated conductor are connected to an ac voltage source from a dual-winding transformer.

426-3. Application of Other Articles. All requirements of this Code shall apply except as specifically amended in this article. Cord- and plug-connected fixed outdoor electric deicing and snow-melting equipment intended for specific use and identified as suitable for this use shall be installed according to Article 422. Fixed outdoor electric deicing and snow-melting equipment for use in hazardous (classified) locations shall comply with Articles 500 through 516.

426-4. Branch-Circuit Sizing. The ampacity of branch-circuit conductors and the rating or setting of overcurrent protective devices supplying fixed outdoor electric deicing and snow-melting equipment shall not be less than 125 percent of the total load of the heaters. The rating or setting of overcurrent protective devices shall be permitted in accordance with Section 240-3(b).

B. Installation

426-10. General. Equipment for outdoor electric deicing and snow melting shall be identified as being suitable for:

   (1) The chemical, thermal, and physical environment, and

   (2) Installation in accordance with the manufacturer's drawings and instructions.

426-11. Use. Electrical heating equipment shall be installed in such a manner as to be afforded protection from physical damage.

426-12. Thermal Protection. External surfaces of outdoor electric deicing and snow-melting equipment that operate at temperatures exceeding 60°C (140°F) shall be physically guarded, isolated, or thermally insulated to protect against contact by personnel in the area.

426-13. Identification. The presence of outdoor electric deicing and snow-melting equipment shall be evident by the posting of appropriate caution signs or markings where clearly visible.

426-14. Special Permission. Fixed outdoor deicing and snow-melting equipment employing methods of construction or installation other than covered by this article shall be permitted only by special permission.

C. Resistance Heating Elements

426-20. Embedded Deicing and Snow-Melting Equipment.

   (a) Watt Density. Panels or units shall not exceed 120 watts per square foot (0.093 sq m) of heated area.

   (b) Spacing. The spacing between adjacent cable runs is dependent upon the rating of the cable, and shall be not less than 1 inch (25.4 mm) on centers.

   (c) Cover. Units, panels, or cables shall be installed:

   (1) On a substantial asphalt or masonry base at least 2 inches (50.8 mm) thick and have at least $1\frac{1}{2}$ inches (38 mm) of asphalt or masonry applied over the units, panels, or cables; or

   (2) They shall be permitted to be installed over other approved bases and embedded within $3\frac{1}{2}$ inches (89 mm) of masonry or asphalt but not less than $1\frac{1}{2}$ inches (38 mm) from the top surface; or

   (3) Equipment that has been specially investigated for other forms of installation shall be installed only in the manner for which it has been investigated.

   (d) Secured. Cables, units, and panels shall be secured in place by frames or spreaders or other approved means while the masonry or asphalt finish is applied.

   (e) Expansion and Contraction. Cables, units, and panels shall not be installed where they bridge expansion joints unless provision is made for expansion and contraction.

426-21. Exposed Deicing and Snow-Melting Equipment.

(a) Secured. Heating element assemblies shall be secured to the surface being heated by approved means.

(b) Overtemperature. Where the heating element is not in direct contact with the surface being heated, the design of the heater assembly shall be such that its temperature limitations shall not be exceeded.

(c) Expansion and Contraction. Heating elements and assemblies shall not be installed where they bridge expansion joints unless provision is made for expansion and contraction.

(d) Flexural Capability. Where installed on flexible structures, the heating elements and assemblies shall have a flexural capability compatible with the structure.

426-22. Installation of Nonheating Leads for Embedded Equipment.

(a) Grounding Sheath or Braid. Nonheating leads having a grounding sheath or braid shall be permitted to be embedded in the masonry or asphalt in the same manner as the heating cable without additional physical protection.

(b) Raceways. All but 1 to 6 inches (25.4 to 152 mm) of nonheating leads of Type TW and other approved types not having a grounding sheath shall be enclosed in a rigid conduit, electrical metallic tubing, intermediate metal conduit, or other raceways within asphalt or masonry; and the distance from the factory splice to raceway shall be not less than 1 inch (25.4 mm) or more than 6 inches (152 mm).

(c) Bushings. Insulating bushings shall be used in the asphalt or masonry where leads enter conduit or tubing.

(d) Expansion and Contraction. Leads shall be protected in expansion joints and where they emerge from masonry or asphalt by rigid conduit, electrical metallic tubing, intermediate metal conduit, other raceways, or other approved means.

(e) Leads in Junction Boxes. Not less than 6 inches (152 mm) of free nonheating lead shall be within the junction box.

426-23. Installation of Nonheating Leads for Exposed Equipment.

(a) Nonheating Leads. Power supply nonheating leads (cold leads) for resistance elements shall be suitable for the temperature encountered. Preassembled nonheating leads on approved heaters shall be permitted to be shortened if the markings specified in Section 426-25 are retained. Not less than 6 inches (152 mm) of nonheating leads shall be provided within the junction box.

(b) Protection. Nonheating power supply leads shall be enclosed in a rigid conduit, intermediate metal conduit, electrical metallic tubing, or other approved means.

426-24. Electrical Connection.

(a) Heating Element Connections. Electrical connections, other than factory connections of heating elements to nonheating elements embedded in masonry or asphalt or on exposed surfaces, shall be made with insulated connectors identified for the use.

(b) Circuit Connections. Splices and terminations at the end of the non-heating leads, other than the heating element end, shall be installed in a box or fitting in accordance with Sections 110-14 and 300-15.

426-25. Marking. Each factory-assembled heating unit shall be legibly marked within 3 inches (76 mm) of each end of the nonheating leads with the permanent identification symbol, catalog number, and ratings in volts and watts, or in volts and amperes.

426-26. Corrosion Protection. Ferrous and nonferrous metal raceways, cable armor, cable sheaths, boxes, fittings, supports, and support hardware shall be permitted to be installed in concrete or in direct contact with the earth, or in areas subject to severe corrosive influences, where made of material suitable for the condition, or where provided with corrosion protection identified as suitable for the condition.

426-27. Grounding.

(a) Metal Parts. Exposed noncurrent-carrying metal parts of equipment likely to become energized shall be bonded together and grounded in a manner specified in Article 250.

(b) Grounding Braid or Sheath. Grounding means, such as copper braid, metal sheath, or other approved means, shall be provided as part of the heated section of the cable, panel, or unit.

D. Impedance Heating

426-30. Personnel Protection. Exposed elements of impedance heating systems shall be physically guarded, isolated, or thermally insulated with weatherproof jacket to protect against contact by personnel in the area.

vs426-31. Voltage Limitations. The impedance heating elements shall not operate at a voltage greater than 30 volts ac.

Exception: The voltage shall be permitted to be greater than 30 volts, but not more than 80 volts, if a ground-fault circuit-interrupter for personnel protection is provided.

426-32. Isolation Transformer. A dual-winding transformer with a grounded shield between the primary and secondary windings shall be used to isolate the distribution system from the heating system.

426-33. Induced Currents. All current-carrying components shall be installed in accordance with Section 300-20.

426-34. Grounding. An impedance heating system that is operating at a voltage greater than 30, but not more than 80, shall be grounded at designated point(s).

E. Skin Effect Heating

426-40. Conductor Ampacity. The current through the electrically insulated conductor inside the ferromagnetic envelope shall be permitted to exceed the ampacity values shown in Article 310, provided it is identified as suitable for this use.

426-41. Pull Boxes. Where pull boxes are used, they shall be accessible without excavation by location in suitable vaults or above grade. Outdoor pull boxes shall be of watertight construction.

426-42. Single Conductor in Enclosure. The provisions of Section 300-20 shall not apply to the installation of a single conductor in a ferromagnetic envelope (metal enclosure).

426-43. Corrosion Protection. Ferromagnetic envelopes, ferrous or nonferrous metal raceways, boxes, fittings, supports, and support hardware shall be permitted to be installed in concrete or in direct contact with the earth, or in areas subjected to severe corrosive influences, where made of material suitable for the condition, or where provided with corrosion protection identified as suitable for the condition. Corrosion protection shall maintain the original wall thickness of the ferromagnetic envelope.

426-44. Grounding. The ferromagnetic envelope shall be grounded at both ends; and, in addition, it shall be permitted to be grounded at intermediate points as required by its design.

The provisions of Section 250-26 shall not apply to the installation of skin effect heating systems.

(FPN): For grounding methods, see Section 250-26(d).

F. Control and Protection

426-50. Disconnecting Means.

(a) Disconnection. All fixed outdoor deicing and snow-melting equipment shall be provided with a means for disconnection from all ungrounded conductors. Where readily accessible to the user of the equipment, the branch-circuit switch or circuit breaker shall be permitted to serve as the disconnecting means. Switches used as the disconnecting means shall be of the indicating type.

(b) Cord- and Plug-Connected Equipment. The factory-installed attachment plug of cord- and plug-connected equipment rated 20 amperes or less and 150 volts or less to ground shall be permitted to be the disconnecting means.

426-51. Controllers.

(a) Temperature Controller with "Off" Position. Temperature controlled switching devices that indicate an "off" position and that interrupt line current shall open all ungrounded conductors when the control device is in the "off" position. These devices shall not be permitted to serve as the disconnecting means unless provided with a positive lockout in the "off" position.

(b) Temperature Controller Without "Off" Position. Temperature controlled switching devices that do not have an "off" position shall not be required to open all ungrounded conductors and shall not be permitted to serve as the disconnecting means.

(c) Remote Temperature Controller. Remote controlled temperature-actuated devices shall not be required to meet the requirements of Section 426-51(a). These devices shall not be permitted to serve as the disconnecting means.

(d) Combined Switching Devices. Switching devices consisting of combined temperature-actuated devices and manually controlled switches that serve both as the controller and the disconnecting means shall comply with all of the following conditions:

(1) Open all ungrounded conductors when manually placed in the "off" position; and

(2) Be so designed that the circuit cannot be energized automatically if the device has been manually placed in the "off" position; and

(3) Be provided with a positive lockout in the "off" position.

426-52. Overcurrent Protection. Fixed outdoor electric deicing and snow-melting equipment shall be permitted to be protected against overcurrent where supplied by a branch circuit as specified in Section 426-4.

426-53. Equipment Protection. Ground-fault protection of equipment shall be provided for branch circuits supplying fixed outdoor electric deicing and snow-melting equipment.

426-54. Cord- and Plug-Connected Deicing and Snow-Melting Equipment. Cord- and plug-connected deicing and snow-melting equipment shall be listed.

ARTICLE 427 — FIXED ELECTRIC HEATING EQUIPMENT FOR PIPELINES AND VESSELS

A. General

427-1. Scope. The requirements of this article shall apply to electrically energized heating systems and the installation of these systems used with pipelines or vessels or both.

427-2. Definitions. For the purpose of this article:

Pipeline: A length of pipe including pumps, valves, flanges, control devices, strainers and/or similar equipment for conveying fluids.

Vessel: A container such as a barrel, drum, or tank for holding fluids or other material.

Integrated Heating System: A complete system consisting of components such as pipelines, vessels, heating elements, heat transfer medium, thermal insulation, moisture barrier, nonheating leads, temperature controllers, safety signs, junction boxes, raceways, and fittings.

Resistance Heating Element: A specific separate element to generate heat that is applied to the pipeline or vessel externally or internally.

(FPN):   Tubular heaters, strip heaters, heating cable, heating tape, heating blankets, and immersion heaters are examples of resistance heaters.

Impedance Heating System: A system in which heat is generated in a pipeline or vessel wall by causing current to flow through the pipeline or vessel wall by direct connection to an ac voltage source from a dual-winding transformer.

Induction Heating System: A system in which heat is generated in a pipeline or vessel wall by inducing current and hysteresis effect in the pipeline or vessel wall from an external isolated ac field source.

Skin Effect Heating System: A system in which heat is generated on the inner surface of a ferromagnetic envelope attached to a pipeline and/or vessel.

(FPN):   Typically, an electrically insulated conductor is routed through and connected to the envelope at the other end. The envelope and the electrically insulated conductor are connected to an ac voltage source from a dual-winding transformer.

427-3. Application of Other Articles.   All requirements of this Code shall apply except as specifically amended in this article. Cord-connected pipe heating assemblies intended for specific use and identified as suitable for this use shall be installed according to Article 422. Fixed electric pipeline and vessel heating equipment for use in hazardous (classified) locations shall comply with Articles 500 through 516.

427-4. Branch-Circuit Sizing.   The ampacity of branch-circuit conductors and the rating or setting of overcurrent protective devices supplying fixed electric heating equipment for pipelines and vessels shall be not less than 125 percent of the total load of the heaters. The rating or setting of overcurrent protective devices shall be permitted in accordance with Section 240-3(b).

B. Installation

427-10. General.   Equipment for pipeline and vessel electrical heating shall be identified as being suitable for (1) the chemical, thermal, and physical environment; and (2) installation in accordance with the manufacturer's drawings and instructions.

427-11. Use.   Electrical heating equipment shall be installed in such a manner as to be afforded protection from physical damage.

427-12. Thermal Protection.   External surfaces of pipeline and vessel heating equipment that operate at temperatures exceeding 60°C (140°F) shall be physically guarded, isolated, or thermally insulated to protect against contact by personnel in the area.

427-13. Identification. The presence of electrically heated pipelines or vessels, or both, shall be evident by the posting of appropriate caution signs or markings at frequent intervals along the pipeline or vessel.

C. Resistance Heating Elements

427-14. Secured. Heating element assemblies shall be secured to the surface being heated by means other than the thermal insulation.

427-15. Not in Direct Contact. Where the heating element is not in direct contact with the pipeline or vessel being heated, means shall be provided to prevent overtemperature of the heating element unless the design of the heater assembly is such that its temperature limitations will not be exceeded.

427-16. Expansion and Contraction. Heating elements and assemblies shall not be installed where they bridge expansion joints unless provisions are made for expansion and contraction.

427-17. Flexural Capability. Where installed on flexible pipelines, the heating elements and assemblies shall have a flexural capability compatible with the pipeline.

427-18. Power Supply Leads.

(a) Nonheating Leads. Power supply nonheating leads (cold leads) for resistance elements shall be suitable for the temperature encountered. Preassembled nonheating leads on approved heaters shall be permitted to be shortened if the markings specified in Section 427-20 are retained. Not less than 6 inches (152 mm) of nonheating leads shall be provided within the junction box.

(b) Power Supply Leads Protection. Nonheating power supply leads shall be protected where they emerge from electrically heated pipeline or vessel heating units by rigid metal conduit, intermediate metal conduit, electrical metallic tubing, or other raceways identified as suitable for the application.

(c) Interconnecting Leads. Interconnecting nonheating leads connecting portions of the heating system shall be permitted to be covered by thermal insulation in the same manner as the heaters.

427-19. Electrical Connections.

(a) Nonheating Interconnections. Nonheating interconnections, where required under thermal insulation, shall be made with insulated connectors identified as suitable for this use.

(b) Circuit Connections. Splices and terminations outside the thermal insulation shall be installed in a box or fitting in accordance with Sections 110-14 and 300-15.

427-20. Marking. Each factory-assembled heating unit shall be legibly marked within 3 inches (76 mm) of each end of the nonheating leads with the permanent identification symbol, catalog number, and ratings in volts and watts, or in volts and amperes.

427-21. Grounding. Exposed noncurrent-carrying metal parts of electric heating equipment that are likely to become energized shall be grounded as required in Article 250.

427-22. Equipment Protection. Ground-fault protection of equipment shall be provided for branch circuits supplying electric heating equipment not having a metal covering.

427-23. Nonmetallic Pipelines. Heating assemblies intended for heating nonmetallic pipelines or vessels shall have an overall grounded metal covering.

D. Impedance Heating

427-25. Personnel Protection. All accessible external surfaces of the pipeline and/or vessel being heated shall be physically guarded, isolated, or thermally insulated (with weatherproof jacket for outside installations) to protect against contact by personnel in the area.

427-26. Voltage Limitations. The secondary winding of the isolation transformer (Section 427-27) connected to the pipeline or vessel being heated shall not have an output voltage greater than 30 volts ac.

Exception: The voltage shall be permitted to be greater than 30 volts but not more than 80 volts if a ground-fault circuit-interrupter for personnel protection is provided.

427-27. Isolation Transformer. A dual-winding transformer with a grounded shield between the primary and secondary windings shall be used to isolate the distribution system from the heating system.

427-28. Induced Currents. All current-carrying components shall be installed in accordance with Section 300-20.

427-29. Grounding. The pipeline or vessel, or both, being heated that is operating at a voltage greater than 30 but not more than 80 shall be grounded at designated points.

427-30. Secondary Conductor Sizing. The ampacity of the conductors connected to the secondary of the transformer shall be at least 100 percent of the total load of the heater.

E. Induction Heating

427-35. Scope. This part covers the installation of line frequency induction heating equipment and accessories for pipelines and vessels.

(FPN): See Article 665 for other applications.

427-36. Personnel Protection. Induction coils that operate or may operate at a voltage greater than 30 volts ac shall be enclosed in a nonmetallic or split metallic enclosure, isolated or made inaccessible by location to protect personnel in the area.

427-37. Induced Current. Induction coils shall be prevented from inducing circulating currents in surrounding metallic equipment, supports, or structures by shielding, isolation, or insulation of the current paths. Stray current paths shall be bonded to prevent arcing.

F. Skin Effect Heating

427-45. Conductor Ampacity. The ampacity of the electrically insulated conductor inside the ferromagnetic envelope shall be permitted to exceed the values given in Article 310, provided it is identified as suitable for this use.

427-46. Pull Boxes. Pull boxes for pulling the electrically insulated conductor in the ferromagnetic envelope shall be permitted to be buried under the thermal insulation, provided their locations are indicated by permanent | markings on the insulation jacket surface and on drawings. For outdoor installations, pull boxes shall be of watertight construction.

427-47. Single Conductor in Enclosure. The provisions of Section 300-20 shall not apply to the installation of a single conductor in a ferromagnetic envelope (metal enclosure).

427-48. Grounding. The ferromagnetic envelope shall be grounded at both ends and, in addition, it shall be permitted to be grounded at intermediate points as required by its design. The ferromagnetic envelope shall be bonded at all joints to assure electrical continuity.

The provisions of Section 250-26 shall not apply to the installation of skin effect heating systems.

(FPN): See Section 250-26(d) for grounding methods. |

G. Control and Protection

427-55. Disconnecting Means.

(a) **Switch or Circuit Breaker.** Means shall be provided to disconnect all fixed electric pipeline or vessel heating equipment from all ungrounded conductors. The branch-circuit switch or circuit breaker, where readily accessible to the user of the equipment, shall be permitted to serve as the disconnecting means. The disconnecting means shall be of the indicating type, and shall be provided with a positive lockout in the "off" position.

(b) **Cord- and Plug-Connected Equipment.** The factory-installed attachment plug of cord- and plug-connected equipment rated 20 amperes or less and 150 volts or less to ground shall be permitted to be the disconnecting means.

427-56. Controls.

(a) Temperature Control with "Off" Position. Temperature controlled switching devices that indicate an "off" position and that interrupt line current shall open all ungrounded conductors when the control device is in this "off" position. These devices shall not be permitted to serve as the disconnecting means unless provided with a positive lockout in the "off" position.

(b) Temperature Control Without "Off" Position. Temperature controlled switching devices that do not have an "off" position shall not be required to open all ungrounded conductors and shall not be permitted to serve as the disconnecting means.

(c) Remote Temperature Controller. Remote controlled temperature-actuated devices shall not be required to meet the requirements of Sections 427-56(a) and (b). These devices shall not be permitted to serve as the disconnecting means.

(d) Combined Switching Devices. Switching devices consisting of combined temperature-actuated devices and manually controlled switches that serve both as the controllers and the disconnecting means shall comply with all the following conditions:

(1) Open all ungrounded conductors when manually placed in the "off" position; and

(2) Be so designed that the circuit cannot be energized automatically if the device has been manually placed in the "off" position; and

(3) Be provided with a positive lockout in the "off" position.

427-57. Overcurrent Protection. Heating equipment shall be considered as protected against overcurrent where supplied by a branch circuit as specified in Section 427-4.

ARTICLE 430 — MOTORS, MOTOR CIRCUITS, AND CONTROLLERS

A. General

430-1. Scope. This article covers motors, motor branch-circuit and feeder conductors and their protection, motor overload protection, motor control circuits, motor controllers, and motor control centers.

Exception No. 1: Installation requirements for motor control centers are covered in Section 384-4.

Exception No. 2: Air-conditioning and refrigerating equipment are covered in Article 440.

(FPN): Diagram 430-1 is for information only.

430-2. Adjustable Speed Drive Systems. The incoming branch circuit or feeder to power conversion equipment included as a part of an adjustable speed drive system shall be based on the rated input to the power conversion equipment. If the power conversion equipment provides overload protection for the motor, additional overload protection is not required.

Part A. General, Sections 430–1 through 430–18
Part B. Motor Circuit Conductors, Sections 430–21 through 430–29.
Part C. Motor and Branch-Circuit Overload Protection, Sections
    430–31 through 430–44.
Part D. Motor Branch-Circuit Short-Circuit and Ground-Fault
    Protection, Sections 430–51 through 430–58
Part E. Motor Feeder Short-Circuit and Ground-Fault Protection,
    Sections 430–61 through 430–63
Part F. Motor Control Circuits, Sections 430–71 through 430–74.
Part G. Motor Controllers, Sections 430–81 through 430–91.
Part H. Motor Control Centers, Sections 430–92 Through 430–98.
Part I. Disconnecting Means, Sections 430–101 through 430–113.
Part J. Over 600 Volts, Nominal, Sections 430–121 through 430–127.
Part K. Protection of Live Parts—All Voltages, Sections 430–131
    through 430–133.
Part L. Grounding—All Voltages, Sections 430–141 through 430–145.
Part M. Tables, Tables 430–147 through 430–152.

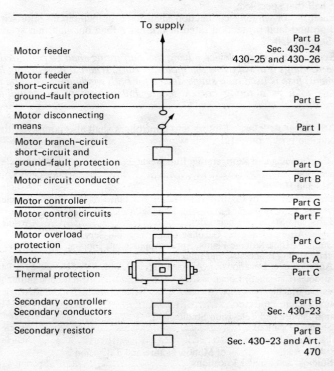

(FPN) Diagram 430-1.

The disconnecting means shall be permitted to be in the incoming line to the conversion equipment and shall have a rating not less than 115 percent of the rated input current of the conversion unit.

430-3. Part-Winding Motors. A part-winding-start induction or synchronous motor is one arranged for starting by first energizing part of its primary (armature) winding and, subsequently, energizing the remainder of this winding in one or more steps. The purpose is to reduce the initial values of the starting current drawn or the starting torque developed by the motor. A standard part-winding-start induction motor is arranged so that one-half of its primary winding can be energized initially, and, subsequently, the remaining half can be energized, both halves then carrying equal current. A hermetic refrigerant compressor motor shall not be considered a standard part-winding-start induction motor.

Where separate overload devices are used with a standard part-winding-start induction motor, each half of the motor winding shall be individually protected in accordance with Sections 430-32 and 430-37 with a trip current one-half that specified.

Each motor-winding connection shall have branch-circuit short-circuit and ground-fault protection rated at not more than one-half that specified by Section 430-52.

Exception: A single device having this half rating shall be permitted for both windings if it will allow the motor to start. Where a time-delay (dual-element) fuse is used as a single device for both windings, it shall be permitted to have a rating not exceeding 150 percent of motor full-load current.

430-5. Other Articles. Motors and controllers shall also comply with the applicable provisions of the following:

Air-Conditioning and Refrigerating Equipment ....Article 440
Capacitors...Sections 460-8, 460-9
Cranes and Hoists...Article 610
Electrically Driven or Controlled Irrigation Machines...................Article 675
Elevators, Dumbwaiters, Escalators, Moving Walks,Wheelchair
    Lifts, and Stairway Chair Lifts..Article 620
Garages, Aircraft Hangars, Gasoline Dispensing and Service
    Stations, Bulk Storage Plants, Spray Application, Dipping,
    and Coating Processes, and Inhalation
    Anesthetizing Locations.....Articles 511, 513, 514, 515, 516, and 517, Part D
Hazardous (Classified) Locations............................Articles 500 through 503
Industrial Machinery..Article 670
Motion Picture Projectors.....................................Sections 540-11, 540-20
Motion Picture and Television Studios and
    Similar Locations...Article 530
Resistors and Reactors...Article 470
Theaters, Audience Areas of Motion Picture and Television
    Studios, and Similar Locations ....Section 520-48
Transformers and Transformer Vaults ....Article 450

430-6. Ampacity and Motor Rating Determination. The size of conductors supplying equipment covered by this article shall be selected from

Tables 310-16 through 310-19 or shall be calculated in accordance with Section 310-15(b). The required ampacity and motor ratings shall be determined as specified in (a), (b), and (c) below.

(a) General Motor Applications. Other than as specified for torque motors in (b) below and for ac adjustable voltage motors in (c) below, where the current rating of a motor is used to determine the ampacity of conductors or ampere ratings of switches, branch-circuit short-circuit and ground-fault protection, etc., the values given in Tables 430-147, 430-148, 430-149, and 430-150, including notes, shall be used instead of the actual current rating marked on the motor nameplate. Separate motor overload protection shall be based on the motor nameplate current rating. Where a motor is marked in amperes, but not horsepower, the horsepower rating shall be assumed to be that corresponding to the value given in Tables 430-147, 430-148, 430-149, and 430-150, interpolated if necessary.

Exception No. 1: Multispeed motors shall be in accordance with Sections 430-22(a) and 430-52.

Exception No. 2: For equipment employing a shaded-pole or permanent-split-capacitor-type fan or blower motor that is marked with the motor type, the full-load current for such motor marked on the nameplate of the equipment in which the fan or blower motor is employed shall be used instead of the horsepower rating to determine the ampacity or rating of the disconnecting means, the branch-circuit conductors, the controller, the branch-circuit short-circuit and ground-fault protection, and the separate overload protection. This marking on the equipment nameplate shall not be less than the current marked on the fan or blower motor nameplate.

(b) Torque Motors. For torque motors, the rated current shall be locked-rotor current, and this nameplate current shall be used to determine the ampacity of the branch-circuit conductors covered in Sections 430-22 and 430-24, the ampere rating of the motor overload protection, and the ampere rating of motor branch-circuit and ground-fault protection in accordance with Section 430-52(b).

(FPN): For motor controllers and disconnecting means, see Section 430-83, Exception No. 3 and Section 430-110.

(c) AC Adjustable Voltage Motors. For motors used in alternating-current, adjustable voltage, variable torque drive systems, the ampacity of conductors, or ampere ratings of switches, branch-circuit short-circuit and ground-fault protection, etc., shall be based on the maximum operating current marked on the motor or control nameplate, or both. If the maximum operating current does not appear on the nameplate, the ampacity determination shall be based on 150 percent of the values given in Tables 430-149 and 430-150.

430-7. Marking on Motors and Multimotor Equipment.

(a) Usual Motor Applications. A motor shall be marked with the following information:

(1) Maker's name.

(2) Rated volts and full-load amperes. For a multispeed motor, full-load amperes for each speed, except shaded-pole and permanent-split capacitor motors where amperes are required only for maximum speed.

(3) Rated frequency and number of phases, if an alternating-current motor.

(4) Rated full-load speed.

(5) Rated temperature rise or the insulation system class and rated ambient temperature.

(6) Time rating. The time rating shall be 5, 15, 30, or 60 minutes, or continuous.

(7) Rated horsepower if ⅛ horsepower or more. For a multispeed motor ⅛ horsepower or more, rated horsepower for each speed, except shaded-pole and permanent-split capacitor motors ⅛ horsepower or more where rated horsepower is required only for maximum speed. Motors of arc welders are not required to be marked with the horsepower rating.

(8) Code letter if an alternating-current motor rated ½ horsepower or more. On polyphase wound-rotor motors the code letter shall be omitted.

(FPN): See (b) below.

(9) Secondary volts and full-load amperes if a wound-rotor induction motor.

(10) Field current and voltage for direct-current excited synchronous motors.

(11) Winding: straight shunt, stabilized shunt, compound, or series, if a direct-current motor. Fractional horsepower dc motors 7 inches (178 mm) or less in diameter shall not be required to be marked.

(12) A motor provided with a thermal protector complying with Sections 430-32(a) (2) or (c) (2) shall be marked "Thermally Protected." Thermally protected motors rated 100 watts or less and complying with Section 430-32(c) (2) shall be permitted to use the abbreviated marking "T.P."

(13) A motor complying with Section 430-32(c) (4) shall be marked "Impedance Protected." Impedance protected motors rated 100 watts or less and complying with Section 430-32(c) (4) shall be permitted to use the abbreviated marking "Z.P."

(b) Locked-Rotor Indicating Code Letters. Code letters marked on motor nameplates to show motor input with locked rotor shall be in accordance with Table 430-7(b).

The code letter indicating motor input with locked rotor shall be in an individual block on the nameplate, properly designated. This code letter shall be used for determining branch-circuit short-circuit and ground-fault protection by reference to Table 430-152, as provided in Section 430-52.

(1) Multispeed motors shall be marked with the code letter designating the locked-rotor kVA per horsepower for the highest speed at which the motor can be started.

Exception: Constant-horsepower multispeed motors shall be marked with the code letter giving the highest locked-rotor kVA per horsepower.

(2) Single-speed motors starting on Y connection and running on delta connections shall be marked with a code letter corresponding to the locked-rotor kVA per horsepower for the Y connection.

(3) Dual-voltage motors that have a different locked-rotor kVA per horsepower on the two voltages shall be marked with the code letter for the voltage giving the highest locked-rotor kVA per horsepower.

(4) Motors with 60- and 50-hertz ratings shall be marked with a code letter designating the locked-rotor kVA per horsepower on 60 hertz.

(5) Part-winding-start motors shall be marked with a code letter designating the locked-rotor kVA per horsepower that is based upon the locked-rotor current for the full winding of the motor.

Table 430-7(b).   Locked-Rotor Indicating Code Letters

Code Letter		Kilovolt-Amperes per Horsepower with Locked Rotor		
A	....	0	—	3.14
B	....	3.15	—	3.54
C	....	3.55	—	3.99
D	....	4.0	—	4.49
E	....	4.5	—	4.99
F	....	5.0	—	5.59
G	....	5.6	—	6.29
H	....	6.3	—	7.09
J	....	7.1	—	7.99
K	....	8.0	—	8.99
L	....	9.0	—	9.99
M	....	10.0	—	11.19
N	....	11.2	—	12.49
P	....	12.5	—	13.99
R	....	14.0	—	15.99
S	....	16.0	—	17.99
T	....	18.0	—	19.99
U	....	20.0	—	22.39
V	....	22.4	—	and up

(c) Torque Motors. Torque motors are rated for operation at standstill and shall be marked in accordance with (a) above.

Exception: Locked-rotor torque shall replace horsepower.

(d) Multimotor and Combination-Load Equipment.

(1) Multimotor and combination-load equipment shall be provided with a visible nameplate marked with the maker's name, the rating in volts, frequency, number of phases, minimum supply circuit conductor ampacity, and the maximum ampere rating of the circuit short-circuit and ground-fault protective device. The conductor ampacity shall be computed in accordance with Section 430-24 and counting all of the motors and other loads that will be operated at the same time. The short-circuit and ground-fault protective device rating shall not exceed the value computed in accordance with Section 430-53. Multimotor equipment for use on two or more circuits shall be marked with the above information for each circuit.

(2) Where the equipment is not factory-wired and the individual nameplates of motors and other loads are visible after assembly of the equipment, the individual nameplates shall be permitted to serve as the required marking.

430-8. Marking on Controllers. A controller shall be marked with the maker's name or identification, the voltage, the current or horsepower rating, and such other necessary data to properly indicate the motors for which it is suitable. A controller that includes motor overload protection suitable for group motor application shall be marked with the motor overload protection and the maximum branch-circuit short-circuit and ground-fault protection for such applications.

Combination controllers employing adjustable instantaneous trip circuit breakers shall be clearly marked to indicate the ampere settings of the adjustable trip element.

Where a controller is built-in as an integral part of a motor or of a motor-generator set, individual marking of the controller shall not be required if the necessary data are on the nameplate. For controllers that are an integral part of equipment approved as a unit, the above marking shall be permitted on the equipment nameplate.

430-9. Terminals.

(a) Markings. Terminals of motors and controllers shall be suitably marked or colored where necessary to indicate the proper connections.

(b) Conductors. Motor controllers and terminals of control circuit devices shall be connected with copper conductors unless identified for use with a different conductor.

(c) Torque Requirements. Control circuit devices with screw-type pressure terminals used with No. 14 or smaller copper conductors shall be torqued to a minimum of 7 pound-inches (0.79 N-m) unless identified for a different torque value.

430-10. Wiring Space in Enclosures.

(a) General. Enclosures for motor controllers and disconnecting means shall not be used as junction boxes, auxiliary gutters, or raceways for conductors feeding through or tapping off to the other apparatus unless designs are employed that provide adequate space for this purpose.

(FPN): See Section 373-8 for switch and overcurrent-device enclosures.

(b) Wire Bending Space in Enclosures. Minimum wire bending space within the enclosures for motor controllers shall be in accordance with Table 430-10(b) where measured in a straight line from the end of the lug or wire connector (in the direction the wire leaves the terminal) to the wall or barrier. Where alternate wire termination means is substituted for that supplied by the manufacturer of the controller, it shall be of a type identified by the manufacturer for use with the controller and shall not reduce the minimum wire bending space.

430-11. Protection Against Liquids. Suitable guards or enclosures shall be provided to protect exposed current-carrying parts of motors and the insulation of motor leads where installed directly under equipment, or in other locations where dripping or spraying oil, water, or other injurious liquid may occur, unless the motor is designed for the existing conditions.

**Table 430-10(b).   Minimum Wire Bending Space
at the Terminals of Enclosed Motor Controllers (in inches)**

Size of Wire AWG or kcmil	Wires per Terminal* 1	2
14-10	Not specified	—
8-6	1½	—
4-3	2	—
2	2½	—
1	3	—
1/0	5	5
2/0	6	6
3/0-4/0	7	7
250	8	8
300	10	10
350-500	12	12
600-700	14	16
750-900	18	19

*Where provision for 3 or more wires per terminal exists, the minimum wire bending space shall be in accordance with the requirements of Article 373.

430-12. Motor Terminal Housings.

(a) Material. Where motors are provided with terminal housings, the housings shall be of metal and of substantial construction.

Exception: In other than hazardous (classified) locations, substantial, nonmetallic, nonburning housings shall be permitted provided an internal grounding means between the motor frame and the equipment grounding connection is incorporated within the housing.

(b) Dimensions and Space — Wire-to-Wire Connections. Where these terminal housings enclose wire-to-wire connections, they shall have minimum dimensions and usable volumes in accordance with Table 430-12(b).

(c) Dimensions and Space — Fixed Terminal Connections. Where these terminal housings enclose rigidly mounted motor terminals, the terminal housing shall be of sufficient size to provide minimum terminal spacings and usable volumes in accordance with Table 430-12(c)(1) and Table 430-12(c)(2).

(d) Large Wire or Factory Connections. For motors with larger ratings, greater number of leads, or larger wire sizes, or where motors are installed as a part of factory-wired equipment, without additional connection being required at the motor terminal housing during equipment installation, the terminal housing shall be of ample size to make connections, but the foregoing provisions for the volumes of terminal housings shall not be considered applicable.

(e) Equipment Grounding Connections. A means for attachment of an equipment grounding conductor termination in accordance with Section

Table 430-12(b). Terminal Housings — Wire-to-Wire Connections
Motors 11 inches in Diameter or Less

HP	Cover Opening, Minimum Dimension, Inches	Usable Volume, Minimum, Cubic Inches
1 and smaller*	1⅝	7½
1½, 2, and 3†	1¾	12
5 and 7½	2	16
10 and 15	2½	26

For SI units: one inch = 25.4 millimeters.

*For motors rated 1 horsepower and smaller and with the terminal housing partially or wholly integral with the frame or end shield, the volume of the terminal housing shall be not less than 0.8 cubic inch per wire-to-wire connection. The minimum cover opening dimension is not specified.

†For motors rated 1½, 2, and 3 horsepower and with the terminal housing partially or wholly integral with the frame or end shield, the volume of the terminal housing shall not be less than 1.0 cubic inch per wire-to-wire connection. The minimum cover opening dimension is not specified.

Motors Over 11 inches in Diameter
Alternating-Current Motors

Max. Full-load Current for Three-phase Motors with Max. of Twelve Leads Amperes	Terminal Box Minimum Dimension Inches	Usable Volume Minimum Cubic Inches	Typical Maximum Horsepower Three Phase	
			230 Volt	460 Volt
45	2.5	26	15	30
70	3.3	55	25	50
110	4.0	100	40	75
160	5.0	180	60	125
250	6.0	330	100	200
400	7.0	600	150	300
600	8.0	1100	250	500

For SI units: one inch = 25.4 millimeters.

250-113 shall be provided at motor terminal housings for wire-to-wire connections or fixed terminal connections. The means for such connections shall be permitted to be located either inside or outside the motor terminal housing.

Exception: Where a motor is installed as a part of factory-wired equipment that is required to be grounded and without additional connection being required at the motor terminal housing during equipment installation, a separate means for motor grounding at the motor terminal housing shall not be required.

430-13. Bushing. Where wires pass through an opening in an enclosure, conduit box, or barrier, a bushing shall be used to protect the conductors from the edges of openings having sharp edges. The bushing shall have

Direct-Current Motors

Maximum Full-Load Current for Motors with Maximum of Six Leads Amperes	Terminal Box Minimum Dimensions Inches	Usable Volume, Minimum Cubic Inches
68	2.5	26
105	3.3	55
165	4.0	100
240	5.0	180
375	6.0	330
600	7.0	600
900	8.0	1100

For SI units: one inch = 25.4 millimeters.

Auxiliary leads for such items as brakes, thermostats, space heaters, exciting fields, etc., shall be permitted to be neglected if their current-carrying area does not exceed 25 percent of the current-carrying area of the machine power leads.

Table 430-12(c)(1).   Terminal Spacings — Fixed Terminals

	Minimum Spacing, Inches	
Nominal Volts	Between Line Terminals	Between Line Terminals and Other Uninsulated Metal Parts
240 or less	1/4	1/4
Over 250 through 600	3/8	3/8

For SI units: one inch = 25.4 millimeters.

Table 430-12(c)(2).   Usable Volumes — Fixed Terminals

Power-Supply Conductor Size, AWG	Minimum Usable Volume per Power-Supply Conductor, Cubic Inches
14	1.0
12 and 10	1 1/4
8 and 6	2 1/4

For SI units: one inch = 25.4 millimeters.

smooth, well-rounded surfaces where it may be in contact with the conductors. If used where oils, greases, or other contaminants may be present, the bushing shall be made of material not deleteriously affected.

(FPN):   For conductors exposed to deteriorating agents, see Section 310-9.

430-14. Location of Motors.

(a) Ventilation and Maintenance. Motors shall be located so that adequate ventilation is provided and so that maintenance, such as lubrication of bearings and replacing of brushes, can be readily accomplished.

(b) Open Motors. Open motors having commutators or collector rings shall be located or protected so that sparks cannot reach adjacent combustible material, but this shall not prohibit the installation of these motors on wooden floors or supports.

430-16. Exposure to Dust Accumulations. In locations where dust or flying material will collect on or in motors in such quantities as to seriously interfere with the ventilation or cooling of motors and thereby cause dangerous temperatures, suitable types of enclosed motors that will not overheat under the prevailing conditions shall be used.

(FPN): Especially severe conditions may require the use of enclosed pipe-ventilated motors, or enclosure in separate dusttight rooms, properly ventilated from a source of clean air.

430-17. Highest Rated or Smallest Rated Motor. In determining compliance with Sections 430-24, 430-53(b), and 430-53(c), the highest rated or smallest rated motor shall be based on the rated full-load current as selected from Tables 430-147, 430-148, 430-149, and 430-150.

430-18. Nominal Voltage of Rectifier Systems. The nominal value of the ac voltage being rectified shall be used to determine the voltage of a rectifier derived system.

Exception: The nominal dc voltage of the rectifier shall be used if it exceeds the peak value of the ac voltage being rectified.

B. Motor Circuit Conductors

430-21. General. Part B specifies sizes of conductors capable of carrying the motor current without overheating under the conditions specified.

Exception: The provisions of Section 430-124 shall apply over 600 volts, nominal.

The provisions of Articles 250, 300, and 310 shall not apply to conductors that form an integral part of approved equipment, or to integral conductors of motors, motor controllers, and the like.

(FPN No. 1): See Sections 300-1(b) and 310-1 for similar requirements.

(FPN No. 2): See Section 430-9(b) for equipment device terminal requirements.

430-22. Single Motor.

(a) General. Branch-circuit conductors supplying a single motor shall have an ampacity not less than 125 percent of the motor full-load current rating.

For a multispeed motor, the selection of branch-circuit conductors on the line side of the controller shall be based on the highest of the full-load current ratings shown on the motor nameplate; the selection of branch-circuit conductors between the controller and the motor shall be based on the current rating of the winding(s) that the conductors energize.

(FPN): See Chapter 9, Example No. 8 and Diagram 430-1.

Exception No. 1: Conductors for a motor used for short-time, intermittent, periodic, or varying duty shall have an ampacity not less than the percentage of the motor nameplate current rating shown in Table 430-22(a), Exception unless the authority having jurisdiction grants special permission for conductors of smaller size.

Exception No. 2: For direct-current motors operating from a rectified single-phase power supply, the conductors between the controller and the motor shall have an ampacity of not less than the following percent of the motor full-load current rating:

a. Where a rectifier bridge of the single-phase half-wave type is used, 190 percent.

b. Where a rectifier bridge of the single-phase full-wave type is used, 150 percent.

Table 430-22(a), Exception.   Duty-Cycle Service

Classification of Service	Percentages of Nameplate Current Rating			
	5-Minute Rated Motor	15-Minute Rated Motor	30 & 60 Minute Rated Motor	Continuous Rated Motor
Short-Time Duty				
Operating valves, raising or lowering rolls, etc.	110	120	150	. . .
Intermittent Duty				
Freight and passenger elevators, tool heads, pumps, drawbridges, turntables, etc. For arc welders, see Section 630-21	85	85	90	140
Periodic Duty				
Rolls, ore- and coal-handling machines, etc.	85	90	95	140
Varying Duty	110	120	150	200

Any motor application shall be considered as continuous duty unless the nature of the apparatus it drives is such that the motor will not operate continuously with load under any condition of use.

(b) Separate Terminal Enclosure. The conductors between a stationary motor rated 1 horsepower or less and the separate terminal enclosure permitted in Section 430-145(b) shall be permitted to be smaller than No. 14 but not smaller than No. 18, provided they have an ampacity as specified in (a) above.

430-23. Wound-Rotor Secondary.

(a) Continuous Duty. For continuous duty, the conductors connecting the secondary of a wound-rotor alternating-current motor to its controller shall have an ampacity not less than 125 percent of the full-load secondary current of the motor.

(b) Other than Continuous Duty. For other than continuous duty, these conductors shall have an ampacity, in percent of full-load secondary current, not less than that specified in Table 430-22(a), Exception.

(c) Resistor Separate from Controller. Where the secondary resistor is separate from the controller, the ampacity of the conductors between controller and resistor shall not be less than that shown in Table 430-23(c).

Table 430-23(c).   Secondary Conductor

Resistor Duty Classification	Ampacity of Conductor in Percent of Full-Load Secondary Current
Light starting duty	35
Heavy starting duty	45
Extra-heavy starting duty	55
Light intermittent duty	65
Medium intermittent duty	75
Heavy intermittent duty	85
Continuous duty	110

430-24. Several Motors or a Motor(s) and Other Load(s).   Conductors supplying several motors, or a motor(s) and other load(s), shall have an ampacity at least equal to the sum of the full-load current rating of all the motors, plus 25 percent of the highest rated motor in the group, plus the ampere rating of other loads determined in accordance with Article 220 and other applicable sections.

Exception No. 1: Where one or more of the motors of the group are used for short-time, intermittent, periodic, or varying duty, the ampere rating of such motors to be used in the summation shall be determined in accordance with Section 430-22(a), Exception No. 1. For the highest rated motor, the greater of either the ampere rating from 430-22(a), Exception No. 1 or the largest continuous duty motor full-load current multiplied by 1.25 shall be used in the summation.

Exception No. 2: The ampacity of conductors supplying motor-operated fixed electric space heating equipment shall conform with Section 424-3(b).

Exception No. 3: Where the circuitry is interlocked so as to prevent operation of selected motors or other loads at the same time, the conductor ampacity shall be permitted to be based on the summation of the currents of the motors and other loads to be operated at the same time that results in the highest total current.

(FPN):   See Chapter 9, Example No. 8.

430-25. Multimotor and Combination-Load Equipment.   The ampacity of the conductors supplying multimotor and combination-load equipment shall be not less than the minimum circuit ampacity marked on the equipment in accordance with Section 430-7(d). Where the equipment is not factory-wired and the individual nameplates are visible in accordance with Section 430-7(d)(2), the conductor ampacity shall be determined in accordance with Section 430-24.

430-26. Feeder Demand Factor.   Where reduced heating of the conductors results from motors operating on duty-cycle, intermittently, or from all motors not operating at one time, the authority having jurisdiction shall be permitted to grant permission for feeder conductors to have an ampacity less than specified in Section 430-24, provided the conductors have sufficient ampacity for the maximum load determined in accordance with the sizes and number of motors supplied and the character of their loads and duties.

430-27. Capacitors with Motors. Where capacitors are installed in motor circuits, conductors shall comply with Sections 460-8 and 460-9.

430-28. Feeder Taps. Feeder tap conductors shall have an ampacity not less than that required by Part B, shall terminate in a branch-circuit protective device and, in addition, shall meet one of the following requirements: (1) be enclosed by either an enclosed controller or by a raceway, be not more than 10 feet (3.05 m) in length, and, for field installation, be protected by an overcurrent device on the line side of the tap conductor, the rating or setting of which shall not exceed 1000 percent of the tap conductor ampacity; or (2) have an ampacity of at least one-third that of the feeder conductors, be protected from physical damage and be not more than 25 feet (7.62 m) in length; or (3) have the same ampacity as the feeder conductors.

Exception: Feeder Taps Over 25 Feet (7.62 m) Long. In high-bay manufacturing buildings [over 35 feet (10.67 m) high at walls] conductors tapped to a feeder shall be permitted to be not over 25 feet (7.62 m) long horizontally and not over 100 feet (30.5) total length where all of the following conditions are met:

a. The ampacity of the tap conductors is not less than one-third that of the feeder conductors.

b. The tap conductors terminate with a single circuit breaker or a single set of fuses conforming with (1) Part D if the tap is a branch circuit or (2) Part E if the tap is a feeder.

c. The tap conductors are suitably protected from physical damage and are installed in raceways.

d. The tap conductors are continuous from end-to-end and contain no splices.

e. The tap conductors shall be No. 6 copper or No. 4 aluminum or larger.

f. The tap conductors shall not penetrate walls, floors, or ceilings.

g. The tap shall not be made less than 30 feet (9.14 m) from the floor.

430-29. Constant Voltage DC Motors — Power Resistors. Conductors connecting the motor controller to separately mounted power accelerating and dynamic braking resistors in the armature circuit shall have an ampacity not less than the value calculated from Table 430-29 using motor full-load current. If an armature shunt resistor is used, the power accelerating resistor conductor ampacity shall be calculated using the total of motor full-load current and armature shunt resistor current.

Armature shunt resistor conductors shall have an ampacity of not less than that calculated from Table 430-29 using rated shunt resistor current as full-load current.

C. Motor and Branch-Circuit Overload Protection

430-31. General. Part C specifies overload devices intended to protect motors, motor-control apparatus, and motor branch-circuit conductors against excessive heating due to motor overloads and failure to start.

Table 430-29. Conductor Rating Factors for Power Resistors

| Time in Seconds | | Ampacity of Conductor in Percent |
On	Off	of Full-Load Current
5	75	35
10	70	45
15	75	55
15	45	65
15	30	75
15	15	85
Continuous Duty		110

Overload in electrical apparatus is an operating overcurrent that, when it persists for a sufficient length of time, would cause damage or dangerous overheating of the apparatus. It does not include short circuits or ground faults.

These provisions shall not be interpreted as requiring overload protection where it might introduce additional or increased hazards, as in the case of fire pumps.

(FPN): See Installation of Centrifugal Fire Pumps, NFPA 20-1990 (ANSI).

The provisions of Part C shall not apply to motor circuits rated over 600 volts, nominal. See Part J.

(FPN): See Chapter 9, Example No. 8.

430-32. Continuous-Duty Motors.

(a) More than 1 Horsepower. Each continuous-duty motor rated more than 1 horsepower shall be protected against overload by one of the following means:

(1) A separate overload device that is responsive to motor current. This device shall be selected to trip or rated at no more than the following percent of the motor nameplate full-load current rating.

Motors with a marked service factor not less than 1.15	125%
Motors with a marked temperature rise not over 40° C................	125%
All other motors	115%

Modification of this value shall be permitted as provided in Section 430-34.

For a multispeed motor, each winding connection shall be considered separately.

Where a separate motor overload device is so connected that it does not carry the total current designated on the motor nameplate, such as for wye-delta starting, the proper percentage of nameplate current applying to the selection or setting of the overload device shall be clearly designated on the equipment, or the manufacturer's selection table shall take this into account.

(2) A thermal protector integral with the motor, approved for use with the motor it protects on the basis that it will prevent dangerous overheating of the motor due to overload and failure to start. The ultimate trip current of a thermally protected motor shall not exceed the following percentage of motor full-load current given in Tables 430-148, 430-149, and 430-150.

Motor full-load current not exceeding 9 amperes........................ 170%
Motor full-load current 9.1 to and including 20 amperes 156%
Motor full-load current greater than 20 amperes........................ 140%

If the motor current-interrupting device is separate from the motor and its control circuit is operated by a protective device integral with the motor, it shall be so arranged that the opening of the control circuit will result in interruption of current to the motor.

(3) A protective device integral with a motor that will protect the motor against damage due to failure to start shall be permitted if the motor is part of an approved assembly that does not normally subject the motor to overloads.

(4) For motors larger than 1500 horsepower, a protective device having embedded temperature detectors that cause current to the motor to be interrupted when the motor attains a temperature rise greater than marked on the nameplate in an ambient of 40°C.

(b) One Horsepower or Less, Nonautomatically Started.

(1) Each continuous-duty motor rated at 1 horsepower or less that is not permanently installed, is nonautomatically started, and is within sight from the controller location shall be permitted to be protected against overload by the branch-circuit short-circuit and ground-fault protective device. This branch-circuit protective device shall not be larger than that specified in Part D of Article 430.

Exception: Any such motor shall be permitted on a nominal 120-volt branch circuit protected at not over 20 amperes.

(2) Any such motor that is not in sight from the controller location shall be protected as specified in Section 430-32(c). Any motor rated at 1 horsepower or less that is permanently installed shall be protected in accordance with Section 430-32(c).

(c) One Horsepower or Less, Automatically Started. Any motor of 1 horsepower or less that is started automatically shall be protected against overload by one of the following means:

(1) A separate overload device that is responsive to motor current. This device shall be selected to trip or rated at no more than the following percentage of the motor nameplate full-load current rating.

Motors with a marked service factor not less than 1.15 .... 125%
Motors with a marked temperature rise not over 40°C................. 125%
All other motors 115%

For a multispeed motor, each winding connection shall be considered separately. Modification of this value shall be permitted as provided in Section 430-34.

(2) A thermal protector integral with the motor, approved for use with the motor that it protects on the basis that it will prevent dangerous overheating of the motor due to overload and failure to start. Where the motor current-interrupting device is separate from the motor and its control circuit is operated by a protective device integral with the motor, it shall be so arranged that the opening of the control circuit will result in interruption of current to the motor.

(3) A protective device integral with a motor that will protect the motor against damage due to failure to start shall be permitted (1) if the motor is part of an approved assembly that does not subject the motor to overloads, or (2) if the assembly is also equipped with other safety controls (such as the safety combustion controls on a domestic oil burner) that protect the motor against damage due to failure to start. Where the assembly has safety controls that protect the motor, it shall be so indicated on the nameplate of the assembly where it will be visible after installation.

(4) In case the impedance of the motor windings is sufficient to prevent overheating due to failure to start, the motor shall be permitted to be protected as specified in Section 430-32(b)(1) for manually started motors if the motor is part of an approved assembly in which the motor will limit itself so that it will not be dangerously overheated.

(FPN):   Many alternating-current motors of less than $\frac{1}{20}$ horsepower, such as clock motors, series motors, etc., and also some larger motors such as torque motors, come within this classification. It does not include split-phase motors having automatic switches that disconnect the starting windings.

(d) Wound-Rotor Secondaries. The secondary circuits of wound-rotor alternating-current motors, including conductors, controllers, resistors, etc., shall be permitted to be protected against overload by the motor-overload device.

430-33. Intermittent and Similar Duty.   A motor used for a condition of service that is inherently short-time, intermittent, periodic, or varying duty, as illustrated by Table 430-22(a), Exception shall be permitted to be protected against overload by the branch-circuit short-circuit and ground-fault protective device, provided the protective device rating or setting does not exceed that specified in Table 430-152.

Any motor application shall be considered to be for continuous duty unless the nature of the apparatus it drives is such that the motor cannot operate continuously with load under any condition of use.

430-34. Selection of Overload Relay.   Where the overload relay selected in accordance with Section 430-32(a)(1) and (c)(1) is not sufficient to start the motor or to carry the load, the next higher size overload relay shall be permitted to be used, provided the trip current of the overload relay does not exceed the following percentage of motor full-load current rating.

Motors with marked service factor not less than 1.15 .... 140%
Motors with a marked temperature rise ot over 40°C.................. 140%
All other motors 130%

If not shunted during the starting period of the motor as provided in Section 430-35, the overload device shall have sufficient time delay to permit the motor to start and accelerate its load.

(FPN): A Class 20 or 30 overload relay will provide a longer motor acceleration time than a Class 10 or 20, respectively. Use of a higher class overload relay may preclude the need for selection of a higher trip current.

430-35. Shunting During Starting Period.

(a) Nonautomatically Started. For a nonautomatically started motor, the overload protection shall be permitted to be shunted or cut out of the circuit during the starting period of the motor if the device by which the overload protection is shunted or cut out cannot be left in the starting position and if fuses or inverse time circuit breakers rated or set at not over 400 percent of the full-load current of the motor are so located in the circuit as to be operative during the starting period of the motor.

(b) Automatically Started. The motor overload protection shall not be shunted or cut out during the starting period if the motor is automatically started.

Exception: The motor overload protection shall be permitted to be shunted or cut out during the starting period on an automatically started motor where:

(1) The motor starting period exceeds the time delay of available motor overload protective devices, and

(2) Listed means are provided to:

a. Sense motor rotation and to automatically prevent the shunting or cutout in the event that the motor fails to start, and

b. Limit the time of overload protection shunting or cutout to less than the locked rotor time rating of the protected motor, and

c. Provide for shutdown and manual restart if motor running condition is not reached.

430-36. Fuses — In Which Conductor.
Where fuses are used for motor overload protection, a fuse shall be inserted in each ungrounded conductor and also in the grounded conductor if the supply system is 3-wire, 3-phase ac with one conductor grounded.

430-37. Devices Other than Fuses — In Which Conductor.
Where devices other than fuses are used for motor overload protection, Table 430-37 shall govern the minimum allowable number and location of overload units such as trip coils or relays.

430-38. Number of Conductors Opened by Overload Device.
Motor overload devices other than fuses or thermal protectors shall simulta-

Table 430-37. Overload Units

Kind of Motor	Supply System	Number and Location of Overload Units, such as trip coils or relays
1-phase ac or dc	2-wire, 1-phase ac or dc ungrounded	1 in either conductor
1-phase ac or dc	2 wire, 1-phase ac or dc, one conductor grounded	1 in ungrounded conductor
1-phase ac or dc	3-wire, 1-phase ac or dc, grounded neutral	1 in either ungrounded conductor
1-phase ac	Any 3-phase	1 in ungrounded conductor
2-phase ac	3-wire, 2-phase ac, ungrounded	2, one in each phase
2-phase ac	3-wire, 2-phase ac, one conductor grounded	2 in ungrounded conductors
2-phase ac	4-wire, 2-phase ac, grounded or ungrounded	2, one per phase in ungrounded conductors
2-phase ac	5-wire, 2-phase ac, grounded neutral or ungrounded	2, one per phase in any ungrounded phase wire
3-phase ac	Any 3-phase	3, one in each phase*

Exception: Where protected by other approved means.

neously open a sufficient number of ungrounded conductors to interrupt current flow to the motor.

430-39. Motor Controller as Overload Protection. A motor controller shall also be permitted to serve as an overload device if the number of overload units complies with Table 430-37 and if these units are operative in both the starting and running position in the case of a direct-current motor, and in the running position in the case of an alternating-current motor.

430-40. Overload Relays. Overload relays and other devices for motor overload protection that are not capable of opening short circuits shall be protected by fuses or circuit breakers with ratings or settings in accordance with Section 430-52 or by a motor short-circuit protector in accordance with Section 430-52.

Exception No. 1: Where approved for group installation and marked to indicate the maximum size of fuse or inverse time circuit breaker by which they must be protected.

Exception No. 2: The fuse or circuit breaker ampere rating shall be permitted to be marked on the nameplate of approved equipment in which overload relay is used.

(FPN): For instantaneous trip circuit breakers or motor short-circuit protectors, see Section 430-52.

430-42. Motors on General-Purpose Branch Circuits. Overload protection for motors used on general-purpose branch circuits as permitted in Article 210 shall be provided as specified in (a), (b), (c), or (d) below.

(a) Not Over 1 Horsepower. One or more motors without individual overload protection shall be permitted to be connected to a general-purpose branch circuit only where the installation complies with the limiting conditions specified in Sections 430-32(b) and (c) and Sections 430-53(a)(1) and (a)(2).

(b) Over 1 Horsepower. Motors of larger ratings than specified in Section 430-53(a) shall be permitted to be connected to general-purpose branch circuits only where each motor is protected by overload protection selected to protect the motor as specified in Section 430-32. Both the controller and the motor overload device shall be approved for group installation with the short-circuit and ground-fault protective device selected in accordance with Section 430-53.

(c) Cord- and Plug-Connected. Where a motor is connected to a branch circuit by means of an attachment plug and receptacle and individual overload protection is omitted as provided in (a) above, the rating of the attachment plug and receptacle shall not exceed 15 amperes at 125 volts, or 10 amperes at 250 volts. Where individual overload protection is required as provided in (b) above for a motor or motor-operated appliance that is attached to the branch circuit through an attachment plug and receptacle, the overload device shall be an integral part of the motor or of the appliance. The rating of the attachment plug and receptacle shall determine the rating of the circuit to which the motor may be connected, as provided in Article 210.

(d) Time Delay. The branch-circuit short-circuit and ground-fault protective device protecting a circuit to which a motor or motor-operated appliance is connected shall have sufficient time delay to permit the motor to start and accelerate its load.

430-43. Automatic Restarting. A motor overload device that can restart a motor automatically after overload tripping shall not be installed unless approved for use with the motor it protects. A motor that can restart automatically after shutdown shall not be installed if its automatic restarting can result in injury to persons.

430-44. Orderly Shutdown. If immediate automatic shutdown of a motor by a motor overload protective device(s) would introduce additional or increased hazard(s) to a person(s) and continued motor operation is necessary for safe shutdown of equipment or process, a motor overload sensing device(s) conforming with the provisions of Part C of this article shall be permitted to be connected to a supervised alarm instead of causing immediate interruption of the motor circuit, so that corrective action or an orderly shutdown can be initiated.

D. Motor Branch-Circuit Short-Circuit and Ground-Fault Protection

430-51. General. Part D specifies devices intended to protect the motor branch-circuit conductors, the motor control apparatus, and the motors against overcurrent due to short circuits or grounds. They add to or amend the provisions of Article 240. The devices specified in Part D do not include the types of devices required by Sections 210-8, 230-95, and 305-6.

The provisions of Part D do not apply to motor circuits rated over 600 volts, nominal. See Part J.

(FPN):   See Chapter 9, Example No. 8.

430-52. Rating or Setting for Individual Motor Circuit.   The motor branch-circuit short-circuit and ground-fault protective device shall be capable of carrying the starting current of the motor.

(a) General Motor Applications.   Other than as specified for torque motors in (b) below, a protective device having a rating or setting not exceeding the value calculated according to the values given in Table 430-152 shall be used.

Exception No. 1:   Where the values for branch-circuit short-circuit and ground-fault protective devices determined by Table 430-152 do not correspond to the standard sizes or ratings of fuses, nonadjustable circuit breakers, thermal protective devices, or possible settings of adjustable circuit breakers, and the next lower standard size, rating, or possible setting is not adequate to carry the load, the next higher standard size, rating, or possible setting shall be permitted.

Exception No. 2:   Where the rating specified in Table 430-152 is not sufficient for the starting current of the motor:

a.   The rating of a nontime-delay fuse not exceeding 600 amperes shall be permitted to be increased but shall in no case exceed 400 percent of the full-load current.

b.   The rating of a time-delay (dual-element) fuse shall be permitted to be increased but shall in no case exceed 225 percent of the full-load current.

c.   The rating of an inverse time circuit breaker shall be permitted to be increased but shall in no case exceed (1) 400 percent for full-load currents of 100 amperes or less or (2) 300 percent for full-load currents greater than 100 amperes.

d.   The rating of a fuse of 601-6000 ampere classification shall be permitted to be increased but shall in no case exceed 300 percent of the full-load current.

(FPN):   See Section 240-6 for standard ratings of fuses and circuit breakers.

An instantaneous trip circuit breaker shall be used only if adjustable, and if part of a listed combination controller having coordinated motor overload and short-circuit and ground-fault protection in each conductor, and if it will operate at not more than 1300 percent of full-load motor current. A motor short-circuit protector shall be permitted in lieu of devices listed in Table 430-152 if the motor short-circuit protector is part of a listed combination controller having coordinated motor overload protection and short-circuit and ground-fault protection in each conductor and if it will operate at not more than 1300 percent of full-load motor current.

(FPN):   For the purpose of this article, instantaneous-trip circuit breakers may include a damping means to accomodate a transient motor inrush current without nuisance tripping of the circuit breaker.

Exception: Where the setting specified in Table 430-152 is not sufficient for the starting current of the motor, the setting of an instantaneous trip circuit breaker shall be permitted to be increased but shall in no case exceed 1300 percent of the motor full-load current. Trip settings above 700 percent shall be permitted where the need has been demonstrated by engineering evaluation. In such cases, it shall not be necessary to first apply an instantaneous trip circuit breaker at 700 percent.

For a multispeed motor, a single short-circuit and ground-fault protective device shall be permitted for two or more windings of the motor, provided the rating of the protective device does not exceed the above applicable percentage of the nameplate rating of the smallest winding protected.

Exception: For a multispeed motor, a single short-circuit and ground-fault protective device shall be permitted to be used and sized according to the full-load current of the highest current winding, provided that each winding is equipped with individual overload protection sized according to its full-load current and that the branch-circuit conductors feeding each winding are sized according to the full-load current of the highest full-load current winding.

Where maximum branch-circuit short-circuit and ground-fault protective device ratings are shown in the manufacturer's overload relay table for use with a motor controller or are otherwise marked on the equipment, they shall not be exceeded even if higher values are allowed as shown above.

(FPN):   See Chapter 9, Example No. 8 and Diagram 430-1.

Suitable fuses shall be permitted in lieu of devices listed in Table 430-152 for an adjustable speed drive system or a solid state motor controller system provided that the marking for replacement fuses is provided adjacent to the fuses.

(b) Torque Motors. Torque motor branch circuits shall be protected at the motor nameplate current rating in accordance with Section 240-3(b).

430-53. Several Motors or Loads on One Branch Circuit.   Two or more motors or one or more motors and other loads shall be permitted to be connected to the same branch circuit under the conditions specified in (a), (b), or (c) below.

(a) Not Over 1 Horsepower. Several motors, each not exceeding 1 horsepower in rating, shall be permitted on a nominal 120-volt branch circuit protected at not over 20 amperes or a branch circuit of 600 volts, nominal, or less, protected at not over 15 amperes, if all of the following conditions are met:

(1) The full-load rating of each motor does not exceed 6 amperes.

(2) The rating of the branch-circuit short-circuit and ground-fault protective device marked on any of the controllers is not exceeded.

(3) Individual overload protection conforms to Section 430-32.

(b) If Smallest Rated Motor Protected. If the branch-circuit short-circuit and ground-fault protective device is selected not to exceed that allowed by Section 430-52 for the smallest rated motor, two or more motors or one or more motors and other load(s), with each motor having individual overload protection, shall be permitted to be connected to a branch circuit where it can be determined that the branch-circuit short-circuit and ground-fault protective device will not open under the most severe normal conditions of service that might be encountered.

(c) Other Group Installations. Two or more motors of any rating or one or more motors and other load(s), with each motor having individual over-load protection, shall be permitted to be connected to one branch circuit where the motor controller(s) and overload device(s) are (1) installed as a listed factory assembly and the motor branch-circuit short-circuit and ground-fault protective device is either provided as part of the assembly or is specified by a marking on the assembly, or (2) the motor branch-circuit short-circuit and ground-fault protective device, the motor controller(s), and overload device(s) are field-installed as separate assemblies listed for such use and provided with manufacturers' instructions for use with each other, and (3) all of the following conditions are complied with:

(1) Each motor overload device is listed for group installation with a specified maximum rating of fuse or inverse time circuit breaker, or both.

(2) Each motor controller is listed for group installation with a specified maximum rating of fuse or circuit breaker, or both.

(3) Each circuit breaker is one of the inverse time type and listed for group installation.

(4) The branch circuit shall be protected by fuses or inverse time circuit breakers having a rating not exceeding that specified in Section 430-52 for the highest rated motor connected to the branch circuit plus an amount equal to the sum of the full-load current ratings of all other motors and the ratings of other loads connected to the circuit. Where this calculation results in a rating less than the ampacity of the supply conductors, it shall be permitted to increase the maximum rating of the fuses or circuit breaker to a value not exceeding that permitted by Section 240-3(b).

(5) The branch-circuit fuses or inverse time circuit breakers are not larger than allowed by Section 430-40 for the overload relay protecting the smallest rated motor of the group.

(FPN): See Section 110-10 for circuit impedance and other characteristics.

(d) Single Motor Taps. For group installations described above, the conductors of any tap supplying a single motor shall not be required to have an individual branch-circuit short-circuit and ground-fault protective device, provided they comply with either of the following: (1) no conductor to the motor shall have an ampacity less than that of the branch-circuit conductors, or (2) no conductor to the motor shall have an ampacity less than one-third that of the branch-circuit conductors, with a minimum in accordance with Section 430-22; the conductors to the motor overload device being not more than 25 feet (7.62 m) long and being protected from physical damage.

430-54. Multimotor and Combination-Load Equipment. The rating of the branch-circuit short-circuit and ground-fault protective device for multim-otor and combination-load equipment shall not exceed the rating marked on the equipment in accordance with Section 430-7(d).

430-55. Combined Overcurrent Protection. Motor branch-circuit short-circuit and ground-fault protection and motor overload protection shall be permitted to be combined in a single protective device where the rating or setting of the device provides the overload protection specified in Section 430-32.

430-56. Branch-Circuit Protective Devices — In Which Conductor. Branch-circuit protective devices shall comply with the provisions of Section 240-20.

430-57. Size of Fuseholder. Where fuses are used for motor branch-circuit short-circuit and ground-fault protection, the fuseholders shall not be of a smaller size than required to accommodate the fuses specified by Table 430-152.

Exception: Where fuses having time delay appropriate for the starting characteristics of the motor are used, fuseholders of smaller size than specified in Table 430-152 shall be permitted.

430-58. Rating of Circuit Breaker. A circuit breaker for motor branch-circuit short-circuit and ground-fault protection shall have a current rating in accordance with Sections 430-52 and 430-110.

E. Motor Feeder Short-Circuit and Ground-Fault Protection

430-61. General. Part E specifies protective devices intended to protect feeder conductors supplying motors against overcurrents due to short circuits or grounds.

(FPN): See Chapter 9, Example No. 8.

430-62. Rating or Setting — Motor Load.

(a) Specific Load. A feeder supplying a specific fixed motor load(s) and consisting of conductor sizes based on Section 430-24 shall be provided with a protective device having a rating or setting not greater than the largest rating or setting of the branch-circuit short-circuit and ground-fault protective device for any motor of the group (based on the maximum permitted value for the specific type of a protective device shown in Table 430-152, or Section 440-22(a) for hermetic refrigerant motor-compressors), plus the sum of the full-load currents of the other motors of the group.

Where the same rating or setting of the branch-circuit short-circuit and ground-fault protective device is used on two or more of the branch circuits of the group, one of the protective devices shall be considered the largest for the above calculations.

Exception: Where one or more instantaneous trip circuit breakers or motor short-circuit protectors are used for motor branch-circuit short-circuit and ground-fault protection as permitted in Section 430-52(a), the procedure provided above for determining the maximum rating of the feeder protective device shall apply with the following provision. For the purpose of the calculation, each instantaneous trip circuit breaker or motor short-circuit protector shall be assumed to have a rating not exceeding the maximum percentage of motor full-load current permitted by Table 430-152 for the type of feeder protective device employed.

(FPN): See Chapter 9, Example No. 8.

(b) Future Additions. For large-capacity installations, where heavy-capacity feeders are installed to provide for future additions or changes, the rating or setting of the feeder protective devices shall be permitted to be based on the ampacity of the feeder conductors.

430-63. Rating or Setting — Power and Light Loads. Where a feeder supplies a motor load and, in addition, a lighting or a lighting and appliance load, the feeder protective device shall be permitted to have a rating or setting sufficient to carry the lighting or the lighting and appliance load as determined in accordance with Articles 210 and 220 plus, for a single motor, the rating permitted by Section 430-52, and, for two or more motors, the rating permitted by Section 430-62.

F. Motor Control Circuits

430-71. General. Part F contains modifications of the general requirements and applies to the particular conditions of motor control circuits.

(FPN): See Section 430-9(b) for equipment device terminal requirements.

Definition of Motor Control Circuit: The circuit of a control apparatus or system that carries the electric signals directing the performance of the controller, but does not carry the main power current.

430-72. Overcurrent Protection.

(a) General. A motor control circuit tapped from the load side of a motor branch-circuit short-circuit and ground-fault protective device(s) and functioning to control the motor(s) connected to that branch circuit shall be protected against overcurrent in accordance with Section 430-72. Such a tapped control circuit shall not be considered to be a branch circuit and shall be permitted to be protected by either a supplementary or branch-circuit overcurrent protective device(s). A motor control circuit other than such a tapped control circuit shall be protected against overcurrent in accordance with Section 725-12 or 725-35, as applicable.

(b) Conductor Protection. The overcurrent protection for conductors shall not exceed the values specified in Column A of Table 430-72(b).

Exception No. 1: Conductors that do not extend beyond the motor control equipment enclosure shall require only short-circuit and ground-fault protection and shall be permitted to be protected by the motor branch-circuit short-circuit and ground-fault protective device(s) where the rating of the protective device(s) is not over the value specified in Column B of Table 430-72(b).

Exception No. 2: Conductors that extend beyond the motor control equipment enclosure shall require only short-circuit and ground-fault protection and shall be permitted to be protected by the motor branch-circuit short-circuit and ground-fault protective device(s) where the rating of the protective device(s) is not over the value specified in Column C of Table 430-72(b).

Exception No. 3: Conductors supplied by the secondary side of a single-phase transformer having only a two-wire (single-voltage) secondary shall be permitted to be protected by overcurrent protection provided on the primary (supply) side of the transformer, provided this protection does not exceed the value determined by multiplying the appropriate maximum rating of the overcurrent device for the secondary conductor from Table 430-72(b) by the secondary-to-primary voltage ratio. Transformer secondary conductors (other than two-wire) are not considered to be protected by the primary overcurrent protection.

Exception No. 4: Conductors of control circuits shall require only short-circuit and ground-fault protection and shall be permitted to be protected by the motor branch-circuit short-circuit and ground-fault protective device(s) where the opening of the control circuit would create a hazard as, for example, the control circuit of a fire pump motor, and the like.

Table 430-72(b). Maximum Rating of Overcurrent Protective Device-Amperes

Control Circuit Conductor Size, AWG	Column A Basic Rule		Column B Exception No. 1		Column C Exception No. 2	
	Copper	Alum. or Copper-Clad Alum.	Copper	Alum. or Copper-Clad Alum.	Copper	Alum. or Copper-Clad Alum.
18	7	—	25	—	7	—
16	10	—	40	—	10	—
14	Note 1	—	100	—	45	—
12	Note 1	Note 1	120	100	60	45
10	Note 1	Note 1	160	140	90	75
larger than 10	Note 1	Note 1	Note 2	Note 2	Note 3	Note 3

Note 1: Value specified in Section 310-15 as applicable.
Note 2: 400 percent of value specified in Table 310-17 for 60°C conductors.
Note 3: 300 percent of value specified in Table 310-16 for 60°C conductors.

(c) Control Circuit Transformer. Where a motor control circuit transformer is provided, the transformer shall be protected in accordance with Article 450.

Exception No. 1: Control circuit transformers rated less than 50 VA and that are an integral part of the motor controller and located within the motor controller enclosure shall be permitted to be protected by primary overcurrent devices, impedance limiting means, or other inherent protective means.

Exception No. 2: Where the control circuit transformer rated primary current is less than 2 amperes, an overcurrent device rated or set at not more than 500 percent of the rated primary current shall be permitted in the primary circuit.

Exception No. 3: Where the transformer supplies a Class 1 power-limited circuit [see Section 725-11(a)], Class 2, or Class 3 remote-control circuit conforming with the requirements of Article 725. See Article 725, Part C.

Exception No. 4: Where protection is provided by other approved means.

Exception No. 5: Overcurrent protection shall be omitted where the opening of the control circuit would create a hazard, as, for example, the control circuit of a fire pump motor and the like.

430-73. Mechanical Protection of Conductor. Where damage to a motor control circuit would constitute a hazard, all conductors of such a remote motor control circuit that are outside the control device itself shall be installed in a raceway or be otherwise suitably protected from physical damage.

Where one side of the motor control circuit is grounded, the motor control circuit shall be so arranged that an accidental ground in the remote-control devices will (1) not start the motor, and (2) not bypass manually operated shutdown devices or automatic safety shutdown devices.

430-74. Disconnection.

(a) General. Motor control circuits shall be so arranged that they will be disconnected from all sources of supply when the disconnecting means is in the open position. The disconnecting means shall be permitted to consist of two or more separate devices, one of which disconnects the motor and the controller from the source(s) of power supply for the motor, and the other(s), the motor control circuit(s) from its power supply. Where separate devices are used, they shall be located immediately adjacent one to each other.

Exception No. 1: Where more than twelve motor control circuit conductors are required to be disconnected, the disconnecting means shall be permitted to be located other than immediately adjacent one to each other where all of the following conditions are complied with:

a. Access to energized parts is limited to qualified persons in accordance with Part K of this article.

b. A warning sign is permanently located on the outside of each equipment enclosure door or cover permitting access to the live parts in the motor control circuit(s), warning that motor control circuit disconnecting means are remotely located and specifying the location and identification of each disconnect. Where energized parts are not in an equipment enclosure as permitted by Sections 430-132 and 430-133, an additional warning sign(s) shall be located where visible to persons who may be working in the area of the energized parts.

Exception No. 2: Where the opening of one or more motor control circuit disconnect means may result in potentially unsafe conditions for personnel or property and the conditions of items a and b of Exception No. 1 above are complied with.

(b) Control Transformer in Controller Enclosure. Where a transformer or other device is used to obtain a reduced voltage for the motor control circuit and is located in the controller enclosure, such transformer or other device shall be connected to the load side of the disconnecting means for the motor control circuit.

G. Motor Controllers

430-81. General.   Part G is intended to require suitable controllers for all motors.

(a) Definition. For definition of "Controller," see Article 100. For the purpose of this article, a controller is any switch or device normally used to start and stop a motor by making and breaking the motor circuit current.

(b) Stationary Motor of ⅛ Horsepower or Less. For a stationary motor rated at ⅛ horsepower or less that is normally left running and is so constructed that it cannot be damaged by overload or failure to start, such as clock motors and the like, the branch-circuit protective device shall be permitted to serve as the controller.

(c) Portable Motor of ⅓ Horsepower or Less. For a portable motor rated at ⅓ horsepower or less, the controller shall be permitted to be an attachment plug and receptacle.

430-82. Controller Design.

(a) Starting and Stopping. Each controller shall be capable of starting and stopping the motor it controls and shall be capable of interrupting the locked-rotor current of the motor.

(b) Autotransformer. An autotransformer starter shall provide an "off" position, a running position, and at least one starting position. It shall be so designed that it cannot rest in the starting position or in any position that will render the overload device in the circuit inoperative.

(c) Rheostats. Rheostats shall be in compliance with the following:

(1) Motor-starting rheostats shall be so designed that the contact arm cannot be left on intermediate segments. The point or plate on which the arm rests when in the starting position shall have no electrical connection with the resistor.

(2) Motor-starting rheostats for direct-current motors operated from a constant voltage supply shall be equipped with automatic devices that will interrupt the supply before the speed of the motor has fallen to less than one-third its normal rate.

430-83. Ratings.

(a) Horsepower Rating at the Application Voltage. The controller shall have a horsepower rating at the application voltage not lower than the horsepower rating of the motor.

Exception No. 1: For a stationary motor rated at 2 horsepower or less, and 300 volts or less, the controller shall be permitted to be a general-use switch having an ampere rating not less than twice the full-load current rating of the motor.

On ac circuits, general-use snap switches suitable only for use on ac (not general-use ac-dc snap switches) shall be permitted to control a motor rated at 2 horsepower or less, and 300 volts or less, having a full-load current rating not more than 80 percent of the ampere rating of the switch.

Exception No. 2: A branch-circuit inverse time circuit breaker rated in amperes only shall be permitted as a controller. Where this circuit breaker is also used for overload protection, it shall conform to the appropriate provisions of this article governing overload protection.

Exception No. 3: The motor controller for a torque motor shall have a continuous-duty, full-load current rating not less than the nameplate current rating of the motor. For a motor controller rated in horsepower but not marked with the foregoing current rating, the equivalent current rating shall be determined from the horsepower rating by using Table 430-147, 430-148, 430-149, or 430-150.

(b) Voltage Rating. A controller with a straight voltage rating, e.g., 240V or 480V, shall be permitted to be applied in a circuit in which the nominal voltage between any two conductors does not exceed the controller's voltage rating. A controller with a slash rating, e.g., 120/240V or 480/277V, shall only be applied in a circuit in which the nominal voltage to ground from any conductor does not exceed the lower of the two values of the controller's voltage rating and the nominal voltage between any two conductors does not exceed the higher value of the controller's voltage rating.

430-84. Need Not Open All Conductors. The controller shall not be required to open all conductors to the motor.

Exception: Where the controller serves also as a disconnecting means, it shall open all ungrounded conductors to the motor as provided in Section 430-111.

430-85. In Grounded Conductors. One pole of the controller shall be permitted to be placed in a permanently grounded conductor, provided the controller is so designed that the pole in the grounded conductor cannot be opened without simultaneously opening all conductors of the circuit.

430-87. Number of Motors Served by Each Controller. Each motor shall be provided with an individual controller.

Exception: For motors rated 600 volts or less, a single controller rated at not less than the sum of the horsepower ratings of all of the motors of the group shall be permitted to serve the group of motors under any one of the following conditions:

a. Where a number of motors drive several parts of a single machine or piece of apparatus, such as metal and woodworking machines, cranes, hoists, and similar apparatus.

b. Where a group of motors is under the protection of one overcurrent device as permitted in Section 430-53(a).

c. Where a group of motors is located in a single room within sight from the controller location.

430-88. Adjustable-Speed Motors. Adjustable-speed motors that are controlled by means of field regulation shall be so equipped and connected that they cannot be started under weakened field.

Exception: Where the motor is designed for such starting.

430-89. Speed Limitation. Machines of the following types shall be provided with speed limiting devices or other speed limiting means:

(a) Separately Excited DC Motors. Separately excited direct-current motors.

(b) Series Motors. Series motors.

(c) Motor-Generators and Converters. Motor-generators and converters that can be driven at excessive speed from the direct-current end, as by a reversal of current or decrease in load.

Exception No. 1: Where the inherent characteristics of the machines, the system, or the load and the mechanical connection thereto are such as to safely limit the speed.

Exception No. 2: Where the machine is always under the manual control of a qualified operator.

430-90. Combination Fuseholder and Switch as Controller. The rating of a combination fuseholder and switch used as a motor controller shall be such that the fuseholder will accommodate the size of the fuse specified in Part C of this article for motor overload protection.

Exception: Where fuses having time delay appropriate for the starting characteristics of the motor are used, fuseholders of smaller size than specified in Part C of this article shall be permitted.

430-91. Motor Controller Enclosure Types. Table 430-91 provides the basis for selecting enclosures for use in specific nonhazardous locations. The enclosures are not intended to protect against conditions such as condensation, icing, corrosion, or contamination that may occur within the enclosure or enter via the conduit or unsealed openings. These internal conditions require special consideration by the installer and user.

H. Motor Control Centers

430-92. General. Part H covers motor control centers installed for the control of motors, lighting, and power circuits.

A motor control center is an assembly of one or more enclosed sections having a common power bus and principally containing motor control units.

430-94. Overcurrent Protection. Motor control centers shall be provided with overcurrent protection in accordance with Article 240 based on the rating of the common power bus. This protection shall be provided by either (1) an upstream overcurrent-protective device or (2) a main overcurrent-protective device located within the motor control center.

Table 430-91. Motor Controller Enclosure Selection Table

Provides a Degree of Protection Against the Following Environmental Conditions	For Outdoor Use Enclosure Type Number †						
	3	3 R	3 S	4	4 X	6	6 P
Incidental contact with the enclosed equipment	X	X	X	X	X	X	X
Rain, snow, and sleet	X	X	X	X	X	X	X
Sleet*	—	—	X	—	—	—	—
Windblown dust	X	—	X	X	X	X	X
Hosedown	—	—	—	X	X	X	X
Corrosive agents	—	—	—	—	X	—	X
Occasional temporary submersion	—	—	—	—	—	X	X
Occasional prolonged submersion	—	—	—	—	—	—	X

*Mechanism shall be operable when ice covered.

Provides a Degree of Protection Against the Following Environmental Conditions	For Indoor Use Enclosure Type Number †									
	1	2	4	4 X	5	6	6 P	12	12K	13
Incidental contact with the enclosed equipment	X	X	X	X	X	X	X	X	X	X
Falling dirt	X	X	X	X	X	X	X	X	X	X
Falling liquids and light splashing	—	X	X	X	X	X	X	X	X	X
Circulating dust, lint, fibers, and flyings	—	—	X	X	—	X	X	X	X	X
Settling airborne dust, lint, fibers, and flyings	—	—	X	X	X	X	X	X	X	X
Hosedown and splasing water	—	—	X	X	—	X	X	—	—	—
Oil and coolant seepage	—	—	—	—	—	—	—	X	X	X
Oil or coolant spraying and splashing	—	—	—	—	—	—	—	—	—	X
Corrosive agents	—	—	—	X	—	—	X	—	—	—
Occasional temporary submersion	—	—	—	—	—	X	X	—	—	—
Occasional prolonged submersion	—	—	—	—	—	—	X	—	—	—

†Enclosure type number shall be marked on the motor controller enclosure.

430-95. Service-Entrance Equipment. Where used as service equipment, each motor control center shall be provided with a single main disconnecting means to disconnect all ungrounded service conductors.

Exception: A second service disconnect shall be permitted to supply additional equipment.

Where a grounded (neutral) conductor is provided, the motor control center shall be provided with a main bonding jumper, sized in accordance with Section 250-79(d), within one of the sections for connecting the grounded (neutral) conductor, on its supply side, to the motor control center equipment ground bus.

430-96. Grounding. Multisection motor control centers shall be bonded together with an equipment grounding conductor or an equivalent grounding bus sized in accordance with Table 250-95. Equipment grounding conductors shall terminate on this grounding bus or to a grounding termination point provided in a single-section motor control center.

430-97. Busbars and Conductors.

(a) **Support and Arrangement.** Busbars shall be protected from physical damage and be held firmly in place. Other than for required interconnections and control wiring, only those conductors that are intended for termination in a vertical section shall be located in that section.

Exception: Conductors shall be permitted to travel horizontally through vertical sections where such conductors are isolated from the busbars by a barrier.

(b) **Phase Arrangement.** The phase arrangement on 3-phase buses shall be A, B, C from front to back, top to bottom, or left to right, as viewed from the front of the motor control center.

Exception: Rear-mounted units connected to a vertical bus that is common to front-mounted units shall be permitted to have a C, B, A phase arrangement where properly identified.

(c) **Minimum Wire Bending Space.** The minimum wire bending space at the motor control center terminals and minimum gutter space shall be as required in Article 373.

(d) **Spacings.** Spacings between motor control center bus terminals and other bare metal parts shall not be less than specified in Table 430-97.

Table 430-97.   Minimum Spacing Between Bare Metal Parts

	Opposite Polarity Where Mounted on the Same Surface	Opposite Polarity Where Held Free in Air	Live Parts to Ground
Not over 125 volts, nominal ...	$3/4$ inch	$1/2$ inch	$1/2$ inch
Not over 250 volts, nominal ...	$1^{1}/4$ inch	$3/4$ inch	$1/2$ inch
Not over 600 volts, nominal ...	2 inches	1 inch	1 inch

For SI units: one inch = 25.4 millimeters.

(e) **Barriers.** Barriers shall be placed in all service-entrance motor control centers to isolate service busbars and terminals from the remainder of the motor control center.

430-98. Marking.

(a) **Motor Control Centers.** Motor control centers shall be marked according to Section 110-21, and such marking shall be plainly visible after installation. Marking shall also include common power bus current rating and motor control center short-circuit rating.

(b) Motor Control Units. Motor control units in a motor control center shall comply with Section 430-8.

I. Disconnecting Means

430-101. General. Part I is intended to require disconnecting means capable of disconnecting motors and controllers from the circuit.

(FPN No. 1): See Diagram 430-1.

(FPN No. 2): See Section 110-22 for identification of disconnecting means.

430-102. Location.

(a) Controller. A disconnecting means shall be located in sight from the controller location and shall disconnect the controller.

Exception No. 1: For motor circuits over 600 volts, nominal, a controller disconnecting means capable of being locked in the open position shall be permitted to be out of sight of the controller, provided the controller is marked with a warning label giving the location of the disconnecting means.

Exception No. 2: A single disconnecting means shall be permitted to be located adjacent to a group of coordinated controllers mounted adjacent one to each other on a multimotor continuous process machine.

(b) Motor. A disconnecting means shall be located in sight from the motor location and the driven machinery location.

Exception: Where the disconnecting means provided in accordance with Section 430-102(a) is capable of being locked in the open position.

430-103. Operation. The disconnecting means shall open all ungrounded supply conductors and shall be so designed that no pole can be operated independently. The disconnecting means shall be permitted in the same enclosure with the controller.

(FPN): See Section 430-113 for equipment receiving energy from more than one source.

430-104. To Be Indicating. The disconnecting means shall plainly indicate whether it is in the open ("off") or closed ("on") position.

430-105. Grounded Conductors. One pole of the disconnecting means shall be permitted to disconnect a permanently grounded conductor, provided the disconnecting means is so designed that the pole in the grounded conductor cannot be opened without simultaneously disconnecting all conductors of the circuit.

430-106. Service Switch as Disconnecting Means. Where an installation consists of a single motor, the service switch shall be permitted to serve as the disconnecting means if it complies with this article and is within sight from the controller location.

430-107. Readily Accessible. One of the disconnecting means shall be readily accessible.

430-108. Every Switch. Every disconnecting means in the motor branch circuit between the point of attachment to the feeder and the point of connection to the motor shall comply with the requirements of Sections 430-109 and 430-110.

430-109. Type. The disconnecting means shall be one of the following types: a motor-circuit switch rated in horsepower, a circuit breaker, or a molded-case switch, and shall be a listed device.

Exception No. 1: For stationary motors of $^1/_8$ horsepower or less, the branch-circuit overcurrent device shall be permitted to serve as the disconnecting means.

Exception No. 2: For stationary motors rated at 2 horsepower or less and 300 volts or less, the disconnecting means shall be permitted to be a general-use switch having an ampere rating not less than twice the full-load current rating of the motor.

On ac circuits, general-use snap switches suitable only for use on ac (not general-use ac-dc snap switches) shall be permitted to disconnect a motor rated 2 horsepower or less and 300 volts or less having a full-load current rating not more than 80 percent of the ampere rating of the switch.

Exception No. 3: For motors of over 2 horsepower to and including 100 horsepower, the separate disconnecting means required for a motor with an autotransformer-type controller shall be permitted to be a general-use switch where all of the following provisions are met:

a. The motor drives a generator that is provided with overload protection.

b. The controller (1) is capable of interrupting the locked-rotor current of the motor; (2) is provided with a no-voltage release; and (3) is provided with running overload protection not exceeding 125 percent of the motor full-load current rating.

c. Separate fuses or an inverse time circuit breaker rated or set at not more than 150 percent of the motor full-load current are provided in the motor branch circuit.

Exception No. 4: For stationary motors rated at more than 40 horsepower direct current or 100 horsepower alternating current, the disconnecting means shall be permitted to be a general-use or isolating switch where plainly marked "Do not operate under load."

Exception No. 5: For a cord- and plug-connected motor, a horsepower-rated attachment plug and receptacle having ratings no less than the motor ratings shall be permitted to serve as the disconnecting means. A horsepower-rated attachment plug and receptacle shall not be required for a cord- and plug-connected appliance in accordance with Section 422-22, a room air conditioner in accordance with Section 440-63, or a portable motor rated $^1/_3$ horsepower or less.

Exception No. 6: For torque motors, the disconnecting means shall be permitted to be a general-use switch.

Exception No. 7: An instantaneous trip circuit breaker that is part of a listed combination motor controller shall be permitted to serve as a disconnecting means.

430-110. Ampere Rating and Interrupting Capacity.

(a) General. The disconnecting means for motor circuits rated 600 volts, nominal, or less, shall have an ampere rating of at least 115 percent of the full-load current rating of the motor.

(b) For Torque Motors. Disconnecting means for a torque motor shall have an ampere rating of at least 115 percent of the motor nameplate current.

(c) For Combination Loads. Where two or more motors are used together or where one or more motors are used in combination with other loads, such as resistance heaters, and where the combined load may be simultaneous on a single disconnecting means, the ampere and horsepower ratings of the combined load shall be determined as follows:

(1) The rating of the disconnecting means shall be determined from the summation of all currents, including resistance loads, at the full-load condition and also at the locked-rotor condition. The combined full-load current and the combined locked-rotor current so obtained shall be considered as a single motor for the purpose of this requirement as follows:

The full-load current equivalent to the horsepower rating of each motor shall be selected from Table 430-148, 430-149, or 430-150. These full-load currents shall be added to the rating in amperes of other loads to obtain an equivalent full-load current for the combined load.

The locked-rotor current equivalent to the horsepower rating of each motor shall be selected from Table 430-151. The locked-rotor currents shall be added to the rating in amperes of other loads to obtain an equivalent locked-rotor current for the combined load. Where two or more motors or other loads cannot be started simultaneously, appropriate combinations of locked-rotor and full-load current shall be permitted to be used to determine the equivalent locked-rotor current for the simultaneous combined loads.

Exception: Where part of the concurrent load is resistance load, and where the disconnecting means is a switch rated in horsepower and amperes, the switch used shall be permitted to have a horsepower rating not less than the combined load of the motor(s), if the ampere rating of the switch is not less than the locked-rotor current of the motor(s) plus the resistance load.

(2) The ampere rating of the disconnecting means shall not be less than 115 percent of the summation of all currents at the full-load condition determined in accordance with (c)(1) above.

(3) For small motors not covered by Tables 430-147, 430-148, 430-149, or 430-150, the locked-rotor current shall be assumed to be six times the full-load current.

430-111. Switch or Circuit Breaker as Both Controller and Disconnecting Means. A switch or circuit breaker complying with Section 430-83 shall be permitted to serve as both controller and disconnecting means if it opens all ungrounded conductors to the motor, if it is protected by an overcurrent device (which shall be permitted to be the branch-circuit fuses) that opens all ungrounded conductors to the switch or circuit breaker, and if it is of one of the types specified in (a), (b), or (c) below:

(a) Air-Break Switch. An air-break switch, operable directly by applying the hand to a lever or handle.

(b) Inverse Time Circuit Breaker. An inverse time circuit breaker operable directly by applying the hand to a lever or handle.

(c) Oil Switch. An oil switch used on a circuit whose rating does not exceed 600 volts or 100 amperes, or by special permission on a circuit exceeding this capacity where under expert supervision.

The oil switch or circuit breaker specified above shall be permitted to be both power and manually operable.

The overcurrent device protecting the controller shall be permitted to be part of the controller assembly or shall be permitted to be separate.

An autotransformer-type controller shall be provided with a separate disconnecting means.

430-112. Motors Served by Single Disconnecting Means. Each motor shall be provided with an individual disconnecting means.

Exception: A single disconnecting means shall be permitted to serve a group of motors under any one of the conditions of a, b, and c below. The single disconnecting means shall be rated in accordance with Section 430-110(c):

a. Where a number of motors drive several parts of a single machine or piece of apparatus, such as metal and woodworking machines, cranes, and hoists.

b. Where a group of motors is under the protection of one set of branch-circuit protective devices as permitted by Section 430-53(a).

c. Where a group of motors is in a single room within sight from the location of the disconnecting means.

430-113. Energy from More than One Source. Motor and motor-operated equipment receiving electrical energy from more than one source shall be provided with disconnecting means from each source of electrical energy immediately adjacent to the equipment served. Each source shall be permitted to have a separate disconnecting means.

Exception No. 1: Where a motor receives electrical energy from more than one source, the disconnecting means for the main power supply to the motor shall not be required to be immediately adjacent to the motor, provided the controller disconnecting means is capable of being locked in the open position.

Exception No. 2: A separate disconnecting means shall not be required for a Class 2 remote-control circuit conforming with Article 725, rated not more than 30 volts, and that is isolated and ungrounded.

J. Over 600 Volts, Nominal

430-121. General. Part J recognizes the additional hazard due to the use of higher voltages. It adds to or amends the other provisions of this article. Other requirements for circuits and equipment operating at over 600 volts, nominal, are in Article 710.

430-122. Marking on Controllers.   In addition to the marking required by Section 430-8, a controller shall be marked with the control voltage.

430-123. Conductor Enclosures Adjacent to Motors.   Flexible metal conduit or liquidtight flexible metal conduit not exceeding 6 feet (1.83 m) in length shall be permitted to be employed for raceway connection to a motor terminal enclosure.

430-124. Size of Conductors.   Conductors supplying motors shall have an ampacity not less than the current at which the motor overload protective device(s) is selected to trip.

430-125. Motor Circuit Overcurrent Protection.

(a) General.   Each motor circuit shall include coordinated protection to automatically interrupt overload and fault currents in the motor, the motor circuit conductors, and the motor control apparatus.

Exception: Where a motor is vital to operation of the plant and the motor should operate to failure if necessary to prevent a greater hazard to persons, the sensing device(s) is permitted to be connected to a supervised annunciator or alarm instead of interrupting the motor circuit.

(b) Overload Protection.

(1) Each motor shall be protected against dangerous heating due to motor overloads and failure to start by a thermal protector integral with the motor or external current-sensing devices, or both.

(2) The secondary circuits of wound-rotor alternating-current motors including conductors, controllers, and resistors rated for the application shall be considered as protected against overcurrent by the motor overload protection means.

(3) Operation of the overload interrupting device shall simultaneously disconnect all ungrounded conductors.

(4) Overload sensing devices shall not automatically reset after trip unless resetting of the overload sensing device does not cause automatic restarting of the motor or there is no hazard to persons created by automatic restarting of the motor and its connected machinery.

(c) Fault-Current Protection.

(1) Fault-current protection shall be provided in each motor circuit by one of the following means:

a. A circuit breaker of suitable type and rating so arranged that it can be serviced without hazard. The circuit breaker shall simultaneously disconnect all ungrounded conductors. The circuit breaker shall be permitted to sense the fault current by means of integral or external sensing elements.

b. Fuses of a suitable type and rating placed in each ungrounded conductor.

Fuses shall be used with suitable disconnecting means or they shall be of a type that can also serve as the disconnecting means. They shall be so arranged that they cannot be serviced while they are energized.

(2) Fault-current interrupting devices shall not reclose the circuit automatically.

Exception: Where circuits are exposed to transient faults and where automatic reclosing of the circuit does not create a hazard to persons.

(3) Overload protection and fault-current protection shall be permitted to be provided by the same device.

430-126. Rating of Motor Control Apparatus. Motor controllers and motor branch-circuit disconnecting means shall have a continuous ampere rating not less than the current at which the overload protective device(s) is selected to trip.

430-127. Disconnecting Means. The controller disconnecting means shall be capable of being locked in the open position.

K. Protection of Live Parts — All Voltages

430-131. General. Part K specifies that live parts shall be protected in a manner judged adequate for the hazard involved.

430-132. Where Required. Exposed live parts of motors and controllers operating at 50 volts or more between terminals shall be guarded against accidental contact by enclosure or by location as follows:

(a) In a Room or Enclosure. By installation in a room or enclosure that is accessible only to qualified persons.

(b) On a Suitable Balcony. By installation on a suitable balcony, gallery, or platform, so elevated and arranged as to exclude unqualified persons.

(c) Elevation. By elevation 8 feet (2.44 m) or more above the floor.

Exception: Stationary motors having commutators, collectors, and brush rigging located inside of motor-end brackets and not conductively connected to supply circuits operating at more than 150 volts to ground.

430-133. Guards for Attendants. Where live parts of motors or controllers operating at over 150 volts to ground are guarded against accidental contact only by location as specified in Section 430-132, and where adjustment or other attendance may be necessary during the operation of the apparatus, suitable insulating mats or platforms shall be provided so that the attendant cannot readily touch live parts unless standing on the mats or platforms.

(FPN):   For working space, see Sections 110-16 and 110-34.

L. Grounding — All Voltages

430-141. General. Part L specifies the grounding of exposed noncurrent-carrying metal parts, likely to become energized, of motor and controller frames to prevent a voltage above ground in the event of accidental contact between energized parts and frames. Insulation, isolation, or guarding are suitable alternatives to grounding of motors under certain conditions.

430-142. Stationary Motors. The frames of stationary motors shall be grounded under any of the following conditions: (1) where supplied by metal-enclosed wiring; (2) where in a wet location and not isolated or guarded; (3) if in a hazardous (classified) location as covered in Articles 500 through 517; (4) if the motor operates with any terminal at over 150 volts to ground.

Where the frame of the motor is not grounded, it shall be permanently and effectively insulated from the ground.

430-143. Portable Motors. The frames of portable motors that operate at over 150 volts to ground shall be guarded or grounded.

(FPN No. 1): See Section 250-45(d) for grounding of portable appliances in other than residential occupancies.

(FPN No. 2): See Section 250-59(b) for color of grounding conductor.

430-144. Controllers. Controller enclosures shall be grounded regardless of voltage. Controller enclosures shall have means for attachment of an equipment grounding conductor termination in accordance with Section 250-113.

Exception No. 1: Enclosures attached to ungrounded portable equipment.

Exception No. 2: Lined covers of snap switches.

430-145. Method of Grounding. Where required, grounding shall be done in the manner specified in Article 250.

(a) Grounding Through Terminal Housings. Where the wiring to fixed motors is metal-enclosed cable or in metal raceways, junction boxes to house motor terminals shall be provided, and the armor of the cable or the metal raceways shall be connected to them in the manner specified in Article 250.

(FPN): See Section 430-12(e) for grounding connection means required at motor terminal housings.

(b) Separation of Junction Box from Motor. The junction box required by (a) above shall be permitted to be separated from the motor by not more than 6 feet (1.83 m), provided the leads to the motor are Type AC cable or armored cord or are stranded leads enclosed in liquidtight flexible metal conduit, flexible metal conduit, intermediate metal conduit, rigid metal conduit, or electrical metallic tubing not smaller than 3/8-inch electrical trade size, the armor or raceway being connected both to the motor and to the box. Where stranded leads are used, protected as specified above, they shall not be larger than No. 10 and shall comply with other requirements of this Code for conductors to be used in raceways.

Liquidtight flexible nonmetallic conduit and rigid nonmetallic conduit shall be permitted to enclose the leads to the motor provided the required grounding conductor is connected to both the motor and to the box.

(c) Grounding of Controller Mounted Devices. Instrument transformer secondaries and exposed noncurrent-carrying metal or other conductive parts or cases of instrument transformers, meters, instruments, and relays shall be grounded as specified in Sections 250-121 through 250-125.

M. Tables

Table 430-147. Full-Load Current in Amperes, Direct-Current Motors

The following values of full-load currents* are for motors running at base speed.

HP	Armature Voltage Rating*					
	90V	**120V**	**180V**	**240V**	**500V**	**550V**
¼	4.0	3.1	2.0	1.6		
⅓	5.2	4.1	2.6	2.0		
½	6.8	5.4	3.4	2.7		
¾	9.6	7.6	4.8	3.8		
1	12.2	9.5	6.1	4.7		
1½		13.2	8.3	6.6		
2		17	10.8	8.5		
3		25	16	12.2		
5		40	27	20		
7½		58		29	13.6	12.2
10		76		38	18	16
15				55	27	24
20				72	34	31
25				89	43	38
30				106	51	46
40				140	67	61
50				173	83	75
60				206	99	90
75				255	123	111
100				341	164	148
125				425	205	185
150				506	246	222
200				675	330	294

*These are average direct-current quantities.

Table 430-148. Full-Load Currents in Amperes
Single-Phase Alternating-Current Motors

The following values of full-load currents are for motors running at usual speeds and motors with normal torque characteristics. Motors built for especially low speeds or high torques may have higher full-load currents, and multispeed motors will have full-load current varying with speed, in which case the namplate current ratings shall be used.

The voltages listed are rated motor voltages. The currents listed shall be permitted for system voltage ranges of 110 to 120 and 220 to 240.

HP	115V	200V	208V	230V
1/6	4.4	2.5	2.4	2.2
1/4	5.8	3.3	3.2	2.9
1/3	7.2	4.1	4.0	3.6
1/2	9.8	5.6	5.4	4.9
3/4	13.8	7.9	7.6	6.9
1	16	9.2	8.8	8
1 1/2	20	11.5	11	10
2	24	13.8	13.2	12
3	34	19.6	18.7	17
5	56	32.2	30.8	28
7 1/2	80	46	44	40
10	100	57.5	55	50

Table 430-149. Full-Load Current
Two-Phase Alternating-Current Motors (4-Wire)

The following values of full-load current are for motors running at speeds usual for belted motors and motors with normal torque characteristics. Motors built for especially low speeds or high torques may require more running current, and multispeed motors will have full-load current varying with speed, in which case the nameplate current rating shall be used. Current in the common conductor of a 2-phase, 3-wire system will be 1.41 times the value given.

The voltages listed are rated motor voltages. The currents listed shall be permitted for system voltage ranges of 110 to 120, 220 to 240, 440 to 480, and 550-600 volts.

HP	Induction Type Squirrel-Cage and Wound-Rotor Amperes				
	115V	230V	460V	575V	2300V
$\frac{1}{2}$	4	2	1	.8	
$\frac{3}{4}$	4.8	2.4	1.2	1.0	
1	6.4	3.2	1.6	1.3	
$1\frac{1}{2}$	9	4.5	2.3	1.8	
2	11.8	5.9	3	2.4	
3		8.3	4.2	3.3	
5		13.2	6.6	5.3	
$7\frac{1}{2}$		19	9	8	
10		24	12	10	
15		36	18	14	
20		47	23	19	
25		59	29	24	
30		69	35	28	
40		90	45	36	
50		113	56	45	
60		133	67	53	14
75		166	83	66	18
100		218	109	87	23
125		270	135	108	28
150		312	156	125	32
200		416	208	167	43

Table 430-150. Full-Load Current*
Three-Phase Alternating-Current Motors

HP	Induction Type Squirrel-Cage and Wound-Rotor Amperes							Synchronous Type Unity Power Factor† Amperes			
	115V	200V	208V	230V	460V	575V	2300V	230V	460V	575V	2300V
½	4	2.3	2.2	2	1	.8					
¾	5.6	3.2	3.1	2.8	1.4	1.1					
1	7.2	4.1	4.0	3.6	1.8	1.4					
1½	10.4	6.0	5.7	5.2	2.6	2.1					
2	13.6	7.8	7.5	6.8	3.4	2.7					
3		11.0	10.6	9.6	4.8	3.9					
5		17.5	16.7	15.2	7.6	6.1					
7½		25.3	24.2	22	11	9					
10		32.2	30.8	28	14	11					
15		48.3	46.2	42	21	17					
20		62.1	59.4	54	27	22					
25		78.2	74.8	68	34	27		53	26	21	
30		92	88	80	40	32		63	32	26	
40		119.6	114.4	104	52	41		83	41	33	
50		149.5	143.0	130	65	52		104	52	42	
60		177.1	169.4	154	77	62	16	123	61	49	12
75		220.8	211.2	192	96	77	20	155	78	62	15
100		285.2	272.8	248	124	99	26	202	101	81	20
125		358.8	343.2	312	156	125	31	253	126	101	25
150		414	396.0	360	180	144	37	302	151	121	30
200		552	528.0	480	240	192	49	400	201	161	40

*These values of full-load current are for motors running at speeds usual for belted motors and motors with normal torque characteristics. Motors built for especially low speeds or high torques may require more runnning current, and multispeed motors will have full-load current varying with speed, in which case the nameplate current rating shall be used.

†For 90 and 80 percent power factor, the above figures shall be multipled by 1.1 and 1.25 respectively.

The voltages listed are rated motor voltages. The currents listed shall be permitted for system voltage ranges of 110 to 120, 220 to 240, 440 to 480, and 550 to 600 volts.

**Table 430-151.   Conversion Table of Locked-Rotor Currents
for Selection of Disconnecting Means and Controllers
as Determined from Horsepower and Voltage Rating**

For use only with Sections 430-110, 440-12, 440-41, and 455-8(c).

Motor Locked-Rotor Current Amperes*							
Single Phase		Two or Three Phase					Max. HP
115V	230V	115V	200V	230V	460V	575V	Rating
58.8	29.4	24	18.8	12	6	4.8	½
82.8	41.4	33.6	19.3	16.8	8.4	6.6	¾
96	48	43.2	24.8	21.6	10.8	8.4	1
120	60	62	35.9	31.2	15.6	12.6	1½
144	72	81	46.9	40.8	20.4	16.2	2
204	102	—	66	58	26.8	23.4	3
336	168	—	105	91	45.6	36.6	5
480	240	—	152	132	66	54	7½
600	300	—	193	168	84	66	10
—	—	—	290	252	126	102	15
—	—	—	373	324	162	132	20
—	—	—	469	408	204	162	25
—	—	—	552	480	240	192	30
—	—	—	718	624	312	246	40
—	—	—	897	780	390	312	50
—	—	—	1063	924	462	372	60
—	—	—	1325	1152	576	462	75
—	—	—	1711	1488	744	594	100
—	—	—	2153	1872	936	750	125
—	—	—	2484	2160	1080	864	150
—	—	—	3312	2880	1440	1152	200

*These values of motor locked-rotor current are approximately six times the full-load current values given in Tables 430-148 and 430-150.

430-152. Maximum Rating or Setting of Motor Branch-Circuit Short-Circuit and Ground-Fault Protective Devices

	Percent of Full-Load Current			
Type of Motor	Nontime Delay Fuse	Dual Element (Time-Delay) Fuse	Instantaneous Trip Breaker	Inverse Time Breaker*
Single-phase, all types				
No code letter	300	175	700	250
All ac single-phase and polyphase squirrel-cage and synchronous motors† with full-voltage, resistor or reactor starting:				
No code letter	300	175	700	250
Code letters F to V	300	175	700	250
Code letters B to E	250	175	700	200
Code letter A	150	150	700	150
All ac squirrel-cage and synchronous motors† with autotransformer starting:				
Not more than 30 amps				
No code letter	250	175	700	200
More than 30 amps				
No code letter	200	175	700	200
Code letters F to V	250	175	700	200
Code letters B to E	200	175	700	200
Code letter A	150	150	700	150
High reactance squirrel-cage				
Not more than 30 amps				
No code letter	250	175	700	250
More than 30 amps				
No code letter	200	175	700	200
Wound-rotor —				
No code letter	150	150	700	150
Direct-current (constant voltage)				
No more than 50 hp				
No code letter	150	150	250	150
More than 50 hp				
No code letter	150	150	175	150

For explanation of code letter marking, see Table 430-7(b).

For certain exceptions to the values specified, see Sections 430-52 through 430-54.

*The values given in the last column also cover the ratings of nonadjustable inverse time types of circuit breakers that may be modified as in Section 430-52.

†Synchronous motors of the low-torque, low-speed type (usually 450 rpm or lower), such as are used to drive reciprocating compressors, pumps, etc., that start unloaded, do not require a fuse rating or circuit-breaker setting in excess of 200 percent of full-load current.

ARTICLE 440 — AIR-CONDITIONING AND REFRIGERATING EQUIPMENT

A. General

440-1. Scope. The provisions of this article apply to electric motor-driven air-conditioning and refrigerating equipment, and to the branch circuits and controllers for such equipment. It provides for the special considerations necessary for circuits supplying hermetic refrigerant motor-compressors and for any air-conditioning or refrigerating equipment that is supplied from an individual branch circuit that supplies a hermetic refrigerant motor-compressor.

440-2. Definitions.

Branch-Circuit Selection Current: Branch-circuit selection current is the value in amperes to be used instead of the rated-load current in determining the ratings of motor branch-circuit conductors, disconnecting means, controllers, and branch-circuit short-circuit and ground-fault protective devices wherever the running overload protective device permits a sustained current greater than the specified percentage of the rated-load current. The value of branch-circuit selection current will always be equal to or greater than the marked rated-load current.

Hermetic Refrigerant Motor-Compressor: A combination consisting of a compressor and motor, both of which are enclosed in the same housing, with no external shaft or shaft seals, the motor operating in the refrigerant.

Rated-Load Current: The rated-load current for a hermetic refrigerant motor-compressor is the current resulting when the motor-compressor is operated at the rated load, rated voltage, and rated frequency of the equipment it serves.

440-3. Other Articles.

(a) Article 430. These provisions are in addition to, or amendatory of, the provisions of Article 430 and other articles in this Code, which apply except as modified in this article.

(b) Article 422, 424, or 430. The rules of Article 422, 424, or 430, as applicable, shall apply to air-conditioning and refrigerating equipment that does not incorporate a hermetic refrigerant motor-compressor. Examples of such equipment are devices that employ refrigeration compressors driven by conventional motors, furnaces with air-conditioning evaporator coils installed, fan-coil units, remote forced air-cooled condensers, remote commercial refrigerators, etc.

(c) Article 422. Devices such as room air conditioners, household refrigerators and freezers, drinking water coolers, and beverage dispensers shall be considered appliances, and the provisions of Article 422 shall also apply.

(d) Other Applicable Articles. Hermetic refrigerant motor-compressors, circuits, controllers, and equipment shall also comply with the applicable provisions of the following:

440-4. Marking on Hermetic Refrigerant Motor-Compressors and Equipment.

(a) Hermetic Refrigerant Motor-Compressor Nameplate. A hermetic refrigerant motor-compressor shall be provided with a nameplate that shall indicate the manufacturer's name, trademark, or symbol; identifying designation; phase; voltage; and frequency. The rated-load current in amperes of the motor-compressor shall be marked by the equipment manufacturer on either or both the motor-compressor nameplate and the nameplate of the equipment in which the motor-compressor is used. The locked-rotor current of each single-phase motor-compressor having a rated-load current of more than 9 amperes at 115 volts, or more than 4.5 amperes at 230 volts, and each polyphase motor-compressor shall be marked on the motor-compressor nameplate. Where a thermal protector complying with Sections 440-52(a)(2) and (b)(2) is used, the motor-compressor nameplate or the equipment nameplate shall be marked with the words "thermally protected." Where a protective system complying with Sections 440-52(a)(4) and (b)(4) is used and is furnished with the equipment, the equipment nameplate shall be marked with the words, "thermally protected system." Where a protective system complying with Sections 440-52(a)(4) and (b)(4) is specified, the equipment nameplate shall be appropriately marked.

(b) Multimotor and Combination-Load Equipment. Multimotor and combination-load equipment shall be provided with a visible nameplate marked with the maker's name, the rating in volts, frequency and number of phases, minimum supply circuit conductor ampacity, and the maximum rating of the branch-circuit short-circuit and ground-fault protective device. The ampacity shall be calculated by using Part D and counting all the motors and other loads that will be operated at the same time. The branch-circuit short-circuit and ground-fault protective device rating shall not exceed the value calculated by using Part C. Multimotor or combination-load equipment for use on two or more circuits shall be marked with the above information for each circuit.

Exception No. 1: Multimotor and combination-load equipment that is suitable under the provisions of this article for connection to a single 15- or 20-ampere, 120-volt, or a 15-ampere, 208- or 240-volt, single-phase branch circuit shall be permitted to be marked as a single load.

Exception No. 2: Room air conditioners as provided in Part G of Article 440.

(c) Branch-Circuit Selection Current. A hermetic refrigerant motor-compressor, or equipment containing such a compressor, having a protection system that is approved for use with the motor-compressor that it protects and that permits continuous current in excess of the specified percentage of nameplate rated-load current given in Section 440-52(b)(2) or (b)(4), shall also be marked with a branch-circuit selection current that complies with Section 440-52(b)(2) or (b)(4). This marking shall be provided by the equipment manufacturer and shall be on the nameplate(s) where the rated-load current(s) appears.   •

440-5. Marking on Controllers. A controller shall be marked with the maker's name, trademark, or symbol; identifying designation; the voltage; phase; full-load and locked-rotor current (or horsepower) rating; and such other data as may be needed to properly indicate the motor-compressor for which it is suitable.

440-6. Ampacity and Rating. The size of conductors for equipment covered by this article shall be selected from Tables 310-16 through 310-19 or calculated in accordance with Section 310-15 as applicable. The required ampacity of conductors and rating of equipment shall be determined as follows.

(a) **Hermetic Refrigerant Motor-Compressor.** For a hermetic refrigerant motor-compressor, the rated-load current marked on the nameplate of the equipment in which the motor-compressor is employed shall be used in determining the rating or ampacity of the disconnecting means, the branch-circuit conductors, the controller, the branch-circuit short-circuit and ground-fault protection, and the separate motor overload protection. Where no rated-load current is shown on the equipment nameplate, the rated-load current shown on the compressor nameplate shall be used. For disconnecting means and controllers, see also Sections 440-12 and 440-41.

Exception No. 1: Where so marked, the branch-circuit selection current shall be used instead of the rated-load current to determine the rating or ampacity of the disconnecting means, the branch-circuit conductors, the controller, and the branch-circuit short-circuit and ground-fault protection.

Exception No. 2: As permitted in Section 440-22(b) for branch-circuit short-circuit and ground-fault protection of cord- and plug-connected equipment.

(b) **Multimotor Equipment.** For multimotor equipment employing a shaded-pole or permanent split-capacitor-type fan or blower motor, the full-load current for such motor marked on the nameplate of the equipment in which the fan or blower motor is employed shall be used instead of the horsepower rating to determine the ampacity or rating of the disconnecting means, the branch-circuit conductors, the controller, the branch-circuit short-circuit and ground-fault protection, and the separate overload protection. This marking on the equipment nameplate shall not be less than the current marked on the fan or blower motor nameplate.

440-7. Highest Rated (Largest) Motor. In determining compliance with this article and with Sections 430-24, 430-53(b) and (c), and 430-62(a), the highest rated (largest) motor shall be considered to be the motor that has the highest rated-load current. Where two or more motors have the same rated-load current, only one of them shall be considered as the highest rated (largest) motor. For other than hermetic refrigerant motor-compressors, and fan or blower motors as covered in Section 440-6(b), the full-load current used to determine the highest rated motor shall be the equivalent value corresponding to the motor horsepower rating selected from Table 430-148, 430-149, or 430-150.

Exception: Where so marked, the branch-circuit selection current shall be used instead of the rated-load current in determining the highest rated (largest) motor-compressor.

440-8. Single Machine. An air-conditioning or refrigerating system shall be considered to be a single machine under the provisions of Section 430-87, Exception and Section 430-112, Exception. The motors shall be permitted to be located remotely from each other.

B. Disconnecting Means

440-11. General. The provisions of Part B are intended to require disconnecting means capable of disconnecting air-conditioning and refrigerating equipment including motor-compressors, and controllers, from the circuit feeder. See Diagram 430-1.

440-12. Rating and Interrupting Capacity.

(a) Hermetic Refrigerant Motor-Compressor. A disconnecting means serving a hermetic refrigerant motor-compressor shall be selected on the basis of the nameplate rated-load current or branch-circuit selection current, whichever is greater, and locked-rotor current, respectively, of the motor-compressor as follows:

(1) The ampere rating shall be at least 115 percent of the nameplate rated-load current or branch-circuit selection current, whichever is greater.

(2) To determine the equivalent horsepower in complying with the requirements of Section 430-109, the horsepower rating shall be selected from Table 430-148, 430-149, or 430-150 corresponding to the rated-load current or branch-circuit selection current, whichever is greater, and also the horsepower rating from Table 430-151 corresponding to the locked-rotor current. In case the nameplate rated-load current or branch-circuit selection current and locked-rotor current do not correspond to the currents shown in Table 430-148, 430-149, 430-150, or 430-151, the horsepower rating corresponding to the next higher value shall be selected. In case different horsepower ratings are obtained when applying these tables, a horsepower rating at least equal to the larger of the values obtained shall be selected.

(b) Combination Loads. Where one or more hermetic refrigerant motor-compressors are used together or are used in combination with other motors or loads, and where the combined load may be simultaneous on a single disconnecting means, the rating for the combined load shall be determined as follows:

(1) The horsepower rating of the disconnecting means shall be determined from the summation of all currents, including resistance loads, at the rated-load condition and also at the locked-rotor condition. The combined rated-load current and the combined locked-rotor current so obtained shall be considered as a single motor for the purpose of this requirement as follows:

a. The full-load current equivalent to the horsepower rating of each motor, other than a hermetic refrigerant motor-compressor, and fan or blower motors as covered in Section 440-6(b) shall be selected from Table 430-148, 430-149, or 430-150. These full-load currents shall be added to the motor-compressor rated-load current(s) or branch-circuit selection current(s), whichever is greater, and to the rating in amperes of other loads to obtain an equivalent full-load current for the combined load.

b. The locked-rotor current equivalent to the horsepower rating of each motor, other than a hermetic refrigerant motor-compressor, shall be selected from Table 430-151, and, for fan and blower motors of the shaded-pole or permanent split-capacitor type marked with the locked-rotor current, the marked value shall be used. The locked-rotor currents shall be added to the motor-compressor locked-rotor current(s) and to the rating in amperes of other loads to obtain an equivalent locked-rotor current for the combined load. Where two or more motors or other loads such as resistance heaters, or both, cannot be started simultaneously, appropriate combinations of locked-rotor and rated-load current or branch-circuit selection current, whichever is greater, shall be an acceptable means of determining the equivalent locked-rotor current for the simultaneous combined load.

Exception: Where part of the concurrent load is a resistance load and the disconnecting means is a switch rated in horsepower and amperes, the switch used shall be permitted to have a horsepower rating not less than the combined load to the motor-compressor(s) and other motor(s) at the locked-rotor condition, if the ampere rating of the switch is not less than this locked-rotor load plus the resistance load.

(2) The ampere rating of the disconnecting means shall be at least 115 percent of the summation of all currents at the rated-load condition determined in accordance with Section 440-12(b)(1).

(c) Small Motor-Compressors. For small motor-compressors not having the locked-rotor current marked on the nameplate, or for small motors not covered by Table 430-147, 430-148, 430-149, or 430-150, the locked-rotor current shall be assumed to be six times the rated-load current. See Section 440-3(a).

(d) Every Switch. Every disconnecting means in the refrigerant motor-compressor circuit between the point of attachment to the feeder and the point of connection to the refrigerant motor-compressor shall comply with the requirements of Section 440-12.

(e) Disconnecting Means Rated in Excess of 100 Horsepower. Where the rated-load or locked-rotor current as determined above would indicate a disconnecting means rated in excess of 100 horsepower, the provisions of Section 430-109, Exception No. 4 shall apply.

440-13. Cord-Connected Equipment. For cord-connected equipment such as room air conditioners, household refrigerators and freezers, drinking water coolers, and beverage dispensers, a separable connector or an attachment plug and receptacle shall be permitted to serve as the disconnecting means. See also Section 440-63.

440-14. Location. Disconnecting means shall be located within sight from and readily accessible from the air-conditioning or refrigerating equipment. The disconnecting means shall be permitted to be installed on or within the air-conditioning or refrigerating equipment.

Exception: Cord- and plug-connected appliances.

(FPN):   See Parts G and I of Article 430 for additional requirements.

C. Branch-Circuit Short-Circuit and Ground-Fault Protection

440-21. General. The provisions of Part C specify devices intended to protect the branch-circuit conductors, control apparatus, and motors in circuits supplying hermetic refrigerant motor-compressors against overcurrent due to short circuits and grounds. They are in addition to or amendatory of the provisions of Article 240.

440-22. Application and Selection.

(a) Rating or Setting for Individual Motor-Compressor. The motor-compressor branch-circuit short-circuit and ground-fault protective device shall be capable of carrying the starting current of the motor. A protective device having a rating or setting not exceeding 175 percent of the motor-compressor rated-load current or branch-circuit selection current, whichever is greater, shall be permitted, provided that, where the protection specified is not sufficient for the starting current of the motor, the rating or setting shall be permitted to be increased, but shall not exceed 225 percent of the motor rated-load current or branch-circuit selection current, whichever is greater.

Exception: The rating of the branch-circuit short-circuit and ground-fault protective device shall not be required to be less than 15 amperes.

(b) Rating or Setting for Equipment. The equipment branch-circuit short-circuit and ground-fault protective device shall be capable of carrying the starting current of the equipment. Where the hermetic refrigerant motor-compressor is the only load on the circuit, the protection shall conform with Section 440-22(a). Where the equipment incorporates more than one hermetic refrigerant motor-compressor or a hermetic refrigerant motor-compressor and other motors or other loads, the equipment short-circuit and ground-fault protection shall conform with Section 430-53 and the following:

(1) Where a hermetic refrigerant motor-compressor is the largest load connected to the circuit, the rating or setting of the branch-circuit short-circuit and ground-fault protective device shall not exceed the value specified in Section 440-22(a) for the largest motor-compressor plus the sum of the rated-load current or branch-circuit selection current, whichever is greater, of the other motor-compressor(s) and the ratings of the other loads supplied.

(2) Where a hermetic refrigerant motor-compressor is not the largest load connected to the circuit, the rating or setting of the branch-circuit short-circuit and ground-fault protective device shall not exceed a value equal to the sum of the rated-load current or branch-circuit selection current, whichever is greater, rating(s) for the motor-compressor(s) plus the value specified in Section 430-53(c)(4) where other motor loads are supplied, or the value specified in Section 240-3 where only nonmotor loads are supplied in addition to the motor-compressor(s).

Exception No. 1: Equipment that will start and operate on a 15- or 20-ampere 120-volt, or 15-ampere 208- or 240-volt single-phase branch circuit, shall be permitted to be protected by the 15- or 20-ampere overcurrent device protecting the branch circuit, but if the maximum branch-circuit short-circuit and ground-fault protective device rating marked on the equipment is less than these values, the circuit protective device shall not exceed the value marked on the equipment nameplate.

Exception No. 2: The nameplate marking of cord- and plug-connected equipment rated not greater than 250 volts, single-phase, such as household refrigerators and freezers, drinking water coolers, and beverage dispensers, shall be used in determining the branch-circuit requirements, and each unit shall be considered as a single motor unless the nameplate is marked otherwise.

(c) Protective Device Rating Not to Exceed the Manufacturer's Values. Where maximum protective device ratings shown on a manufacturer's heater table for use with a motor controller are less than the rating or setting selected in accordance with Sections 440-22(a) and (b), the protective device rating shall not exceed the manufacturer's values marked on the equipment.

D. Branch-Circuit Conductors

440-31. General. The provisions of Part D and Articles 300 and 310 specify sizes of conductors required to carry the motor current without overheating under the conditions specified, except as modified in Section 440-6(a), Exception No. 1.

The provisions of these articles shall not apply to integral conductors of motors, motor controllers and the like, or to conductors that form an integral part of approved equipment.

(FPN): See Sections 300-1(b) and 310-1 for similar requirements.

440-32. Single Motor-Compressor. Branch-circuit conductors supplying a single motor-compressor shall have an ampacity not less than 125 percent of either the motor-compressor rated-load current or the branch-circuit selection current, whichever is greater.

440-33. Motor-Compressor(s) With or Without Additional Motor Loads. Conductors supplying one or more motor-compressor(s) with or without additional load(s) shall have an ampacity not less than the sum of the rated-load or branch-circuit selection current ratings, whichever is larger, of all the motor-compressor(s) plus the full-load currents of the other motor(s), plus 25 percent of the highest motor or motor-compressor rating in the group.

Exception No. 1: Where the circuitry is so interlocked as to prevent the starting and running of a second motor-compressor or group of motor-compressors, the conductor size shall be determined from the largest motor-compressor or group of motor-compressors that is to be operated at a given time.

Exception No. 2: Room air conditioners as provided in Part G of Article 440.

440-34. Combination Load. Conductors supplying a motor-compressor load in addition to a lighting or appliance load as computed from Article 220 and other applicable articles shall have an ampacity sufficient for the lighting or appliance load plus the required ampacity for the motor-compressor load determined in accordance with Section 440-33, or, for a single motor-compressor, in accordance with Section 440-32.

Exception: Where the circuitry is so interlocked as to prevent simultaneous operation of the motor-compressor(s) and all other loads connected, the conductor size shall be determined from the largest size required for the motor-compressor(s) and other loads to be operated at a given time.

440-35. Multimotor and Combination-Load Equipment. The ampacity of the conductors supplying multimotor and combination-load equipment shall not be less than the minimum circuit ampacity marked on the equipment in accordance with Section 440-4(b).

E. Controllers for Motor-Compressors

440-41. Rating.

(a) Motor-Compressor Controller. A motor-compressor controller shall have both a continuous-duty full-load current rating, and a locked-rotor current rating, not less than the nameplate rated-load current or branch-circuit selection current, whichever is greater, and locked-rotor current, respectively (see Sections 440-6 and 440-7) of the compressor. In case the motor controller is rated in horsepower, but is without one or both of the foregoing current ratings, equivalent currents shall be determined from the ratings as follows: use Table 430-148, 430-149, or 430-150 to determine the equivalent full-load current rating. Use Table 430-151 to determine the equivalent locked-rotor current ratings.

(b) Controller Serving More than One Load. A controller, serving more than one motor-compressor or a motor-compressor and other loads, shall have a continuous-duty full-load current rating, and a locked-rotor current rating not less than the combined load as determined in accordance with Section 440-12(b).

F. Motor-Compressor and Branch-Circuit
Overload Protection

440-51. General. The provisions of Part F specify devices intended to protect the motor-compressor, the motor-control apparatus, and the branch-circuit conductors against excessive heating due to motor overload and failure to start. See Sections 240-3(e) through (h).

(FPN): Note: Overload in electrically driven apparatus is an operating overcurrent that, when it persists for a sufficient length of time, would cause damage or dangerous overheating. It does not include short circuits or ground faults.

440-52. Application and Selection.

(a) **Protection of Motor-Compressor.** Each motor-compressor shall be protected against overload and failure to start by one of the following means:

(1) A separate overload relay that is responsive to motor-compressor current. This device shall be selected to trip at not more than 140 percent of the motor-compressor rated-load current.

(2) A thermal protector integral with the motor-compressor, approved for use with the motor-compressor that it protects on the basis that it will prevent dangerous overheating of the motor-compressor due to overload and failure to start. If the current-interrupting device is separate from the motor-compressor and its control circuit is operated by a protective device integral with the motor-compressor, it shall be so arranged that the opening of the control circuit will result in interruption of current to the motor-compressor.

(3) A fuse or inverse time circuit breaker responsive to motor current, which shall also be permitted to serve as the branch-circuit short-circuit and ground-fault protective device. This device shall be rated at not more than 125 percent of the motor-compressor rated-load current. It shall have sufficient time delay to permit the motor-compressor to start and accelerate its load. The equipment or the motor-compressor shall be marked with this maximum branch-circuit fuse or inverse time circuit breaker rating.

(4) A protective system, furnished or specified and approved for use with the motor-compressor that it protects on the basis that it will prevent dangerous overheating of the motor-compressor due to overload and failure to start. If the current interrupting device is separate from the motor-compressor and its control circuit is operated by a protective device that is not integral with the current-interrupting device, it shall be so arranged that the opening of the control circuit will result in interruption of current to the motor-compressor.

(b) **Protection of Motor-Compressor Control Apparatus and Branch-Circuit Conductors.** The motor-compressor controller(s), the disconnecting means, and branch-circuit conductors shall be protected against overcurrent due to motor overload and failure to start by one of the following means, which shall be permitted to be the same device or system protecting the motor-compressor in accordance with Section 440-52(a).

Exception: For motor-compressors and equipment on 15- or 20-ampere single-phase branch circuits as provided in Sections 440-54 and 440-55.

(1) An overload relay selected in accordance with Section 440-52(a)(1).

(2) A thermal protector applied in accordance with Section 440-52(a)(2) and that will not permit a continuous current in excess of 156 percent of the marked rated-load current or branch-circuit selection current.

(3) A fuse or inverse time circuit breaker selected in accordance with Section 440-52(a)(3).

(4) A protective system in accordance with Section 440-52(a)(4) and that will not permit a continuous current in excess of 156 percent of the marked rated-load current or branch-circuit selection current.

440-53. Overload Relays. Overload relays and other devices for motor overload protection, that are not capable of opening short circuits, shall be protected by fuses or inverse time circuit breakers with ratings or settings in accordance with Part C unless approved for group installation or for part-winding motors and marked to indicate the maximum size of fuse or inverse time circuit breaker by which they shall be protected.

Exception: The fuse or inverse time circuit breaker size marking shall be permitted on the nameplate of approved equipment in which the overload relay or other overload device is used.

440-54. Motor-Compressors and Equipment on 15- or 20-Ampere Branch Circuits — Not Cord- and Attachment Plug-Connected. Overload protection for motor-compressors and equipment used on 15- or 20-ampere 120-volt, or 15-ampere 208- or 240-volt single-phase branch circuits as permitted in Article 210 shall be permitted as indicated in (a) and (b) below.

(a) Overload Protection. The motor-compressor shall be provided with overload protection selected as specified in Section 440-52(a). Both the controller and motor overload protective device shall be approved for installation with the short-circuit and ground-fault protective device for the branch circuit to which the equipment is connected.

(b) Time Delay. The short-circuit and ground-fault protective device protecting the branch circuit shall have sufficient time delay to permit the motor-compressor and other motors to start and accelerate their loads.

440-55. Cord- and Attachment Plug-Connected Motor-Compressors and Equipment on 15- or 20-Ampere Branch Circuits. Overload protection for motor-compressors and equipment that are cord- and attachment plug-connected and used on 15- or 20-ampere 120-volt, or 15-ampere 208- or 240-volt single-phase branch circuits as permitted in Article 210 shall be permitted as indicated in (a), (b), and (c) below.

(a) Overload Protection. The motor-compressor shall be provided with overload protection as specified in Section 440-52(a). Both the controller and the motor overload protective device shall be approved for installation with the short-circuit and ground-fault protective device for the branch circuit to which the equipment is connected.

(b) Attachment Plug and Receptacle Rating. The rating of the attachment plug and receptacle shall not exceed 20 amperes at 125 volts or 15 amperes at 250 volts.

(c) Time Delay. The short-circuit and ground-fault protective device protecting the branch circuit shall have sufficient time delay to permit the motor-compressor and other motors to start and accelerate their loads.

G. Provisions for Room Air Conditioners

440-60. General. The provisions of Part G shall apply to electrically energized room air conditioners that control temperature and humidity. For the purpose of Part G, a room air conditioner (with or without provisions for heating) shall be considered as an alternating-current appliance of the air-cooled window, console, or in-wall type that is installed in the conditioned room and that incorporates a hermetic refrigerant motor-compressor(s). The provisions of Part G cover equipment rated not over 250 volts, single phase, and such equipment shall be permitted to be cord- and attachment plug-connected.

A room air conditioner that is rated three phase or rated over 250 volts shall be directly connected to a wiring method recognized in Chapter 3, and provisions of Part G shall not apply.

440-61. Grounding. Room air conditioners shall be grounded in accordance with Sections 250-42, 250-43, and 250-45.

440-62. Branch-Circuit Requirements.

(a) Room Air Conditioner as a Single Motor Unit. A room air conditioner shall be considered as a single motor unit in determining its branch-circuit requirements where all the following conditions are met:

(1) It is cord- and attachment plug-connected.

(2) Its rating is not more than 40 amperes and 250 volts, single phase.

(3) Total rated-load current is shown on the room air-conditioner nameplate rather than individual motor currents, and

(4) The rating of the branch-circuit short-circuit and ground-fault protective device does not exceed the ampacity of the branch-circuit conductors or the rating of the receptacle, whichever is less.

(b) Where No Other Loads Are Supplied. The total marked rating of a cord- and attachment plug-connected room air conditioner shall not exceed 80 percent of the rating of a branch circuit where no other loads are supplied.

(c) Where Lighting Units or Other Appliances Are also Supplied. The total marked rating of a cord-and attachment plug-connected room air conditioner shall not exceed 50 percent of the rating of a branch circuit where lighting units or other appliances are also supplied.

440-63. Disconnecting Means. An attachment plug and receptacle shall be permitted to serve as the disconnecting means for a single-phase room air conditioner rated 250 volts or less if (1) the manual controls on the room air conditioner are readily accessible and located within 6 feet (1.83 m) of the floor or (2) an approved manually operable switch is installed in a readily accessible location within sight from the room air conditioner.

440-64. Supply Cords. Where a flexible cord is used to supply a room air conditioner, the length of such cord shall not exceed (1) 10 feet (3.05 m) for a nominal, 120-volt rating or (2) 6 feet (1.83 m) for a nominal, 208- or 240-volt rating.

ARTICLE 445 — GENERATORS

445-1. General. Generators and their associated wiring and equipment shall also comply with the applicable provisions of Articles 230, 250, 700, 701, 702, and 705.

445-2. Location. Generators shall be of a type suitable for the locations in which they are installed. They shall also meet the requirements for motors in Section 430-14. Generators installed in hazardous (classified) locations as described in Articles 500 through 503, or in other locations as described in Articles 510 through 517, and in Articles 520, 530, and 665 shall also comply with the applicable provisions of those articles.

445-3. Marking. Each generator shall be provided with a nameplate giving the maker's name, the rated frequency, power factor, number of phases if of alternating current, the rating in kilowatts or kilovolt amperes, the normal volts and amperes corresponding to the rating, rated revolutions per minute, insulation system class and rated ambient temperature or rated temperature rise, and time rating.

445-4. Overcurrent Protection.

(a) **Constant-Voltage Generators.** Constant-voltage generators, except alternating-current generator exciters, shall be protected from overloads by inherent design, circuit breakers, fuses, or other acceptable overcurrent protective means suitable for the conditions of use.

(b) **Two-Wire Generators.** Two-wire, direct-current generators shall be permitted to have overcurrent protection in one conductor only if the overcurrent device is actuated by the entire current generated other than the current in the shunt field. The overcurrent device shall not open the shunt field.

(c) **65 Volts or Less.** Generators operating at 65 volts or less and driven by individual motors shall be considered as protected by the overcurrent device protecting the motor if these devices will operate when the generators are delivering not more than 150 percent of their full-load rated current.

(d) **Balancer Sets.** Two-wire, direct-current generators used in conjunction with balancer sets to obtain neutrals for 3-wire systems shall be equipped with overcurrent devices that will disconnect the 3-wire system in case of excessive unbalancing of voltages or currents.

(e) **Three-Wire, Direct-Current Generators.** Three-wire, direct-current generators, whether compound or shunt wound, shall be equipped with overcurrent devices, one in each armature lead, and so connected as to be actuated by the entire current from the armature. Such overcurrent devices shall consist either of a double-pole, double-coil circuit breaker, or of a 4-pole circuit breaker connected in the main and equalizer leads and tripped by two overcurrent devices, one in each armature lead. Such protective devices shall be so interlocked that no one pole can be opened without simultaneously disconnecting both leads of the armature from the system.

Exception to (a) through (e): Where deemed by the authority having jurisdiction, a generator is vital to the operation of an electrical system and the generator should operate to failure to prevent a greater hazard to persons, the overload sensing device(s) shall be permitted to be connected to an annunciator or alarm supervised by authorized personnel instead of interrupting the generator circuit.

445-5. Ampacity of Conductors. The ampacity of the phase conductors from the generator terminals to the first overcurrent device shall not be less than 115 percent of the nameplate current rating of the generator. It shall be permitted to size the neutral conductors in accordance with Section 220-22. Conductors that must carry ground-fault currents shall not be smaller than required by Section 250-23(b).

Exception No. 1: Where the design and operation of the generator prevent overloading, the ampacity of the conductors shall not be less than 100 percent of the nameplate current rating of the generator.

Exception No. 2: Where the generator manufacturer's leads are connected directly to an overcurrent device that is an integral part of the generator set assembly.

445-6. Protection of Live Parts. Live parts of generators operated at more than 50 volts to ground shall not be exposed to accidental contact where accessible to unqualified persons.

445-7. Guards for Attendants. Where necessary for the safety of attendants, the requirements of Section 430-133 shall apply.

445-8. Bushings. Where wires pass through an opening in an enclosure, conduit box, or barrier, a bushing shall be used to protect the conductors from the edges of an opening having sharp edges. The bushing shall have smooth, well-rounded surfaces where it may be in contact with the conductors. If used where oils, grease, or other contaminants may be present, the bushing shall be made of a material not deleteriously affected.

ARTICLE 450 — TRANSFORMERS AND TRANSFORMER VAULTS

(Including Secondary Ties)

450-1. Scope. This article covers the installation of all transformers.

Exception No. 1: Current transformers.

Exception No. 2: Dry-type transformers that constitute a component part of other apparatus and comply with the requirements for such apparatus.

Exception No. 3: Transformers that are an integral part of an X-ray, high-frequency, or electrostatic-coating apparatus.

Exception No. 4: Transformers used with Class 2 and Class 3 circuits that comply with Article 725.

Exception No. 5: Transformers for sign and outline lighting that comply with Article 600.

Exception No. 6: Transformers for electric-discharge lighting that comply with Article 410.

Exception No. 7: Transformers used for power-limited fire protective signaling circuits that comply with Part C of Article 760.

Exception No. 8: Transformers used for research, development, or testing, where effective arrangements are provided to safeguard persons from contacting energized parts.

This article also covers the installation of transformers in hazardous (classified) locations as modified by Articles 501 through 504.

A. General Provisions

450-2. Definitions.   For the purpose of this article:

Transformer: The word transformer is intended to mean an individual transformer, single- or polyphase, identified by a single nameplate, unless otherwise indicated in this article.

450-3. Overcurrent Protection.   Overcurrent protection of transformers shall comply with (a), (b), (c), or (d) below. The secondary overcurrent device shall be permitted to consist of not more than six circuit breakers or six sets of fuses grouped in one location. Where multiple overcurrent devices are utilized, the total of all the device ratings shall not exceed the allowed value of a single overcurrent device. If both breakers and fuses are utilized as the overcurrent device, the total of the device ratings shall not exceed that allowed for fuses. As used in this section, the word "transformer" shall mean a transformer or polyphase bank of two or more single-phase transformers operating as a unit.

(FPN No. 1):   See Sections 240-3, 240-21, and 240-100 for overcurrent protection of conductors.

(FPN No. 2):   Nonlinear currents caused by loads such as electric-discharge lighting, electronic computer/data processing, or similar equipment can increase heat in a transformer without operating its overcurrent protective device.

(a) Transformers Over 600 Volts, Nominal.

(1) Primary and Secondary.   Each transformer over 600 volts, nominal, shall have primary and secondary protective devices rated or set to open at no more than the values of transformer rated currents as noted in Table 450-3(a)(1) Electronically-actuated fuses that may be set to open at a specific current shall be set in accordance with settings for circuit breakers.

Exception No. 1: Where the required fuse rating or circuit breaker setting does not correspond to a standard rating or setting, the next higher standard rating or setting shall be permitted.

Exception No. 2: As provided in (a)(2) below.

(2) Supervised Installations.   Where conditions of maintenance and supervision assure that only qualified persons will monitor and service the transformer installation, overcurrent protection as provided in (a)(2)a. shall be permitted.

a. Primary. Each transformer over 600 volts, nominal, shall be protected by an individual overcurrent device on the primary side.

Where fuses are used, their continuous current rating shall not exceed 250 percent of the rated primary current of the transformer. Where circuit breakers or electronically-actuated fuses are used, they shall be set at not more than 300 percent of the rated primary current of the transformer.

Table 450-3(a)(1).
Transformers Over 600 Volts

	Maximum Rating or Setting for Overcurrent Device				
	Primary		Secondary		
	Over 600 Volts		Over 600 Volts		600 Volts or Below
Transformer Rated Impedance	Circuit Breaker Setting	Fuse Rating	Circuit Breaker Setting	Fuse Rating	Circuit Breaker Setting or Fuse Rating
Not more than 6%	600%	300%	300%	250%	125%
More than 6% and not more than 10%	400%	300%	250%	225%	125%

Exception No. 1: Where the required fuse rating or circuit breaker setting does not correspond to a standard rating or setting, the next higher standard rating or setting shall be permitted.

Exception No. 2: An individual overcurrent device shall not be required where the primary circuit overcurrent device provides the protection specified in this section.

Exception No. 3: As provided in (a)(2)b below:

b. Primary and Secondary. A transformer over 600 volts, nominal, having an overcurrent device on the secondary side rated or set to open at not more than the values noted in Table 450-3(a)(2)b, or a transformer equipped with a coordinated thermal overload protection by the manufacturer, shall not be required to have an individual overcurrent device in the primary connection, provided the primary feeder overcurrent device is rated or set to open at not more than the values noted in Table 450-3(a)(2)b.

Table 450-3(a)(2)b.
Transformers Over 600 Volts in Supervised Locations

	Maximum Rating or Setting for Overcurrent Device				
	Primary		Secondary		
	Over 600 Volts		Over 600 Volts		600 Volts or Below
Transformer Rated Impedance	Circuit Breaker Setting	Fuse Rating	Circuit Breaker Setting	Fuse Rating	Circuit Breaker Setting or Fuse Rating
Not more than 6%	600%	300%	300%	250%	250%
More than 6% and not more than 10%	400%	300%	250%	225%	250%

(b) Transformers 600 Volts, Nominal, or Less. Overcurrent protection of transformers rated 600 volts, nominal, or less, shall comply with (1) or (2) below.

(1) Primary. Each transformer 600 volts, nominal, or less, shall be protected by an individual overcurrent device on the primary side, rated or set at not more than 125 percent of the rated primary current of the transformer.

Exception No. 1: Where the rated primary current of a transformer is 9 amperes or more and 125 percent of this current does not correspond to a standard rating of a fuse or nonadjustable circuit breaker, the next higher standard rating described in Section 240-6 shall be permitted. Where the rated primary current is less than 9 amperes, an overcurrent device rated or set at not more than 167 percent of the primary current shall be permitted.
Where the rated primary current is less than 2 amperes, an overcurrent device rated or set at not more than 300 percent shall be permitted.

Exception No. 2: An individual overcurrent device shall not be required where the primary circuit overcurrent device provides the protection specified in this section.

Exception No. 3: As provided in (b)(2) below.

(2) Primary and Secondary. A transformer 600 volts, nominal, or less, having an overcurrent device on the secondary side rated or set at not more than 125 percent of the rated secondary current of the transformer shall not be required to have an individual overcurrent device on the primary side if the primary feeder overcurrent device is rated or set at a current value not more than 250 percent of the rated primary current of the transformer.

A transformer 600 volts, nominal, or less, equipped with coordinated thermal overload protection by the manufacturer and arranged to interrupt the primary current, shall not be required to have an individual overcurrent device on the primary side if the primary feeder overcurrent device is rated or set at a current value not more than six times the rated current of the transformer for transformers having not more than 6 percent impedance, and not more than four times the rated current of the transformer for transformers having more than 6 percent but not more than 10 percent impedance.

Exception: Where the rated secondary current of a transformer is 9 amperes or more and 125 percent of this current does not correspond to a standard rating of a fuse or nonadjustable circuit breaker, the next higher standard rating described in Section 240-6 shall be permitted.

Where the rated secondary current is less than 9 amperes, an overcurrent device rated or set at not more than 167 percent of the rated secondary current shall be permitted.

(c) Voltage Transformers. Voltage transformers installed indoors or enclosed shall be protected with primary fuses.

(FPN): For protection of instrument circuits including voltage transformers, see Section 384-32.

(d) Fire Pump Installations. Where the transformer is dedicated to supplying power to a fire pump installation, it shall not require secondary overcurrent protection. Primary overcurrent protection shall be provided in accordance with Section 450-3(a) or 450-3(b). The primary rating or setting shall be sufficient to carry the equivalent of the transformer secondary current sum of the locked-rotor currents of the fire pump motor(s) and associated fire pump accessory equipment indefinitely.

450-4. Autotransformers 600 Volts, Nominal, or Less.

(a) Overcurrent Protection. Each autotransformer 600 volts, nominal, or less, shall be protected by an individual overcurrent device installed in series with each ungrounded input conductor. Such overcurrent device shall be rated or set at not more than 125 percent of the rated full-load input current of the autotransformer. An overcurrent device shall not be installed in series with the shunt winding (the winding common to both the input and the output circuits) of the autotransformer between Points A and B as shown in Diagram 450-4.

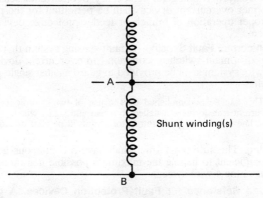

Shunt winding(s)

Diagram 450-4

Exception: Where the rated input current of an autotransformer is 9 amperes or more and 125 percent of this current does not correspond to a standard rating of a fuse or nonadjustable circuit breaker, the next higher standard rating described in Section 240-6 shall be permitted. Where the rated input current is less than 9 amperes, an overcurrent device rated or set at not more than 167 percent of the input current shall be permitted.

(b) Transformer Field-Connected as an Autotransformer. A transformer field-connected as an autotransformer shall be identified for use at elevated voltage.

(FPN): For information on permitted uses of autotransformers, see Section 210-9.

450-5. Grounding Autotransformers. Grounding autotransformers covered in this section are zig-zag or T-connected transformers connected to 3-phase, 3-wire ungrounded systems for the purpose of creating a 3-phase, 4-wire distribution system or to provide a neutral reference for grounding purposes. Such transformers shall have a continuous per phase current rating and a continuous neutral current rating.

(FPN):   The phase current in a grounding autotransformer is one-third the neutral current.

(a) Three-Phase, 4-Wire System. A grounding autotransformer used to create a 3-phase, 4-wire distribution system from a 3-phase, 3-wire ungrounded system shall conform to the following:

(1) Connections. The transformer shall be directly connected to the ungrounded phase conductors and shall not be switched or provided with overcurrent protection that is independent of the main switch and common-trip overcurrent protection for the 3-phase, 4-wire system.

(2) Overcurrent Protection. An overcurrent sensing device shall be provided that will cause the main switch or common-trip overcurrent protection referred to in (a)(1) above to open if the load on the autotransformer reaches or exceeds 125 percent of its continuous current per phase or neutral rating. Delayed tripping for temporary overcurrents sensed at the autotransformer overcurrent device shall be permitted for the purpose of allowing proper operation of branch or feeder protective devices on the 4-wire system.

(3) Transformer Fault Sensing. A fault sensing system that will cause the opening of a main switch or common-trip overcurrent device for the 3-phase, 4-wire system shall be provided to guard against single-phasing or internal faults.

(FPN):   This can be accomplished by the use of two subtractive-connected donut-type current transformers installed to sense and signal when an unbalance occurs in the line current to the autotransformer of 50 percent or more of rated current.

(4) Rating. The autotransformer shall have a continuous neutral current rating sufficient to handle the maximum possible neutral unbalanced load current of the 4-wire system.

(b) Ground Reference for Fault Protection Devices. A grounding autotransformer used to make available a specified magnitude of ground-fault current for operation of a ground responsive protective device on a 3-phase, 3-wire ungrounded system shall conform to the following requirements:

(1) Rating. The autotransformer shall have a continuous neutral current rating sufficient for the specified ground-fault current.

(2) Overcurrent Protection. An overcurrent protective device of adequate short-circuit rating that will open simultaneously all ungrounded conductors when it operates shall be applied in the grounding autotransformer branch circuit and rated or set at a current not exceeding 125 percent of the autotransformer continuous per phase current rating or 42 percent of the continuous current rating of any series connected devices in the autotransformer neutral connection. Delayed tripping for temporary overcurrents to permit the proper operation of ground responsive tripping devices on the main system shall be permitted, but shall not exceed values that would be more than the short-time current rating of the grounding autotransformer or any series connected devices in the neutral connection thereto.

(c) Ground Reference for Damping Transitory Overvoltages. A grounding autotransformer used to limit transitory overvoltages shall be of suitable rating and connected in accordance with (a)(1) above.

450-6. Secondary Ties. A secondary tie is a circuit operating at 600 volts, nominal, or less, between phases that connects two power sources or power supply points, such as the secondaries of two transformers. The tie shall be permitted to consist of one or more conductors per phase.

As used in this section, the word "transformer" means a transformer or a bank of transformers operating as a unit.

(a) Tie Circuits. Tie circuits shall be provided with overcurrent protection at each end as required in Article 240.

Exception: Under the conditions described in (a)(1) and (a)(2) below, the overcurrent protection shall be permitted to be in accordance with (a)(3) below.

(1) Loads at Transformer Supply Points Only. Where all loads are connected at the transformer supply points at each end of the tie and overcurrent protection is not provided in accordance with Article 240, the rated ampacity of the tie shall not be less than 67 percent of the rated secondary current of the largest transformer connected to the secondary tie system.

(2) Loads Connected Between Transformer Supply Points. Where load is connected to the tie at any point between transformer supply points and overcurrent protection is not provided in accordance with Article 240, the rated ampacity of the tie shall not be less than 100 percent of the rated secondary current of the largest transformer connected to the secondary tie system.

Exception: As otherwise provided in (a)(4) below.

(3) Tie Circuit Protection. Under the conditions described in (a)(1) and (a)(2) above, both ends of each tie conductor shall be equipped with a protective device that will open at a predetermined temperature of the tie conductor under short-circuit conditions. This protection shall consist of one of the following: (1) a fusible link cable connector, terminal, or lug, commonly known as a limiter, each being of a size corresponding with that of the conductor and of construction and characteristics according to the operating voltage and the type of insulation on the tie conductors or (2) automatic circuit breakers actuated by devices having comparable current-time characteristics.

(4) Interconnection of Phase Conductors Between Transformer Supply Points. Where the tie consists of more than one conductor per phase, the conductors of each phase shall be interconnected in order to establish a load supply point, and the protection specified in (a)(3) above shall be provided in each tie conductor at this point.

Exception: Loads shall be permitted to be connected to the individual conductors of a paralleled conductor tie without interconnecting the conductors of each phase and without the protection specified in (a)(3) above at load connection points, provided the tie conductors of each phase have a combined capacity of not less than 133 percent of the rated secondary current of the largest transformer connected to the secondary tie system; the total load of such taps does not exceed the rated secondary current of the largest transformer; and the loads are equally divided on each phase and on the individual conductors of each phase as far as practicable.

(5) Tie Circuit Control. Where the operating voltage exceeds 150 volts to ground, secondary ties provided with limiters shall have a switch at each end that, when open, will deenergize the associated tie conductors and limiters. The current rating of the switch shall not be less than the rated current of the conductors connected to the switch. It shall be capable of opening its rated current, and it shall be constructed so that it will not open under the magnetic forces resulting from short-circuit current.

(b) Overcurrent Protection for Secondary Connections. Where secondary ties are used, an overcurrent device rated or set at not more than 250 percent of the rated secondary current of the transformers shall be provided in the secondary connections of each transformer. In addition, an automatic circuit breaker actuated by a reverse-current relay set to open the circuit at not more than the rated secondary current of the transformer shall be provided in the secondary connection of each transformer.

450-7. Parallel Operation. Transformers shall be permitted to be operated in parallel and switched as a unit provided that the overcurrent protection for each transformer meets the requirements of Section 450-3 (a)(1) or (b)(2).

450-8. Guarding. Transformers shall be guarded as specified in (a) through (d) below.

(a) Mechanical Protection. Appropriate provisions shall be made to minimize the possibility of damage to transformers from external causes where the transformers are exposed to physical damage.

(b) Case or Enclosure. Dry-type transformers shall be provided with a noncombustible moisture-resistant case or enclosure that will provide reasonable protection against the accidental insertion of foreign objects.

(c) Exposed Energized Parts. Switches or other equipment operating at 600 volts, nominal, or less, and serving only equipment within a transformer enclosure shall be permitted to be installed in the transformer enclosure if accessible to qualified persons only. All energized parts shall be guarded in accordance with Sections 110-17 and 110-34.

(d) Voltage Warning. The operating voltage of exposed live parts of transformer installations shall be indicated by signs or visible markings on the equipment or structures.

450-9. Ventilation. The ventilation shall be adequate to dispose of the transformer full-load losses without creating temperature rise that is in excess of the transformer rating.

(FPN): See General Requirements for Liquid-Immersed Distribution, Power, and Regulating Transformers, ANSI/IEEE C57.12.00-1987, and General Requirements for Dry-Type Distribution and Power Transformers, ANSI/IEEE C57.12.01-1979.

Transformers with ventilating openings shall be installed so that the ventilating openings are not blocked by walls or other obstructions. The required clearances shall be clearly marked on the transformer.

450-10. Grounding. Exposed noncurrent-carrying metal parts of transformer installations, including fences, guards, etc., shall be grounded where required under the conditions and in the manner specified for electric equipment and other exposed metal parts in Article 250.

450-11. Marking. Each transformer shall be provided with a nameplate giving the name of the manufacturer; rated kilovolt-amperes; frequency; primary and secondary voltage; impedance of transformers 25 kVA and larger; required clearances for transformers with ventilating openings; and the amount and kind of insulating liquid where used. In addition, the nameplate of each dry-type transformer shall include the temperature class for the insulation system.

450-12. Terminal Wiring Space. The minimum wire bending space at fixed, 600 volts and below terminals of transformer line and load connections shall be as required in Section 373-6. Wiring space for pigtail connections shall conform to Table 370-16(b).

450-13. Location. Transformers and transformer vaults shall be readily accessible to qualified personnel for inspection and maintenance.

Exception No. 1: Dry-type transformers 600 volts, nominal, or less, located in the open on walls, columns, or structures, shall not be required to be readily accessible.

Exception No. 2: Dry-type transformers not exceeding 600 volts, nominal, and 50 kVA shall be permitted in fire-resistant hollow spaces of buildings not permanently closed in by structure and provided they meet the ventilation requirements of Section 450-9.

Unless specified otherwise in this article, the term "fire-resistant" means a construction having a minimum fire rating of 1 hour.

(FPN No. 1): See Method for Fire Tests of Building Construction and Materials, ANSI/ASTM E119-83, and Methods of Fire Tests of Building Construction and Materials, NFPA 251-1990.

(FPN No. 2): The location of different types of transformers is covered in Part B of Article 450. The location of transformer vaults is covered in Section 450-41.

B. Specific Provisions Applicable to Different Types of Transformers

450-21. Dry-type Transformers Installed Indoors.

(a) Not Over 112½ kVA. Dry-type transformers installed indoors and rated 112½ kVA or less shall have a separation of at least 12 inches (305 mm) from combustible material.

Exception No. 1: Where separated from the combustible material by a fire-resistant, heat-insulating barrier.

Exception No. 2: Transformers 600 volts, nominal, or less completely enclosed, with or without ventilating openings.

(b) Over 112½ kVA. Individual dry-type transformers of more than 112½ kVA rating shall be installed in a transformer room of fire-resistant construction.

Exception No. 1: Transformers with 80°C rise or higher ratings and separated from combustible material by a fire-resistant, heat-insulating barrier or by not less than 6 feet (1.83 m) horizontally and 12 feet (3.66 m) vertically.

Exception No. 2: Transformers with 80°C rise or higher ratings and completely enclosed except for ventilating openings.

(c) Over 35,000 Volts. Dry-type transformers rated over 35,000 volts shall be installed in a vault complying with Part C of this article.

450-22. Dry-type Transformers Installed Outdoors. Dry-type transformers installed outdoors shall have a weatherproof enclosure.

Transformers exceeding 112½ kVA shall not be located within 12 inches (305 mm) of combustible materials of buildings.

Exception: Transformers with 80°C rise or higher ratings and completely enclosed except for ventilating openings.

450-23. Less-Flammable Liquid-Insulated Transformers. Transformers insulated with listed less-flammable liquids shall be permitted to be installed without a vault in Type I and Type II buildings in areas in which no combustible materials are stored, provided there is a liquid confinement area, the liquid has a fire point of not less than 300°C, and the installation complies with all restrictions provided for in the listing of the liquid. Such indoor transformer installations not meeting the restrictions of the liquid listing, or installed in other than Type I or Type II buildings, or in areas where combustible materials are stored, shall (1) be provided with an automatic fire extinguishing system and a liquid confinement area or (2) be installed in a vault complying with Part C of this article.

Transformers installed indoors and rated over 35,000 volts shall be installed in a vault.

Less-flammable liquid-insulated transformers shall be permitted to be installed outdoors attached to, adjacent to, or on the roof of Type I or Type II buildings.

Such installations attached to, adjacent to, or on the roof of other than Type I or Type II buildings, adjacent to combustible material, fire escapes, or door and window openings, shall be provided with one or more safeguards according to the degree of hazard. Fire barriers, space separation, and compliance with all of the requirements provided for in the listing of the liquid are recognized safeguards.

(FPN No. 1): As used in this section, "noncombustible" refers to Type I and Type II building construction and noncombustible materials as defined in Types of Building Construction, NFPA 220-1992 (ANSI).

(FPN No. 2): See definition of "Listed" in Article 100.

450-24. Nonflammable Fluid-Insulated Transformers. Transformers insulated with a dielectric fluid identified as nonflammable shall be permitted to be installed indoors or outdoors. Such transformers installed indoors and rated over 35,000 volts shall be installed in a vault. Such transformers installed indoors shall be furnished with a liquid confinement area and a pressure relief vent. The transformers shall be furnished with a means for absorbing any gases generated by arcing inside the tank, or the pressure relief vent shall be connected to a chimney or flue that will carry such gases to an environmentally safe area.

(FPN): Safety will be increased if fire hazard analyses are performed for such transformer installations.

For the purposes of this section, a nonflammable dielectric fluid is one that does not have a flash point or fire point, and is not flammable in air.

450-25. Askarel-Insulated Transformers Installed Indoors. Askarel-insulated transformers installed indoors and rated over 25 kVA shall be furnished with a pressure-relief vent. Where installed in a poorly ventilated place, they shall be furnished with a means for absorbing any gases generated by arcing inside the case, or the pressure-relief vent shall be connected to a chimney or flue that will carry such gases outside the building. Askarel-insulated transformers rated over 35,000 volts shall be installed in a vault.

450-26. Oil-Insulated Transformers Installed Indoors. Oil-insulated transformers installed indoors shall be installed in a vault constructed as specified in Part C of this article.

Exception No. 1: Where the total capacity does not exceed 112½ kVA, the vault specified in Part C of this article shall be permitted to be constructed of reinforced concrete not less than 4 inches (102 mm) thick.

Exception No. 2: Where the nominal voltage does not exceed 600, a vault shall not be required if suitable arrangements are made to prevent a transformer oil fire from igniting other materials, and the total capacity in one location does not exceed 10 kVA in a section of the building classified as combustible, or 75 kVA where the surrounding structure is classified as fire-resistant construction.

Exception No. 3: Electric furnace transformers having a total rating not exceeding 75 kVA shall be permitted to be installed without a vault in a building or room of fire-resistant construction, provided suitable arrangements are made to prevent a transformer oil fire from spreading to other combustible material.

Exception No. 4: Transformers shall be permitted to be installed in a detached building that does not comply with Part C of this article if neither the building nor its contents presents a fire hazard to any other building or property, and if the building is used only in supplying electric service and the interior is accessible only to qualified persons.

Exception No. 5: Oil-insulated transformers shall be permitted to be used without a vault in portable and mobile surface mining equipment (such as electric excavators) if each of the following conditions is met:

a. Provision is made for draining leaking fluid to the ground.

b. Safe egress is provided for personnel.

c. A minimum ¼-inch (6.35-mm) steel barrier is provided for personnel protection.

450-27. Oil-Insulated Transformers Installed Outdoors. Combustible material, combustible buildings, and parts of buildings, fire escapes, and door and window openings shall be safeguarded from fires originating in oil-insulated transformers installed on roofs, attached to, or adjacent to a building or combustible material.

Space separations, fire-resistant barriers, automatic water spray systems, and enclosures that confine the oil of a ruptured transformer tank are recognized safeguards. One or more of these safeguards shall be applied according to the degree of hazard involved in cases where the transformer installation presents a fire hazard.

Oil enclosures shall be permitted to consist of fire-resistant dikes, curbed areas or basins, or trenches filled with coarse, crushed stone. Oil enclosures shall be provided with trapped drains where the exposure and the quantity of oil involved are such that removal of oil is important.

(FPN): For additional information on transformers installed on poles or structures or underground, see National Electrical Safety Code, ANSI C2-1990.

450-28. Modification of Transformers. When modifications are made to a transformer in an existing installation that change the type of the transformer with respect to Part B of this article, such transformer shall be marked to show the type of insulating liquid installed, and the modified transformer installation shall comply with the applicable requirements for that type of transformer.

C. Transformer Vaults

450-41. Location. Vaults shall be located where they can be ventilated to the outside air without using flues or ducts wherever such an arrangement is practicable.

450-42. Walls, Roof, and Floor. The walls and roofs of vaults shall be constructed of materials that have adequate structural strength for the conditions with a minimum fire resistance of 3 hours. The floors of vaults in contact with the earth shall be of concrete not less than 4 inches (102 mm) thick, but where the vault is constructed with a vacant space or other stories below it, the floor shall have adequate structural strength for the load imposed thereon and a minimum fire resistance of 3 hours. For the purposes of this section, studs and wall board construction shall not be acceptable.

(FPN No. 1): For additional information, see Method for Fire Tests of Building Construction and Materials, ANSI/ASTM E119-83, and Methods of Fire Tests of Building Construction and Materials, NFPA 251-1990.

(FPN No. 2): Six-inch (152-mm) thick reinforced concrete is a typical 3-hour construction.

Exception: Where transformers are protected with automatic sprinkler, water spray, carbon dioxide, or halon, construction of 1-hour rating shall be permitted.

450-43. Doorways. Vault doorways shall be protected as follows:

(a) Type of Door. Each doorway leading into a vault from the building interior shall be provided with a tight-fitting door having a minimum fire rating of 3 hours. The authority having jurisdiction shall be permitted to require such a door for an exterior wall opening where conditions warrant.

Exception: Where transformers are protected with automatic sprinkler, water spray, carbon dioxide, or halon, construction of 1-hour rating shall be permitted.

(FPN): For additional information, see Fire Doors and Fire Windows, NFPA 80-1992 (ANSI).

(b) Sills. A door sill or curb of sufficient height to confine within the vault the oil from the largest transformer shall be provided, and in no case shall the height be less than 4 inches (102 mm).

(c) Locks. Doors shall be equipped with locks, and doors shall be kept locked, access being allowed only to qualified persons. Personnel doors shall swing out and be equipped with panic bars, pressure plates, or other devices that are normally latched but open under simple pressure.

450-45. Ventilation Openings. Where required by Section 450-9, openings for ventilation shall be provided in accordance with (a) through (f) below.

(a) Location. Ventilation openings shall be located as far away as possible from doors, windows, fire escapes, and combustible material.

(b) Arrangement. A vault ventilated by natural circulation of air shall be permitted to have roughly half of the total area of openings required for ventilation in one or more openings near the floor and the remainder in one or more openings in the roof or in the sidewalls near the roof, or all of the area required for ventilation shall be permitted in one or more openings in or near the roof.

(c) Size. For a vault ventilated by natural circulation of air to an outdoor area, the combined net area of all ventilating openings, after deducting the area occupied by screens, gratings, or louvers, shall not be less than 3 square inches (1936 sq mm) per kVA of transformer capacity in service, and in no case shall the net area be less than 1 square foot (0.093 sq m) for any capacity under 50 kVA.

(d) Covering. Ventilation openings shall be covered with durable gratings, screens, or louvers, according to the treatment required in order to avoid unsafe conditions.

(e) Dampers. All ventilation openings to the indoors shall be provided with automatic closing fire dampers that operate in response to a vault fire. Such dampers shall possess a standard fire rating of not less than $1\frac{1}{2}$ hours.

(FPN): See Standard for Fire Dampers, ANSI/UL 555-1972.

(f) Ducts. Ventilating ducts shall be constructed of fire-resistant material.

450-46. Drainage. Where practicable, vaults containing more than 100 kVA transformer capacity shall be provided with a drain or other means that will carry off any accumulation of oil or water in the vault unless local conditions make this impracticable. The floor shall be pitched to the drain where provided.

450-47. Water Pipes and Accessories. Any pipe or duct system foreign to the electrical installation shall not enter or pass through a transformer vault. Piping or other facilities provided for vault fire protection, or for transformer cooling, shall not be considered foreign to the electrical installation.

450-48. Storage in Vaults. Materials shall not be stored in transformer vaults.

ARTICLE 455 — PHASE CONVERTERS

A. General

455-1. Scope. This article covers the installation and use of phase converters.

455-2. Definition. A phase converter is an electrical device that converts single-phase power to 3-phase electrical power. Phase converters are of two types: static and rotary.

(FPN): Phase converters have characteristics that modify the starting torque and locked-rotor current of motors served, and consideration is required in selecting a phase converter for a specific load.

455-3. Other Articles. All requirements of this Code shall apply to phase converters except as amended by this article.

455-4. Marking. Each phase converter shall be provided with a permanent nameplate indicating (1) manufacturer's name; (2) rated input and output voltages; (3) frequency; (4) rated single-phase input full-load amperes; (5) rated minimum and maximum single load in kVA or horsepower; and (6) maximum total load in kVA or horsepower.

455-5. Equipment Grounding Connection. A means for attachment of an equipment grounding conductor termination in accordance with Section 250-113 shall be provided.

455-6. Conductor Ampacity. The ampacity of the single-phase supply conductors shall not be less than 125 percent of the phase converter nameplate single-phase input full-load amperes.

Exception: Where the phase converter supplies specific fixed loads, the conductors shall have an ampacity not less than 250 percent of the sum of the full-load, 3-phase current rating of the motors and other loads served where the input and output voltages of the phase converter are identical.

Where the input and output voltages of the phase converter are different, the current as determined by this section shall be multiplied by the ratio of output to input voltage.

(FPN): Single-phase conductors sized to prevent a voltage drop not exceeding 3 percent from the source of supply to the phase converter will help ensure proper starting and operation of motor loads.

455-7. Overcurrent Protection. The single-phase supply conductor and phase converter shall be protected from overcurrent at not more than 125 percent of the phase converter nameplate single-phase input full-load amperes.

Exception No. 1: Where the phase converter supplies specific fixed loads, the overcurrent protection shall have a rating not greater than 250 percent of the sum of the full-load, 3-phase current rating of the motors and other loads served where the input and output voltages of the phase converter are identical.

Where the input and output voltages of the phase converter are different, the current as determined by this section shall be multiplied by the ratio of output to input voltage.

Exception No. 2: Where the required fuse rating or circuit breaker setting does not correspond to a standard rating or setting, the next higher standard rating or setting shall be permitted.

455-8. Disconnecting Means. Means shall be provided to disconnect simultaneously all ungrounded single-phase supply conductors to the phase converter.

(a) Location. The disconnecting means shall be readily accessible and located in sight from the phase converter.

(b) Type. The disconnecting means shall be a switch rated in horsepower, a circuit breaker, or a molded-case switch.

Exception: Where only nonmotor loads are served, an ampere-rated switch shall be permitted.

(c) Rating. The ampere rating of the disconnecting means shall not be less than 115 percent of the rated maximum single-phase input full-load amperes.

Exception No. 1: Where the phase converter supplies specific fixed loads, and the input and output voltages of the phase converter are identical, the disconnecting means shall be permitted to be a circuit breaker or molded-case switch with an ampere rating not less than 250 percent of the sum of the following:

a. Full-load 3-phase current ratings of the motors, and

b. Other loads served.

Where the input and output voltages of the phase converter are different, the current shall be multiplied by the ratio of the output to input voltage.

Exception No. 2: Where the phase converter supplies specific fixed loads, and the input and output voltages of the phase converter are identical, the disconnecting means shall be permitted to be a switch with a horsepower rating. The horsepower rating shall be equivalent to 200 percent of the sum of the following:

a. Nonmotor loads,

b. The 3-phase locked-rotor current of the largest motor as determined from Table 430-151, and

c. *The full-load current of all other 3-phase motors operating at the same time.*

Where the input and output voltages of the phase converter are different, the current shall be multiplied by the ratio of the output to input voltage.

455-9. Connection of Single-Phase Loads. Where single-phase loads are connected on the load side of a phase converter, they shall not be connected to the derived phase.

455-10. Terminal Housings. A terminal housing shall be provided on a phase converter, and the terminal housing shall be in accordance with the provisions of Section 430-12.

B. Specific Provisions Applicable to Different Types of Phase Converters

455-20. Disconnecting Means. The single-phase disconnecting means for the input of a static phase converter shall be permitted to serve as the disconnecting means for the phase converter and a single load if the load is within sight of the disconnecting means.

455-21. Start-Up. Power to the utilization equipment shall not be supplied until the rotary phase converter has been started.

455-22. Power Interruption. Utilization equipment supplied by a rotary phase converter shall be controlled in such a manner that power to the equipment will be disconnected in the event of a power interruption.

(FPN): Magnetic motor starters, magnetic contactors, and similar devices, with manual or time delay restarting for the load, will provide restarting after power interruption.

455-23. Capacitors. Capacitors that are not an integral part of the rotary phase conversion system but are installed for a motor load shall be connected to the line side of that motor overload protective device.

ARTICLE 460 — CAPACITORS

460-1. Scope. This article covers the installation of capacitors on electric circuits.

Surge capacitors or capacitors included as a component part of other apparatus and conforming with the requirements of such apparatus are excluded from these requirements.

This article also covers the installation of capacitors in hazardous (classified) locations as modified by Articles 501 through 503.

460-2. Enclosing and Guarding.

(a) Containing More than 3 Gallons (11.36 L) of Flammable Liquid. Capacitors containing more than 3 gallons (11.36 L) of flammable liquid shall be enclosed in vaults or outdoor fenced enclosures complying with Article 710. This limit shall apply to any single unit in an installation of capacitors.

(b) Accidental Contact. Capacitors shall be enclosed, located, or guarded so that persons cannot come into accidental contact or bring conducting materials into accidental contact with exposed energized parts, terminals, or buses associated with them.

Exception: No additional guarding is required for enclosures accessible only to authorized and qualified persons.

A. 600 Volts, Nominal, and Under

460-6. Discharge of Stored Energy. Capacitors shall be provided with a means of discharging stored energy.

(a) Time of Discharge. The residual voltage of a capacitor shall be reduced to 50 volts, nominal, or less, within 1 minute after the capacitor is disconnected from the source of supply.

(b) Means of Discharge. The discharge circuit shall be either permanently connected to the terminals of the capacitor or capacitor bank, or provided with automatic means of connecting it to the terminals of the capacitor bank on removal of voltage from the line. Manual means of switching or connecting the discharge circuit shall not be used.

460-8. Conductors.

(a) Ampacity. The ampacity of capacitor circuit conductors shall not be less than 135 percent of the rated current of the capacitor. The ampacity of conductors that connect a capacitor to the terminals of a motor or to motor circuit conductors shall not be less than one third the ampacity of the motor circuit conductors and in no case less than 135 percent of the rated current of the capacitor.

(b) Overcurrent Protection.

(1) An overcurrent device shall be provided in each ungrounded conductor for each capacitor bank.

Exception: A separate overcurrent device shall not be required for a capacitor connected on the load side of a motor overload protective device.

(2) The rating or setting of the overcurrent device shall be as low as practicable.

(c) Disconnecting Means.

(1) A disconnecting means shall be provided in each ungrounded conductor for each capacitor bank.

Exception: Where a capacitor is connected on the load side of a motor overload protective device.

(2) The disconnecting means shall open all ungrounded conductors simultaneously.

(3) The disconnecting means shall be permitted to disconnect the capacitor from the line as a regular operating procedure.

(4) The rating of the disconnecting means shall not be less than 135 percent of the rated current of the capacitor.

460-9. Rating or Setting of Motor Overload Device. Where a motor installation includes a capacitor connected on the load side of the motor overload device, the rating or setting of the motor overload device shall be based on the improved power factor of the motor circuit.

The effect of the capacitor shall be disregarded in determining the motor circuit conductor rating in accordance with Section 430-22.

460-10. Grounding. Capacitor cases shall be grounded in accordance with Article 250.

Exception: Where the capacitor units are supported on a structure that is designed to operate at other than ground potential.

460-12. Marking. Each capacitor shall be provided with a nameplate giving the name of the manufacturer, rated voltage, frequency, kilovar or amperes, number of phases, and, if filled with a combustible liquid, the amount of liquid in gallons. Where filled with a nonflammable liquid, the nameplate shall so state. The nameplate shall also indicate if a capacitor has a discharge device inside the case.

B. Over 600 Volts, Nominal

460-24. Switching.

(a) Load Current. Group-operated switches shall be used for capacitor switching and shall be capable of (1) carrying continuously not less than 135 percent of the rated current of the capacitor installation; (2) interrupting the maximum continuous load current of each capacitor, capacitor bank, or capacitor installation that will be switched as a unit; (3) withstanding the maximum inrush current, including contributions from adjacent capacitor installations; (4) carrying currents due to faults on capacitor side of switch.

(b) Isolation.

(1) A means shall be installed to isolate from all sources of voltage each capacitor, capacitor bank, or capacitor installation that will be removed from service as a unit.

(2) The isolating means shall provide a visible gap in the electrical circuit adequate for the operating voltage.

(3) Isolating or disconnecting switches (with no interrupting rating) shall be interlocked with the load-interrupting device or shall be provided with prominently displayed caution signs in accordance with Section 710-22 to prevent switching load current.

(c) **Additional Requirements for Series Capacitors.** The proper switching sequence shall be assured by use of one of the following: (1) mechanically sequenced isolating and bypass switches; (2) interlocks; or (3) switching procedure prominently displayed at the switching location.

460-25. Overcurrent Protection.

(a) **Provided to Detect and Interrupt Fault Current.** A means shall be provided to detect and interrupt fault current likely to cause dangerous pressure within an individual capacitor.

(b) **Single-Phase or Multiphase Devices.** Single-phase or multiphase devices shall be permitted for this purpose.

(c) **Protected Individually or in Groups.** Capacitors shall be permitted to be protected individually or in groups.

(d) **Protective Devices Rated or Adjusted.** Protective devices for capacitors or capacitor equipment shall be rated or adjusted to operate within the limits of the safe zone for individual capacitors.

Exception: If the protective devices are rated or adjusted to operate within the limits for Zone 1 or Zone 2, the capacitors shall be enclosed or isolated.

In no event shall the rating or adjustment of the protective devices exceed the maximum limit of Zone 2.

(FPN): For definitions of the Safe Zone, Zone 1, and Zone 2, see Shunt Power Capacitors, ANSI/IEEE 18-1980.

460-26. Identification.
Each capacitor shall be provided with a permanent nameplate giving the maker's name, rated voltage, frequency, kilovar or amperes, number of phases, and the amount of liquid in gallons identified as flammable, if such is the case.

460-27. Grounding.
Capacitor neutrals and cases, if grounded, shall be grounded in accordance with Article 250.

Exception: Where the capacitor units are supported on a structure that is designed to operate at other than ground potential.

460-28. Means for Discharge.

(a) **Means to Reduce the Residual Voltage.** A means shall be provided to reduce the residual voltage of a capacitor to 50 volts or less within 5 minutes after the capacitor is disconnected from the source of supply.

(b) **Connection to Terminals.** A discharge circuit shall be either permanently connected to the terminals of the capacitor or provided with automatic means of connecting it to the terminals of the capacitor bank after disconnection of the capacitor from the source of supply. The windings of motors, or transformers, or of other equipment directly connected to capacitors without a switch or overcurrent device interposed shall meet the requirements of (a) above.

ARTICLE 470 — RESISTORS AND REACTORS
For Rheostats, see Section 430-82.

A. 600 Volts, Nominal, and Under

470-1. Scope. This article covers the installation of separate resistors and reactors on electric circuits.

Exception: Resistors and reactors that are component parts of other apparatus.

This article also covers the installation of resistors and reactors in hazardous (classified) locations as modified by Articles 501 through 504.

470-2. Location. Resistors and reactors shall not be placed where exposed to physical damage.

470-3. Space Separation. A thermal barrier shall be required if the space between the resistors and reactors and any combustible material is less than 12 inches (305 mm).

470-4. Conductor Insulation. Insulated conductors used for connections between resistance elements and controllers shall be suitable for an operating temperature of not less than 90°C (194°F).

Exception: Other conductor insulations shall be permitted for motor starting service.

B. Over 600 Volts, Nominal

470-18. General.

   (a) Protected Against Physical Damage. Resistors and reactors shall be protected against physical damage.

   (b) Isolated by Enclosure or Elevation. Resistors and reactors shall be isolated by enclosure or elevation to protect personnel from accidental contact with energized parts.

   (c) Combustible Materials. Resistors and reactors shall not be installed in close enough proximity to combustible materials to constitute a fire hazard and shall have a clearance of not less than 1 foot (305 mm) from combustible materials.

   (d) Clearances. Clearances from resistors and reactors to grounded surfaces shall be adequate for the voltage involved.

   (FPN):   See Article 710.

   (e) Temperature Rise from Induced Circulating Currents. Metallic enclosures of reactors and adjacent metal parts shall be installed so that the temperature rise from induced circulating currents will not be hazardous to personnel or constitute a fire hazard.

470-19. Grounding. Resistor and reactor cases or enclosures shall be grounded in accordance with Article 250.

470-20. Oil-Filled Reactors. Installation of oil-filled reactors, in addition to the above requirements, shall comply with applicable requirements of Article 450.

ARTICLE 480 — STORAGE BATTERIES

480-1. Scope. The provisions of this article shall apply to all stationary installations of storage batteries.

480-2. Definitions.

Storage Battery: A battery comprised of one or more rechargeable cells of the lead-acid, nickel-cadmium, or other rechargeable electrochemical types.

Sealed Cell or Battery: A sealed cell or battery is one that has no provision for the addition of water or electrolyte or for external measurement of electrolyte specific gravity. The individual cells shall be permitted to contain a venting arrangement as described in Section 480-9(b).

Nominal Battery Voltage: The voltage computed on the basis of 2.0 volts per cell for the lead-acid type and 1.2 volts per cell for the alkali type.

480-3. Wiring and Equipment Supplied from Batteries. Wiring and equipment supplied from storage batteries shall be subject to the requirements of this Code applying to wiring and equipment operating at the same voltage.

Exception: As otherwise provided for communication systems in Article 800.

480-4. Grounding. The requirements of Article 250 shall apply.

480-5. Insulation of Batteries of Not Over 250 Volts. This section shall apply to storage batteries having cells so connected as to operate at a nominal battery voltage of not over 250 volts.

(a) Vented Lead-Acid Batteries. Cells and multicompartment batteries with covers sealed to containers of nonconductive, heat-resistant material shall not require additional insulating support.

(b) Vented Alkaline-type Batteries. Cells with covers sealed to jars of nonconductive, heat-resistant material shall require no additional insulation support. Cells in jars of conductive material shall be installed in trays of nonconductive material with not more than 20 cells (24 volts, nominal) in the series circuit in any one tray.

(c) Rubber Jars. Cells in rubber or composition containers shall require no additional insulating support where the total nominal voltage of all cells in series does not exceed 150 volts. Where the total voltage exceeds 150 volts, batteries shall be sectionalized into groups of 150 volts or less, and each group shall have the individual cells installed in trays or on racks.

(d) Sealed Cells or Batteries. Sealed cells and multicompartment sealed batteries constructed of nonconductive, heat-resistant material shall not require additional insulating support. Batteries constructed of a conducting container shall have insulating support if a voltage is present between the container and ground.

480-6. Insulation of Batteries of Over 250 Volts. The provisions of Section 480-5 shall apply to storage batteries having the cells so connected as to operate at a nominal voltage exceeding 250 volts, and, in addition, the provisions of this section shall also apply to such batteries. Cells shall be installed in groups having a total nominal voltage of not over 250 volts. Insulation, which can be air, shall be provided between groups and shall have a minimum separation between live battery parts of opposite polarity of 2 inches (50.8 mm) for battery voltages not exceeding 600 volts.

480-7. Racks and Trays. Racks and trays shall comply with (a) and (b) below.

(a) Racks. Racks, as required in this article, are rigid frames designed to support cells or trays. They shall be substantial and made of:

(1) Metal, so treated as to be resistant to deteriorating action by the electrolyte and provided with nonconducting members directly supporting the cells or with continuous insulating material other than paint or conducting members; or

(2) Other construction such as fiberglass or other suitable nonconductive materials.

(b) Trays. Trays are frames, such as crates or shallow boxes usually of wood or other nonconductive material, so constructed or treated as to be resistant to deteriorating action by the electrolyte.

480-8. Battery Locations. Battery locations shall conform to (a) and (b) below.

(a) Ventilation. Provisions shall be made for sufficient diffusion and ventilation of the gases from the battery to prevent the accumulation of an explosive mixture.

(b) Live Parts. Guarding of live parts shall comply with Section 110-17.

480-9. Vents.

(a) Vented Cells. Each vented cell shall be equipped with a flame arrestor designed to prevent destruction of the cell due to ignition of gases within the cell by an external spark or flame under normal operating conditions.

(b) Sealed Cells. Sealed battery/cells shall be equipped with a pressure-release vent to prevent excessive accumulation of gas pressure, or the battery/cell shall be designed to prevent scatter of cell parts in event of a cell explosion.

Chapter 5. Special Occupancies

ARTICLE 500 — HAZARDOUS (CLASSIFIED) LOCATIONS

500-1. Scope — Articles 500 Through 504. Articles 500 through 504 cover the requirements for electrical equipment and wiring for all voltages in locations where fire or explosion hazards may exist due to flammable gases or vapors, flammable liquids, combustible dust, or ignitible fibers or flyings.

500-2. Location and General Requirements. Locations shall be classified depending on the properties of the flammable vapors, liquids or gases, or combustible dusts or fibers that may be present and the likelihood that a flammable or combustible concentration or quantity is present. Where pyrophoric materials are the only materials used or handled, these locations shall not be classified.

Each room, section, or area shall be considered individually in determining its classification.

Exception: Except as modified in Articles 500 through 504, all other applicable rules contained in this Code shall apply to electric equipment and wiring installed in hazardous (classified) locations.

(FPN No. 1): For definitions of "approved" and "explosionproof" as used in these articles, see Article 100; "dust-ignition-proof" is defined in Section 502-1.

Intrinsically safe apparatus and wiring shall be permitted in any hazardous (classified) location for which it is approved, and the provisions of Articles 501 through 503 and 510 through 516 shall not be considered applicable to such installations except as required by Article 504.

Wiring of intrinsically safe circuits shall be physically separated from wiring of all other circuits that are not intrinsically safe. Means shall be provided to minimize the passage of gases and vapors.

Installation of intrinsically safe apparatus and wiring shall be in accordance with the requirements of Article 504.

(FPN No. 2): Through the exercise of ingenuity in the layout of electrical installations for hazardous (classified) locations, it is frequently possible to locate much of the equipment in less hazardous or in nonhazardous locations and, thus, to reduce the amount of special equipment required. In some cases, hazards may be reduced or hazardous (classified) locations limited or eliminated by adequate positive-pressure ventilation from a source of clean air in conjunction with effective safeguards against ventilation failure. For further information, see Purged and Pressurized Enclosures for Electrical Equipment, NFPA 496-1989 (ANSI), Flammable and Combustible Liquids Code, NFPA 30-1990 (ANSI), Section 5-3.3, and American Petroleum Institute, API RP 500-1991, Recommended Practice for Classification of Locations for Electrical Installations at Petroleum Facilities, Section 4.6.

(FPN No. 3): It is important that the authority having jurisdiction be familiar with such recorded industrial experience as well as with such standards of the National Fire Protection Association as may be of use in the classification of various areas with respect to hazard.

(FPN No. 4): For further information, see Flammable and Combustible Liquids Code, NFPA 30-1990; Drycleaning Plants, NFPA 32-1990; Manufacture of Organic Coatings, NFPA 35-1987 (ANSI); Solvent Extraction Plants, NFPA 36-1988 (ANSI); Storage and Handling of Liquefied Petroleum Gases, NFPA 58-1992; Storage and Handling of Liquefied Petroleum Gases at Utility Gas Plants, NFPA 59-1992; Classification of Class I Hazardous (Classified) Locations for Electrical Installations in Chemical Process Areas, NFPA 497A-1992 (ANSI); and Fire Protection in Wastewater Treatment and Collection Facilities, NFPA 820-1992 (ANSI).

(FPN No. 5): For protection against static electricity hazards, see Recommended Practice on Static Electricity, NFPA 77-1988 (ANSI).

(FPN No. 6): For electrical classification of laboratory areas, see Fire Protection for Laboratories Using Chemicals, NFPA 45-1991.

All conduit referred to herein shall be threaded with an NPT standard conduit cutting die that provides $3/4$-inch taper per foot. Such conduit shall be made up wrenchtight to minimize sparking when fault current flows through the conduit system. Where it is impractical to make a threaded joint tight, a bonding jumper shall be utilized.

500-3. Special Precaution. Articles 500 through 504 require equipment construction and installation that will ensure safe performance under conditions of proper use and maintenance.

(FPN No. 1): It is important that inspection authorities and users exercise more than ordinary care with regard to installation and maintenance.

(FPN No. 2): Low ambient conditions require special consideration. Explosionproof or dust-ignitionproof equipment may not be suitable for use at temperatures lower than -25°C (-13°F) unless they are approved for low-temperature service. However, at low ambient temperatures, flammable concentrations of vapors may not exist in a location classified Class I, Division 1 at normal ambient temperature.

For purposes of testing, approval, and area classification, various air mixtures (not oxygen-enriched) shall be grouped in accordance with Sections 500-3(a) and 500-3(b).

Exception: Equipment approved for a specific gas, vapor, or dust.

(FPN): This grouping is based on the characteristics of the materials. Facilities have been made available for testing and approving equipment for use in the various atmospheric groups.

(a) Class I Group Classifications. Class I groups shall be as follows:

(1) Group A. Atmospheres containing acetylene.

(2) Group B: Atmospheres containing hydrogen, fuel and combustible process gases containing more than 30 percent hydrogen by volume, or gases or vapors of equivalent hazard such as butadiene, ethylene oxide, propylene oxide, and acrolein.

Exception No. 1: Group D equipment shall be permitted to be used for atmospheres containing butadiene if such equipment is isolated in accordance with Section 501-5(a) by sealing all conduit $1/2$-inch size or larger.

Exception No. 2: Group C equipment shall be permitted to be used for atmospheres containing ethylene oxide, propylene oxide, and acrolein if such equipment is isolated in accordance with Section 501-5(a) by sealing all conduit $1/2$-inch size or larger.

(3) Group C: Atmospheres such as ethyl ether, ethylene, or gases or vapors of equivalent hazard.

(4) Group D: Atmospheres such as acetone, ammonia, benzene, butane, cyclopropane, ethanol, gasoline, hexane, methanol, methane, natural gas, naphtha, propane, or gases or vapors of equivalent hazard.

Exception: For atmospheres containing ammonia, the authority having jurisdiction for enforcement of this Code shall be permitted to reclassify the location to a less hazardous location or a nonhazardous location.

(FPN No. 1): For additional information on the properties and group classification of Class I materials, see Classification of Gases, Vapors, and Dusts for Electrical Equipment in Hazardous (Classified) Locations, NFPA 497M-1991, and Fire Hazard Properties of Flammable Liquids, Gases, and Volatile Solids, NFPA 325M-1991.

(FPN No. 2): The explosion characteristics of air mixtures of gases or vapors vary with the specific material involved. For Class I locations, Groups A, B, C, and D, the classification involves determinations of maximum explosion pressure and maximum safe clearance between parts of a clamped joint in an enclosure. It is necessary, therefore, that equipment be approved not only for class but also for the specific group of the gas or vapor that may be present.

(FPN No. 3): Certain chemical atmospheres may have characteristics that require safeguards beyond those required for any of the above groups. Carbon disulfide is one of these chemicals because of its low ignition temperature [100°C (212°F)] and the small joint clearance permitted to arrest its flame.

(FPN No. 4): For classification of areas involving ammonia atmosphere, see Safety Code for Mechanical Refrigeration, ANSI/ASHRAE 15-1978, and Safety Requirements for the Storage and Handling of Anhydrous Ammonia, ANSI/CGA G2.1-1981.

(b) Class II Group Classifications. Class II groups shall be as follows:

(1)^x Group E: Atmospheres containing combustible metal dusts, including aluminum, magnesium, and their commercial alloys, or other combustible dusts whose particle size, abrasiveness, and conductivity present similar hazards in the use of electrical equipment.

(FPN): Certain metal dusts may have characteristics that require safeguards beyond those required for atmospheres containing the dusts of aluminum, magnesium, and their commercial alloys. For example, zirconium, thorium, and uranium dusts have extremely low ignition temperatures [as low as 20°C (68°F)] and minimum ignition energies lower than any material classified in any of the Class I or Class II Groups.

(2)^x Group F: Atmospheres containing combustible carbonaceous dusts, including carbon black, charcoal, coal, or coke dusts that have more than 8 percent total entrapped voltatiles, or dusts that have been sensitized by other materials so that they present an explosion hazard.

(FPN): See ASTM D 3175-89, *Standard Test Method for Volatile Material in the Analysis Sample for Coal and Coke.*

(3) Group G: Atmospheres containing combustible dusts not included in Group E or F, including flour, grain, wood, plastic, and chemicals.

(FPN No. 1): For additional information on group classification of Class II materials, see Classification of Gases, Vapors, and Dusts for Electrical Equipment in Hazardous (Classified) Locations, NFPA 497M-1991.

(FPN No. 2): The explosion characteristics of air mixtures of dust vary with the materials involved. For Class II locations, Groups E, F, and G, the classification involves the tightness of the joints of assembly and shaft openings to prevent the entrance of dust in the dust-ignitionproof enclosure, the blanketing effect of layers of dust on the equipment that may cause overheating, and the ignition temperature of the dust. It is necessary, therefore, that equipment be approved not only for the class, but also for the specific group of dust that will be present.

(FPN No. 3):   Certain dusts may require additional precautions due to chemical phenomena that can result in the generation of ignitible gases. See National Electrical Safety Code, ANSI C2-1990, Section 127A-Coal Handling Areas.

(c) Approval for Class and Properties. Equipment shall be approved not only for the class of location but also for the explosive, combustible, or ignitible properties of the specific gas, vapor, dust, fiber, or flyings that will be present. In addition, Class I equipment shall not have any exposed surface that operates at a temperature in excess of the ignition temperature of the specific gas or vapor. Class II equipment shall not have an external temperature higher than that specified in Section 500-3(f). Class III equipment shall not exceed the maximum surface temperatures specified in Section 503-1.

Equipment that has been approved for a Division 1 location shall be permitted in a Division 2 location of the same class and group.

Where specifically permitted in Articles 501 through 503, general-purpose equipment or equipment in general-purpose enclosures shall be permitted to be installed in Division 2 locations if the equipment does not constitute a source of ignition under normal operating conditions.

Unless otherwise specified, normal operating conditions for motors shall be assumed to be rated full-load steady conditions.

Where flammable gases or combustible dusts are or may be present at the same time, the simultaneous presence of both shall be considered when determining the safe operating temperature of the electrical equipment.

(FPN):   The characteristics of various atmospheric mixtures of gases, vapors, and dusts depend on the specific material involved.

(d) Marking. Approved equipment shall be marked to show the Class, Group, and operating temperature or temperature range referenced to a 40°C ambient.

The temperature range, if provided, shall be indicated in identification numbers, as shown in Table 500-3(d).

Identification numbers marked on equipment nameplates shall be in accordance with Table 500-3(d).

Equipment that is approved for Class I and Class II shall be marked with the maximum safe operating temperature, as determined by simultaneous exposure to the combinations of Class I and Class II conditions.

Exception No. 1: Equipment of the nonheat-producing type, such as junction boxes, conduit, and fittings and equipment of the heat-producing type having a maximum temperature not more than 100°C (212°F) shall not be required to have a marked operating temperature or temperature range.

Exception No. 2: Fixed lighting fixtures marked for use in Class I, Division 2 or Class II, Division 2 locations only shall not be required to be marked to indicate the group.

Exception No. 3: Fixed general-purpose equipment in Class I locations, other than fixed lighting fixtures, that is acceptable for use in Class I, Division 2 locations shall not be required to be marked with the Class, Group, Division, or operating temperature.

Exception No. 4: Fixed dusttight equipment other than fixed lighting fixtures that are acceptable for use in Class II, Division 2 and Class III locations shall not be required to be marked with the Class, Group, Division, or operating temperature.

Table 500-3(d).   Identification Numbers

Degrees C	Maximum Temperature Degrees F	Identification Number
450	842	T1
300	572	T2
280	536	T2A
260	500	T2B
230	446	T2C
215	419	T2D
200	392	T3
180	356	T3A
165	329	T3B
160	320	T3C
135	275	T4
120	248	T4A
100	212	T5
85	185	T6

(FPN): Since there is no consistent relationship between explosion properties and ignition temperature, the two are independent requirements.

(e) Class I Temperature. The temperature marking specified in (d) above shall not exceed the ignition temperature of the specific gas or vapor to be encountered.

(FPN):   For information regarding ignition temperatures of gases and vapors, see Classification of Gases, Vapors, and Dusts for Electrical Equipment in Hazardous (Classified) Locations, NFPA 497M-1991, and Fire Hazard Properties of Flammable Liquids, Gases, and Volatile Solids, NFPA 325M-1991.

Formerly, the temperature limit of each group was assumed to be the lowest ignition temperature of any material in the group, i.e., 280°C (536°F) for Group D and 180°C (356°F) for Group C.

(FPN):   To avoid revising this limit as new gases are added (see hexane in Group D and acetaldehyde in Group C), temperature will be specified in future markings.

The ignition temperature for which equipment was approved prior to this requirement shall be assumed to be as follows:

Group A – 280°C (536°F)   Group C – 180°C (356°F)
Group B – 280°C (536°F)   Group D – 280°C (536°F)

(f) Class II Temperature. The temperature marking specified in (d) above shall be less than the ignition temperature of the specific dust to be encountered. For organic dusts that may dehydrate or carbonize, the temperature marking shall not exceed the lower of either the ignition temperature or 165°C (329°F). See Classification of Gases, Vapors, and Dusts for Electrical Equipment in Hazardous (Classified) Locations, NFPA 497M-1991, for minimum ignition temperatures of specific dusts.

The ignition temperature for which equipment was approved prior to this requirement shall be assumed to be as shown in Table 500-3(f).

Table 500-3(f)

Class II Group	Equipment that is Not Subject to Overloading		Equipment (such as Motors or Power Transformers) that May Be Overloaded			
			Normal Operation		Abnormal Operation	
	Degrees C	Degrees F	Degrees C	Degrees F	Degrees C	Degrees F
E	200	392	200	392	200	392
F	200	392	150	302	200	392
G	165	329	120	248	165	329

500-4. Specific Occupancies. Articles 510 through 517 cover garages, aircraft hangars, gasoline dispensing and service stations, bulk storage plants, spray application, dipping and coating processes, and health care facilities.

500-5. Class I Locations. Class I locations are those in which flammable gases or vapors are or may be present in the air in quantities sufficient to produce explosive or ignitible mixtures. Class I locations shall include those specified in (a) and (b) below.

(a) Class I, Division 1. A Class I, Division 1 location is a location (1) in which ignitible concentrations of flammable gases or vapors can exist under normal operating conditions; or (2) in which ignitible concentrations of such gases or vapors may exist frequently because of repair or maintenance operations or because of leakage; or (3) in which breakdown or faulty operation of equipment or processes might release ignitible concentrations of flammable gases or vapors, and might also cause simultaneous failure of electric equipment.

(FPN): This classification usually includes locations where volatile flammable liquids or liquefied flammable gases are transferred from one container to another; interiors of spray booths and areas in the vicinity of spraying and painting operations where volatile flammable solvents are used; locations containing open tanks or vats of volatile flammable liquids; drying rooms or compartments for the evaporation of flammable solvents; locations containing fat and oil extraction equipment using volatile flammable solvents; portions of cleaning and dyeing plants where flammable liquids are used; gas generator rooms and other portions of gas manufacturing plants where flammable gas may escape; inadequately ventilated pump rooms for flammable gas or for volatile flammable liquids; the interiors of refrigerators and freezers in which volatile flammmable materials are stored in open, lightly stoppered, or easily ruptured containers; and all other locations where ignitible concentrations of flammable vapors or gases are likely to occur in the course of normal operations.

(b) Class I, Division 2. A Class I, Division 2 location is a location (1) in which volatile flammable liquids or flammable gases are handled, processed, or used, but in which the liquids, vapors, or gases will normally be confined within closed containers or closed systems from which they can escape only in case of accidental rupture or breakdown of such containers or systems, or in case of abnormal operation of equipment; or (2) in which ignitible concentrations of gases or vapors are normally prevented by positive mechanical ventilation, and which might become hazardous through failure or abnormal operation of the ventilating equipment; or (3) that is adjacent to a Class I, Division 1 location, and to which ignitible concentrations of gases or vapors might occasionally be communicated unless such communication is prevented by adequate positive-pressure ventilation from a source of clean air, and effective safeguards against ventilation failure are provided.

(FPN No. 1): This classification usually includes locations where volatile flammable liquids or flammable gases or vapors are used but that, in the judgment of the authority having jurisdiction, would become hazardous only in case of an accident or of some unusual operating condition. The quantity of flammable material that might escape in case of accident, the adequacy of ventilating equipment, the total area involved, and the record of the industry or business with respect to explosions or fires are all factors that merit consideration in determining the classification and extent of each location.

(FPN No. 2): Piping without valves, checks, meters, and similar devices would not ordinarily introduce a hazardous condition even though used for flammable liquids or gases. Locations used for the storage of flammable liquids or of liquefied or compressed gases in sealed containers would not normally be considered hazardous unless subject to other hazardous conditions also.

Electrical conduits and their associated enclosures separated from process fluids by a single seal or barrier shall be classed as a Division 2 location if the outside of the conduit and enclosures is an unclassified location.

500-6. Class II Locations. Class II locations are those that are hazardous because of the presence of combustible dust. Class II locations shall include those specified in (a) and (b) below.

(a) Class II, Division 1. A Class II, Division 1 location is a location (1) in which combustible dust is in the air under normal operating conditions in quantities sufficient to produce explosive or ignitible mixtures; or (2) where mechanical failure or abnormal operation of machinery or equipment might cause such explosive or ignitible mixtures to be produced, and might also provide a source of ignition through simultaneous failure of electric equipment, operation of protection devices, or from other causes; or (3) in which combustible dusts of an electrically conductive nature may be present in hazardous quantities.

(FPN): Combustible dusts that are electrically nonconductive include dusts produced in the handling and processing of grain and grain products, pulverized sugar and cocoa, dried egg and milk powders, pulverized spices, starch and pastes, potato and woodflour, oil meal from beans and seed, dried hay, and other organic materials that may produce combustible dusts when processed or handled. Only Group E dusts are considered to be electrically conductive for classification purposes. Dusts containing magnesium or aluminum are particularly hazardous, and the use of extreme precaution will be necessary to avoid ignition and explosion.

(b) Class II, Division 2. A Class II, Division 2 location is a location where combustible dust is not normally in the air in quantities sufficient to produce explosive or ignitible mixtures, and dust accumulations are normally insufficient to interfere with the normal operation of electrical equipment or other apparatus, but combustible dust may be in suspension in the air as a result of infrequent malfunctioning of handling or processing equipment and where combustible dust accumulations on, in, or in the vicinity of the electrical equipment may be sufficient to interfere with the safe dissipation of heat from electrical equipment or may be ignitible by abnormal operation or failure of electrical equipment.

(FPN No. 1):   The quantity of combustible dust that may be present and the adequacy of dust removal systems are factors that merit consideration in determining the classification and may result in an unclassified area.

(FPN No. 2):   Where products such as seed are handled in a manner that produces low quantities of dust, the amount of dust deposited may not warrant classification.

500-7. Class III Locations.   Class III locations are those that are hazardous because of the presence of easily ignitible fibers or flyings, but in which such fibers or flyings are not likely to be in suspension in the air in quantities sufficient to produce ignitible mixtures. Class III locations shall include those specified in (a) and (b) below.

(a) Class III, Division 1. A Class III, Division 1 location is a location in which easily ignitible fibers or materials producing combustible flyings are handled, manufactured, or used.

(FPN No. 1):   Such locations usually include some parts of rayon, cotton, and other textile mills; combustible fiber manufacturing and processing plants; cotton gins and cotton-seed mills; flax-processing plants; clothing manufacturing plants; woodworking plants; and establishments and industries involving similar hazardous processes or conditions.

(FPN No. 2):   Easily ignitible fibers and flyings include rayon, cotton (including cotton linters and cotton waste), sisal or henequen, istle, jute, hemp, tow, cocoa fiber, oakum, baled waste kapok, Spanish moss, excelsior, and other materials of similar nature.

(b) Class III, Division 2. A Class III, Division 2 location is a location in which easily ignitible fibers are stored or handled.

Exception: In process of manufacture.

ARTICLE 501 — CLASS I LOCATIONS

501-1. General.   The general rules of this Code shall apply to the electric wiring and equipment in locations classified as Class I in Section 500-5.

Exception: As modified by this article.

501-2. Transformers and Capacitors.

(a) Class I, Division 1. In Class I, Division 1 locations, transformers and capacitors shall comply with the following:

(1) Containing Liquid that Will Burn. Transformers and capacitors containing a liquid that will burn shall be installed only in approved vaults that comply with Sections 450-41 through 450-48, and, in addition, (1) there shall be no door or other communicating opening between the vault and the Division 1 location; and (2) ample ventilation shall be provided for the continuous removal of flammable gases or vapors; and (3) vent openings or ducts shall lead to a safe location outside of buildings; and (4) vent ducts and openings shall be of sufficient area to relieve explosion pressures within the vault, and all portions of vent ducts within the buildings shall be of reinforced concrete construction.

(2) Not Containing Liquid that Will Burn. Transformers and capacitors that do not contain a liquid that will burn shall (1) be installed in vaults complying with (a)(1) above or (2) be approved for Class I locations.

(b) Class I, Division 2. In Class I, Division 2 locations, transformers and capacitors shall comply with Sections 450-21 through 450-27.

501-3. Meters, Instruments, and Relays.

(a) Class I, Division 1. In Class I, Division 1 locations, meters, instruments, and relays, including kilowatt-hour meters, instrument transformers, resistors, rectifiers, and thermionic tubes, shall be provided with enclosures approved for Class I, Division 1 locations.

Enclosures approved for Class I, Division 1 locations include (1) explosionproof enclosures and (2) purged and pressurized enclosures.

(FPN): See Purged and Pressurized Enclosures for Electrical Equipment, NFPA 496-1989 (ANSI).

(b) Class I, Division 2. In Class I, Division 2 locations, meters, instruments, and relays shall comply with the following:

(1) Contacts. Switches, circuit breakers, and make-and-break contacts of pushbuttons, relays, alarm bells, and horns shall have enclosures approved for Class I, Division 1 locations in accordance with (a) above.

Exception: General-purpose enclosures shall be permitted, if current-interrupting contacts are:

a. Immersed in oil; or

b. Enclosed within a chamber hermetically sealed against the entrance of gases or vapors; or

c. In nonincendive circuits that under normal conditions do not release sufficient energy to ignite a specific ignitible atmospheric mixture.

(2) Resistors and Similar Equipment. Resistors, resistance devices, thermionic tubes, rectifiers, and similar equipment that are used in or in connection with meters, instruments, and relays shall comply with (a) above.

Exception: General-purpose-type enclosures shall be permitted if such equipment is without make-and-break or sliding contacts [other than as provided in (b)(1) above] and if the maximum operating temperature of any exposed surface will not exceed 80 percent of the ignition temperature in degrees Celsius of the gas or vapor involved or has been tested and found incapable of igniting the gas or vapor.

(3) Without Make-or-Break Contacts. Transformer windings, impedance coils, solenoids, and other windings that do not incorporate sliding or make-or-break contacts shall be provided with enclosures. General-purpose-type enclosures shall be permitted.

(4) General-Purpose Assemblies. Where an assembly is made up of components for which general-purpose enclosures are acceptable as provided in (b)(1), (b)(2), and (b)(3) above, a single general-purpose enclosure shall be acceptable for the assembly. Where such an assembly includes any of the equipment described in (b)(2) above, the maximum obtainable surface temperature of any component of the assembly shall be clearly and permanently indicated on the outside of the enclosure. Alternatively, approved equipment shall be permitted to be marked to indicate the temperature range for which it is suitable, using the identification numbers of Table 500-3(d).

(5) Fuses. Where general-purpose enclosures are permitted in (b)(1), (b)(2), (b)(3), and (b)(4) above, fuses for overcurrent protection of instrument circuits not subject to overloading in normal use shall be permitted to be mounted in general-purpose enclosures if each such fuse is preceded by a switch complying with (b)(1) above.

(6) Connections. To facilitate replacements, process control instruments shall be permitted to be connected through flexible cord, attachment plug, and receptacle, provided (1) a switch complying with (b)(1) above is provided so that the attachment plug is not depended on to interrupt current; and (2) the current does not exceed 3 amperes at 120 volts, nominal; and (3) the power-supply cord does not exceed 3 feet (914 mm), is of a type approved for extra-hard usage or for hard usage if protected by location, and is supplied through an attachment plug and receptacle of the locking and grounding type; and (4) only necessary receptacles are provided; and (5) the receptacle carries a label warning against unplugging under load.

501-4. Wiring Methods. Wiring methods shall comply with (a) and (b) below.

(a) Class I, Division 1. In Class I, Division 1 locations, threaded rigid metal conduit, threaded steel intermediate metal conduit, or Type MI cable with termination fittings approved for the location shall be the wiring method employed. All boxes, fittings, and joints shall be threaded for connection to conduit or cable terminations and shall be explosionproof. Threaded joints shall be made up with at least five threads fully engaged. Type MI cable shall be installed and supported in a manner to avoid tensile stress at the termination fittings. Where necessary to employ flexible connections, as at motor terminals, flexible fittings approved for Class I locations shall be used.

Exception: As provided in Section 501-11.

(b) Class I, Division 2. In Class I, Division 2 locations, threaded rigid metal conduit, threaded steel intermediate metal conduit, enclosed gasketed busways, enclosed gasketed wireways, or Type PLTC cable in accordance with the provisions of Article 725, Type MI, MC, MV, TC, or

SNM cable with approved termination fittings shall be the wiring method employed. Type PLTC, MI, MC, MV, TC, or SNM cable shall be permitted to be installed in cable tray systems and shall be installed in a manner to avoid tensile stress at the termination fittings. Boxes, fittings, and joints shall not be required to be explosionproof except as required by Sections 501-3(b)(1), 501-6(b)(1), and 501-14(b)(1). Where provision must be made for limited flexibility, as at motor terminals, flexible metal fittings, flexible metal conduit with approved fittings, liquidtight flexible metal conduit with approved fittings, liquidtight flexible nonmetallic conduit with approved fittings, or flexible cord approved for extra-hard usage and provided with approved bushed fittings shall be used. An additional conductor for grounding shall be included in the flexible cord.

(FPN):   See Section 501-16(b) for grounding requirements where flexible conduit is used.

Exception: Wiring in nonincendive circuits shall be permitted using any of the methods suitable for wiring in ordinary locations.

501-5. Sealing and Drainage.   Seals in conduit and cable systems shall comply with (a) through (f) below. Sealing compound shall be of a type approved for the conditions and use. Sealing compound shall be used in Type MI cable termination fittings to exclude moisture and other fluids from the cable insulation.

(FPN No. 1):   Seals are provided in conduit and cable systems to minimize the passage of gases and vapors and prevent the passage of flames from one portion of the electrical installation to another through the conduit. Such communication through Type MI cable is inherently prevented by construction of the cable. Unless specifically designed and tested for the purpose, conduit and cable seals are not intended to prevent the passage of liquids, gases, or vapors at a continuous pressure differential across the seal. Even at differences in pressure across the seal equivalent to a few inches of water, there may be a slow passage of gas or vapor through a seal, and through conductors passing through the seal. See Section 501-5(e)(2). Temperature extremes and highly corrosive liquids and vapors can affect the ability of seals to perform their intended function. See Section 501-5(c)(2).

(FPN No. 2):   Gas or vapor leakage and propagation of flames may occur through the interstices between the strands of standard stranded conductors larger than No. 2. Special conductor constructions, e.g., compacted strands or sealing of the individual strands, are means of reducing leakage and preventing the propagation of flames.

(a) Conduit Seals, Class I, Division 1.   In Class I, Division 1 locations, conduit seals shall be located as follows:

(1) In each conduit run entering an enclosure for switches, circuit breakers, fuses, relays, resistors, or other apparatus that may produce arcs, sparks, or high temperatures. Seals shall be installed within 18 inches (457 mm) from such enclosures. Explosionproof unions, couplings, reducers, elbows, capped elbows, and conduit bodies similar to "L," "T," and "Cross" types shall be the only enclosures or fittings permitted between the sealing fitting and the enclosure. The conduit bodies shall not be larger than the largest trade size of the conduits.

Exception: Conduit 1½ inches and smaller entering an explosionproof enclosure for switches, circuit breakers, fuses, relays, or other apparatus that may produce arcs or sparks shall not be required to be sealed if the current-interrupting contacts are:

  a. Enclosed within a chamber hermetically sealed against the entrance of gases or vapors, or

  b. Immersed in oil in accordance with Section 501-6(b)(1)(2).

  (2) In each conduit of 2-inch size or larger entering the enclosure or fitting housing terminals, splices, or taps and within 18 inches (457 mm) of such enclosure or fitting.

  (FPN):   See Section 500-3(c)(2).

  (3) Where two or more enclosures for which seals are required under (a)(1) and (a)(2) above are connected by nipples or by runs of conduit not more than 36 inches (914 mm) long, a single seal in each such nipple connectic..1 or run of conduit shall be considered sufficient if located not more than 18 inches (457 mm) from either enclosure.

  (4) In each conduit run leaving the Class I, Division 1 location. The sealing fitting shall be permitted on either side of the boundary of such location but shall be so designed and installed to minimize the amount of gas or vapor that may have entered the conduit system within the Division 1 location from being communicated to the conduit beyond the seal. There shall be no union, coupling, box, or fitting in the conduit between the sealing fitting and the point at which the conduit leaves the Division 1 location.

  Exception: Metal conduit containing no unions, couplings, boxes, or fittings that passes completely through a Class I, Division 1 location with no fittings less than 12 inches (305 mm) beyond each boundary shall not be required to be sealed if the termination points of the unbroken conduit are in unclassified locations.

  (b) Conduit Seals, Class I, Division 2. In Class I, Division 2 locations, conduit seals shall be located as follows:

  (1) For connections to explosionproof enclosures that are required to be approved for Class I locations, seals shall be provided in accordance with (a)(1), (a)(2), and (a)(3) above. All portions of the conduit run or nipple between the seal and such enclosure shall comply with Section 501-4(a).

  (2) In each conduit run passing from a Class I, Division 2 location into an unclassified location. The sealing fitting shall be permitted on either side of the boundary of such location but shall be so designed and installed to minimize the amount of gas or vapor that may have entered the conduit system within the Division 2 location from being communicated to the conduit beyond the seal. Rigid metal conduit or threaded steel intermediate metal conduit shall be used between the sealing fitting and the point at which the conduit leaves the Division 2 location, and a threaded connection shall be used at the sealing fitting. There shall be no union, coupling, box, or fitting in the conduit between the sealing fitting and the point at which the conduit leaves the Division 2 location.

  Exception No. 1: Metal conduit containing no unions, couplings, boxes, or fittings that passes completely through a Class I, Division 2 location with no fittings less than 12 inches (305 mm) beyond each boundary shall not be required to be sealed if t¹ᴸ termination points of the unbroken conduit are in unclassified locations.

Exception No. 2: Conduit systems terminating at an outdoor unclassi-fied location where a wiring method transition is made to cable tray, cable-bus, ventilated busway, Type MI cable, or open wiring shall not be required to be sealed where passing from the Class I, Division 2 location into the unclassified area. The conduits shall not terminate at an enclosure containing an ignition source.

(c) Class I, Divisions 1 and 2. Where required, seals in Class I, Division 1 and 2 locations shall comply with the following:

(1) Fittings. Enclosures for connections or equipment shall be pro-vided with an approved integral means for sealing, or sealing fittings approved for Class I locations shall be used. Sealing fittings shall be accessible.

(2) Compound. Sealing compound shall be approved and shall pro-vide a seal against passage of gas or vapors through the seal fitting, shall not be affected by the surrounding atmosphere or liquids, and shall not have a melting point of less than 93°C (200°F).

(3) Thickness of Compounds. In a completed seal, the minimum thickness of the sealing compound shall not be less than the trade size of the conduit and, in no case, less than ⅝ inch (16 mm).

(4) Splices and Taps. Splices and taps shall not be made in fittings intended only for sealing with compound, nor shall other fittings in which splices or taps are made be filled with compound.

(5) Assemblies. In an assembly where equipment that may produce arcs, sparks, or high temperatures is located in a compartment separate from the compartment containing splices or taps, and an integral seal is provided where conductors pass from one compartment to the other, the entire assembly shall be approved for Class I locations. Seals in conduit connections to the compartment containing splices or taps shall be pro-vided in Class I, Division 1 locations where required by (a)(2) above.

(d) Cable Seals, Class I, Division 1. In Class I, Division 1 locations, each multiconductor cable in conduit shall be considered as a single con-ductor if the cable is incapable of transmitting gases or vapors through the cable core. These cables shall be sealed in accordance with (a) above.

Cables with a gas/vapor-tight continuous sheath capable of transmitting gases or vapors through the cable core shall be sealed in the Division 1 location after removing the jacket and any other coverings so that the seal-ing compound will surround each individual insulated conductor and the outer jacket.

Exception: Multiconductor cables with a gas/vapor-tight continuous sheath capable of transmitting gases or vapors through the cable core shall be permitted to be considered as a single conductor by sealing the cable in the conduit within 18 inches (457 mm) of the enclosure and the cable end within the enclosure by an approved means to prevent the entrance of gases or vapors or propagation of flame into the cable core, or by other approved methods.

(e) Cable Seals, Class I, Division 2. In Class I, Division 2 locations, cable seals shall be located as follows:

(1) Cables entering enclosures that are required to be approved for Class I locations shall be sealed at the point of entrance. The sealing fitting shall comply with (b)(1) above. Multiconductor cables with a gas/vapor-tight continuous sheath capable of transmitting gases or vapors through the cable core shall be sealed in an approved fitting in the Division 2 location after removing the jacket and any other coverings so that the sealing compound will surround each individual insulated conductor in such a manner as to minimize the passage of gases and vapors. Multiconductor cables in conduit shall be sealed as described in (d) above.

(2) Cables with a gas/vapor-tight continuous sheath and that will not transmit gases or vapors through the cable core in excess of the quantity permitted for seal fittings shall not be required to be sealed except as required in (e)(1) above. The minimum length of such cable run shall not be less than that length that limits gas or vapor flow through the cable core to the rate permitted for seal fittings [0.007 cubic feet per hour (198 cubic centimeters per hour) of air at a pressure of 6 inches of water (1493 pascals)].

(FPN No. 1):   See Outlet Boxes and Fittings for Use in Hazardous Locations, ANSI/UL 886-1985.

(FPN No. 2):   The cable core does not include the interstices of the conductor strands.

(3) Cables with a gas/vapor-tight continuous sheath capable of transmitting gases or vapors through the cable core shall not be required to be sealed except as required in (e)(1) above, unless the cable is attached to process equipment or devices that may cause a pressure in excess of 6 inches (1493 pascals) of water to be exerted at a cable end, in which case a seal, barrier, or other means shall be provided to prevent migration of flammables into an unclassified area.

Exception: Cables with an unbroken gas/vapor-tight continuous sheath shall be permitted to pass through a Class I, Division 2 location without seals.

(4) Cables that do not have gas/vapor-tight continuous sheath shall be sealed at the boundary of the Division 2 and unclassified location in such a manner as to minimize the passage of gases or vapors into an unclassified location.

(FPN):   The sheath mentioned in (d) and (e) above may be either metal or a nonmetallic material.

(f) Drainage.

(1) Control Equipment. Where there is a probability that liquid or other condensed vapor may be trapped within enclosures for control equipment or at any point in the raceway system, approved means shall be provided to prevent accumulation or to permit periodic draining of such liquid or condensed vapor.

(2) Motors and Generators. Where the authority having jurisdiction judges that there is a probability that liquid or condensed vapor may accumulate within motors or generators, joints and conduit systems shall be arranged to minimize entrance of liquid. If means to prevent accumulation or to permit periodic draining are judged necessary, such means shall be provided at the time of manufacture and shall be considered an integral part of the machine.

(3) Canned Pumps, Process or Service Connections, Etc. For canned pumps, process or service connections for flow, pressure, or analysis measurement, etc., that depend upon a single compression seal, diaphragm, or tube to prevent flammable or combustible fluids from entering the electrical conduit system, an additional approved seal, barrier, or other means shall be provided to prevent the flammable or combustible fluid from entering the conduit system beyond the additional devices or means, if the primary seal fails.

The additional approved seal or barrier and the interconnecting enclosure shall meet the temperature and pressure conditions to which they will be subjected upon failure of the primary seal, unless other approved means are provided to accomplish the purpose above.

Drains, vents, or other devices shall be provided so that primary seal leakage will be obvious.

(FPN): See also the last paragraph of Section 500-5(b) and Fine Print Notes to Section 501-5.

501-6. Switches, Circuit Breakers, Motor Controllers, and Fuses.

(a) Class I, Division 1. In Class I, Division 1 locations, switches, circuit breakers, motor controllers, and fuses, including pushbuttons, relays, and similar devices, shall be provided with enclosures, and the enclosure in each case, together with the enclosed apparatus, shall be approved as a complete assembly for use in Class I locations.

(b) Class I, Division 2. Switches, circuit breakers, motor controllers, and fuses in Class I, Division 2 locations shall comply with the following:

(1) Type Required. Circuit breakers, motor controllers, and switches intended to interrupt current in the normal performance of the function for which they are installed shall be provided with enclosures approved for Class I, Division 1 locations in accordance with Section 501-3(a), unless general-purpose enclosures are provided and (1) the interruption of current occurs within a chamber hermetically sealed against the entrance of gases and vapors, or (2) the current make-and-break contacts are oil-immersed and of the general-purpose type having a 2-inch (50.8-mm) minimum immersion for power contacts and a 1-inch (25.4-mm) minimum immersion for control contacts, or (3) the interruption of current occurs within a factory-sealed explosionproof chamber approved for the location.

(2) Isolating Switches. Fused or unfused disconnect and isolating switches for transformers or capacitor banks that are not intended to interrupt current in the normal performance of the function for which they are installed shall be permitted to be installed in general-purpose enclosures.

(3) Fuses. For the protection of motors, appliances, and lamps, other than as provided in (b)(4) below, standard plug or cartridge fuses shall be permitted, provided they are placed within enclosures approved for the location; or fuses shall be permitted if they are within general-purpose enclosures, and if they are of a type in which the operating element is immersed in oil or other approved liquid, or the operating element is enclosed within a chamber hermetically sealed against the entrance of gases and vapors.

(4) Fuses or Circuit Breakers for Overcurrent Protection. Where not more than ten sets of approved enclosed fuses or not more than ten circuit breakers that are not intended to be used as switches for the interruption of current are installed for branch-circuit or feeder protection in any one room, area, or section of the Class I, Division 2 location, general-purpose-type enclosures for such fuses or circuit breakers shall be permitted if the fuses or circuit breakers are for the protection of circuits or feeders supplying lamps in fixed positions only.

(FPN): A set of fuses is all the fuses required to protect all the ungrounded conductors of a circuit. For example, a group of three fuses protecting an ungrounded 3-phase circuit and a single fuse protecting the ungrounded conductor of an identified 2-wire, single-phase circuit is a set of fuses in each instance.

Fuses complying with (b)(3) above shall not be required to be included in counting the ten sets of fuses permitted in general-purpose enclosures.

(5) Fuses Internal to Lighting Fixtures. Approved cartridge fuses shall be permitted as supplementary protection within lighting fixtures.

501-7. Control Transformers and Resistors. Transformers, impedance coils, and resistors used as, or in conjunction with, control equipment for motors, generators, and appliances shall comply with (a) and (b) below.

(a) Class I, Division 1. In Class I, Division 1 locations, transformers, impedance coils, and resistors, together with any switching mechanism associated with them, shall be provided with enclosures approved for Class I, Division 1 locations in accordance with Section 501-3(a).

(b) Class I, Division 2. In Class I, Division 2 locations, control transformers and resistors shall comply with the following:

(1) Switching Mechanisms. Switching mechanisms used in conjunction with transformers, impedance coils, and resistors shall comply with Section 501-6(b).

(2) Coils and Windings. Enclosures for windings of transformers, solenoids, or impedance coils shall be permitted to be of the general-purpose type.

(3) Resistors. Resistors shall be provided with enclosures; and the assembly shall be approved for Class I locations, unless resistance is non-variable and maximum operating temperature, in degrees Celsius, will not exceed 80 percent of the ignition temperature of the gas or vapor involved, or has been tested and found incapable of igniting the gas or vapor.

501-8. Motors and Generators.

(a) Class I, Division 1. In Class I, Division 1 locations, motors, generators, and other rotating electric machinery shall be (1) approved for Class I, Division 1 locations; or (2) of the totally enclosed type supplied with positive-pressure ventilation from a source of clean air with discharge to a safe area, so arranged to prevent energizing of the machine until ventilation has been established and the enclosure has been purged with at least 10 volumes of air, and also arranged to automatically deenergize the equipment when the air supply fails; or (3) of the totally enclosed inert gas-filled type supplied with a suitable reliable source of inert gas for pressuring the enclosure, with devices provided to ensure a positive pressure in the enclosure and arranged to automatically deenergize the equipment when the gas supply fails; or (4) of a type designed to be submerged in a liquid that is flammable only when vaporized and mixed with air, or in a gas or vapor at a pressure greater than atmospheric and that is flammable only when mixed with air; and the machine is so arranged to prevent energizing it until it has been purged with the liquid or gas to exclude air, and also arranged to automatically deenergize the equipment when the supply of liquid or gas or vapor fails or the pressure is reduced to atmospheric.

Totally enclosed motors of Types (2) or (3) shall have no external surface with an operating temperature in degrees Celsius in excess of 80 percent of the ignition temperature of the gas or vapor involved. Appropriate devices shall be provided to detect and automatically deenergize the motor or provide an adequate alarm if there is any increase in temperature of the motor beyond designed limits. Auxiliary equipment shall be of a type approved for the location in which it is installed.

(FPN):   See ASTM Test Procedure (Designation D 2155-69).

(b) Class I, Division 2. In Class I, Division 2 locations, motors, generators, and other rotating electric machinery in which are employed sliding contacts, centrifugal or other types of switching mechanism (including motor overcurrent, overloading, and overtemperature devices), or integral resistance devices, either while starting or while running, shall be approved for Class I, Division 1 locations, unless such sliding contacts, switching mechanisms, and resistance devices are provided with enclosures approved for Class I, Division 2 locations in accordance with Section 501-3(b). The exposed surface of space heaters used to prevent condensation of moisture during shutdown periods shall not exceed 80 percent of the ignition temperature in degrees Celsius of the gas or vapor involved when operated at rated voltage, and the maximum surface temperature [based upon a 40°C (104°F) ambient] shall be permanently marked on a visible nameplate mounted on the motor. Otherwise, space heaters shall be approved for Class I, Division 2 locations.

In Class I, Division 2 locations, the installation of open or nonexplosion-proof enclosed motors, such as squirrel-cage induction motors without brushes, switching mechanisms, or similar arc-producing devices, shall be permitted.

(FPN No. 1):   It is important to consider the temperature of internal and external surfaces that may be exposed to the flammable atmosphere.

(FPN No. 2):   It is important to consider the risk of ignition due to currents arcing across discontinuities and overheating of parts in multisection enclosures of large motors and generators. Such motors and generators may require equipotential bonding jumpers across joints in the enclosures and from enclosure to ground and clean-air purging immediately prior to and during start-up periods.

501-9. Lighting Fixtures.   Lighting fixtures shall comply with (a) or (b) below.

(a) Class I, Division 1. In Class I, Division 1 locations, lighting fixtures shall comply with the following:

(1) Approved Fixtures. Each fixture shall be approved as a complete assembly for the Class I, Division 1 location and shall be clearly marked to indicate the maximum wattage of lamps for which it is approved. Fixtures intended for portable use shall be specifically approved as a complete assembly for that use.

(2) Physical Damage. Each fixture shall be protected against physical damage by a suitable guard or by location.

(3) Pendant Fixtures. Pendant fixtures shall be suspended by and supplied through threaded rigid metal conduit stems or threaded steel intermediate conduit stems, and threaded joints shall be provided with set-screws or other effective means to prevent loosening. For stems longer than 12 inches (305 mm), permanent and effective bracing against lateral displacement shall be provided at a level not more than 12 inches (305 mm) above the lower end of the stem, or flexibility in the form of a fitting or flexible connector approved for the Class I, Division 1 location shall be provided not more than 12 inches (305 mm) from the point of attachment to the supporting box or fitting.

(4) Supports. Boxes, box assemblies, or fittings used for the support of lighting fixtures shall be approved for Class I locations.

(b) Class I, Division 2. In Class I, Division 2 locations, lighting fixtures shall comply with the following:

(1) Portable Lighting Equipment. Portable lighting equipment shall comply with (a)(1) above.

Exception: Where portable lighting equipment are mounted on movable stands and are connected by flexible cords, as covered in Section 501-11, they shall be permitted, where mounted in any position, provided that they conform to Section 501-9(b)(2) below.

(2) Fixed Lighting. Lighting fixtures for fixed lighting shall be protected from physical damage by suitable guards or by location. Where there is danger that falling sparks or hot metal from lamps or fixtures might ignite localized concentrations of flammable vapors or gases, suitable enclosures or other effective protective means shall be provided. Where lamps are of a size or type that may, under normal operating conditions, reach surface temperatures exceeding 80 percent of the ignition temperature in degrees Celsius of the gas or vapor involved, fixtures shall comply with (a)(1) above or shall be of a type that has been tested in order to determine the marked operating temperature or temperature range.

(3) Pendant Fixtures. Pendant fixtures shall be suspended by threaded rigid metal conduit stems, threaded steel intermediate metal conduit stems, or by other approved means. For rigid stems longer than 12 inches (305 mm), permanent and effective bracing against lateral displacement shall be provided at a level not more than 12 inches (305 mm) above the lower end of the stem, or flexibility in the form of an approved fitting or flexible connector shall be provided not more than 12 inches (305 mm) from the point of attachment to the supporting box or fitting.

(4) Switches. Switches that are a part of an assembled fixture or of an individual lampholder shall comply with Section 501-6(b)(1).

(5) Starting Equipment. Starting and control equipment for electric-discharge lamps shall comply with Section 501-7(b).

Exception: A thermal protector potted into a thermally protected fluorescent lamp ballast if the lighting fixture is approved for locations of this Class and Division.

501-10. Utilization Equipment.

(a) Class I, Division 1. In Class I, Division 1 locations, all utilization equipment shall be approved for Class I, Division 1 locations.

(b) Class I, Division 2. In Class I, Division 2 locations, all utilization equipment shall comply with the following:

(1) Heaters. Electrically heated utilization equipment shall conform with either item a or b below.

a. The heater shall not exceed 80 percent of the ignition temperature in degrees Celsius of the gas or vapor involved on any surface that is exposed to the gas or vapor when continuously energized at the maximum rated ambient temperature. If a temperature controller is not provided, these conditions shall apply when the heater is operated at 120 percent of rated voltage.

Exception No. 1: For motor-mounted anticondensation space heaters, see Section 501-8(b).

Exception No. 2: A current-limiting device is applied to the circuit serving the heater that will limit the current in the heater to a value less than that required to raise the heater surface temperature to 80 percent of the ignition temperature.

b. The heater shall be approved for Class I, Division 1 locations.

Exception: Electrical resistance heat tracing approved for Class 1, Division 2 locations.

(2) Motors. Motors of motor-driven utilization equipment shall comply with Section 501-8(b).

(3) Switches, Circuit Breakers, and Fuses. Switches, circuit breakers, and fuses shall comply with Section 501-6(b).

501-11. Flexible Cords, Class I, Divisions 1 and 2. A flexible cord shall be permitted only for connection between portable lighting equipment or other portable utilization equipment and the fixed portion of its supply circuit; and, where used, shall (1) be of a type approved for extra-hard usage; (2) contain, in addition to the conductors of the circuit, a grounding conductor complying with Section 400-23; (3) be connected to terminals or to supply conductors in an approved manner; (4) be supported by clamps or by other suitable means in such a manner that there will be no tension on the terminal connections; and (5) be provided with suitable seals where the flexible cord enters boxes, fittings, or enclosures of the explosionproof type.

Exception: As provided in Sections 501-3(b)(6) and 501-4(b).

Electric submersible pumps with means for removal without entering the wet-pit shall be considered portable utilization equipment. The extension of the flexible cord within a suitable raceway between the wet-pit and the power source shall be permitted.

Electric mixers intended for travel into and out of open-type mixing tanks or vats shall be considered portable utilization equipment.

(FPN): See Section 501-13 for flexible cords exposed to liquids having a deleterious effect on the conductor insulation.

501-12. Receptacles and Attachment Plugs, Class I, Divisions 1 and 2. Receptacles and attachment plugs shall be of the type providing for connection to the grounding conductor of a flexible cord and shall be approved for the location.

Exception: As provided in Section 501-3(b)(6).

501-13. Conductor Insulation, Class I, Divisions 1 and 2. Where condensed vapors or liquids may collect on, or come in contact with, the insulation on conductors, such insulation shall be of a type approved for use under such conditions; or the insulation shall be protected by a sheath of lead or by other approved means.

501-14. Signaling, Alarm, Remote-Control, and Communication Systems.

(a) Class I, Division 1. In Class I, Division 1 locations, all apparatus and equipment of signaling, alarm, remote-control, and communication systems, regardless of voltage, shall be approved for Class I, Division 1 locations, and all wiring shall comply with Sections 501-4(a) and 501-5(a) and (c).

(b) Class I, Division 2. In Class I, Division 2 locations, signaling, alarm, remote-control, and communication systems shall comply with the following:

(1) Contacts. Switches, circuit breakers, and make-and-break contacts of pushbuttons, relays, alarm bells, and horns shall have enclosures approved for Class I, Division 1 locations in accordance with Section 501-3(a).

Exception: General-purpose enclosures shall be permitted if current-interrupting contacts are:

a. Immersed in oil; or

b. Enclosed within a chamber hermetically sealed against the entrance of gases or vapors; or

c. In nonincendive circuits that under normal conditions do not release sufficient energy to ignite a specific ignitible atmospheric mixture.

(2) Resistors and Similar Equipment. Resistors, resistance devices, thermionic tubes, rectifiers, and similar equipment shall comply with Section 501-3(b)(2).

(3) Protectors. Enclosures shall be provided for lightning protective devices and for fuses. Such enclosures shall be permitted to be of the general-purpose type.

(4) Wiring and Sealing. All wiring shall comply with Sections 501-4(b) and 501-5(b) and (c).

501-15. Live Parts, Class I, Divisions 1 and 2. There shall be no exposed live parts.

501-16. Grounding, Class I, Divisions 1 and 2. Wiring and equipment in Class I, Division 1 and 2 locations shall be grounded as specified in Article 250 and with the following additional requirements:

(a) Bonding. The locknut-bushing and double-locknut types of contacts shall not be depended upon for bonding purposes, but bonding jumpers with proper fittings or other approved means of bonding shall be used. Such means of bonding shall apply to all intervening raceways, fittings, boxes, enclosures, etc., between Class I locations and the point of grounding for service equipment or point of grounding of a separately derived system.

Exception: The specific bonding means shall only be required to the point of grounding of a building disconnecting means as specified in Sections 250-24(a), (b), and (c), provided that branch-circuit overcurrent protection is located on the load side of the disconnecting means.

(FPN): See Section 250-78 for additional bonding requirements in hazardous (classified) locations.

(b) Types of Equipment Grounding Conductors. Where flexible metal conduit or liquidtight flexible metal conduit is used as permitted in Section 501-4(b) and is to be relied upon to complete a sole equipment grounding path, it shall be installed with internal or external bonding jumpers in parallel with each conduit and complying with Section 250-79.

Exception: In Class I, Division 2 areas, the bonding jumper shall be permitted to be deleted where all the following conditions are met:

a. Approved liquidtight flexible metal conduit 6 feet (1.83 m) or less in length, with approved fittings, is used;

b. Overcurrent protection in the circuit is limited to 10 amperes or less;

c. The load is not a power utilization load.

501-17. Surge Protection.

(a) Class I, Division 1. Surge arresters, including their installation and connection, shall comply with Article 280. The surge arresters and capacitors shall be installed in enclosures approved for Class I, Division 1 locations. Surge-protective capacitors shall be of a type designed for specific duty.

(b) Class I, Division 2. Surge arresters shall be nonarcing, such as metal-oxide varistor (MOV), sealed type, and surge-protective capacitors shall be of a type designed for specific duty. Installation and connection shall comply with Article 280.

Enclosures shall be permitted to be of the general-purpose type. Surge protection of types other than described above shall be installed in enclosures approved for Class I, Division 1 locations.

501-18. Multiwire Branch Circuits. In a Class I, Division 1 location, a separate grounded conductor shall be installed in each single-phase branch circuit that is part of a multiwire branch circuit.

Exception: Where the disconnect device(s) for the circuit opens all ungrounded conductors of the multiwire circuit simultaneously.

ARTICLE 502 — CLASS II LOCATIONS

502-1. General.   The general rules of this Code shall apply to the electric wiring and equipment in locations classified as Class II locations in Section 500-6.

Exception: As modified by this article.

"Dust-ignitionproof," as used in this article, shall mean enclosed in a manner that will exclude dusts and, where installed and protected in accordance with this Code, will not permit arcs, sparks, or heat otherwise generated or liberated inside of the enclosure to cause ignition of exterior accumulations or atmospheric suspensions of a specified dust on or in the vicinity of the enclosure.

(FPN:) For further information on dust-ignitionproof enclosures, see Type 9 enclosure in Enclosures for Electrical Equipment, ANSI/NEMA 250-1985, and Explosionproof and Dust-Ignitionproof Electrical Equipment for Hazardous (Classified) Locations, ANSI/UL 1203-1988.

Equipment installed in Class II locations shall be able to function at full rating without developing surface temperatures high enough to cause excessive dehydration or gradual carbonization of any organic dust deposits that may occur.

(FPN):   Dust that is carbonized or excessively dry is highly susceptible to spontaneous ignition.

Equipment and wiring of the type defined in Article 100 as explosionproof shall not be required and shall not be acceptable in Class II locations unless approved for such locations.

Where Class II, Group E dusts are present in hazardous quantities, there are only Division 1 locations.

502-2. Transformers and Capacitors.

(a) Class II, Division 1. In Class II, Division 1 locations, transformers and capacitors shall comply with the following:

(1) Containing Liquid that Will Burn. Transformers and capacitors containing a liquid that will burn shall be installed only in approved vaults complying with Sections 450-41 through 450-48, and, in addition, (1) doors or other openings communicating with the Division 1 location shall have self-closing fire doors on both sides of the wall, and the doors shall be carefully fitted and provided with suitable seals (such as weather stripping) to minimize the entrance of dust into the vault; (2) vent openings and ducts shall communicate only with the outside air; and (3) suitable pressure-relief openings communicating with the outside air shall be provided.

(2) Not Containing Liquid that Will Burn. Transformers and capacitors that do not contain a liquid that will burn shall (1) be installed in vaults complying with Sections 450-41 through 450-48 or (2) be approved as a complete assembly, including terminal connections for Class II locations.

(3) Metal Dusts. No transformer or capacitor shall be installed in a location where dust from magnesium, aluminum, aluminum bronze powders, or other metals of similarly hazardous characteristics may be present.

(b) Class II, Division 2. In Class II, Division 2 locations, transformers and capacitors shall comply with the following:

(1) Containing Liquid that Will Burn. Transformers and capacitors containing a liquid that will burn shall be installed in vaults complying with Sections 450-41 through 450-48.

(2) Containing Askarel. Transformers containing askarel and rated in excess of 25 kVA shall (1) be provided with pressure-relief vents; (2) be provided with a means for absorbing any gases generated by arcing inside the case, or the pressure-relief vents shall be connected to a chimney or flue that will carry such gases outside the building; and (3) have an air space of not less than 6 inches (152 mm) between the transformer cases and any adjacent combustible material.

(3) Dry-type Transformers. Dry-type transformers shall be installed in vaults or shall (1) have their windings and terminal connections enclosed in tight metal housings without ventilating or other openings and (2) operate at not over 600 volts, nominal.

502-4. Wiring Methods. Wiring methods shall comply with (a) and (b) below.

(a) Class II, Division 1. In Class II, Division 1 locations, threaded rigid metal conduit, threaded steel intermediate metal conduit, or Type MI cable with termination fittings approved for the location shall be the wiring method employed. Type MI cable shall be installed and supported in a manner to avoid tensile stress at the termination fittings.

(1) Fittings and Boxes. Fittings and boxes shall be provided with threaded bosses for connection to conduit or cable terminations, shall have close-fitting covers, and shall have no openings (such as holes for attachment screws) through which dust might enter or through which sparks or burning material might escape. Fittings and boxes in which taps, joints, or terminal connections are made, or that are used in locations where dusts are of a combustible, electrically conductive nature, shall be approved for Class II locations.

(2) Flexible Connections. Where necessary to employ flexible connections, dusttight flexible connectors, liquidtight flexible metal conduit with approved fittings, liquidtight flexible nonmetallic conduit with approved fittings, or flexible cord approved for extra-hard usage and provided with bushed fittings shall be used. Where flexible cords are used, they shall comply with Section 502-12. Where flexible connections are subject to oil or other corrosive conditions, the insulation of the conductors shall be of a type approved for the condition or shall be protected by means of a suitable sheath.

(FPN): See Section 502-16(b) for grounding requirements where flexible conduit is used.

(b) Class II, Division 2. In Class II, Division 2 locations, rigid metal conduit, intermediate metal conduit, electrical metallic tubing, dusttight wireways, or Type MC, MI, or SNM cable with approved termination fittings, or Type PLTC in cable trays, or Type MC or TC cable installed in ladder, ventilated trough, or ventilated channel cable trays in a single layer, with a space not less than the larger cable diameter between the two adjacent cables, shall be the wiring method employed.

Exception: Wiring in nonincendive circuits shall be permitted using any of the methods suitable for wiring in ordinary locations.

(1) Wireways, Fittings, and Boxes. Wireways, fittings, and boxes in which taps, joints, or terminal connections are made shall be designed to minimize the entrance of dust and (1) shall be provided with telescoping or close-fitting covers or other effective means to prevent the escape of sparks or burning material and (2) shall have no openings (such as holes for attachment screws) through which, after installation, sparks or burning material might escape or through which adjacent combustible material might be ignited.

(2) Flexible Connections. Where flexible connections are necessary, (a)(2) above shall apply.

502-5. Sealing, Class II, Divisions 1 and 2. Where a raceway provides communication between an enclosure that is required to be dust-ignitionproof and one that is not, suitable means shall be provided to prevent the entrance of dust into the dust-ignitionproof enclosure through the raceway. One of the following means shall be permitted: (1) a permanent and effective seal; (2) a horizontal raceway not less than 10 feet (3.05 m) long; or (3) a vertical raceway not less than 5 feet (1.52 m) long and extending downward from the dust-ignitionproof enclosure.

Where a raceway provides communication between an enclosure that is required to be dust-ignitionproof and an enclosure in an unclassified location, seals shall not be required.

Sealing fittings shall be accessible.

502-6. Switches, Circuit Breakers, Motor Controllers, and Fuses.

(a) Class II, Division 1. In Class II, Division 1 locations, switches, circuit breakers, motor controllers, and fuses shall comply with the following:

(1) Type Required. Switches, circuit breakers, motor controllers, and fuses, including pushbuttons, relays, and similar devices that are intended to interrupt current during normal operation, or that are installed where combustible dusts of an electrically conductive nature may be present, shall be provided with dust-ignitionproof enclosures, which, together with the enclosed equipment in each case, shall be approved as a complete assembly for Class II locations.

(2) Isolating Switches. Disconnecting and isolating switches containing no fuses and not intended to interrupt current and not installed where dusts may be of an electrically conductive nature shall be provided with tight metal enclosures that shall be designed to minimize the entrance of dust and that shall (1) be equipped with telescoping or close-fitting covers or with other effective means to prevent the escape of sparks or burning material and (2) have no openings (such as holes for attachment screws) through which, after installation, sparks or burning material might escape or through which exterior accumulations of dust or adjacent combustible material might be ignited.

(3) Metal Dusts. In locations where dust from magnesium, aluminum, aluminum bronze powders, or other metals of similarly hazardous characteristics may be present, fuses, switches, motor controllers, and circuit breakers shall have enclosures specifically approved for such locations.

(b) Class II, Division 2. In Class II, Division 2 locations, enclosures for fuses, switches, circuit breakers, and motor controllers, including pushbuttons, relays, and similar devices, shall be dusttight.

502-7. Control Transformers and Resistors.

(a) Class II, Division 1. In Class II, Division 1 locations, control transformers, solenoids, impedance coils, resistors, and any overcurrent devices or switching mechanisms associated with them shall have dust-ignitionproof enclosures approved for Class II locations. No control transformer, impedance coil, or resistor shall be installed in a location where dust from magnesium, aluminum, aluminum bronze powders, or other metals of similarly hazardous characteristics may be present unless provided with an enclosure approved for the specific location.

(b) Class II, Division 2. In Class II, Division 2 locations, transformers and resistors shall comply with the following:

(1) Switching Mechanisms. Switching mechanisms (including overcurrent devices) associated with control transformers, solenoids, impedance coils, and resistors shall be provided with dusttight enclosures.

(2) Coils and Windings. Where not located in the same enclosure with switching mechanisms, control transformers, solenoids, and impedance coils shall be provided with tight metal housings without ventilating openings.

(3) Resistors. Resistors and resistance devices shall have dust-ignition-proof enclosures approved for Class II locations.

Exception: Where the maximum normal operating temperature of the resistor will not exceed 120°C (248°F), nonadjustable resistors or resistors that are part of an automatically timed starting sequence shall be permitted to have enclosures complying with (b)(2) above.

502-8. Motors and Generators.

(a) Class II, Division 1. In Class II, Division 1 locations, motors, generators, and other rotating electrical machinery shall be:

(1) Approved for Class II, Division 1 locations, or

(2) Totally enclosed pipe-ventilated, meeting temperature limitations in Section 502-1.

(b) Class II, Division 2. In Class II, Division 2 locations, motors, generators, and other rotating electrical equipment shall be totally enclosed nonventilated, totally enclosed pipe-ventilated, totally enclosed fan-cooled or dust-ignitionproof for which maximum full-load external temperature shall be in accordance with Section 500-3(f) for normal operation when operating in free air (not dust blanketed) and shall have no external openings.

Exception: If the authority having jurisdiction believes accumulations of nonconductive, nonabrasive dust will be moderate and if machines can be easily reached for routine cleaning and maintenance, the following shall be permitted to be installed:

a. Standard open-type machines without sliding contacts, centrifugal or other types of switching mechanism (including motor overcurrent, overloading, and overtemperature devices), or integral resistance devices.

b. Standard open-type machines with such contacts, switching mechanisms, or resistance devices enclosed within dusttight housings without ventilating or other openings.

c. Self-cleaning textile motors of the squirrel-cage type.

502-9. Ventilating Piping. Ventilating pipes for motors, generators, or other rotating electric machinery, or for enclosures for electric equipment, shall be of metal not less than 0.021 inch (533 micrometers) in thickness, or of equally substantial noncombustible material, and shall comply with the following: (1) lead directly to a source of clean air outside of buildings; (2) be screened at the outer ends to prevent the entrance of small animals or birds; and (3) be protected against physical damage and against rusting or other corrosive influences.

Ventilating pipes shall also comply with (a) and (b) below.

(a) Class II, Division 1. In Class II, Division 1 locations, ventilating pipes, including their connections to motors or to the dust-ignitionproof enclosures for other equipment, shall be dusttight throughout their length. For metal pipes, seams and joints shall comply with one of the following: (1) be riveted and soldered; (2) be bolted and soldered; (3) be welded; or (4) be rendered dusttight by some other equally effective means.

(b) Class II, Division 2. In Class II, Division 2 locations, ventilating pipes and their connections shall be sufficiently tight to prevent the entrance of appreciable quantities of dust into the ventilated equipment or enclosure and to prevent the escape of sparks, flame, or burning material that might ignite dust accumulations or combustible material in the vicinity. For metal pipes, lock seams and riveted or welded joints shall be permitted; and tight-fitting slip joints shall be permitted where some flexibility is necessary, as at connections to motors.

502-10. Utilization Equipment.

(a) Class II, Division 1. In Class II, Division 1 locations, all utilization equipment shall be approved for Class II locations. Where dust from magnesium, aluminum, aluminum bronze powders, or other metals of similarly hazardous characteristics may be present, such equipment shall be approved for the specific location.

(b) Class II, Division 2. In Class II, Division 2 locations, all utilization equipment shall comply with the following:

(1) Heaters. Electrically heated utilization equipment shall be approved for Class II locations.

Exception: Metal-enclosed radiant heating panel equipment shall be dusttight and marked in accordance with Section 500-3(d).

(2) Motors. Motors of motor-driven utilization equipment shall comply with Section 502-8(b).

(3) Switches, Circuit Breakers, and Fuses. Enclosures for switches, circuit breakers, and fuses shall be dusttight.

(4) Transformers, Impedance Coils, and Resistors. Transformers, solenoids, impedance coils, and resistors shall comply with Section 502-7(b).

502-11. Lighting Fixtures. Lighting fixtures shall comply with (a) and (b) below.

(a) Class II, Division 1. In Class II, Division 1 locations, lighting fixtures for fixed and portable lighting shall comply with the following:

(1) Approved Fixtures. Each fixture shall be approved for Class II locations and shall be clearly marked to indicate the maximum wattage of the lamp for which it is approved. In locations where dust from magnesium, aluminum, aluminum bronze powders, or other metals of similarly hazardous characteristics may be present, fixtures for fixed or portable lighting and all auxiliary equipment shall be approved for the specific location.

(2) Physical Damage. Each fixture shall be protected against physical damage by a suitable guard or by location.

(3) Pendant Fixtures. Pendant fixtures shall be suspended by threaded rigid metal conduit stems, threaded steel intermediate metal conduit stems, by chains with approved fittings, or by other approved means. For rigid stems longer than 12 inches (305 mm), permanent and effective bracing against lateral displacement shall be provided at a level not more than 12 inches (305 mm) above the lower end of the stem, or flexibility in the form of a fitting or a flexible connector approved for the location shall be provided not more than 12 inches (305 mm) from the point of attachment to the supporting box or fitting. Threaded joints shall be provided with set-screws or other effective means to prevent loosening. Where wiring between an outlet box or fitting and a pendant fixture is not enclosed in conduit, flexible cord approved for hard usage shall be used, and suitable seals shall be provided where the cord enters the fixture and the outlet box or fitting. Flexible cord shall not serve as the supporting means for a fixture.

(4) Supports. Boxes, box assemblies, or fittings used for the support of lighting fixtures shall be approved for Class II locations.

(b) Class II, Division 2. In Class II, Division 2 locations, lighting fixtures shall comply with the following:

(1) Portable Lighting Equipment. Portable lighting equipment shall be approved for Class II locations. They shall be clearly marked to indicate the maximum wattage of lamps for which they are approved.

(2) Fixed Lighting. Lighting fixtures for fixed lighting, where not of a type approved for Class II locations, shall provide enclosures for lamps and lampholders that shall be designed to minimize the deposit of dust on lamps and to prevent the escape of sparks, burning material, or hot metal. Each fixture shall be clearly marked to indicate the maximum wattage of the lamp that shall be permitted without exceeding an exposed surface temperature in accordance with Section 500-3(f) under normal conditions of use.

(3) Physical Damage. Lighting fixtures for fixed lighting shall be protected from physical damage by suitable guards or by location.

(4) Pendant Fixtures. Pendant fixtures shall be suspended by threaded rigid metal conduit stems, threaded steel intermediate metal conduit stems, by chains with approved fittings, or by other approved means. For rigid stems longer than 12 inches (305 mm), permanent and effective bracing against lateral displacement shall be provided at a level not more than 12 inches (305 mm) above the lower end of the stem, or flexibility in the form of an approved fitting or a flexible connector shall be provided not more than 12 inches (305 mm) from the point of attachment to the supporting box or fitting. Where wiring between an outlet box or fitting and a pendant fixture is not enclosed in conduit, flexible cord approved for hard usage shall be used. Flexible cord shall not serve as the supporting means for a fixture.

(5) Electric-Discharge Lamps. Starting and control equipment for electric-discharge lamps shall comply with the requirements of Section 502-7(b).

502-12. Flexible Cords, Class II, Divisions 1 and 2. Flexible cords used in Class II locations shall comply with the following: (1) be of a type approved for extra-hard usage; (2) contain, in addition to the conductors of the circuit, a grounding conductor complying with Section 400-23; (3) be connected to terminals or to supply conductors in an approved manner; (4) be supported by clamps or by other suitable means in such a manner that there will be no tension on the terminal connections; and (5) be provided with suitable seals to prevent the entrance of dust where the flexible cord enters boxes or fittings that are required to be dust-ignitionproof.

502-13. Receptacles and Attachment Plugs.

(a) Class II, Division 1. In Class II, Division 1 locations, receptacles and attachment plugs shall be of the type providing for connection to the grounding conductor of the flexible cord and shall be approved for Class II locations.

(b) Class II, Division 2. In Class II, Division 2 locations, receptacles and attachment plugs shall be of the type providing for connection to the grounding conductor of the flexible cord and shall be so designed that connection to the supply circuit cannot be made or broken while live parts are exposed.

502-14. Signaling, Alarm, Remote-Control, and Communication Systems, Meters, Instruments, and Relays.

(FPN): See Article 800 for rules governing the installation of communication circuits.

(a) Class II, Division 1. In Class II, Division 1 locations, signaling, alarm, remote-control, and communication systems; and meters, instruments, and relays shall comply with the following:

(1) Wiring Methods. The wiring method shall comply with Section 502-4(a).

(2) Contacts. Switches, circuit breakers, relays, contactors, fuses and current-breaking contacts for bells, horns, howlers, sirens, and other devices in which sparks or arcs may be produced shall be provided with enclosures approved for a Class II location.

Exception: Where current-breaking contacts are immersed in oil or where the interruption of current occurs within a chamber sealed against the entrance of dust, enclosures shall be permitted to be of the general-purpose type.

(3) Resistors and Similar Equipment. Resistors, transformers, choke coils, rectifiers, thermionic tubes, and other heat-generating equipment shall be provided with enclosures approved for Class II locations.

Exception: Where resistors or similar equipment are immersed in oil or enclosed in a chamber sealed against the entrance of dust, enclosures shall be permitted to be of the general-purpose type.

(4) Rotating Machinery. Motors, generators, and other rotating electric machinery shall comply with Section 502-8(a).

(5) Combustible, Electrically Conductive Dusts. Where dusts are of a combustible, electrically conductive nature, all wiring and equipment shall be approved for Class II locations.

(6) Metal Dusts. Where dust from magnesium, aluminum, aluminum bronze powders, or other metals of similarly hazardous characteristics may be present, all apparatus and equipment shall be approved for the specific conditions.

(b) Class II, Division 2. In Class II, Division 2 locations, signaling, alarm, remote-control, and communication systems; and meters, instruments, and relays shall comply with the following:

(1) Contacts. Enclosures shall comply with (a)(2) above; or contacts shall have tight metal enclosures designed to minimize the entrance of dust and shall have telescoping or tight-fitting covers and no openings through which, after installation, sparks or burning material might escape.

Exception: In nonincendive circuits that under normal conditions do not release sufficient energy to ignite a dust layer, enclosures shall be permitted to be of the general-purpose type.

(2) Transformers and Similar Equipment. The windings and terminal connections of transformers, choke coils, and similar equipment shall be provided with tight metal enclosures without ventilating openings.

(3) Resistors and Similar Equipment. Resistors, resistance devices, thermionic tubes, rectifiers, and similar equipment shall comply with (a)(3) above.

Exception: Enclosures for thermionic tubes, nonadjustable resistors, or rectifiers for which maximum operating temperature will not exceed 120°C (248°F) shall be permitted to be of the general-purpose type.

(4) Rotating Machinery. Motors, generators, and other rotating electric machinery shall comply with Section 502-8(b).

(5) Wiring Methods. The wiring method shall comply with Section 502-4(b).

502-15. Live Parts, Class II, Divisions 1 and 2. Live parts shall not be exposed.

502-16. Grounding, Class II, Divisions 1 and 2. Wiring and equipment in Class II, Division 1 and 2 locations shall be grounded as specified in Article 250 and with the following additional requirements:

(a) Bonding. The locknut-bushing and double-locknut types of contact shall not be depended upon for bonding purposes, but bonding jumpers with proper fittings or other approved means of bonding shall be used. Such means of bonding shall apply to all intervening raceways, fittings, boxes, enclosures, etc., between Class II locations and the point of grounding for service equipment or point of grounding of a separately derived system.

Exception: The specified bonding means shall only be required to the point of grounding of a building disconnecting means as specified in Sections 250-24 (a), (b), and (c), provided that branch-circuit protection is located on the load side of the disconnecting means.

(FPN): See Section 250-78 for additional bonding requirements in hazardous (classified) locations.

(b) Types of Equipment Grounding Conductors. Where flexible conduit is used as permitted in Section 502-4, it shall be installed with internal or external bonding jumpers in parallel with each conduit and complying with Section 250-79.

Exception: In Class II, Division 2 areas, the bonding jumper shall be permitted to be deleted where all the following conditions are met:

*   a. Approved liquidtight flexible metal conduit 6 feet (1.83 m) or less in length, with approved fittings, is used;*

*   b. Overcurrent protection in the circuit is limited to 10 amperes or less;*

*   c. The load is not a power utilization load.*

502-17. Surge Protection, Class II, Divisions 1 and 2. Surge arresters, including their installation and connection, shall comply with Article 280. In addition, surge arresters, if installed in a Class II, Division 1 location, shall be in suitable enclosures.

Surge-protective capacitors shall be of a type designed for specific duty.

502-18. Multiwire Branch Circuits. In a Class II, Division 1 location, a separate grounded conductor shall be installed in each single-phase branch circuit that is part of a multiwire branch circuit.

Exception: Where the disconnect device(s) for the circuit opens all ungrounded conductors of the multiwire circuit simultaneously.

ARTICLE 503 — CLASS III LOCATIONS

503-1. General. The general rules of this Code shall apply to electric wiring and equipment in locations classified as Class III locations in Section 500-7.

Exception: As modified by this article.

Equipment installed in Class III locations shall be able to function at full rating without developing surface temperatures high enough to cause excessive dehydration or gradual carbonization of accumulated fibers or flyings. Organic material that is carbonized or excessively dry is highly susceptible to spontaneous ignition. The maximum surface temperatures under operating conditions shall not exceed 165°C (329°F) for equipment that is not subject to overloading, and 120°C (248°F) for equipment (such as motors or power transformers) that may be overloaded.

(FPN):   For electric trucks, see Powered Industrial Trucks Including Type Designations, Areas of Use, Maintenance, and Operation, NFPA 505-1992 (ANSI).

503-2. Transformers and Capacitors, Class III, Divisions 1 and 2.   Transformers and capacitors shall comply with Section 502-2(b).

503-3. Wiring Methods.   Wiring methods shall comply with (a) and (b) below.

(a) Class III, Division 1. In Class III, Division 1 locations, the wiring method shall be rigid metal conduit, rigid nonmetallic conduit, intermediate metal conduit, electrical metallic tubing, dusttight wireways, or Type MC, MI, or SNM cable with approved termination fittings.

(1) Boxes and Fittings. All boxes and fittings shall be dusttight.

(2) Flexible Connections. Where necessary to employ flexible connections, dusttight flexible connectors, liquidtight flexible metal conduit with approved fittings, liquidtight flexible nonmetallic conduit with approved fittings, or flexible cord in conformance with Section 503-10 shall be used.

(FPN):   See Section 503-16(b) for grounding requirements where flexible conduit is used.

(b) Class III, Division 2. In Class III, Division 2 locations, the wiring method shall comply with (a) above.

Exception: In sections, compartments, or areas used solely for storage and containing no machinery, open wiring on insulators shall be permitted where installed in accordance with Article 320, but only on condition that protection as required by Section 320-14 be provided where conductors are not run in roof spaces and are well out of reach of sources of physical damage.

503-4. Switches, Circuit Breakers, Motor Controllers, and Fuses, Class III, Divisions 1 and 2.   Switches, circuit breakers, motor controllers, and fuses, including pushbuttons, relays, and similar devices, shall be provided with dusttight enclosures.

503-5. Control Transformers and Resistors, Class III, Divisions 1 and 2. Transformers, impedance coils, and resistors used as or in conjunction with control equipment for motors, generators, and appliances shall be provided with dusttight enclosures complying with the temperature limitations in Section 503-1.

503-6. Motors and Generators, Class III, Divisions 1 and 2. In Class III, Division 1 and 2 locations, motors, generators, and other rotating machinery shall be totally enclosed nonventilated, totally enclosed pipe-ventilated, or totally enclosed fan-cooled.

Exception: In locations where, in the judgment of the authority having jurisdiction, only moderate accumulations of lint or flyings will be likely to collect on, in, or in the vicinity of a rotating electric machine and where such machine is readily accessible for routine cleaning and maintenance, one of the following shall be permitted:

a. Self-cleaning textile motors of the squirrel-cage types;

b. Standard open-type machines without sliding contacts, centrifugal or other types of switching mechanism, including motor overload devices; or

c. Standard open-type machines having such contacts, switching mechanisms, or resistance devices enclosed within tight housings without ventilating or other openings.

503-7. Ventilating Piping, Class III, Divisions 1 and 2. Ventilating pipes for motors, generators, or other rotating electric machinery, or for enclosures for electric equipment, shall be of metal not less than 0.021 inch (533 micrometers) in thickness, or of equally substantial noncombustible material, and shall comply with the following: (1) lead directly to a source of clean air outside of buildings; (2) be screened at the outer ends to prevent the entrance of small animals or birds; and (3) be protected against physical damage and against rusting or other corrosive influences.

Ventilating pipes shall be sufficiently tight, including their connections, to prevent the entrance of appreciable quantities of fibers or flyings into the ventilated equipment or enclosure and to prevent the escape of sparks, flame, or burning material that might ignite accumulations of fibers or flyings or combustible material in the vicinity. For metal pipes, lock seams and riveted or welded joints shall be permitted; and tight-fitting slip joints shall be permitted where some flexibility is necessary, as at connections to motors.

503-8. Utilization Equipment, Class III, Divisions 1 and 2.

(a) Heaters. Electrically heated utilization equipment shall be approved for Class III locations.

(b) Motors. Motors of motor-driven utilization equipment shall comply with Section 503-6.

(c) Switches, Circuit Breakers, Motor Controllers, and Fuses. Switches, circuit breakers, motor controllers, and fuses shall comply with Section 503-4.

503-9. Lighting Fixtures, Class III, Divisions 1 and 2.

(a) Fixed Lighting. Lighting fixtures for fixed lighting shall provide enclosures for lamps and lampholders that are designed to minimize entrance of fibers and flyings and to prevent the escape of sparks, burning material, or hot metal. Each fixture shall be clearly marked to show the maximum wattage of the lamps that shall be permitted without exceeding an exposed surface temperature of 165°C (329°F) under normal conditions of use.

(b) Physical Damage. A fixture that may be exposed to physical damage shall be protected by a suitable guard.

(c) Pendant Fixtures. Pendant fixtures shall be suspended by stems of threaded rigid metal conduit, threaded intermediate metal conduit, threaded metal tubing of equivalent thickness, or by chains with approved fittings. For stems longer than 12 inches (305 mm), permanent and effective bracing against lateral displacement shall be provided at a level not more than 12 inches (305 mm) above the lower end of the stem, or flexibility in the form of an approved fitting or a flexible connector shall be provided not more than 12 inches (305 mm) from the point of attachment to the supporting box or fitting.

(d) Portable Lighting Equipment. Portable lighting equipment shall be equipped with handles and protected with substantial guards. Lampholders shall be of the unswitched type with no provision for receiving attachment plugs. There shall be no exposed current-carrying metal parts, and all exposed noncurrent-carrying metal parts shall be grounded. In all other respects, portable lighting equipment shall comply with (a) above.

503-10. Flexible Cords, Class III, Divisions 1 and 2.

Flexible cords shall comply with the following: (1) be of a type approved for extra-hard usage; (2) contain, in addition to the conductors of the circuit, a grounding conductor complying with Section 400-23; (3) be connected to terminals or to supply conductors in an approved manner; (4) be supported by clamps or other suitable means in such a manner that there will be no tension on the terminal connections; and (5) be provided with suitable means to prevent the entrance of fibers or flyings where the cord enters boxes or fittings.

503-11. Receptacles and Attachment Plugs, Class III, Divisions 1 and 2.

Receptacles and attachment plugs shall be of the grounding type and shall be so designed to minimize the accumulation or the entry of fibers or flyings, and shall prevent the escape of sparks or molten particles.

Exception: In locations where, in the judgment of the authority having jurisdiction, only moderate accumulations of lint or flyings will be likely to collect in the vicinity of a receptacle, and where such receptacle is readily accessible for routine cleaning, general-purpose grounding-type receptacles mounted so as to minimize the entry of fibers or flyings shall be permitted.

503-12. Signaling, Alarm, Remote-Control, and Local Loudspeaker Intercommunication Systems, Class III, Divisions 1 and 2.

Signaling, alarm, remote-control, and local loudspeaker intercommunication systems shall comply with the requirements of Article 503 regarding wiring methods, switches, transformers, resistors, motors, lighting fixtures, and related components.

503-13. Electric Cranes, Hoists, and Similar Equipment, Class III, Divisions 1 and 2. Where installed for operation over combustible fibers or accumulations of flyings, traveling cranes and hoists for material handling, traveling cleaners for textile machinery, and similar equipment shall comply with (a) through (d) below.

(a) **Power Supply.** Power supply to contact conductors shall be isolated from all other systems and shall be equipped with an acceptable ground detector that will give an alarm and automatically deenergize the contact conductors in case of a fault to ground or will give a visual and audible alarm as long as power is supplied to the contact conductors and the ground fault remains.

(b) **Contact Conductors.** Contact conductors shall be so located or guarded as to be inaccessible to other than authorized persons and shall be protected against accidental contact with foreign objects.

(c) **Current Collectors.** Current collectors shall be so arranged or guarded as to confine normal sparking and prevent escape of sparks or hot particles. To reduce sparking, two or more separate surfaces of contact shall be provided for each contact conductor. Reliable means shall be provided to keep contact conductors and current collectors free of accumulations of lint or flyings.

(d) **Control Equipment.** Control equipment shall comply with Sections 503-4 and 503-5.

503-14. Storage-Battery Charging Equipment, Class III, Divisions 1 and 2. Storage-battery charging equipment shall be located in separate rooms built or lined with substantial noncombustible materials so constructed as to adequately exclude flyings or lint and shall be well ventilated.

503-15. Live Parts, Class III, Divisions 1 and 2. Live parts shall not be exposed.

Exception: As provided in Section 503-13.

503-16. Grounding, Class III, Divisions 1 and 2. Wiring and equipment in Class III, Divisions 1 and 2 shall be grounded as specified in Article 250 and with the following additional requirements:

(a) **Bonding.** The locknut-bushing and double-locknut types of contacts shall not be depended upon for bonding purposes, but bonding jumpers with proper fittings or other approved means of bonding shall be used. Such means of bonding shall apply to all intervening raceways, fittings, boxes, enclosures, etc., between Class III locations and the point of grounding for service equipment or point of grounding of a separately derived system.

Exception: The specified bonding means shall only be required to the point of grounding of a building disconnecting means as specified in Sections 250-24(a), (b), and (c), provided that branch-circuit overcurrent protection is located on the load side of the disconnecting means.

(FPN): See Section 250-78 for additional bonding requirements in hazardous (classified) locations.

(b) Types of Equipment Grounding Conductors. Where flexible conduit is used as permitted in Section 503-3, it shall be installed with internal or external bonding jumpers in parallel with each conduit and complying with Section 250-79.

Exception: In Class III, Division 1 and 2 areas, the bonding jumper shall be permitted to be deleted where all the following conditions are met:

a. Approved liquidtight flexible metal 6 feet (1.83 m) or less in length, with approved fittings, is used;

b. Overcurrent protection in the circuit is limited to 10 amperes or less;

c. The load is not a power utilization load.

ARTICLE 504 — INTRINSICALLY SAFE SYSTEMS

504-1. Scope. This article covers the installation of intrinsically safe (I.S.) apparatus, wiring, and systems for Class I, II, and III locations.

(FPN): For further information, see Installation of Intrinsically Safe Instrument Systems in Class I Hazardous Locations, ANSI/ISA RP 12.6-1987.

504-2. Definitions. For the purpose of this article:

Associated Apparatus: Apparatus in which the circuits are not necessarily intrinsically safe themselves, but that affect the energy in the intrinsically safe circuits and are relied upon to maintain intrinsic safety. Associated apparatus may be either:

1. Electrical apparatus that has an alternative type protection for use in the appropriate hazardous (classified) location, or

2. Electrical apparatus not so protected that shall not be used within a hazardous (classified) location.

(FPN): Associated apparatus has identified intrinsically safe connections for intrinsically safe apparatus and also may have connections for nonintrinsically safe apparatus.

Control Drawing: A drawing or other document provided by the manufacturer of the intrinsically safe or associated apparatus that details the allowed interconnections between the intrinsically safe and associated apparatus.

Different Intrinsically Safe Circuits: Different intrinsically safe circuits are intrinsically safe circuits in which the possible interconnections have not been evaluated and approved as intrinsically safe.

Intrinsically Safe Apparatus: Apparatus in which all the circuits are intrinsically safe.

Intrinsically Safe Circuit: A circuit in which any spark or thermal effect is incapable of causing ignition of a mixture of flammable or combustible material in air under prescribed test conditions.

(FPN): Test conditions are described in Standard for Safety, Intrinsically Safe Apparatus and Associated Apparatus for Use in Class I, II, and III, Division 1, Hazardous (Classified) Locations, ANSI/UL 913-1988.

Intrinsically Safe System: An assembly of interconnected intrinsically safe apparatus, associated apparatus, and interconnecting cables in that those parts of the system that may be used in hazardous (classified) locations are intrinsically safe circuits.

(FPN): An intrinsically safe system may include more than one intrinsically safe circuit.

504-3. Application of Other Articles. Except as modified by this article, all applicable articles of this Code shall apply.

504-4. Equipment Approval. All intrinsically safe apparatus and associated apparatus shall be approved.

504-10. Equipment Installation.

(a) Control Drawing. Intrinsically safe apparatus, associated apparatus, and other equipment shall be installed in accordance with the control drawing(s).

(FPN): The control drawing identification is marked on the apparatus.

(b) Location. Intrinsically safe and associated apparatus shall be permitted to be installed in any hazardous (classified) location for which it has been approved.

(FPN): Associated apparatus may be installed in hazardous (classified) locations if protected by other means permitted by Articles 501 through 503.

General-purpose enclosures shall be permitted for intrinsically safe apparatus.

504-20. Wiring Methods. Intrinsically safe apparatus and wiring shall be permitted to be installed using any of the wiring methods suitable for unclassified locations. Sealing shall be as provided in Section 504-70, and separation shall be as provided in Section 504-30.

504-30. Separation of Intrinsically Safe Conductors.

(a) From Nonintrinsically Safe Circuit Conductors.

(1) Open Wiring. Conductors and cables of intrinsically safe circuits not in raceways or cable trays shall be separated at least 2 inches (50.8 mm) and secured from conductors and cables of any nonintrinsically safe circuits.

Exception: Where either (1) all of the intrinsically safe circuit conductors are in Type MI, MC, or SNM cables or (2) all of the nonintrinsically safe circuit conductors are in raceways or Type MI, MC, or SNM cables where the sheathing or cladding is capable of carrying fault current to ground.

(2) In Raceways, Cable Trays, and Cables. Conductors of intrinsically safe circuits shall not be placed in any raceway, cable tray, or cable with conductors of any nonintrinsically safe circuit.

Exception No. 1: Where conductors of intrinsically safe circuits are separated from conductors of nonintrinsically safe circuits by a distance of at least 2 inches (50.8 mm) and secured, or by a grounded metal partition or an approved insulating partition.

(FPN):   No. 20 gauge sheet metal partitions 0.0359 inch (912 micrometers) or thicker are generally considered acceptable.

Exception No. 2: Where either (1) all of the intrinsically safe circuit conductors or (2) all of the nonintrinsically safe circuit conductors are in grounded metal-sheathed or metal-clad cables where the sheathing or cladding is capable of carrying fault current to ground.

(FPN):   Cables meeting the requirements of Articles 330, 334, and 337 are typical of those considered acceptable.

(3) Within Enclosures.

a. Conductors of intrinsically safe circuits shall be separated at least 2 inches (50.8 mm) from conductors of any nonintrinsically safe circuits, or as specified in Section 504-30(a)(2).

b. All conductors shall be secured so that any conductor that might come loose from a terminal cannot come in contact with another terminal.

(FPN No. 1):   The use of separate wiring compartments for the intrinsically safe and nonintrinsically safe terminals is the preferred method of complying with this requirement.

(FPN No. 2):   Physical barriers such as grounded metal partitions or approved insulating partitions or approved restricted access wiring ducts separated from other such ducts by at least $\frac{3}{4}$ in. (19 mm) can be used to help assure the required separation of the wiring.

(b) From Different Intrinsically Safe Circuit Conductors. Different intrinsically safe circuits shall be in separate cables or shall be separated from each other by one of the following means:

(1) The conductors of each circuit are within a grounded metal shield;

(2) The conductors of each circuit have insulation with a minimum thickness of 0.01 inches (254 micrometers).

Exception: Unless otherwise approved.

504-50. Grounding.

(a) Intrinsically Safe Apparatus, Associated Apparatus, and Raceways. Intrinsically safe apparatus, associated apparatus, cable shields, enclosures, and raceways, if of metal, shall be grounded.

(FPN):   Supplementary bonding to the grounding electrode may be needed for some associated apparatus, e.g., zener diode barriers, if specified in the control drawing. See Installation of Intrinsically Safe Instrument Systems in Class I Hazardous Locations, ANSI/ISA RP 12.6-1987.

(b) Connection to Grounding Electrodes. Where connection to a grounding electrode is required, the grounding electrode shall be as specified in Sections 250-81(a), (b), (c), (d) and shall comply with Section 250-26(c). Section 250-83 shall not be used if electrodes specified in Section 250-81 are available.

(c) Shields. Where shielded conductors or cables are used, shields shall be grounded.

Exception: Where a shield is part of an intrinsically safe circuit.

504-60. Bonding.

(a) Hazardous Locations. In hazardous (classified) locations, intrinsically safe apparatus shall be bonded in the hazardous (classified) location in accordance with Section 250-78.

(b) Nonhazardous Locations. In nonhazardous locations, where metal raceways are used for intrinsically safe system wiring in hazardous locations, associated apparatus shall be bonded in accordance with Sections 501-16(a), 502-16(a), or 503-16(a), as applicable.

504-70. Sealing. Conduits and cables that are required to be sealed by Sections 501-5 and 502-5 shall be sealed to minimize the passage of gases, vapors, or dust.

Exception: Seals are not required for enclosures that contain only intrinsically safe apparatus except as required by Section 501-5(f)(3).

(FPN): It is not the intent of this section to require an explosionproof seal.

504-80. Identification. Labels required by this section shall be suitable for the environment where they are installed with consideration given to exposure to chemicals and sunlight.

(a) Terminals. Intrinsically safe circuits shall be identified at terminal and junction locations in a manner that will prevent unintentional interference with the circuits during testing and servicing.

(b) Wiring. Raceways, cable trays, and open wiring for intrinsically safe system wiring shall be identified with permanently affixed labels with the wording "Intrinsic Safety Wiring" or equivalent. The labels shall be so located as to be visible after installation and placed so that they may be readily traced through the entire length of the installation. Spacing between labels shall not be more than 25 feet (7.62 m).

Exception: Circuits run underground shall be permitted to be identified where they become accessible after emergence from the ground.

(FPN No. 1): Wiring methods permitted in nonhazardous locations may be used for intrinsically safe systems in hazardous (classified) locations. Without lables to identify the application of the wiring, enforcement authorities cannot determine that an installation is in compliance with the Code.

(FPN No. 2): In nonhazardous locations, the identification is necessary to assure that nonintrinsically safe wire will not be inadvertently added to existing raceways at a later date.

(c) Color Coding. Color coding shall be permitted to identify intrinsically safe conductors where they are colored light blue and where no other conductors colored light blue are used.

ARTICLE 510 — HAZARDOUS (CLASSIFIED) LOCATIONS — SPECIFIC

510-1. Scope. Articles 511 through 517 cover occupancies or parts of occupancies that are or may be hazardous because of atmospheric concentrations of flammable liquids, gases, or vapors, or because of deposits or accumulations of materials that may be readily ignitible.

510-2. General. The general rules of this Code shall apply to electric wiring and equipment in occupancies within the scope of Articles 511 through 517, except as such rules are modified in those articles. Where unusual conditions exist in a specific occupancy, the authority having jurisdiction shall judge with respect to the application of specific rules.

ARTICLE 511 — COMMERCIAL GARAGES, REPAIR AND STORAGE

511-1. Scope. These occupancies shall include locations used for service and repair operations in connection with self-propelled vehicles (including passenger automobiles, buses, trucks, tractors, etc.) in which volatile flammable liquids are used for fuel or power.

511-2. Locations. Areas in which flammable fuel is transferred to vehicle fuel tanks shall conform to Article 514. Parking garages used for parking or storage and where no repair work is done except exchange of parts and routine maintenance requiring no use of electrical equipment, open flame, welding, or the use of volatile flammable liquids are not classified, but they shall be adequately ventilated to carry off the exhaust fumes of the engines.

(FPN): For further information, see Parking Structures, NFPA 88A-1991, and Repair Garages, NFPA 88B-1991.

511-3. Class I Locations. Classification under Article 500.

(a) Up to a Level of 18 Inches (457 mm) Above the Floor. For each floor, the entire area up to a level of 18 inches (457 mm) above the floor shall be considered to be a Class I, Division 2 location except where the enforcing agency determines that there is mechanical ventilation providing a minimum of four air changes per hour.

(b) Any Pit or Depression Below Floor Level. Any pit or depression below floor level shall be considered to be a Class I, Division 1 location which shall extend up to said floor level, except that any pit or depression in which six air changes per hour are exhausted at the floor level of the pit shall be permitted to be judged by the enforcing agency to be a Class I, Division 2 location.

Exception: Lubrication and service rooms without dispensing shall be classified in accordance with Table 514-2.

(c) Areas Adjacent to Defined Locations or with Positive Pressure Ventilation. Areas adjacent to defined locations in which flammable vapors are not likely to be released, such as stock rooms, switchboard rooms, and other similar locations, shall not be classified where mechanically ventilated at a rate of four or more air changes per hour or where effectively cut off by walls or partitions.

(d) Adjacent Areas by Special Permission. Adjacent areas that by reason of ventilation, air pressure differentials, or physical spacing are such that, in the opinion of the authority enforcing this Code, no ignition hazard exists shall be classified as nonhazardous.

(e) Fuel Dispensing Units. Where fuel dispensing units (other than liquid petroleum gas, which is prohibited) are located within buildings, the requirements of Article 514 shall govern.

Where mechanical ventilation is provided in the dispensing area, the controls shall be interlocked so that the dispenser cannot operate without ventilation as prescribed in Section 500-5(b).

(f) Portable Lighting Equipment. Portable lighting equipment shall be equipped with handle, lampholder, hook, and substantial guard attached to the lampholder or handle. All exterior surfaces that might come in contact with battery terminals, wiring terminals, or other objects shall be of nonconducting material or shall be effectively protected with insulation. Lampholders shall be of unswitched type and shall not provide means for plug-in of attachment plugs. The outer shell shall be of molded composition or other suitable material. Unless the lamp and its cord are supported or arranged in such a manner that they cannot be used in the locations classified in Section 511-3, they shall be of a type approved for Class I, Division 1 locations.

511-4. Wiring and Equipment in Class I Locations. Within Class I locations as defined in Section 511-3, wiring and equipment shall conform to applicable provisions of Article 501. Raceways embedded in a masonry wall or buried beneath a floor shall be considered to be within the Class I location above the floor if any connections or extensions lead into or through such areas.

511-5. Sealing. Approved seals conforming to the requirements of Section 501-5 shall be provided, and Section 501-5(b)(2) shall apply to horizontal as well as vertical boundaries of the defined Class I locations.

511-6. Wiring in Spaces Above Class I Locations.

(a) Fixed Wiring Above Class I Locations. All fixed wiring above Class I locations shall be in metal raceways, rigid nonmetallic conduit, electrical nonmetallic tubing, flexible metal conduit, liquidtight flexible metal conduit, or liquidtight flexible nonmetallic conduit or shall be Type MC, MI, manufactured wiring systems, SNM, or PLTC cable in accordance with Article 725, or TC cable. Cellular metal floor raceways or cellular concrete floor raceways shall be permitted to be used only for supplying ceiling outlets or extensions to the area below the floor, but such raceways shall have no connections leading into or through any Class I location above the floor.

(b) Pendants. For pendants, flexible cord suitable for the type of service and approved for hard usage shall be used.

(c) **Grounded Conductor.** Where a circuit that supplies portables or pendants includes a grounded conductor as provided in Article 200, receptacles, attachment plugs, connectors, and similar devices shall be of polarized type, and the grounded conductor of the flexible cord shall be connected to the screw shell of any lampholder or to the grounded terminal of any utilization equipment supplied.

(d) **Attachment Plug Receptacles.** Attachment plug receptacles in fixed position shall be located above the level of any defined Class I location or be approved for the location.

511-7. Equipment Above Class I Locations.

(a) **Arcing Equipment.** Equipment that is less than 12 feet (3.66 m) above the floor level and that may produce arcs, sparks, or particles of hot metal, such as cutouts, switches, charging panels, generators, motors, or other equipment (excluding receptacles, lamps, and lampholders) having make-and-break or sliding contacts, shall be of the totally enclosed type or so constructed as to prevent escape of sparks or hot metal particles.

(b) **Fixed Lighting.** Lamps and lampholders for fixed lighting that is located over lanes through which vehicles are commonly driven or that may otherwise be exposed to physical damage shall be located not less than 12 feet (3.66 m) above floor level, unless of the totally enclosed type or so constructed as to prevent escape of sparks or hot metal particles.

511-8. Battery Charging Equipment.

Battery chargers and their control equipment, and batteries being charged, shall not be located within locations classified in Section 511-3.

511-9. Electric Vehicle Charging.

(a) **Connections.** Flexible cords and connectors used for charging shall be suitable for the type of service and approved for extra-hard usage. Their ampacity shall be adequate for the charging current.

(b) **Connector Design and Location.** Connectors shall be so designed and installed that they will disconnect readily at any position of the charging cable, and live parts shall be guarded from accidental contact. No connector shall be located within a Class I location as defined in Section 511-3.

(c) **Plug Connections to Vehicles.** Where plugs are provided for direct connection to vehicles, the point of connection shall not be within a Class I location as defined in Section 511-3, and, where the cord is suspended from overhead, it shall be so arranged that the lowest point of sag is at least 6 inches (152 mm) above the floor. Where the vehicle is equipped with an approved plug that will disconnect readily and where an automatic arrangement is provided to pull both cord and plug beyond the range of physical damage, no additional connector shall be required in the cable or at the outlet.

511-10. Ground-Fault Circuit-Interrupter Protection for Personnel.

All 125-volt single-phase 15- and 20-ampere receptacles installed in areas where electrical diagnostic equipment, electrical hand tools, or portable lighting devices are to be used shall provide ground-fault circuit-interrupter protection for personnel.

ARTICLE 513 — AIRCRAFT HANGARS

513-1. Definition. An aircraft hangar is a location used for storage or servicing of aircraft in which gasoline, jet fuels, or other volatile flammable liquids or flammable gases are used. It shall not include locations used exclusively for aircraft that have never contained such liquids or gases, or that have been drained and properly purged.

513-2. Classification of Locations.

(a) Below Floor Level. Any pit or depression below the level of the hangar floor shall be classified as a Class I, Division 1 location that shall extend up to said floor level.

(b) Areas Not Cut Off or Ventilated. The entire area of the hangar, including any adjacent and communicating areas not suitably cut off from the hangar, shall be classified as a Class I, Division 2 location up to a level 18 inches (457 mm) above the floor.

(c) Vicinity of Aircraft. The area within 5 feet (1.52 m) horizontally from aircraft power plants or aircraft fuel tanks shall be classified as a Class I, Division 2 location that shall extend upward from the floor to a level 5 feet (1.52 m) above the upper surface of wings and of engine enclosures.

(d) Areas Suitably Cut Off and Ventilated. Adjacent areas in which flammable liquids or vapors are not likely to be released, such as stock rooms, electrical control rooms, and other similar locations, shall not be classified where adequately ventilated and where effectively cut off from the hangar itself by walls or partitions.

513-3. Wiring and Equipment in Class I Locations. All wiring and equipment that is or may be installed or operated within any of the Class I locations defined in Section 513-2 shall comply with the applicable provisions of Article 501. All wiring installed in or under the hangar floor shall comply with the requirements for Class I, Division 1 locations. Where such wiring is located in vaults, pits, or ducts, adequate drainage shall be provided; and the wiring shall not be placed within the same compartment with any service other than piped compressed air.

Attachment plugs and receptacles in Class I locations shall be approved for Class I locations or shall be so designed that they cannot be energized while the connections are being made or broken.

513-4. Wiring Not Within Class I Locations.

(a) Fixed Wiring. All fixed wiring in a hangar, but not within a Class I location as defined in Section 513-2, shall be installed in metal raceways or shall be Type MI, TC, SNM, or Type MC cable.

Exception: Wiring in unclassified locations as defined in Section 513-2(d) shall be of a type recognized in Chapter 3.

(b) Pendants. For pendants, flexible cord suitable for the type of service and approved for hard usage shall be used. Each such cord shall include a separate equipment grounding conductor.

(c) **Portable Equipment.** For portable utilization equipment and lamps, flexible cord suitable for the type of service and approved for extra-hard usage shall be used. Each such cord shall include a separate equipment | grounding conductor.

(d) **Grounded and Grounding Conductors.** Where a circuit supplies portables or pendants and includes a grounded conductor as provided in Article 200, receptacles, attachment plugs, connectors, and similar devices shall be of the grounding type, and the grounded conductor of the flexible cord shall be connected to the screw shell of any lampholder or to the grounded terminal of any utilization equipment supplied. Approved means shall be provided for maintaining continuity of the grounding conductor between the fixed wiring system and the noncurrent-carrying metal | portions of pendant fixtures, portable lamps, and portable utilization equipment.

513-5. Equipment Not Within Class I Locations.

(a) **Arcing Equipment.** In locations other than those described in Section 513-2, equipment that is less than 10 feet (3.05 m) above wings and engine enclosures of aircraft and that may produce arcs, sparks, or particles of hot metal, such as lamps and lampholders for fixed lighting, cutouts, switches, receptacles, charging panels, generators, motors, or other equipment having make-and-break or sliding contacts, shall be of the totally enclosed type or so constructed as to prevent escape of sparks or hot metal particles.

Exception: Equipment in areas described in Section 513-2(d) shall be permitted to be of the general-purpose type.

(b) **Lampholders.** Lampholders of metal-shell, fiber-lined types shall not be used for fixed incandescent lighting.

(c) **Portable Lighting Equipment.** Portable lighting equipment that are used within a hangar shall be approved for the location in which they are used.

(d) **Portable Equipment.** Portable utilization equipment that is or may be used within a hangar shall be of a type suitable for use in Class I, Division 2 locations.

513-6. Stanchions, Rostrums, and Docks.

(a) **In Class I Location.** Electric wiring, outlets, and equipment (including lamps) on or attached to stanchions, rostrums, or docks that are located or likely to be located in a Class I location as defined in Section 513-2(c) shall comply with the requirements for Class I, Division 2 locations.

(b) **Not in Class I Location.** Where stanchions, rostrums, or docks are not located or likely to be located in a Class I location as defined in Section 513-2(c), wiring and equipment shall comply with Sections 513-4 and 513-5, except that such wiring and equipment not more than 18 inches (457 mm) above the floor in any position shall comply with (a) above. Receptacles and attachment plugs shall be of a locking type that will not readily disconnect.

(c) Mobile Type. Mobile stanchions with electric equipment complying with (b) above shall carry at least one permanently affixed warning sign to read: "WARNING — KEEP 5 FEET CLEAR OF AIRCRAFT ENGINES AND FUEL TANK AREAS."

513-7. Sealing. Approved seals shall be provided in accordance with Section 501-5. Sealing requirements specified in Sections 501-5(a)(4) and (b)(2) shall apply to horizontal as well as to vertical boundaries of the defined Class I locations. Raceways embedded in a concrete floor or buried beneath a floor shall be considered to be within the Class I location above the floor where any connections or extensions lead into or through such location.

513-8. Aircraft Electrical Systems. Aircraft electrical systems shall be deenergized when the aircraft is stored in a hangar, and, whenever possible, while the aircraft is undergoing maintenance.

513-9. Aircraft Battery — Charging and Equipment. Aircraft batteries shall not be charged where installed in an aircraft located inside or partially inside a hangar.

Battery chargers and their control equipment shall not be located or operated within any of the Class I locations defined in Section 513-2 and shall preferably be located in a separate building or in an area such as defined in Section 513-2(d). Mobile chargers shall carry at least one permanently affixed warning sign to read: "WARNING — KEEP 5 FEET CLEAR OF AIRCRAFT ENGINES AND FUEL TANK AREAS." Tables, racks, trays, and wiring shall not be located within a Class I location and, in addition, shall comply with Article 480.

513-10. External Power Sources for Energizing Aircraft.

(a) Not Less than 18 Inches (457 mm) Above Floor. Aircraft energizers shall be so designed and mounted that all electric equipment and fixed wiring will be at least 18 inches (457 mm) above floor level and shall not be operated in a Class I location as defined in Section 513-2(c).

(b) Marking for Mobile Units. Mobile energizers shall carry at least one permanently affixed warning sign to read: "WARNING — KEEP 5 FEET CLEAR OF AIRCRAFT ENGINES AND FUEL TANK AREAS."

(c) Cords. Flexible cords for aircraft energizers and ground support equipment shall be approved for the type of service and extra-hard usage and shall include an equipment grounding conductor.

513-11. Mobile Servicing Equipment with Electric Components.

(a) General. Mobile servicing equipment (such as vacuum cleaners, air compressors, air movers, etc.) having electric wiring and equipment not suitable for Class I, Division 2 locations shall be so designed and mounted that all such fixed wiring and equipment will be at least 18 inches (457 mm) above the floor. Such mobile equipment shall not be operated within the Class I location defined in Section 513-2(c) and shall carry at least one permanently affixed warning sign to read: "WARNING — KEEP 5 FEET CLEAR OF AIRCRAFT ENGINES AND FUEL TANK AREAS."

(b) **Cords and Connectors.** Flexible cords for mobile equipment shall be suitable for the type of service and approved for extra-hard usage and shall include an equipment grounding conductor. Attachment plugs and receptacles shall be approved for the location in which they are installed and shall provide for connection of the equipment grounding conductor.

(c) **Restricted Use.** Equipment that is not identified as suitable for Class I, Division 2 locations shall not be operated in locations where maintenance operations likely to release flammable liquids or vapors are in progress.

513-12. Grounding. All metal raceways, metal-jacketed cables, and all noncurrent-carrying metal portions of fixed or portable equipment, regardless of voltage, shall be grounded as provided in Article 250.

ARTICLE 514 — GASOLINE DISPENSING AND SERVICE STATIONS

514-1. Definition. A gasoline dispensing and service station is a location where gasoline or other volatile flammable liquids or liquefied flammable gases are transferred to the fuel tanks (including auxiliary fuel tanks) of self-propelled vehicles or approved containers.

Other areas used as lubritoriums, service rooms, repair rooms, offices, salesrooms, compressor rooms, and similar locations shall comply with Articles 510 and 511 with respect to electric wiring and equipment.

Where the authority having jurisdiction can satisfactorily determine that flammable liquids having a flash point below 38°C (100°F), such as gasoline, will not be handled, such location shall not be required to be classified.

(FPN No. 1): For further information regarding safeguards for gasoline dispensing and service stations, see Automotive and Marine Service Station Code, NFPA 30A-1990.

(FPN No. 2): For information on classified areas pertaining to LP-Gas Systems other than residential or commercial, see Storage and Handling of Liquefied Petroleum Gases, NFPA 58-1992 and Storage and Handling of Liquefied Petroleum Gases at Utility Gas Plants, NFPA 59-1992.

(FPN No. 3): See Section 555-9 for gasoline dispensing stations in marinas and boatyards.

514-2.ˣ Class I Locations. Table 514-2 shall be applied where Class I liquids are stored, handled, or dispensed and shall be used to delineate and classify service stations. A Class I location shall not extend beyond an unpierced wall, roof, or other solid partition.

514-3. Wiring and Equipment Within Class I Locations. All electric equipment and wiring within Class I locations defined in Section 514-2 shall comply with the applicable provisions of Article 501.

Exception: As permitted in Section 514-8.

(FPN):   For special requirements for conductor insulation, see Section 501-13.

514-4. Wiring and Equipment Above Class I Locations. Wiring and equipment above the Class I locations defined in Section 514-2 shall comply with Sections 511-6 and 511-7.

514-5. Circuit Disconnects.

(a) General. Each circuit leading to or through dispensing equipment, including equipment for remote pumping systems, shall be provided with a switch or other acceptable means to disconnect simultaneously from the source of supply all conductors of the circuit, including the grounded conductor, if any.

Single-pole breakers utilizing handle ties shall not be permitted.

(b) Attended Self-Service Stations.ˣ Emergency controls as specified in (a) above shall be installed at a location acceptable to the authority having jurisdiction, but controls shall not be more than 100 feet (30 m) from dispensers.

(c) Unattended Self-Service Stations.ˣ Emergency controls as specified in (a) above shall be installed at a location acceptable to the authority having jurisdiction, but the controls shall be more than 20 feet (7 m) but less than 100 feet (30 m) from the dispensers. Additional emergency controls shall be installed on each group of dispensers or the outdoor equipment used to control the dispensers. Emergency controls shall shut off all power to all dispensing equipment at the station. Controls shall be manually reset only in a manner approved by the authority having jurisdiction.

(FPN):   For additional information, see Sections 9-4.5 and 9-5.3 of Automotive and Marine Service Station Code, NFPA 30A-1990.

514-6. Sealing.

(a) At Dispenser. An approved seal shall be provided in each conduit run entering or leaving a dispenser or any cavities or enclosures in direct communication therewith. The sealing fitting shall be the first fitting after the conduit emerges from the earth or concrete.

(b) At Boundary. Additional seals shall be provided in accordance with Section 501-5. Sections 501-5(a)(4) and (b)(2) shall apply to horizontal as well as to vertical boundaries of the defined Class I locations.

514-7. Grounding. Metal portions of dispensing pumps, metal raceways, metal-jacketed cables, and all noncurrent-carrying metal parts of electric equipment, regardless of voltage, shall be grounded as provided in Article 250.

514-8. Underground Wiring. Underground wiring shall be installed in threaded rigid metal conduit or threaded steel intermediate metal conduit. Any portion of electrical wiring or equipment that is below the surface of a Class I, Division 1 or Division 2 location (as defined in Table 514-2) shall be considered to be in a Class I, Division 1 location, which shall extend at least to the point of emergence above grade. Refer to Table 300-5.

Table 514-2.[x] **Class I Locations — Service Stations**

Location	Class I, Group D Division	Extent of Classified Location
Underground Tank		
Fill Opening	1	Any pit, box, or space below grade level, any part of which is within the Division 1 or 2 classified location.
	2	Up to 18 inches above grade level within a horizontal radius of 10 feet from a loose fill connection and within a horizontal radius of 5 feet from a tight fill connection.
Vent — Discharging Upward	1	Within 3 feet of open end of vent, extending in all directions.
	2	Space between 3 feet and 5 feet of open end of vent, extending in all directions.
Dispensing Device (except overhead type)*		
Pits	1	Any pit, box, or space below grade level, any part of which is within the Division 1 or 2 classified location.
Dispenser		(FPN): Space classification inside the dispenser enclosure is covered in Power Operated Dispensing Devices for Petroleum Products, ANSI/UL 87-1987.
	2	Within 18 inches horizontally in all directions extending to grade from (1) the dispenser enclosure or (2) that portion of the dispenser enclosure containing liquid handling components.
		(FPN): Space classification inside the dispenser enclosure is covered in Power Operated Dispensing Devices for Petroleum Products, ANSI/UL 87-1987.
Outdoor	2	Up to 18 inches above grade level within 20 feet horizontally of any edge of enclosure.
Indoor with Mechanical Ventilation	2	Up to 18 inches above grade or floor level within 20 feet horizontally of any edge of enclosure.
with Gravity Ventilation	2	Up to 18 inches above grade or floor level within 25 feet horizontally of any edge of enclosure.

Table 514-2.   Class I Locations — Service Stations (cont.)

Location	Class I, Group D Division	Extent of Classified Location
Dispensing Device[1]		
Overhead Type*	1	The space within the dispenser enclosure, and all electrical equipment integral with the dispensing hose or nozzle.
	2	A space extending 18 inches horizontally in all directions beyond the enclosure and extending to grade.
	2	Up to 18 inches above grade level within 20 feet horizontally measured from a point vertically below the edge of any dispenser enclosure.
Remote Pump — Outdoor	1	Any pit, box, or space below grade level if any part is within a horizontal distance of 10 feet from any edge of pump.
	2	Within 3 feet of any edge of pump, extending in all directions. Also up to 18 inches above grade level within 10 feet horizontally from any edge of pump.
Remote Pump — Indoor	1	Entire space within any pit.
	2	Within 5 feet of any edge of pump, extending in all directions. Also up to 3 feet above grade level within 25 feet horizontally from any edge of pump.
Lubrication or Service Room — with Dispensing	1	Any pit within any unventilated space.
	2	Any pit with ventilation.
	2	Space up to 18 inches above floor or grade level and 3 feet horizontally from a lubrication pit.
Dispenser for Class I Liquids	2	Within 3 feet of any fill or dispensing point, extending in all directions.
Lubrication or Service Room — without Dispensing	2	Entire area within any pit used for lubrication or similar services where Class I liquids may be released.
	2	Area up to 18 inches above any such pit, and extending a distance of 3 ft horizontally from any edge of the pit.
	2	Entire unventilated area within any pit, below-grade area, or subfloor area.

Table 514-2.   Class I Locations — Service Stations (cont.)

Location	Class I, Group D Division	Extent of Classified Location
	2	Area up to 18 inches above any such unventilated pit, below-grade work area, or subfloor work area and extending a distance of 3 feet horizontally from the edge of any such pit, below-grade work area, or subfloor work area.
	Non-classified	Any pit, below-grade work area, or subfloor work area that is ventilated in accordance with 5-1.3.
Special Enclosure Inside Building**	1	Entire enclosure
Sales, Storage, and Rest Rooms	Non-classified	If there is any opening to these rooms within the extent of a Division 1 location, the entire room shall be classified as Division 1.
Vapor Processing Systems Pits	1	Any pit, box, or space below grade level, any part of which is within a Division 1 or 2 classified location or that houses any equipment used to transport or process vapors.
Vapor Processing Equipment Located within Protective Enclosures (FPN): See Automotive and Marine Service Station Code, NFPA 30A-1990, Section 4-5.7	2	Within any protective enclosure housing vapor processing equipment
Vapor Processing Equipment Not within Protective Enclosures (excluding piping and combustion devices)	2	The space within 18 inches in all directions of equipment containing flammable vapor or liquid extending to grade level. Up to 18 inches above grade level within 10 feet horizontally of the vapor processing equipment.
Equipment Enclosures	1	Any space within the enclosure where vapor or liquid is present under normal operating conditions.
Vacuum-Assist Blowers	2	The space within 18 inches in all directions extending to grade level. Up to 18 inches above grade level within 10 feet horizontally.

For SI units: one inch = 2.5 cm; one foot = 0.3048 meter.
[1]Refer to Figure 1 for an illustration of classified location around dispensing devices.
*Ceiling mounted hose reel.
**(FPN): See Automotive and Marine Service Station Code, NFPA 30A-1990, Section 2-2.

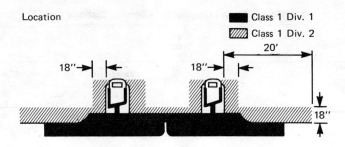

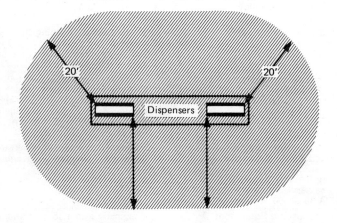

Figure 1.   Classified Locations Adjacent to Dispensers as Detailed in Table 514-2.

Exception No. 1: Type MI cable shall be permitted where it is installed in accordance with Article 330.

Exception No. 2: Rigid nonmetallic conduit complying with Article 347 shall be permitted where buried under not less than 2 feet (610 mm) of earth. Where rigid nonmetallic conduit is used, threaded rigid metal conduit or threaded steel intermediate metal conduit shall be used for the last 2 feet (610 mm) of the underground run to emergence or to the point of connection to the aboveground raceway; an equipment grounding conductor shall be included to provide electrical continuity of the raceway system and for grounding of noncurrent-carrying metal parts.

ARTICLE 515 — BULK STORAGE PLANTS

515-1. Definition. A bulk storage plant is that portion of a property where flammable liquids are received by tank vessel, pipelines, tank car, or tank vehicle, and are stored or blended in bulk for the purpose of distributing such liquids by tank vessel, pipeline, tank car, tank vehicle, portable tank, or container.

(FPN): For further information, see Flammable and Combustible Liquids Code, NFPA 30-1990 (ANSI).

515-2.ˣ Class I Locations. Table 515-2 shall be applied where Class I liquids are stored, handled, or dispensed and shall be used to delineate and classify bulk storage plants. The Class I location shall not extend beyond a floor, wall, roof, or other solid partition that has no communicating openings.

(FPN): The area classifications listed in Table 515-2 are based on the premise that the installation meets the applicable requirements of Flammable and Combustible Liquids Code, NFPA 30-1990 (ANSI), Chapter 5 in all respects. Should this not be the case, the authority having jurisdiction has the authority to classify the extent of the classified space.

515-3. Wiring and Equipment Within Class I Locations. All electric wiring and equipment within the Class I locations defined in Section 515-2 shall comply with the applicable provisions of Article 501.

Exception: As permitted in Section 515-5.

515-4. Wiring and Equipment Above Class I Locations. All fixed wiring above Class I locations shall be in metal raceways or PVC schedule 80 rigid nonmetallic conduit, or equivalent, or be Type MI, TC, SNM, or Type MC cable. Fixed equipment that may produce arcs, sparks, or particles of hot metal, such as lamps and lampholders for fixed lighting, cutouts, switches, receptacles, motors, or other equipment having make-and-break or sliding contacts, shall be of the totally enclosed type or be so constructed as to prevent escape of sparks or hot metal particles. Portable lamps or other utilization equipment and their flexible cords shall comply with the provisions of Article 501 for the class of location above which they are connected or used.

515-5. Underground Wiring.

(a) **Wiring Method.** Underground wiring shall be installed in threaded rigid metal conduit or threaded steel intermediate metal conduit or, where buried under not less than 2 feet (610 mm) of earth, shall be permitted in rigid nonmetallic conduit or an approved cable. Where rigid nonmetallic conduit is used, threaded rigid metal conduit or threaded steel intermediate metal conduit shall be used for the last 2 feet (610 mm) of the conduit run to emergence or to the point of connection to the aboveground raceway. Where cable is used, it shall be enclosed in threaded rigid metal conduit or threaded steel intermediate metal conduit from the point of lowest buried cable level to the point of connection to the aboveground raceway.

(b) **Insulation.** Conductor insulation shall comply with Section 501-13.

Table 515-2.[x]   Class I Locations — Bulk Plants

Location	Class I, Division	Extent of Classified Location
Indoor equipment installed in accordance with Flammable and Combustible Liquids Code, NFPA 30-1990 (ANSI), Section 5-3.3.2 where flammable vapor-air mixtures may exist under normal operation.	1	Space within 5 feet of any edge of such equipment, extending in all directions.
	2	Space between 5 feet and 8 feet of any edge of such equipment, extending in all directions. Also, space up to 3 feet above floor or grade level within 5 feet to 25 feet horizontally from any edge of such equipment.*
Outdoor equipment of the type covered in Flammable and Combustible Liquids Code, NFPA 30-1990 (ANSI), Section 5-3.3.2 where flammable vapor-air mixtures may exist under normal operation.	1	Space within 3 feet of any edge of such equipment, extending in all directions.
	2	Space between 3 feet and 8 feet of any edge of such equipment, extending in all directions. Also, space up to 3 feet above floor or grade level within 3 feet to 10 feet horizontally from any edge of such equipment.
Tank — Aboveground**	1	Space inside dike where dike height is greater than the distance from the tank to the dike for more than 50 percent of the tank circumference.
Shell, Ends, or Roof and Dike Space	2	Within 10 feet from shell, ends, or roof of tank. Space inside dikes to level of top of dike.
Vent	1	Within 5 feet of open end of vent, extending in all directions.
	2	Space between 5 feet and 10 feet from open end of vent, extending in all directions.
Floating Roof	1	Space above the roof and within the shell.
Underground Tank Fill Opening	1	Any pit, box, or space below grade level, if any part is within a Division 1 or 2 classified location.
	2	Up to 18 inches above grade level within a horizontal radius of 10 feet from a loose fill connection, and within a horizontal radius of 5 feet from a tight fill connection.
Vent — Discharging Upward	1	Within 3 feet of open end of vent, extending in all directions.
	2	Space between 3 feet and 5 feet of open end of vent, extending in all directions.

*The release of Class I liquids may generate vapors to the extent that the entire building, and possibly a zone surrounding it, should be considered a Class I, Division 2 location.

**For Tanks — Underground, see Section 514-2.

Table 515-2. Class I Locations — Bulk Plants (cont.)

Location	Class I, Division	Extent of Classified Location
Drum and Container Filling		
Outdoors, or Indoors with Adequate Ventilation	1	Within 3 feet of vent and fill openings, extending in all directions.
	2	Space between 3 feet and 5 feet from vent or fill opening, extending in all directions. Also, up to 18 inches above floor or grade level within a horizontal radius of 10 feet from vent or fill openings.
Pumps, Bleeders, Withdrawal Fittings, Meters and Similar Devices		
Indoors	2	Within 5 feet of any edge of such devices, extending in all directions. Also up to 3 feet above floor or grade level within 25 feet horizontally from any edge of such devices.
Outdoors	2	Within 3 feet of any edge of such devices, extending in all directions. Also up to 18 inches above grade level within 10 feet horizontally from any edge of such devices.
Pits		
Without Mechanical Ventilation	1	Entire space within pit if any part is within a Division 1 or 2 classified location.
With Adequate Mechanical Ventilation	2	Entire space within pit if any part is within a Division 1 or 2 classified location.
Containing Valves, Fittings, or Piping, and not within a Division 1 or 2 Classified Location	2	Entire pit.
Drainage Ditches, Separators, Impounding Basins		
Outdoor	2	Space up to 18 inches above ditch, separator, or basin. Also up to 18 inches above grade within 15 feet horizontally from any edge.
Indoor		Same as pits.
Tank Vehicle and Tank Car*		
Loading Through Open Dome	1	Within 3 feet of edge of dome, extending in all directions.
	2	Space between 3 feet and 15 feet from edge of dome, extending in all directions.

*When classifying extent of space, consideration shall be given to fact that tank cars or tank vehicles may be spotted at varying points. Therefore, the extremities of the loading or unloading positions shall be used.

Table 515-2. Class I Locations — Bulk Plants (cont.)

Location	Class I, Division	Extent of Classified Location
Loading through Bottom Connections with Atmospheric Venting	1	Within 3 feet of point of venting to atmosphere, extending in all directions.
	2	Space between 3 feet and 15 feet from point of venting to atmosphere, extending in all directions. Also up to 18 inches above grade within a horizontal radius of 10 feet from point of loading connection.
Office and Rest Rooms	Ordinary	If there is any opening to these rooms within the extent of an indoor classified location, the room shall be classified the same as if the wall, curb, or partition did not exist.
Loading through Closed Dome with Atmospheric Venting	1	Within 3 feet of open end of vent, extending in all directions.
	2	Space between 3 feet and 15 feet from open end of vent, extending in all directions. Also within 3 feet of edge of dome, extending in all directions.
Loading through Closed Dome with Vapor Control	2	Within 3 feet of point of connection of both fill and vapor lines, extending in all directions.
Bottom Loading with Vapor Control Any Bottom Unloading	2	Within 3 feet of point of connections, extending in all directions. Also up to 18 inches above grade within a horizontal radius of 10 feet from point of connections.
Storage and Repair Garage for Tank Vehicles	1	All pits or spaces below floor level.
	2	Space up to 18 inches above floor or grade level for entire storage or repair garage.
Garages for other than Tank Vehicles	Ordinary	If there is any opening to these rooms within the extent of an outdoor classified location, the entire room shall be classified the same as the space classification at the point of the opening.
Outdoor Drum Storage	Ordinary	
Indoor Warehousing Where There is No Flammable Liquid Transfer	Ordinary	If there is any opening to these rooms within the extent of an indoor classified location, the room shall be classified the same as if the wall, curb, or partition did not exist.
Piers and Wharves		See Figure 1.

For SI units: one inch = 25.4 millimeters; one foot = 0.3048 meter.

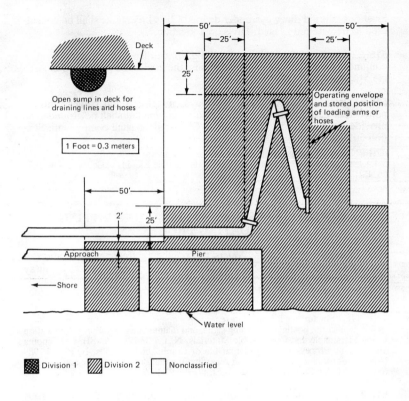

Notes:
(1) The "source of vapor" shall be the operating envelope and stored position of the outboard flange connection of the loading arm (or hose).
(2) The berth area adjacent to tanker and barge cargo tanks is to be Division 2 to the following extent:
    a. 25 ft (7.6 m) horizontally in all directions on the pier side from that portion of the hull containing cargo tanks.
    b. From the water level to 25 ft (7.6 m) above the cargo tanks at their highest position.
(3) Additional locations may have to be classified as required by the presence of other sources of flammable liquids on the berth, or by Coast Guard or other regulations.

Figure 1.[X]   Marine Terminal Handling Flammable Liquids.

(c) Nonmetallic Wiring. Where rigid nonmetallic conduit or cable with a nonmetallic sheath is used, an equipment grounding conductor shall be included to provide for electrical continuity of the raceway system and for grounding of noncurrent-carrying metal parts.

515-6. Sealing. Approved seals shall be provided in accordance with Section 501-5. Sealing requirements in Sections 501-5(a)(4) and (b)(2) shall apply to horizontal as well as to vertical boundaries of the defined Class I locations. Buried raceways under defined Class I locations shall be considered to be within a Class I, Division 1 location.

515-7. Gasoline Dispensing. Where gasoline dispensing is carried on in conjunction with bulk station operations, the applicable provisions of Article 514 shall apply.

515-8. Grounding. All metal raceways, metal-jacketed cables, and all noncurrent-carrying metal parts of electric equipment shall be grounded as provided in Article 250. Grounding in Class I areas shall comply with Section 501-16.

(FPN): For information on grounding for static protection, see Sections 5-4.4.1.2 and 5-4.4.1.7 of Flammable and Combustible Liquids Code, NFPA 30-1990 (ANSI).

ARTICLE 516 — SPRAY APPLICATION, DIPPING, AND COATING PROCESSES

516-1. Scope. This article covers the regular or frequent application of flammable liquids, combustible liquids, and combustible powders by spray operations and the application of flammable liquids, or combustible liquids at temperatures above their flashpoint, by dipping, coating, or other means.

(FPN): For further information regarding safeguards for these processes, such as fire protection, posting of warning signs, and maintenance, see Spray Application Using Flammable and Combustible Materials, NFPA 33-1989 (ANSI), and Dipping and Coating Processes Using Flammable or Combustible Liquids, NFPA 34-1989 (ANSI). For additional information regarding ventilation, see Blower and Exhaust Systems for Air Conveying of Materials, NFPA 91-1992.

516-2. Classification of Locations. Classification is based on dangerous quantities of flammable vapors, combustible mists, residues, dusts or deposits.

(a) Class I or Class II, Division 1 Locations. The following spaces shall be considered Class I or Class II, Division 1 locations, as applicable.

(1)[x] The interiors of spray booths and rooms except as specifically provided in Section 516-3(d).

(2)[x] The interior of exhaust ducts.

(3)[x] Any area in the direct path of spray operations.

(4)[x] For dipping and coating operations, all space within a 5-foot (1.52-m) radial distance from the vapor sources extending from these surfaces to the floor. The vapor source shall be the liquid exposed in the process and the drainboard, and any dipped or coated object from which it is possible to measure vapor concentrations exceeding 25 percent of the lower flammable limit at a distance of 1 foot (305 mm), in any direction, from the object.

(5)[x] Pits within 25 feet (7.62 m) horizontally of the vapor source. If pits extend beyond 25 feet (7.62 m) of the vapor source, a vapor stop shall be provided, or the entire pit shall be classified as Class I, Division 1.

(6)[x] The interior of any enclosed coating or dipping process.

(b) Class I or Class II, Division 2 Locations. The following spaces shall be considered Class I or Class II, Division 2 as applicable.

(1)[x] For open spraying, all space outside of but within 20 feet (6.10 m) horizontally and 10 feet (3.05 m) vertically of the Class I, Division 1 location as defined in Section 516-2(a), and not separated from it by partitions. See Figure 1[x] on following page.

(2)[x] For spraying operations conducted within a closed top, open face, or front spray booth or room, the space shown in Figure 2, and the space within 3 feet (914 mm) in all directions from openings other than the open face or front.

The Class I or Class II, Division 2 location shown in Figure 2 shall extend from the open face or front of the spray booth or room in accordance with the following:

   a. If the ventilation system is interlocked with the spraying equipment so as to make the spraying equipment inoperable when the ventilation system is not in operation, the space shall extend 5 feet (1.52 m) from the open face or front of the spray booth or room, and as otherwise shown in Figure 2A.

   b. If the ventilation system is not interlocked with the spraying equipment so as to make the spraying equipment inoperable when the ventilation system is not in operation, the space shall extend 10 feet (3.05 m) from the open face or front of the spray booth or room, and as otherwise shown in Figure 2B.

(3)[x] For spraying operations conducted within an open top spray booth, the space 3 feet (914 mm) vertically above the booth and within 3 feet (914 mm) of other booth openings shall be considered Class I or Class II, Division 2.

(4)[x] For spraying operations confined to an enclosed spray booth or room, the space within 3 feet (914 mm) in all directions from any openings shall be considered Class I or Class II, Division 2 as shown in Figure 3.

(5)[x] For dip tanks and drain boards, the 3-foot (914-mm) space surrounding the Class I, Division 1 location as defined in Section 516-2(a)(4) above and as shown in Figure 4.

(6) For dip tanks and drain boards, the space 3 feet (914 mm) above the floor and extending 20 feet (6.1 m) horizontally in all directions from the Class I, Division 1 location.

Exception: This space shall not be required to be considered a hazardous (classified) location where the vapor source area is 5 square feet (0.46 sq m) or less, and where the contents of the open tank, trough, or container do not exceed 5 gallons (18.9 L). In addition, the vapor concentration during operation and shutdown periods shall not exceed 25 percent of the lower flammable limit outside the Class I location specified in Section 516-2(a)(4) above.

(c)[x] **Enclosed Coating and Dipping Operations.** The space adjacent to enclosed coating and dipping operations shall be considered nonclassified.

Exception: The space within 3 feet (914 mm) in all directions from any opening in the enclosure shall be classified as Class I, Division 2.

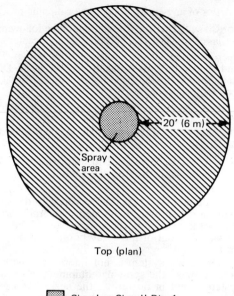

Top (plan)

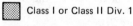

Class I or Class II Div. 1

Class I or Class II Div. 2

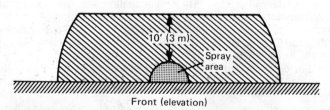

Front (elevation)

For SI units:
one inch = 25.4 millimeters;
one foot = 0.3048 meter.

Figure 1.^x Class I or Class II, Division 2 Location Adjacent to an Unenclosed Spray Operation.

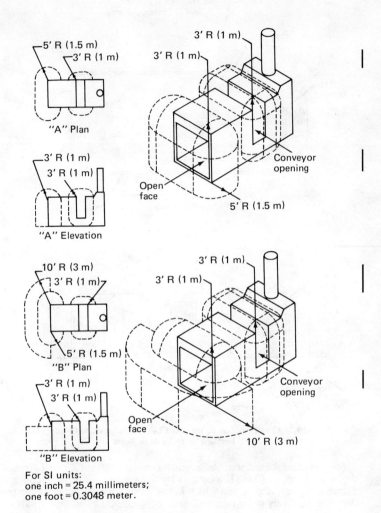

For SI units:
one inch = 25.4 millimeters;
one foot = 0.3048 meter.

Figure 2.[X]   Class I or Class II, Division 2 Locations Adjacent to a Closed Top,
Open Face or Open Front Spray Booth or Room.

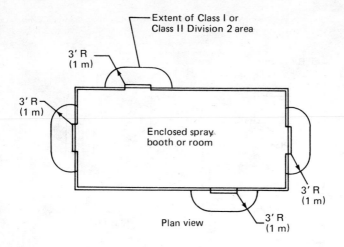

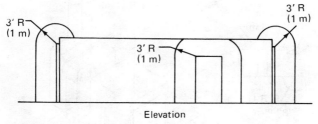

For SI units:
one inch = 25.4 millimeters;
one foot = 0.3048 meter.

Figure 3.^{X} Class I or Class II, Division 2 Location Adjacent to Openings in an Enclosed Spray Booth or Room

(d) Adjacent Locations. Adjacent locations that are cut off from the defined Class I or Class II locations by tight partitions without communicating openings, and within which hazardous vapors or combustible powders are not likely to be released, shall be classified as nonhazardous.

(e) Nonhazardous Locations. Locations utilizing drying, curing, or fusion apparatus and provided with positive mechanical ventilation adequate to prevent accumulation of flammable concentrations of vapors, and provided with effective interlocks to deenergize all electric equipment (other than equipment approved for Class I locations) in case the ventilating equipment is inoperative, shall be permitted to be classified as nonhazardous where the authority having jurisdiction so judges.

(FPN): For further information regarding safeguards, see Ovens and Furnaces, NFPA 86-1990 (ANSI).

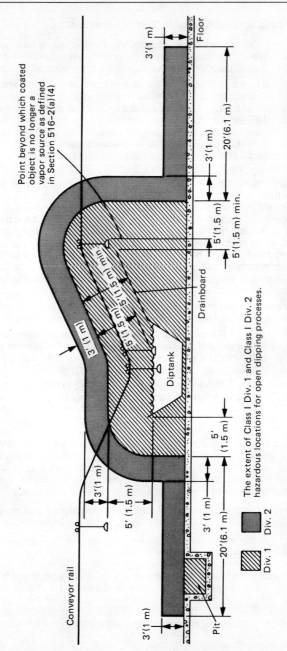

Conveyor rail

Point beyond which coated object is no longer a vapor source as defined in Section 516-2(a)(4)

3' (1 m)

3' (1 m)

5' (1.5 m)

5' (1.5 m) min.

5' (1.5 m)

Floor

3' (1 m)

20' (6.1 m)

5' (1.5 m) min.

Drainboard

Diptank

5' (1.5 m)

3' (1 m)

3' (1 m)

20' (6.1 m)

Pit

The extent of Class I, Division I and Class I, Div. 2 hazardous locations for open dipping processes.

Div. 2

Div. 1

Figure 4.^x   The Extent of Class I, Division I and Class I, Division 2 Hazardous (Classified) Locations for Open Dipping Processes.

516-3. Wiring and Equipment in Class I Locations.

(a) Wiring and Equipment — Vapors. All electric wiring and equipment within the Class I location (containing vapor only—not residues) defined in Section 516-2 shall comply with the applicable provisions of Article 501.

(b)ˣ Wiring and Equipment — Vapors and Residues. Unless specifically listed for locations containing deposits of dangerous quantities of flammable or combustible vapors, mists, residues, dusts or deposits (as applicable), there shall be no electrical equipment in any spray area as herein defined whereon deposits of combustible residue may readily accumulate, except wiring in rigid metal conduit, intermediate metal conduit, Type MI cable, or in metal boxes or fittings containing no taps, splices or terminal connections.

(c) Illumination. Illumination of readily ignitible areas through panels of glass or other transparent or translucent material shall be permitted only if it complies with the following: (1) fixed lighting units are used as the source of illumination; (2) the panel effectively isolates the Class I location from the area in which the lighting unit is located; (3) the lighting unit is approved for its specific location; (4) the panel is of a material or is so protected that breakage will be unlikely; and (5) the arrangement is such that normal accumulations of hazardous residue on the surface of the panel will not be raised to a dangerous temperature by radiation or conduction from the source of illumination.

(d)ˣ Portable Equipment. Portable electric lamps or other utilization equipment shall not be used in a spray area during spray operations.

Exception No. 1: Where portable electric lamps are required for operations in spaces not readily illuminated by fixed lighting within the spraying area, they shall be of the type approved for Class I, Division 1 locations where readily ignitible residues may be present.

Exception No. 2: Where portable electric drying apparatus are used in automobile refinishing spray booths and the following requirements are met: (1) the apparatus and its electrical connections are not located within the spray enclosure during spray operations; (2) electrical equipment within 18 inches (45.7 cm) of the floor is approved for Class I, Division 2 locations; (3) all metallic parts of the drying apparatus are electrically bonded and grounded; and (4) interlocks are provided to prevent the operation of spray equipment while drying apparatus is within the spray enclosure, to allow for a 3-minute purge of the enclosure before energizing the drying apparatus, and to shut off drying apparatus on failure of ventilation system.

(e) Electrostatic Equipment. Electrostatic spraying or detearing equipment shall be installed and used only as provided in Section 516-4.

(FPN): For further information, see Spray Application Using Flammable and Combustible Materials, NFPA 33-1989 (ANSI).

516-4.ˣ Fixed Electrostatic Equipment.

This section shall apply to any equipment using electrostatically charged elements for the atomization, charging, and (or) precipitation of hazardous materials for coatings on articles or for other similar purposes in which the charging or atomizing device is attached to a mechanical support or manipulator. This includes robotic devices. This section shall not apply to devices that are held or manipulated by hand. Where robotic programming procedures involve manual manipulation of the robot arm while spraying with the high voltage on, the provisions of Section 516-5 shall apply. The installation of electrostatic spraying equipment shall comply with (a) through (j) below. Spray equipment shall be listed or approved.

All automatic electrostatic equipment systems shall comply with (a) through (i) below.

(a) Power and Control Equipment. Transformers, high-voltage supplies, control apparatus, and all other electric portions of the equipment shall be installed outside of the Class I location as defined in Section 516-2 or be of a type approved for the location.

Exception: High-voltage grids, electrodes, electrostatic atomizing heads, and their connections shall be permitted within the Class I location.

(b) Electrostatic Equipment. Electrodes and electrostatic atomizing heads shall be adequately supported in permanent locations and shall be effectively insulated from ground. Electrodes and electrostatic atomizing heads that are permanently attached to their bases, supports, reciprocators, or robots shall be deemed to comply with this section.

(c) High-Voltage Leads. High-voltage leads shall be properly insulated and protected from mechanical damage or exposure to destructive chemicals. Any exposed element at high voltage shall be effectively and permanently supported on suitable insulators and shall be effectively guarded against accidental contact or grounding.

(d) Support of Goods. Goods being coated using this process shall be supported on conveyors or hangers. The conveyors or hangers shall be arranged to (1) assure that the parts being coated are electrically connected to ground with a resistance of 1 megohm or less and (2) prevent parts from swinging.

(e) Automatic Controls. Electrostatic apparatus shall be equipped with automatic means that will rapidly deenergize the high-voltage elements under any of the following conditions: (1) stoppage of ventilating fans or failure of ventilating equipment from any cause; (2) stoppage of the conveyor carrying goods through the high-voltage field unless stoppage is required by the spray process; (3) occurrence of excessive current leakage at any point in the high-voltage system; (4) deenergizing the primary voltage input to the power supply.

(f) Grounding. All electrically conductive objects in the spray area, except those objects required by the process to be at high voltage, shall be adequately grounded. This requirement shall apply to paint containers, wash cans, guards, hose connectors, brackets, and any other electrically conductive objects or devices in the area.

(g) Isolation. Safeguards such as adequate booths, fencing, railings, interlocks, or other means shall be placed about the equipment or incorporated therein so that they, either by their location or character, or both, assure that a safe separation of the process is maintained.

(h) Signs. Signs shall be conspicuously posted to (1) designate the process zone as dangerous with regard to fire and accident; (2) identify the grounding requirements for all electrically conductive objects in the spray area; (3) restrict access to qualified personnel only.

(i) Insulators. All insulators shall be kept clean and dry.

(j) Other Than Nonincendive Equipment. Spray equipment that cannot be classified as nonincendive shall comply with (1) and (2) below.

(1) Conveyors or hangers shall be so arranged to maintain a safe distance of at least twice the sparking distance between goods being painted and electrodes, electrostatic atomizing heads, or charged conductors. Warnings defining this safe distance shall be posted.

(2) The equipment shall provide an automatic means of rapidly deenergizing the high-voltage elements in the event the distance between the goods being painted and the eletrodes or electrostatic atomizing heads falls below that specified in item (1) above.

516-5.ˣ Electrostatic Hand-Spraying Equipment. This section shall apply to any equipment using electrostatically charged elements for the atomization, charging, and/or precipitation of materials for coatings on articles, or for other similar purposes in which the atomizing device is hand held or manipulated during the spraying operation. Electrostatic hand-spraying equipment and devices used in connection with paint-spraying operations shall be of approved types and shall comply with (a) through (e) below.

(a) General. The high-voltage circuits shall be designed so as not to produce a spark of sufficient intensity to ignite the most readily ignitible of those vapor-air mixtures likely to be encountered, nor result in appreciable shock hazard upon coming in contact with a grounded object under all normal operating conditions. The electrostatically charged exposed elements of the hand gun shall be capable of being energized only by an actuator that also controls the coating material supply.

(b) Power Equipment. Transformers, power packs, control apparatus, and all other electric portions of the equipment shall be located outside of the Class I location or be approved for the location.

Exception: The hand gun itself and its connections to the power supply shall be permitted within the Class I location.

(c) Handle. The handle of the spraying gun shall be electrically connected to ground by a metallic connection and be so constructed that the operator in normal operating position is in intimate electrical contact with the grounded handle to prevent buildup of a static charge on the operator's body. Signs indicating the necessity for grounding other persons entering the spray area shall be conspicuously posted.

(d) Electrostatic Equipment. All electrically conductive objects in the spraying area shall be adequately grounded. This requirement shall apply to paint containers, wash cans, and any other electrically conductive objects or devices in the area. The equipment shall carry a prominent, permanently installed warning regarding the necessity for this grounding feature.

(e) Support of Objects. Objects being painted shall be maintained in metallic contact with the conveyor or other grounded support. Hooks shall be regularly cleaned to ensure adequate grounding of 1 megohm or less.

Areas of contact shall be sharp points or knife edges where possible. Points of support of the object shall be concealed from random spray where feasible; and, where the objects being sprayed are supported from a conveyor, the point of attachment to the conveyor shall be so located as to not collect spray material during normal operation.

516-6.[x] **Powder Coating.** This section shall apply to processes in which combustible dry powders are applied. The hazards associated with combustible dusts are present in such a process to a degree, depending upon the chemical composition of the material, particle size, shape, and distribution.

(FPN): The hazards associated with combustible dusts are inherent in this process. Generally speaking, the hazard rating of the powders employed is dependent upon the chemical composition of the material, particle size, shape, and distribution.

(a) Electric Equipment and Sources of Ignition. Electric equipment and other sources of ignition shall comply with the requirements of Article 502. Portable electric lamps and other utilization equipment shall not be used within a Class II location during operation of the finishing processes. When such lamps or utilization equipment are used during cleaning or repairing operations, they shall be of a type approved for Class II, Division 1 locations, and all exposed metal parts shall be effectively grounded.

Exception: Where portable electric lamps are required for operations in spaces not readily illuminated by fixed lighting within the spraying area, they shall be of the type approved for Class II, Division 1 locations where readily ignitible residues may be present.

(b) Fixed Electrostatic Spraying Equipment. The provisions of Sections 516-4 and (a) above shall apply to fixed electrostatic spraying equipment.

(c) Electrostatic Hand-Spraying Equipment. The provisions of Sections 516-5 and (a) above shall apply to electrostatic hand-spraying equipment.

(d) Electrostatic Fluidized Beds. Electrostatic fluidized beds and associated equipment shall be of approved types. The high-voltage circuits shall be so designed that any discharge produced when the charging electrodes of the bed are approached or contacted by a grounded object shall not be of sufficient intensity to ignite any powder-air mixture likely to be encountered nor to result in an appreciable shock hazard.

(1) Transformers, power packs, control apparatus, and all other electric portions of the equipment shall be located outside the powder-coating area or shall otherwise comply with the requirements of (a) above.

Exception: The charging electrodes and their connections to the power supply shall be permitted within the powder-coating area.

(2) All electrically conductive objects within the powder-coating area shall be adequately grounded. The powder-coating equipment shall carry a prominent, permanently installed warning regarding the necessity for grounding these objects.

(3) Objects being coated shall be maintained in electrical contact (less than 1 megohm) with the conveyor or other support in order to ensure proper grounding. Hangers shall be regularly cleaned to ensure effective electrical contact. Areas of electrical contact shall be sharp points or knife edges where possible.

(4) The electric equipment and compressed-air supplies shall be inter-locked with a ventilation system so that the equipment cannot be operated unless the ventilating fans are in operation.

516-7. Wiring and Equipment Above Class I and II Locations.

(a) Wiring. All fixed wiring above the Class I and II locations shall be in metal raceways, rigid nonmetallic conduit, or electrical nonmetallic tubing, or shall be Type MI, TC, SNM, or Type MC cable. Cellular metal floor raceways shall be permitted only for supplying ceiling outlets or extensions to the area below the floor of a Class I or II location, but such raceways shall have no connections leading into or through the Class I or II location above the floor unless suitable seals are provided.

(b) Equipment. Equipment that may produce arcs, sparks, or particles of hot metal, such as lamps and lampholders for fixed lighting, cutouts, switches, receptacles, motors, or other equipment having make-and-break or sliding contacts, where installed above a Class I or II location or above a location where freshly finished goods are handled, shall be of the totally enclosed type or be so constructed as to prevent escape of sparks or hot metal particles.

516-8. Grounding.
All metal raceways, metal-jacketed cables, and all noncurrent-carrying metal parts of fixed or portable equipment, regardless of voltage, shall be grounded as provided in Article 250.

ARTICLE 517 — HEALTH CARE FACILITIES

A. General

517-1. Scope.

The provisions of this article shall apply to electrical construction and installation criteria in health care facilities.

(FPN No. 1):   This article is not intended to apply to veterinary facilities.

(FPN No. 2):   For information concerning performance, maintenance, and test-ing criteria, refer to the appropriate health care facilities documents.

517-2. General.
The requirements in Parts B and C apply not only to single-function buildings, but are also intended to be individually applied to their respective forms of occupancy within a multifunction building (e.g., a doctor's examining room located within a limited care facility would be required to meet the provisions of Section 517-10).

517-3. Definitions.

Alternate Power Source: One or more generator sets, or battery sys-tems where permitted, intended to provide power during the interruption of the normal electrical services or the public utility electrical service intended to provide power during interruption of service normally provided by the generating facilities on the premises.

Anesthetizing Location: Any area of a health care facility that has been designated to be used for the administration of any flammable or nonflammable inhalation anesthetic agent in the course of examination or treatment, including the use of such agents for relative analgesia.

Critical Branch: A subsystem of the emergency system consisting of feeders and branch circuits supplying energy to task illumination, special power circuits, and selected receptacles serving areas and functions related to patient care, and which are connected to alternate power sources by one or more transfer switches during interruption of the normal power source.

Electrical Life-Support Equipment: Electrically powered equipment whose continuous operation is necessary to maintain a patient's life.

Emergency System: A system of feeders and branch circuits meeting the requirements of Article 700, and intended to supply alternate power to a limited number of prescribed functions vital to the protection of life and patient safety, with automatic restoration of electrical power within 10 seconds of power interruption.

Equipment System: A system of feeders and branch circuits arranged for delayed, automatic or manual connection to the alternate power source and which serves primarily 3-phase power equipment.

Essential Electrical System: A system comprised of alternate sources of power and all connected distribution systems and ancillary equipment, designed to assure continuity of electrical power to designated areas and functions of a health care facility during disruption of normal power sources, and also designed to minimize disruption within the internal wiring system.

Exposed Conductive Surfaces: Those surfaces which are capable of carrying electric current and which are unprotected, unenclosed, or unguarded, permitting personal contact. Paint, anodizing, and similar coatings are not considered suitable insulation, unless they are listed for use.

Flammable Anesthetics: Gases or vapors, such as fluroxene, cyclopropane, divinyl ether, ethyl chloride, ethyl ether, and ethylene, which may form flammable or explosive mixtures with air, oxygen, or reducing gases such as nitrous oxide.

Flammable Anesthetizing Location: Any area of the facility which has been designated to be used for the administration of any flammable inhalation anesthetic agents in the normal course of examination or treatment.

Hazard Current: For a given set of connections in an isloated power system, the total current that would flow through a low impedance if it were connected between either isolated conductor and ground.

FAULT HAZARD CURRENT: The hazard current of a given isolated system with all devices connected except the line isolation monitor.

MONITOR HAZARD CURRENT: The hazard current of the line isolation monitor alone.

TOTAL HAZARD CURRENT: The hazard current of a given isolated system with all devices, including the line isolation monitor, connected.

Health Care Facilities: Buildings or portions of buildings that contain, but are not limited to, occupancies such as: hospitals; nursing homes; limited care; supervisory care; clinics; medical and dental offices; and ambulatory care, whether permanent or movable.

Hospital: A building or part thereof used for the medical, psychiatric, obstetrical, or surgical care, on a 24-hour basis, of four or more inpatients. Hospital, wherever used in this Code, shall include general hospitals, mental hospitals, tuberculosis hospitals, children's hospitals, and any such facilites providing inpatient care.

Isolated Power System: A system comprising an isolating transformer or its equivalent, a line isolation monitor, and its ungrounded circuit conductors.

Isolation Transformer: A transformer of the multiple-winding type, with the primary and secondary windings physically separated, which inductively couples its secondary winding to the grounded feeder systems that energize its primary winding.

Life Safety Branch: A subsystem of the emergency system consisting of feeders and branch circuits, meeting the requirements of Article 700 and intended to provide adequate power needs to ensure safety to patients and personnel, and which are automatically connected to alternate power sources during interruption of the normal power source.

Limited Care Facility: A building or part thereof used on a 24-hour basis for the housing of four or more persons who are incapable of self-preservation because of age, physical limitation due to accident or illness, or mental limitations, such as mental retardation/developmental disability, mental illness, or chemical dependency.

Line Isolation Monitor: A test instrument designed to continually check the balanced and unbalanced impedance from each line of an isolated circuit to ground and equipped with a built-in test circuit to exercise the alarm without adding to the leakage current hazard.

(FPN): "Line isolation monitor" was formerly known as "ground contact indicator."

Nursing Home: A building or part thereof used for the lodging, boarding and nursing care, on a 24-hour basis, of four or more persons who, because of mental or physical incapacity, may be unable to provide for their own needs and safety without the assistance of another person. Nursing home, wherever used in this Code, shall include nursing and convalescent homes, skilled nursing facilities, intermediate care facilities, and infirmaries of homes for the aged.

Nurses' Stations: Areas intended to provide a center of nursing activity for a group of nurses serving bed patients, where the patient calls are received, nurses are dispatched, nurses' notes written, inpatient charts prepared, and medications prepared for distribution to patients. Where such activities are carried on in more that one location within a nursing unit, all such separate areas are considered a part of the nurses' station.

Patient Bed Location: The location of an inpatient sleeping bed or the bed or procedure table used in a critical patient care area.

Patient Care Area: Any portion of a health care facility wherein patients are intended to be examined or treated.

(FPN): Business offices, corridors, lounges, day rooms, dining rooms, or similar areas typically are not classified as patient care areas.

Patient Care Areas of a Hospital: Areas of a hospital in which patient care is administered are classified as general care areas, or critical care areas, either of which may be classified as a wet location. The governing body of the facility shall designate these areas in accordance with the type of patient care anticipated and with the following definitions of the area classification.

(1) General care areas are patient bedrooms, examining rooms, treatment rooms, clinics, and similar areas in which it is intended that the patient shall come in contact with ordinary appliances such as a nurse call system, electrical beds, examining lamps, telephone, and entertainment devices. In such areas, it may also be intended that patients be connected to electromedical devices (such as heating pads, electrocardiographs, drainage pumps, monitors, otoscopes, ophthalmoscopes, intravenous lines, etc.).

(2) Critical care areas are those special care units, intensive care units, coronary care units, angiography laboratories, cardiac catheterization laboratories, delivery rooms, operating rooms, and similar areas in which patients are intended to be subjected to invasive procedures and connected to line-operated, electromedical devices.

(3) Wet locations are those patient care areas that are normally subject to wet conditions while patients are present. These include standing fluids on the floor or drenching of the work area, either of which condition is intimate to the patient or staff. Routine housekeeping procedures and incidental spillage of liquids do not define a wet location.

Patient Equipment Grounding Point: A jack or terminal bus which serves as the collection point for redundant grounding of electric appliances serving a patient vicinity or for grounding other items in order to eliminate electromagnetic interference problems.

Patient Vicinity: In an area in which patients are normally cared for, the patient vicinity is the space with surfaces likely to be contacted by the patient or an attendant who can touch the patient. Typically in a patient room, this encloses a space within the room not less than 6 feet (1.83 m) beyond the perimeter of the bed in its nominal location, and extending vertically not less than 7½ feet (2.29 m) above the floor.

Psychiatric Hospital: A building used exclusively for the psychiatric care, on a 24-hour basis, of four or more inpatients.

Reference Grounding Point: The ground bus of the panelboard or isolated power system panel supplying the patient care area.

Selected Receptacles: A minimum number of electric receptacles to accommodate appliances ordinarily required for local tasks or likely to be used in patient care emergencies.

Task Illumination: Provision for the minimum lighting required to carry out necessary tasks in the described areas, including safe access to supplies and equipment, and access to exits.

Therapeutic High-Frequency Diathermy Equipment: Therapeutic high-frequency diathermy equipment is therapeutic induction and dielectric heating equipment.

X-Ray Installations (Long-Time Rating):   A rating based on an operating interval of 5 minutes or longer.

X-Ray Installations (Mobile):   X-ray equipment mounted on a permanent base with wheels, casters, or a combination of both to facilitate moving the equipment while completely assembled.

X-Ray Installations (Momentary Rating):   A rating based on an operating interval that does not exceed 5 seconds.

X-Ray Installations (Portable):   X-ray equipment designed to be hand carried.

X-Ray Installations (Transportable):   X-ray equipment to be installed in a vehicle or that may be readily disassembled for transport in a vehicle.

B. Wiring and Protection

517-10. Applicability.   Part B shall apply to all health care facilities.

Exception No. 1: Part B shall not apply to business offices, corridors, waiting rooms, and the like in clinics, medical and dental offices, and outpatient facilities.

Exception No. 2: Part B shall not apply to patient sleeping areas in nursing homes and residential care facilities wired in accordance with Chapters 1 through 4 of this Code.

Exception No. 3:ˣ Part B shall not apply to freestanding buildings used as nursing homes and residential care facilities provided:

a. It maintains admitting and discharge policies that preclude the provision of care for any patient or resident who may need to be sustained by electrical life-support equipment, and

b. Offers no surgical treatment requiring general anesthesia, and

c. Provides an automatic battery-operated system(s) or equipment that shall be effective for at least 1¹/₂ hours and is otherwise in accordance with Section 700-12 and that shall be capable of supplying lighting for exit lights, exit corridors, stairways, nursing stations, medical preparation areas, boiler rooms, and communication areas. This system shall also supply power to operate all alarm systems.

(FPN):   See Life Safety Code, NFPA 101-1991 (ANSI).

517-11. General Installation/Construction Criteria. It is the purpose of this article to specify the installation criteria and wiring methods that will minimize electrical hazards by the maintenance of adequately low-potential differences only between exposed conductive surfaces that are likely to become energized and could be contacted by a patient.

(FPN): In a health care facility, it is difficult to prevent the occurrence of a conductive or capacitive path from the patient's body to some grounded object, because that path may be established accidentally or through instrumentation directly connected to the patient. Other electrically conductive surfaces that may make an additional contact with the patient, or instruments that may be connected to the patient, then become possible sources of electric currents that can traverse the patient's body. The hazard is increased as more apparatus is associated with the patient, and, therefore, more intensive precautions are needed. Control of electric shock hazard requires the limitation of electric current that might flow in an electric circuit involving the patient's body by raising the resistance of the conductive circuit that includes the patient, or by insulating exposed surfaces that might become energized, in addition to reducing the potential difference that can appear between exposed conductive surfaces in the patient vicinity, or by combinations of these methods. A special problem is presented by the patient with an externalized direct conductive path to the heart muscle. The patient may be electrocuted at current levels so low that additional protection in the design of appliances, insulation of the catheter, and control of medical practice are required.

517-12. Wiring Methods. Except as modified in this article, wiring methods shall comply with the applicable requirements of Chapters 1 through 4 of this Code.

517-13. Grounding of Receptacles and Fixed Electric Equipment.

(a) Patient Care Area. In an area used for patient care, the grounding terminals of all receptacles and all noncurrent-carrying conductive surfaces of fixed electric equipment likely to become energized that are subject to personal contact, operating at over 100 volts, shall be grounded by an insulated copper conductor. The grounding conductor shall be sized in accordance with Table 250-95 and installed in metal raceways with the branch-circuit conductors supplying these receptacles or fixed equipment.

Exception No. 1: Metal raceways shall not be required where Type MI cable, and Types MC and AC cables where the outer metal jacket is an approved grounding means of a listed cable assembly.

Exception No. 2: Metal faceplates shall be permitted to be grounded by means of a metal mounting screw(s) securing the faceplate to a grounded outlet box or grounded wiring device.

(b) Methods. In addition to the requirements of Section 517-13(a), all branch circuits serving patient care areas shall be provided with a ground path for fault current by installation in a metal raceway system or cable assembly. The metal raceway system, or cable armor or sheath assembly, shall qualify as an equipment grounding return path in accordance with Section 250-91(b). Type MC and Type MI cable shall have an outer metal armor or sheath that is identified as an acceptable grounding return path.

517-14. Panelboard Bonding. The equipment grounding terminal busses of the normal and essential branch-circuit panelboards serving the same individual patient vicinity shall be bonded together with an insulated continuous copper conductor not smaller than No. 10. Where more than two

(2) panels serve the same location, this conductor shall be continuous from panel to panel, but shall be permitted to be broken in order to terminate on the ground bus in each panel.

517-16. Receptacles with Insulated Grounding Terminals. Receptacles with insulated grounding terminals as permitted in Section 250-74, Exception No. 4 shall be identified; such identification shall be visible after installation.

(FPN): Caution is important in specifying such a system with receptacles having insulated grounding terminals, since the grounding impedance is controlled only by the grounding conductors and does not benefit functionally from any parallel grounding paths.

517-17. Ground-Fault Protection.

(a) Feeders. Where ground-fault protection is provided for operation of the service disconnecting means or feeder disconnecting means as specified by Sections 230-95 or 215-10, an additional step of ground-fault protection shall be provided in the next level of feeder disconnecting means downstream toward the load. Such protection shall consist of overcurrent devices and current transformers or other equivalent protective equipment that shall cause the feeder disconnecting means to open.

(FPN): The additional levels of ground-fault protection required by Section 517-17(a) are not intended:

a. On the load side of an essential electrical system transfer switch, or

b. Between the on-site generating unit(s) described in Section 517-35(b) and the essential electrical system transfer switch(es), or

c. On electrical systems that are not solidly grounded wye systems with greater than 150 volts to ground, but not exceeding 600 volts phase-to-phase.

(b) Selectivity. Ground-fault protection for operation of the service and feeder disconnecting means shall be fully selective such that the feeder device and not the service device shall open on ground faults on the load side of the feeder device. A six-cycle minimum separation between the service and feeder ground-fault tripping bands shall be provided. Operating time of the disconnecting devices shall be considered in selecting the time spread between these two bands to achieve 100 percent selectivity.

(FPN): See Section 230-95, Fine Print Note, for transfer of alternate source where ground-fault protection is applied.

(c) Testing. When equipment ground-fault protection is first installed, each level shall be performance tested to ensure compliance with (b) above.

517-18. General Care Areas.

(a) Patient Bed Location Branch Circuits. Each patient bed location shall be supplied by at least two branch circuits, at least one of which originates in a normal system panelboard; all branch circuits from the normal system shall originate in the same panelboard.

Exception No. 1: Branch circuits serving only special-purpose outlets or receptacles, such as portable X-ray outlets, shall not be required to be served from the same distribution panel or panels.

Exception No. 2: Clinics, medical and dental offices, and outpatient facilities; psychiatric, substance abuse, and rehabilitation hospitals; nursing homes and residential custodial care facilities meeting the requirements of the exceptions to Section 517-10.

(b) Patient Bed Location Receptacles. Each patient bed location shall be provided with a minimum of four receptacles. They shall be permitted to be of the single or duplex types or a combination of both. All receptacles, whether four or more, shall be listed "hospital grade" and so identified. The requirement that the receptacles be listed "hospital grade" and so identified shall become effective January 1, 1993. Each receptacle shall | be grounded by means of an insulated copper conductor sized in accordance with Table 250-95.

Exception No. 1: Psychiatric, substance abuse, and rehabilitation hospitals meeting the requirements of the exceptions to Section 517-10.

Exception No. 2: Psychiatric security rooms shall not be required to have receptacle outlets installed in the room.

(FPN): It is not intended that there be a total, immediate replacement of existing non-"hospital grade" receptacles. It is intended, however, that non-"hospital grade" receptacles be replaced with "hospital grade" receptacles upon modification of use, renovation, or as existing receptacles need replacement.

(c) Pediatric Locations. Fifteen- and 20-ampere, 125-volt receptacles | intended to supply patient care areas of pediatric wards, rooms, or areas | shall be tamper resistant. For the purpose of this section, a tamper resistant receptacle is a receptacle that, by its construction, limits improper access to its energized contacts.

517-19. Critical Care Areas.

(a) Patient Bed Location Branch Circuits. Each patient bed location shall be supplied by at least two branch circuits, one or more from the emergency system and one or more circuits from the normal system. At least one branch circuit from the emergency system shall supply an outlet(s) only at that bed location. All branch circuits from the normal system shall be from a single panelboard. Emergency system receptacles shall be | identified and shall also indicate the panelboard and circuit number supplying them. |

Exception: Branch circuits serving only special-purpose receptacles or equipment in critical care areas shall be permitted to be served by other panelboards.

(b) Patient Bed Location Receptacles. Each patient bed location shall be provided with a minimum of six receptacles. They may be of the single or duplex types or a combination of both. All receptacles, whether six or more, shall be listed "hospital grade" and so identified; each receptacle shall be grounded to the reference grounding point by means of an insulated copper equipment grounding conductor.

(c) Patient Vicinity Grounding and Bonding (Optional). A patient vicinity shall be permitted to have a patient equipment grounding point. The patient equipment grounding point, where supplied, shall be permitted to contain one or more jacks listed for the purpose. An equipment bonding jumper, not smaller than No. 10, shall be used to connect the grounding terminal of all grounding-type receptacles to the patient equipment grounding point. The bonding conductor shall be permitted to be arranged centrically or looped as convenient.

(FPN): Where there is no patient equipment grounding point, it is important that the distance between the reference grounding point and the patient vicinity be as short as possible to minimize any potential differences.

(d) Panelboard Grounding. Where a grounded electrical distribution system is used, and metal feeder raceway or Type MC or MI cable is installed, grounding of a panelboard or switchboard shall be assured by one of the following means at each termination or junction point of the raceway or Type MC or MI cable:

(1) A grounding bushing and a continuous copper bonding jumper, sized in accordance with Section 250-95, with the bonding jumper connected to the junction enclosure or the ground bus of the panel.

(2) Connection of feeder raceways or Type MC or MI cable to threaded hubs or bosses on terminating enclosures.

(3) Other approved devices, such as bonding-type locknuts or bushings.

(e) Additional Protective Techniques in Critical Care Areas (Optional). Isolated power systems shall be permitted to be used for critical care areas, and, if used, the isolated power system equipment shall be listed for the purpose and the system so designed and installed that it meets the provisions and is in accordance with Section 517-160.

Exception: The audible and visual indicators of the line isolation monitor shall be permitted to be located at the nursing station for the area being served.

(f) Isolated Power System Grounding. Where an isolated ungrounded power source is used and limits the first-fault current to a low magnitude, the grounding conductor associated with the secondary circuit shall be permitted to be run outside of the enclosure of the power conductors in the same circuit.

(FPN): Although it is permitted to run the grounding conductor outside of the conduit, it is safer to run it with the power conductors to provide better protection in case of a second ground fault.

(g) Special-Purpose Receptacle Grounding. The equipment grounding conductor for special-purpose receptacles such as the operation of mobile X-ray equipment shall be extended to the reference grounding points of branch circuits for all locations likely to be served from such receptacles. Where such a circuit is served from an isolated ungrounded system, the grounding conductor shall not be required to be run with the power conductors; however, the equipment grounding terminal of the special-purpose receptacle shall be connected to the reference grounding point.

517-20. Wet Locations.

(a) All receptacles and fixed equipment within the area of the wet location shall have ground-fault circuit-interrupter protection for personnel if interruption of power under fault conditions can be tolerated, or be served by an isolated power system if such interruption cannot be tolerated.

Exception: Branch circuits supplying only listed, fixed, therapeutic and diagnostic equipment shall be permitted to be supplied from a normal grounded service, single- or 3-phase system, provided:

a. Wiring for grounded and isolated circuits does not occupy the same raceway, and

b. All conductive surfaces of the equipment are grounded.

(b) Where an isolated power system is utilized, the equipment shall be listed for the purpose and installed so that it meets the provisions of, and is in accordance with, Section 517-160.

(FPN):   For requirements for installation of therapeutic pools and tubs, see Part F of Article 680.

C. Essential Electrical System

517-25.[x] **Scope.**   The essential electrical system for these facilities shall comprise a system capable of supplying a limited amount of lighting and power service which is considered essential for life safety and orderly cessation of procedures during the time normal electrical service is interrupted for any reason. This includes clinics, medical and dental offices, outpatient facilities, nursing homes, limited care facilities, hospitals, and other health care facilities serving patients.

(FPN):   For information as to the need for an essential electrical system, see Health Care Facilities, NFPA 99-1990 (ANSI).

517-30. Essential Electrical Systems For Hospitals.

(a) Applicability. The requirements of Part C, Sections 517-30 through 517-35, shall apply to hospitals where an essential electrical system is required.

(FPN No. 1):   For performance, maintenance, and testing requirements of essential electrical systems in hospitals, see Health Care Facilities, NFPA 99-1990 (ANSI). For installation of centrifugal fire pumps, see Installation of Centrifugal Fire Pumps, NFPA 20-1990 (ANSI).

(FPN No. 2):   For additional information, see Health Care Facilities, NFPA 99-1990 (ANSI).

(b) General.

(1)[x] Essential electrical systems for hospitals shall be comprised of two separate systems capable of supplying a limited amount of lighting and power service which is considered essential for life safety and effective hospital operation during the time the normal electrical service is interrupted for any reason. These two systems shall be the emergency system and the equipment system.

(2)[x] The emergency system shall be limited to circuits essential to life safety and critical patient care. These are designated the life safety branch and the critical branch.

(3)[x] The equipment system shall supply major electric equipment necessary for patient care and basic hospital operation.

(4)[x] The number of transfer switches to be used shall be based upon reliability, design, and load considerations. Each branch of the essential electrical system shall be served by one or more transfer switches as shown in Diagrams 517-30(1) and 517-30(2). One transfer switch shall be permitted to serve one or more branches or systems in a facility with a maximum demand on the essential electrical system of 150 kVA as shown in Diagram 517-30(3).

(FPN): See Health Care Facilities, NFPA 99-1990 (ANSI): Section 3-5.1.2.2(a), Transfer Switch Operation Type I; Section 3-4.2.1.4, Automatic Transfer Switch Features; and Section 3-4.2.1.6, Nonautomatic Transfer Device Features.

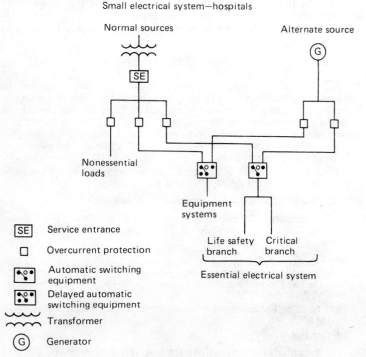

Diagram 517-30(1)

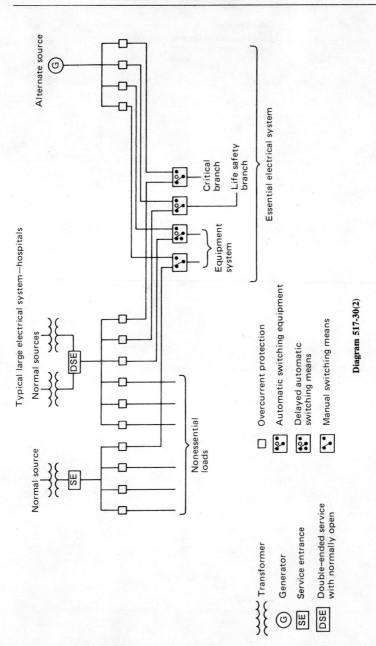

Diagram 517-30(2)

Small electrical system—hospitals
(single transfer switch)

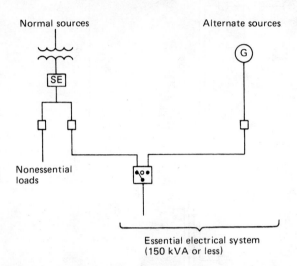

Normal sources   Alternate sources

Nonessential
loads

Essential electrical system
(150 kVA or less)

SE	Service entrance
	Overcurrent protection
	Automatic switching equipment
	Transformer
G	Generator

Diagram 517-30 (3)

(c) Wiring Requirements.

(1) Separation from Other Circuits. The life safety branch and critical branch of the emergency system shall be kept entirely independent of all other wiring and equipment and shall not enter the same raceways, boxes, or cabinets with each other or other wiring, except as follows:

a. In transfer switches,

b. In exit or emergency lighting fixtures supplied from two sources, or

c. In a common junction box attached to exit or emergency lighting fixtures supplied from two sources.

The wiring of the equipment system shall be permitted to occupy the same raceways, boxes, or cabinets of other circuits that are not part of the emergency system.

(2) Isolated Power Systems. Where isolated power systems are installed in any of the areas in Sections 517-33(a)(1) and (a)(2), each system shall be supplied by an individual circuit serving no other load.

(3) Mechanical Protection of the Emergency System. The wiring of the emergency system of a hospital shall be mechanically protected by installation in metal raceways.

Exception No. 1: Flexible power cords of appliances, or other utilization equipment, connected to the emergency system shall not be required to be enclosed in raceways.

Exception No. 2: Secondary circuits of transformer-powered communication or signaling systems shall not be required to be enclosed in raceways unless otherwise specified by Chapter 7 or 8.

Exception No. 3: Schedule 80 rigid nonmetallic conduit shall be permitted except for branch circuits serving patient care areas.

Exception No. 4: Where encased in not less than 2 inches (50.8 mm) of concrete, Schedule 40 PVC shall be permitted except for branch circuits serving patient care areas.

(FPN): See Section 517-13(b) for additional grounding requirements in patient care areas.

(d) Capacity of Systems. The essential electrical system shall have adequate capacity to meet the demand for the operation of all functions and equipment to be served by each system and branch.

517-31.^x Emergency System. Those functions of patient care depending on lighting or appliances that are connected to the emergency system shall be divided into two mandatory branches: the life safety branch and the critical branch, described in Sections 517-32 and 517-33.

The branches of the emergency system shall be installed and connected to the alternate power source so that all functions specified herein for the emergency system shall be automatically restored to operation within 10 seconds after interruption of the normal source.

517-32.^x Life Safety Branch. No function other than those listed in (a) through (f) shall be connected to the life safety branch. The life safety branch of the emergency system shall supply power for the following lighting, receptacles, and equipment:

(a) Illumination of Means of Egress. Illumination of means of egress, such as lighting required for corridors, passageways, stairways and landings at exit doors, and all necessary ways of approach to exits. Switching arrangements to transfer patient corridor lighting in hospitals from general illumination circuits to night illumination circuits shall be permitted provided only one of two circuits can be selected, and both circuits cannot be extinguished at the same time.

(FPN):   See Life Safety Code, NFPA 101-1991 (ANSI), Section 5-10.

(b) Exit Signs. Exit signs and exit directional signs.

(FPN):   See Life Safety Code, NFPA 101-1991 (ANSI), Section 5-11.

(c) Alarm and Alerting Systems. Alarm and alerting systems including:

(1) Fire alarms.

(FPN):   See Life Safety Code, NFPA 101-1991 (ANSI), Sections 12-1 and 12-2.

(4) Alarms required for systems used for the piping of nonflammable medical gases.

(FPN):   See Health Care Facilities, NFPA 99-1990 (ANSI), Section 12-3.4.1.

(d) Communication Systems. Hospital communication systems, where used for issuing instructions during emergency conditions.

(e) Generator Set Location. Task illumination battery charger for emergency battery-powered lighting unit(s) and selected receptacles at the generator set location.

(f) Elevators. Elevator cab lighting, control, communication, and signal systems.

517-33.ˣ Critical Branch.

(a) Task Illumination and Selected Receptacles. The critical branch of the emergency system shall supply power for task illumination, fixed equipment, selected receptacles, and special power circuits serving the following areas and functions related to patient care.

(1) Anesthetizing locations — task illumination, all receptacles and fixed equipment.

(2) The isolated power systems in special environments.

(3) Patient care areas — task illumination and selected receptacles in:

  a. Infant nurseries,

  b. Medication preparation areas,

  c. Pharmacy dispensing areas,

  d. Selected acute nursing areas,

  e. Psychiatric bed areas (omit receptacles),

  f. Ward treatment rooms, and

  g. Nurses' stations (unless adequately lighted by corridor luminaires).

(4) Additional specialized patient care task illumination and receptacles, where needed.

(5) Nurse call systems.

(6) Blood, bone, and tissue banks.

(7) Telephone equipment room and closets.

(8) Task illumination, selected receptacles, and selected power circuits for:

a. General care beds (at least one duplex receptacle per patient bedroom),

b. Angiographic labs,

c. Cardiac catheterization labs,

d. Coronary care units,

e. Hemodialysis rooms or areas,

f. Emergency room treatment areas (selected),

g. Human physiology labs,

h. Intensive care units, and

i. Postoperative recovery rooms (selected).

(9) Additional task illumination, receptacles, and selected power circuits needed for effective hospital operation. Single-phase fractional horsepower exhaust fan motors that are interlocked with 3-phase motors on the equipment shall be permitted to be connected to the critical branch.

(b) Subdivision of the Critical Branch. It shall be permitted to subdivide the critical branch into two or more branches.

(FPN): It is important to analyze the consequences of supplying an area with only critical care branch power when failure occurs between the area and the transfer switch. Some proportion of normal and critical power, or critical power from separate transfer switches, may be appropriate.

517-34.ˣ Equipment System Connection to Alternate Power Source. The equipment system shall be installed and connected to the alternate source, such that the equipment described in Section 517-34(a) is automatically restored to operation at appropriate time-lag intervals following the energizing of the emergency system. Its arrangement shall also provide for the subsequent connection of equipment described in Section 517-34(b).

(a)ˣ Equipment for Delayed Automatic Connection. The following equipment shall be arranged for delayed automatic connection to the alternate power source:

(1) Central suction systems serving medical and surgical functions, including controls. Such suction systems shall be permitted on the critical branch.

(2) Sump pumps and other equipment required to operate for the safety of major apparatus, including associated control systems and alarms.

(3) Compressed air systems serving medical and surgical functions, including controls.

(FPN): The above equipment may be arranged for sequential delayed automatic connection to the alternate power source to prevent overloading the generator where engineering studies indicate it is necessary.

(b)ˣ Equipment for Delayed Automatic or Manual Connection. The following equipment shall be arranged for either delayed automatic or manual connection to the alternate power source:

(1) Heating equipment to provide heating for operating, delivery, labor, recovery, intensive care, coronary care, nurseries, infection/isolation rooms, emergency treatment spaces, and general patient rooms.

Exception: Heating of general patient rooms and infection/isolation rooms during disruption of the normal source shall not be required under any of the following conditions:

a. The outside design temperature is higher than +20°F (-6.7°C), or

b. The outside design temperature is lower than +20°F (-6.7°C) and where a selected room(s) is provided for the needs of all confined patients, then only such room(s) need be heated, or

c. The facility is served by a dual source of normal power as described in Section 517-35(c), Fine Print Note.

(FPN): The design temperature is based on the 97½ percent design value as shown in Chapter 24 of the ASHRAE Handbook of Fundamentals (1985).

(2) Elevator(s) selected to provide service to patient, surgical, obstetrical, and ground floors during interruption of normal power.

In instances where interruption of normal power would result in other elevators stopping between floors, throw-over facilities shall be provided to allow the temporary operation of any elevator for the release of patients or other persons who may be confined between floors.

(3) Supply, return, and exhaust ventilating systems for surgical and obstetrical delivery suites, intensive care, coronary care, nurseries, infection/isolation rooms, emergency treatment spaces, and exhaust fans for laboratory fume hoods, nuclear medicine areas where radioactive material is used, ethylene oxide evacuation, and anesthesia evacuation.

(4) Hyperbaric facilities.

(5) Hypobaric facilities.

(6) Automatically operated doors.

(7) Minimal electrically heated autoclaving equipment shall be permitted to be arranged for either automatic or manual connection to the alternate source.

(8) Other selected equipment shall be permitted to be served by the equipment system.

517-35. Sources of Power.

(a)ˣ Two Independent Sources of Power. Essential electrical systems shall have a minimum of two independent sources of power: a normal source generally supplying the entire electrical system, and one or more alternate sources for use when the normal source is interrupted.

(b)ˣ Alternate Source of Power. The alternate source of power shall be a generator(s) driven by some form of prime mover(s), and located on the premises.

Exception: Where the normal source consists of generating units on the premises, the alternate source shall be either another generating set, or an external utility service.

(c) Location of Essential Electrical System Components. Careful consideration shall be given to the location of the spaces housing the components of the essential electrical system to minimize interruptions caused by natural forces common to the area (e.g., storms, floods, earthquakes, or hazards created by adjoining structures or activities). Consideration shall also be given to the possible interruption of normal electrical services resulting from similar causes as well as possible disruption of normal electrical service due to internal wiring and equipment failures.

(FPN): Facilities whose normal source of power is supplied by two or more separate central station-fed services experience greater than normal electrical service reliability than those with only a single feed. Such a dual source of normal power consists of two or more electrical services fed from separate generator sets or a utility distribution network having multiple power input sources and arranged to provide mechanical and electrical separation so that a fault between the facility and the generating sources will not likely cause an interruption of more than one of the facility service feeders.

517-40. Essential Electrical Systems for Nursing Homes and Limited | Care Facilities.

(a) **Applicability.** The requirements of Part C, Sections 517-40(c) through 517-44, shall apply to nursing homes and limited care facilities. | Facilities that fulfill Section 517-10, Exception No. 3 shall be exempt from these requirements.

(b) **Inpatient Hospital Care Facilities.** Nursing homes and limited care | facilities that provide inpatient hospital care shall comply with the requirements of Part C, Sections 517-30 through 517-35.

(c) **Facilities Contiguous with Hospitals.** Nursing homes and limited | care facilities that are contiguous with a hospital shall be permitted to have their essential electrical systems supplied by that of the hospital.

(FPN): For performance, maintenance, and testing requirements of essential electrical systems in nursing homes and limited care facilities, see Health Care | Facilities, NFPA 99-1990 (ANSI).

517-41. Essential Electrical Systems.

(a)[x] **General.** Essential electrical systems for nursing homes and limited | care facilities shall be comprised of two separate branches capable of supplying a limited amount of lighting and power service which is considered essential for the protection of life safety and effective operation of the institution during the time normal electrical service is interrupted for any reason. These two separate branches shall be the life safety branch and the critical branch.

(b)[x] **Transfer Switches.** The number of transfer switches to be used shall be based upon reliability, design, and load considerations. Each branch of the essential electrical system shall be served by one or more transfer switches as shown in Diagrams 517-41(1) and 517-41(2). One transfer switch shall be permitted to serve one or more branches or systems in a facility with a maximum demand on the essential electrical system of 150 kVa as shown in Diagram 517-41(3).

(FPN): See Health Care Facilities, NFPA 99-1990 (ANSI): Section 3-5.1.2.2(b), | Transfer Switch Operation Type II; Section 3-4.2.1.4, Automatic Transfer Switch Features; and Section 3-4.2.1.6, Nonautomatic Transfer Device Features.

Small electrical system—Nursing homes
                        and residential custodial
                        care facilities

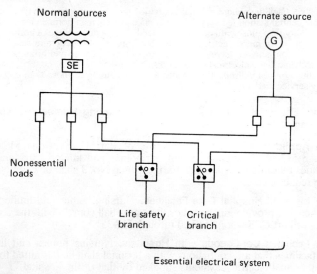

SE   Service entrance

☐   Overcurrent
         protection

[⟋°•]   Automatic switching
         equipment

[⁝°•]   Delayed automatic
         switching equipment

〰〰〰   Transformer

Ⓖ   Generator

Diagram 517-41(1)

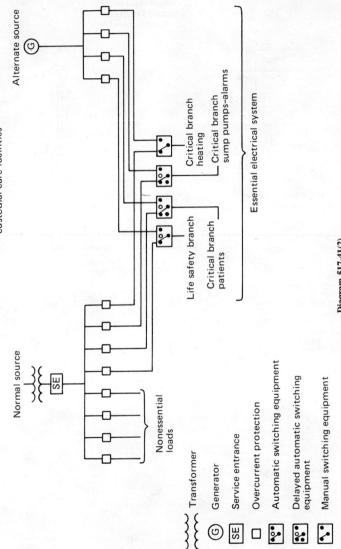

Typical large electrical system—Nursing homes and residential custodial care facilities

Alternate source

Normal source

Nonessential loads

Critical branch heating

Critical branch sump pumps–alarms

Life safety branch

Critical branch patients

Essential electrical system

Transformer

Generator

Service entrance

Overcurrent protection

Automatic switching equipment

Delayed automatic switching equipment

Manual switching equipment

Diagram 517-41(2)

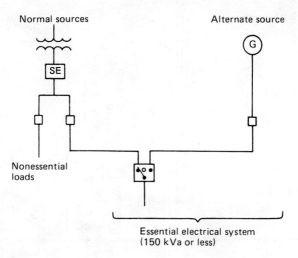

Small electrical system—Nursing homes
and residential custodial
care facilities
(single transfer switch)

Normal sources   Alternate source

SE

Nonessential
loads

Essential electrical system
(150 kVa or less)

SE	Service entrance
□	Overcurrent protection
	Automatic switching equipment
	Transformer
G	Generator

Diagram 517-41(3)

(c) Capacity of System. The essential electrical system shall have adequate capacity to meet the demand for the operation of all functions and equipment to be served by each branch at one time.

(d) Separation from Other Circuits. The life safety branch shall be kept entirely independent of all other wiring and equipment and shall not enter the same raceways, boxes, or cabinets with other wiring except as follows:

(1) In transfer switches,

(2) In exit or emergency lighting fixtures supplied from two sources, or

(3) In a common junction box attached to exit or emergency lighting fixtures supplied from two sources.

The wiring of the critical branch shall be permitted to occupy the same raceways, boxes, or cabinets of other circuits that are not part of the life safety branch.

517-42.ˣ Automatic Connection to Life Safety Branch. The life safety branch shall be so installed and connected to the alternate source of power that all functions specified herein shall be automatically restored to operation within 10 seconds after the interruption of the normal source. No function other than those listed in (a) through (g) shall be connected to the life safety branch. The life safety branch shall supply power for the following lighting, receptacles, and equipment:

(FPN): The life safety branch is called the emergency system in Health Care Facilities, NFPA 99-1990 (ANSI).

(a) Illumination of Means of Egress. Illumination of means of egress as is necessary for corridors, passageways, stairways, landings, and exit doors and all ways of approach to exits. Switching arrangement to transfer patient corridor lighting from general illumination circuits shall be permitted providing only one of two circuits can be selected, and both circuits cannot be extinguished at the same time.

(FPN): See Life Safety Code, NFPA 101-1991 (ANSI), Section 5-10.

(b) Exit Signs. Exit signs and exit directional signs.

(FPN): See Life Safety Code, NFPA 101-1991 (ANSI), Section 5-11.

(c) Alarm and Alerting Systems. Alarm and alerting systems, including:

(1) Fire alarms.

(FPN): See Life Safety Code, NFPA 101-1991 (ANSI), Sections 7-6, 12-3.4, and 12-3.5.

(2) Alarms required for systems used for the piping of nonflammable medical gases.

(FPN): See Health Care Facilities, NFPA 99-1990 (ANSI), Section 16-3.4.1.

(d) Communications Systems. Communications systems, where used for issuing instructions during emergency conditions.

(e) Dining and Recreation Areas. Sufficient lighting in dining and recreation areas to provide illumination to exit ways.

(f) Generator Set Location. Task illumination and selected receptacles in the generator set location.

(g) Elevators. Elevator cab lighting, control, communication, and signal systems.

517-43.[x] **Connection to Critical Branch.** The critical branch shall be so installed and connected to the alternate power source that the equipment listed in Section 517-43(a) shall be automatically restored to operation at appropriate time-lag intervals following the restoration of the life safety branch to operation. Its arrangement shall also provide for the additional connection of equipment listed in Section 517-43(b) by either delayed automatic or manual operation.

(a) Delayed Automatic Connection. The following equipment shall be connected to the critical branch and shall be arranged for delayed automatic connection to the alternate power source:

(1) Patient care areas — task illumination and selected receptacles in:

a. Medication preparation areas.

b. Pharmacy dispensing areas.

c. Nurses' stations (unless adequately lighted by corridor luminaires).

(2) Sump pumps and other equipment required to operate for the safety of major apparatus and associated control systems and alarms.

(b) Delayed Automatic or Manual Connection. The following equipment shall be connected to the critical branch and shall be arranged for either delayed automatic or manual connection to the alternate power source:

(1) Heating equipment to provide heating for patient rooms.

Exception: Heating of general patient rooms during disruption of the normal source shall not be required under any of the following conditions:

a. The outside design temperature is higher than +20°F (-6.7°C), or

b. The outside design temperature is lower than +20°F (-6.7°C) and where a selected room(s) is provided for the needs of all confined patients, then only such room(s) need be heated, or

c. The facility is served by a dual source of normal power as described in Section 517-44(c), Fine Print Note.

(FPN): The outside design temperature is based on the 97½ percent design values as shown in Chapter 24 of the ASHRAE Handbook of Fundamentals (1985).

(2) Elevator Service. In instances where disruption of power would result in elevators stopping between floors, throw-over facilities shall be provided to allow the temporary operation of any elevator for the release of passengers. For elevator cab lighting, control, and signal system requirements, see Section 517-42(g).

(3) Additional illumination, receptacles, and equipment shall be permitted to be connected only to the critical branch.

517-44. Sources of Power.

(a)[x] **Two Independent Sources of Power.** Essential electrical systems shall have a minimum of two independent sources of power: a normal source generally supplying the entire electrical system, and one or more alternate sources for use when the normal source is interrupted.

(b)[x] **Alternate Source of Power.** The alternate source of power shall be a generator(s) driven by some form of prime mover(s), and located on the premises.

Exception No. 1: Where the normal source consists of generating units on the premises, the alternate source shall be either another generator set, or an external utility service.

Exception No. 2: Nursing homes or residential custodial care facilities meeting the requirements of Section 517-10, Exception No. 3 shall be permitted to use a battery system or self-contained battery integral with the equipment.

(c) Location of Essential Electrical System Components. Careful consideration shall be given to the location of the spaces housing the components of the essential electrical system to minimize interruptions caused by natural forces common to the area (e.g., storms, floods, earthquakes, or hazards created by adjoining structures or activities). Consideration shall also be given to the possible interruption of normal electrical services resulting from similar causes as well as possible disruption of normal electrical service due to internal wiring and equipment failures.

(FPN): Facilities whose normal source of power is supplied by two or more separate central station-fed services experience greater than normal electrical service reliability than those with only a single feed. Such a dual source of normal power consists of two or more electrical services fed from separate generator sets or a utility distribution network having multiple power input sources and arranged to provide mechanical and electrical separation so that a fault between the facility and the generating sources will not likely cause an interruption of more than one of the facility service feeders.

517-50. Essential Electrical Systems for Clinics, Medical and Dental Offices, Outpatient Facilities, and Other Health Care Facilities Not Covered in Sections 517-30 and 517-40.

(a)[x] Applicability. The requirements of this section shall apply to those health care facilities described in Section 517-50, and in which:

(1) Inhalation anesthetics are administered in any concentration to patients, or

(2) Patients require electrical life-support equipment.

(b)[x] Connections. The essential electrical system shall supply power for:

(1) Task illumination which is related to the safety of life and which is necessary for the safe cessation of procedures in progress.

(2) All anesthesia and resuscitative equipment used in areas where inhalation anesthetics are administered to patients, including alarm and alerting devices.

(FPN): See Health Care Facilities, NFPA 99-1990 (ANSI), Sections 13-3.4.1, 14-3.4.1, and 15-3.4.1.

(3) All electrical life-support equipment in areas where procedures are performed that require such equipment for the support of the patient's life.

(c) Alternate Source of Power.

(1)[x] Power Source. The alternate source of power for the system shall be specifically designed for this purpose and shall be either a generator, battery system, or self-contained battery integral with the equipment.

(2)^x **System Capacity.** The alternate source of power shall be separate and independent of the normal source and shall have a capacity to sustain its connected loads for a minimum of $1\frac{1}{2}$ hours after loss of the normal source.

(3)^x **System Operation.** The system shall be so arranged that, in the event of a failure of the normal power source, the alternate source of power shall be automatically connected to the load within 10 seconds.

(FPN): See Health Care Facilities, NFPA 99-1990 (ANSI), Sections 3-5.1.2.2(c), Transfer Switch Operation for Type III with Generator Sets, and 3-5.1.2.2(d), Transfer Switch Operation for Type III with Battery Systems.

D. Inhalation Anesthetizing Locations

(FPN): For further information regarding safeguards for anesthetizing locations, see Health Care Facilities, NFPA 99-1990 (ANSI).

517-60. Anesthetizing Location Classification.

(FPN): If either of the following anesthetizing locations is designated a wet location, refer to Section 517-20.

(a) Hazardous (Classified) Location.

(1)^x In a location where flammable anesthetics are employed, the entire area shall be considered to be a Class I, Division 1 location which shall extend upward to a level 5 feet (1.52 m) above the floor. The remaining volume up to the structural ceiling is considered to be above a hazardous (classified) location.

(2) Any room or location in which flammable anesthetics or volatile flammable disinfecting agents are stored shall be considered to be a Class I, Division 1 location from floor to ceiling.

(b) Other-than-Hazardous (Classified) Location.
Any inhalation anesthetizing location designated for the exclusive use of nonflammable anesthetizing agents shall be considered to be an other-than-hazardous (classified) location.

517-61. Wiring and Equipment.

(a) Within Hazardous Anesthetizing Locations.

(1)^x Except as permitted in Section 517-160, each power circuit within, or partially within, a flammable anesthetizing location as referred to in Section 517-60 shall be isolated from any distribution system by the use of an isolated power system.

(2) Isolated power system equipment shall be listed for the purpose and the system so designed and installed that it meets the provisions and is in accordance with Part G.

(3)^x In hazardous (classified) location(s) referred to in Section 517-60, all fixed wiring and equipment, and all portable equipment, including lamps and other utilization equipment, operating at more than 10 volts between conductors shall comply with the requirements of Sections 501-1 through 501-15 and Sections 501-16(a) and (b) for Class I, Division 1 locations. All such equipment shall be specifically approved for the hazardous atmospheres involved.

(4) Where a box, fitting, or enclosure is partially, but not entirely, within a hazardous (classified) location(s), the hazardous (classified) location(s) shall be considered to be extended to include the entire box, fitting, or enclosure.

(5) Receptacles and attachment plugs in hazardous (classified) location(s) shall be listed for use in Class I, Group C hazardous (classified) locations and shall have provision for the connection of a grounding conductor.

(6) Flexible cords used in hazardous areas for connection to portable utilization equipment, including lamps operating at more than 8 volts between conductors, shall be of a type approved for extra-hard usage in accordance with Table 400-4 and shall include an additional conductor for grounding.

(7) A storage device for the flexible cord shall be provided and shall not subject the cord to bending at a radius of less than 3 inches (76 mm).

(b) Above Hazardous Anesthetizing Locations.

(1) Wiring above a hazardous area referred to in Section 517-60 shall be installed in rigid metal conduit, electrical metallic tubing, intermediate metal conduit, Type MI cable, or Type MC cable that employs a continuous, gas/vapor-tight metal sheath.

(2) Installed equipment that may produce arcs, sparks, or particles of hot metal, such as lamps and lampholders for fixed lighting, cutouts, switches, generators, motors, or other equipment having make-and-break or sliding contacts, shall be of the totally enclosed type or so constructed as to prevent escape of sparks or hot metal particles.

Exception: Wall-mounted receptacles installed above the hazardous area in flammable anesthetizing locations shall not be required to be totally enclosed or have openings guarded or screened to prevent dispersion of particles.

(3) Surgical and other lighting fixtures shall conform to Section 501-9(b).

Exception No. 1: The surface temperature limitations set forth in Section 501-9(b)(2) shall not apply.

Exception No. 2: Integral or pendant switches that are located above and cannot be lowered into the hazardous (classified) location(s) shall not be required to be explosionproof.

(4) Approved seals shall be provided in conformance with Section 501-5, and Section 501-5(a)(4) shall apply to horizontal as well as to vertical boundaries of the defined hazardous (classified) locations.

(5) Receptacles and attachment plugs located above hazardous anesthetizing locations shall be listed for hospital use for services of prescribed voltage, frequency, rating, and number of conductors with provision for the connection of the grounding conductor. This requirement shall apply to attachment plugs and receptacles of the 2-pole, 3-wire grounding type for single-phase 120-volt, nominal, ac service.

(6) Plugs and receptacles rated 250-volt, for connection of 50-ampere, and 60-ampere ac medical equipment for use above hazardous (classified) locations shall be so arranged that the 60-ampere receptacle will accept either the 50-ampere or the 60-ampere plug. Fifty-ampere receptacles shall be designed so as not to accept the 60-ampere attachment plug. The plugs shall be of the 2-pole, 3-wire design with a third contact connecting to the insulated (green or green with yellow stripe) equipment grounding conductor of the electrical system.

(c) Other-than-Hazardous Anesthetizing Locations.

(1) Wiring serving other-than-hazardous (classified) locations, as defined in Section 517-60, shall be installed in a metal raceway system or cable assembly. The metal raceway system, or cable armor or sheath assembly, shall qualify as an equipment grounding return path in accordance with Section 250-91(b). Type MC and Type MI cable shall have an outer metal armor or sheath that is identified as an acceptable grounding return path.

Exception: Pendant receptacle constructions employing at least Type SJO or equivalent flexible cords suspended not less than 6 feet (1.83 m) from the floor.

(2) Receptacles and attachment plugs installed and used in other-than-hazardous (classified) locations shall be listed for hospital use for services of prescribed voltage, frequency, rating, and number of conductors with provision for connection of the grounding conductor. This requirement shall apply to 2-pole, 3-wire grounding type for single-phase 120-, 208-, or 240-volt, nominal, ac service.

(3) Plugs and receptacles rated 250-volt, for connection of 50-ampere, and 60-ampere ac medical equipment for use in other-than-hazardous (classified) locations shall be so arranged that the 60-ampere receptacle will accept either the 50-ampere or the 60-ampere plug. Fifty-ampere receptacles shall be designed so as not to accept the 60-ampere attachment plug. The plugs shall be of the 2-pole, 3-wire design with a third contact connecting to the insulated (green or green with yellow stripe) equipment grounding conductor of the electrical system.

517-62. Grounding. In any anesthetizing area, all metal raceways and metal-sheathed cables, and all noncurrent-carrying conductive portions of fixed electric equipment, shall be grounded.

Exception: Equipment operating at not more than 10 volts between conductors shall not be required to be grounded.

517-63. Grounded Power Systems in Anesthetizing Locations.

(a) General-Purpose Lighting Circuit. A general-purpose lighting circuit connected to the normal grounded service shall be installed in each operating room.

Exception: Where connected to any alternate source permitted in Section 700-12 that is separate from the source serving the emergency system.

(b) Branch-Circuit Wiring. Branch circuits supplying only listed, fixed, therapeutic and diagnostic equipment, permanently installed above the hazardous (classified) location and in other-than-hazardous (classified) locations, shall be permitted to be supplied from a normal grounded service, single- or three-phase system, provided:

(1) Wiring for grounded and isolated circuits does not occupy the same raceway,

(2) All conductive surfaces of the equipment are grounded,

(3) Equipment (except enclosed X-ray tubes and the leads to the tubes) are located at least 8 feet (2.44 m) above the floor or outside the anesthetizing location, and

(4) Switches for the grounded branch circuit are located outside the hazardous (classified) location.

Exception: Sections 517-63 (b)(3) and (b)(4) shall not apply in other- | *than-hazardous (classified) locations.*

(c) **Fixed Lighting Branch Circuits.** Branch circuits supplying only fixed lighting shall be permitted to be supplied by a normal grounded service, provided:

(1) Such fixtures are located at least 8 feet (2.44 m) above the floor,

(2) All conductive surfaces of fixtures are grounded,

(3) Wiring for circuits supplying power to fixtures does not occupy the same raceway for circuits supplying isolated power, and

(4) Switches are wall-mounted and located above hazardous (classified) locations.

(d) **Remote-Control Stations.** Wall-mounted remote-control stations for remote-control switches operating at 24 volts or less shall be permitted to be installed in any anesthetizing location.

(e) **Location of Isolated Power Systems.** An isolated power center listed for the purpose and its grounded primary feeder shall be permitted to be located in an anesthetizing location, provided it is installed above a hazardous (classified) location or in an other-than-hazardous (classified) location.

(f) **Circuits in Anesthetizing Locations.** Except as permitted above, each power circuit within, or partially within, a flammable anesthetizing location as referred to in Section 517-60 shall be isolated from any distribution system supplying other-than-anesthetizing locations.

517-64.[x] Low-Voltage Equipment and Instruments.

(a) **Equipment Requirements.** Low-voltage equipment that is frequently in contact with the bodies of persons or has exposed current-carrying elements shall:

(1) Operate on an electrical potential of 10 volts or less, or

(2) Be approved as intrinsically safe or double-insulated equipment.

(3) Be moisture-resistant.

(b) **Power Supplies.** Power shall be supplied to low-voltage equipment from:

(1) An individual portable isolating transformer (autotransformers shall not be used) connected to an isolated power circuit receptacle by means of an appropriate cord and attachment plug, or

(2) A common low-voltage isolating transformer installed in a nonhazardous location, or

(3) Individual dry-cell batteries, or

(4) Common batteries made up of storage cells located in a nonhazardous location.

(c) Isolated Circuits. Isolating-type transformers for supplying low-voltage circuits shall:

(1) Have approved means for insulating the secondary circuit from the primary circuit, and

(2) Have the core and case grounded.

(d) Controls. Resistance or impedance devices shall be permitted to control low-voltage equipment but shall not be used to limit the maximum available voltage to the equipment.

(e) Battery-Powered Appliances. Battery-powered appliances shall not be capable of being charged while in operation unless their charging circuitry incorporates an integral isolating-type transformer.

(f) Receptacles or Attachment Plugs. Any receptacle or attachment plug used on low-voltage circuits shall be of a type that does not permit interchangeable connection with circuits of higher voltage.

(FPN): Any interruption of the circuit, even circuits as low as 10 volts, either by any switch, or loose or defective connections anywhere in the circuit, may produce a spark sufficient to ignite flammable anesthetic agents. See Section 7-5.1.2.3 of Health Care Facilities, NFPA 99-1990 (ANSI).

E. X-Ray Installations

Nothing in this part shall be construed as specifying safeguards against the useful beam or stray X-ray radiation.

(FPN No. 1): Radiation safety and performance requirements of several classes of X-ray equipment are regulated under Public Law 90-602 and are enforced by the Department of Health and Human Services.

(FPN No. 2): In addition, information on radiation protection by the National Council on Radiation Protection and Measurements is published as Reports of the National Council on Radiation Protection and Measurement. These reports are obtainable from NCRP Publications, P.O. Box 30175, Washington, D.C. 20014.

517-71. Connection to Supply Circuit.

(a) Fixed and Stationary Equipment. Fixed and stationary X-ray equipment shall be connected to the power supply by means of a wiring method meeting the general requirements of this Code.

Exception: Equipment properly supplied by a branch circuit rated at not over 30 amperes shall be permitted to be supplied through a suitable attachment plug and hard-service cable or cord.

(b) Portable, Mobile, and Transportable Equipment. Individual branch circuits shall not be required for portable, mobile, and transportable medical X-ray equipment requiring a capacity of not over 60 amperes.

(c) Over 600-Volt Supply. Circuits and equipment operated on a supply circuit of over 600 volts shall comply with Article 710.

517-72. Disconnecting Means.

(a) Capacity. A disconnecting means of adequate capacity for at least 50 percent of the input required for the momentary rating or 100 percent of the input required for the long-time rating of the X-ray equipment, whichever is greater, shall be provided in the supply circuit.

(b) Location. The disconnecting means shall be operable from a location readily accessible from the X-ray control.

(c) Portable Equipment. For equipment connected to a 120-volt branch circuit of 30 amperes or less, a grounding-type attachment plug and receptacle of proper rating shall be permitted to serve as a disconnecting means.

517-73. Rating of Supply Conductors and Overcurrent Protection.

(a) Diagnostic Equipment.

(1) The ampacity of supply branch-circuit conductors and the current rating of overcurrent protective devices shall not be less than 50 percent of the momentary rating or 100 percent of the long-time rating, whichever is greater.

(2) The ampacity of supply feeders and the current rating of overcurrent protective devices supplying two or more branch circuits supplying X-ray units shall not be less than 50 percent of the momentary demand rating of the largest unit plus 25 percent of the momentary demand rating of the next largest unit plus 10 percent of the momentary demand rating of each additional unit. Where simultaneous biplane examinations are undertaken with the X-ray units, the supply conductors and overcurrent protective devices shall be 100 percent of the momentary demand rating of each X-ray unit.

(FPN):   The minimum conductor size for branch and feeder circuits is also governed by voltage regulation requirements. For a specific installation, the manufacturer usually specifies minimum distribution transformer and conductor sizes, rating of disconnecting means, and overcurrent protection.

(b) Therapeutic Equipment. The ampacity of conductors and rating of overcurrent protective devices shall not be less than 100 percent of the current rating of medical X-ray therapy equipment.

(FPN):   The ampacity of the branch-circuit conductors and the ratings of disconnecting means and overcurrent protection for X-ray equipment are usually designated by the manufacturer for the specific installation.

517-74. Control Circuit Conductors.

(a) Number of Conductors in Raceway. The number of control circuit conductors installed in a raceway shall be determined in accordance with Section 300-17.

(b) Minimum Size of Conductors. Size No. 18 or No. 16 fixture wires as specified in Section 725-16 and flexible cords shall be permitted for the control and operating circuits of X-ray and auxiliary equipment where protected by not larger than 20-ampere overcurrent devices.

517-75. Equipment Installations. All equipment for new X-ray installations and all used or reconditioned X-ray equipment moved to and reinstalled at a new location shall be of an approved type.

517-76. Transformers and Capacitors. Transformers and capacitors that are part of an X-ray equipment shall not be required to comply with Articles 450 and 460.

Capacitors shall be mounted within enclosures of insulating material or grounded metal.

517-77. Installation of High Tension X-Ray Cables. Cables with grounded shields connecting X-ray tubes and image intensifiers shall be permitted to be installed in cable trays or cable troughs along with X-ray equipment control and power supply conductors without the need for barriers to separate the wiring.

517-78. Guarding and Grounding.

(a) High-Voltage Parts. All high-voltage parts, including X-ray tubes, shall be mounted within grounded enclosures. Air, oil, gas, or other suitable insulating media shall be used to insulate the high-voltage from the grounded enclosure. The connection from the high-voltage equipment to X-ray tubes and other high-voltage components shall be made with high-voltage shielded cables.

(b) Low-Voltage Cables. Low-voltage cables connecting to oil-filled units that are not completely sealed, such as transformers, condensers, oil coolers, and high-voltage switches, shall have insulation of the oil-resistant type.

(c) Noncurrent-Carrying Metal Parts. Noncurrent-carrying metal parts of X-ray and associated equipment (controls, tables, X-ray tube supports, transformer tanks, shielded cables, X-ray tube heads, etc.) shall be grounded in the manner specified in Article 250, as modified by Sections 517-13(a) and (b).

Exception: Battery-operated equipment.

F. Communications, Signaling Systems, Data Systems, Fire Protective Signaling Systems, and Systems Less than 120 Volts, Nominal

517-80. Patient Care Areas. Equivalent insulation and isolation to that required for the electrical distribution systems in patient care areas shall be provided for communications, signaling systems, data system circuits, fire protective signaling systems, and systems less than 120 volts, nominal.

(FPN): An acceptable alternate means of providing isolation for patient/nurse call systems is by the use of nonelectrified signaling, communication, or control devices held by the patient or within reach of the patient.

517-81. Other-than-Patient-Care Areas. In other than patient care areas, installations shall be in accordance with the appropriate provisions of Articles 725, 760, and 800.

517-82. Signal Transmission Between Appliances.

(a) General. Permanently installed signal cabling from an appliance in a patient location to remote appliances shall employ a signal transmission system that prevents hazardous grounding interconnection of the appliances.

(FPN):   See Sections 517-13(b) and 517-15.

(b) Common Signal Grounding Wire. Common signal grounding wires (i.e., the chassis ground for single-ended transmission) shall be permitted to be used between appliances all located within the patient vicinity, provided the appliances are served from the same reference grounding point.

G. Isolated Power Systems

517-160. Isolated Power Systems.

(a) Installations.

(1) Each isolated power circuit shall be controlled by a switch having a disconnecting pole in each isolated circuit conductor to simultaneously disconnect all power. Such isolation shall be accomplished by means of one or more transformers having no electrical connection between primary and secondary windings, by means of motor generator sets, or by means of suitably isolated batteries.

(2)[x] Circuits supplying primaries of isolating transformers shall operate at not more than 600 volts between conductors and shall be provided with proper overcurrent protection. The secondary voltage of such transformers shall not exceed 600 volts between conductors of each circuit. All circuits supplied from such secondaries shall be ungrounded, and shall have an approved overcurrent device of proper ratings in each conductor. Circuits supplied directly from batteries or from motor generator sets shall be ungrounded, and shall be protected against overcurrent in the same manner as transformer-fed secondary circuits. If an electrostatic shield is present, it shall be connected to the reference grounding point.

(3) The isolating transformers, motor generator sets, batteries and battery chargers, and associated primary or secondary overcurrent devices shall not be installed in hazardous (classified) locations. The isolated secondary circuit wiring extending into a hazardous anesthetizing location shall be installed in accordance with Section 501-4.

(4) An isolated branch circuit supplying an anesthetizing location shall supply no other location.

(5) The isolated circuit conductors shall be identified as follows:

Isolated Conductor No. 1 — Orange
Isolated Conductor No. 2 — Brown

For 3-phase systems, the third conductor shall be identified as yellow.

(6) Wire-pulling compounds that increase the dielectric constant shall not be used on the secondary conductors of the isolated power supply.

(FPN No. 1):   It is desirable to limit the size of the isolation transformer to 10 kVA or less and to use conductor insulation with low leakage to meet impedance requirements.

(FPN No. 2):   Minimizing the length of branch-circuit conductors and utilizing conductor insulations with a dielectric constant less than 3.5 and insulation resistance constant greater than 6100 megohm-meters (20,000 megohm-feet) at 60°F (16°C) reduces leakage from line to ground reducing the monitor hazard current.

(b)[x] Line Isolation Monitor.

(1) In addition to the usual control and overcurrent protective devices, each isolated power system shall be provided with a continually operating line isolation monitor that indicates possible leakage or fault currents from either isolated conductor to ground. The monitor shall be designed so that a green signal lamp, conspicuously visible to persons in the anesthetizing location, remains lighted when the system is adequately isolated from ground; an adjacent red signal lamp and an audible warning signal (remote if desired) shall be energized when the total hazard current (consisting of possible resistive and capacitive leakage currents) from either isolated conductor to ground reaches a threshold value of 5 milliamperes under nominal line voltage conditions. The line monitor shall not alarm for a fault hazard of less than 3.7 milliamperes or for a total hazard current of less than 5.0 milliamperes.

Exception: A system shall be permitted to be designed to operate at a lower threshold value of total hazard current. A line isolation monitor for such a system shall be permitted to be approved with the provision that the fault hazard current shall be permitted to be reduced but not to less than 35 percent of the corresponding threshold value of the total hazard current, and the monitor hazard current is to be correspondingly reduced to no more than 50 percent of the alarm threshold value of the total hazard current.

(2) The line isolation monitor shall be designed to have sufficient internal impedance such that, when properly connected to the isolated system, the maximum internal current that can flow through the line isolation monitor, when any point of the isolated system is grounded, shall be 1 milliampere.

Exception: The line isolation monitor shall be permitted to be of the low-impedance type such that the current through the line isolation monitor, when any point of the isolated system is grounded, will not exceed twice the alarm threshold value for a period not exceeding 5 milliseconds.

(FPN): Reduction of the monitor hazard current, provided this reduction results in an increased "not alarm" threshold value for the fault hazard current, will increase circuit capacity.

(3) An ammeter calibrated in the total hazard current of the system (contribution of the fault hazard current plus monitor hazard current) shall be mounted in a plainly visible place on the line isolation monitor with the "alarm on" zone at approximately the center of the scale.

Exception: The line isolation monitor may be a composite unit, with a sensing section cabled to a separate display panel section on which the alarm or test functions are located.

(FPN): It is desirable to locate the ammeter so that it is conspicuously visible to persons in the anesthetizing location.

ARTICLE 518 — PLACES OF ASSEMBLY

518-1. Scope. This article covers all buildings or portions of buildings or structures designed or intended for the assembly of 100 or more persons.

518-2. General Classifications. Places of assembly shall include, but are not limited to:

Assembly Halls
Armories
Restaurants
Dance Halls
Museums
Gymnasiums
Bowling Lanes
Club Rooms
Court Rooms
Conference Rooms
Auditoriums within:
    Schools
    Mercantile Establishments
    Business Establishments
    Other Occupancies.

Exhibition Halls
Dining Facilities
Church Chapels
Mortuary Chapels
Skating Rinks
Multipurpose Rooms
Pool Rooms
Places of Awaiting Transportation
Auditoriums

Occupancy of any room or space for assembly purposes by less than 100 persons in a building of other occupancy, and incidental to such other occupancy, shall be classed as part of the other occupancy and subject to the provisions applicable thereto.

Where any such building structure, or portion thereof, contains a projection booth or stage platform or area for the presentation of theatrical or musical productions, either fixed or portable, the wiring for that area and all equipment that is used in the referenced area, and portable equipment and wiring for use in the production that will not be connected to permanently installed wiring, shall comply with Article 520.

(FPN): For methods of determining population capacity, see local building code or in its absence Life Safety Code, NFPA 101-1991 (ANSI).

518-3. Other Articles.

(a) **Hazardous (Classified) Areas.** Hazardous (classified) areas located in any assemblage occupancy shall be installed in accordance with Article 500, Hazardous (Classified) Locations.

(b) **Temporary Wiring.** In exhibition halls used for display booths, as in trade shows, the temporary wiring shall be installed in accordance with Article 305, Temporary Wiring, except that approved flexible cables and cords shall be permitted to be laid on floors where protected from contact by the general public.

(c) **Emergency Systems.** Control of emergency systems shall comply with Article 700, Emergency Systems.

518-4. Wiring Methods.
The fixed wiring methods shall be metal raceways, nonmetallic raceways encased in not less than 2 inches (50.8 mm) of concrete, Type MI cable, or Type MC cable.

Exception No. 1: Nonmetallic-sheathed cable, type AC cable, electrical nonmetallic tubing, and rigid nonmetallic conduit shall be permitted to be installed in those buildings or portions thereof that are not required to be fire-rated construction by the applicable building code.

Exception No. 2: As provided in Article 640, Sound-Recording and Similar Equipment, in Article 800, Communication Circuits, and in Article 725 for Class 2 and Class 3 remote-control and signaling circuits, and in Article 760 for fire protective signaling circuits.

(FPN): Fire-rated construction is the fire-resistive classification used in building codes.

518-5. Supply. Portable switchboards and portable power distribution equipment shall be supplied only from listed power outlets of sufficient voltage and ampere rating. Such power outlets shall be protected by overcurrent devices. Such overcurrent devices and power outlets shall not be accessible to the general public. Provisions for connection of an equipment grounding conductor shall be provided. The neutral of feeders supplying solid-state, 3-phase, 4-wire dimmer systems shall be considered a current-carrying conductor.

ARTICLE 520 — THEATERS, AUDIENCE AREAS OF MOTION PICTURE AND TELEVISION STUDIOS, AND SIMILAR LOCATIONS

A. General

520-1. Scope. This article covers all buildings or that part of a building or structure designed or used for presentation, dramatic, musical, motion picture projection, or similar purposes and to specific audience seating areas within motion picture or television studios.

520-2. Motion Picture Projectors. Motion picture equipment and its installation and use shall comply with Article 540.

520-3. Sound Reproduction. Sound-reproducing equipment and its installation shall comply with Article 640.

520-4. Wiring Methods. The fixed wiring method shall be metal raceways, nonmetallic raceways encased in at least 2 inches (50.8 mm) of concrete, Type MI cable, or Type MC cable.

Exception No. 1: The wiring for portable switchboards, stage set lighting, stage effects, and other wiring not fixed as to location shall be permitted with approved flexible cords and cables as provided elsewhere in Article 520. Fastening such cables and cords by uninsulated staples or nailing shall not be permitted.

Exception No. 2: As provided in Article 640 for sound recording, in Article 800 for communication circuits, in Article 725 for Class 2 and Class 3 remote-control and signaling circuits, and in Article 760 for fire protective signaling circuits.

Exception No. 3: Nonmetallic-sheathed cable, Type AC cable, electrical nonmetallic tubing, and rigid nonmetallic conduit shall be permitted to be installed in those buildings or portions thereof that are not required to be of fire-rated construction by the applicable building code.

520-5. Number of Conductors in Raceway. The number of conductors permitted in any metal conduit, rigid nonmetallic conduit as permitted in this article, or electrical metallic tubing for border or stage pocket circuits or for remote-control conductors shall not exceed the percentage fill shown in Table 1 of Chapter 9. Where contained within an auxiliary gutter or a wireway, the sum of the cross-sectional areas of all contained conductors at any cross section shall not exceed 20 percent of the interior cross-sectional area of the auxiliary gutter or wireway. The thirty-conductor limitation of Sections 362-5 and 374-5 shall not apply.

520-6. Enclosing and Guarding Live Parts. Live parts shall be enclosed or guarded to prevent accidental contact by persons and objects. All switches shall be of the externally operable type. Dimmers, including rheostats, shall be placed in cases or cabinets that enclose all live parts.

520-7. Emergency Systems. Control of emergency systems shall comply with Article 700, Emergency Systems.

520-8. Branch Circuits. A branch circuit of any size supplying one or more receptacles shall be permitted to supply stage set lighting. The voltage rating of the receptacles shall not be less than the circuit voltage. Receptacle ampere ratings and branch-circuit conductor ampacity shall not be less than the branch-circuit overcurrent device ampere rating. Table 210-21(b)(2) shall not apply.

B. Fixed Stage Switchboard

520-21. Dead Front. Stage switchboards shall be of the dead-front type and shall comply with Part C of Article 384 unless approved based on suitability as a stage switchboard as determined by a qualified testing laboratory and recognized test standards and principles.

520-22. Guarding Back of Switchboard. Stage switchboards having exposed live parts on the back of such boards shall be enclosed by the building walls, wire mesh grills, or by other approved methods. The entrance to this enclosure shall be by means of a self-closing door.

520-23. Control and Overcurrent Protection of Receptacle Circuits. Means shall be provided at a stage lighting switchboard to which load circuits are connected for overcurrent protection of stage lighting branch circuits, including branch circuits supplying stage and auditorium receptacles used for cord- and plug-connected stage equipment. Where the stage switchboard contains dimmers to control nonstage lighting, the locating of the overcurrent protective devices for these branch circuits at the stage switchboard shall be permitted.

520-24. Metal Hood. A stage switchboard that is not completely enclosed dead-front and dead-rear or recessed into a wall shall be provided with a metal hood extending the full length of the board to protect all equipment on the board from falling objects.

520-25. Dimmers. Dimmers shall comply with (a) through (d) below.

  (a) Disconnection and Overcurrent Protection. Where dimmers are installed in ungrounded conductors, each dimmer shall have overcurrent protection not greater than 125 percent of the dimmer rating, and shall be disconnected from all ungrounded conductors when the master or individual switch or circuit breaker supplying such dimmer is in the open position.

(b) Resistance- or Reactor-type Dimmers. Resistance- or series reactor-type dimmers shall be permitted to be placed in either the grounded or the ungrounded conductor of the circuit. Where designed to open either the supply circuit to the dimmer or the circuit controlled by it, the dimmer shall then comply with Section 380-1. Resistance- or reactor-type dimmers placed in the grounded neutral conductor of the circuit shall not open the circuit.

(c) Autotransformer-type Dimmers. The circuit supplying an autotransformer-type dimmer shall not exceed 150 volts between conductors. The grounded conductor shall be common to the input and output circuits.

(d) Solid-State-type Dimmers. The circuit supplying a solid-state dimmer shall not exceed 150 volts between conductors unless the dimmer is specifically approved for higher voltage operation. Where a grounded conductor supplies a dimmer, it shall be common to the input and output circuits. Dimmer chassis shall be connected to the equipment grounding conductor.

(FPN): See Section 210-9 for circuits derived from autotransformers.

520-26. Type of Switchboard. Stage switchboard shall be either one or a combination of the following types:

(a) Manual. Dimmers and switches are operated by handles mechanically linked to the control devices.

(b) Remotely Controlled. Devices are operated electrically from a pilot-type control console or panel. Pilot control panels shall either be part of the switchboard or shall be permitted to be at another location.

(c) Intermediate. A stage switchboard with circuit interconnections is a secondary switchboard (patch panel) or panelboard remote to the primary stage switchboard. It shall contain overcurrent protection. Where the required branch circuit overcurrent protection is provided in the dimmer panel, it shall be permitted to be omitted from the intermediate switchboard.

520-27. Stage Switchboard Feeders. Feeders supplying stage switchboards shall be one of the following:

(a) Single Feeder. A single feeder disconnected by a single disconnect device. The neutral of feeders supplying solid-state, 3-phase, 4-wire dimming systems shall be considered a current-carrying conductor.

(b) Multiple Feeders to Intermediate Stage Switchboard (Patch Panel). Multiple feeders of unlimited quantity shall be permitted, provided that all multiple feeders are part of a single system. Where combined, neutral conductors in a given raceway shall be of sufficient ampacity to carry the maximum unbalanced current supplied by multiple feeder conductors in the same raceway, but need not be greater than the ampacity of the neutral supplying the primary stage switchboard. Parallel neutral conductors shall comply with Section 310-4. The neutral of feeders supplying solid-state, 3-phase, 4-wire dimming systems shall be considered a current-carrying conductor.

(c) Separate Feeders to Single Primary Stage Switchboard (Dimmer Bank). Installations with separate feeders to a single primary stage switchboard shall have a disconnecting means for each feeder. The primary stage switchboard shall have a permanent and obvious label stating the number and location of disconnecting means. If the disconnecting means are located in more than one distribution switchboard, the primary stage switchboard shall be provided with barriers to correspond with these multiple locations. The neutral of feeders supplying solid-state, 3-phase, 4-wire dimming systems shall be considered a current-carrying conductor.

For purposes of computing supply capacity to switchboards, it shall be permissible to consider the maximum load that the switchboard is intended to control in a given installation, provided that all feeders supplying the switchboard shall be protected by an overcurrent device with a rating not greater than the ampacity of the feeder, and that the opening of the overcurrent device does not have any effect on the egress or emergency lighting systems.

(FPN):   For computation of stage switchboard feeder loads, see Section 220-10.

C. Stage Equipment—Fixed

520-41. Circuit Loads.   Footlights, border lights, and proscenium side lights shall be so arranged that no branch circuit supplying such equipment will carry a load exceeding 20 amperes.

Exception: Where heavy-duty lampholders only are used, such circuits shall be permitted to comply with Article 210 for circuits supplying heavy-duty lampholders.

520-42. Conductor Insulation.   Foot, border, proscenium, or portable strip light fixtures and connector strips shall be wired with conductors having insulation suitable for the temperature at which the conductors will be operated, but not less than 125°C (257°F). The ampacity of the 125°C (257°F) conductors shall be that of 60°C (140°F) conductors. All drops from connector strips shall be 90°C (194°F) wire with no more than 6 inches (152 mm) of conductor extending into the connector strip. Article 310, Note 8(a), Notes to Ampacity Tables of 0 to 2000 Volts, shall not apply.

(FPN):   See Table 310-13 for conductor types.

520-43. Footlights.

(a) Metal Trough Construction.   Where metal trough construction is employed for footlights, the trough containing the circuit conductors shall be made of sheet metal not lighter than No. 20 MSG treated to prevent oxidation. Lampholder terminals shall be kept at least 1/2 inch (12.7 mm) from the metal of the trough. The circuit conductors shall be soldered to the lampholder terminals.

(b) Other-than-Metal-Trough Construction.   Where the metal trough construction specified in Section 520-43(a) is not used, footlights shall consist of individual outlets with lampholders wired with rigid metal conduit, intermediate metal conduit, or flexible metal conduit, Type MC cable, or mineral-insulated, metal-sheathed cable. The circuit conductors shall be soldered to the lampholder terminals.

(c) Disappearing Footlights. Disappearing footlights shall be so arranged that the current supply will be automatically disconnected when the footlights are replaced in the storage recesses designed for them.

520-44. Borders and Proscenium Sidelights.

(a) General. Borders and proscenium sidelights shall be (1) constructed as specified in Section 520-43; (2) suitably stayed and supported; and (3) so designed that the flanges of the reflectors or other adequate guards will protect the lamps from mechanical damage and from accidental contact with scenery or other combustible material.

(b) Cables for Border Lights. Cables for supply to border lights shall be listed for extra-hard usage. The cables shall be suitably supported. Such cables shall be employed only where flexible conductors are necessary. Ampacity of the conductors shall be as provided in Section 400-5.

Exception: Listed multiconductor extra-hard usage type cords not in direct contact with equipment containing heat-producing elements shall be permitted to have their ampacity determined by Table 520-44. Maximum load current in any conductor shall not exceed the values in Table 520-44.

Table 520-44.   Allowable Ampacity of Listed Extra-Hard Usage Cords with Temperature Ratings of 75°C (167°F) and 90°C (194°F)*
[Based on Ambient Temperature of 30°C (86°F)]

| Size | Temperature Rating of Cord | | Maximum |
AWG	75° (167°F)	90°C (194°F)	Rating of Over-current Device
14	24	28	15
12	32	35	20
10	41	47	25
8	57	65	35
6	77	87	45
4	101	114	60
2	133	152	80

*Ampacity shown is the allowable ampacity for multiconductor cords where only three copper conductors are current-carrying. If the number of current-carrying conductors in a cord exceeds three and the load diversity factor is a minimum of 50 percent, the allowable ampacity of each conductor shall be reduced as shown in the following table.

Number of Conductors	Percent of Allowable Ampacity
4 thru 6	80
7 thru 24	70
25 thru 42	60
43 and above	50

Note: Ultimate Insulation Temperature. In no case shall conductors be associated together in such a way with respect to the kind of circuit, the wiring method used, or the number of conductors such that the temperature limit of the conductors will be exceeded.

520-45. Receptacles. Receptacles for electrical equipment or fixtures on stages shall be rated in amperes.

Conductors supplying receptacles shall be in accordance with Articles 310 and 400.

520-46. Connector Strips, Drop Boxes, and Stage Pockets. Receptacles for the connection of portable stage lighting equipment shall be pendant or mounted in suitable pockets or enclosures and comply with Section 520-45. Supply cables for connector strips and drop boxes shall be as specified in Section 520-44(b).

520-47. Lamps in Scene Docks. Lamps installed in scene docks shall be so located and guarded as to be free from physical damage and shall provide an air space of not less than 2 inches (50.8 mm) between such lamps and any combustible material.

520-48. Curtain Machines. Curtain machines shall be listed.

520-49. Flue Damper Control. Where stage flue dampers are released by an electrical device, the circuit operating the device shall be normally closed and shall be controlled by at least two externally operable switches, one switch being placed on stage at either the electrician's or stage manager's station and the other where designated by the authority having jurisdiction. The device shall be designed for the full voltage of the circuit to which it is connected, no resistance being inserted. The device shall be located in the loft above the scenery and shall be enclosed in a suitable metal box having a tight, self-closing door.

D. Portable Switchboards on Stage

520-50. Road-Show Connection Panel (A Type of Patch Panel). A panel designed to allow for road-show connection of portable stage switchboards to fixed lighting outlets by means of permanently installed supplementary circuits. The panel, supplementary circuits, and outlets shall comply with (a) through (d) below.

(a) Load Circuits. Circuits shall terminate in grounded polarized inlets of current and voltage rating that match the fixed load receptacle.

(b) Circuit Transfer. Circuits that are transferred between fixed and portable switchboards shall have both the line and neutral transferred simultaneously.

(c) Overcurrent Protection. The supply devices of these supplementary circuits shall be protected by branch-circuit overcurrent protective devices. The individual supplementary circuit, within the road-show connection panel and theater, shall be protected by branch-circuit overcurrent protective devices of suitable ampacity installed within the road-show connection panel.

(d) Enclosure. Panel construction shall be in accordance with Article 384.

520-51. Supply. Portable switchboards shall be supplied only from power outlets of sufficient voltage and ampere rating. Such power outlets shall include only externally operable, enclosed fused switches or circuit breakers mounted on stage or at the permanent switchboard in locations readily accessible from the stage floor. Provisions for connection of an equipment grounding conductor shall be provided. The neutral of feeders supplying solid-state, 3-phase, 4-wire dimmer systems shall be considered a current-carrying conductor.

520-52. Overcurrent Protection. Circuits from portable switchboards directly supplying equipment containing incandescent lamps of not over 300 watts shall be protected by overcurrent protective devices having a rating or setting of not over 20 amperes. Circuits for lampholders over 300 watts shall be permitted where overcurrent protection complies with Article 210.

520-53. Construction and Feeders. Portable switchboards and feeders for use on stages shall comply with (a) through (p) below.

(a) Enclosure. Portable switchboards shall be placed within an enclosure of substantial construction, which shall be permitted to be so arranged that the enclosure is open during operation. Enclosures of wood shall be completely lined with sheet metal of not less than No. 24 MSG and shall be well-galvanized, enameled, or otherwise properly coated to prevent corrosion or be of a corrosion-resistant material.

(b) Energized Parts. There shall not be exposed energized parts within the enclosure.

Exception: For dimmer faceplates as provided in (e) below.

(c) Switches and Circuit Breakers. All switches and circuit breakers shall be of the externally operable, enclosed type.

(d) Circuit Protection. Overcurrent devices shall be provided in each ungrounded conductor of every circuit supplied through the switchboard. Enclosures shall be provided for all overcurrent devices in addition to the switchboard enclosure.

(e) Dimmers. The terminals of dimmers shall be provided with enclosures, and dimmer faceplates shall be so arranged that accidental contact cannot be readily made with the faceplate contacts.

(f) Interior Conductors. All conductors other than busbars within the switchboard enclosure shall be stranded. Conductors shall be approved for an operating temperature at least equal to the approved operating temperature of the dimming devices used in the switchboard and in no case less than the following: (1) resistance-type dimmers, 200°C (392°F); or (2) reactor-type, autotransformer, and solid-state dimmers, 125°C (257°F). All control wiring shall comply with Article 725.

Each conductor shall have an ampacity not less than the rating of the circuit breaker, switch, or fuse that it supplies. Circuit interrupting and bus bracing shall be in accordance with Sections 110-9 and 110-10. Switchboards with inadequate short-circuit withstand rating shall be protected on the line side by current-limiting devices. The short-circuit withstand rating shall be marked on the switchboard.

Exception: Conductors for pilot light circuits having overcurrent protection of not over 20 amperes.

Conductors shall be enclosed in metal wireways or be securely fastened in position and shall be bushed where they pass through metal.

(g) Pilot Light. A pilot light shall be provided within the enclosure and shall be so connected to the circuit supplying the board that the opening of the master switch will not cut off the supply to the lamp. This lamp shall be on an individual branch circuit having overcurrent protection rated or set at not over 15 amperes.

(h) Supply Conductors. The supply to a portable switchboard shall be by means of listed extra-hard usage cords or cables. The supply cords or cable shall terminate within the switchboard enclosure, in an externally operable fused master switch or circuit breaker, or in a connector assembly identified for the purpose. The supply cords or cable (and connector assembly) shall have sufficient ampacity to carry the total load connected to the switchboard and shall be protected by overcurrent devices.

Single-conductor portable supply cable sets shall not be smaller than No. 2 conductors. The equipment grounding conductor shall not be smaller than No. 6 conductor. The single-conductor cables for a supply shall be of the same length, type, size, and be grouped together, but not bundled. The equipment grounding conductor shall be permitted to be of a different type, provided it meets the other requirements of this section, and it shall be permitted to be reduced in size as permitted by Section 250-95. Grounded (neutral) and equipment grounding conductors shall be identified in accordance with Sections 200-6, 250-57(b), and 310-12. Grounded conductors shall be permitted to be identified by marking at least the first 6 inches (152.4 mm) from both ends of each length of conductor with white or natural gray. Equipment grounding conductors shall be permitted to be identified by marking at least the first 6 inches (152.4 mm) from both ends of each length of conductor with green or green with yellow stripes. Where more than one nominal voltage exists within the same premises, each ungrounded system conductor shall be identified by system.

Exception No. 1: Supply Conductors Not Over 10 Feet (3.05 m) Long. In cases where supply conductors do not exceed 10 feet (3.05 m) in length between supply and switchboard or supply and a subsequent overcurrent device, the ampacity of the supply conductors shall be at least one-quarter of the ampacity of the supply overcurrent protection device where all of the following conditions are met:

a. The supply conductors shall terminate in a single overcurrent protection device that will limit the load to the ampacity of the supply conductors. This single overcurrent device shall be permitted to supply additional overcurrent devices on its load side.

b. The supply conductors shall not penetrate walls, floors, or ceilings or be run through doors or traffic areas. The supply conductors shall be adequately protected from physical damage.

c. The supply conductors shall be suitably terminated in an approved manner.

d. Conductors shall be continuous without splices or connectors.

e. Conductors shall not be bundled.

f. Conductors shall be supported above the floor in an approved manner.

Exception No. 2: Supply Conductors Not Over 20 Feet (6.1 m) Long. In cases where supply conductors do not exceed 20 feet (6.1 m) in length between supply and switchboard or supply and a subsequent overcurrent protection device, the ampacity of the supply conductors shall be at least one-half the rating of the supply overcurrent protection device where all of the following conditions are met:

a. The supply conductors shall terminate in a single overcurrent protection device that will limit the load to the ampacity of the supply conductors. This single overcurrent device shall be permitted to supply additional overcurrent devices on its load side.

b. The supply conductors shall not penetrate walls, floors, or ceilings or be run through doors or traffic areas. The supply conductors shall be adequately protected from physical damage.

c. The supply conductors shall be suitably terminated in an approved manner.

d. The supply conductors shall be supported in an approved manner at least 7 feet (2.13 m) above the floor except at terminations.

e. The supply conductors shall not be bundled.

f. Tap conductors shall be in unbroken lengths.

(i) Cable Arrangement. Cables shall be protected by bushings where they pass through enclosures and shall be so arranged that tension on the cable will not be transmitted to the connections. Where power conductors pass through metal, the requirements of Section 300-20 shall apply.

(j) Number of Supply Interconnections. Where connectors are used in a supply conductor, there shall be a maximum number of three interconnections (mated connector pairs) where the total length from supply to switchboard does not exceed 100 feet (30.5 m). In cases where the total length from supply to switchboard exceeds 100 feet (30.5 m), one additional interconnection shall be permitted for each additional 100 feet (30.5 m) of supply conductor.

(k) Single-Pole Separable Connectors. Where single-pole portable cable connectors are used, they shall be listed and of the locking type. The use of such connectors shall comply with at least one of the following conditions:

(1) Connection and disconnection of connectors are only possible where the supply connectors are interlocked to the source and it is not possible to connect or disconnect connectors when the supply is energized.

(2) Line connectors are of the listed sequential-interlocking type so that load connectors shall be connected in the following sequence:

a. Equipment grounding conductor connection.

b. Grounded circuit conductor connection, if provided.

c. Ungrounded conductor connection,

and that disconnection shall be in the reverse order.

(3) A caution notice shall be provided adjacent to the line connectors indicating that plug connection shall be in the following order:

  a. Equipment grounding conductor connectors.

  b. Grounded circuit conductor connectors, if provided.

  c. Ungrounded conductor connectors,

and that disconnection shall be in the reverse order.

(l) Protection of Supply Conductors and Connectors. All supply conductors and connectors shall be protected against physical damage by an approved means. This protection shall not be required to be raceways.

(m) Flanged Surface Inlets. Flanged surface inlets (recessed plugs) that are used to accept the power shall be rated in amperes.

(n) Terminals. Terminals to which stage cables are connected shall be so located as to permit convenient access to the terminals.

(o) Supply Neutral Terminal. In portable switchboard equipment designed for use with 3-phase, 4-wire with ground supply, the supply neutral terminal, its associated busbar, or equivalent wiring, or both, shall have an ampacity equal to twice the ampacity of the largest ungrounded supply terminal. The power supply lines for portable switchboards shall be sized considering the neutral as a current-carrying conductor. Where single-conductor feeder cables, not installed in raceways, are used on multiphase circuits, the grounded neutral conductor shall have an ampacity of 130 percent of the ungrounded circuit conductors feeding the portable switchboard.

Exception: Where portable switchboard equipment is specifically constructed and identified to be internally converted in the field, in an approved manner, from use with a balanced 3-phase, 4-wire with ground supply to a balanced single-phase, 3-wire with ground supply, the supply neutral terminal and its associated busbar, equivalent wiring, or both, shall have an ampacity equal to that of the largest ungrounded single-phase supply terminal.

(p) Qualified Personnel. The routing of portable supply conductors, the making and breaking of supply connectors and other supply connections, and the energization and deenergization of supply services shall be performed by qualified personnel, and portable switchboards shall be so marked, indicating this requirement in a permanent and conspicuous manner.

Exception: Connection of a portable switchboard to a permanently installed supply receptacle outlet, that supply receptacle outlet being protected for its rated ampacity by an overcurrent device of not greater than 150 amps, and where the receptacle outlet, interconnection, and switchboard further:

  a. Employ listed multipole connectors suitable for the purpose for every supply interconnection, and

  b. Prevent access to all supply connections by the general public, and

  c. Employ listed extra-hard usage multiconductor cords or cables with an ampacity suitable for the type of load and not less than the ampere rating of the connectors.

E. Stage Equipment — Portable

520-61. Arc Lamp Fixtures. Arc lamp fixtures, including enclosed arc lamp fixtures and associated ballasts, shall be listed. Interconnecting cord sets and interconnecting cords and cables shall be extra-hard usage type and listed.

520-62. Portable Plugging Boxes. Portable plugging boxes shall comply with (a) through (e) below.

(a) Enclosure. The construction shall be such that no current-carrying part will be exposed.

(b) Receptacles and Overcurrent Protection. Receptacles shall comply with Section 520-45 and shall have branch-circuit overcurrent protection in the box. Fuses and circuit breakers shall be protected against physical damage. Cords or cables supplying pendant receptacles shall be listed for extra-hard usage.

(c) Busbars and Terminals. Busbars shall have an ampacity equal to the sum of the ampere ratings of all the circuits connected to the busbar. Lugs shall be provided for the connection of the master cable.

(d) Flanged Surface Inlets. Flanged surface inlets (recessed plugs) that are used to accept the power shall be rated in amperes.

(e) Cable Arrangement. Cables shall be adequately protected where they pass through enclosures and be so arranged that tension on the cable will not be transmitted to the terminations.

520-63. Bracket Fixture Wiring.

(a) Bracket Wiring. Brackets for use on scenery shall be wired internally, and the fixture stem shall be carried through to the back of the scenery where a bushing shall be placed on the end of the stem.

Exception: Externally wired brackets or other fixtures shall be permitted where wired with cords designed for hard usage that extend through scenery and without joint or splice in canopy of fixture back and terminate in an approved-type stage connector located, where practical, within 18 inches (457 mm) of the fixture.

(b) Mounting. Fixtures shall be securely fastened in place.

520-64. Portable Strips. Portable strips shall be constructed in accordance with the requirements for border lights and proscenium side lights in Section 520-44(a). The supply cable shall be protected by bushings where it passes through metal and shall be so arranged that tension on the cable will not be transmitted to the connections.

(FPN No. 1): See Section 520-42 for wiring of portable strips.

(FPN No. 2): See Section 520-68(a), Exception No. 2 for insulation types required on single conductors.

520-65. Festoons. Joints in festoon wiring shall be staggered. Lamps enclosed in lanterns or similar devices of combustible material shall be equipped with guards.

520-66. Special Effects. Electrical devices used for simulating lightning, waterfalls, and the like shall be so constructed and located that flames, sparks, or hot particles cannot come in contact with combustible material.

520-67. Cable Connectors. Cable connectors, male and female, for flexible conductors shall be constructed so that tension on the cord or cable will not be transmitted to the connections. The female half shall be attached to the load end of the power supply cord or cable. The connector shall be rated in amperes and designed so that differently rated devices cannot be connected together. AC multipole connectors shall be polarized and comply with Section 410-56(f) and Section 410-58.

(FPN): See Section 400-10 for pull at terminals.

520-68. Conductors for Portables.

(a) **Conductor Type.** Flexible conductors, including cable extensions, used to supply portable stage equipment shall be listed, extra-hard usage cords or cables.

Exception No. 1: Reinforced cord shall be permitted to supply stand lamps where the cord is not subject to severe physical damage and is protected by an overcurrent device rated at not over 20 amperes.

Exception No. 2: A special assembly of conductors in sleeving no longer than 3.3 feet (1.0 m) shall be permitted to be employed in lieu of flexible cord if the individual wires are stranded and rated not less than 125°C (257°F) and the outer sleeve is glass fiber with a wall thickness of at least 0.025 inches (635 micrometers).

Exception No. 3: Portable stage equipment requiring flexible supply conductors with a higher temperature rating where one end is permanently attached to the equipment shall be permitted to employ alternate, suitable conductors as determined by a qualified testing laboratory and recognized test standards.

(b) **Conductor Ampacity.** The ampacity of conductors shall be as given in Section 400-5, except multiconductor listed extra-hard usage portable cords, that are not in direct contact with equipment containing heat-producing elements, shall be permitted to have their ampacity determined by Table 520-44. Maximum load current in any conductor shall not exceed the values in Table 520-44.

Exception: Where alternate conductors are allowed in Exception Nos. 2 and 3 of Section 520-68(a), their ampacity shall be as given in the appropriate table in this Code for the types of conductors employed.

520-69. Adapters. Adapters, two-fers, and other single and multiple circuit outlet devices shall comply with (a) and (b) below.

(a) No Reduction in Current Rating. Each receptacle and its corresponding cable shall have the same current and voltage rating as the plug supplying it. It shall not be utilized in a stage circuit with a greater current rating.

(b) Connectors. All connectors shall be wired in accordance with Sections 520-67 and 520-68(a).

F. Dressing Rooms

520-71. Pendant Lampholders. Pendant lampholders shall not be installed in dressing rooms.

520-72. Lamp Guards. All exposed incandescent lamps in dressing rooms, where less than 8 feet (2.44 m) from the floor, shall be equipped with open-end guards riveted to the outlet box cover or otherwise sealed or locked in place.

520-73. Switches Required. All lights and receptacles in dressing rooms shall be controlled by wall switches installed in the dressing rooms. Each switch controlling receptacles shall be provided with a pilot light to indicate when the receptacles are energized.

G. Grounding

520-81. Grounding. All metal raceways and metal-sheathed cables shall be grounded. The metal frames and enclosures of all equipment, including border lights and portable lighting fixtures, shall be grounded. Grounding, where used, shall be in accordance with Article 250.

ARTICLE 530 — MOTION PICTURE AND TELEVISION STUDIOS AND SIMILAR LOCATIONS

A. General

530-1. Scope. The requirements of this article shall apply to television studios and motion picture studios using either film or electronic cameras, except as provided in Section 520-1, and exchanges, factories, laboratories, stages, or a portion of the building in which film or tape more than $7/8$ inch (22 mm) in width is exposed, developed, printed, cut, edited, rewound, repaired, or stored.

(FPN): For methods of protecting against cellulose nitrate film hazards, see Storage and Handling of Cellulose Nitrate Motion Picture Film, NFPA 40-1988 (ANSI).

530-2. Definitions.

Portable Equipment: Equipment intended to be moved from one place to another.

Plugging Box: A dc device consisting of one or more 2-pole, 2-wire, nonpolarized, ungrounded receptacles intended to be used on dc circuits only.

AC Power Distribution Box (AC Plugging Box, Scatter Box): An ac distribution center or box that contains one or more grounded, polarized receptacles, that may contain overcurrent protection devices.

B. Stage or Set

530-11. Permanent Wiring. The permanent wiring shall be Type MC cable, Type MI cable, or in approved raceways.

Exception: Communications circuits, sound-recording and reproducing circuits, Class 2 and Class 3 remote-control or signaling circuits and power-limited fire protective signaling circuits shall be permitted to be wired in accordance with Articles 640, 725, 760, and 800.

530-12. Portable Wiring. The wiring for stage set lighting, stage effects, electric equipment used as stage properties, and other wiring not fixed as to location shall be done with approved flexible cords and cables. Splices or taps shall be permitted in flexible cords used to supply stage properties where such are made with approved devices and the circuit is protected at not more than 20 amperes. Such cables and cords shall not be fastened by staples or nailing.

530-13. Stage Lighting and Effects Control. Switches used for studio stage set lighting and effects (on the stages and lots and on location) shall be of the externally operable type. Where contactors are used as the disconnecting means for fuses, an individual externally operable switch, such as a tumbler switch, for the control of each contactor shall be located at a distance of not more than 6 feet (1.83 m) from the contactor, in addition to remote-control switches.

Exception: A single externally operable switch shall be permitted to simultaneously disconnect all the contactors on any one location board, where located at a distance of not more than 6 feet (1.83 m) from the location board.

530-14. Plugging Boxes. Each receptacle of dc plugging boxes shall be rated at not less than 30 amperes.

530-15. Enclosing and Guarding Live Parts.

(a) **Live Parts.** Live parts shall be enclosed or guarded to prevent accidental contact by persons and objects.

(b) **Switches.** All switches shall be of the externally operable type.

(c) Rheostats. Rheostats shall be placed in approved cases or cabinets that enclose all live parts, having only the operating handles exposed.

(d) Current-Carrying Parts. Current-carrying parts of bull-switches, location boards, spiders, and plugging boxes shall be so enclosed, guarded, or located that persons cannot accidentally come into contact with them or bring conductive material into contact with them.

530-16. Portable Lamps. Portable lamps and work lights shall be equipped with flexible cords, composition or metal-sheathed porcelain sockets, and substantial guards.

Exception: Portable lamps used as properties in a motion picture set or television stage set, on a studio stage or lot, or on location.

530-17. Portable Arc Lamp Fixtures.

(a) Portable Carbon Arc Lamps. Portable carbon arc lamps shall be substantially constructed. The arc shall be provided with an enclosure designed to retain sparks and carbons and to prevent persons or materials from coming into contact with the arc or bare live parts. The enclosures shall be ventilated. All switches shall be of the externally-operable type.

(b) Portable Noncarbon Arc Electric-Discharge Lamp Fixtures. Portable noncarbon arc lamp fixtures, including enclosed arc lamp fixtures, and associated ballasts shall be listed. Interconnecting cord sets, and interconnecting cords and cables, shall be extra-hard usage type and listed.

530-18. Overcurrent Protection — Short-Time Rating.*

General. Automatic overcurrent protective devices (circuit breakers or fuses) for motion picture studio stage set lighting and the stage cables for such stage set lighting shall be as given in (a) through (f) below.

(FPN): Note: * Special consideration is given to motion picture studios and similar locations, because filming periods are of short duration of 20 minutes.

(a) Stage Cables. Stage cables for stage set lighting shall be protected by means of overcurrent devices set at not more than 400 percent of the ampacity given in applicable tables of Articles 310 and 400.

(b) Feeders. In buildings used primarily for motion picture production, the feeders from the substations to the stages shall be protected by means of overcurrent devices (generally located in the substation) having suitable ampere rating. The overcurrent devices shall be permitted to be multipole or single-pole gang-operated. No pole or overcurrent device shall be required in the neutral conductor. The overcurrent device setting for each feeder shall not exceed 400 percent of the ampacity of the feeder, as given in applicable tables of Article 310. Short-term ratings shall not be permitted when equipment is operated for periods in excess of 20 minutes.

(c) Location Boards. Overcurrent protection (fuses or circuit breakers) shall be provided at the "location boards." Fuses in the "location boards" shall have an ampere rating of not over 400 percent of the ampacity of the cables between the "location boards" and the plugging boxes.

(d) **Plugging Boxes.** Cables and cords supplied through plugging boxes shall be of copper. Cables and cords smaller than No. 8 shall be attached to the plugging box by means of a plug containing two cartridge fuses or a 2-pole circuit breaker. The rating of the fuses or the setting of the circuit breaker shall not be over 400 percent of the rated ampacity of the cables or cords as given in applicable tables of Articles 310 and 400. Plugging boxes shall not be permitted on ac systems.

(e) **AC Power Distribution Boxes.** AC power distribution boxes used on sound stages and shooting locations shall contain connection receptacles of a polarized, grounding type.

(f) **Lighting.** Work lights, stand lamps, and fixtures, where connected to dc plugging boxes, shall be by means of plugs containing two cartridge fuses not larger than 20 amperes, or they shall be permitted to be connected to special outlets on circuits protected by fuses or circuit breakers rated at not over 20 amperes. Plug fuses shall not be used unless they are on the load side of the fuse or circuit breakers on the "location boards."

530-19. Sizing of Feeder Conductors for Television Studio Sets.

(a) **General.** It shall be permissible to apply the demand factors listed in Table 530-19(a) to that portion of the maximum possible connected load for studio or stage set lighting for all permanently installed feeders between substations and stages and to all permanently installed feeders between the main stage switchboard and stage distribution centers or location boards.

Table 530-19(a). Demand Factors for Stage Set Lighting

Portion of Stage Set Lighting Load to Which Demand Factor Applied (volt-amperes)	Feeder Demand Factor
First 50,000 or less at	100%
From 50,001 to 100,000 at	75%
From 100,001 to 200,000 at	60%
Remaining over 200,000 at	50%

(b) **Portable Feeders.** A demand factor of 50 percent of maximum possible connected load shall be permitted for all portable feeders.

530-20. Grounding. Type MC cable, Type MI cable, metal raceways, and all noncurrent-carrying metal parts of appliances, devices, and equipment shall be grounded as specified in Article 250. This shall not apply to pendant and portable lamps, to stage lighting and stage sound equipment, nor to other portable and special stage equipment operating at not over 150 volts dc to ground.

530-21. Single-Pole Separable Connectors. Where single-pole portable cable connectors are used, they shall be listed and of the locking type. The use of such connectors shall comply with at least one of the following conditions:

(1) Connection and disconnection of connectors are only possible where the supply connectors are interlocked to the source and it is not possible to connect or disconnect connectors when the supply is energized.

(2) Line connectors are of the listed sequential-interlocking type so that load connectors shall be connected in the following sequence:

a. Equipment grounding conductor connection.

b. Grounded circuit conductor connection, if provided.

c. Ungrounded conductor connection, and that disconnection shall be in the reverse order.

(3) A caution notice shall be provided adjacent to the line connectors indicating that plug connection shall be in the following order:

a. Equipment grounding conductor connectors.

b. Grounded circuit conductor connectors, if provided.

c. Ungrounded conductor connectors, and that disconnection shall be in the reverse order.

530-22. Branch Circuits. A branch circuit of any size supplying one or more receptacles shall be permitted to supply stage set lighting loads.

C. Dressing Rooms

530-31. Dressing Rooms. Fixed wiring in dressing rooms shall be installed in accordance with wiring methods covered in Chapter 3. Wiring for portable dressing rooms shall be approved.

D. Viewing, Cutting, and Patching Tables

530-41. Lamps at Tables. Only composition or metal-sheathed, porcelain, keyless lampholders equipped with suitable means to guard lamps from physical damage and from film and film scrap shall be used at patching, viewing, and cutting tables.

E. Cellulose Nitrate Film Storage Vaults

530-51. Lamps in Cellulose Nitrate Film Storage Vaults. Lamps in cellulose nitrate film storage vaults shall be installed in rigid fixtures of the glass enclosed and gasketed type. Lamps shall be controlled by a switch having a pole in each ungrounded conductor. This switch shall be located outside of the vault and provided with a pilot light to indicate whether the switch is on or off. This switch shall disconnect from all sources of supply all ungrounded conductors terminating in any outlet in the vault.

530-52. Motors and Other Equipment in Cellulose Nitrate Film Storage Vaults. Except as permitted in Section 530-51, no receptacles, outlets, electric motors, heaters, portable lights, or other portable electric equipment shall be located in cellulose nitrate film storage vaults.

F. Substations

530-61. Substations. Wiring and equipment of over 600 volts, nominal, shall comply with Article 710.

530-62. Low-Voltage Switchboards. On 600 volts, nominal, or less, switchboards shall comply with Article 384.

530-63. Overcurrent Protection of DC Generators. Three-wire dc generators shall have protection consisting of overcurrent devices having an ampere rating or setting in accordance with the generator ampere rating. Single-pole or double-pole overcurrent devices shall be permitted, and no pole or overcurrent coil shall be required in the neutral lead (whether it is grounded or ungrounded).

530-64. Working Space and Guarding. Working space and guarding in permanent fixed substations shall comply with Sections 110-16 and 110-17.

(FPN): For guarding of live parts on motors and generators, see Sections 430-11 and 430-14.

Exception: Switchboards of not over 250 volts dc between conductors, where located in substations or switchboard rooms accessible to qualified persons only, shall not be required to be dead-front.

530-65. Portable Substations. Wiring and equipment in portable substations shall conform to the sections applying to installations in permanently fixed substations, but, due to the limited space available, the working spaces shall be permitted to be reduced, provided that the equipment shall be so arranged that the operator can work safely and so that other persons in the vicinity cannot accidentally come into contact with current-carrying parts or bring conducting objects into contact with them while they are energized.

530-66. Grounding at Substations. Noncurrent-carrying metal parts shall be grounded in accordance with Article 250.

Exception: Frames of dc circuit breakers installed on switchboards.

ARTICLE 540 — MOTION PICTURE PROJECTORS

A. General

540-1. Scope. The provisions of this article apply to motion picture projection rooms, motion picture projectors, and associated equipment of the professional and nonprofessional types using incandescent, carbon arc, xenon, or other light source equipment that develops hazardous gases, dust, or radiation.

(FPN): For further information, see Storage and Handling of Cellulose Nitrate Motion Picture Film, NFPA 40-1988 (ANSI).

B. Definitions

540-2. Professional Projector: The professional projector is a type using 35-or 70-millimeter film that has a minimum width of 1⅜ inches (35 mm) and has on each edge 5.4 perforations per inch, or a type using carbon arc, xenon, or other light source equipment that develops hazardous gases, dust, or radiation.

540-3. Nonprofessional Projector: Nonprofessional projectors are those types other than described in Section 540-2.

C. Equipment and Projectors of the Professional Type

540-10. Motion Picture Projection Room Required.   Every professional-type projector shall be located within a projection room. Every projection room shall be of permanent construction, approved for the type of building in which the projection room is located. All projection ports, spotlight ports, viewing ports, and similar openings shall be provided with glass or other approved material so as to completely close the opening. Such rooms shall not be considered as hazardous (classified) locations as defined in Article 500.

(FPN):   For further information on protecting openings in projection rooms handling cellulose nitrate motion picture film, see Life Safety Code, NFPA 101-1991 (ANSI).

540-11. Location of Associated Electrical Equipment.

(a) Motor Generator Sets, Transformers, Rectifiers, Rheostats, and Similar Equipment. Motor generator sets, transformers, rectifiers, rheostats, and similar equipment for the supply or control of current to projection or spotlight equipment shall, if practicable, be located in a separate room. Where placed in the projection room, they shall be so located or guarded that arcs or sparks cannot come in contact with film, and motor generator sets shall have the commutator end or ends protected as provided in Section 520-48.

(b) Switches, Overcurrent Devices, or Other Equipment. Switches, overcurrent devices, or other equipment not normally required or used for projectors, sound reproduction, flood or other special effect lamps, or other equipment shall not be installed in projection rooms.

Exception No. 1: In projection rooms approved for use only with cellulose acetate (safety) film, the installation of appurtenant electrical equipment used in conjunction with the operation of the projection equipment and the control of lights, curtains, and audio equipment, etc., shall be permitted. In such projection rooms, a sign reading "Safety Film Only Permitted in This Room" shall be posted on the outside of each projection room door and within the projection room itself in a conspicuous location.

Exception No. 2: Remote-control switches for the control of auditorium lights or switches for the control of motors operating curtains and masking of the motion picture screen.

(c) Emergency Systems. Control of emergency systems shall comply with Article 700, Emergency Systems.

540-12. Work Space.   Each motion picture projector, floodlight, spotlight, or similar equipment shall have clear working space not less than 30 inches (762 mm) wide on each side and at the rear thereof.

Exception: One such space shall be permitted between adjacent pieces of equipment.

540-13. Conductor Size. Conductors supplying outlets for arc and xenon projectors of the professional type shall not be smaller than No. 8 and shall be of sufficient size for the projector employed. Conductors for incandescent-type projectors shall conform to normal wiring standards as provided in Section 210-24.

540-14. Conductors on Lamps and Hot Equipment. Insulated conductors having a rated operating temperature of not less than 200°C (392°F) shall be used on all lamps or other equipment where the ambient temperature at the conductors as installed will exceed 50°C (122°F).

540-15. Flexible Cords. Cords approved for hard usage as provided in Table 400-4 shall be used on portable equipment.

540-20. Approval. Projectors and enclosures for arc, xenon and incandescent lamps and rectifiers, transformers, rheostats, and similar equipment shall be listed.

540-21. Marking. Projectors and other equipment shall be marked with the maker's name or trademark and with the voltage and current for which they are designed in accordance with Section 110-21.

D. Nonprofessional Projectors

540-31. Motion Picture Projection Room Not Required. Projectors of the nonprofessional or miniature type, where employing cellulose acetate (safety) film, shall be permitted to be operated without a projection room.

540-32. Approval. Projection equipment shall be listed.

E. Sound Recording and Reproduction

540-50. Sound Recording and Reproduction. Sound-recording and reproduction equipment shall be installed as provided in Article 640.

ARTICLE 545 — MANUFACTURED BUILDING

A. General

545-1. Scope. This article covers requirements for a manufactured building and building components as herein defined.

545-2. Other Articles. Wherever the requirements of other articles of this Code and Article 545 differ, the requirements of Article 545 shall apply.

545-3. Definitions.

Manufactured Building: "Manufactured Building" means any building that is of closed construction and is made or assembled in manufacturing facilities on or off the building site for installation, or assembly and installation on the building site, other than mobile homes or recreational vehicles.

Building Component: "Building Component" means any subsystem, subassembly, or other system designed for use in or integral with or as part of a structure, which can include structural, electrical, mechanical, plumbing, and fire protection systems, and other systems affecting health and safety.

Building System: "Building System" means plans, specifications, and documentation for a system of manufactured building or for a type or a system of building components, which can include structural, electrical, mechanical, plumbing, and fire protection systems, and other systems affecting health and safety, and including such variations thereof as are specifically permitted by regulation, and which variations are submitted as part of the building system or amendment thereto.

Closed Construction: "Closed Construction" means any building, building component, assembly, or system manufactured in such a manner that all concealed parts of processes of manufacture cannot be inspected before installation at the building site without disassembly, damage, or destruction.

545-4. Wiring Methods.

(a) Methods Permitted. All raceway and cable wiring methods included in this Code and such other wiring systems specifically intended and listed for use in manufactured buildings shall be permitted with listed fittings and with fittings listed and identified for manufactured buildings.

(b) Securing Cables. In closed construction, cables shall be permitted to be secured only at cabinets, boxes, or fittings where No. 10 or smaller conductors are used and protection against physical damage is provided as required by Section 300-4.

545-5. Service-Entrance Conductors.

Service-entrance conductors shall meet the requirements of Article 230. Provisions shall be made to route the service-entrance conductors from the service equipment to the point of attachment of the service.

(FPN):   See Section 310-10 for temperature limitation of conductors.

545-6. Installation of Service-Entrance Conductors.

Service-entrance conductors shall be installed after erection at the building site.

Exception: Where point of attachment is known prior to manufacture.

545-7. Service Equipment Location.

Service equipment shall be installed in accordance with Section 230-70(a).

545-8. Protection of Conductors and Equipment.

Protection shall be provided for exposed conductors and equipment during processes of manufacturing, packaging, in transit, and erection at the building site.

545-9. Boxes.

(a) Other Dimensions. Boxes of dimensions other than those required in Table 370-16(a) shall be permitted to be installed when tested, identified, and listed to applicable standards.

(b) Not Over 100 Cubic Inches. Any box not over 100 cubic inches in size, intended for mounting in closed construction, shall be affixed with anchors or clamps so as to provide a rigid and secure installation.

545-10. Receptacle or Switch with Integral Enclosure. A receptacle or switch with integral enclosure and mounting means, when tested, identified, and listed to applicable standards, shall be permitted to be installed.

545-11. Bonding and Grounding. Prewired panels and building components shall provide for the bonding, or bonding and grounding, of all exposed metals likely to become energized, in accordance with Article 250, Parts E, F, and G.

545-12. Grounding Electrode Conductor. The grounding electrode conductor shall meet the requirements of Article 250, Part J. Provisions shall be made to route the grounding electrode conductor from the service equipment to the point of attachment to the grounding electrode.

545-13. Component Interconnections. Fittings and connectors that are intended to be concealed at the time of on-site assembly, when tested, identified, and listed to applicable standards, shall be permitted for on-site interconnection of modules or other building components. Such fittings and connectors shall be equal to the wiring method employed in insulation, temperature rise, and fault-current withstand and shall be capable of enduring the vibration and minor relative motions occurring in the components of manufactured building.

ARTICLE 547 — AGRICULTURAL BUILDINGS

547-1. Scope. The provisions of this article shall apply to the following agricultural buildings or that part of a building or adjacent area as specified in (a) and (b) below.

(a) Excessive Dust and Dust with Water. Agricultural buildings where excessive dust and dust with water may accumulate, including all areas of poultry, livestock, and fish confinement systems, where litter dust, or feed dust, including mineral feed particles, may accumulate, and adjacent areas of similar or like nature.

(b) Corrosive Atmosphere. Agricultural buildings where a corrosive atmosphere exists. Such buildings include areas where (1) poultry and animal excrement may cause corrosive vapors; (2) corrosive particles may combine with water; (3) the area is damp and wet by reason of periodic washing for cleaning and sanitizing with water and cleansing agents; (4) similar conditions exist.

547-2. Other Articles. For agricultural buildings not having conditions as specified in Section 547-1, the electrical installations shall be made in accordance with the applicable articles in this Code.

547-3. Surface Temperatures. Electrical equipment or devices installed in accordance with the provisions of this article shall be installed in a manner such that they will function at full rating without developing surface temperatures in excess of the specified normal safe operating range of the equipment or device.

547-4. Wiring Methods. In agricultural buildings as described in Sections 547-1(a) and (b), Types UF, NMC, SNM, copper SE, or other cables or raceways suitable for the location, with approved termination fittings, shall be the wiring methods employed. Article 320 and Article 502 wiring methods shall be permitted for Section 547-1(a). Article 347 and Article 351, Part B wiring methods shall be permitted. All cables shall be secured within 8 inches (203 mm) of each cabinet, box, or fitting. The 1/4-inch (6.35-mm) air space required for nonmetallic boxes, fittings, conduit, and cables in Section 300-6(c) shall not be required in buildings covered by this article.

(FPN): See Sections 300-7 and 347-9 for installation of raceway systems exposed to widely different temperatures.

(a) Boxes, Fittings, and Wiring Devices. All boxes and fittings shall comply with Section 547-5.

(b) Flexible Connections. Where necessary to employ flexible connections, dusttight flexible connectors, liquidtight flexible conduit, or flexible cord listed and identified for hard usage shall be used. All shall be used with listed and identified fittings.

547-5. Switches, Circuit Breakers, Controllers, and Fuses. Switches, circuit breakers, controllers, and fuses, including pushbuttons, relays, and similar devices, used in buildings described in Sections 547-1(a) and (b), shall be provided with enclosures as specified in (a) and (b) below.

(a) Excessive Dust and Dust with Water. Buildings described in Section 547-1(a) shall utilize dustproof and weatherproof enclosures.

(b) Corrosive Atmosphere. Buildings described in Section 547-1(b) shall utilize watertight, corrosion-resistant enclosures.

(FPN): Cast aluminum and magnetic steels will corrode in agricultural environments.

547-6. Motors. Motors and other rotating electrical machinery shall be totally enclosed or so designed as to minimize the entrance of dust, moisture, or corrosive particles.

547-7. Lighting Fixtures. Lighting fixtures installed in agricultural buildings described in Section 547-1 shall comply with the following:

(a) Minimize the Entrance of Dust. Lighting fixtures shall be installed to minimize the entrance of dust, foreign matter, moisture, and corrosive material.

(b) **Exposed to Physical Damage.** Any lighting fixture that may be exposed to physical damage shall be protected by a suitable guard.

(c) **Exposed to Water.** A fixture that may be exposed to water from condensation, building cleansing water, or solution shall be watertight.

547-8. Grounding, Bonding, and Equipotential Plane.

(a) **Grounding and Bonding.** Grounding and bonding shall comply with Article 250.

(FPN): See Section 250-21 for objectionable current over grounding conductors.

Exception No. 1: The main bonding jumper shall not be required at the distribution panelboard in or on buildings housing livestock or poultry where all the following conditions are met:

a. All buildings and premises wiring are under the same ownership.

b. An equipment grounding conductor is run with the supply conductors and is of the same size as the largest supply conductor, if of the same material, or is adjusted in size in accordance with the equivalent size columns of Table 250-95 if of different materials.

c. Service disconnecting means is provided at the distribution point for supply to those buildings.

d. The equipment grounding conductor is bonded to the grounded circuit conductor at the service equipment or the source of a separately derived system.

e. A grounding electrode is provided and connected to the equipment grounding conductor in the distribution panelboard.

Exception No. 2: A metal interior water piping system or other interior metal or metallic piping system of an agricultural building to which electrical equipment requiring grounding is not attached or in electrical contact shall be permitted to be bonded to the service equipment enclosure, the grounded conductor at the service, or to the equipment grounding terminal bar in a panelboard that serves the building by means of an impedance device listed for the purpose, provided that all the following conditions are complied with:

a. The device shall have the short-circuit withstand rating identified where other than the minimum of 10,000 amperes.

b. The bonding conductor shall be an insulated copper conductor not smaller than No. 8 and shall be installed without splice other than where connected to the impedance device.

c. The bonding conductor shall be installed in a raceway suitable for the conditions.

d. The bonding conductor shall be connected to the metal piping system or other metal by means of a listed pressure connector suitable for the conditions or by exothermic welding.

(b) **Concrete Embedded Elements.** Wire mesh or other conductive elements where provided in the concrete floor of animal confinement areas to provide an equipotential plane shall be bonded to the building grounding electrode system. The bonding conductor shall be copper, insulated, covered or bare, and not smaller than No. 8. The means of bonding to wire mesh or conductive elements shall be by pressure connectors or clamps of brass, copper, copper alloy, or an equally substantial approved means.

Equipotential Plane: An equipotential plane is an area where a wire mesh or other conductive elements are embedded in concrete, bonded to all adjacent conductive equipment, structures, or surfaces, and connected to the electrical grounding system to prevent a difference in voltage from developing within the plane.

(FPN): If a wire mesh or other conductive grid is embedded in a concrete floor or platform, and, if this grid is bonded to the electrical system grounding bus, live-stock making contact between the concrete floor or platform and the equipment or metal structure will be less likely to be exposed to a level of voltage that may alter animal behavior or productivity.

(c) Separate Equipment Grounding Conductor. In agricultural buildings as described in Sections 547-1(a) and (b), noncurrent-carrying metal parts of equipment, raceways, and other enclosures, where required to be grounded, shall be grounded by a copper equipment grounding conductor installed between the equipment and the building disconnecting means. If installed underground, the equipment grounding conductor shall be insulated or covered.

(FPN): Grounding electrode system resistances lower than required by Article 250, Part H, may reduce potential differences in livestock facilities.

(d) Water Pumps and Metal Well Casings. The frame of the motor of any water pump shall be grounded as required for Section 430-142. Where a submersible pump is used in metal well casing, the well casing shall be bonded to the pump circuit equipment grounding conductor, or to the equipment grounding bus of the panelboard supplying the submersible pump.

ARTICLE 550 — MOBILE HOMES
AND MOBILE HOME PARKS

A. General

550-1. Scope. The provisions of this article cover the electrical conductors and equipment installed within or on mobile homes, the conductors that connect mobile homes to a supply of electricity, and the installation of electrical wiring, fixtures, equipment, and appurtenances related to electrical installations within a mobile home park up to the mobile home service-entrance conductors or, if none, the mobile home service equipment.

550-2. Definitions.

Appliance, Fixed: An appliance that is fastened or otherwise secured at a specific location.

Appliance, Portable: An appliance that is actually moved or can easily be moved from one place to another in normal use.

(FPN): For the purpose of this article, the following major appliances, other than built-in, are considered portable if cord-connected: refrigerators, gas range equipment, clothes washers, dishwashers without booster heaters, or other similar appliances.

Appliance, Stationary: An appliance that is not easily moved from one place to another in normal use.

Distribution Panelboard: See definition of panelboard in Article 100.

Feeder Assembly: The overhead or under-chassis feeder conductors, including the grounding conductor, together with the necessary fittings and equipment or a power-supply cord listed for mobile home use, designed for the purpose of delivering energy from the source of electrical supply to the distribution panelboard within the mobile home.

Laundry Area: An area containing or designed to contain a laundry tray, clothes washer, or a clothes dryer.

Mobile Home: A factory-assembled structure or structures transportable in one or more sections, that is built on a permanent chassis and designed to be used as a dwelling without a permanent foundation where connected to the required utilities, and includes the plumbing, heating, air-conditioning, and electric systems contained therein.

Unless otherwise indicated, the term "mobile home" includes manufactured homes that are similar structure(s) designed to be used with a permanent foundation.

(FPN): The phrase "with a permanent foundation" indicates that the home is permanently attached to a permanent foundation acceptable to the authority having jurisdiction, such that moving the structure is not likely to occur.

Mobile Home Accessory Building or Structure: Any awning, cabana, ramada, storage cabinet, carport, fence, windbreak, or porch established for the use of the occupant of the mobile home upon a mobile home lot.

Mobile Home Lot: A designated portion of a mobile home park designed for the accommodation of one mobile home and its accessory buildings or structures for the exclusive use of its occupants.

Mobile Home Park: A contiguous parcel of land that is used for the accommodation of occupied mobile homes.

Mobile Home Service Equipment: The equipment containing the disconnecting means, overcurrent protective devices, and receptacles or other means for connecting a mobile home feeder assembly.

Park Electrical Wiring Systems: All of the electrical wiring, fixtures, equipment, and appurtenances related to electrical installations within a mobile home park, including the mobile home service equipment.

550-3. Other Articles. Wherever the requirements of other articles of this Code and Article 550 differ, the requirements of Article 550 shall apply.

550-4. General Requirements.

(a) Mobile Home Not Intended as a Dwelling Unit. A mobile home not intended as a dwelling unit, as for example equipped for sleeping purposes only, contractor's on-site offices, construction job dormitories, mobile studio dressing rooms, banks, clinics, mobile stores, or intended for the display or demonstration of merchandise or machinery, shall not be required to meet the provisions of this article pertaining to the number or capacity of circuits required. It shall, however, meet all other applicable requirements of this article if provided with an electrical installation intended to be energized from a 120-volt or 120/240-volt ac power supply system. Where different voltage is required by either design or available power supply system, adjustment shall be made in accordance with other articles and sections for the voltage used.

(b) In Other than Mobile Home Parks. Mobile homes installed in other than mobile home parks shall comply with the provisions of this article.

(c) Connection to Wiring System. The provisions of this article apply to mobile homes intended for connection to a wiring system rated 120/240 volts, nominal, 3-wire ac, with grounded neutral.

(d) Listed or Labeled. All electrical materials, devices, appliances, fittings, and other equipment shall be listed or labeled by a qualified testing agency and shall be connected in an approved manner when installed.

B. Mobile Homes

550-5. Power Supply.

(a) Feeder. The power supply to the mobile home shall be a feeder assembly consisting of not more than one listed 50-ampere mobile home power-supply cord with integral molded cap, or a permanently installed feeder.

Exception No. 1: A mobile home that is factory-equipped with gas or oil-fired central heating equipment and cooking appliances shall be permitted to be provided with a listed mobile home power-supply cord rated 40 amperes.

Exception No. 2: Manufactured homes constructed in accordance with Section 550-23(a), Exception No. 2.

(b) Power-Supply Cord. If the mobile home has a power-supply cord, it shall be permanently attached to the distribution panelboard or to a junction box permanently connected to the distribution panelboard, with the free end terminating in an attachment plug cap.

Cords with adapters and pigtail ends, extension cords, and similar items shall not be attached to, or shipped with, a mobile home.

A suitable clamp or the equivalent shall be provided at the distribution panelboard knockout to afford strain relief for the cord to prevent strain from being transmitted to the terminals when the power-supply cord is handled in its intended manner.

The cord shall be a listed type with four conductors, one of which shall be identified by a continuous green color or a continuous green color with one or more yellow stripes for use as the grounding conductor.

(c) Attachment Plug Cap. The attachment plug cap shall be a 3-pole, 4-wire, grounding type, rated 50 amperes, 125/250 volts with a configuration as shown in Figure 550-5(c) and intended for use with the 50-ampere, 125/250 receptacle configuration shown in Figure 550-5(c). It shall be molded of butyl rubber, neoprene, or other materials that have been found suitable for the purpose and shall be molded to the flexible cord so that it adheres tightly to the cord at the point where the cord enters the attachment plug cap. If a right-angle cap is used, the configuration shall be so oriented that the grounding member is farthest from the cord.

(FPN): Complete details of the 50-ampere plug and receptacle shown in Figure 550-5(c) can be found in ANSI Standard Dimensions of Caps, Plugs and Receptacles, C73.17-1972.

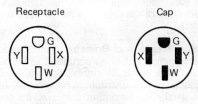

125/250-volt, 50-amp,
3-pole, 4-wire, grounding type

Figure 550-5(c). 50-ampere, 125/250-Volt Receptacle and Attachment-plug-cap Configurations, 3-pole, 4-wire, Grounding-types, Used for Mobile Homes Supply Cords and Mobile Home Parks.

(d) Overall Length of a Power-Supply Cord. The overall length of a power-supply cord, measured from the end of the cord, including bared leads, to the face of the attachment plug cap shall not be less than 21 feet (6.4 m) and shall not exceed 36½ feet (11.13 m). The length of the cord from the face of the attachment plug cap to the point where the cord enters the mobile home shall not be less than 20 feet (6.1 m).

(e) Marking. The power-supply cord shall bear the following marking: "For use with mobile homes — 40 amperes" or "For use with mobile homes — 50 amperes."

(f) Point of Entrance. The point of entrance of the feeder assembly to the mobile home shall be in the exterior wall, floor, or roof.

(FPN No. 1): For location of distribution panelboard, see Section 550-6(a).

(FPN No. 2): For location of attachment of feeder assembly, see Section 550-23(e).

(g) Protected. Where the cord passes through walls or floors, it shall be protected by means of conduits and bushings or equivalent. The cord shall be permitted to be installed within the mobile home walls, provided a continuous raceway having a maximum size of 1¼ inches (31.8 mm) is installed from the branch-circuit panelboard to the underside of the mobile home floor.

(h) Protection Against Corrosion and Mechanical Damage. Permanent provisions shall be made for the protection of the attachment plug cap of the power-supply cord and any connector cord assembly or receptacle against corrosion and mechanical damage if such devices are in an exterior location while the mobile home is in transit.

(i) Mast Weatherhead or Raceway. Where the calculated load exceeds 50 amperes or where a permanent feeder is used, the supply shall be by means of:

(1) One mast weatherhead installation, installed in accordance with Article 230, containing four continuous, insulated, color-coded feeder conductors, one of which shall be an equipment grounding conductor; or

(2) A metal raceway or rigid nonmetallic conduit from the disconnecting means in the mobile home to the underside of the mobile home, with provisions for the attachment to a suitable junction box or fitting to the raceway on the underside of the mobile home (with or without conductors as in Section 550-5(i)(1)).

550-6. Disconnecting Means and Branch-Circuit Protective Equipment. The branch-circuit equipment shall be permitted to be combined with the disconnecting means as a single assembly. Such a combination shall be permitted to be designated as a distribution panelboard. If a fused distribution panelboard is used, the maximum fuse size for the mains shall be plainly marked, with lettering at least ¼ inch (6.4 mm) high and visible when fuses are changed.

Where plug fuses and fuseholders are used, they shall be tamper-resistant Type S, enclosed in dead-front fuse panelboards. Electrical distribution panelboards containing circuit breakers shall also be dead-front type.

(FPN): See Section 110-22 concerning identification of each disconnecting means and each service, feeder, or branch circuit at the point where it originated and the type marking needed.

(a) Disconnecting Means. A single disconnecting means shall be provided in each mobile home consisting of a circuit breaker, or a switch and fuses and its accessories installed in a readily accessible location near the point of entrance of the supply cord or conductors into the mobile home. The main circuit breakers or fuses shall be plainly marked "Main." This equipment shall contain a solderless type of grounding connector or bar for the purposes of grounding, with sufficient terminals for all grounding conductors. The neutral bar termination of the grounded circuit conductors shall be insulated in accordance with Section 550-11(a). The disconnecting equipment shall have a rating suitable for the connected load. The distribution equipment, either circuit breaker or fused type, shall be located a minimum of 24 inches (610 mm) from the bottom of such equipment to the floor level of the mobile home.

(FPN): See Section 550-15(b) for information on disconnecting means for branch circuits designed to energize heating or air-conditioning equipment, or both, located outside the mobile home, other than room air conditioners.

A distribution panelboard shall be rated not less than 50 amperes and employ a 2-pole circuit breaker rated 40 amperes for a 40-ampere supply cord, or 50 amperes for a 50-ampere supply cord. A distribution panelboard employing a disconnect switch and fuses shall be rated 60 amperes and shall employ a single 2-pole, 60-ampere fuseholder with 40- or 50-ampere main fuses for 40- or 50-ampere supply cords, respectively. The outside of the distribution panelboard shall be plainly marked with the fuse size.

The distribution panelboard shall be located in an accessible location but shall not be located in a bathroom or a clothes closet. A clear working space at least 30 inches (762 mm) wide and 30 inches (762 mm) in front of the distribution panelboard shall be provided. This space shall extend from floor to the top of the distribution panelboard.

(b) Branch-Circuit Protective Equipment. Branch-circuit distribution equipment shall be installed in each mobile home and shall include over-current protection for each branch circuit consisting of either circuit breakers or fuses.

The branch-circuit overcurrent devices shall be rated (1) not more than the circuit conductors; and (2) not more than 150 percent of the rating of a single appliance rated 13.3 amperes or more that is supplied by an individual branch circuit; but (3) not more than the overcurrent protection size and of the type marked on the air conditioner or other motor-operated appliance.

A 15-ampere multiple receptacle shall be permitted where connected to a 20-ampere laundry circuit.

(c) Two-Pole Circuit Breakers. Where circuit breakers are provided for branch-circuit protection, 240-volt circuits shall be protected by a 2-pole common or companion trip, or handle-tied paired circuit breakers.

(d) Electrical Nameplates. A metal nameplate on the outside adjacent to the feeder assembly entrance shall read: "This Connection for 120/240-Volt, 3-Pole, 4-Wire, 60 Hertz, . . . Ampere Supply." The correct ampere rating shall be marked in the blank space.

550-7. Branch Circuits. The number of branch circuits required shall be determined in accordance with (a) through (c) below.

(a) Lighting. Based on 3 volt-amperes per square foot (32.26 VA/sq m) times outside dimensions of the mobile home (coupler excluded) divided by 120 volts to determine the number of 15- or 20-ampere lighting area circuits, e.g.,

$$\frac{3 \times \text{Length} \times \text{Width}}{120 \times 15 \text{ (or 20)}} = \text{No. of 15- (or 20-) ampere circuits}$$

The lighting circuits shall be permitted to serve built-in gas ovens with electric service only for lights, clocks or timers, or listed cord-connected garbage disposal units.

(b) Small Appliances. Small appliance branch circuits shall be installed in accordance with Section 210-52(b).

(c) General Appliances. (Including furnace, water heater, range, and central or room air conditioner, etc.) There shall be one or more circuits of adequate rating in accordance with the following:

(1) Ampere rating of fixed appliances not over 50 percent of circuit rating if lighting outlets (receptacles, other than kitchen, dining area, and laundry, considered as lighting outlets) are on the same circuit;

(2) For fixed appliances on a circuit without lighting outlets, the sum of rated amperes shall not exceed the branch-circuit rating. Motor loads or other continuous duty loads shall not exceed 80 percent of the branch-circuit rating;

(3) The rating of a single cord- and plug-connected appliance on a circuit having no other outlets shall not exceed 80 percent of the circuit rating;

(4) The rating of a range branch circuit shall be based on the range demand as specified for ranges in Section 550-13(b)(5).

(FPN No. 1):   For the laundry branch circuit, see Section 220-4(c).

(FPN No. 2):   For central air conditioning, see Article 440.

550-8. Receptacle Outlets.

(a) Grounding-type Receptacle Outlets. All receptacle outlets (1) shall be of grounding type; (2) shall be installed according to Section 210-7; and (3), except where supplying specific appliances, receptacles shall be 15- or 20-ampere, 125-volt, either single or duplex, and shall accept parallel-blade attachment plugs.

(b) Ground-Fault Circuit-Interrupters. All 120-volt, single-phase, 15- and 20-ampere receptacle outlets installed outdoors and in bathrooms, including receptacles in light fixtures, shall have ground-fault circuit-interrupter protection for personnel. Ground-fault circuit-interrupter protection for personnel shall be provided for receptacle outlets located within 6 feet (1.83 m) of any lavatory or sink.

Exception: Receptacles installed for appliances in dedicated spaces, such as for dishwashers, disposals, refrigerators, freezers, and laundry equipment.

No receptacle shall be required in the area occupied by a toilet, shower, tub, or any combination thereof. If a receptacle is installed in such an area, it shall have ground-fault circuit-interrupter protection for personnel.

Feeders supplying branch circuits shall be permitted to be protected by a ground-fault circuit-interrupter in lieu of the provision for such interrupters specified herein.

(c) Cord-Connected Fixed Appliance. A grounding-type receptacle outlet shall be provided for each cord-connected fixed appliance installed.

(d) Required Receptacle Outlets. Receptacle outlets shall be provided in all rooms other than the bath, closet, and hall areas, and shall be installed so that no point along the floor line is more than 6 feet (1.83 m) measured horizontally from an outlet in that space. Counter tops shall have receptacles located every 6 feet (1.83 m). The contiguous measurement of counter top and floor line shall be permitted where measured from the required receptacle in rooms requiring small appliance circuits. Receptacle outlets on small appliance circuits shall not be included in determining the spacing for receptacle outlets of other circuits.

Exception No. 1: Where the measured distance is interrupted by an interior doorway, sink, refrigerator, range, oven, or cooktop, an additional receptacle outlet shall be provided where the interrupted space is at least 2 feet (610 mm) wide at the floor line and at least 12 inches (305 mm) wide at the counter top.

Exception No. 2: Receptacles rendered not readily accessible by stationary appliances shall not be considered as the required outlets.

Exception No. 3: The distance along a floor line occupied by a door opened fully against that space shall not be required to be included in establishing the horizontal measurement if the door swing is limited to 90 degrees nominal by that wall space.

Exception No. 4: Receptacle requirements for bar-type counters and for fixed room dividers shall be permitted to be provided by a receptacle outlet in the wall at the nearest point where the counter or room divider attaches to the wall provided:

    a. The divider does not exceed 8 feet (2.44 m) in length; and

    b. The divider does not exceed 4 feet (1.22 m) in height; and

    c. The divider is attached to a wall at one end only.

(e) Outdoor Receptacle Outlets. At least one receptacle outlet shall be installed outdoors. A receptacle outlet located in a compartment accessible from the outside of the mobile home shall be considered an outdoor receptacle. Outdoor receptacle outlets shall be protected as required in Section 550-8(b).

(f) Receptacle Outlets Not Permitted.

    (1) Shower or Bathtub Space. Receptacle outlets shall not be installed in or within reach [30 inches (762 mm)] of a shower or bathtub space.

    (2) Face-Up Position. A receptacle shall not be installed in a face-up position in any counter top.

(g) Heat Tape Outlet. A heat tape outlet, if installed, and if located on the underside of the mobile home at least 3 feet (914 mm) from the outside edge, shall not be considered an outdoor receptacle outlet. A heat tape outlet, if installed, shall be located within 2 feet (610 mm) of the cold water inlet.

550-9. Fixtures and Appliances.

(a) Fasten Appliances in Transit. Means shall be provided to securely fasten appliances when the mobile home is in transit. (See Section 550-11 for provisions on grounding.)

(b) Accessibility. Every appliance shall be accessible for inspection, service, repair, or replacement without removal of permanent construction.

(c) Pendants. Pendant-type fixtures or pendant cords shall be listed and identified for the interconnection of building components.

(d) Bathtub and Shower Fixtures. Where a lighting fixture is installed over a bathtub or in a shower stall, it shall be of the enclosed and gasketed type listed for wet locations.

(e) Location of Switches. The switch for shower lighting fixtures and exhaust fans located over a tub or in a shower stall shall be located outside the tub or shower space.

550-10. Wiring Methods and Materials.   Except as specifically limited in this section, the wiring methods and materials included in this Code shall be used in mobile homes.

(a) Nonmetallic Boxes. Nonmetallic boxes shall be permitted only with nonmetallic cable or nonmetallic raceways.

(b) Nonmetallic Cable Protection. Nonmetallic cable located 15 inches (381 mm) or less above the floor, if exposed, shall be protected from physical damage by covering boards, guard strips, or raceways. Cable likely to be damaged by stowage shall be so protected in all cases.

(c) Metal-Covered and Nonmetallic Cable Protection. Metal-covered and nonmetallic cables shall be permitted to pass through the centers of the wide side of 2-inch by 4-inch studs. However, they shall be protected where they pass through 2-inch by 2-inch studs or at other studs or frames where the cable or armor would be less than 1½ inches (38 mm) from the inside or outside surface of the studs where the wall covering materials are in contact with the studs. Steel plates on each side of the cable, or a tube, with not less than No. 16 MSG wall thickness shall be required to protect the cable. These plates or tubes shall be securely held in place.

(d) Metal Faceplates. Where metal faceplates are used, they shall be effectively grounded.

(e) Installation Requirements. If a range, clothes dryer, or similar appliance is connected by metal-covered cable or flexible metal conduit, a length of not less than 3 feet (914 mm) of free cable or conduit shall be provided to permit moving the appliance. The cable or flexible metal conduit shall be secured to the wall. Type NM or Type SE cable shall not be used to connect a range or dryer. This shall not prohibit the use of Type NM or Type SE cable between the branch-circuit overcurrent-protective device and a junction box or range or dryer receptacle.

(f) Raceways. Where rigid metal conduit or intermediate metal conduit is terminated at an enclosure with a locknut and bushing connection, two locknuts shall be provided, one inside and one outside of the enclosure. Rigid nonmetallic conduit or electrical nonmetallic tubing shall be permitted. All cut ends of conduit and tubing shall be reamed or otherwise finished to remove rough edges.

(g) Switches. Switches shall be rated as follows:

(1) For lighting circuits, switches shall be rated not less than 10 amperes, 120-125 volts, and in no case less than the connected load.

(2) For motors or other loads, switches shall have ampere or horsepower ratings, or both, adequate for loads controlled. (An "ac general-use" snap switch shall be permitted to control a motor 2 horsepower or less with full-load current not over 80 percent of the switch ampere rating.)

(h) Free Conductor at Each Box. At least 4 inches (102 mm) of free conductor shall be left at each box except where conductors are intended to loop without joints.

(i) Under-Chassis Wiring. (Exposed to weather.)

(1) Where outdoor or under-chassis line-voltage (120 volts, nominal, or higher) wiring is exposed to moisture or physical damage, it shall be protected by rigid metal conduit or intermediate metal conduit. The conductors shall be suitable for wet locations.

Exception: Electrical metallic tubing or rigid nonmetallic conduit shall be permitted where closely routed against frames and equipment enclosures.

(2) The cables or conductors shall be Type NMC, TW, or equivalent.

(j) Boxes, Fittings, and Cabinets. Boxes, fittings, and cabinets shall be securely fastened in place and shall be supported from a structural member of the home, either directly or by using a substantial brace.

Exception: Snap-in type boxes. Boxes provided with special wall or ceiling brackets and wiring devices with integral enclosures, that securely fasten to walls or ceilings and are identified for the use, shall be permitted without support from a structural member or brace. The testing and approval shall include the wall and ceiling construction systems for which the boxes and devices are intended to be used.

(k) Appliance Terminal Connections. Appliances having branch-circuit terminal connections that operate at temperatures higher than 60°C (140°F) shall have circuit conductors as described in (1) or (2) below.

(1) Branch-circuit conductors having an insulation suitable for the temperature encountered shall be permitted to be run directly to the appliance.

(2) Conductors having an insulation suitable for the temperature encountered shall be run from the appliance terminal connection to a readily accessible outlet box placed at least 1 foot (305 mm) from the appliance. These conductors shall be in a suitable raceway that shall extend for at least 4 feet (1.22 m).

(l) Component Interconnections. Fittings and connectors that are intended to be concealed at the time of assembly shall be listed and identified for the interconnection of building components. Such fittings and connectors shall be equal to the wiring method employed in insulation, temperature rise, and fault-current withstanding and shall be capable of enduring the vibration and shock occurring in mobile home transportation.

550-11. Grounding. Grounding of both electrical and nonelectrical metal parts in a mobile home shall be through connection to a grounding bus in the mobile home distribution panelboard. The grounding bus shall be grounded through the green-colored insulated conductor in the supply cord or the feeder wiring to the service ground in the service-entrance equipment located adjacent to the mobile home location. Neither the frame of the mobile home nor the frame of any appliance shall be connected to the grounded circuit conductor (neutral) in the mobile home.

(a) Insulated Neutral.

(1) The grounded circuit conductor (neutral) shall be insulated from the grounding conductors and from equipment enclosures and other grounded parts. The grounded (neutral) circuit terminals in the distribution panelboard and in ranges, clothes dryers, counter-mounted cooking units, and wall-mounted ovens shall be insulated from the equipment enclosure. Bonding screws, straps, or buses in the distribution panelboard or in appliances shall be removed and discarded.

(2) Connections of ranges and clothes dryers with 120/240-volt, 3-wire ratings shall be made with 4-conductor cord and 3-pole, 4-wire, grounding-type plugs or by Type AC cable, Type MC cable, or conductors enclosed in flexible metal conduit.

(b) Equipment Grounding Means.

(1) The green-colored insulated grounding wire in the supply cord or permanent feeder wiring shall be connected to the grounding bus in the distribution panelboard or disconnecting means.

(2) In the electrical system, all exposed metal parts, enclosures, frames, lamp fixture canopies, etc., shall be effectively bonded to the grounding terminal or enclosure of the distribution panelboard.

(3) Cord-connected appliances, such as washing machines, clothes dryers, refrigerators, and the electrical system of gas ranges, etc., shall be grounded by means of a cord with grounding conductor and grounding-type attachment plug.

(c) Bonding of Noncurrent-Carrying Metal Parts.

(1) All exposed noncurrent-carrying metal parts that may become energized shall be effectively bonded to the grounding terminal or enclosure of the distribution panelboard. A bonding conductor shall be connected between the distribution panelboard and accessible terminal on the chassis.

(2) Grounding terminals shall be of the solderless type and listed as pressure-terminal connectors recognized for the wire size used. The bonding conductor shall be solid or stranded, insulated or bare, and shall be No. 8 copper minimum, or equal. The bonding conductor shall be routed so as not to be exposed to physical damage.

(3) Metallic gas, water, and waste pipes and metallic air-circulating ducts shall be considered bonded if they are connected to the terminal on the chassis [see Section 550-11(c)(1)] by clamps, solderless connectors, or by suitable grounding-type straps.

(4) Any metallic roof and exterior covering shall be considered bonded if (a) the metal panels overlap one another and are securely attached to the wood or metal frame parts by metallic fasteners and (b) if the lower panel of the metallic exterior covering is secured by metallic fasteners at a cross member of the chassis by two metal straps per mobile home unit or section at opposite ends.

The bonding strap material shall be a minimum of 4 inches (102 mm) in width of material equivalent to the skin or a material of equal or better electrical conductivity. The straps shall be fastened with paint-penetrating fittings, such as screws and starwashers or equivalent.

550-12. Testing.

(a) Dielectric Strength Test. The wiring of each mobile home shall be subjected to a 1-minute, 900-volt, dielectric strength test (with all switches closed) between live parts (including neutral) and the mobile home ground. Alternatively, the test shall be permitted to be performed at 1,080 volts for 1 second. This test shall be performed after branch circuits are complete and after fixtures or appliances are installed.

Exception: Listed fixtures or appliances shall not be required to withstand the dielectric strength test.

(b) Continuity and Operational Tests and Polarity Checks. Each mobile home shall be subjected to:

(1) An electrical continuity test to ensure that all exposed electrically conductive parts are properly bonded;

(2) An electrical operational test to demonstrate that all equipment, except water heaters and electric furnaces, is connected and in working order; and

(3) Electrical polarity checks of permanently wired equipment and receptacle outlets to determine that connections have been properly made.

550-13. Calculations. The following method shall be employed in computing the supply-cord and distribution-panelboard load for each feeder assembly for each mobile home in lieu of the procedure shown in Article 220 and shall be based on a 3-wire, 120/240-volt supply with 120-volt loads balanced between the two legs of the 3-wire system.

(a) Lighting and Small Appliance Load:

Lighting Volt-Amperes: Length times width of mobile home floor (outside dimensions) times 3 volt-amperes per square foot; e.g.,

Length × width × 3 = ... lighting volt-amperes.

Small Appliance Volt-Amperes: Number of circuits times 1,500 volt-amperes for each 20-ampere appliance receptacle circuit (see definition of "Appliance, Portable" with note) including 1,500 volt-amperes for laundry circuit; e.g.,

Number of circuits × 1,500 = ... small appliance volt-amperes.

Total: Lighting volt-amperes plus small appliance = ... total volt-amperes.

First 3,000 total volt-amperes at 100 percent plus remainder at 35 percent = ... volt-amperes to be divided by 240 volts to obtain current (amperes) per leg.

(b) Total Load for Determining Power Supply. Total load for determining power supply is the summation of:

(1) Lighting and small appliance load as calculated in Section 550-13(a).

(2) Nameplate amperes for motors and heater loads (exhaust fans, air conditioners, electric, gas, or oil heating).

Omit smaller of the heating and cooling loads, except include blower motor if used as air-conditioner evaporator motor. Where an air conditioner is not installed and a 40-ampere power-supply cord is provided, allow 15 amperes per leg for air conditioning.

(3) 25 percent of current of largest motor in (2).

(4) Total of nameplate amperes for: disposal, dishwasher, water heater, clothes dryer, wall-mounted oven, cooking units.

Where number of these appliances exceeds three, use 75 percent of total.

(5) Derive amperes for freestanding range (as distinguished from separate ovens and cooking units) by dividing the following values by 240 volts.

Nameplate Rating	Use
0 thru 10,000 watts	80 percent of rating
Over 10,000 thru 12,500 watts	8,000 volt-amperes
Over 12,500 thru 13,500 watts	8,400 volt-amperes
Over 13,500 thru 14,500 watta	8,800 volt-amperes
Over 14,500 thru 15,500 watts	9,200 volt-amperes
Over 15,500 thru 16,500 watts	9,600 volt-amperes
Over 16,500 thru 17,500 watts	10,000 volt-amperes

(6) If outlets or circuits are provided for other than factory-installed appliances, include the anticipated load.

See following Example for illustration of application of this calculation.

Example

A mobile home floor is 70 feet × 10 feet and has two small appliance circuits, a 1000-volt-ampere, 240-volt heater, a 200-volt-ampere, 120-volt exhaust fan, a 400-volt-ampere, 120-volt dishwasher, and a 7000-volt-ampere electric range.

Lighting and small appliance load

Lighting 70 × 10 × 3 VA/sq ft = ...	2100 volt-amperes
Small appliance 1500 × 2=	3000 volt-amperes
Laundry 1500 × 1=	1500 volt-amperes
	6600 volt-amperes
1st 3000 volt-amperes at 100 percent	3000 volt-amperes
Remainder (6600 - 3000) at 35 percent	1260 volt-amperes
	4260 volt-amperes

$$\frac{4260 \text{ volt-amperes}}{240 \text{ volts}} = 17.75 \text{ amperes per leg}$$

	Amperes per leg	
	A	**B**
Lighting and appliances	17.75	17.75
Heater, 1000 VA ÷ 240 volt =	4.2	4.2
Fan, 200 VA ÷ 120 volt =	1.7	
Dishwasher, 400VA ÷ 120 =		3.3
Range, 7000 VA × .8 ÷ 240 =	23.3	23.3
Totals	46.95	48.55

Based on the higher current calculated for either leg, a minimum 50-ampere supply cord is required.

For SI units: one square foot = 0.093 square meter; one foot = 0.3048 meter.

(c) Optional Method of Calculation for Lighting and Appliance Load. For mobile homes, the optional method for calculating lighting and appliance load shown in Section 220-30 and Table 220-30 shall be permitted.

550-14. Interconnection of Multiple Section Mobile Home Units.

(a) Fixed-type Wiring. Approved and listed fixed-type wiring methods shall be used to join portions of a circuit that must be electrically joined that are located in adjacent sections of mobile homes after the home is installed on its support foundation. The circuit's junction shall be accessible for disassembly when the home is prepared for relocation.

(b) Disconnecting Means. Multiple section mobile homes not having permanently installed feeders, and that are to be moved from one location to another, shall be permitted to have disconnecting means with branch-circuit protective equipment in each unit where so located that, after assembly or joining together of units, they shall not be interconnected on either the line side or the load side, except that the grounding means shall be electrically interconnected.

(FPN): Subsection (b) above applies to connection of previously constructed mobile homes where multiple feeder assemblies were allowed. The present Code does not permit more than one cord or feeder to a mobile home.

550-15. Outdoor Outlets, Fixtures, Air-Cooling Equipment, Etc.

(a) Listed for Outdoor Use. Outdoor fixtures and equipment shall be listed for outdoor use. Outdoor receptacle or convenience outlets shall be of a gasketed-cover type for use in wet locations.

(b) Outside Heating Equipment, Air-Conditioning Equipment, or Both. A mobile home provided with a branch circuit designed to energize outside heating equipment or air-conditioning equipment, or both, located outside the mobile home, other than room air conditioners, shall have such branch-circuit conductors terminate in a listed outlet box, or disconnecting means, located on the outside of the mobile home. A label shall be permanently affixed adjacent to the outlet box and contain the following information:

> This connection is for heating and/or air-conditioning equipment. The branch circuit is rated at not more than _____ amperes, at _____ volts, 60-Hertz, _____ conductor ampacity. A disconnecting means shall be located within sight of the equipment.

The correct voltage and ampere rating shall be given. The tag shall be not less than 0.020-inch (508-micrometer) thick etched brass, stainless steel, anodized or alclad aluminum, or equivalent. The tag shall not be less than 3 inches (76 mm) by 1¾ inches (44.5 mm) minimum size.

C. Services and Feeders

550-21. Distribution System.
The mobile home park secondary electrical distribution system to mobile home lots shall be single-phase, 120/240 volts, nominal. For the purpose of Part C, where the park service exceeds 240 volts, nominal, transformers and secondary distribution panelboards shall be treated as services.

(FPN No. 1): See Table 550-22 for calculation of load.

(FPN No. 2): See Section 550-4(b) for mobile homes not located in mobile home parks.

550-22. Minimum Allowable Demand Factors.
Park electrical wiring systems shall be calculated (at 120/240 volts) on the larger of (1) 16,000 volt-amperes for each mobile home lot or (2) the load calculated in accordance with Section 550-13 for the largest typical mobile home that each lot will accept. It shall be permissible to compute the feeder or service load in accordance with Table 550-22. No demand factor shall be allowed for any other load, except as provided in this Code.

Table 550-22.
Demand Factors for Feeders and Service-Entrance Conductors

Number of Mobile Homes	Demand Factor (percent)
1	100
2	55
3	44
4	39
5	33
6	29
7-9	28
10-12	27
13-15	26
16-21	25
22-40	24
41-60	23
61 and over	22

Service and feeder conductors to a mobile home in compliance with Article 310, Note 3 of Notes to Ampacity Tables of 0 to 2000 Volts shall be permitted.

550-23. Mobile Home Service Equipment.

(a) Service Equipment. The mobile home service equipment shall be located adjacent to the mobile home and not mounted in or on the mobile home. The service equipment shall be located in sight from and not more than 30 feet (9.14 m) from the exterior wall of the mobile home it serves.

Exception No. 1: The service equipment shall be permitted to be located elsewhere on the premises, provided that all of the following conditions are met:

a. A disconnecting means suitable for service equipment is located in sight from and not more than 30 feet (9.14 m) from the exterior wall of the mobile home it serves.

b. Grounding at the disconnecting means shall be in accordance with Section 250-24.

Exception No. 2: Service equipment shall be permitted to be installed in or on a manufactured home, provided that all of the following conditions are met:

a. The service equipment is completely installed by the manufacturer of the structure.

b. The installation of the service equipment complies with Article 230.

c. Means are provided for the connection of a grounding electrode conductor to the service equipment and routing it outside the structure.

• **(b) Rating.** Mobile home service equipment shall be rated at not less than 100 amperes, and provision shall be made for connecting a mobile home feeder assembly by a permanent wiring method. Power outlets used as mobile home service equipment shall also be permitted to contain receptacles rated up to 50 amperes with appropriate overcurrent protection. Fifty-ampere receptacles shall conform to the configuration shown in Figure 550-5(c).

(FPN): Complete details on the 50-ampere attachment plug cap configuration can be found in American National Standard Dimensions of Caps, Plugs and Receptacles, ANSI C73.17-1972.

(c) Additional Outside Electrical Equipment. Mobile home service equipment shall also contain a means for connecting a mobile home accessory building or structure or additional electrical equipment located outside a mobile home by a fixed wiring method.

(d) Additional Receptacles. Additional receptacles shall be permitted for connection of electrical equipment located outside the mobile home, and all such 125-volt, single-phase, 15- and 20-ampere receptacles shall be protected by a listed ground-fault circuit-interrupter protection.

(e) Mounting Height. Outdoor mobile home disconnecting means shall be installed so the bottom of the enclosure containing the disconnecting means is not less than 2 feet (610 mm) above finished grade or working platform. The disconnecting means shall be so installed that the center of the grip of the operating handle, when in its highest position, will not be more than 6½ feet (1.98 m) above the finished grade or working platform.

(f) Grounded. Each mobile home service equipment shall be grounded in accordance with Article 250 for service equipment.

(g) Marking. Where a 125/250-volt receptacle is used in mobile home service equipment, the service equipment shall be marked as follows:

"Turn disconnecting switch or circuit breaker off before inserting or removing plug. Plug must be fully inserted or removed."

The marking shall be located on the service equipment adjacent to the receptacle outlet.

550-24. Feeder.

(a) Feeder Conductors. Mobile home feeder conductors shall consist of either a listed cord, factory-installed in accordance with Section 550-5(b), or a permanently installed feeder consisting of four, insulated, color-coded conductors that shall be identified by the factory or field marking of the conductors in compliance with Section 310-12. Equipment grounding conductors shall not be identified by stripping the insulation.

Exception: A mobile home feeder located between service equipment and a mobile home disconnecting means as covered in Section 550-23(a), Exception No. 1 shall be permitted to omit the equipment grounding conductor where the grounded circuit conductor is grounded at the disconnecting means as permitted in Section 250-24(a).

(b) Adequate Feeder Capacity. Mobile home lot feeder circuit conductors shall have adequate capacity for the loads supplied and shall be rated at not less than 100 amperes at 120/240 volts.

ARTICLE 551 — RECREATIONAL VEHICLES AND RECREATIONAL VEHICLE PARKS

A. General

551-1. Scope. The provisions of this article cover the electrical conductors and equipment installed within or on recreational vehicles, the conductors that connect recreational vehicles to a supply of electricity, and the installation of equipment and devices related to electrical installations within a recreational vehicle park.

(FPN): For requirements on the installation of plumbing and heating systems in recreational vehicles, refer to the Standard on Recreational Vehicles, NFPA 501C-1990 (ANSI).

551-2. Definitions. (See Article 100 for other definitions.)

Air-Conditioning or Comfort-Cooling Equipment: All of that equipment intended or installed for the purpose of processing the treatment of air so as to control simultaneously its temperature, humidity, cleanliness, and distribution to meet the requirements of the conditioned space.

Appliance, Fixed: An appliance that is fastened or otherwise secured at a specific location.

Appliance, Portable: An appliance that is actually moved or can easily be moved from one place to another in normal use.

(FPN): For the purpose of this article, the following major appliances, other than built-in, are considered portable if cord-connected: refrigerators, gas range equipment, clothes washers, dishwashers without booster heaters, or other similar appliances.

Appliance, Stationary: An appliance that is not easily moved from one place to another in normal use.

Camping Trailer: A vehicular portable unit mounted on wheels and constructed with collapsible partial side walls that fold for towing by another vehicle and unfold at the campsite to provide temporary living quarters for recreational, camping, or travel use. (See "Recreational Vehicle.")

Converter: A device that changes electrical energy from one form to another, as from alternating current to direct current.

Dead Front: (As applied to switches, circuit breakers, switchboards, and distribution panelboards.) So designed, constructed, and installed that no current-carrying parts are normally exposed on the front.

Disconnecting Means: The necessary equipment usually consisting of a circuit breaker or switch and fuses, and their accessories, located near the point of entrance of supply conductors in a recreational vehicle and intended to constitute the means of cutoff for the supply to that recreational vehicle.

Distribution Panelboard: A single panel or group of panel units designed for assembly in the form of a single panel; including buses, and with or without switches and/or automatic overcurrent-protective devices for the control of light, heat, or power circuits of small individual as well as aggregate capacity; designed to be placed in a cabinet or cutout box placed in or against a wall or partition and accessible only from the front.

Frame: Chassis rail and any welded addition thereto of metal thickness of 16 MSG or greater.

Low Voltage: An electromotive force rated 24 volts, nominal, or less, supplied from a transformer, converter, or battery.

Motor Home: A vehicular unit designed to provide temporary living quarters for recreational, camping, or travel use built on or permanently attached to a self-propelled motor vehicle chassis or on a chassis cab or van that is an integral part of the completed vehicle. (See "Recreational Vehicle.")

Power-Supply Assembly: The conductors, including ungrounded, grounded, and equipment grounding conductors, the connectors, attachment plug caps, and all other fittings, grommets, or devices installed for the purpose of delivering energy from the source of electrical supply to the distribution panel within the recreational vehicle.

Recreational Vehicle: A vehicular-type unit primarily designed as temporary living quarters for recreational, camping, or travel use, which either has its own motive power or is mounted on or drawn by another vehicle. The basic entities are: travel trailer, camping trailer, truck camper, and motor home.

Recreational Vehicle Park. A plot of land upon which two or more recreational vehicle sites are located, established, or maintained for occupancy by recreational vehicles of the general public as temporary living quarters for recreation or vacation purposes.

Recreational Vehicle Site. A plot of ground within a recreational vehicle park intended for the accommodation of either a recreational vehicle, tent, or other individual camping unit on a temporary basis.

Recreational Vehicle Site Feeder Circuit Conductors. The conductors from the park service equipment to the recreational vehicle site supply equipment.

Recreational Vehicle Site Supply Equipment. The necessary equipment, usually a power outlet, consisting of a circuit breaker or switch and fuse and their accessories, located near the point of entrance of supply conductors to a recreational vehicle site and intended to constitute the disconnecting means for the supply to that site.

Recreational Vehicle Stand. That area of a recreational vehicle site intended for the placement of a recreational vehicle.

Transformer: A device that, when used, will raise or lower the voltage of alternating current of the original source.

Travel Trailer: A vehicular unit mounted on wheels, designed to provide temporary living quarters for recreational, camping, or travel use, of such size or weight as not to require special highway movement permits when towed by a motorized vehicle, and of gross trailer area less than 320 square feet (29.77 sq m). (See "Recreational Vehicle.")

Truck Camper: A portable unit constructed to provide temporary living quarters for recreational, travel, or camping use, consisting of a roof, floor, and sides, designed to be loaded onto and unloaded from the bed of a pick-up truck. (See "Recreational Vehicle.")

551-3. Other Articles. Wherever the requirements of other articles of this Code and Article 551 differ, the requirements of Article 551 shall apply.

551-4. General Requirements.

(a) Not Covered. A recreational vehicle not used for the purposes as defined in Section 551-2 shall not be required to meet the provisions of Part A pertaining to the number or capacity of circuits required. It shall, however, meet all other applicable requirements of this article if the recreational vehicle is provided with an electrical installation intended to be energized from a 120- or 120/240-volt, nominal, ac power-supply system.

(b) Systems. This article covers battery and direct-current power (24-volts or less) systems, combination electrical systems, generator installations, and 120- or 120/240-volt, nominal, systems.

B. Low-Voltage Systems

551-10. Low-Voltage Systems.

(a) Low-Voltage Circuits. Low-voltage circuits furnished and installed by the recreational vehicle manufacturer, other than those related to braking and cranking/ignition, are subject to this Code. Circuits supplying lights subject to federal or state regulations shall comply with applicable government regulations and this Code.

(b) Low-Voltage Wiring.

(1) Copper conductors shall be used for low-voltage circuits.

Exception: Metal chassis or frame shall be permitted as the return path to the source of supply. Connections to the chassis or frame shall be made (1) in an accessible location, (2) by means of copper conductors and copper or copper alloy terminals of the solderless type identified for the size of wire used, and (3) mechanically secure.

(2) Conductors shall conform to the requirements for Type HDT, SGT, SGR, or Type SXL or shall have insulation in accordance with Table 310-13 or the equivalent. Conductor sizes No. 6 through 18 or SAE shall be listed.

(FPN): See SAE Standard J1128-1975 for Types HDT and SXL and SAE Standard J1127-1980 for Types SGT and SGR.

(3) Single-wire, low-voltage conductors shall be of the stranded type.

(4) All insulated low-voltage conductors shall be surface marked at intervals no greater than 4 feet (1.22 m) as follows:

a. Listed conductors shall be marked as required by the listing agency.

b. SAE conductors shall be marked with the name or logo of the manufacturer, specification designation, and wire gage.

c. Other conductors shall be marked with the name or logo of the manufacturer, temperature rating, wire gage, conductor material, and insulation thickness.

(c) Low-Voltage Wiring Methods.

(1) Conductors shall be protected against physical damage and shall be secured. Where insulated conductors are clamped to the structure, the conductor insulation shall be supplemented by an additional wrap or layer of equivalent material, except that jacketed cables need not be so protected. Wiring shall be routed away from sharp edges, moving parts, or heat sources.

(2) Conductors shall be spliced or joined with splicing devices that provide a secure connection or by brazing, welding, or soldering with a fusible metal or alloy. Soldered splices shall first be so spliced or joined as to be mechanically and electrically secure without solder and then soldered. All splices, joints, and free ends of conductors shall be covered with an insulation equivalent to that on the conductors.

(3) Battery and direct-current circuits shall be physically separated by at least a $\frac{1}{2}$-inch (12.7-mm) gap or other approved means from circuits of a different power source. Acceptable methods shall be by clamping, routing, or equivalent means that ensure permanent total separation. Where circuits of different power sources cross, the external jacket of the nonmetallic-sheathed cables shall be deemed adequate separation.

(4) Ground terminals shall be accessible for service. The surface on which ground terminals make contact shall be cleaned and free from oxide or paint, or shall be electrically connected through use of a cadmium, tin, or zinc-plated internal-external toothed lockwasher or lockring terminals. Ground terminal attaching screws, rivets or bolts, nuts, and lockwashers shall be cadmium, tin, or zinc-plated, except rivets shall be permitted to be unanodized aluminum when attaching to aluminum structures.

(5) The chassis-grounding terminal of the battery shall be bonded to the vehicle chassis with a minimum No. 8 copper conductor or equivalent. In the event the power lead from the battery exceeds No. 8, then the bonding conductor shall be of an equal size.

(d) Battery Installations.
Storage batteries subject to the provisions of this Code shall be securely attached to the vehicle and installed in an area vapor-tight to the interior and ventilated directly to the exterior of the vehicle. Where batteries are installed in a compartment, the compartment shall be ventilated with openings having a minimum area of 1.7 square inches (1100 sq mm) at both the top and at the bottom. Where compartment doors are equipped for ventilation, the openings shall be within 1 inch (25.4 mm) of the top and bottom. Batteries shall not be installed in a compartment containing spark- or flame-producing equipment, except that they shall be permitted to be installed in the engine generator compartment if the only charging source is from the engine generator.

(e) Overcurrent Protection.

(1) Low-voltage circuit wiring shall be protected by overcurrent-protective devices rated not in excess of the ampacity of copper conductors, as follows:

551-10(e)1

Wire Size	Ampacity	Wire Type
18	6	Stranded only
16	8	Stranded only
14	15	Stranded or Solid
12	20	Stranded or Solid
10	30	Stranded or Solid

(2) Circuit breakers or fuses shall be of an approved type, including automotive types. Fuseholders shall be clearly marked with maximum fuse size and shall be protected against shorting and physical damage by a cover or equivalent means.

(FPN): For further information, see Society of Automotive Engineers (SAE) Standard for Electric Fuses (Cartridge Type), ANSI/SAE J554(b)-1981; Standard for Blade Type Electric Fuses, SAE J1284; and Underwriters Laboratories Inc. Standard for Automotive Glass Tube Fuses, UL 275-1986.

(3) Higher current-consuming, direct-current appliances such as pumps, compressors, heater blowers, and similar motor-driven appliances shall be installed in accordance with themanufacturer's instructions.

Motors that are controlled by automatic switching or by latching-type manual switches shall be protected in accordance with Section 430-32(c).

(4) The overcurrent-protective device shall be installed in an accessible location on the vehicle within 18 inches (457 mm) of the point where the power supply connects to the vehicle circuits. If located outside the recreational vehicle, the device shall be protected against weather and physical damage.

Exception: External low-voltage supply shall be permitted to be fused within 18 inches (457 mm) after entering the vehicle or after leaving a metal raceway.

(f) Switches. Switches shall have a direct-current rating not less than the connected load.

(g) Lighting Fixtures. All low-voltage interior lighting fixtures shall be listed.

Exception: Fixtures rated 4 watts or less, employing lamps rated 1.2 watts or less.

(h) Cigarette Lighter Receptacles. Twelve-volt receptacles that will accept and energize cigarette lighters shall be installed in a noncombustible outlet box, or the assembly shall be identified by the manufacturer of the product as thermally protected.

C. Combination Electrical Systems

551-20. Combination Electrical Systems.

(a) General. Vehicle wiring suitable for connection to a battery or direct-current supply source shall be permitted to be connected to a 120-volt source, provided that the entire wiring system and equipment are rated and installed in full conformity with Parts A, C, D, E, and F requirements covering 120-volt electrical systems. Circuits fed from alternating-current transformers shall not supply direct-current appliances.

(b) Voltage Converters (120-Volt Alternating Current to Low-Voltage Direct Current). The 120-volt alternating current side of the voltage converter shall be wired in full conformity with Parts A, C, D, E, and F requirements for 120-volt electrical systems.

Exception: Converters supplied as an integral part of a listed appliance shall not be subject to the above.

All converters and transformers shall be listed for use in recreation vehicles and designed or equipped to provide over-temperature protection. To determine the converter rating, the following formula shall be applied to the total connected load, including average battery charging rate, of all 12-volt equipment:

The first 20 amperes of load at 100 percent; plus
The second 20 amperes of load at 50 percent; plus
All load above 40 amperes at 25 percent.

Exception: A low-voltage appliance that is controlled by a momentary switch (normally "open") that has no means for holding in the "closed" position shall not be considered as a "connected load" when determining the required converter rating. Momentarily energized appliances shall be limited to those used to prepare the vehicle for occupancy or travel.

(c) Bonding Voltage Converter Enclosures. The noncurrent-carrying metal enclosure of the voltage converter shall be bonded to the frame of the vehicle with a No. 8 copper conductor minimum or equivalent. The grounding conductor for the battery and the metal enclosure shall be permitted to be the same conductor.

(d) Dual-Voltage Fixtures or Appliances. Fixtures or appliances having both 120-volt and low-voltage connections shall be listed for dual voltage.

(e) Autotransformers. Autotransformers shall not be used.

(f) Receptacles and Plug Caps. Where a recreational vehicle is equipped with a 120-volt or 120/240-volt alternating-current system, a low-voltage system, or both, receptacles and plug caps of the low-voltage system shall differ in configuration from those of the 120- or 120/240-volt system. Where a vehicle equipped with a battery or direct-current system has an external connection for low-voltage power, the connector shall have a configuration that will not accept 120-volt power.

D. Other Power Sources

551-30. Generator Installations.

(a) Mounting. Generators shall be mounted in such a manner as to be effectively bonded to the recreational vehicle chassis.

(b) Generator Protection. Equipment shall be installed to ensure that the current-carrying conductors from the engine generator and from an outside source are not connected to a vehicle circuit at the same time.

Receptacles used as disconnecting means shall be accessible (as applied to wiring methods) and capable of interrupting their rated current without hazard to the operator.

(c) Installation of Storage Batteries and Generators. Storage batteries and internal-combustion-driven generator units (subject to the provisions of this Code) shall be secured in place to avoid displacement from vibration and road shock.

(d) Ventilation of Generator Compartments. Compartments accommodating internal-combustion-driven generator units shall be provided with ventilation in accordance with instructions provided by the manufacturer of the generator unit.

(FPN):   For generator compartment construction requirements, see the Standard on Recreational Vehicles, NFPA 501C-1990.

(e) Supply Conductors. The supply conductors from the engine generator to the first termination on the vehicle shall be of the stranded type and be installed in listed flexible conduit. The point of first termination shall be in a (1) panelboard, (2) junction box with a blank cover, (3) junction box with a receptacle, (4) enclosed transfer switch, or (5) receptacle assembly listed in conjunction with the generator.

The panelboard or junction box with a receptacle shall be installed within the vehicle's interior and within 18 inches (457 mm) of the compartment wall but not inside the compartment. A junction box with a blank cover shall be mounted on the compartment wall and shall be permitted inside or outside the compartment. A receptacle assembly listed in conjunction with the generator shall be mounted in accordance with its listing. Overcurrent protection in accordance with Section 240-3 shall be provided for supply conductors as an integral part of a listed generator or shall be located within 18 inches (457 mm) of their point of entry into the vehicle.

551-31. Multiple Supply Source.

(a) Multiple Supply Sources. Where a multiple supply system consisting of an alternate power source and a power-supply cord is installed, the feeder from the alternate power source shall be protected by an overcurrent-protective device. Installation shall be in accordance with Sections 551-30(a) and (b) and 551-40.

(b) Calculation of Loads. Calculation of loads shall be in accordance with Section 551-42.

(c) Multiple Supply Sources Capacity. The multiple supply sources shall not be required to be of the same capacity.

(d) Alternate Power Sources Exceeding 30 Amperes. If an alternate power source exceeds 30 amperes, 120 volts, nominal, it shall be permissible to wire it as a 120-volt, nominal, system or a 120/240-volt, nominal, system, provided an overcurrent-protective device of the proper rating is installed in the feeder.

(e) Power-Supply Assembly Not Less than 30 Amperes. The external power-supply assembly shall be permitted to be less than the calculated load but not less than 30 amperes and shall have overcurrent protection not greater than the capacity of the external power-supply assembly.

551-32. Other Sources. Other sources of ac power such as inverters or motor generators shall be listed for use in recreational vehicles and shall be installed in accordance with the terms of the listing. Other sources of ac power shall be wired in full conformity with the requirements in Parts A, C, D, E, and F of this article covering 120-volt electrical systems.

551-33. Alternate Source Restriction. Transfer equipment, if not integral with the listed power source, shall be installed to ensure that the current-carrying conductors from other sources of ac power and from an outside source are not connected to the vehicle circuit at the same time.

E. Nominal 120- or 120/240-Volt Systems

551-40. 120- or 120/240-Volt, Nominal, Systems.

(a) General Requirements. The electrical equipment and material of recreational vehicles indicated for connection to a wiring system rated 120 volts, nominal, 2-wire with ground, or a wiring system rated 120/240 volts, nominal, 3-wire with ground, shall be listed and installed in accordance with the requirements of Parts A, C, D, E, and F.

(b) Materials and Equipment. Electrical materials, devices, appliances, fittings, and other equipment installed, intended for use in, or attached to the recreational vehicle shall be listed. All products shall be used only in the manner in which they have been tested and found suitable for the intended use.

(c) Ground-Fault Circuit-Interrupter Protection. The internal wiring of a recreational vehicle having only one 15- or 20-ampere branch circuit as permitted in Sections 551-42(a) and (b) shall have ground-fault circuit-interrupter protection for personnel. The ground-fault circuit-interrupter shall be installed at the point where the power supply assembly terminates within the recreational vehicle. Where a separable cord set is not employed, the ground-fault circuit-interrupter shall be permitted to be an integral part of the attachment plug of the power supply assembly. The ground-fault circuit-interrupter shall provide protection also under the conditions of an open grounded circuit conductor, interchanged circuit conductors, or both.

551-41. Receptacle Outlets Required.

(a) Spacing. Receptacle outlets shall be installed at wall spaces 2 feet (610 mm) wide or more so that no point along the floor line is more than 6 feet (1.83 m), measured horizontally, from an outlet in that space.

Exception No. 1: Bath and hall areas.

Exception No. 2: Wall spaces occupied by kitchen cabinets, wardrobe cabinets, built-in furniture, behind doors that may open fully against a wall surface, or similar facilities.

(b) Location. Receptacle outlets shall be installed:

(1) Adjacent to counter tops in the kitchen [at least one on each side of the sink if counter tops are on each side and are 12 inches (305 mm) or over in width].

(2) Adjacent to the refrigerator and gas range space, except where a gas-fired refrigerator or cooking appliance, requiring no external electrical connection, is factory-installed.

(3) Adjacent to counter top spaces of 12 inches (305 mm) or more in width that cannot be reached from a receptacle required in Section 551-41(b)(1) by a cord of 6 feet (1.83 m) without crossing a traffic area, cooking appliance, or sink.

(c) Ground-Fault Circuit-Interrupter Protection. Where provided, each 120-volt, single-phase, 15- or 20-ampere receptacle outlet shall have ground-fault circuit-interrupter protection for personnel in the following locations:

(1) Adjacent to a bathroom lavatory. [The receptacle outlet shall be a minimum of 24 inches (610 mm) from the compartment floor.]

(2) Within 6 feet (1.83 m) of any lavatory or sink.

Exception: Receptacles installed for appliances in dedicated spaces, such as for dishwashers, disposals, refrigerators, freezers, and laundry equipment.

(3) In the area occupied by a toilet, shower, tub, or any combination thereof.

(4) On the exterior of the vehicle.

Exception: Receptacles that are located inside of an access panel that is installed on the exterior of the vehicle to supply power for an installed appliance shall not be required to have ground-fault circuit-interrupter protection.

The receptacle outlet shall be permitted in a listed lighting fixture. A receptacle outlet shall not be installed in a tub or combination tub-shower compartment.

Exception: Where ground-fault circuit-interrupter protection is provided in accordance with Section 551-40(c).

(d) Face-Up Position. A receptacle shall not be installed in a face-up position in any counter top or similar horizontal surfaces within the living area.

551-42. Branch Circuits Required. Each recreational vehicle containing a 120-volt electrical system shall contain one of the following:

(a) One 15-Ampere Circuit. One 15-ampere circuit to supply lights, receptacle outlets, and fixed appliances. Such recreational vehicles shall be equipped with one 15-ampere switch and fuse or 15-ampere circuit breaker.

(b) One 20-Ampere Circuit. One 20-ampere circuit to supply lights, receptacle outlets, and fixed appliances. Such recreational vehicles shall be equipped with one 20-ampere switch and fuse or 20-ampere circuit breaker.

(c) Two or More 15- or 20-Ampere Circuits. Two or more 15- or 20-ampere circuits to supply lights, receptacle outlets, and fixed appliances. Such recreational vehicles shall be equipped with a 30-ampere rated main power supply assembly.

(FPN): See Section 210-23(a) for permissible loads. See Section 551-45(c) for main disconnect and overcurrent protection requirements.

(d) Power-Supply Assembly. A 40- or 50-ampere power-supply assembly that shall be calculated in accordance with the following method:

(1) Lighting. If electric lighting is provided either directly or indirectly (through a voltage converter) by the 120-volt or 120/240-volt system, calculate lighting wattage at 3 volt-amperes per square foot using exterior dimensions (exclusive of hitch and cab) as follows:

Length (feet) × width (feet) × 3 = _____ lighting volt-amperes.

(2) Small Appliance. Number of circuits times 1,500 volt-amperes for each 20-ampere appliance receptacle circuit, e.g.,

Number of Circuits × 1,500 =_____ small appliance volt-amperes.

(3) Total. Lighting volt-amperes plus small appliance volt-amperes = _____ total volt-amperes.

(4) First 3,000 total volt-amperes at 100 percent plus remainder at 35 percent = _____ volt-amperes to be divided by voltage to obtain current (amperes) per leg.

	Amperes per Leg	
	A	*B*
Lighting and small appliance current (amperes) per leg [from (d) above] =		_____

(5) Add nameplate amperes for motors and heater loads (exhaust fans, air conditioners,* electric, gas, or oil heating*). Also include anticipated loads in above categories where prewired outlets or circuits are installed for other than factory-installed major appliances. _____

*Omit smaller of heating or air-conditioning load, except include any motor common to both functions.

(6) Add 25 percent of amperes of largest motor in E = _____

(7) Add nameplate amperes of the following appliances. Include anticipated loads where prewired outlets or circuits are installed for other than factory-installed major appliances. Where the number of appliances is four or more, use 75 percent of total.

	A	B	
Disposal	_____	_____	
Water Heater	_____	_____	
Wall-Mounted Ovens	_____	_____	
Cooking Units	_____	_____	
TOTAL	_____	_____ =	_____

(8) Add amperes for freestanding range as distinguished from separate ovens and cooking units. Derive from following table by dividing volt-amperes by 240 volts.

551-42(8)

Range	Nameplate Rating (watts)	Use (volt-amperes)
(Freestanding range as	0 thru 10,000	80 percent of rating
distinguished from	Over 10,000 thru 12,500	8,000
separate oven and	Over 12,500 thru 13,500	8,400
cooking units	Over 13,500 thru 14,500	8,800
	Over 14,500 thru 15,500	9,200
	Over 15,500 thru 16,500	9,600
	Over 16,500 thru 17,500	10,000

551-43. Branch-Circuit Protection.

(a) Rating. The branch-circuit overcurrent devices shall be rated:

(1) Not more than the circuit conductors; and

(2) Not more than 150 percent of the rating of a single appliance rated 13.3 amperes or more and supplied by an individual branch circuit; but

(3) Not more than the overcurrent protection size marked on an air conditioner or other motor-operated appliances.

(b) Protection for Smaller Conductors. A 20-ampere fuse or circuit breaker shall be permitted for protection for fixture leads, cords, or small appliances, and No. 14 tap conductors, not over 6 feet (1.83 m) long for recessed lighting fixtures.

(c) 15-Ampere Receptacle Considered Protected by 20 Amperes. If more than one outlet or load is on a branch circuit, a 15-ampere receptacle shall be permitted to be protected by a 20-ampere fuse or circuit breaker.

551-44. Power-Supply Assembly.

(a) 15-Ampere Main Power-Supply Assembly. Recreational vehicles wired in accordance with Section 551-42(a) shall use a listed 15-ampere, or larger, main power-supply assembly.

(b) 20-Ampere Main Power-Supply Assembly. Recreational vehicles wired in accordance with Section 551-42(b) shall use a listed 20-ampere, or larger, main power-supply assembly.

(c) 30-Ampere Main Power-Supply Assembly. Recreational vehicles wired in accordance with Section 551-42(c) shall use a listed 30-ampere, or larger, main power-supply assembly.

(d) 40- or 50-Ampere Power-Supply Assembly. In accordance with Section 551-42(d), any recreational vehicle with a rating in excess of 30 amperes, 120 volts, shall use a listed 40-ampere or 50-ampere, 120/240-volt power-supply assembly.

Exception No. 1: Where the calculated load of the recreational vehicle exceeds 30 amperes, 120 volts, a second power-supply cord shall be permitted. Where a two-cord supply system is installed, they shall not be interconnected on either the line side or the load side. The grounding circuits and grounding means shall be electrically interconnected.

Exception No. 2: For a dual-supply source consisting of a generator and a power-supply cord, see Section 551-31.

551-45. Distribution Panelboard.

(a) Listed and Appropriately Rated. A listed and appropriately rated distribution panelboard or other equipment specifically listed for the purpose shall be used. The grounded conductor termination bar shall be insulated from the enclosure as provided in Section 551-54(c). An equipment grounding terminal bar shall be attached inside the metal enclosure of the panelboard.

(b) Location. The distribution panelboard shall be installed in a readily accessible location. Working clearance for the panelboard shall be no less than 24 inches (610 mm) wide and 30 inches (762 mm) deep.

Exception No. 1: Where the panelboard cover is exposed to the inside aisle space, then one of the working clearance dimensions shall be permitted to be reduced to a minimum of 22 inches (559 mm). A panelboard is considered exposed where the panelboard cover is within 2 inches (50.8 mm) of the aisle's finished surface.

Exception No. 2: Compartment doors used for access to a generator shall be permitted to be equipped with a locking system.

(c) Dead-Front Type. The distribution panelboard shall be of the dead-front type and shall consist of one or more circuit breakers or Type S fuseholders. A main disconnecting means shall be provided where fuses are used or where more than two circuit breakers are employed. A main overcurrent protective device not exceeding the power-supply assembly rating shall be provided where more than two branch circuits are employed.

551-46. Means for Connecting to Power Supply.

(a) Assembly. The power-supply assembly or assemblies shall be factory-supplied or factory-installed and be of one of the types as specified herein:

(1) Separable. Where a separable power-supply assembly consisting of a cord with a female connector and molded attachment plug cap is provided, the vehicle shall be equipped with a permanently mounted, flanged surface inlet (male, recessed-type motor-base attachment plug) wired directly to the distribution panelboard by an approved wiring method. The attachment plug cap shall be of a listed type.

(2) Permanently Connected. Each power-supply assembly shall be connected directly to the terminals of the distribution panelboard or conductors within a junction box and provided with means to prevent strain from being transmitted to the terminals. The ampacity of the conductors between each junction box and the terminals of each distribution panelboard shall be at least equal to the ampacity of the power-supply cord. The supply end of the assembly shall be equipped with an attachment plug of the type described in Section 551-46(c). Where the cord passes through the walls or floors, it shall be protected by means of conduit and bushings or equivalent. The cord assembly shall have permanent provisions for protection against corrosion and mechanical damage while the vehicle is in transit.

(b) Cord. The cord exposed usable length shall be measured from the point of entrance to the recreational vehicle or the face of the flanged surface inlet (motor-base attachment plug) to the face of the attachment plug at the supply end.

The cord exposed usable length, measured to the point of entry on the vehicle exterior, shall be a minimum of 23 feet (7.0 m) where the point of entrance is at the side of the vehicle, or shall be a minimum 28 feet (8.5 m) where the point of entrance is at the rear of the vehicle.

Where the cord entrance into the vehicle is more than 3 feet (0.9 m) above the ground, the minimum cord lengths above shall be increased by the vertical distance of the cord entrance heights above 3 feet (0.9 m).

(FPN):   See Section 551-46(e).

(c) Attachment Plugs.

(1) Recreational vehicles having only one 15-ampere branch circuit as permitted by Section 551-42(a) shall have an attachment plug that shall be 2-pole, 3-wire, grounding type, rated 15 amperes, 125 volts, conforming to the configuration shown in Figure 551-46(c).

(FPN):   Complete details of this configuration can be found in American National Standard ANSI C73.11-1972.

(2) Recreational vehicles having only one 20-ampere branch circuit as permitted in Section 551-42(b) shall have an attachment plug that shall be 2-pole, 3-wire, grounding type, rated 20 amperes, 125 volts, conforming to the configuration shown in Figure 551-46(c).

(FPN):   Complete details of this configuration can be found in American National Standard ANSI C73.12-1972.

(3) Recreational vehicles wired in accordance with Section 551-42(c) shall have an attachment plug that shall be 2-pole, 3-wire, grounding type, rated 30 amperes, 125 volts, conforming to the configuration shown in Figure 551-46(c) intended for use with units rated at 30 amperes, 125 volts.

(FPN):   Complete details of this configuration can be found in American National Standard Dimensions of Caps, Plugs and Receptacles, ANSI C73.13-1972.

(4) Recreational vehicles having a power-supply assembly rated 40 amperes or 50 amperes as permitted by Section 551-42(d) shall have a 3-pole, 4-wire, grounding-type attachment plug rated 50 amperes, 125/250 volts, conforming to the configuration shown in Figure 551-46(c).

(FPN):   Complete details of this configuration can be found in American National Standard Dimensions of Caps, Plugs and Receptacles, ANSI C73.17-1972.

(d) Labeling at Electrical Entrance. Each recreational vehicle shall have permanently affixed to the exterior skin, at or near the point of entrance of the power-supply cord(s), a label 3 inches (76 mm) by 1¾ inches (44.5 mm) minimum size, made of etched, metal-stamped or embossed brass, stainless steel, or anodized or alclad aluminum not less than 0.020 inch

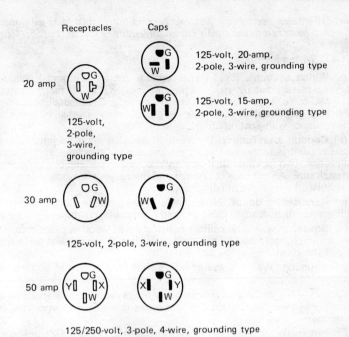

Receptacles   Caps

20 amp

125-volt,
2-pole,
3-wire,
grounding type

125-volt, 20-amp,
2-pole, 3-wire, grounding type

125-volt, 15-amp,
2-pole, 3-wire, grounding type

30 amp

125-volt, 2-pole, 3-wire, grounding type

50 amp

125/250-volt, 3-pole, 4-wire, grounding type

Figure 551-46(c).   Configurations for Grounding-type Receptacles and Attachment Plug Caps Used for Recreational Vehicle Supply Cords and Recreatonal Vehicle Lots.

(508 micrometers) thick, or other suitable material [e.g., 0.005-inch (127-micrometer) thick plastic laminate], that reads, as appropriate, either:

"This connection is for 110-125 volt ac, 60 Hz ____ ampere supply," or

"This connection is for 120/240 volt ac, 3-pole, 4-wire 60 Hz____ ampere supply."

The correct ampere rating shall be marked in the blank space.

(e) Location. The point of entrance of a power-supply assembly shall be located within 15 feet (4.57 m) of the rear, on the left (road) side or at the rear, left of the longitudinal center of the vehicle, within 18 inches (457 mm) of the outside wall.

Exception No. 1: A recreational vehicle equipped with only a listed flex- ible drain system or a side-vent drain system shall be permitted to have the electrical point of entrance located on either side, provided the drain(s) for the plumbing system is (are) located on the same side.

Exception No. 2: A recreational vehicle shall be permitted to have the electrical point of entrance located more than 15 feet (4.57 m) from the rear. Where this occurs, the distance beyond the 15-foot (4.57-m) dimension shall be added to the cord's minimum length as specified in 551-46(b).

551-47. Wiring Methods.

(a) Wiring Systems. Rigid metal conduit, intermediate metal conduit, electrical metallic tubing, rigid nonmetallic conduit, flexible metal conduit, Type MC cable, Type MI cable, Type AC cable, and nonmetallic-sheathed cable shall be permitted. An equipment grounding means shall be provided in accordance with Section 250-91.

(b) Conduit and Tubing. Where rigid metal conduit or intermediate metal conduit is terminated at an enclosure with a locknut and bushing connection, two locknuts shall be provided, one inside and one outside of the enclosure. All cut ends of conduit and tubing shall be reamed or otherwise finished to remove rough edges.

(c) Nonmetallic Boxes. Nonmetallic boxes shall be acceptable only with nonmetallic-sheathed cable or rigid nonmetallic conduit.

(d) Boxes. In walls and ceilings constructed of wood or other combustible material, boxes and fittings shall be flush with the finished surface or project therefrom.

(e) Mounting. Wall and ceiling boxes shall be mounted in accordance with Article 370.

Exception No. 1: Snap-in type boxes or boxes provided with special wall or ceiling brackets that securely fasten boxes in walls or ceilings shall be permitted.

Exception No. 2: A wooden plate providing a 1½-inch (38-mm) minimum width backing around the box and of a thickness of ½ inch (12.7 mm) or greater (actual) glued to the wall panel shall be considered as approved means for mounting outlet boxes.

(f) Sheath Armor. The sheath of nonmetallic-sheathed cable, metal-clad cable, and Type AC cable shall be continuous between outlet boxes and other enclosures.

(g) Protected. Metal-clad, Type AC, or nonmetallic-sheathed cables shall be permitted to pass through the centers of the wide side of 2-inch by 4-inch wood studs. However, they shall be protected where they pass through 2-inch by 2-inch wood studs or at other wood studs or frames where the cable would be less than 1½ inches (38 mm) from the inside or outside surface. Steel plates on each side of the cable, or a steel tube, with not less than No. 16 MSG wall thickness, shall be installed to protect the cable. These plates or tubes shall be securely held in place. Where nonmetallic-sheathed cables pass through punched, cut, or drilled slots or holes in metal members, the cable shall be protected by bushings or grommets securely fastened in the opening prior to installation of the cable.

(h) Bends. No bend shall have a radius of less than five times the cable diameter.

(i) Cable Supports. Where connected with cable connectors or clamps, cables shall be supported within 12 inches (305 mm) of outlet boxes, distribution panelboards, and splice boxes on appliances. Supports shall be provided every 4½ feet (1.37 m) at other places.

(j) Nonmetallic Box Without Cable Clamps. Nonmetallic-sheathed cables shall be supported within 8 inches (203 mm) of a nonmetallic outlet box without cable clamps.

Exception: Where wiring devices with integral enclosures are employed with a loop of extra cable to permit future replacement of the device, the cable loop shall be considered as an integral portion of the device.

(k) Physical Damage. Where subject to physical damage, exposed nonmetallic cable shall be protected by covering boards, guard strips, raceways, or other means.

(l) Metal Faceplates. Metal faceplates shall be of ferrous metal not less than 0.030 inch (762 micrometers) in thickness or of nonferrous metal not less than 0.040 inch (1.02 mm) in thickness. Nonmetallic faceplates shall be listed.

(m) Metal Faceplates Effectively Grounded. Where metal faceplates are used, they shall be effectively grounded.

(n) Moisture or Physical Damage. Where outdoor or underchassis wiring is 120 volts, nominal, or over and is exposed to moisture or physical damage, the wiring shall be protected by rigid metal conduit, intermediate metal conduit, or by electrical metallic tubing or rigid nonmetallic conduit that is closely routed against frames and equipment enclosures or other raceway or cable identified for the application.

(o) Component Interconnections. Fittings and connectors that are intended to be concealed at the time of assembly shall be listed and identified for the interconnection of building components. Such fittings and connectors shall be equal to the wiring method employed in insulation, temperature rise, fault-current withstanding, and shall be capable of enduring the vibration and shock occurring in recreational vehicles.

(p) Method of Connecting Expandable Units.

(1) That portion of a branch circuit that is installed in an expandable unit shall be permitted to be connected to the portion of the branch circuit in the main body of the vehicle by means of an attachment plug and cord listed for hard usage. The cord and its connections shall conform to all provisions of Article 400 and shall be considered as a permitted use under Section 400-7.

Exception: Where the attachment plug and cord are located within the vehicle's interior, use of plastic thermoset or elastomer parallel cord Type SPT-3, SP-3, or SPE shall be permitted.

(2) If the receptacle provided for connection of the cord to the main circuit is located on the outside of the vehicle, it shall be protected with a ground-fault circuit-interrupter for personnel and be listed for wet locations. A cord located on the outside of a vehicle shall be identified for outdoor use.

(3) Unless removable or stored within the vehicle interior, the cord assembly shall have permanent provisions for protection against corrosion and mechanical damage while the vehicle is in transit.

(4) The cord shall be installed so as not to permit exposed live attachment plug pins.

(q) Prewiring for Air Conditioning Installation. Prewiring installed for the purpose of facilitating future air-conditioning installation shall conform to the following and other applicable portions of this article. The circuit shall serve no other purpose.

(1) An overcurrent-protective device with a rating compatible with the circuit conductors shall be installed in the distribution panelboard and wiring connections completed.

(2) The load end of the circuit shall terminate in a junction box with a blank cover or a device listed for the purpose. Where a junction box with a blank cover is used, the free ends of the conductors shall be adequately capped or taped.

(3) A label conforming to Section 551-46(d) shall be placed on or adjacent to the junction box and shall read:

AIR-CONDITIONING CIRCUIT. THIS
CONNECTION IS FOR AIR CONDITIONERS
RATED 110-125 VOLT AC, 60 HZ _____
AMPERES MAXIMUM. DO NOT EXCEED
CIRCUIT RATING.

An ampere rating, not to exceed 80 percent of the circuit rating, shall be legibly marked in the blank space.

(r) Prewiring for Generator Installation. Prewiring installed for the purpose of facilitating future generator installation shall conform to the following and other applicable portions of this article.

(1) Circuit conductors shall be appropriately sized in relation to the anticipated load and shall be protected by an overcurrent device in accordance with their ampacities.

(2) Where junction boxes are utilized at either of the circuit originating or terminus points, free ends of the conductors shall be adequately capped or taped.

(3) Where devices such as receptacle outlet, transfer switch, etc., are installed, the installation shall be complete, including circuit conductor connections. All devices shall be listed and appropriately rated.

(4) A label conforming to Section 551-46(d) shall be placed on the cover of each junction box containing incomplete circuitry and shall read, as appropriate, either:

GENERATOR CIRCUIT. THIS CONNECTION
IS FOR GENERATORS RATED 110-125 VOLT
AC, 60 HZ _____ AMPERES MAXIMUM.

OR

GENERATOR CIRCUIT. THIS CONNECTION
IS FOR GENERATORS RATED 120/240 VOLT
AC, 60 HZ _____ AMPERES MAXIMUM.

The correct ampere rating shall be legibly marked in the blank space.

551-48. Conductors and Boxes.

(a) Maximum Number of Conductors. The maximum number of conductors permitted in boxes shall be in accordance with Section 370-16.

(b) Free Conductor at Each Box. At least 4 inches (102 mm) of free conductor shall be left at each box except where conductors are intended to loop without joints.

551-49. Grounded Conductors. The identification of grounded conductors shall be in accordance with Section 200-6.

551-50. Connection of Terminals and Splices. Conductor splices and connections at terminals shall be in accordance with Section 110-14. If splices of the equipment grounding conductor in nonmetallic-sheathed cable are made in boxes containing devices, the splices shall be insulated.

551-51. Switches. Switches shall be rated as follows:

(a) Lighting Circuits. For lighting circuits, switches shall be rated not less than 10 amperes, 120-125 volts and in no case less than the connected load.

(b) Motors or Other Loads. For motors or other loads, switches shall have ampere or horsepower ratings, or both, adequate for loads controlled. (An ac general-use snap switch shall be permitted to control a motor 2 horsepower or less with full-load current not over 80 percent of the switch ampere rating.)

551-52. Receptacles. All receptacle outlets shall be (1) of the grounding type and (2) installed in accordance with Sections 210-7 and 210-21.

551-53. Lighting Fixtures.

(a) General. Any combustible wall or ceiling finish exposed between the edge of a fixture canopy, or pan and the outlet box, shall be covered with noncombustible material or a material identified for the purpose.

(b) Shower Fixtures. If a lighting fixture is provided over a bathtub or in a shower stall, it shall be of the enclosed and gasketed type and listed for the type of installation, and it shall be ground-fault circuit-interrupter protected if rated at 120 volts, nominal.
The switch for shower lighting fixtures and exhaust fans, located over a tub or in a shower stall, shall be located outside the tub or shower space.

(c) Outdoor Outlets, Fixtures, Air-Cooling Equipment, Etc. Outdoor fixtures and other equipment shall be listed for outdoor use.

551-54. Grounding. (See also Section 551-56 on bonding of noncurrent-carrying metal parts.)

(a) Power-Supply Grounding. The grounding conductor in the supply cord or feeder shall be connected to the grounding bus or other approved grounding means in the distribution panelboard.

(b) Distribution Panelboard. The distribution panelboard shall have a grounding bus with sufficient terminals for all grounding conductors or other approved grounding means.

(c) Insulated Neutral.

(1) The grounded circuit conductor (neutral) shall be insulated from the equipment grounding conductors and from equipment enclosures and other grounded parts. The grounded (neutral) circuit terminals in the distribution panelboard and in ranges, clothes dryers, counter-mounted cooking units, and wall-mounted ovens shall be insulated from the equipment enclosure. Bonding screws, straps, or buses in the distribution panelboard or in appliances shall be removed and discarded.

(2) Connection of electric ranges and electric clothes dryers utilizing a grounded (neutral) conductor, if cord-connected, shall be made with 4-conductor cord and 3-pole, 4-wire, grounding-type plug caps and receptacles.

551-55. Interior Equipment Grounding.

(a) Exposed Metal Parts. In the electrical system, all exposed metal parts, enclosures, frames, lighting fixture canopies, etc., shall be effectively bonded to the grounding terminals or enclosure of the distribution panelboard.

(b) Equipment Grounding Conductors. Bare wires, green-colored wires, or green wires with yellow stripe(s) shall be used for equipment grounding conductors only.

(c) Grounding of Electrical Equipment. Where grounding of electrical equipment is specified, it shall be permitted as follows:

(1) Connection of metal raceway (conduit or electrical metallic tubing), the sheath of Type MC and Type MI cable where the sheath is identified for grounding, or the armor of Type AC cable to metal enclosures.

(2) A connection between the one or more equipment grounding conductors and a metal box by means of a grounding screw, which shall be used for no other purpose, or a listed grounding device.

(3) The equipment grounding conductor in nonmetallic-sheathed cable shall be permitted to be secured under a screw threaded into the fixture canopy other than a mounting screw or cover screw, or attached to a listed grounding means (plate) in a nonmetallic outlet box for fixture mounting (grounding means shall also be permitted for fixture attachment screws).

(d) Grounding Connection in Nonmetallic Box. A connection between the one or more grounding conductors brought into a nonmetallic outlet box shall be so arranged that a connection can be made to any fitting or device in that box that requires grounding.

(e) Grounding Continuity. Where more than one equipment grounding conductor of a branch circuit enters a box, all such conductors shall be in good electrical contact with each other, and the arrangement shall be such that the disconnection or removal of a receptacle, fixture, or other device fed from the box will not interfere with or interrupt the grounding continuity.

(f) Cord-Connected Appliances. Cord-connected appliances, such as washing machines, clothes dryers, refrigerators, and the electrical system of gas ranges, etc., shall be grounded by means of an approved cord with equipment grounding conductor and grounding-type attachment plug.

551-56. Bonding of Noncurrent-Carrying Metal Parts.

(a) Required Bonding. All exposed noncurrent-carrying metal parts that may become energized shall be effectively bonded to the grounding terminal or enclosure of the distribution panelboard.

(b) Bonding Chassis. A bonding conductor shall be connected between any distribution panelboard and an accessible terminal on the chassis. Aluminum or copper-clad aluminum conductors shall not be used for bonding if such conductors or their terminals are exposed to corrosive elements.

Exception: Any recreational vehicle that employs a unitized metal chassis-frame construction to which the distribution panelboard is securely fastened with a bolt(s) and nut(s) or by welding or riveting shall be considered to be bonded.

(c) Bonding Conductor Requirements. Grounding terminals shall be of the solderless type and listed as pressure terminal connectors recognized for the wire size used. The bonding conductor shall be solid or stranded, insulated or bare, and shall be No. 8 copper minimum, or equal.

(d) Metallic Roof and Exterior Bonding. The metal roof and exterior covering shall be considered bonded where:

(1) The metal panels overlap one another and are securely attached to the wood or metal frame parts by metal fasteners, and

(2) The lower panel of the metal exterior covering is secured by metal fasteners at each cross member of the chassis, or the lower panel is bonded to the chassis by a metal strap.

(e) Gas, Water, and Waste Pipe Bonding. The gas, water, and waste pipes shall be considered grounded if they are bonded to the chassis.

(FPN): See Section 551-56(b) for chassis bonding.

(f) Furnace and Metal Air Duct Bonding. Furnace and metal circulating air ducts shall be bonded.

551-57. Appliance Accessibility and Fastening. Every appliance shall be accessible for inspection, service, repair, and replacement without removal of permanent construction. Means shall be provided to securely fasten appliances in place when the recreational vehicle is in transit.

F. Factory Tests

551-60. Factory Tests (Electrical). Each recreational vehicle shall be subjected to the following tests:

(a) Circuits of 120 Volts or 120/240 Volts. Each recreational vehicle designed with a 120-volt or a 120/240-volt electrical system shall withstand the applied potential without electrical breakdown of a 1-minute, 900-volt dielectric strength test, or a 1-second, 1080-volt dielectric strength test, with all switches closed, between ungrounded and grounded conductors | and the recreational vehicle ground. During the test, all switches and other controls shall be in the "on" position. Fixtures and permanently installed appliances shall not be required to withstand this test.

Each recreational vehicle shall be subjected to (1) a continuity test to | assure that all metal parts are properly bonded; (2) operational tests to demonstrate that all equipment is properly connected and in working order; and (3) polarity checks to determine that connections have been properly made.

(b) Low-Voltage Circuits. Low-voltage circuit conductors in each recreational vehicle shall withstand the applied potential without electrical breakdown of a 1-minute, 500-volt or a 1-second, 600-volt dielectric strength test. The potential shall be applied between ungrounded and grounded conductors.

The test shall be permitted on running light circuits before the lights are installed provided the vehicle's outer covering and interior cabinetry has been secured. The braking circuit shall be permitted to be tested before being connected to the brakes, provided the wiring has been completely secured.

G. Recreational Vehicle Parks

551-71. Type Receptacles Provided.   Every recreational vehicle site with electrical supply shall be equipped with at least one 20-ampere, 125-volt receptacle. A minimum of 75 percent of all recreational vehicle sites with electrical supply shall each be equipped with a 30-ampere, 125-volt receptacle conforming to Figure 551-46(c). This supply shall be permitted to include additional receptacle configurations conforming to Section 551-81. The remainder of all recreational vehicle sites with electrical supply shall be equipped with one or more of the receptacle configurations conforming to Section 551-81. All 20-ampere, 125-volt receptacles shall have listed ground-fault circuit-interrupter protection for personnel.

Additional receptacles shall be permitted for the connection of electrical equipment outside the recreational vehicle within the recreational vehicle park. All such 125-volt, single-phase, 15- and 20-ampere receptacles shall have ground-fault circuit-interrupter protection for personnel.

551-72. Distribution System.   The recreational vehicle park secondary electrical distribution system to recreational vehicle sites shall be derived from a single-phase 120/240-volt, 3-wire system.

(FPN): On a 120/240-volt, 3-wire, single-phase service or feeder, the neutral cannot be reduced in size (below the size of the ungrounded conductors) if there are no 240-volt loads, because, under the most severe conditions of unbalance, the neutral will carry the same current as the ungrounded conductor supplying the load.

551-73. Calculated Load.

(a) Basis of Calculations. Electrical service and feeders shall be calculated on the basis of not less than 9,600 volt-amperes per site equipped with 50-ampere 120/240-volt supply facilities, 3,600 volt-amperes per site equipped with both 20-ampere and 30-ampere supply facilities, and 2,400 volt-amperes per site equipped with only 20-ampere supply facilities. The demand factors set forth in Table 551-73 shall be the minimum allowable demand factors that shall be permitted in calculating load for service and feeders.

Loads for other amenities such as, but not limited to, service buildings, recreational buildings, and swimming pools shall be sized separately and then be added to the value calculated for the recreational vehicle sites where they are all supplied by one service.

Table 551-73.
Demand Factors for Site Feeders and Service-Entrance
Conductors for Park Sites

Number of Recreational Vehicle Sites	Demand Factor (percent)	Number of Recreational Vehicle Sites	Demand Factor (percent)
1	100	10-12	50
2	90	13-15	48
3	80	16-18	47
4	75	19-21	45
5	65	22-24	43
6	60	25-35	42
7-9	55	36 plus	41

(b) Transformers and Secondary Distribution Panelboards. For the purpose of this Code, where the park service exceeds 240 volts, transformers and secondary distribution panelboards shall be treated as services.

(c) Demand Factors. The demand factor for a given number of sites shall apply to all sites indicated. For example: twenty sites calculated at 42 percent of 3,600 volt-amperes result in a permissible demand of 1512 volt-amperes per site or a total of 30,240 volt-amperes for twenty sites.

(FPN): These demand factors may be inadequate in areas of extreme hot or cold temperature with loaded circuits for heating or air conditioning.

(d) Feeder Circuit Capacity. Recreational vehicle site feeder circuit conductors shall have adequate ampacity for the loads supplied and shall be rated at not less than 30 amperes. The grounded conductors shall have the same ampacity as the ungrounded conductors.

(FPN): Due to the long circuit lengths typical in most recreational vehicle parks, feeder conductor sizes found in ampacity tables of Article 310 may be inadequate to maintain the voltage regulation suggested in the Fine Print Note to Section 210-19. Total circuit voltage drop is a summation of the voltage drops of each serial circuit segment, where the load for each segment is calculated using the load that segment sees and the demand factors of Section 551-73(a).

551-74. Overcurrent Protection. Overcurrent protection shall be provided in accordance with Article 240.

551-75. Grounding. All electrical equipment and installations in recreational vehicle parks shall be grounded as required by Article 250.

551-76. Recreational Vehicle Site Supply Equipment.

(a) Location. Where provided, the recreational vehicle site electrical supply equipment shall be located on the left (road) side of the parked vehicle, on a line that is 9 feet (2.74 m), ± 1 foot (0.3 m), from the longitudinal centerline of the stand and shall be located at any point on this line from the rear of the stand to 15 feet (4.57 m) forward of the rear of the stand.

Exception: For pull-through sites, it shall be permitted to locate the electrical supply equipment at any point along the line from 16 feet (4.88 m) forward of the rear of the stand to 32 feet (9.75 m) forward of the rear of the stand.

(b) Disconnecting Means. A disconnecting switch or circuit breaker shall be provided in the site supply equipment for disconnecting the power supply to the recreational vehicle.

(c) Access. All site supply equipment shall be accessible by an unobstructed entrance or passageway not less than 2 feet (610 mm) wide and 6½ feet (1.98 m) high.

(d) Mounting Height. Site supply equipment shall be located not less than 2 feet (610 mm) nor more than 6½ feet (1.98 m) above the ground.

(e) Working Space. Sufficient space shall be provided and maintained about all electric equipment to permit ready and safe operation, in accordance with Section 110-16.

551-77. Grounding, Recreational Vehicle Site Supply Equipment.

(a) Exposed Noncurrent-Carrying Metal Parts. Exposed noncurrent-carrying metal parts of fixed equipment, metal boxes, cabinets, and fittings, that are not electrically connected to grounded equipment, shall be grounded by a continuous equipment grounding conductor run with the circuit conductors from the service equipment or from the transformer of a secondary distribution system. Equipment grounding conductors shall be sized in accordance with Section 250-95.

The arrangement of equipment grounding connnections shall be such that the disconnection or removal of a receptacle or other device will not interfere with, or interrupt, the grounding continuity.

(b) Secondary Distribution System. Each secondary distribution system shall be grounded at the transformer.

(c) Neutral Conductor Not to Be Used as an Equipment Ground. The neutral conductor shall not be used as an equipment ground for recreational vehicles or equipment within the recreational vehicle park.

(d) No Connection on the Load Side. No connection to a grounding electrode shall be made to the neutral conductor on the load side of the service disconnecting means or tranformer distribution panelboard.

551-78. Protection of Outdoor Equipment.

(a) Wet Locations. All switches, circuit breakers, receptacles, control equipment, and metering devices located in wet locations or outside of a building shall be rainproof equipment.

(b) Meters. If secondary meters are installed, meter sockets without meters installed shall be blanked-off with an approved blanking plate.

551-79. Clearance for Overhead Conductors.
Open conductors of not over 600 volts, nominal, shall have a vertical clearance of not less than 18 feet (5.49 m) and a horizontal clearance of not less than 3 feet (914 mm) in all areas subject to recreational vehicle movement. In all other areas, clearances shall conform to Sections 225-18 and 225-19.

(FPN): For clearances of conductors over 600 volts, nominal, see National Electrical Safety Code, ANSI C2-1990.

551-80. Underground Service, Feeder, Branch-Circuit and Recreational Vehicle Site Feeder Circuit Conductors.

(a) General. All direct-burial conductors, including the equipment grounding conductor if of aluminum, shall be insulated and identified for the use. All conductors shall be continuous from equipment to equipment. All splices and taps shall be made in approved junction boxes or by use of material listed and identified for the purpose.

(b) Protection Against Physical Damage. Direct-buried conductors and cables entering or leaving a trench shall be protected by rigid metal conduit, intermediate metal conduit, electrical metallic tubing with supplementary corrosion protection, rigid nonmetallic conduit, or other approved raceways or enclosures. Where subject to physical damage, the conductors or cables shall be protected by rigid metal conduit, intermediate conduit, or Schedule 80 rigid nonmetallic conduit. All such protection shall extend at least 18 inches (457 mm) into the trench from finished grade.

(FPN): See Section 300-5 and Article 339 for conductors or Type UF cable used underground or in direct burial in earth.

551-81. Receptacles. A receptacle to supply electric power to a recreational vehicle shall be one of the configurations shown in Figure 551-46(c) in the following ratings:

(a) 50-Ampere. 125/250 volts, 50-ampere, 3-pole, 4-wire, grounding type for 120/240-volt systems.

(b) 30-Ampere. 125-volt, 30-ampere, 2-pole, 3-wire, grounding type for 120-volt systems.

(c) 20-Ampere. 125-volt, 20-ampere, 2-pole, 3-wire, grounding type for 120-volt systems.

(FPN): Complete details of these configurations can be found in American National Standard Dimensions of Caps, Plugs and Receptacles, ANSI C73.17-1972; ANSI C73.13-1972; and C73.12-1972.

ARTICLE 553 — FLOATING BUILDINGS

A. General

553-1. Scope. This article covers wiring, services, feeders, and grounding for floating buildings.

553-2. Definition.

Floating Building. A building unit as defined in Article 100 that floats on water, is moored in a permanent location, and has a premises wiring system served through connection by permanent wiring to an electricity supply system not located on the premises.

553-3. Application of Other Articles. Wiring for floating buildings shall comply with the applicable provisions of other articles of this Code, except as modified by this article.

B. Services and Feeders

553-4. Location of Service Equipment. The service equipment for a floating building shall be located adjacent to, but not in or on, the building.

553-5. Service Conductors. One set of service conductors shall be permitted to serve more than one set of service equipment.

553-6. Feeder Conductors. Each floating building shall be supplied by a single set of feeder conductors from its service equipment.

Exception: Where the floating building has multiple occupancy, each occupant shall be permitted to be supplied by a single set of feeder conductors extended from the occupant's service equipment to the occupant's panelboard.

553-7. Installation of Services and Feeders.

(a) Flexibility. Flexibility of the wiring system shall be maintained between floating buildings and the supply conductors. All wiring shall be so installed that motion of the water surface and changes in the water level will not result in unsafe conditions.

(b) Wiring Methods. Liquidtight flexible metal conduit or liquidtight flexible nonmetallic conduit with approved fittings shall be permitted for feeders and where flexible connections are required for services. Extra-hard usage portable power cable listed for both wet locations and sunlight resistance shall be permitted for a feeder to a floating building where flexibility is required.

(FPN): See Sections 555-1 and 555-6.

C. Grounding

553-8. General Requirements. Grounding of both electrical and nonelectrical parts in a floating building shall be through connection to a grounding bus in the building panelboard. The grounding bus shall be grounded through a green-colored insulated equipment grounding conductor run with the feeder conductors and connected to a grounding terminal in the service equipment. The grounding terminal in the service equipment shall be grounded by connection through a green insulated grounding electrode conductor to a grounding electrode on shore.

553-9. Insulated Neutral. The grounded circuit conductor (neutral) shall be a white insulated conductor. The neutral conductor shall be connected to the equipment grounding terminal in the service equipment, and, except for that connection, it shall be insulated from the equipment grounding conductors, equipment enclosures, and all other grounded parts. The neutral circuit terminals in the panelboard and in ranges, clothes dryers, counter-mounted cooking units, and the like shall be insulated from the enclosures.

553-10. Equipment Grounding.

(a) Electrical Systems. All enclosures and exposed metal parts of electrical systems shall be bonded to the grounding bus.

(b) Cord-Connected Appliances. Where required to be grounded, cord-connected appliances shall be grounded by means of an equipment grounding conductor in the cord and a grounding-type attachment plug.

553-11. Bonding of Noncurrent-Carrying Metal Parts. All metal parts in contact with the water, all metal piping, and all noncurrent-carrying metal parts that may become energized, shall be bonded to the grounding bus in the panelboard.

ARTICLE 555 — MARINAS AND BOATYARDS

555-1. Scope. This article covers the installation of wiring and equipment in the areas comprising fixed or floating piers, wharfs, docks, and other areas in marinas, boatyards, boat basins, and similar establishments that are used, or intended for use, for the purpose of repair, berthing, launching, storage, or fueling of small craft and the moorage of floating buildings.

555-2. Application of Other Articles. Wiring and equipment for marinas and boatyards shall comply with this article and also with the applicable provisions of other articles of this Code.

(FPN No. 1): See fine print notes following Sections 210-19(a) and 215-2(b) for voltage drop on branch circuits and feeders, respectively.

(FPN No. 2): For disconnection of auxiliary power from boats, see Pleasure and Commercial Motor Craft, NFPA 302-1989 (ANSI).

555-3. Receptacles. Where shore power is supplied, those accommodations for boats 20 feet (6.1 m) or less in length shall be equipped with shore-power receptacles of a locking and grounding type rated at not less than 20 amperes.

Where shore power is supplied to accommodations for boats longer than 20 feet (6.1 m) in length, shore-power receptacles of a locking and grounding type rated at 30 amperes or more shall be provided.

Fifteen- and 20-ampere, single-phase, 125-volt receptacles other than those supplying shore power to boats located at piers, wharfs, and other locations shall be protected by ground-fault circuit-interrupters.

(FPN No. 1): For various configurations and ratings of locking- and grounding-type receptacles and caps, see Dimensions of Caps, Plugs, and Receptacles, ANSI C73-1972, and Supplement ANSI C73a-1980.

(FPN No. 2): For locking- and grounding-type receptacles for auxiliary power to boats, see Marinas and Boatyards, NFPA 303-1990 (ANSI).

(FPN No. 3): In locating receptacles, consideration should be given to the maximum tide level and wave action. See Marinas and Boatyards, NFPA 303-1990 (ANSI), for establishment of datum plane.

555-4. Branch Circuits. Each single receptacle that supplies shore power to boats shall be supplied from a power outlet or panelboard by an individual or multiwire branch circuit of the voltage class and rating corresponding to the rating of the receptacle.

(FPN): Supplying receptacles at voltages other than the voltages marked on the receptacle may cause overheating or malfunctioning of connected equipment; for example, supplying single-phase, 120/240V, 3-wire loads for a 208Y/120V, 3-wire source.

555-5. Feeders and Services. The load for each ungrounded feeder and service conductor supplying receptacles that supply shore power for boats shall be calculated as follows:

For 1-4 receptacles	100%	of the sum of the rating of the receptacles
For 5-8	90%	"	"	"	"
For 9-14	80%	"	"	"	"
For 15-30	70%	"	"	"	"
For 31-40	60%	"	"	"	"
For 41-50	50%	"	"	"	"
For 51-70	40%	"	"	"	"
For 71-100	30%	"	"	"	"
For 101 plus	20%	"	"	"	"

(FPN): These demand factors may be inadequate in areas of extreme hot or cold temperature with loaded circuits for heating, air-conditioning, or refrigerating equipment.

555-6. Wiring Methods. The wiring method shall be of a type identified for use in wet locations. Extra-hard usage portable power cable listed for both wet locations and sunlight resistance shall be permitted for a feeder where flexibility is required.

Open wiring shall be permitted only by special permission.

(FPN No. 1): In granting special permission, major factors include possible contact of open wires with masts, cranes, or similar structures or equipment.

(FPN No. 2): For further information on wiring methods for various locations and for establishment of datum plane, see Marinas and Boatyards, NFPA 303-1990 (ANSI).

555-7. Grounding.

(a) Equipment to Be Grounded. The following items shall be connected to an equipment grounding conductor run with the circuit conductors in a raceway or cable:

(1) Boxes, cabinets, and all other metal enclosures.

(2) Metal frames of utilization equipment.

(3) Grounding terminals of grounding-type receptacles.

(b) Type of Equipment Grounding Conductor. The equipment grounding conductor shall be an insulated copper conductor with a continuous outer finish that is either green or green with one or more yellow stripes.

Exception: The equipment grounding conductor of Type MI cable shall be permitted to be identified at terminations.

(c) Size of Equipment Grounding Conductor. The insulated copper equipment grounding conductor shall be sized in accordance with Section 250-95 but not smaller than No. 12.

(d) Branch-Circuit Equipment Grounding Conductor. The insulated equipment grounding conductor for branch circuits shall terminate at a grounding terminal in a remote panelboard or the grounding terminal in the main service equipment.

(e) Feeder Equipment Grounding Conductors. Where a feeder supplies a remote panelboard, an insulated equipment grounding conductor shall extend from a grounding terminal in the service equipment to a grounding terminal in the remote panelboard.

555-8. Wiring Over and Under Navigable Water. Wiring over and under navigable water shall be subject to approval by the authority having jurisdiction.

555-9. Gasoline Dispensing Stations — Hazardous (Classified) Locations. Electrical equipment and wiring located in gasoline dispensing stations shall comply with Article 514.

(FPN): For further information, see Automotive and Marine Service Station Code, NFPA 30A-1990, and Marinas and Boatyards, NFPA 303-1990 (ANSI).

555-10. Location of Service Equipment. The service equipment for floating docks or marinas shall be located adjacent to, but not on or in, the floating structure.

Chapter 6.   Special Equipment

ARTICLE 600 — ELECTRIC SIGNS AND OUTLINE LIGHTING

A. General

600-1. Scope.   This article covers the installation of conductors and equipment for electric signs and outline lighting as defined in Article 100.

600-2. Disconnect Required.   Each outline lighting installation, and each sign of other than the portable type, shall be controlled by an externally operable switch or breaker that will open all ungrounded conductors.

(a) In Sight of Sign.   The disconnecting means shall be within sight of the sign or outline lighting that it controls.

Exception No. 1: A disconnecting means shall not be required for an exit directional sign connected to a circuit within the scope of Article 700.

Exception No. 2: Signs or outline lighting operated by electronic or electromechanical controllers located external to the sign shall have a disconnecting means located within sight from the controller location. The disconnecting means shall disconnect the sign and the controller from all ungrounded supply conductors and shall be so designed that no pole can be operated independently. The disconnecting means shall be permitted to be in the same enclosure with the controller. The disconnecting means shall be capable of being locked in the open position.

(b) Control Switch Rating.   Switches, flashers, and similar devices controlling transformers shall be either rated for controlling inductive load(s) or have an ampere rating not less than twice the ampere rating of the transformer.

(FPN):   See Section 380-14 for rating of snap switches.

600-3. Enclosures as Pull Boxes.   The wiring method used to supply signs and outline lighting shall terminate in the sign or transformer enclosures.

Exception: Such signs and transformer boxes shall be permitted to be used as pull or junction boxes for conductors supplying other adjacent signs, outline lighting systems, and floodlights that are part of signs, provided the conductors extending from the equipment are protected by an overcurrent device rated 20 amperes or less.

600-4. Listing Required.   Every electric sign of any type, fixed or portable, shall be listed and installed in conformance with that listing, unless otherwise permitted by special permission.

600-5. Grounding.   Signs, troughs, tube terminal boxes, and other metal frames shall be grounded in the manner specified in Article 250.

Exception: Isolated Parts. Isolated noncurrent-carrying metal parts of outline lighting shall be permitted to be bonded by No. 14 conductors, protected from physical damage, and grounded in accordance with Article 250.

600-6. Branch Circuits.

(a) Rating. Circuits that supply lamps, ballasts, and transformers, or combinations, shall be rated not to exceed 20 amperes. Circuits containing electric-discharge lighting transformers exclusively shall not be rated in excess of 30 amperes.

(b) Required Branch Circuit. Each commercial building and each commercial occupancy accessible to pedestrians shall be provided at an accessible location outside the entrance to each tenant space, with at least one outlet for sign or outline lighting use. The outlet(s) shall be supplied by a 20-ampere branch circuit that supplies no other load.

Exception: Interior service hallways or corridors shall not be considered outside the occupancy.

(c) Computed Load. The load for the required branch circuit installed for the supply of exterior signs or outline lighting shall be computed at a minimum of 1200 volt-amperes.

600-7. Marking.

(a) Signs. Signs shall be marked with the maker's name; and, for incandescent lamp signs, with the number of lampholders; and, for electric-discharge-lamp signs, with input amperes at full load and input voltage. The marking of the sign shall be visible after installation.

(b) Transformers. Transformers shall be marked with the maker's name; and transformers for electric-discharge-lamp signs shall be marked with the input rating in amperes or volt-amperes, the input voltage, and the open-circuit output voltage.

600-8. Enclosures.

(a) Conductors and Terminals. Conductors and terminals in sign boxes, cabinets, and outline troughs shall be enclosed in metal or other noncombustible material.

Exception: The supply leads shall not be required to be enclosed.

(b) Cutouts, Flashers, Etc. Cutouts, flashers, and similar devices shall be enclosed in metal boxes, the doors of which shall be arranged so they can be opened without removing obstructions or finished parts of the enclosure.

(c) Strength. Enclosures shall have ample strength and rigidity.

(d) Material. Signs and outline lighting shall be constructed of metal or other noncombustible material. Wood shall be permitted for external decoration if placed not less than 2 inches (50.8 mm) from the nearest lampholder or current-carrying part.

Exception: Portable signs of the indoor type shall not be required to meet this requirement.

(e) Minimum Thickness — Enclosure Metal. Sheet copper or aluminum shall be at least 0.020 inches (508 micrometers) thick. Sheet steel shall be of No. 28 MSG.

Exception: For outline lighting and for electric-discharge signs, sheet steel shall be of No. 24 MSG if not ribbed, corrugated, or embossed over its entire surface and of No. 26 MSG if it is so ribbed, corrugated, or embossed.

(f) Protection of Metal. All steel parts of enclosures shall be galvanized or otherwise protected from corrosion.

(g) Enclosures Exposed to Weather. Enclosures for outdoor use shall be weatherproof and shall have at least two drain holes, each not larger than $\frac{1}{2}$ inch (12.7 mm) or smaller than $\frac{1}{4}$ inch (6.35 mm).

600-9. Portable Signs or Sections. Portable signs or sections, letters, fixtures, symbols, and similar displays used in conjunction with fixed outdoor signs shall only be used where in compliance with all applicable provisions of this Code and, in addition, shall meet all of the following requirements:

(a) Weatherproof Receptacle and Attachment Plug. A weatherproof receptacle and attachment plug having one pole for grounding shall be provided for each individual letter, fixture, or sign.

(b) Cords. All cords shall be of the junior hard service or hard service types as designated in Table 400-4 and shall be 3-conductor with one conductor grounded as provided in Section 600-9(a).

(c) Cord from Ground Level. No cord shall be less than 10 feet (3.05 m) from the ground level directly underneath.

600-10. Clearances.

(a) Vertical and Horizontal. Signs and outline system enclosures shall have not less than the vertical and horizontal clearances from open conductors specified in Article 225.

(b) Elevation. The bottom of sign and outline lighting enclosures shall not be less than 16 feet (4.88 m) above areas accessible to vehicles.

Exception: The bottom of such enclosures shall be permitted to be less than 16 feet (4.88 m) above areas accessible to vehicles where such enclosures are protected from physical damage.

600-11. Outdoor Portable Signs. The wiring of an outdoor sign that is portable or mobile and is readily accessible shall be provided with factory-installed ground-fault circuit-interrupter protection for personnel. The ground-fault circuit-interrupter shall be an integral part of the attachment plug or shall be located in the power-supply cord within 12 inches (305 mm) of the attachment plug. Conductive supports of a sign specified in this section shall be considered part of the sign.

B. 1000 Volts, Nominal, or Less

600-21. Installation of Conductors.

(a) Wiring Method. Conductors shall be installed using any wiring method included in Chapter 3 suitable for the conditions that provides a means for equipment grounding, or in metal poles complying with all the requirements of Section 410-15(b).

(b) Insulation and Size. Conductors shall be of a type listed for general use and shall not be smaller than No. 14.

Exception No. 1: Conductors not smaller than No. 18 of a type listed in Table 402-3 shall be permitted:

a. In portable signs.

b. As short leads permanently attached to lampholders or electric-discharge ballasts.

c. As leads not more than 8 feet (2.44 m) long permanently attached to electric-discharge lampholders or electric-discharge ballasts, if the leads are enclosed in wiring channels.

d. For signs with multiple incandescent lamps requiring one conductor from a control to one or more lamps whose total load does not exceed 250 watts, if in an approved cable assembly of two or more conductors.

Exception No. 2: Conductors not smaller than No. 20 shall be permitted as short leads permanently attached to synchronous motors.

(c) Exposed to Weather. Conductors in raceways, metal-clad cable, or enclosures exposed to the weather shall be of a type listed for the conditions.

Exception: This shall not apply where rigid metal conduit, intermediate metal conduit, rigid nonmetallic conduit, electrical metallic tubing, or enclosures are made raintight and arranged to drain.

(d) Number of Conductors in Raceway. The number of conductors in a raceway for sign fixtures shall be in accordance with Table 1 of Chapter 9.

(e) Conductors Soldered to Terminals. Where the conductors are fastened to lampholders other than of the pin type, they shall be soldered to the terminals or made with wire connectors, and the exposed parts of conductors and terminals shall be treated to prevent corrosion. Where the conductors are fastened to pin-type lampholders that protect the terminals from the entrance of water, and that have been found acceptable for sign use, the conductors shall be of the stranded type but shall not be required to be soldered to the terminals.

600-22. Lampholders.
Lampholders shall be of the unswitched type having bodies of suitable insulating material and shall be so constructed and installed as to prevent turning. The screw-shell of all sign lampholders shall be connected to the grounded conductor of the circuit. Lampholders for outdoor signs shall be suitable for the application.

600-23. Conductors Within Signs and Troughs.
Wires within the sign and outline lighting troughs shall be installed as to be mechanically secure.

600-24. Protection of Leads. Bushings shall be employed to protect wires feeding through enclosures.

C. Over 1000 Volts, Nominal |

600-31. Installation of Conductors.

(a) **Wiring Method.** Conductors shall be installed as concealed conductors on insulators, in rigid metal conduit, in intermediate metal conduit, in rigid nonmetallic conduit, in flexible metal conduit, in liquidtight flexible metal conduit, or in electrical metallic tubing, or as Type MC cable.

(FPN):   See Section 600-5 for grounding requirements. |

Exception: Liquidtight flexible nonmetallic conduit shall be permitted where requirements for flexibility exist and where weather conditions require corrosion protection.

(b) **Insulation and Size.** Conductors shall be of a type identified for voltage not less than the voltage of the circuit and shall not be smaller than No. 14.

Exception: Conductors not smaller than No. 18 shall be permitted:

a. As leads not more than 8 feet (2.44 m) long permanently attached to electric-discharge lampholders or electric-discharge ballasts, if the leads are enclosed in wiring channels.

b. In show window displays or small portable signs, as leads not more than 8 feet (2.44 m) long that run from the line ends of the tubing to the secondary windings of transformers, if the leads are permanently attached within the transformer enclosure.

(c) **Bends in Conductors.** Sharp bends in the conductors shall be avoided.

(d) **Concealed Conductors on Insulators — Indoors.** Concealed conductors on insulators shall be separated from each other and from all objects other than the insulators on which they are mounted by a spacing of not less than 1½ inches (38 mm) for voltages above 10,000 and not less than 1 inch (25.4 mm) for voltages of 10,000 or less. They shall be installed in channels lined with noncombustible material and used for no other purpose, except that the primary circuit conductors shall be permitted to be in the same channel. The insulators shall be of noncombustible, nonabsorbent material. Concealed conductors on insulators shall not be allowed outside the sign enclosure.

(e) **Conductors in Raceways.** Where conductors are covered with metal sheathing, the covering shall extend beyond the end of the raceway, and the surface of the cable shall not be damaged where the covering terminates.

(1) In damp or wet locations, the insulation on all conductors shall extend beyond the metal covering or raceway not less than 4 inches (102 mm) for voltages over 10,000, 3 inches (76 mm) for voltages over 5000 but not exceeding 10,000, and 2 inches (50.8 mm) for voltages of 5000 or less.

(2) In dry locations, the insulation shall extend beyond the end of the metal covering or raceways not less than 2½ inches (64 mm) for voltages

over 10,000, 2 inches (50.8 mm) for voltages over 5000 but not exceeding 10,000, and 1½ inches (38 mm) for voltages of 5000 or less.

(3) For conductors at grounded midpoint terminals, no spacing shall be required.

(4) A metal raceway containing a single conductor from one secondary terminal of a transformer shall not exceed 20 feet (6.1 m) in length.

(f) Show Windows and Similar Locations. Conductors that hang freely in the air, away from combustible material, and where not subject to physical damage, as in some show window displays, shall be required to be insulated for the voltage but not otherwise protected.

(g) Between Tubing and Grounded Midpoint. Conductors shall be permitted to be run from the ends of tubing to the grounded midpoint of transformers specifically designed for the purpose and provided with terminals at the midpoint. Where such connections are made to the transformer grounded midpoint, the connections between the high-voltage terminals of the transformer and the line ends of the tubing shall be as short as possible.

600-32. Transformers.

(a) Voltage. The transformer secondary open-circuit voltage shall not exceed 15,000 volts with an allowance on test of 1000 volts additional. For end-grounded transformers, the secondary open-circuit voltage shall not exceed 7500 volts with an allowance on test of 500 volts additional.

(b) Type. Transformers shall be of a type identified for use with electrical-discharge tubing and shall be limited in rating to a maximum of 4500 volt-amperes.

Open core-and-coil-type transformers shall be limited to 5000 volts with an allowance on test of 500 volts and to indoor applications in small portable signs.

Transformers for outline lighting installations shall have secondary current ratings not more than 60 milliamperes.

Exception: Where the transformers and all wiring connected to them are installed in accordance with Article 410 for electric-discharge lighting of the same voltage.

(c) Exposed to Weather. Transformers used outdoors shall be of the weatherproof type or shall be protected from the weather by enclosure in the sign body or in a separate metal box.

(d) Transformer Secondary Connections. The secondary windings of transformers shall not be connected in parallel or in series.

Exception No. 1: Two transformers, each having one end of its secondary winding connected to the metal enclosure, shall be permitted to have their secondary windings connected in series to form the equivalent of a midpoint-grounded transformer. The grounded ends shall be connected by insulated conductors not smaller than No. 14.

Exception No. 2: Transformers for small portable signs in show windows and similar locations, listed for the purpose, shall be permitted to be connected in series where they are equipped with secondary leads permanently attached to the secondary winding within the transformer enclosure.

The secondary leads shall not extend more than 8 feet (2.44 m) beyond the enclosure for attachment to the electrode ends of the tubing. The grounded ends shall be connected by insulated conductors not smaller than No. 18.

(e) Accessibility. Transformers shall be located where accessible and shall be securely fastened in place.

(f) Working Space. A work space at least 3 feet (914 mm) high and measuring at least 3 feet (914 mm) by 3 feet (914 mm) horizontally shall be provided about each transformer or its enclosure where not installed in a sign.

(g) Attic Locations. Transformers shall be permitted to be located in attics, provided there is a passageway at least 3 feet (914 mm) in height and at least 2 feet (610 mm) in width, provided with a suitable, permanent, fixed walkway or catwalk at least 12 inches (305 mm) in width extending from the point of entry into the attic to each transformer.

600-33. Electric-Discharge Tubing.

(a) Design. The tubing shall be of such length and design as not to cause a continuous overvoltage on the transformer.

(b) Support. Tubing shall be adequately supported on noncombustible, nonabsorbent supports. Tubing supports shall, where practicable, be adjustable.

(c) Contact with Flammable Material and Other Surfaces. The tubing shall be free from contact with flammable material and shall be located where not normally exposed to physical damage. Where operating at over 7500 volts, the tubing shall be supported on noncombustible, nonabsorbent insulating supports that maintain a spacing of not less than ¼ inch (6.35 mm) between the tubing and the nearest surface.

600-34. Terminals and Electrode Receptacles for Electric-Discharge Tubing.

(a) Terminals. Terminals of the tubing shall be inaccessible to unqualified persons and isolated from combustible material and grounded metal or shall be enclosed. Where enclosed, they shall be separated from grounded metal and combustible material by noncombustible, nonabsorbent insulating material or by not less than 1½ inches (38 mm) of air. Terminals shall be relieved from stress by the independent support of the tubing.

(b) Tube Connections Other than with Receptacles. Where tubes do not terminate in receptacles designed for the purpose, all energized parts of tube terminals and conductors shall be separated from grounded metal and combustible material by noncombustible, nonabsorbent insulating material or be supported so as to maintain a separation of not less than 1½ inches (38 mm) between conductors or between conductors and any grounded metal. Soldering or connection devices shall not be required for the connection of tube terminals to the conductors where they are spliced or joined so that they are mechanically and electrically secure.

(c) Receptacles. Electrode receptacles for the tubing shall be of noncombustible, nonabsorbent insulating material.

(d) Bushings. Where electrodes enter the enclosure of outdoor signs or of an indoor sign operating at a voltage in excess of 7500 volts, bushings shall be used unless receptacles are provided. Electrode terminal assemblies shall be supported not more than 6 inches (152 mm) from the electrode terminals. All electric-discharge tubing terminals shall be insulated with a material suitable for the voltages and environmental conditions expected during use.

(e) Show Windows. In the exposed type of show-window signs, terminals shall be enclosed by receptacles.

(f) Receptacles and Bushing Seals. A flexible, nonconducting seal shall be permitted to close the opening between the tubing and the receptacle or bushing against the entrance of dust or moisture. This seal shall not be in contact with grounded conductive material and shall not be depended upon for the insulation of the tubing.

(g) Enclosures of Metal. Enclosures of metal for electrodes shall not be less than No. 24 MSG sheet metal.

(h) Enclosures of Insulating Material. Enclosures of insulating material shall be noncombustible, nonabsorbent, and suitable for the voltage of the circuit.

(i) Energized Parts. Energized parts shall be enclosed or suitably guarded to prevent contact.

600-35. Switches on Doors. Doors or covers giving access to uninsulated parts of indoor signs or outline lighting exceeding 600 volts, nominal, and accessible to unqualified persons shall either be provided with interlock switches that on the opening of the doors or covers disconnect the primary circuit, or shall be so fastened that the use of other than ordinary tools will be necessary to open them.

600-36. Fixed Outline Lighting and Skeleton-type Signs for Interior Use.

(a) Tube Support. Gas tubing shall be supported independently of the conductors by means of insulators of noncombustible, nonabsorptive materials such as glass or porcelain or by suspension from suitable wires or chains.

(b) Transformers. Transformers shall be installed in metal enclosures and as near as practicable to the gas tubing system.

(c) Secondary Conductors. Secondary conductors shall be insulated for the voltage of the circuit and shall be enclosed in grounded metal raceway.

Exception: Conductors not exceeding 4 feet (1.22 m) in length between gas tubing and adjacent metal enclosures shall be permitted to be enclosed in continuous glass or other insulating sleeves.

600-37. Portable Gas Tube Signs for Show Windows and Interior Use. This section shall apply to the installation and use of portable gas tube signs.

(a) Location. Portable gas tube signs shall be for indoor use only.

(b) Transformer. The transformer shall be of the window type or shall be within a metal enclosure.

(c) Supply Conductors. Supply conductors shall consist of hard or extra-hard usage-type cord containing an equipment grounding conductor. The cord shall not exceed 10 feet (3.05 m) in length.

(d) Secondary Conductors. Secondary conductors shall not be more than 6 feet (1.83 m) long and shall be located where not subject to physical damage, and shall be insulated for the voltage of the circuit and be protected by continuous glass or other insulating sleeves or tubing.

(e) Grounding. Transformers and attached noncurrent-carrying metal parts shall be grounded in the manner specified in Article 250.

(f) Support. Portable indoor signs shall be held in place by not more than two open hooks attached to the transformer case.

ARTICLE 604 — MANUFACTURED WIRING SYSTEMS

604-1. Scope. The provisions of this article apply to field-installed wiring using off-site manufactured subassemblies for branch circuits, remote-control circuits, signaling circuits, and communication circuits in accessible areas.

604-2. Definition.

Manufactured Wiring System: A system containing component parts that are assembled in the process of manufacture and cannot be inspected at the building site without damage or destruction to the assembly.

604-3. Other Articles. Except as modified by the requirements of this article, all other applicable articles of this Code shall apply.

604-4. Uses Permitted. The manufactured wiring systems shall be permitted in accessible and dry locations and in plenums and spaces used for environmental air, where listed for this application, and installed in accordance with Section 300-22.

Exception: In concealed spaces, one end of tapped cable shall be permitted to extend into hollow walls for direct termination at switch and outlet points.

604-5. Uses Not Permitted. Where conductors or cables are limited by the provisions in Articles 333 and 334.

604-6. Construction.

(a) Cable or Conduit Types.

(1) Cable shall be listed armored cable or metal-clad cable containing nominal 600-volt No. 10 or 12 copper insulated conductors with a bare or insulated copper equipment grounding conductor equivalent in size to the ungrounded conductor.

(2) Conduit shall be listed flexible metal conduit containing nominal 600-volt No. 10 or 12 copper insulated conductors with a bare or insulated copper equipment grounding conductor equivalent in size to the ungrounded conductor.

Exception No. 1 to (1) and (2): A fixture tap maximum 6 feet (1.83 m) long intended for connection to a single fixture shall be permitted to contain conductors smaller than No. 12 but not smaller than No. 18.

Exception No. 2 to (1) and (2): Conductors smaller than No. 12 shall be permitted for remote-control, signaling, or communications circuits. The assembly shall be listed for the purpose.

(3) Each section shall be marked to identify the type of cable or conduit.

(b) Receptacles and Connectors. Receptacles and connectors shall be locking type, uniquely polarized and identified for the purpose, and shall be part of a listed assembly for the appropriate system.

(c) Other Component Parts. Other component parts shall be listed for the appropriate system.

604-7. Unused Outlets. All unused outlets shall be capped to effectively close the connector openings.

ARTICLE 605 — OFFICE FURNISHINGS

(Consisting of Lighting Accessories and Wired Partitions)

605-1. Scope. This article covers electrical equipment, lighting accessories, and wiring systems used to connect, or contained within, or installed on relocatable wired partitions.

605-2. General. Wiring systems shall be identified as suitable for providing power for lighting accessories and appliances in wired partitions. These partitions shall not extend from floor to ceiling.

Exception: Where permitted by the authority having jurisdiction, these relocatable wired partitions shall be permitted to extend to the ceiling but shall not penetrate the ceiling.

(a) Use. These assemblies shall be installed and used only as provided for by this article.

(b) Other Articles. Except as modified by the requirements of this article, all other articles of this Code shall apply.

(c) Hazardous (Classified) Locations. Where used in hazardous (classified) locations, these assemblies shall conform with Articles 500 through 517 in addition to this article.

605-3. Wireways. All conductors and connections shall be contained within wiring channels of metal or other material identified as suitable for the conditions of use. Wiring channels shall be free of projections or other conditions that may damage conductor insulation.

(FPN):   Conductors as used in this section do not include flexible cord.

605-4. Partition Interconnections. The electrical connection between partitions shall be a flexible assembly identified for use with wired partitions.

Exception: Flexible cord shall be permitted for the connection between partitions, provided all of the following conditions are met:

   a. The cord is extra-hard usage type.

   b. The partitions are mechanically contiguous.

   c. The cord is not longer than necessary for maximum positioning of the partitions but is in no case to exceed 2 feet (610 mm).

   d. The cord is terminated at an attachment plug and cord-connector with strain relief.

605-5. Lighting Accessories. Lighting equipment listed and identified for use with wired partitions shall comply with all of the following:

   (a) Support. A means for secure attachment or support shall be provided.

   (b) Connection. Where cord- and plug-connection is provided, the cord length shall be suitable for the intended application but shall not exceed 9 feet (2.74 m) in length. The cord shall not be smaller than No. 18, shall | contain an equipment grounding conductor, and shall be of the hard usage type. Connection by other means shall be identified as suitable for the condition of use.

   (c) Receptacle Outlet. Convenience receptacles shall not be permitted in lighting accessories.

605-6. Fixed-type Partitions. Wired partitions that are fixed (secured to building surfaces) shall be permanently connected to the building electrical system by one of the wiring methods of Chapter 3.

605-7. Freestanding-Type Partitions. Partitions of the freestanding type (not fixed) shall be permitted to be permanently connected to the building electrical system by one of the wiring methods of Chapter 3.

605-8. Freestanding-Type Partitions, Cord- and Plug-Connected. Individual partitions of the freestanding type, or groups of individual partitions that are electrically connected, mechanically contiguous, and do not exceed 30 feet (9.14 m) when assembled, shall be permitted to be connected to the building electrical system by a single flexible cord and plug, provided all of the following conditions are met:

   (a) Flexible Power-Supply Cord. The flexible power-supply cord shall be extra-hard-usage type with No. 12 or larger conductors with an insu- | lated grounding conductor and not exceeding 2 feet (610 mm) in length.

   (b) Receptacle Supplying Power. The receptacle(s) supplying power shall be on a separate circuit serving only panels and no other loads and shall be located not more than 12 inches (305 mm) from the partition that is connected to it.

   (c) Receptacle Outlets, Maximum. Individual partitions or groups of interconnected individual partitions shall not contain more than thirteen 15-ampere, 125-volt receptacle outlets.

(d) Multiwire Circuits, Not Permitted. Individual partitions or groups of interconnected individual partitions shall not contain multiwire circuits.

(FPN):   See Section 210-4 for circuits supplying partitions in Sections 605-6 and 605-7.

ARTICLE 610 — CRANES AND HOISTS

A. General

610-1. Scope.   This article covers the installation of electric equipment and wiring used in connection with cranes, monorail hoists, hoists, and all runways.

(FPN):   For further information, see Safety Code for Cranes, Derricks, Hoists, Jacks, and Slings, (ANSI B-30).

610-2. Special Requirements for Particular Locations.

(a) Hazardous (Classified) Locations.   All equipment that operates in a hazardous (classified) location shall conform to Article 500.

(1) Equipment used in locations that are hazardous because of the presence of flammable gases or vapors shall conform to Article 501.

(2) Equipment used in locations that are hazardous because of combustible dust shall conform to Article 502.

(3) Equipment used in locations that are hazardous because of the presence of easily ignitible fibers or flyings shall conform to Article 503.

(b) Combustible Materials.   Where a crane, hoist, or monorail hoist operates over readily combustible material, the resistors shall be placed in a well-ventilated cabinet composed of noncombustible material so constructed that it will not emit flames or molten metal.

Exception: Resistors shall be permitted to be located in a cage or cab constructed of noncombustible material that encloses the sides of the cage or cab from the floor to a point at least 6 inches (152 mm) above the top of the resistors.

(c) Electrolytic Cell Lines.   See Section 668-32.

B. Wiring

610-11. Wiring Method.   Conductors shall be enclosed in raceways or be Type MC cable or Type MI cable.

Exception No. 1: Contact conductors.

Exception No. 2: Short lengths of open conductors at resistors, collectors, and other equipment.

Exception No. 3: Where flexible connections are necessary to motors and similar equipment, flexible stranded conductors shall be installed in flexible metal conduit, liquidtight flexible metal conduit, multiconductor cable, or an approved nonmetallic enclosure.

Exception No. 4: Where multiconductor cable is used with a suspended pushbutton station, the station shall be supported in some satisfactory manner that protects the electric conductors against strain.

Exception No. 5: Where flexibility is required for power or control to moving parts, a cord suitable for the purpose shall be permitted provided:

a. Suitable strain relief and protection from physical damage is provided; and

b. In Class 1, Division 2 hazardous locations, cord shall be approved for extra-hard usage.

610-12. Raceway or Cable Terminal Fittings. Conductors leaving raceways or cables shall comply with one of the following:

(a) Separately Bushed Hole. A box or terminal fitting having a separately bushed hole for each conductor shall be used wherever a change is made from rigid metal conduit, intermediate metal conduit, electrical metallic tubing, metal-clad cable, mineral-insulated cable, or surface raceway wiring to open wiring. A fitting used for this purpose shall not contain taps or splices and shall not be used at fixture outlets.

(b) Bushing in Lieu of a Box. A bushing shall be permitted to be used in lieu of a box at the end of a rigid metal conduit, intermediate metal conduit, or electrical metallic tubing where the raceway terminates at unenclosed controls or similar equipment including contact conductors, collectors, resistors, brakes, power circuit limit switches, and dc split frame motors.

610-13. Types of Conductors. Conductors shall comply with Table 310-13.

Exception No. 1: Conductor(s) exposed to external heat or connected to resistors shall have a flame-resistant outer covering or be covered with flame-resistant tape individually or as a group.

Exception No. 2: Contact conductors along runways, crane bridges, and monorails shall be permitted to be bare, and shall be copper, aluminum, steel, or other alloys or combinations thereof in the form of hard drawn wire, tees, angles, tee rails, or other stiff shapes.

Exception No. 3: Flexible conductors shall be permitted to be used to conduct current and, where practicable, cable reels or take-up devices shall be used.

610-14. Rating and Size of Conductors.

(a) Ampacity. The allowable ampacities of conductors shall be as shown in Table 610-14(a).

(FPN): For the ampacities of conductors between controllers and resistors, see Section 430-23.

(b) Secondary Resistor Conductors. Where the secondary resistor is separate from the controller, the minimum size of the conductors between controller and resistor shall be calculated by multiplying the motor secondary current by the appropriate factor from Table 610-14(b) and selecting a wire from Table 610-14(a).

Table 610-14(a). Ampacities of Insulated Copper Conductors Used with Short-Time Rated Crane and Hoist Motors. Based on Ambient Temperature of 30°C (86°F). Up to Four Conductors in Raceway or Cable* Up to 3 ac** or 4 dc*Conductors in Raceway or Cable.

Maximum Operating Temp.	75°C (167°F)		90°C (194°F)		125°C (257°F)		Maximum Operating Temp.
Size AWG (kcmil)	Types MTW, RH, RHW, THW, THWN, XHHW, USE, ZW		Types TA, TBS, SA, SIS, PFA, FEP, FEPB, RHH, THHN, XHHW, Z, ZW		Types FEP, FEPB, PFA, PFAH, SA, TFE, Z, ZW		Size AWG (kcmil)
	60 Min	30 Min	60 Min	30 Min	60 Min	30 Min	
16	10	12	. . .	. . .	. . . .	. . . .	16
14	25	26	31	32	38	40	14
12	30	33	36	40	45	50	12
10	40	43	49	52	60	65	10
8	55	60	63	69	73	80	8
6	76	86	83	94	101	119	6
5	85	95	95	106	115	134	5
4	100	117	111	130	133	157	4
3	120	141	131	153	153	183	3
2	137	160	148	173	178	214	2
1	143	175	158	192	210	253	1
1/0	190	233	211	259	253	304	1/0
2/0	222	267	245	294	303	369	2/0
3/0	280	341	305	372	370	452	3/0
4/0	300	369	319	399	451	555	4/0
250	364	420	400	461	510	635	250
300	455	582	497	636	587	737	300
350	486	646	542	716	663	837	350
400	538	688	593	760	742	941	400
450	600	765	660	836	818	1042	450
500	660	847	726	914	896	1143	500

AMPACITY CORRECTION FACTORS							
Ambient Temp. °C	For ambient temperatures other than 30°C (86°F), multiply the ampacities shown above by the appropriate factor shown below.						Ambient Temp. °F
21-25	1.05	1.05	1.04	1.04	1.02	1.02	70-77
26-30	1.00	1.00	1.00	1.00	1.00	1.00	79-86
31-35	.94	.94	.96	.96	.97	.97	88-95
36-40	.88	.88	.91	.91	.95	.95	97-104
41-45	.82	.82	.87	.87	.92	.92	106-113
46-50	.75	.75	.82	.82	.89	.89	115-122
51-55	.67	.67	.76	.76	.86	.86	124-131
56-60	.58	.58	.71	.71	.83	.83	133-140
61-70	.33	.33	.58	.58	.76	.76	142-158
71-80	. . .	. . .	.41	.41	.69	.69	160-176
81-90	. . .	. . .	. . .	. . .	.61	.61	177-194
91-100	. . .	. . .	. . .	. . .	.51	.51	195-212
101-120	. . .	. . .	. . .	. . .	.40	.40	213-248

Other insulations shown in Table 310-13 and approved for the temperature and location shall be permitted to be substituted for those shown in Table 610-14(a). The allowable ampacities of conductors used with 15-minute motors shall be the 30-minute ratings increased by 12%.

* For 5 to 8 simultaneously energized power conductors in raceway or cable, the ampacity of each power conductor shall be reduced to a value of 80% of that shown in the table.

** For 4 to 6 simultaneously energized 125°C (257°F) ac power conductors in raceway or cable, the ampacity of each power conductor shall be reduced to a value of 80% of that shown in the table.

Table 610-14(b). Secondary Conductor Rating Factors

Time in Seconds		Ampacity of Wire in Percent of Full-Load Secondary Current
On	Off	
5	75	35
10	70	45
15	75	55
15	45	65
15	30	75
15	15	85
Continuous Duty		110

(c) Minimum Size. Conductors external to motors and controls shall not be smaller than No. 16.

Exception No. 1: No. 18 wire in multiple conductor cord shall be permitted for control circuits at not over 7 amperes.

Exception No. 2: Wires not smaller than No. 20 shall be permitted for electronic circuits.

(d) Contact Conductors. Contact wires shall have an ampacity not less than that required by Table 610-14(a) for 75°C (167°F) wire, and in no case shall they be smaller than the following:

Distance Between End Strain Insulators or Clamp-type Intermediate Supports	Size of Wire
0-30 feet	No. 6
30-60 feet	No. 4
Over 60 feet	No. 2

For SI units: one foot = 0.3048 meter.

(e) Calculation of Motor Load.

(1) For one motor, use 100 percent of motor nameplate full-load ampere rating.

(2) For multiple motors on a single crane or hoist, the minimum ampacity of the power supply conductors shall be the nameplate full-load ampere rating of the largest motor or group of motors for any single crane motion, plus 50 percent of the nameplate full-load ampere rating of the next largest motor or group of motors, using that column of Table 610-14(a) that applies to the longest time-rated motor.

(3) For multiple cranes and/or hoists supplied by a common conductor system, compute the motor minimum ampacity for each crane as defined in Section 610-14(e), add them together, and multiply the sum by the appropriate demand factor from Table 610-14(e).

Table 610-14(e). Demand Factors

Number of Cranes of Hoists	Demand Factor
2	0.95
3	0.91
4	0.87
5	0.84
6	0.81
7	0.78

(f) Other Loads. Additional loads, such as heating, lighting, and air conditioning, shall be provided for by application of the appropriate sections of this Code.

(g) Nameplate. Each crane, monorail, or hoist shall be provided with a visible nameplate marked with the maker's name, the rating in volts, frequency, number of phases, and circuit amperes as calculated in Sections 610-14(e) and (f).

610-15. Common Return. Where a crane or hoist is operated by more than one motor, a common-return conductor of proper ampacity shall be permitted.

C. Contact Conductors

610-21. Installation of Contact Conductors. Contact conductors shall comply with (a) through (h) below.

(a) Locating or Guarding Contact Conductors. Runway contact conductors shall be guarded, and bridge contact conductors shall be located or guarded in a manner that persons cannot inadvertently touch energized current-carrying parts.

(b) Contact Wires. Wires that are used as contact conductors shall be secured at the ends by means of approved strain insulators and shall be so mounted on approved insulators that the extreme limit of displacement of the wire will not bring the latter within less than $1\frac{1}{2}$ inches (38 mm) from the surface wired over.

(c) Supports Along Runways. Main contact conductors carried along runways shall be supported on insulating supports placed at intervals not exceeding 20 feet (6.1 m).

Exception: Supports for grounded rail conductors as provided in (f) below shall not be required to be of the insulating type.

Such conductors shall be separated not less than 6 inches (152 mm) other than for monorail hoists where a spacing of not less than 3 inches (76 mm) shall be permitted. Where necessary, intervals between insulating supports shall be permitted to be increased up to 40 feet (12.2 m), the separation between conductors being increased proportionately.

(d) Supports on Bridges. Bridge wire contact conductors shall be kept at least 2½ inches (64 mm) apart, and, where the span exceeds 80 feet (24.4 m), insulating saddles shall be placed at intervals not exceeding 50 feet (15.2 m).

(e) Supports for Rigid Conductors. Conductors along runways and crane bridges, that are of the rigid type specified in Section 610-13, Exception No. 2 and not contained within an approved enclosed assembly, shall be carried on insulating supports spaced at intervals of not more than eighty times the vertical dimension of the conductor, but in no case greater than 15 feet (4.57 m), and spaced apart sufficiently to give a clear electrical separation of conductors or adjacent collectors of not less than 1 inch (25.4 mm).

(f) Track as Circuit Conductor. Monorail, tramrail, or crane-runway tracks shall be permitted as a conductor of current for one phase of a 3-phase, alternating-current system furnishing power to the carrier, crane, or trolley, provided all of the following conditions are met:

(1) The conductors supplying the other two phases of the power supply are insulated.

(2) The power for all phases is obtained from an insulating transformer.

(3) The voltage does not exceed 300 volts.

(4) The rail serving as a conductor is effectively grounded at the transformer and also shall be permitted to be grounded by the fittings used for the suspension or attachment of the rail to a building or structure.

(g) Electrical Continuity of Contact Conductors. All sections of contact conductors shall be mechanically joined to provide a continuous electrical connection.

(h) Not to Supply Other Equipment. Contact conductors shall not be used as feeders for any equipment other than the crane or cranes that they are primarily designed to serve.

610-22. Collectors. Collectors shall be so designed as to reduce to a minimum sparking between them and the contact conductor; and, where operated in rooms used for the storage of easily ignitible combustible fibers and materials, they shall comply with Section 503-13.

D. Disconnecting Means

610-31. Runway Conductor Disconnecting Means. A disconnecting means having a continuous ampere rating not less than that computed in Sections 610-14(e) and (f) shall be provided between the runway contact conductors and the power supply. Such disconnecting means shall consist of a motor circuit switch or circuit breaker. This disconnecting means shall:

(1) Be readily accessible and operable from the ground or floor level.

(2) Be arranged to be locked in the open position.

(3) Open all ungrounded conductors simultaneously.

(4) Be placed within view of the crane or hoist and the runway contact conductors.

610-32. Disconnecting Means for Cranes and Monorail Hoists. A motor circuit switch or circuit breaker arranged to be locked in the open position shall be provided in the leads from the runway contact conductors or other power supply on all cranes and monorail hoists.

Exception: Where a monorail hoist or hand-propelled crane bridge installation meets all of the following, the disconnect shall be permitted to be omitted.

    a. The unit is floor controlled.

    b. The unit is within view of the power supply disconnecting means.

    c. No fixed work platform has been provided for servicing the unit.

Where the disconnecting means is not readily accessible from the crane or monorail hoist operating station, means shall be provided at the operating station to open the power circuit to all motors of the crane or monorail hoist.

610-33. Rating of Disconnecting Means. The continuous ampere rating of the switch or circuit breaker required by Section 610-32 shall not be less than 50 percent of the combined short-time ampere rating of the motors, nor less than 75 percent of the sum of the short-time ampere rating of the motors required for any single motion.

E. Overcurrent Protection

610-41. Feeders, Runway Conductors. The runway supply conductors and main contact conductors of a crane or monorail shall be protected by an overcurrent device(s) that shall not be greater than the largest rating or setting of any branch-circuit protective device, plus the sum of the nameplate ratings of all the other loads with application of the demand factors from Table 610-14(e).

610-42. Branch-Circuit Short-Circuit and Ground-Fault Protection. Branch circuits shall be protected as follows:

    (a) Fuse or Circuit Breaker Rating. Crane, hoist, and monorail hoist motor branch circuits shall be protected by fuses or inverse-time circuit breakers having a rating in accordance with Table 430-152. Taps to control circuits shall be permitted to be taken from the load side of a branch-circuit protective device, provided each tap and piece of equipment is properly protected.

    Exception No. 1: When two or more motors operate a single motion, the sum of their nameplate current ratings shall be considered as a single motor current in the above calculations.

    Exception No. 2: Two or more motors shall be permitted to be connected to the same branch circuit if no tap conductor to an individual motor has an ampacity less than one-third that of the branch circuit and if each motor is protected from overload according to Section 610-43.

    (b) Taps to Brake Coils. Taps to brake coils do not require separate overcurrent protection.

610-43. Motor and Branch-Circuit Overload Protection. Each motor, motor control, and branch-circuit conductor shall be protected from overload by one of the following means:

(1) A single motor shall be considered as protected where the branch-circuit overcurrent device meets the rating requirements of Section 610-42.

(2) Overload relay elements in each ungrounded circuit conductor, with all relay elements protected from short circuit by the branch-circuit protection.

(3) Thermal sensing device(s), sensitive to motor temperature or to temperature and current that are thermally in contact with the motor winding(s). A hoist or trolley is considered to be protected if the sensing device is connected in the hoist's upper limit switch circuit so as to prevent further hoisting during an overload condition of either motor.

Exception No. 1: If the motor is manually controlled, with spring return controls, the overload protective device shall not be required to protect the motor against stalled rotor conditions.

Exception No. 2: Where two or more motors drive a single trolley, truck, or bridge and are controlled as a unit by a single set of overload devices with a rating equal to the sum of their rated full-load currents. A hoist or trolley shall be considered to be protected if the sensing device is connected in the hoist's upper limit switch circuit so as to prevent further hoisting during an overtemperature condition of either motor.

Exception No. 3: Hoists and monorail hoists and their trolleys that are not used as part of an overhead traveling crane shall not require individual motor overload protection, provided the largest motor does not exceed $7^1/_2$ horsepower and all motors are under manual control of the operator.

F. Control

610-51. Separate Controllers. Each motor shall be provided with an individual controller.

Exception No. 1: Where two or more motors drive a single hoist, carriage, truck, or bridge, they shall be permitted to be controlled by a single controller.

Exception No. 2: One controller shall be permitted to be switched between motors, provided:

a. The controller shall have a horsepower rating that shall not be lower than the horsepower rating of the largest motor.

b. Only one motor is operated at one time.

610-53. Overcurrent Protection. Conductors of control circuits shall be protected against overcurrent. Control circuits shall be considered as protected by overcurrent devices that are rated or set at not more than 300 percent of the ampacity of the control conductors.

Exception No. 1: Taps to control transformers shall be considered as protected where the secondary circuit is protected by a device rated or set at not more than 200 percent of the rated secondary current of the transformer and not more than 200 percent of the ampacity of the control circuit conductors.

Exception No. 2: Such conductors shall be considered as being properly protected by the branch-circuit overcurrent devices where the opening of the control circuit would create a hazard, as for example, the control circuit of a hot metal crane.

610-55. Limit Switch. A limit switch or other device shall be provided to prevent the load block from passing the safe upper limit of travel of all hoisting mechanisms.

610-57. Clearance. The dimension of the working space in the direction of access to live parts that are likely to require examination, adjustment, servicing, or maintenance while energized shall be a minimum of 2½ feet (762 mm). Where controls are enclosed in cabinets, the door(s) shall either open at least 90 degrees or be removable.

G. Grounding

610-61. Grounding. All exposed noncurrent-carrying metal parts of cranes, monorail hoists, hoists, and accessories including pendant controls shall be metallically joined together into a continuous electrical conductor so that the entire crane or hoist will be grounded in accordance with Article 250. Moving parts, other than removable accessories or attachments, having metal-to-metal bearing surfaces shall be considered to be electrically connected to each other through the bearing surfaces for grounding purposes. The trolley frame and bridge frame shall be considered as electrically grounded through the bridge and trolley wheels and its respective tracks unless local conditions, such as paint or other insulating material, prevent reliable metal-to-metal contact. In this case, a separate bonding conductor shall be provided.

ARTICLE 620 — ELEVATORS, DUMBWAITERS, ESCALATORS, MOVING WALKS, WHEELCHAIR LIFTS, AND STAIRWAY CHAIR LIFTS

A. General

620-1. Scope. This article covers the installation of electric equipment and wiring used in connection with elevators, dumbwaiters, escalators, moving walks, wheelchair lifts, and stairway chair lifts.

(FPN): For further information, see Safety Code for Elevators and Escalators, ASME/ANSI A17.1-1990.

620-2. Voltage Limitations. The nominal voltage used for elevator, dumbwaiter, escalator, moving walk, wheelchair lift, and stairway chair lift operating control and signaling circuits, operating equipment, driving machine motors, machine brakes, and motor-generator sets shall not exceed the following:

(a) 300 Volts. For operating control and signaling circuits and related equipment, including door operator motors.

Exception: Higher voltages shall be permitted for frequencies of 25-through 60-hertz alternating current or for direct current, provided the current in the system cannot, under any conditions, exceed 8 milliamperes for alternating current or 30 milliamperes for direct current.

(b) 600 Volts. Driving machine motors, machine brakes, and motor-generator sets.

Exception: Higher voltages shall be permitted for driving motors of motor-generator sets.

620-3. Live Parts Enclosed. All live parts of electric apparatus in the hoistways, at the landings, or in or on the cars of elevators and dumbwaiters or in the wellways or the landings of escalators or moving walks, or in the runways and machinery spaces of wheelchair lifts and stairway chair lifts shall be enclosed to protect against accidental contact.

B. Conductors

620-11. Insulation of Conductors. The insulation of conductors installed in connection with elevators, dumbwaiters, escalators, moving walks, wheelchair lifts, and stairway chair lifts shall comply with (a) through (d) below.

(a) Hoistway Door Interlock Wiring. The conductors to the hoistway door interlocks from the hoistway riser shall be flame-retardant and suitable for a temperature of not less than 200°C (392°F). Conductors shall be Type SF or equivalent.

(b) Traveling Cables. Traveling cables used as flexible connections between the elevator or dumbwaiter car and the raceway shall be of the types of elevator cable listed in Table 400-4 or other approved types.

(c) Other Wiring. All conductors in raceways; in or on the cars of elevators and dumbwaiters; in the wellways of escalators and moving walks; in the runways and machinery spaces of wheelchair lifts and stairway chair lifts; and in the machine room of elevators, dumbwaiters, escalators, and moving walks shall have flame-retardant insulation.

(d) Insulation. All conductors shall have an insulation voltage rating equal to at least the maximum nominal circuit voltage rating of any conductor within the enclosure, cable, or raceway.

Conductors shall be Type MTW, TF, TFF, TFN, TFFN, THHN, THW, THWN, TW, XHHW, hoistway cable, or any other conductor with insulation designated as flame-retardant. Shielded conductors shall be permitted, provided such conductors are insulated for the maximum voltage found in the cable or raceway system.

620-12. Minimum Size of Conductors. The minimum size of conductors used for elevator, dumbwaiter, escalator, moving walk, wheelchair lift, and stairway chair lift wiring, other than conductors that form an integral part of control equipment, shall be as follows:

(a) Traveling Cables.

(1) For lighting circuits: No. 14.

Exception: No. 20 or larger conductors shall be permitted in parallel, provided the ampacity is equivalent to at least that of No. 14 wire.

(2) Operating control and signaling circuits: No. 20.

(b) Other Wiring. All operating control and signaling circuits: No. 24.

620-13. Motor Circuit Conductors. Conductors supplying elevator, dumbwaiter, escalator, moving walk, wheelchair lift, and stairway chair lift motors shall have an ampacity in accordance with (a) and (b) below based on the nameplate current rating of the motors. With generator field control, the ampacity shall be based on the nameplate current rating of the driving motor of the motor-generator set that supplies power to the elevator motor.

(FPN): The heating of conductors depends on root-mean-square current values, which, with generator field control, are reflected by the nameplate current rating of the motor-generator set driving motor rather than by the rating of the elevator motor, which represents actual but short-time and intermittent full-load current values.

(a) Conductors Supplying Single Motor. Conductors supplying a single motor shall have an ampacity in conformance with Section 430-22 and Table 430-22(a), Exception.

(b) Conductors Supplying Several Motors. Conductors supplying two or more motors shall have an ampacity of not less than 125 percent of the nameplate current rating of the highest rated motor in the group plus the sum of the nameplate current ratings of the remainder of the motors in the group.

620-14. Adjustable Speed Drive Systems. Conductors supplying elevators, dumbwaiters, escalators, moving walks, wheelchair lifts, or stairway chair lifts shall have an ampacity in accordance with the following:

(a) Drive Power Transformer Integral with the Power Conversion Equipment. Conductor ampacity shall be based on the nameplate current rating of the power conversion equipment.

(b) Drive Power Transformer Not Integral with the Power Conversion Equipment. Conductor ampacity shall be based on the nameplate current rating of the drive power transformer and all other connected loads, or the nameplate current rating of the power conversion equipment reflected to the primary side of the transformer, plus all other connected loads, whichever is the larger.

(FPN): Diagram 620-14 is for information only.

620-15. Feeder Demand Factor. Feeder conductors of less ampacity than required by Sections 620-13 and 620-14 shall be permitted subject to the requirements of Section 430-26 and Table 620-15.

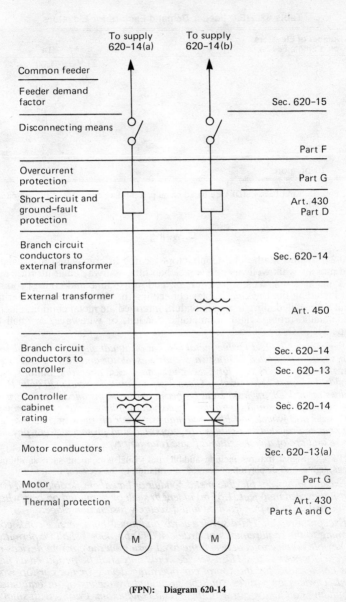

(FPN): Diagram 620-14

Table 620-15.　Feeder Demand Factors for Elevators

Number of Elevators on a Single Feeder	DF
1	1.00
2	.95
3	.90
4	.85
5	.82
6	.79
7	.77
8	.75
9	.73
10 or more	.72

(FPN): Demand Factors (DF) are based on 50 percent duty cycle (i.e., half time on and half time off).

C. Wiring

620-21. Wiring Methods.　Conductors located in hoistways, in escalator and moving walk wellways, in wheelchair lifts, stairway chair lift runways, and machinery spaces, in or on cars, and in machine and control rooms, not including the traveling cables connecting the car and hoistway wiring, shall be installed in rigid metal conduit, intermediate metal conduit, electrical metallic tubing, rigid nonmetallic conduit, or wireways, or shall be Type MC, MI, or AC cable.

Exception No. 1: Flexible metal conduit or liquidtight flexible metal conduit shall be permitted in hoistways and in escalator and moving walk wellways between risers and limit switches, interlocks, operating buttons, and similar devices. Cables used in Class 2 power-limited circuits (30 volts RMS or less or 42 VDC or less) shall be permitted to be installed between risers and signal fixtures and operating devices and within escalators and moving walkways and wheelchair lifts, and stairway chair lift runways and machinery spaces provided the cables are supported and protected from physical abuse and are of a jacketed and flame-retardant type.

(FPN):　Signal fixtures include audible and visual equipment such as chimes, gongs, lights, and displays that convey information to the elevator user.

Exception No. 2: Flexible metal conduit or liquidtight flexible metal conduit not exceeding 6 feet (1.83 m) in length shall be permitted on cars where so located as to be free from oil and if securely fastened in place.

Exception No. 3: Hard-service cords and junior hard-service cords conforming to the requirements of Article 400 (Table 400-4) shall be permitted as flexible connections between the fixed wiring on the car and devices on the car doors or gates. Hard-service cords only shall be permitted as flexible connections for the top-of-car operating device or the car-top work light. Devices or fixtures shall be grounded by means of an equipment grounding conductor run with the circuit conductors. Cables with smaller conductors and other types and thicknesses of insulation and jackets shall be permitted as flexible connections between the fixed wiring on the car and devices on the car doors or gates, if listed for this use.

Exception No. 4: Flexible metal conduit or liquidtight flexible metal conduit, not exceeding 6 feet (1.83 m) in length, shall be permitted between control panels and machine motors, machine brakes, motor-generator sets, disconnecting means, and pumping unit motors and valves. Conductors shall also be permitted to be grouped together and taped or corded without being installed in a raceway. Such cable groups shall be supported at intervals not over 3 feet (914 mm) and so located as to be free from physical damage.

Exception No. 5: Flexible metal conduit or liquidtight flexible metal conduit, of ³/₈-inch nominal trade size, shall be permitted in lengths not in excess of 6 feet (1.83 m).

Exception No. 6: Flexible metal conduit, conforming to Article 350, or liquidtight flexible metal conduit, or hard-service cords conforming to the requirements of Article 400 (Table 400-4) shall be permitted as flexible connections on escalators, moving walk or elevator control panels and disconnecting means where the entire control panel and disconnecting means are arranged for removal from machine spaces as permitted in Section 620-72, Exception Nos. 1 and 2.

Exception No. 7: Flexible metal conduit, liquidtight flexible metal conduit, or flexible cords and cables, or conductors grouped together and taped or corded that are part of listed equipment, a driving machine, or a driving machine brake shall be permitted on the counterweight assembly, in lengths not to exceed 6 feet (1.83 mm), without being installed in a raceway and where located to be free from physical damage.

Where motor-generators, machine motors, or pumping unit motors and valves are located adjacent to or underneath control equipment and are provided with extra-length terminal leads not exceeding 6 feet (1.83 m) in length, such leads shall be permitted to be extended to connect directly to controller terminal studs without regard to the carrying-capacity requirements of Articles 430 and 445. Auxiliary gutters shall be permitted in machine and control rooms between controllers, starters, and similar apparatus.

620-22. Branch Circuits for Car Lighting, Accessories, Heating, and Air Conditioning.

(a) **Car Light Source.** A dedicated branch circuit shall supply the car lights and accessories on each elevator car.

(b) **Air Conditioning and Heating Source.** A dedicated branch circuit shall supply the air-conditioning and heating units on each elevator car.

D. Installation of Conductors

620-31. Raceway Terminal Fittings.
Conductors shall comply with Section 300-16(b). In locations where raceways project from the floor and terminate in other than a wiring enclosure, they shall extend at least 6 inches (152 mm) above the floor.

620-32. Wireways.
Section 362-5 shall not apply to wireways. The sum of the cross-sectional area of the individual conductors in a wireway shall not be more than 50 percent of the interior cross-sectional area of the wireway.

Vertical runs of wireways shall be securely supported at intervals not exceeding 15 feet (4.57 m) and shall have not more than one joint between supports. Adjoining wireway sections shall be securely fastened together to provide a rigid joint.

620-33. Number of Conductors in Raceways. The sum of the cross-sectional area of the operating and control circuit conductors in raceways shall not exceed 40 percent of the interior cross-sectional area of the raceway.

Exception: In wireways as permitted in Section 620-32.

620-34. Supports. Supports for cables or raceways in a hoistway or in an escalator or moving walk wellway or wheelchair lift and stairway chair lift runway shall be securely fastened to the guide rail, escalator or moving walk truss, or to the hoistway, wellway, or runway construction.

620-35. Auxiliary Gutters (Wiring Troughs). Auxiliary gutters shall not be subject to the restrictions of Section 374-2 as to length or of Section 374-5 as to number of conductors.

620-36. Different Systems in One Raceway or Traveling Cable. Conductors for operating, control, power, signaling, and lighting circuits of 600 volts or less shall be permitted to be run in the same traveling cable or raceway system if all conductors are insulated for the maximum voltage found in the cables or raceway system and if all live parts of the equipment are insulated from ground for this maximum voltage. Such a traveling cable or raceway shall also be permitted to include shielded conductors and/or one or more coaxial cables, if such conductors are insulated for the maximum voltage found in the cable or raceway system. Conductors shall be permitted to be covered with suitable shielding for telephone, audio, video, or higher frequency communication circuits.

620-37. Wiring in Hoistways and Machine Rooms. Only such electric wiring, raceways, and cables used directly in connection with the elevator or dumbwaiter, including wiring for signals, for communication with the car, for lighting, heating, air conditioning, and ventilating the car, for fire detecting systems, for pit sump pumps, and for heating and lighting the hoistway, shall be permitted inside the hoistway and the machine room. Main feeders for supplying power to elevators and dumbwaiters shall be installed outside the hoistway.

Exception No. 1: By special permission, feeders for elevators shall be permitted within an existing hoistway if no conductors are spliced within the hoistway.

Exception No. 2: Feeders shall be permitted inside the hoistway for elevators with driving machine motors located in the hoistway or on the car or counterweight.

620-38. Electric Equipment in Garages and Similar Occupancies. Electric equipment and wiring used for elevators, dumbwaiters, escalators, moving walks, and wheelchair lifts and stairway chair lifts in garages shall comply with the requirements of Article 511. Wiring and equipment located on the underside of the car platform shall be considered as being located in the hazardous area.

(FPN):   Garages used for parking or storage and where no repair work is done in accordance with Section 511-2 are not classified.

620-39. Sidewalk Elevators.   Sidewalk elevators with sidewalk doors located exterior to the building shall have all electric wiring in rigid metal conduit, rigid nonmetallic conduit, intermediate metal conduit, liquidtight flexible metal conduit, liquidtight flexible nonmetallic conduit, or electrical metallic tubing, and all electrical outlets, switches, junction boxes, and fittings shall be weatherproof.

E. Traveling Cables

620-41. Suspension of Traveling Cables.   Traveling cables shall be so suspended at the car and hoistways' ends, or counterweight end where applicable, as to reduce the strain on the individual copper conductors to a minimum.

Traveling cables shall be supported by one of the following means: (1) by its steel supporting member(s); (2) by looping the cables around supports for unsupported lengths less than 100 feet (30.5 m); (3) by suspending from the supports by a means that automatically tightens around the cable when tension is increased for unsupported lengths up to 200 feet (61 m).

(FPN):   Unsupported length for the hoistway suspension means is that length of cable as measured from the point of suspension in the hoistway to the bottom of the loop, with the elevator car located at the bottom landing. Unsupported length for the car suspension means is that length of cable as measured from the point of suspension on the car to the bottom of the loop, with the elevator car located at the top landing.

620-42. Hazardous (Classified) Locations.   In hazardous (classified) locations, traveling cables shall be of a type approved for hazardous (classified) locations and shall comply with Section 501-11, 502-12, or 503-10, as applicable.

620-43. Location of and Protection for Cables.   Traveling cable supports shall be so located as to reduce to a minimum the possibility of damage due to the cables coming in contact with the hoistway construction or equipment in the hoistway. Where necessary, suitable guards shall be provided to protect the cables against damage.

620-44. Installation of Traveling Cables.   Traveling cable shall be permitted to be run without the use of a raceway for a distance not exceeding 6 feet (1.83 m) in length as measured from the first point of support on the elevator car or hoistway wall, or counterweight where applicable, provided the conductors are grouped together and taped or corded, or in the original sheath.

Traveling cables shall be permitted to be continued to elevator control panels and to elevator car and machine room connections, as fixed wiring, provided they are suitably supported and protected from physical damage.

F. Disconnecting Means and Control

620-51. Disconnecting Means. Elevators, dumbwaiters, escalators, moving walks, wheelchair lifts, and stairway chair lifts shall have a single means for disconnecting all ungrounded main power supply conductors for each unit. Where multiple driving machines are connected to a single elevator, escalator, moving walk, or pumping unit, there shall be one disconnecting means to disconnect the motor(s) and control valve operating magnets.

Where there is more than one driving machine in a machine room, disconnecting means shall be numbered to correspond to the number of the driving machine that they control.

(a) Type. The disconnecting means shall be an enclosed externally operable fused motor circuit switch or circuit breaker arranged to be locked in the open position. No provision shall be made to open or close this disconnecting means from outside of the hoistway, machine room, or machinery spaces.

The fuses or circuit breakers provided for in the disconnecting means shall be selectively coordinated with any and all other supply side overcurrent-protective devices.

(FPN):   For additional information, see Safety Code for Elevators and Escalators, ASME/ANSI A17.1-1990.

(b) Location. The disconnecting means shall be located where it is readily accessible to qualified persons.

(1) On elevators without generator field control, the disconnecting means shall be located within sight of the power converter or motor starter. Where the disconnecting means is not within sight of the hoist machine or control panel, an additional manually operated switch shall be installed adjacent to the remote equipment, connected in the control circuit to prevent starting.

(2) On elevators with generator field control, the disconnecting means shall be located within sight of the motor starter for the driver motor of the motor-generator set. Where the disconnecting means is not within sight of the hoist machine, the control panel, or the motor-generator set, an additional manually operated switch shall be installed adjacent to the remote equipment, connected in the control circuit to prevent starting.

(3) On escalators and moving walks, the disconnecting means shall be installed in the space where the controller is located.

(4) On wheelchair lifts and stairway chair lifts, the disconnecting means shall be located within sight of the motor controller. The disconnecting means shall be permitted in the same enclosure with the motor controller.

620-52. Power from More than One Source.

(a) Single- and Multi-Car Installations. On single- and multi-car installations, equipment receiving electrical power from more than one source shall be provided with a disconnecting means for each source of electrical power. The disconnecting means shall be within sight of the equipment served.

(b) **Warning Sign for Multiple Disconnecting Means.** Where multiple disconnecting means are used and parts of the control panel remain energized from a source other than the one disconnected, a warning sign shall be mounted on or adjacent to the disconnecting means. The sign shall be clearly legible and shall read "Warning — Parts of the Control Panel Are Not Deenergized by This Switch."

(c) **Interconnection Multi-Car Control Panels.** Where interconnections between control panels are necessary for the operation of the system on multi-car installations that remain energized from a source other than the one disconnected, a warning sign in accordance with Section 620-52(b) shall be mounted on or adjacent to the disconnecting means.

620-53. Car Light and Accessories Disconnecting Means. Elevators shall have a single means for disconnecting all ungrounded car light and accessories power-supply conductors for each unit.

Where there is equipment for more than one car in the machine room, disconnecting means shall be numbered to correspond to the number of the elevator car whose light source they control.

The disconnecting means shall be arranged to be locked in the open position and shall be located in the machine room for that car.

620-54. Heating and Air-Conditioning Disconnecting Means. Elevators shall have a single means for disconnecting all ungrounded car heating and air-conditioning power-supply conductors for each unit.

Where there is equipment for more than one car in the machine room, disconnecting means shall be numbered to correspond to the number of the elevator car whose heating and air-conditioning source they control. The disconnecting means shall be arranged to be locked in the open position and shall be located in the machine room for that car.

G. Overcurrent Protection

620-61. Overcurrent Protection. Overcurrent protection shall be provided as follows:

(a) **Control and Operating Circuits.** Control and operating circuits and signaling circuits shall be protected against overcurrent in accordance with the requirements of Section 725-12.

(b) **Overload Protection for Motors.**

(1) Duty on elevator and dumbwaiter driving machine motors and driving motors of motor-generators used with generator field control shall be classed as intermittent. Such motors shall be protected against overload in accordance with Section 430-33.

(2) Duty on escalator and moving walk driving machine motors shall be classed as continuous. Such motors shall be protected against overload in accordance with Section 430-32.

(3) Escalator and moving walk driving machine motors and driving motors of motor-generator sets shall be protected against running overload as provided in Table 430-37.

(4) Duty on wheelchair lift and stairway chair lift driving machine motors shall be classed as intermittent. Such motors shall be protected against overload in accordance with Section 430-33.

H. Machine Room

620-71. Guarding Equipment. Elevator, dumbwaiter, escalator, and moving walk driving machines, motor-generator sets, motor controllers, and disconnecting means shall be installed in a room or enclosure set aside for that purpose. The room or enclosure shall be secured against unauthorized access.

Exception No. 1: Dumbwaiter, escalator, moving walk, or wheelchair lift and stairway chair lift motor controllers shall be permitted outside the spaces herein specified, provided they are enclosed in cabinets with doors or removable panels capable of being locked in the closed position and the disconnecting means is located adjacent to or is an integral part of the motor controller. Motor controller cabinets for escalator or moving walks shall be permitted in the balustrading on the side located away from the moving steps or moving treadway. If the disconnecting means is an integral part of the controller, it must be operable without opening the cabinet.

Exception No. 2: Elevator motor controllers and driving machines shall be permitted outside the spaces herein specified, provided the motor controllers are enclosed in cabinets with doors or removable panels capable of being locked in the closed position and the disconnecting means is located adjacent to or is an integral part of the controller. If the disconnecting means is an integral part of the controller, it must be operable without opening the cabinet.

620-72. Clearance Around Control Panels and Disconnecting Means. Sufficient clear working space shall be provided around control panels and disconnecting means to provide safe and convenient access to all live parts of the equipment necessary for maintenance and adjustment. The minimum clear working space about live parts on control panels and disconnecting means shall not be less than specified in Section 110-16.

Exception No. 1: Where an escalator or moving walk control panel and disconnecting means are mounted in the same space as the escalator or moving walk drive machine and the clearances specified cannot be provided, the clearance requirements of Section 110-16 shall be permitted to be waived where the entire panel and disconnecting means are arranged so that they can be readily removed from the machine space and are provided with flexible leads to all external connections.

Exception No. 2: Where an elevator motor control panel and disconnecting means are mounted in the hoistway or on the car, and the clearances specified cannot be provided, the clearance requirements of Section 110-16 shall be permitted to be waived where the entire panel and disconnecting means are arranged so that they can be readily removed from the machine space and are provided with flexible leads to all external connections.

Where control panels are not located in the same space as the drive machine, they shall be located in cabinets with doors or removable panels capable of being locked in the closed position. Such cabinets shall be permitted in the balustrading on the side away from the moving steps or moving treadway.

J. Grounding

620-81. Metal Raceways Attached to Cars. Metal raceways, Type MC cable, Type MI cable, or Type AC cable attached to elevator cars shall be bonded to grounded metal parts of the car that they contact.

620-82. Electric Elevators. For electric elevators, the frames of all motors, elevator machines, controllers, and the metal enclosures for all electric devices in or on the car or in the hoistway shall be grounded in accordance with Article 250.

620-83. Nonelectric Elevators. For elevators other than electric having any electric conductors attached to the car, the metal frame of the car, where normally accessible to persons, shall be grounded in accordance with Article 250.

620-84. Escalators, Moving Walks, Wheelchair Lifts, and Stairway Lifts. Escalators, moving walks, wheelchair lifts, and stairway chair lifts shall comply with Article 250.

620-85. Ground-Fault Circuit-Interrupter Protection for Personnel. All 125-volt, single-phase, 15- and 20-ampere receptacles installed in machine rooms, machinery spaces, pits, and elevator car tops shall have ground-fault circuit-interrupter protection.

K. Overspeed

620-91. Overspeed Protection for Elevators. Under overhauling load conditions, a means shall be provided on the load side of each elevator power disconnecting means to prevent the elevator from attaining a speed equal to the governor tripping speed or a speed in excess of 125 percent of the elevator rated speed, whichever is the lesser.

Overhauling load conditions shall include all loads up to rated elevator loads for freight elevators and all loads up to 125 percent of rated elevator loads for passenger elevators.

620-92. Motor-Generator Overspeed Device. Motor-generators driven by direct-current motors and used to supply direct current for the operation of elevator machine motors shall be provided with speed-limiting devices as required by Section 430-89(c) that will prevent the elevator from attaining at any time a speed of more than 125 percent of its rated speed.

620-101. Emergency Power. An elevator can be powered by an emergency power system, provided that when operating on such emergency power there is conformance with Section 620-91.

Exception: Where the emergency power system is designed to operate only one elevator at a time, the energy absorption means, if required, shall be permitted on the line side of the disconnecting means, provided all other requirements of Section 620-91 are conformed to when operating any of the elevators the system might serve.

(a) Other Building Loads. Other building loads, such as power and light that can be supplied by the emergency power system, shall not be considered as means of absorbing the regenerated energy for the purpose of conforming to Section 620-91, unless such loads are using their normal power from the emergency power system when it is activated.

(b) Disconnecting Means. The disconnecting means required by Section 620-51 shall disconnect the emergency power service and the normal power service.

ARTICLE 630 — ELECTRIC WELDERS

A. General

630-1. Scope. This article covers electric arc welding, resistance welding apparatus, and other similar welding equipment that is connected to an electric supply system.

B. AC Transformer and DC Rectifier Arc Welders

630-11. Ampacity of Supply Conductors. The ampacity of conductors for ac transformer and dc rectifier arc welders shall be as follows:

(a) Individual Welders. The ampacity of the supply conductors shall not be less than the current values determined by multiplying the rated primary current in amperes given on the welder nameplate and the following factor based upon the duty cycle or time rating of the welder.

Duty Cycle (percent)	100	90	80	70	60	50	40	30	20 or less
Multiplier	1.00	.95	.89	.84	.78	.71	.63	.55	.45

For a welder having a time rating of 1 hour, the multiplying factor shall be 0.75.

(b) Group of Welders. The ampacity of conductors that supply a group of welders shall be permitted to be less than the sum of the currents, as determined in accordance with (a) above, of the welders supplied. The conductor rating shall be determined in each case according to the welder loading based on the use to be made of each welder and the allowance permissible in the event that all the welders supplied by the conductors will not be in use at the same time. The load value used for each welder shall take into account both the magnitude and the duration of the load while the welder is in use.

(FPN): Conductor ratings based on 100 percent of the current, as determined in accordance with (a) above, of the two largest welders, 85 percent for the third largest welder, 70 percent for the fourth largest welder, and 60 percent for all the remaining welders, can be assumed to provide an ample margin of safety under high-production conditions with respect to the maximum permissible temperature of the conductors. Percentage values lower than those given are permissible in cases where the work is such that a high-operating duty cycle for individual welders is impossible.

630-12. Overcurrent Protection. Overcurrent protection for ac transformer and dc rectifier arc welders shall be as provided in (a) and (b) below. Where the nearest standard rating of the overcurrent device used is under the value specified in this section, or where the rating or setting specified results in unnecessary opening of the overcurrent device, the next higher standard rating or setting shall be permitted.

(a) For Welders. Each welder shall have overcurrent protection rated or set at not more than 200 percent of the rated primary current of the welder.

Exception: An overcurrent device shall not be required for a welder having supply conductors protected by an overcurrent device rated or set at not more than 200 percent of the rated primary current of the welder.

(b) For Conductors. Conductors that supply one or more welders shall be protected by an overcurrent device rated or set at not more than 200 percent of the conductor rating.

630-13. Disconnecting Means. A disconnecting means shall be provided in the supply circuit for each ac transformer and dc rectifier arc welder that is not equipped with a disconnect mounted as an integral part of the welder.

The disconnecting means shall be a switch or circuit breaker, and its rating shall not be less than that necessary to accommodate overcurrent protection as specified under Section 630-12.

630-14. Marking. A nameplate shall be provided for ac transformer and dc rectifier arc welders giving the following information:　name of manufacturer; frequency; number of phases; primary voltage; rated primary current; maximum open-circuit voltage; rated secondary current; basis of rating, such as the duty cycle or time rating.

C. Motor-Generator Arc Welders

630-21. Ampacity of Supply Conductors. The ampacity of conductors for motor-generator arc welders shall be as follows:

(a) Individual Welders. The ampacity of the supply conductors shall not be less than the current values determined by multiplying the rated primary current in amperes given on the welder nameplate and the following factor based upon the duty cycle or time rating of the welder.

Duty Cycle (percent)	100	90	80	70	60	50	40	30	20 or less
Multiplier	1.00	.96	91	.86	. 81	.75	.69	.62	.55

For a welder having a time rating of 1 hour, the multiplying factor shall be 0.80.

(b) Group of Welders. The ampacity of conductors that supply a group of welders shall be permitted to be less than the sum of the currents, as determined in accordance with (a) above, of the welders supplied. The conductor rating shall be determined in each case according to the welder loading based on the use to be made of each welder and the allowance permissible in the event that all the welders supplied by the conductors will not be in use at the same time. The load value used for each welder shall take into account both the magnitude and the duration of the load while the welder is in use.

(FPN): Conductor ratings based on 100 percent of the current, as determined in accordance with (a) above, of the two largest welders, 85 percent for the third largest welder, 70 percent for the fourth largest welder, and 60 percent for all the remaining welders, can be assumed to provide an ample margin of safety under high-production conditions with respect to the maximum permissible temperature of the conductors. Percentage values lower than those given are permissible in cases where the work is such that a high-operating duty cycle for individual welders is impossible.

630-22. Overcurrent Protection. Overcurrent protection for motor-generator arc welders shall be as provided in (a) and (b) below. Where the nearest standard rating of the overcurrent device used is under the value specified in this section, or where the rating or setting specified results in unnecessary opening of the overcurrent device, the next higher standard rating or setting shall be permitted.

(a) For Welders. Each welder shall have overcurrent protection rated or set at not more than 200 percent of the rated primary current of the welder.

Exception: An overcurrent device shall not be required for a welder having supply conductors protected by an overcurrent device rated or set at not more than 200 percent of the rated primary current of the welder.

(b) For Conductors. Conductors that supply one or more welders shall be protected by an overcurrent device rated or set at not more than 200 percent of the conductor rating.

630-23. Disconnecting Means. A disconnecting means shall be provided in the supply connection of each motor-generator arc welder.

The disconnecting means shall be a circuit breaker or motor-circuit switch, and its rating shall not be less than that necessary to accommodate overcurrent protection as specified under Section 630-22.

630-24. Marking. A nameplate shall be provided for each motor-generator arc welder giving the following information: name of manufacturer; rated frequency; number of phases; input voltage; input current; maximum open-circuit voltage; rated output current; basis of rating, such as duty cycle or time rating.

D. Resistance Welders

630-31. Ampacity of Supply Conductors. The ampacity of the supply conductors for resistance welders necessary to limit the voltage drop to a value permissible for the satisfactory performance of the welder is usually greater than that required to prevent overheating as prescribed in (a) and (b) below.

(a) Individual Welders. The rated ampacity for conductors for individual welders shall comply with the following:

(1) The ampacity of the supply conductors for a welder that may be | operated at different times at different values of primary current or duty cycle shall not be less than 70 percent of the rated primary current for seam and automatically fed welders, and 50 percent of the rated primary current for manually operated nonautomatic welders.

(2) The ampacity of the supply conductors for a welder wired for a | specific operation for which the actual primary current and duty cycle are known and remain unchanged shall not be less than the product of the actual primary current and the multiplier given below for the duty cycle at which the welder will be operated.

Duty Cycle
(percent)........ 50 40 30 25 20 15 10 7.5 5.0 or less
Multiplier ...... .71 .63 .55 .50 .45 .39 .32 .27 .22

| **(b) Groups of Welders.** The ampacity of conductors that supply two or more welders shall not be less than the sum of the value obtained in accordance with (a) above for the largest welder supplied and 60 percent of the values obtained for all the other welders supplied.

(FPN): Explanation of Terms. (1) The rated primary current is the rated kVA multiplied by 1000 and divided by the rated primary voltage, using values given on the nameplate. (2) The actual primary current is the current drawn from the supply circuit during each welder operation at the particular heat tap and control setting used. (3) The duty cycle is the percentage of the time during which the welder is loaded. For instance, a spot welder supplied by a 60-hertz system (216,000 cycles per hour) making four hundred 15-cycle welds per hour would have a duty cycle of 2.8 percent (400 multiplied by 15, divided by 216,000, multiplied by 100). A seam welder operating 2 cycles "on" and 2 cycles "off" would have a duty cycle of 50 percent.

630-32. Overcurrent Protection. Overcurrent protection for resistance welders shall be as provided in (a) and (b) below. Where the nearest standard rating of the overcurrent device used is under the value specified in this section, or where the rating or setting specified results in unnecessary opening of the overcurrent device, the next higher standard rating or setting shall be permitted.

(a) For Welders. Each welder shall have an overcurrent device rated or set at not more than 300 percent of the rated primary current of the welder.

Exception: An overcurrent device shall not be required for a welder having a supply circuit protected by an overcurrent device rated or set at not more than 300 percent of the rated primary current of the welder.

(b) For Conductors. Conductors that supply one or more welders shall be protected by an overcurrent device rated or set at not more than 300 percent of the conductor rating.

630-33. Disconnecting Means. A switch or circuit breaker shall be provided by which each resistance welder and its control equipment can be disconnected from the supply circuit. The ampere rating of this disconnecting means shall not be less than the supply conductor ampacity determined in accordance with Section 630-31. The supply circuit switch shall be permitted as the welder disconnecting means where the circuit supplies only one welder.

630-34. Marking. A nameplate shall be provided for each resistance welder giving the following information: name of manufacturer; frequency; primary voltage rated kVA at 50 percent duty cycle; maximum and minimum open-circuit secondary voltage; short-circuit secondary current at maximum secondary voltage; and specified throat and gap setting.

E. Welding Cable

630-41. Conductors. Insulation of conductors intended for use in the secondary circuit of electric welders shall be flame-retardant.

630-42. Installation. Cables shall be permitted to be installed in a dedicated cable tray as provided in (a), (b), and (c) below.

(a) Cable Support. The cable tray shall provide support at not greater than 6-inch (152-mm) intervals.

(b) Spread of Fire and Products of Combustion. The installation shall comply with Section 300-21.

(c) Signs. A permanent sign shall be attached to the cable tray at intervals not greater than 20 feet (6.1 m). The sign shall read "Cable tray for welding cables only."

ARTICLE 640 — SOUND-RECORDING AND SIMILAR EQUIPMENT

640-1. Scope. This article covers equipment and wiring for sound-recording and reproduction, centralized distribution of sound, public address, speech-input systems, and electronic organs.

640-2. Application of Other Articles.

(a) Wiring to and Between Devices. Wiring and equipment from source of power to and between devices connected to the interior wiring systems shall comply with the requirements of Chapters 1 through 4, except as modified by this article.

(b) Wiring and Equipment. Wiring and equipment for public-address, speech-input, radio-frequency and audio-frequency systems, and amplifying equipment associated with radio receiving stations in centralized distribution systems shall comply with Article 725.

640-3. Number of Conductors in Conduit or Tubing. The number of conductors in a conduit or tubing shall comply with Tables 1 through 6 of Chapter 9.

Exception No. 1: Special permission may be granted for the installation of two 2-conductor lead-covered cables in ³/₄-inch conduit or tubing, provided the cross-sectional area of each cable does not exceed .11 square inch (70.97 mm ²).

Exception No. 2: Special permission may be granted for the installation of two 2-conductor No. 19 lead-covered cables in ¹/₂-inch conduit or tubing, provided the sum of the cross-sectional areas of the cables does not exceed 32 percent of the internal cross-sectional area of the conduit or tubing.

640-4. Wireways and Auxiliary Gutters. Wireways shall comply with the requirements of Article 362, and auxiliary gutters shall comply with the requirements of Article 374.

Exception: Where used for sound-recording and reproduction, the following shall be complied with:

a. Conductors in wireways or gutters shall not fill the raceway to more than 75 percent of its depth.

b. Where the cover of auxiliary gutters is flush with the flooring and is subject to the moving of heavy objects, it shall be of steel at least ¹/₄ inch (6.35 mm) in thickness; where not subject to moving of heavy objects, as in the rear of patch or other equipment panels, the cover shall be at least No. 10 MSG.

c. Wireways shall be permitted in concealed places, provided they are run in a straight line between outlets or junction boxes. Covers of boxes shall be accessible. Edges of metal shall be rounded at outlet or junction boxes and all rough projections smoothed to prevent abrasion of insulation or conductors.

d. Wireways and auxiliary gutters shall be grounded and bonded in accordance with the requirements of Article 250. Where the wireway or auxiliary gutter does not contain power-supply wires, the equipment grounding conductor shall not be required to be larger than No. 14 copper or its equivalent. Where the wireway or auxiliary gutter contains power-supply wires, the equipment grounding conductor shall not be smaller than specified in Section 250-95.

640-5. Conductors. Amplifier output circuits carrying audio-program signals of 70 volts or less and whose open-circuit voltage will not exceed 100 volts shall be permitted to employ Class 2 or Class 3 wiring as covered in Article 725.

(FPN): The above is based on amplifiers whose open-circuit voltage will not exceed 100 volts when driven with a signal at any frequency from 60 to 100 hertz sufficient to produce rated output (70.7 volts) into its rated load. This also accepts the known fact that the average program material is 12 db below the amplifier rating; thus, the average rms voltage for an open-circuit 70-volt output would be only 25 volts.

640-6. Grouping of Conductors. Conductors of different systems grouped in the same raceway or other enclosure or in portable cords or cables shall comply with (a) through (c) below.

(a) Power-Supply Conductors. Power-supply conductors shall be properly identified and shall be used solely for supplying power to the equipment to which the other conductors are connected.

(b) Leads to Motor-Generator or Rotary Converter. Input leads to a motor-generator or rotary converter shall be run separately from the output leads.

(c) Conductor Insulation. The conductors shall be insulated individually, or collectively in groups, by insulation at least equivalent to that on the power supply and other conductors.

Exception: Where the power supply and other conductors are separated by a lead sheath or other continuous metallic covering.

640-7. Flexible Cords. Flexible cords and cables shall be of Type S, SJ, ST, SJO, or SJT or other approved type. The conductors of flexible cords, other than power-supply conductors, shall be permitted to be of a size not smaller than No. 26, provided such conductors are not in direct electrical connection with the power-supply conductors and are equipped with a current-limiting means so that the maximum power under any condition will not exceed 150 watts.

640-8. Terminals. Terminals shall be marked to show their proper connections. Terminals for conductors other than power-supply conductors shall be separated from the terminals of the power-supply conductors by a spacing at least as great as the spacing between power-supply terminals of opposite polarity.

640-9. Storage Batteries. Storage batteries shall comply with (a) and (b) below.

(a) Installation. Storage batteries shall be installed in accordance with Article 480.

(b) Conductor Insulation. Storage-battery leads shall be rubber-covered or thermoplastic-covered.

640-10. Circuit Overcurrent Protection. Overcurrent protection shall be provided as follows:

(a) Heater or Filament (Cathode). Circuits to the heater or filament (cathode) of an electronic tube shall have overcurrent protection not exceeding 15 amperes where supplied by lighting branch circuits or by storage batteries exceeding 20 ampere-hour capacity.

(b) Plate (Anode-Positive). Circuits to the plate (anode-positive) and to the screen grid of an electronic tube shall have overcurrent protection not exceeding 1.0 ampere.

(c) Control Grid. Circuits to the control grid of an electronic tube shall have overcurrent protection not exceeding 1.0 ampere where supplied by lighting branch circuits or by storage batteries exceeding 20 ampere-hour capacity.

(d) Location. Overcurrent devices shall be located as near as practicable to the source of power supply.

640-11. Amplifiers and Rectifiers — Type.

(a) Approved Type. Amplifiers and rectifiers shall be of an approved type and shall be suitably housed.

(b) Readily Accessible. Amplifiers and rectifiers shall be so located as to be readily accessible.

(c) Ventilation. Amplifiers and rectifiers shall be so located as to provide sufficient ventilation to prevent undue temperature rise within the housing.

640-12. Hazardous (Classified) Locations. Equipment used in hazardous (classified) locations shall comply with Article 500.

640-13. Protection Against Physical Damage. Amplifiers, rectifiers, loudspeakers, and other equipment shall be so located or protected as to guard against physical damage, such as might result in fire or personal hazard.

ARTICLE 645 — ELECTRONIC COMPUTER/DATA PROCESSING EQUIPMENT

645-1. Scope. This article covers equipment, power-supply wiring, equipment interconnecting wiring, and grounding of electronic computer/data processing equipment and systems, including terminal units, in an electronic computer/data processing room.

(FPN): For further information, see Protection of Electronic Computer/Data Processing Equipment, NFPA 75-1992 (ANSI).

645-2. Special Requirements for Electronic Computer/Data Processing Equipment Room. This article applies provided all the following conditions are met:

(1) Disconnecting means complying with Section 645-10 are provided.

(2) A separate heating/ventilating/air conditioning (HVAC) system is provided that is dedicated for electronic computer/data processing equipment use and is separated from other areas of occupancy. Any HVAC system that serves other occupancies may also serve the electronic computer/data processing equipment room if fire/smoke dampers are provided at the point of penetration of the room boundary. Such dampers shall operate on activation of smoke detectors and also by operation of the disconnecting means required by Section 645-10.

(FPN): For further information, see Protection of Electronic Computer/Data Processing Equipment, NFPA 75-1992 (ANSI).

(3) Listed electronic computer/data processing equipment is installed.

(FPN): For further information, see Protection of Electronic Computer/Data Processing Equipment, NFPA 75-1992 (ANSI), Section 5-1.1.

(4) Occupied only by those personnel needed for the maintenance and functional operation of the installed electronic computer/data processing equipment.

(FPN): The computer room is not to be used for the storage of combustibles beyond that necessary for the day-to-day operation of the equipment. For further information, see Protection of Electronic Computer/Data Processing Equipment, NFPA 75-1992 (ANSI).

(5) The room is separated from other occupancies by fire-resistant-rated walls, floors, and ceilings with protected openings.

(6) The building construction, rooms, or areas and occupancy comply with the applicable building code.

645-5. Supply Circuits and Interconnecting Cables.

(a) Branch Circuit Conductors. The branch circuit conductors supplying one or more units of a data processing system shall have an ampacity not less than 125 percent of the total connected load.

(b) Connecting Cables. The data processing system shall be permitted to be connected to a branch circuit by any of the following means listed for the purpose:

(1) Computer/data processing cable and attachment plug cap.

(2) Flexible cord and an attachment plug cap.

(3) Cord-set assembly. Where run on the surface of the floor, they shall be protected against physical damage.

(c) Interconnecting Cables. Separate data processing units shall be permitted to be interconnected by means of cables and cable assemblies listed for the purpose. Where run on the surface of the floor, they shall be protected against physical damage.

(d) Under Raised Floors. Power cables, communications cables, connecting cables, interconnecting cables, and receptacles associated with the data processing equipment shall be permitted under a raised floor, provided:

(1) The raised floor is of suitable construction and the area under the floor is accessible.

(FPN): See Protection of Electronic Computer/Data Processing Equipment, NFPA 75-1992 (ANSI).

(2) The branch-circuit supply conductors to receptacles or field-wired equipment are in rigid metal conduit, rigid nonmetallic conduit, intermediate metal conduit, electrical metallic tubing, metal wireway, surface metal raceway with metal cover, flexible metal conduit, liquidtight flexible metal or nonmetallic conduit, Type MI cable, Type MC cable, or Type AC cable. These supply conductors shall be installed in accordance with the requirements of Section 300-11.

(3) Ventilation in the underfloor area is used for the data processing equipment and data processing area only.

(4) Openings in raised floors for cables protect cables against abrasions and minimize the entrance of debris beneath the floor.

(5) Cables, other than those covered in (2) above, shall be listed as Type DP cable having adequate fire-resistance characteristics suitable for use under raised floors of a computer room. This listing requirement shall become effective July 1, 1994.

Exception No. 1: Where the interconnecting cables are enclosed in conduit or raceway.

Exception No. 2: Interconnecting cables listed with equipment manufactured prior to July 1, 1994 shall be permitted to be reinstalled with that equipment.

Exception No. 3: Other cable type designations that satisfy the above requirement are Type TC (Article 340); Types CL2, CL3, and PLTC (Article 725); Type FPL (Article 760); Types OFC and OFN (Article 770); Types CM and MP (Article 800); Type CATV (Article 820). These designations shall be permitted to have an additional letter P or R.

(FPN): One method of defining fire resistance is by establishing that the cables do not spread fire to the top of the tray in the "Vertical Tray Flame Test" referenced in the Standard for Electrical Wires, Cables, and Flexible Cords, ANSI/UL 1581-1985.

Another method of defining fire resistance is for the damage (char length) not to exceed 4 feet 11 inches (1.5 m) when performing the CSA "Vertical Flame Test — Cables in Cable Trays," as described in Test Methods for Electrical Wires and Cables, CSA C22.2 No. 0.3-M-1985.

(e) Securing in Place. Power cables, communications cables, connecting cables, interconnecting cables, and associated boxes, connectors, plugs, and receptacles that are listed as part of, or for, electronic computer/data processing equipment shall not be required to be secured in place.

645-6. Cables not in Computer Room. Cables extending beyond the computer room shall be subject to the applicable requirements of this Code.

(FPN): For signaling circuits, refer to Article 725; for fiber optic circuits, refer to Article 770; and for communication circuits, refer to Article 800. For fire protective signaling systems, refer to Article 760.

645-7. Penetrations. Penetrations of the fire resistant room boundary shall be in accordance with Section 300-21.

645-10. Disconnecting Means. A means shall be provided to disconnect power to all electronic equipment in the electronic computer/data processing equipment room. There shall also be a similar means to disconnect the power to all dedicated HVAC systems serving the room and cause all required fire/smoke dampers to close. The control for these disconnecting means shall be grouped and identified and shall be readily accessible at the principal exit doors. A single means to control both the electronic equipment and HVAC systems shall be permitted.

Exception: Installations qualifying under the provisions of Article 685.

645-11. Uninterruptible Power Supplies (UPS). UPS systems installed within the electronic computer/data processing room, and their supply and output circuits, shall comply with Section 645-10. The disconnecting means shall also disconnect the battery from its load.

Exception No. 1: Installations qualifying under the provisions of Article 685.

Exception No. 2: A disconnecting means complying with Section 645-10 shall not be required for power sources capable of supplying 750 volt-amperes or less derived from UPS equipment or from battery circuits integral to electronic equipment, provided all other requirements of Section 645-11 are met.

645-15. Grounding. All exposed noncurrent-carrying metal parts of an electronic computer/data processing system shall be grounded in accordance with Article 250 or shall be double-insulated. Power systems derived within listed electronic computer/data processing equipment that supply electronic computer/data processing systems through receptacles or cable assemblies supplied as part of this equipment shall not be considered separately derived for the purpose of applying Section 250-5(d).

(FPN No. 1): This listed equipment provides the bonding and grounding requirements in accordance with the intent of Article 250.

(FPN No. 2): Where isolated grounding-type receptacles are used, see Section 250-74, Exception No. 4.

645-16. Marking. Each unit of an electronic computer/data processing system supplied by a branch-circuit shall be provided with a manufacturer's nameplate, which shall also include the input power requirements for voltage, frequency, and maximum rated load in amperes.

ARTICLE 650 — PIPE ORGANS

650-1. Scope. This article covers those electrical circuits and parts of electrically operated pipe organs that are employed for the control of the sounding apparatus and keyboards.

650-2. Other Articles. Electronic organs shall comply with the appropriate provisions of Article 640.

650-3. Source of Energy. The source of power shall be a transformer-type rectifier, the dc potential of which shall not exceed 30 volts dc.

650-4. Grounding. The rectifier shall be grounded according to the provisions in Article 250.

650-5. Conductors. Conductors shall comply with (a) through (d) below:

(a) **Size.** Not less than No. 28 for electronic signal circuits and not less than No. 26 for electromagnetic valve supply and the like. A main common-return conductor in the electromagnetic supply shall not be less than No. 14.

(b) **Insulation.** Conductors shall have thermoplastic or thermosetting insulation.

(c) **Conductors to Be Cabled.** Except for the common-return conductor and conductors inside the organ proper, the organ sections and the organ console conductors shall be cabled. The common-return conductors shall be permitted under an additional covering enclosing both cable and return conductor, or shall be permitted as a separate conductor and shall be permitted to be in contact with the cable.

(d) Cable Covering. Each cable shall be provided with an outer covering, either overall or on each of any subassemblies of grouped conductors. Tape shall be permitted in place of a covering. Where not installed in metal raceway, the covering shall be flame-retardant or the cable or each cable subassembly shall be covered with a closely wound fireproof tape.

650-6. Installation of Conductors. Cables shall be securely fastened in place and shall be permitted to be attached directly to the organ structure without insulating supports. Cables shall not be placed in contact with other conductors.

650-7. Overcurrent Protection. Circuits shall be so arranged that all conductors shall be protected from overcurrent by an overcurrent device rated at not more than 6 amperes.

Exception: The main supply conductors and the common-return conductors.

ARTICLE 660 — X-RAY EQUIPMENT

A. General

660-1. Scope. This article covers all X-ray equipment operating at any frequency or voltage for industrial or other nonmedical or nondental use.

(FPN):   See Article 517, Part E, for X-ray installations in health care facilities. |

Nothing in this article shall be construed as specifying safeguards against the useful beam or stray X-ray radiation.

(FPN No. 1):   Radiation safety and performance requirements of several classes of X-ray equipment are regulated under Public Law 90-602 and are enforced by the Department of Health and Human Services.

(FPN No. 2):   In addition, information on radiation protection by the National Council on Radiation Protection and Measurements is published as Reports of the National Council on Radiation Protection and Measurement. These reports are obtainable from NCRP Publications, 7910 Woodmont Ave., Suite 1016, Bethesda, MD 20814.

660-2. Definitions.

Long-Time Rating: A rating based on an operating interval of 5 minutes or longer.

Mobile: X-ray equipment mounted on a permanent base with wheels and/or casters for moving while completely assembled.

Momentary Rating: A rating based on an operating interval that does not exceed 5 seconds.

Portable: X-ray equipment designed to be hand-carried.

Transportable: X-ray equipment to be installed in a vehicle or that may be readily disassembled for transport in a vehicle.

660-3. Hazardous (Classified) Locations. Unless approved for the location, X-ray and related equipment shall not be installed or operated in hazardous (classified) locations.

(FPN): See Article 517, Part D.

660-4. Connection to Supply Circuit.

(a) Fixed and Stationary Equipment. Fixed and stationary X-ray equipment shall be connected to the power supply by means of a wiring method meeting the general requirements of this Code.

Exception: Equipment properly supplied by a branch circuit rated at not over 30 amperes shall be permitted to be supplied through a suitable attachment plug cap and hard-service cable or cord.

(b) Portable, Mobile, and Transportable Equipment. Individual branch circuits shall not be required for portable, mobile, and transportable X-ray equipment requiring a capacity of not over 60 amperes. Portable and mobile types of X-ray equipment of any capacity shall be supplied through a suitable hard-service cable or cord. Transportable X-ray equipment of any capacity shall be permitted to be connected to its power supply by suitable connections and hard-service cable or cord.

(c) Over 600 Volts, Nominal. Circuits and equipment operated at more than 600 volts, nominal, shall comply with Article 710.

660-5. Disconnecting Means. A disconnecting means of adequate capacity for at least 50 percent of the input required for the momentary rating or 100 percent of the input required for the long-time rating of the X-ray equipment, whichever is greater, shall be provided in the supply circuit. The disconnecting means shall be operable from a location readily accessible from the X-ray control. For equipment connected to a 120-volt, nominal, branch circuit of 30 amperes or less, a grounding-type attachment plug cap and receptacle of proper rating shall be permitted to serve as a disconnecting means.

660-6. Rating of Supply Conductors and Overcurrent Protection.

(a) Branch-Circuit Conductors. The ampacity of supply branch-circuit conductors and the overcurrent protective devices shall not be less than 50 percent of the momentary rating or 100 percent of the long-time rating, whichever is the greater.

(b) Feeder Conductors. The rated ampacity of conductors and overcurrent devices of a feeder for two or more branch circuits supplying X-ray units shall not be less than 100 percent of the momentary demand rating [as determined by (a)] of the two largest X-ray apparatus plus 20 percent of the momentary ratings of other X-ray apparatus.

(FPN): The minimum conductor size for branch and feeder circuits is also governed by voltage regulation requirements. For a specific installation, the manufacturer usually specifies minimum distribution transformer and conductor sizes, rating of disconnect means, and overcurrent protection.

660-7. Wiring Terminals. X-ray equipment shall be provided with suitable wiring terminals or leads for the connection of power-supply conductors of the size required by the rating of the branch circuit for the equipment.

Exception: Where provided with a permanently attached cord or a cord set.

660-8. Number of Conductors in Raceway. The number of control circuit conductors installed in a raceway shall be determined in accordance with Section 300-17.

660-9. Minimum Size of Conductors. Size No. 18 or 16 fixture wires, as specified in Section 725-16, and flexible cords shall be permitted for the control and operating circuits of X-ray and auxiliary equipment where protected by not larger than 20-ampere overcurrent devices.

660-10. Equipment Installations. All equipment for new X-ray installations and all used or reconditioned X-ray equipment moved to and reinstalled at a new location shall be of an approved type.

B. Control

660-20. Fixed and Stationary Equipment.

(a) **Separate Control Device.** A separate control device, in addition to the disconnecting means, shall be incorporated in the X-ray control supply or in the primary circuit to the high-voltage transformer. This device shall be a part of the X-ray equipment but shall be permitted in a separate enclosure immediately adjacent to the X-ray control unit.

(b) **Protective Device.** A protective device, which shall be permitted to be incorporated into the separate control device, shall be provided to control the load resulting from failures in the high-voltage circuit.

660-21. Portable and Mobile Equipment. Portable and mobile equipment shall comply with Section 660-20, but the manually controlled device shall be located in or on the equipment.

660-23. Industrial and Commercial Laboratory Equipment.

(a) **Radiographic and Fluoroscopic Types.** All radiographic- and fluoroscopic-type equipment shall be effectively enclosed or shall have interlocks that deenergize the equipment automatically to prevent ready access to live current-carrying parts.

(b) **Diffraction and Irradiation Types.** Diffraction- and irradiation-type equipment shall be provided with a positive means to indicate when it is energized. The indicator shall be a pilot light, readable meter deflection, or equivalent means.

Exception: Equipment or installations effectively enclosed or provided with interlocks to prevent access to live current-carrying parts during operation.

660-24. Independent Control. Where more than one piece of equipment is operated from the same high-voltage circuit, each piece or each group of equipment as a unit shall be provided with a high-voltage switch or equivalent disconnecting means. This disconnecting means shall be constructed, enclosed, or located so as to avoid contact by persons with its live parts.

C. Transformers and Capacitors

660-35. General. Transformers and capacitors that are part of an X-ray equipment shall not be required to comply with Articles 450 and 460.

660-36. Capacitors. Capacitors shall be mounted within enclosures of insulating material or grounded metal.

D. Guarding and Grounding

660-47. General.

(a) High-Voltage Parts. All high-voltage parts, including X-ray tubes, shall be mounted within grounded enclosures. Air, oil, gas, or other suitable insulating media shall be used to insulate the high voltage from the grounded enclosure. The connection from the high-voltage equipment to X-ray tubes and other high-voltage components shall be made with high-voltage shielded cables.

(b) Low-Voltage Cables. Low-voltage cables connecting to oil-filled units that are not completely sealed, such as transformers, condensers, oil coolers, and high-voltage switches, shall have insulation of the oil-resistant type.

660-48. Grounding. Noncurrent-carrying metal parts of X-ray and associated equipment (controls, tables, X-ray tube supports, transformer tanks, shielded cables, X-ray tube heads, etc.) shall be grounded in the manner specified in Article 250. Portable and mobile equipment shall be provided with an approved grounding-type attachment plug cap.

Exception: Battery-operated equipment.

ARTICLE 665 — INDUCTION AND DIELECTRIC HEATING EQUIPMENT

A. General

665-1. Scope. This article covers the construction and installation of induction and dielectric heating equipment and accessories for industrial and scientific applications, but not for medical or dental applications, appliances, or line frequency pipelines and vessels heating.

(FPN No. 1): See Article 422 for appliances.

(FPN No. 2): See Article 427, Part E, for line frequency pipelines and vessels heating.

665-2. Definitions.

Dielectric Heating: Dielectric heating is the heating of a nominally insulating material due to its own dielectric losses when the material is placed in a varying electric field.

Heating Equipment: The term "heating equipment" as used in this article includes any equipment used for heating purposes whose heat is generated by induction or dielectric methods.

Induction Heating: Induction heating is the heating of a nominally conductive material due to its own I^2R losses when the material is placed in a varying electromagnetic field.

665-3. Other Articles. Wiring from the source of power to the heating equipment shall comply with Chapters 1 through 4. Circuits and equipment operated at more than 600 volts, nominal, shall comply with Article 710.

665-4. Hazardous (Classified) Locations. Induction and dielectric heating equipment shall not be installed in hazardous (classified) locations as defined in Article 500.

Exception: Where the equipment and wiring are designed and approved for the hazardous (classified) locations.

B. Guarding, Grounding, and Labeling

665-20. Enclosures. The converting apparatus (including the dc line) and high-frequency electric circuits (excluding the output circuits and remote-control circuits) shall be completely contained within an enclosure or enclosures of noncombustible material.

665-21. Panel Controls. All panel controls shall be of dead-front construction.

665-22. Access to Internal Equipment. Doors or detachable panels shall be employed for internal access. Where doors are used giving access to voltages from 500 to 1000 volts ac or dc, either door locks shall be provided or interlocking shall be installed. Where doors are used giving access to voltages of over 1000 volts ac or dc, either mechanical lockouts with a disconnecting means to prevent access until voltage is removed from the cubicle, or both door interlocking and mechanical door locks, shall be provided. Detachable panels not normally used for access to such parts shall be fastened in a manner that will make them inconvenient to remove.

665-23. Warning Labels or Signs. Warning labels or signs that read "DANGER — HIGH VOLTAGE — KEEP OUT" shall be attached to the equipment and shall be plainly visible where unauthorized persons might come in contact with energized parts, even when doors are open or when panels are removed from compartments containing over 250 volts ac or dc.

665-24. Capacitors. Where capacitors in excess of 0.1 microfarad are used in dc circuits, either as rectifier filter components or suppressors, etc., having circuit voltages of over 240 volts to ground, bleeder resistors or grounding switches shall be used as grounding devices. The time of discharge shall be in accordance with Section 460-6(a).

Where capacitors are individually switched out of a circuit, a bleeder resistor or automatic switch shall be used as a discharge means.

Where auxiliary rectifiers are used with filter capacitors in the output for bias supplies, tube keyers, etc., bleeder resistors shall be used, even though the dc voltage may not exceed 240 volts.

665-25. Work Applicator Shielding. Protective cages or adequate shielding shall be used to guard work applicators other than induction heating coils. Induction heating coils shall be permitted to be protected by insulation and/or refractory materials. Interlock switches shall be used on all hinged access doors, sliding panels, or other easy means of access to the applicator. All interlock switches shall be connected in such a manner as to remove all power from the applicator when any one of the access doors or panels is open. Interlocks on access doors or panels shall not be required if the applicator is an induction heating coil at dc ground potential or operating at less than 150 volts ac.

665-26. Grounding and Bonding. Grounding and/or inter-unit bonding shall be used wherever required for circuit operation, for limiting to a safe value radio frequency potentials between all exposed noncurrent-carrying parts of the equipment and earth ground, between all equipment parts and surrounding objects, and between such objects and earth ground. Such grounding and bonding shall be installed in accordance with Article 250.

665-27. Marking. Each heating equipment shall be provided with a nameplate giving the manufacturer's name and model identification and the following input data: line volts, frequency, number of phases, maximum current, full-load kVA, and full-load power factor.

665-28. Control Enclosures. Direct current or low-frequency ac shall be permitted in the control portion of the heating equipment. This shall be limited to not over 150 volts. Solid or stranded wire No. 18 or larger shall be used. A step-down transformer with proper overcurrent protection shall be permitted in the control enclosure to obtain an ac voltage of less than 150 volts. The higher-voltage terminals shall be guarded to prevent accidental contact. 60-hertz components shall be permitted to control high frequency where properly rated by the induction heating equipment manufacturer. Electronic circuits utilizing solid-state devices and tubes shall be permitted printed circuits or wires smaller than No. 18.

C. Motor-Generator Equipment

665-40. General. Motor-generator equipment shall include all rotating equipment designed to operate from an ac or dc motor or by mechanical drive from a prime mover, producing an alternating current of any frequency for induction and/or dielectric heating.

665-41. Ampacity of Supply Conductors. The ampacity of supply conductors shall be determined in accordance with Article 430.

665-42. Overcurrent Protection. Overcurrent protection shall be provided as specified in Article 430 for the electric supply circuit.

665-43. Disconnecting Means. The disconnecting means shall be provided as specified in Article 430.

A readily accessible disconnecting means shall be provided by which each heating equipment can be isolated from its supply circuit. The ampere rating of this disconnecting means shall not be less than the nameplate current rating of the equipment. The supply circuit disconnecting means shall be permitted as a heating equipment disconnecting means where the circuit supplies only one equipment.

665-44. Output Circuit. The output circuit shall include all output components external to the generator, including contactors, transformers, busbars, and other conductors, and shall comply with (a) and (b) below.

(a) Generator Output. The output circuit shall be isolated from ground.

Exception No. 1: Where the capacitive coupling inherent in the generator causes the generator terminals to have voltages from terminal to ground that are equal.

Exception No. 2: Where a vacuum or controlled atmosphere is used with a coil in a tank or chamber, the center point of the coil shall be grounded to maintain an equal potential between each terminal and ground.

Where rated at over 500 volts, the output circuit shall incorporate a dc ground protector unit. The dc impressed on the output circuit shall not exceed 30 volts and shall not exceed a current capability of 5 milliamperes.

An isolating transformer for matching the load and the source shall be permitted in the output circuit if the output secondary is not at dc ground potential.

(b) Component Interconnections. The various components required for a complete induction heating equipment installation shall be connected by properly protected multiconductor cable, busbar, or coaxial cable. Cables shall be installed in nonferrous raceways. Busbars shall be protected, where required, by nonferrous enclosures.

665-47. Remote Control.

(a) Selector Switch. Where remote controls are used for applying power, a selector switch shall be provided and interlocked to provide power from only one control point at a time.

(b) Foot Switches. Switches operated by foot pressure shall be provided with a shield over the contact button to avoid accidental closing of a switch.

D. Equipment Other than Motor-Generator

665-60. General. Equipment other than motor-generators shall consist of all static multipliers and oscillator-type units utilizing vacuum tubes and/or solid-state devices. The equipment shall be capable of converting ac or dc to an ac frequency suitable for induction and/or dielectric heating.

665-61. Ampacity of Supply Conductors. The ampacity of supply conductors shall be determined in accordance with (a) and (b) below.

(a) Nameplate Rating. The ampacity of the circuit conductors shall not be less than the nameplate current rating of the equipment.

(b) Two or More. The ampacity of conductors supplying two or more equipments shall not be less than the sum of the nameplate current ratings on all equipments.

Exception: If simultaneous operation of two or more equipments supplied from the same feeder is not possible, the ampacity of the feeder shall not be less than the sum of the nameplate ratings for the largest group of machines capable of simultaneous operation, plus 100 percent of the stand-by currents of the remaining machines supplied.

665-62. Overcurrent Protection. Overcurrent protection shall be provided as specified in Article 240 for the equipment as a whole. This overcurrent protection shall be provided separately or as a part of the equipment.

665-63. Disconnecting Means. A readily accessible disconnecting means shall be provided by which each heating equipment can be isolated from its supply circuit. The rating of this disconnecting means shall not be less than the nameplate rating of the equipment. The supply circuit disconnecting means shall be permitted for disconnecting the heating equipment where the circuit supplies only one equipment.

665-64. Output Circuit. The output circuit shall include all output components external to the converting device, including contactors, transformers, busbars, and other conductors, and shall comply with (a) and (b) below.

(a) Converter Output. The output circuit shall be isolated from ground.

Exception: Where a dc voltage can exist at the terminals because of an internal component failure, then the output circuit (direct or coupled) shall be at dc ground potential.

(b) Converter and Applicator Connection. Where the connections between the converter and the work applicator exceed 2 feet (610 mm) in length, the connections shall be enclosed or guarded with nonferrous, noncombustible material.

665-66. Line Frequency in Converter Equipment Output. Commercial frequencies of 25- to 60-hertz alternating-current output shall be permitted to be coupled for control purposes, but shall be limited to not over 150 volts during periods of circuit operation.

665-67. Keying. Where high-speed keying circuits dependent on the effect of "oscillator blocking" are employed, the peak radio-frequency output voltage during the blocked portion of the cycle shall not exceed 100 volts in units employing radio-frequency converters.

665-68. Remote Control.

(a) Selector Switch. Where remote controls are used for applying power, a selector switch shall be provided and interlocked to provide power from only one control point at a time.

(b) Foot Switches. Switches operated by foot pressure shall be provided with a shield over the contact button to avoid accidental closing of the switch.

ARTICLE 668 — ELECTROLYTIC CELLS

668-1. Scope. The provisions of this article apply to the installation of the electrical components and accessory equipment of electrolytic cells, electrolytic cell lines and process power supply for the production of aluminum, cadmium, chlorine, copper, fluorine, hydrogen peroxide, magnesium, sodium, sodium chlorate, and zinc.

Not covered by this article are cells used as a source of electric energy and for electroplating processes and cells used for the production of hydrogen.

(FPN No. 1): In general, any cell line or group of cell lines operated as a unit for the production of a particular metal, gas, or chemical compound may differ from any other cell lines producing the same product because of variations in the particular raw materials used, output capacity, use of proprietary methods or process practices, or other modifying factors to the extent that detailed Code requirements become overly restrictive and do not accomplish the stated purpose of this Code.

(FPN No. 2): For further information, see IEEE Standard for Electrical Safety Practices in Electrolytic Cell Line Working Zones: IEEE Std. 463-1977.

668-2. Definitions.

Cell Line: An assembly of electrically interconnected electrolytic cells supplied by a source of direct-current power.

Cell Line Attachments and Auxiliary Equipment: As applied to Article 668, cell line attachments and auxiliary equipment include, but are not limited to, auxiliary tanks; process piping; duct work; structural supports; exposed cell line conductors; conduits and other raceways; pumps, positioning equipment, and cell cutout or bypass electrical devices. Auxiliary equipment includes tools, welding machines, crucibles, and other portable equipment used for operation and maintenance within the electrolytic cell line working zone.

In the cell line working zone, auxiliary equipment includes the exposed conductive surfaces of ungrounded cranes and crane-mounted cell-servicing equipment.

Electrolytic Cell: A receptacle or vessel in which electrochemical reactions are caused by applying electrical energy for the purpose of refining or producing usable materials.

Electrolytic Cell Line Working Zone: The cell line working zone is the space envelope wherein operation or maintenance is normally performed on or in the vicinity of exposed energized surfaces of electrolytic cell lines or their attachments.

668-3. Other Articles.

(a) Lighting, Ventilating, Material Handling. Chapters 1 through 4 shall apply to services, feeders, branch circuits, and apparatus for supplying lighting, ventilating, material handling, and the like, that are outside the electrolytic cell line working zone.

(b) Systems Not Electrically Connected. Those elements of a cell line power-supply system that are not electrically connected to the cell supply system, such as the primary winding of a two-winding transformer, the motor of a motor-generator set, feeders, branch circuits, disconnecting means, motor controllers, and overload protective equipment shall be required to comply with all applicable provisions of this Code.

(FPN):   For the purpose of this section, "electrically connected" means connection capable of carrying current as distinguished from connection through electromagnetic induction.

(c) Electrolytic Cell Lines. Electrolytic cell lines shall comply with the provisions of Chapters 1, 2, 3, and 4.

Exception No. 1: The electrolytic cell line conductors shall not be required to comply with the provisions of Articles 110, 210, 215, 220, and 225. See Section 668-11.

Exception No. 2: Overcurrent protection of electrolytic cell dc process power circuits shall not be required to comply with the requirements of Article 240.

Exception No. 3: Equipment located or used within the electrolytic cell line working zone or associated with the cell line dc power circuits shall not be required to comply with the provisions of Article 250.

Exception No. 4: The electrolytic cells, cell line attachments, and the wiring of auxiliary equipments and devices within the cell line working zone shall not be required to comply with the provisions of Articles 110, 210, 215, 220, and 225. See Section 668-30.

(FPN):   See Section 668-15 on equipment, apparatus, and structural component grounding.

668-10. Cell Line Working Zone.

(a) Area Covered. The space envelope of the cell line working zone shall encompass any space:

(1) Within 96 inches (2.44 m) above energized surfaces of electrolytic cell lines or their energized attachments.

(2) Below energized surfaces of electrolytic cell lines or their energized attachments, provided the head room in the space beneath is less than 96 inches (2.44 m).

(3) Within 42 inches (1.07 m) horizontally from energized surfaces of electrolytic cell lines or their energized attachments or from the space envelope described in Section 668-10(a)(1) or (a)(2).

(b) Area Not Covered. The cell line working zone shall not be required to extend through or beyond walls, floors, roofs, partitions, barriers, or the like.

668-11. DC Cell Line Process Power Supply.

(a) **Not Grounded.** The dc cell line process power supply conductors shall not be required to be grounded.

(b) **Metal Enclosures Grounded.** All metal enclosures of dc cell line process power supply apparatus operating at a power supply potential between terminals of over 50 volts shall be grounded:

(1) Through protective relaying equipment, or

(2) By No. 2/0 minimum copper grounding conductor or a conductor | of equal ampacity.

(c) **Grounding Requirements.** The grounding connections required by Section 668-11(b) shall be installed in accordance with Sections 250-112, 250-113, 250-115, 250-117, and 250-118.

668-12. Cell Line Conductors.

(a) **Insulation and Material.** Cell line conductors shall be either bare, covered, or insulated and of copper, aluminum, copper-clad aluminum, steel, or other suitable material.

(b) **Size.** Cell line conductors shall be of such cross-sectional area that the temperature rise under maximum load conditions and at maximum ambient shall not exceed the safe operating temperature of the conductor insulation or the material of the conductor supports.

(c) **Connections.** Cell line conductors shall be joined by bolted, welded, clamped, or compression connectors.

668-13. Disconnecting Means.

(a) **More than One Process Power Supply.** Where more than one dc cell line process power supply serves the same cell line, a disconnecting means shall be provided on the cell line circuit side of each power supply to disconnect it from the cell line circuit.

(b) **Removable Links or Conductors.** Removable links or removable conductors shall be permitted to be used as the disconnecting means.

668-14. Shunting Means.

(a) **Partial or Total Shunting.** Partial or total shunting of cell line circuit current around one or more cells shall be permitted.

(b) **Shunting One or More Cells.** The conductors, switches, or combination of conductors and switches used for shunting one or more cells shall comply with the applicable requirements of Section 668-12.

668-15. Grounding. For equipment, apparatus, and structural components that are required to be grounded by provisions of Article 668, the provisions of Article 250 shall apply.

Exception No. 1: A water pipe electrode shall not be required to be used.

Exception No. 2: Any electrode or combination of electrodes described in Sections 250-81 and 250-83 shall be permitted.

668-20. Portable Electrical Equipment.

(a) Portable Electrical Equipment Not to Be Grounded. The frames and enclosures of portable electrical equipment used within the cell line working zone shall not be grounded.

Exception No. 1: Where the cell line circuit voltage does not exceed 200 volts dc, these frames and enclosures shall be permitted to be grounded.

Exception No. 2: These frames and enclosures shall be permitted to be grounded where guarded.

(b) Isolating Transformers. Electrically powered, hand-held, cord-connected portable equipment with ungrounded frames or enclosures used within the cell line working zone shall be connected to receptacle circuits having only ungrounded conductors such as a branch circuit supplied by an isolating transformer with an ungrounded secondary.

(c) Marking. Ungrounded portable electrical equipment shall be distinctively marked and shall employ plugs and receptacles of a configuration that prevents connection of this equipment to grounding receptacles and that prevents inadvertent interchange of ungrounded and grounded portable electrical equipments.

668-21. Power Supply Circuits and Receptacles for Portable Electrical Equipment.

(a) Isolated Circuits. Circuits supplying power to ungrounded receptacles for hand-held, cord-connected equipments shall be electrically isolated from any distribution system supplying areas other than the cell line working zone and shall be ungrounded. Power for these circuits shall be supplied through isolating transformers. Primaries of such transformers shall operate at not more than 600 volts between conductors and shall be provided with proper overcurrent protection. The secondary voltage of such transformers shall not exceed 300 volts between conductors, and all circuits supplied from such secondaries shall be ungrounded and shall have an approved overcurrent device of proper rating in each conductor.

(b) Noninterchangeability. Receptacles and their mating plugs for ungrounded equipment shall not have provision for a grounding conductor and shall be of a configuration that prevents their use for equipment required to be grounded.

(c) Marking. Receptacles on circuits supplied by an isolating transformer with an ungrounded secondary shall be a distinctive configuration, distinctively marked, and shall not be used in any other location in the plant.

668-30. Fixed and Portable Electrical Equipment.

(a) Electrical Equipment Not Required to Be Grounded. AC systems supplying fixed and portable electrical equipments within the cell line working zone shall not be required to be grounded.

(b) Exposed Conductive Surfaces Not Required to Be Grounded. Exposed conductive surfaces, such as electrical equipment housings, cabinets, boxes, motors, raceways, and the like that are within the cell line working zone shall not be required to be grounded.

(c) Wiring Methods. Auxiliary electrical devices such as motors, transducers, sensors, control devices, and alarms, mounted on an electrolytic cell or other energized surface, shall be connected to premises wiring systems by any of the following means:

(1) Multiconductor hard usage cord;

(2) Wire or cable in suitable raceways or metal or nonmetallic cable trays. If metal conduit, cable tray, armored cable, or similar metallic systems are used, they shall be installed with insulating breaks such that they will not cause a potentially hazardous electrical condition.

(d) Circuit Protection. Circuit protection shall not be required for control and instrumentation that are totally within the cell line working zone.

(e) Bonding. Bonding of fixed electrical equipment to the energized conductive surfaces of the cell line, its attachments, or auxiliaries shall be permitted. Where fixed electrical equipment is mounted on an energized conductive surface, it shall be bonded to that surface.

668-31. Auxiliary Nonelectric Connections. Auxiliary nonelectric connections, such as air hoses, water hoses, and the like, to an electrolytic cell, its attachments, or auxiliary equipments shall not have continuous conductive reinforcing wire, armor, braids, and the like. Hoses shall be of a nonconductive material.

668-32. Cranes and Hoists.

(a) Conductive Surfaces to Be Insulated from Ground. The conductive surfaces of cranes and hoists that enter the cell line working zone shall not be required to be grounded. The portion of an overhead crane or hoist that contacts an energized electrolytic cell or energized attachments shall be insulated from ground.

(b) Hazardous Electrical Conditions. Remote crane or hoist controls that may introduce hazardous electrical conditions into the cell line working zone shall employ one or more of the following systems:

(1) Insulated and ungrounded control circuit in accordance with Section 668-21(a);

(2) Nonconductive rope operator;

(3) Pendant pushbutton with nonconductive supporting means and having nonconductive surfaces or ungrounded exposed conductive surfaces;

(4) Radio.

668-40. Enclosures. General-purpose electrical equipment enclosures shall be permitted where a natural draft ventilation system prevents the accumulation of gases.

ARTICLE 669 — ELECTROPLATING

669-1. Scope. The provisions of this article apply to the installation of the electrical components and accessory equipment that supply the power and controls for electroplating, anodizing, electropolishing, and electrostripping. For purposes of this article, the term electroplating shall be used to identify any or all of these processes.

669-2. Other Articles. Except as modified by this article, wiring and equipment used for electroplating processes shall comply with the applicable requirements of Chapters 1 through 4.

669-3. General. Equipment for use in electroplating processes shall be identified for such service.

669-5. Branch-Circuit Conductors. Branch-circuit conductors supplying one or more units of equipment shall have an ampacity of not less than 125 percent of the total connected load. The ampacities for busbars shall be in accordance with Section 374-6.

669-6. Wiring Methods. Conductors connecting the electrolyte tank equipment to the conversion equipment shall be as follows:

(a) Systems Not Exceeding 50 Volts DC. Insulated conductors shall be permitted to be run without insulated support provided they are protected from physical damage. Bare copper or aluminum conductors shall be permitted where supported on insulators.

(b) Systems Exceeding 50 Volts DC. Insulated conductors shall be permitted to be run on insulated supports, provided they are protected from physical damage. Bare copper or aluminum conductors shall be permitted where supported on insulators and guarded against accidental contact in accordance with Section 110-17.

Exception: Unguarded bare conductors shall be permitted at the terminals.

669-7. Warning Signs. Warning signs shall be posted to indicate the presence of bare conductors.

669-8. Disconnecting Means.

(a) More than One Power Supply. Where more than one power supply serves the same dc system, a disconnecting means shall be provided on the dc side of each power supply.

(b) Removable Links or Conductors. Removable links or removable conductors shall be permitted to be used as the disconnecting means.

669-9. Overcurrent Protection. DC conductors shall be protected from overcurrent by one or more of the following: (1) fuses or circuit breakers; (2) a current-sensing device that operates a disconnecting means; or (3) other approved means.

ARTICLE 670 — INDUSTRIAL MACHINERY

670-1. Scope. This article covers the definition of, the size and overcurrent protection of supply conductors to, and the nameplate data required on industrial metalworking machine tools, woodworking machinery, plastics machinery, and mass production equipment, not portable by hand.

(FPN): For further information, see Electrical Standard for Industrial Machinery, NFPA 79-1991 (ANSI).

670-2. Definition of Industrial Machinery. For the purposes of this article, a machine tool is defined as a power-driven machine not portable by hand, used to shape or form metal or plastic by cutting, impact, pressure, electrical techniques, or a combination of these processes. Plastics machinery is defined as a power-driven machine not portable by hand, used to shape or form plastic by application of thermal and/or mechanical energy, by cutting, impact, pressure, or a combination of these processes.

Mass production industrial equipment is defined as a systematic array of one or more machine tools, plastics machinery, or assembly machines that is not portable by hand and that includes any associated material handling, manipulating, gaging, measuring, or inspection equipment.

670-3. Machine Nameplate Data.

(a) Permanent Nameplate. A permanent nameplate that lists supply voltage, phase, frequency, full-load current, the maximum ampere rating of the short-circuit and ground-fault protective device, ampere rating of largest motor or load, short-circuit interrupting capacity of the machine overcurrent-protective device, if furnished, and diagram number shall be attached to the control equipment enclosure or machine where plainly visible after installation.

The full-load current shown on the nameplate shall not be less than the sum of the full-load currents required for all motors and other equipment that may be in operation at the same time under normal conditions of use. Where unusual type loads, duty cycles, etc., require oversized conductors, the required capacity shall be included in the marked "full-load current."

Where more than one incoming supply circuit is to be provided, the nameplate shall state the above information for each circuit.

(b) Overcurrent Protection. Where overcurrent protection is provided in accordance with Section 670-4(b), the machine shall be marked "overcurrent protection provided at machine supply terminals."

670-4. Supply Conductors and Overcurrent Protection.

(a) Size. The size of the supply conductor shall be such as to have an ampacity not less than 125 percent of the full-load current rating of all resistance heating loads plus 125 percent of the full-load current rating of highest rated motor plus the sum of the full-load current ratings of all other connected motors and apparatus that may be in operation at the same time.

(FPN): See the 0-2000 volt ampacity tables of Article 310 for ampacity of conductors rated 600 volts and below.

(b) Overcurrent Protection. A machine shall be considered as an individual unit and therefore shall be provided with a disconnecting means. The disconnecting means shall be permitted to be supplied by branch circuits protected by either fuses or circuit breakers. The disconnecting means shall not be required to incorporate overcurrent protection. Where furnished as part of the machine, overcurrent protection shall consist of a single circuit breaker or set of fuses, the machine shall bear the marking required in Section 670-3, and the supply conductors shall be considered either as feeders or taps as covered by Section 240-21.

The rating or setting of the overcurrent-protective device for the circuit supplying the machine shall not be greater than the sum of the largest rating or setting of the branch-circuit short-circuit and ground-fault protective device provided with the machine, plus 125 percent of the full-load current rating of all resistance heating loads, plus the sum of the full-load currents of all other motors and apparatus that may be in operation at the same time.

Exception: Where one or more instantaneous trip circuit breakers or motor short-circuit protectors are used for motor branch-circuit short-circuit and ground-fault protection as permitted by Section 430-52(a), the procedure specified above for determining the maximum rating of the protective device for the circuit supplying the machine shall apply with the following provision. For the purpose of the calculation, each instantaneous trip circuit breaker or motor short-circuit protector shall be assumed to have a rating not exceeding the maximum percentage of motor full-load current permitted by Table 430-152 for the type of machine supply circuit protective device employed.

Where no branch-circuit short-circuit and ground-fault protective device is provided with the machine, the rating or setting of the overcurrent-protective device shall be based on Sections 430-52 and 430-53 as applicable.

670-5. Clearance. Where the conditions of maintenance and supervision assure that only qualified persons will service the installation, the dimensions of the working space in the direction of access to live parts operating at not over 150 volts that are likely to require examination, adjustment, servicing, or maintenance while energized shall be a minimum of 2½ feet (762 mm). Where controls are enclosed in cabinets, the door(s) shall open at least 90 degrees or be removable.

Exception: Where the enclosure requires a tool to open, and where only diagnostic and troubleshooting testing is involved on live parts operating at not over 150 volts, line-to-line, the clearances shall be permitted to be less than 2½ feet (762 mm).

ARTICLE 675 — ELECTRICALLY DRIVEN OR CONTROLLED IRRIGATION MACHINES

A. General

675-1. Scope. The provisions of this article apply to electrically driven or controlled irrigation machines, and to the branch circuits and controllers for such equipment.

675-2. Definitions.

Center Pivot Irrigation Machines: A center pivot irrigation machine is a multimotored irrigation machine that revolves around a central pivot and employs alignment switches or similar devices to control individual motors.

Collector Rings: A collector ring is an assembly of slip rings for transferring electrical energy from a stationary to a rotating member.

Irrigation Machines: An irrigation machine is an electrically driven or controlled machine, with one or more motors, not hand portable, and used primarily to transport and distribute water for agricultural purposes.

675-3. Other Articles. These provisions are in addition to, or amendatory of, the provisions of Article 430 and other articles in this Code that apply except as modified in this article.

675-4. Irrigation Cable.

(a) Construction. The cable used to interconnect enclosures on the structure of an irrigation machine shall be an assembly of stranded, insulated conductors with nonhygroscopic and nonwicking filler in a core of moisture- and flame-resistant nonmetallic material overlaid with a metallic covering and jacketed with a moisture-, corrosion-, and sunlight-resistant nonmetallic material.

The conductor insulation shall be of a type listed in Table 310-13 for an operating temperature of 75°C (167°F) and for use in wet locations. The core insulating material thickness shall not be less than 30 mils (762 micrometers), and the metallic overlay thickness shall not be less than 8 mils (203 micrometers). The jacketing material thickness shall not be less than 50 mils (1.27 mm).

A composite of power, control, and grounding conductors in the cable shall be permitted.

(b) Alternate Wiring Methods. Other cables listed for the purpose.

(c) Supports. Irrigation cable shall be secured by straps, hangers, or similar fittings identified for the purpose and installed as not to damage the | cable. Cable shall be supported at intervals not exceeding 4 feet (1.22 m).

(d) Fittings. Fittings shall be used at all points where irrigation cable terminates. The fittings shall be designed for use with the cable and shall be suitable for the conditions of service.

675-5. More than Three Conductors in a Raceway or Cable. The signal and control conductors of a raceway or cable shall not be counted for the purpose of derating the conductors as required in Article 310, Note 8 (a) of Notes to Ampacity Tables of 0 to 2000 Volts.

675-6. Marking on Main Control Panel. The main control panel shall be provided with a nameplate that shall give the following information: (1) the manufacturer's name, the rated voltage, the phase, and the frequency; (2) the current rating of the machine; and (3) the rating of the main disconnecting means and size of overcurrent protection required.

675-7. Equivalent Current Ratings. Where intermittent duty is not involved, the provisions of Article 430 shall be used for determining ratings for controllers, disconnecting means, conductors, and the like. Where irrigation machines have inherent intermittent duty, the following determinations of equivalent current ratings shall be used.

(a) Continuous-Current Rating. The equivalent continuous-current rating for the selection of branch-circuit conductors and overcurrent protection shall be equal to 125 percent of the motor nameplate full-load current rating of the largest motor plus a quantity equal to the sum of each of the motor nameplate full-load current ratings of all remaining motors on the circuit multiplied by the maximum percent duty cycle at which they can continuously operate.

(b) Locked-Rotor Current. The equivalent locked-rotor current rating shall be equal to the numerical sum of the locked-rotor current of the two largest motors plus 100 percent of the sum of the motor nameplate full-load current ratings of all the remaining motors on the circuit.

675-8. Disconnecting Means.

(a) Main Controller. A controller that is used to start and stop the complete machine shall meet all of the following requirements:

(1) An equivalent continuous current rating not less than specified in Section 675-7(a) or 675-22(a).

(2) A horsepower rating not less than the value from Table 430-151 based on the equivalent locked-rotor current specified in Section 675-7(b) or 675-22(b).

(b) Main Disconnecting Means. The main disconnecting means for the machine shall be at the point of connection of electrical power to the machine or shall be visible and not more than 50 feet (15.2 m) from the machine and shall be readily accessible and capable of being locked in the open position. This disconnecting means shall have a horsepower and current rating not less than required for the main controller.

Exception: Circuit breakers without marked horsepower ratings shall be permitted in accordance with Section 430-109.

(c) Disconnecting Means for Individual Motors and Controllers. A disconnecting means shall be provided to simultaneously disconnect all ungrounded conductors for each motor and controller and shall be located as required by Article 430, Part I. The disconnecting means shall not be required to be readily accessible.

675-9. Branch-Circuit Conductors. The branch-circuit conductors shall have an ampacity not less than specified in Section 675-7(a) or 675-22(a).

675-10. Several Motors on One Branch Circuit.

(a) Protection Required. Several motors, each not exceeding 2-horsepower rating, shall be permitted to be used on an irrigation machine circuit protected at not more than 30 amperes at 600 volts, nominal, or less, provided all of the following conditions are met:

(1) The full-load rating of any motor in the circuit shall not exceed 6 amperes.

(2) Each motor in the circuit shall have individual overload protection in accordance with Section 430-32.

(3) Taps to individual motors shall not be smaller than No. 14 copper and not more than 25 feet (7.62 m) in length.

(b) Individual Protection Not Required. Individual branch-circuit short-circuit protection for motors and motor controllers shall not be required where the requirements of Section 675-10(a) are met.

675-11. Collector Rings.

(a) Transmitting Current for Power Purposes. Collector rings shall have a current rating not less than 125 percent of the full-load current of the | largest device served plus the full-load current of all other devices served, or as determined from Section 675-7(a) or 675-22(a).

(b) Control and Signal Purposes. Collector rings for control and signal purposes shall have a current rating not less than 125 percent of the full- | load current of the largest device served plus the full-load current of all other devices served.

(c) Grounding. The collector ring used for grounding shall have a current rating of not less than that sized in accordance with Section 675-11(a). |

(d) Protection. Collector rings shall be protected from the expected environment and from accidental contact by means of a suitable enclosure.

675-12. Grounding. The following equipment shall be grounded: (1) All electrical equipment on the irrigation machine; (2) all electrical equipment associated with the irrigation machine; (3) metal junction boxes and enclosures; and (4) control panels or control equipment that supply or control electrical equipment to the irrigation machine.

Exception: Grounding shall not be required on machines where all of the following provisions are met:

a. The machine is electrically controlled but not electrically driven.

b. The control voltage is 30 volts or less.

c. The control or signal circuits are current-limited as specified in Section 725-31.

675-13. Methods of Grounding. Machines that require grounding shall have a noncurrent-carrying equipment grounding conductor provided as an integral part of each cord, cable, or raceway. This grounding conductor shall be sized not less than the largest supply conductor in each cord, cable, or | raceway. Feeder circuits supplying power to irrigation machines shall have | an equipment grounding conductor sized according to Table 250-95.

675-14. Bonding. Where electrical grounding is required on an irrigation machine, the metallic structure of the machine, metallic conduit, or metallic sheath of cable shall be bonded to the grounding conductor. Metal-to-metal contact with a part that is bonded to the grounding conductor and the noncurrent-carrying parts of the machine shall be considered as an acceptable bonding path.

675-15. Lightning Protection. If an irrigation machine has a stationary point, a grounding electrode system in accordance with Article 250, Part H, shall be connected to the machine at the stationary point for lightning protection.

675-16. Energy from More than One Source. Equipment within an enclosure receiving electrical energy from more than one source shall not be required to have a disconnecting means for the additional source, provided that its voltage is 30 volts or less and meets the requirements of Section 725-31.

675-17. Connectors. External plugs and connectors on the equipment shall be of the weatherproof type.

Unless provided solely for the connection of circuits meeting the requirements of Section 725-31, external plugs and connectors shall be constructed as specified in Section 250-99(a).

B. Center Pivot Irrigation Machines

675-21. General. The provisions of Part B are intended to cover additional special requirements that are peculiar to center pivot irrigation machines. See Section 675-2 for definition of "Center Pivot Irrigation Machines."

675-22. Equivalent Current Ratings. In order to establish ratings of controllers, disconnecting means, conductors, and the like, for the inherent intermittent duty of center pivot irrigation machines, the following determination shall be used:

(a) Continuous-Current Rating. The equivalent continuous-current rating for the selection of branch-circuit conductors and branch-circuit devices shall be equal to 125 percent of the motor nameplate full-load current rating of the largest motor plus 60 percent of the sum of the motor nameplate full-load current ratings of all remaining motors on the circuit.

(b) Locked-Rotor Current. The equivalent locked-rotor current rating shall be equal to the numerical sum of two times the locked-rotor current of the largest motor plus 80 percent of the sum of the motor nameplate full-load current ratings of all the remaining motors on the circuit.

ARTICLE 680 — SWIMMING POOLS, FOUNTAINS, AND SIMILAR INSTALLATIONS

A. General

680-1. Scope. The provisions of this article apply to the construction and installation of electric wiring for and equipment in or adjacent to all swimming, wading, therapeutic, and decorative pools, fountains, hot tubs, spas, and hydromassage bathtubs, whether permanently installed or storable, and to metallic auxiliary equipment, such as pumps, filters, and similar equipment.

(FPN): The term "pool" as used in the balance of this article includes swimming, wading, and permanently installed therapeutic pools. The term "fountain" as used in the balance of this article includes fountains, ornamental pools, display pools, and reflection pools. The term is not intended to include drinking water fountains.

680-2. Approval of Equipment. All electric equipment installed in the water, walls, or decks of pools, fountains, and similar installations shall comply with the provisions of this article.

680-3. Other Articles. Except as modified by this article, wiring and equipment in or adjacent to pools and fountains shall comply with the applicable requirements of Chapters 1 through 4.

(FPN): See Section 370-23 for junction boxes and Section 347-3 for rigid non-metallic conduit.

680-4. Definitions.

Cord- and Plug-Connected Lighting Assembly: A lighting assembly consisting of a lighting fixture intended for installation in the wall of a spa, hot tub, or storable pool, and a cord- and plug-connected transformer.

Dry-Niche Lighting Fixture: A lighting fixture intended for installation in the wall of a pool or fountain in a niche that is sealed against the entry of pool water.

Forming Shell: A structure designed to support a wet-niche lighting fixture assembly and intended for mounting in a pool or fountain structure.

Hydromassage Bathtub: A permanently installed bathtub equipped with a recirculating piping system, pump, and associated equipment. It is designed so it can accept, circulate, and discharge water upon each use.

No-Niche Lighting Fixture: A lighting fixture intended for installation above or below the water without a niche.

Permanently Installed Decorative Fountains and Reflection Pools: Those that are constructed in the ground, on the ground, or in a building in such a manner that the fountain cannot be readily disassembled for storage, whether or not served by electrical circuits of any nature. These units are primarily constructed for their aesthetic value and not intended for swimming or wading.

Permanently Installed Swimming, Wading, and Therapeutic Pools: Those that are constructed in the ground, on the ground, or in a building in such a manner that the pool cannot be readily disassembled for storage, whether or not served by electrical circuits of any nature.

Pool Cover, Electrically Operated: Motor-driven equipment designed to cover and uncover the water surface of a pool by means of a flexible sheet or rigid frame.

Spa or Hot Tub: A hydromassage pool, or tub for recreational or therapeutic use, not located in health care facilities, designed for immersion of users and usually having a filter, heater, and motor-driven blower. It may be installed indoors or outdoors, on the ground or supporting structure, or in the ground or supporting structure. Generally, a spa or hot tub is not designed or intended to have its contents drained or discharged after each use.

Storable Swimming or Wading Pool: A pool with a maximum dimension of 18 feet (5.49 m) and a maximum wall height of 42 inches (1.07 m) and so constructed that it may be readily disassembled for storage and reassembled to its original integrity. A pool with nonmetallic inflatable walls regardless of dimensions is considered to be a storable pool.

Wet-Niche Lighting Fixture: A lighting fixture intended for installation in a forming shell mounted in a pool or fountain structure where the fixture will be completely surrounded by water.

680-5. Transformers and Ground-Fault Circuit-Interrupters.

(a) Transformers. Transformers used for the supply of underwater fixtures, together with the transformer enclosure, shall be identified for the purpose. The transformer shall be an isolated winding type having a grounded metal barrier between the primary and secondary windings.

(b) Ground-Fault Circuit-Interrupters. Ground-fault circuit-interrupters shall be self-contained units, circuit-breaker types, receptacle types, or other approved types.

(c) Wiring. Conductors on the load side of a ground-fault circuit-interrupter or of a transformer, used to comply with provisions of Section 680-20(a)(1), shall not occupy raceway, boxes, or enclosures containing other conductors.

Exception No. 1: Ground-fault circuit-interrupters shall be permitted in a panelboard that contains circuits protected by other than ground-fault circuit-interrupters.

Exception No. 2: Supply conductors to a feed-through, receptacle-type, ground-fault circuit-interrupter shall be permitted in the same enclosure.

Exception No. 3: Conductors on the load side of a ground-fault circuit-interrupter shall be permitted to occupy raceway, boxes, or enclosures containing only conductors protected by ground-fault circuit-interrupters.

Exception No. 4: Grounding conductors.

680-6. Receptacles, Lighting Fixtures, Lighting Outlets, Switching Devices, and Ceiling Fans.

(a) Receptacles.

(1) Receptacles on the property shall be located at least 10 feet (3.05 m) from the inside walls of a pool or fountain.

Exception: Receptacle(s) that provide power for water-pump motor(s) for a permanently installed pool or fountain, as permitted in Section 680-7, shall be permitted between 5 and 10 feet (1.52 and 3.05 m) from the inside walls of the pool or fountain, and, where so located, shall be single and of the locking and grounding types and shall be protected by ground-fault circuit-interrupter(s).

(2) Where a permanently installed pool is installed at a dwelling unit(s), at least one 125-volt receptacle shall be located a minimum of 10 feet (3.05 m) from and not more than 20 feet (6.08 m) from the inside wall of the pool.

(3) All 125-volt receptacles located within 20 feet (6.08 m) of the inside walls of a pool shall be protected by a ground-fault circuit-interrupter. See Section 210-8(a)(3).

(FPN): In determining the above dimensions, the distance to be measured is the shortest path the supply cord of an appliance connected to the receptacle would follow without piercing a floor, wall, ceiling, doorway with hinged or sliding door, window opening, or other effective permanent barrier.

(b) Lighting Fixtures, Lighting Outlets, and Ceiling Fans.

(1) Lighting fixtures, lighting outlets, and ceiling fans shall not be installed over the pool or over the area extending 5 feet (1.52 m) horizontally from the inside walls of a pool unless no part of the lighting fixture or ceiling fan is less than 12 feet (3.66 m) above the maximum water level.

Exception No. 1: Existing lighting fixtures and lighting outlets located less than 5 feet (1.52 m) measured horizontally from the inside walls of a pool shall be at least 5 feet (1.52 m) above the surface of the maximum water level and shall be rigidly attached to the existing structure.

Exception No. 2: In indoor pool areas, the limitations of Section 680-6(b)(1) shall not apply if all of the following conditions are complied with: (1) fixtures are of totally enclosed type; (2) a ground-fault circuit-interrupter is installed in the branch circuit supplying the fixture(s) or ceiling fans; and (3) the distance from the bottom of the fixture or ceiling fan to the maximum water level is not less than 7.5 feet (2.29 m).

(2) Lighting fixtures and lighting outlets installed in the area extending between 5 feet (1.52 m) and 10 feet (3.05 m) horizontally from the inside walls of a pool shall be protected by a ground-fault circuit-interrupter unless installed 5 feet (1.52 m) above the maximum water level and rigidly attached to the structure adjacent to or enclosing the pool.

(3) Cord-connected lighting fixtures shall meet the same specifications as other cord- and plug-connected equipment as set forth in Section 680-7 where installed within 16 feet (4.88 m) of any point on the water surface, measured radially.

(c) Switching Devices. Switching devices on the property shall be located at least 5 feet (1.52 m) horizontally from the inside walls of a pool unless separated from the pool by a solid fence, wall, or other permanent barrier.

680-7. Cord- and Plug-Connected Equipment. Fixed or stationary equipment rated 20 amperes or less, other than an underwater lighting fixture for a permanently installed pool, shall be permitted to be connected with a flexible cord to facilitate the removal or disconnection for maintenance or repair. For other than storable pools, the flexible cord shall not exceed 3 feet (914 mm) in length and shall have a copper equipment grounding conductor not smaller than No. 12 with a grounding-type attachment plug.

(FPN): See Section 680-25(e) for connection with flexible cords.

680-8. Overhead Conductor Clearances. The following parts of pools shall not be placed under existing service-drop conductors or any other open overhead wiring; nor shall such wiring be installed above the following: (1) pools and the area extending 10 feet (3.05 m) horizontally from the inside of the walls of the pool; (2) diving structure; or (3) observation stands, towers, or platforms.

Exception No. 1: Structures listed in items (1), (2), and (3) above shall be permitted under supply lines or service drops where such installations provide the following clearances:

	Insulated supply or service drop cables, 0-750 volts to ground, supported on and cabled together with an effectively grounded bare messenger or effectively grounded neutral conductor	All other supply or service drop conductors	
		Voltage to Ground	
		0-15 kV	greater than 15 to 50kV
A. Clearance in any direction to the water level, edge of water surface, base of diving platform, or permanently-anchored raft	18 feet (5.49 m)	25 feet (7.62 m)	27 feet (8.23 m)
B. Clearance in any direction to the diving platform or tower	14 feet (4.27 m)	16 feet (4.88 m)	18 feet (5.49 m)
C. Horizontal limit of clearance measured from inside wall of the pool	This limit shall extend to the outer edge of the structures listed in (1) and (2) above but not less than 10 feet (3.05 m).		

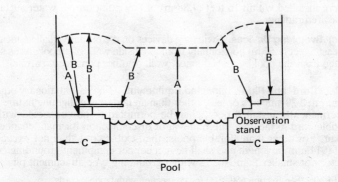

Figure 680-8, Exception No. 1

Exception No. 2: Utility-owned, -operated, and -maintained communication conductors, community antenna system coaxial cables complying with Article 820, and the supporting messengers shall be permitted at a height of not less than 10 feet (3.05 m) above swimming and wading pools, diving structures, and observation stands, towers, or platforms.

(FPN): See Sections 225-18 and 225-19 for clearances for conductors not covered by this section.

680-9. Electric Pool Water Heaters.   All electric pool water heaters shall have the heating elements subdivided into loads not exceeding 48 amperes and protected at not more than 60 amperes.

The ampacity of the branch-circuit conductors and the rating or setting of overcurrent protective devices shall not be less than 125 percent of the total load of the nameplate rating.

680-10. Underground Wiring Location.   Underground wiring shall not be permitted under the pool or within the area extending 5 feet (1.52 m) horizontally from the inside wall of the pool.

Exception No. 1: Wiring necessary to supply pool equipment permitted by this article shall be allowed within this area.

Exception No. 2: Where space limitations prevent wiring from being routed 5 feet (1.52 m) or more from the pool, such wiring shall be permitted where installed in rigid metal conduit, intermediate metal conduit, or a non-metallic raceway system. All metal conduit shall be corrosion-resistant and suitable for the location. The minimum burial depth shall be as follows:

Wiring Method	Minimum Burial (inches)
Rigid Metal Conduit	6
Intermediate Metal Conduit	6
Rigid Nonmetallic Conduit Approved for Direct Burial without Concrete Encasement	18
Other Approved Raceways*	18

For SI units: one inch = 25.4 millimeters
*Note: Raceways approved for burial only where concrete-encased shall require a concrete envelope not less than 2 inches (50.8 mm) thick.

680-11. Equipment Rooms and Pits.   Electric equipment shall not be installed in rooms or pits that do not have adequate drainage to prevent water accumulation during normal operation or filter maintenance.

B. Permanently Installed Pools

680-20. Underwater Lighting Fixtures.   Paragraphs (a) through (d) of this section apply to all lighting fixtures installed below the normal water level of the pool.

(a) General.

(1) The design of an underwater lighting fixture supplied from a branch circuit either directly or by way of a transformer meeting the requirements of Section 680-5(a) shall be such that, where the fixture is properly installed without a ground-fault circuit-interrupter, there is no shock hazard with any likely combination of fault conditions during normal use (not relamping).

In addition, a ground-fault circuit-interrupter shall be installed in the branch circuit supplying fixtures operating at more than 15 volts, so that there is no shock hazard during relamping. The installation of the ground-fault circuit-interrupter shall be such that there is no shock hazard with any likely fault-condition combination that involves a person in a conductive path from any ungrounded part of the branch circuit or the fixture to ground.

Compliance with this requirement shall be obtained by the use of an approved underwater lighting fixture and by installation of an approved ground-fault circuit-interrupter in the branch circuit.

(2) No lighting fixtures shall be installed for operation on supply circuits over 150 volts between conductors.

(3) Lighting fixtures mounted in walls shall be installed with the top of the fixture lens at least 18 inches (457 mm) below the normal water level of the pool. A lighting fixture facing upward shall have the lens adequately guarded to prevent contact by any person.

Exception: Lighting fixtures identified for use at a depth of not less than 4 inches (102 mm) below the normal water level of the pool shall be permitted.

(4) Fixtures that depend on submersion for safe operation shall be inherently protected against the hazards of overheating when not submerged.

(b) Wet-Niche Fixtures.

(1) Forming shells shall be installed for the mounting of all wet-niche underwater fixtures and shall be equipped with provisions for conduit entries.

Conduit shall extend from the forming shell to a suitable junction box or other enclosure located as provided in Section 680-21. Conduit shall be rigid metal, intermediate metal, or rigid nonmetallic.

Metal conduit shall be of brass or other approved corrosion-resistant metal.

Where rigid nonmetallic conduit is used, a No. 8 insulated copper conductor shall be installed in this conduit with provisions for terminating in the forming shell, junction box or transformer enclosure, or ground-fault circuit-interrupter enclosure. The termination of the No. 8 conductor in the forming shell shall be covered with, or encapsulated in, a listed potting compound to protect such connection from the possible deteriorating effect of pool water. Metal parts of the fixture and forming shell in contact with the pool water shall be of brass or other approved corrosion-resistant metal.

(2) The end of the flexible-cord jacket and the flexible-cord conductor terminations within a fixture shall be covered with, or encapsulated in, a suitable potting compound to prevent the entry of water into the fixture through the cord or its conductors. In addition, the grounding connection within a fixture shall be similarly treated to protect such connection from the deteriorating effect of pool water in the event of water entry into the fixture.

(3) The fixture shall be bonded to and secured to the forming shell by a positive locking device that assures a low-resistance contact and requires a tool to remove the fixture from the forming shell.

(c) Dry-Niche Fixtures. A dry-niche lighting fixture shall be provided with (1) provision for drainage of water and (2) means for accommodating one equipment grounding conductor for each conduit entry.

Approved rigid metal conduit, intermediate metal conduit, or rigid non-metallic conduit shall be installed from the fixture to the service equipment or panelboard. A junction box shall not be required but, if used, shall not be required to be elevated or located as specified in Section 680-21(a)(4) if the fixture is specifically identified for the purpose.

Exception: Electrical metallic tubing shall be permitted to be used to protect conductors where installed on or within buildings.

(d) No-Niche Fixtures. A no-niche fixture shall:

(1) Be listed for the purpose.

(2) Be installed in accordance with the requirements of Section 680-20(b).

Where connection to a forming shell is specified, the connection shall be to the mounting bracket.

680-21. Junction Boxes and Enclosures for Transformers or Ground-Fault Circuit-Interrupters.

(a) Junction Boxes. A junction box connected to a conduit that extends directly to a forming shell or mounting bracket of a no-niche fixture shall be:

(1) Equipped with threaded hubs or bosses or a nonmetallic hub listed for the purpose; and

(2) Of copper, brass, suitable plastic, or other approved corrosion-resistant material; and

(3) Provided with electrical continuity between every connected metal conduit and the grounding terminals by means of copper, brass, or other approved corrosion-resistant metal that is integral with the box; and

(4) Located not less than 4 inches (102 mm), measured from the inside of the bottom of the box, above the ground level, or pool deck, or not less than 8 inches (203 mm) above the maximum pool water level, whichever provides the greatest elevation, and located not less than 4 feet (1.22 m) from the inside wall of the pool, unless separated from the pool by a solid fence, wall, or other permanent barrier.

Exception: On lighting systems of 15 volts or less, a flush deck box shall be permitted provided:

a. An approved potting compound is used to fill the box to prevent the entrance of moisture; and

b. The flush deck box is located not less than 4 feet (1.22 m) from the inside wall of the pool.

(b) Other Enclosures. An enclosure for a transformer, ground-fault circuit-interrupter, or a similar device connected to a conduit that extends directly to a forming shell or mounting bracket of a no-niche fixture shall be:

(1) Equipped with threaded hubs or bosses or a nonmetallic hub listed for the purpose; and

(2) Provided with an approved seal, such as duct seal at the conduit connection, that prevents circulation of air between the conduit and the enclosures; and

(3) Provided with electrical continuity between every connected metal conduit and the grounding terminals by means of copper, brass, or other approved corrosion-resistant metal that is integral with the enclosures; and

(4) Located not less than 4 inches (102 mm), measured from the inside bottom of the enclosure, above the ground level or pool deck, or not less than 8 inches (203 mm) above the maximum pool water level, whichever provides the greater elevation, and located not less than 4 feet (1.22 mm) from the inside wall of the pool, unless separated from the pool by a solid fence, wall, or other permanent barrier.

(c) Protection. Junction boxes and enclosures mounted above the grade of the finished walkway around the pool shall not be located in the walkway unless afforded additional protection, such as by location under diving boards, adjacent to fixed structures, and the like.

(d) Grounding Terminals. Junction boxes, transformer enclosures, and ground-fault circuit-interrupter enclosures connected to a conduit that extends directly to a forming shell or mounting bracket of a no-niche fixture shall be provided with a number of grounding terminals that shall be at least one more than the number of conduit entries.

(e) Strain Relief. The termination of a flexible cord of an underwater lighting fixture within a junction box, transformer enclosure, ground-fault circuit-interrupter, or other enclosure shall be provided with a strain relief.

680-22. Bonding.

(FPN): It is not the intent of this subsection to require that the No. 8 or larger solid copper bonding conductor be extended or attached to any remote panelboard, service equipment, or any electrode, but only that it be employed to eliminate voltage gradients in the pool area as prescribed.

(a) Bonded Parts. The following parts shall be bonded together:

(1) All metallic parts of the pool structure, including the reinforcing metal of the pool shell, coping stones, and deck.

(2) All forming shells and mounting brackets of a no-niche fixture.

(3) All metal fittings within or attached to the pool structure.

(4) Metal parts of electric equipment associated with the pool water circulating system, including pump motors.

(5) Metal parts of equipment associated with pool covers, including electric motors.

(6) Metal-sheathed cables and raceways, metal piping, and all fixed metal parts that are within 5 feet (1.52 m) horizontally of the inside walls of the pool, and within 12 feet (3.66 m) above the maximum water level of the pool, or any observation stands, towers, or platforms, or from any diving structures, and that are not separated from the pool by a permanent barrier.

Exception No. 1: The usual steel tie wires shall be considered suitable for bonding the reinforcing steel together, and welding or special clamping shall not be required. These tie wires shall be made tight.

Exception No. 2: Isolated parts that are not over 4 inches (102 mm) in any dimension and do not penetrate into the pool structure more than 1 inch (25.4 mm) shall not require bonding.

Exception No. 3: Structural reinforcing steel or the walls of bolted or welded metal pool structures shall be permitted as a common bonding grid for nonelectrical parts where connections can be made in accordance with Section 250-113.

(b) Common Bonding Grid. These parts shall be connected to a common bonding grid with a solid, copper conductor, insulated, covered, or bare, not smaller than No. 8. Connection shall be made by pressure connectors or clamps of stainless steel, brass, copper, or copper alloy. The common bonding grid shall be permitted to be any of the following:

(1) The structural reinforcing steel of a concrete pool where the reinforcing rods are bonded together by the usual steel tie wires or the equivalent; or

(2) The wall of a bolted or welded metal pool; or

(3) A solid, copper conductor, insulated, covered, or bare, not smaller than No. 8.

(c) Pool Water Heaters. For pool water heaters rated at more than 50 amperes that have specific instructions regarding bonding and grounding, only those parts designated to be bonded shall be bonded, and only those parts designated to be grounded shall be grounded.

680-23. Underwater Audio Equipment.   All underwater audio equipment shall be identified for the purpose.

(a) Speakers. Each speaker shall be mounted in an approved metal forming shell, the front of which is enclosed by a captive metal screen, or equivalent, that is bonded to and secured to the forming shell by a positive locking device that assures a low-resistance contact and requires a tool to open for installation or servicing of the speaker. The forming shell shall be installed in a recess in the wall or floor of the pool.

(b) Wiring Methods. Rigid metal conduit or intermediate metal conduit of brass or other identified corrosion-resistant metal or rigid nonmetallic conduit shall extend from the forming shell to a suitable junction box or other enclosure as provided in Section 680-21. Where rigid nonmetallic conduit is used, a No. 8 insulated copper conductor shall be installed in this conduit with provisions for terminating in the forming shell and the junction box. The termination of the No. 8 conductor in the forming shell shall be covered with, or encapsulated in, a suitable potting compound to protect such connection from the possible deteriorating effect of pool water.

(c) **Forming Shell and Metal Screen.** The forming shell and metal screen shall be of brass or other approved corrosion-resistant metal.

680-24. Grounding.   The following equipment shall be grounded: (1) wet-niche and no-niche underwater lighting fixtures; (2) dry-niche underwater lighting fixtures; (3) all electric equipment located within 5 feet (1.52 m) of the inside wall of the pool; (4) all electric equipment associated with the recirculating system of the pool; (5) junction boxes; (6) transformer enclosures; (7) ground-fault circuit-interrupters; (8) panelboards that are not part of the service equipment and that supply any electric equipment associated with the pool.

680-25. Methods of Grounding.

(a) **General.** The following provisions shall apply to the grounding of underwater lighting fixtures, junction boxes, metal transformer enclosures, panelboards, motors, and other electrical enclosures and equipment.

(b) **Pool Lighting Fixtures and Related Equipment.**

(1) Wet-niche, dry-niche, or no-niche lighting fixtures shall be connected to an equipment grounding conductor sized in accordance with Table 250-95 but not smaller than No. 12. It shall be an insulated copper conductor and shall be installed with the circuit conductors in rigid metal conduit, intermediate metal conduit, or rigid nonmetallic conduit.

Exception No. 1: Electrical metallic tubing shall be permitted to be used to protect conductors where installed on or within buildings.

Exception No. 2: The equipment grounding conductor between the wiring chamber of the secondary winding of a transformer and a junction box shall be sized in accordance with the overcurrent device in this circuit.

(2) The junction box, transformer enclosure, or other enclosure in the supply circuit to a wet-niche or no-niche lighting fixture and the field-wiring chamber of a dry-niche lighting fixture shall be grounded to the equipment grounding terminal of the panelboard. This terminal shall be directly connected to the panelboard enclosure. The equipment grounding conductor shall be installed without joint or splice.

Exception No. 1: Where more than one underwater lighting fixture is supplied by the same branch circuit, the equipment grounding conductor, installed between the junction boxes, transformer enclosures, or other enclosures in the supply circuit to wet-niche fixtures, or between the field-wiring compartments of dry-niche fixtures, shall be permitted to be terminated on grounding terminals.

Exception No. 2: Where the underwater lighting fixture is supplied from a transformer, ground-fault circuit-interrupter, clock-operated switch, or a manual snap switch that is located between the panelboard and a junction box connected to the conduit that extends directly to the underwater lighting fixture, the equipment grounding conductor shall be permitted to terminate on grounding terminals on the transformer, ground-fault circuit-interrupter, clock-operated switch enclosure, or an outlet box used to enclose a snap switch.

(3) Wet-niche or no-niche lighting fixtures that are supplied by a flexible cord or cable shall have all exposed noncurrent-carrying metal parts grounded by an insulated copper equipment grounding conductor that is an integral part of the cord or cable. This grounding conductor shall be connected to a grounding terminal in the supply junction box, transformer enclosure, or other enclosure. The grounding conductor shall not be smaller than the supply conductors and not smaller than No. 16.

(c) Motors. Pool-associated motors shall be connected to an equipment grounding conductor sized in accordance with Table 250-95 but not smaller than No. 12. It shall be an insulated copper conductor and shall be installed with the circuit conductors in rigid metal conduit, intermediate metal conduit, rigid nonmetallic conduit, or Type MC cable listed for the application.

Exception No. 1: Electrical metallic tubing shall be permitted to be used to protect conductors where installed on or within buildings.

Exception No. 2: Where necessary to employ flexible connections at or adjacent to the motor, liquidtight flexible metal or nonmetallic conduit with approved fittings shall be permitted.

Exception No. 3: Any of the wiring methods recognized in Chapter 3 of this Code that contain an insulated or covered equipment grounding conductor not smaller than No. 12 shall be permitted to be used in the interior of a one-family dwelling.

Exception No. 4: Flexible cord shall be permitted in accordance with Section 680-7.

(d) Panelboards. A panelboard, not part of the service equipment, shall have an equipment grounding conductor installed between its grounding terminal and the grounding terminal of the service equipment. This conductor shall be sized in accordance with Table 250-95 but not smaller than No. 12. It shall be an insulated conductor and shall be installed with the feeder conductors in rigid metal conduit, intermediate metal conduit, or rigid nonmetallic conduit. The equipment grounding conductor shall be connected to an equipment grounding terminal of the panelboard.

Exception No. 1: The equipment grounding conductor between an existing remote panelboard and the service equipment shall not be required to be in one of the conduits listed in paragraph (d) if the interconnection is by means of a flexible metal conduit or an approved cable assembly with an insulated or covered equipment grounding conductor.

Exception No. 2: Electrical metallic tubing shall be permitted to be used to protect conductors where installed on or within the building.

(FPN):   For permitted use of electrical metallic tubing, see Section 348-1.

(e) Cord-Connected Equipment. Where fixed or stationary equipment is connected with a flexible cord to facilitate removal or disconnection for maintenance, repair, or storage as provided in Section 680-7, the equipment grounding conductors shall be connected to a fixed metal part of the assembly. The removable part shall be mounted on or bonded to the fixed metal part.

(f) Other Equipment. Other electrical equipment shall be grounded in accordance with Article 250 and connected by wiring methods of Chapter 3.

680-26. Electrically Operated Pool Covers.

(a) Motors and Controllers. The electric motors, controllers, and wiring shall be located at least 5 feet (1.52 m) from the inside wall of the pool unless separated from the pool by a wall, cover, or other permanent barrier. Electric motors installed below grade level shall be of the totally enclosed type.

(FPN No. 1): For cabinets installed in damp and wet locations, see Section 373-2(a).

(FPN No. 2): For switches or circuit breakers installed in a wet location, see Section 380-4.

(FPN No. 3): For protection against liquids, see Section 430-11.

(b) Wiring Methods. The electric motor and controller shall be connected to a circuit protected by a ground-fault circuit-interrupter.

680-27. Deck Area Heating.

The provisions of this section apply to all pool deck areas, including a covered pool, where electrically operated comfort heating units are installed within 20 feet (6.1 m) of the inside wall of the pool.

(a) Unit Heaters. Unit heaters shall be rigidly mounted to the structure and shall be of the totally enclosed or guarded types. Unit heaters shall not be mounted over the pool or within the area extending 5 feet (1.52 m) horizontally from the inside walls of a pool.

(b) Permanently Wired Radiant Heaters. Radiant electric heaters shall be suitably guarded and securely fastened to their mounting device(s). Heaters shall not be installed over a pool or within the area extending 5 feet (1.52 m) horizontally from the inside walls of the pool and shall be mounted at least 12 feet (3.66 m) vertically above the pool deck unless otherwise approved.

(c) Radiant Heat Cables Not Permitted. Radiant heating cables embedded in or below the deck shall not be permitted.

C. Storable Pools

680-30. Pumps.

A cord-connected pool filter pump shall incorporate an approved system of double insulation or its equivalent and shall be provided with means for grounding only the internal and nonaccessible noncurrent-carrying metal parts of the appliance.

The means for grounding shall be an equipment grounding conductor run with the power-supply conductors in the flexible cord that is properly terminated in a grounding-type attachment plug having a fixed grounding contact member.

680-31. Ground-Fault Circuit-Interrupters Required. All electric equipment, including power-supply cords, used with storable pools shall be protected by ground-fault circuit-interrupters.

(FPN): Where flexible cords are used, see Section 400-4.

680-32. Lighting Fixtures. A lighting fixture installed in or on the wall of a storable pool shall be part of a cord- and plug-connected lighting assembly. This assembly shall:

(1) Have no exposed metal parts;

(2) Have a fixture lamp that operates at 15 volts or less;

(3) Have an impact-resistant polymeric lens, fixture body, and transformer enclosure;

(4) Have a transformer meeting the requirements of Section 680-5(a) with a primary rating not over 150 volts; and

(5) Be listed as an assembly for the purpose.

Exception: A lighting assembly without a transformer, and with the fixture lamp(s) operating at not over 150 volts, shall be permitted to be cord-and plug-connected where the assembly complies with all of the following:

a. It has no exposed metal parts.

b. It has an impact-resistant polymeric lens and fixture body.

c. A ground-fault circuit-interrupter with open neutral protection is provided as an integral part of the assembly.

d. The fixture lamp is permanently connected to the ground-fault circuit-interrupter with open-neutral protection.

e. It complies with the requirements of Section 680-20(a).

f. It is listed as an assembly for the purpose.

D. Spas and Hot Tubs

680-40. Outdoor Installations. A spa or hot tub installed outdoors shall comply with the provisions of Parts A and B.

Exception No. 1: Metal bands or hoops used to secure wooden staves shall be exempt from Section 680-22.

Exception No. 2: Listed packaged units shall be permitted to be cord-and plug-connected with a cord no longer than 15 feet (4.57 m) if protected by a ground-fault circuit-interrupter.

Exception No. 3: Bonding by metal-to-metal mounting on a common frame or base shall be permitted.

Exception No. 4: Listed packaged units utilizing a factory-installed remote panelboard shall be permitted to be connected with not more than 3 feet (914 mm) of liquidtight flexible conduit.

680-41. Indoor Installations. A spa or hot tub installed indoors shall conform to the requirements of this part and shall be connected by wiring methods of Chapter 3.

Exception: Listed packaged units rated 20 amperes or less shall be permitted to be cord- and plug-connected to facilitate the removal or disconnection for maintenance and repair.

(a) Receptacles. At least one convenience receptacle shall be located a minimum of 5 feet (1.52 m) from and not more than 10 feet (3.05 m) from the inside wall of the spa or hot tub.

(1) Receptacles on the property shall be located at least 5 feet (1.52 m) from the inside walls of the spa or hot tub.

(2) 125-volt receptacles located within 10 feet (3.05 m) of the inside walls of a spa or hot tub shall be protected by a ground-fault circuit-interrupter.

(FPN):   In determining the above dimensions, the distance to be measured is the shortest path the supply cord of an appliance connected to the receptacle would follow without piercing a floor, wall, or ceiling of a building or other effective permanent barrier.

(3) Receptacles that provide power for a spa or hot tub shall be ground-fault circuit-interrupter protected.

(b) Lighting Fixtures, Lighting Outlets, and Ceiling Fans.

(1) Lighting fixtures, lighting outlets, and ceiling fans located over the spa or hot tub or within 5 feet (1.52 m) from the inside walls of the spa or hot tub shall be a minimum of 7 feet 6 inches (2.29 m) above the maximum water level and shall be protected by a ground-fault circuit-interrupter.

Exception No. 1: Lighting fixtures, lighting outlets, and ceiling fans located 12 feet (3.66 m) or more above the maximum water level shall not require protection by a ground-fault circuit-interrupter.

Exception No. 2: Lighting fixtures meeting the requirements of items a or b below and protected by a ground-fault circuit-interrupter shall be permitted to be installed less than 7 feet 6 inches (2.29 m) over a spa or hot tub.

a. Recessed fixtures with a glass or plastic lens and nonmetallic or electrically isolated metal trim, suitable for use in damp locations.

b. Surface-mounted fixtures with a glass or plastic globe and a nonmetallic body or a metallic body isolated from contact. Such fixtures shall be suitable for use in damp locations.

(2) Underwater lighting fixtures shall comply with the provisions of Part B or Part C of this article.

(c) Wall Switches. Switches shall be located at least 5 feet (1.52 m), measured horizontally, from the inside walls of the spa or hot tub.

(d) Bonding. The following parts shall be bonded together:

(1) All metal fittings within or attached to the spa or hot tub structure.
(2) Metal parts of electric equipment associated with the spa or hot tub water circulating system, including pump motors.
(3) Metal conduit and metal piping within 5 feet (1.52 m) of the inside walls of the spa or hot tub and that are not separated from the spa or hot tub by a permanent barrier.

(4) All metal surfaces that are within 5 feet (1.52 m) of the inside walls of the spa or hot tub and not separated from the spa or hot tub area by a permanent barrier.

Exception: Small conductive surfaces not likely to become energized, such as air and water jets and drain fittings, where not connected to metallic piping, towel bars, mirror frames, and similar nonelectric equipment, need not be bonded.

(5) Electrical devices and controls not associated with the spas or hot tubs shall be located a minimum of 5 feet (1.52 m) away from such units or be bonded to the spa or hot tub system.

(e) Methods of Bonding. All metal parts associated with the spa or hot tub shall be bonded by any of the following methods: the interconnection of threaded metal piping and fittings; metal-to-metal mounting on a common frame or base; or by the provisions of a copper bonding jumper, insulated, covered, or bare, not smaller than No. 8 solid.

(f) Grounding. The following equipment shall be grounded:

(1) All electric equipment located within 5 feet (1.52 m) of the inside wall of the spa or hot tub.

(2) All electric equipment associated with the circulating system of the spa or hot tub.

(g) Methods of Grounding.

(1) All electric equipment shall be grounded in accordance with Article 250 and connected by the wiring methods of Chapter 3.

(2) Where equipment is connected with a flexible cord, the equipment grounding conductor shall be connected to a fixed metal part of the assembly.

(h) Electric Water Heaters. All electric spa or hot tub water heaters shall be listed and have the heating elements subdivided into loads not exceeding 48 amperes and protected at not more than 60 amperes.

The ampacity of the branch-circuit conductors, and the rating or setting of overcurrent protective devices, shall not be less than 125 percent of the total load of the nameplate rating.

(i) Underwater Audio Equipment. Underwater audio equipment shall comply with the provisions of Part B or Part C of this article.

680-42. Protection. Spas and hot tubs and associated electrical components shall be protected by ground-fault circuit-interrupters. This requirement shall become effective on January 1, 1994.

E. Fountains

680-50. General. The provisions of Part E shall apply to all fountains as defined in Section 680-4. Fountains that have water common to a pool shall comply with the pool requirements of this article.

Exception: Self-contained, portable fountains no larger than 5 feet (1.52 m) in any dimension are not covered by Part E.

680-51. Lighting Fixtures, Submersible Pumps, and Other Submersible Equipment.

(a) Ground-Fault Circuit-Interrupter. A ground-fault circuit-interrupter shall be installed in the branch circuit supplying fountain equipment.

Exception: Ground-fault circuit-interrupters shall not be required for equipment operating at 15 volts or less and supplied by a transformer complying with Section 680-5(a).

(b) Operating Voltage. All lighting fixtures shall be installed for operation at 150 volts or less between conductors. Submersible pumps and other submersible equipment shall operate at 300 volts or less between conductors.

(c) Lighting Fixture Lenses. Lighting fixtures shall be installed with the top of the fixture lens below the normal water level of the fountain unless approved for above-water locations. A lighting fixture facing upward shall have the lens adequately guarded to prevent contact by any person.

(d) Overheating Protection. Electric equipment that depends on submersion for safe operation shall be protected against overheating by a low-water cutoff or other approved means when not submerged.

(e) Wiring. Equipment shall be equipped with provisions for threaded conduit entries or be provided with a suitable flexible cord. The maximum length of exposed cord in the fountain shall be limited to 10 feet (3.05 m). Cords extending beyond the fountain perimeter shall be enclosed in approved wiring enclosures. Metal parts of equipment in contact with water shall be of brass or other approved corrosion-resistant metal.

(f) Servicing. All equipment shall be removable from the water for relamping or normal maintenance. Fixtures shall not be permanently imbedded into the fountain structure so that the water level must be reduced or the fountain drained for relamping, maintenance, or inspection.

(g) Stability. Equipment shall be inherently stable or be securely fastened in place.

680-52. Junction Boxes and Other Enclosures.

(a) General. Junction boxes and other enclosures used for other than underwater installation shall comply with Sections 680-21(a) (1), (2), and (3); and (b), (c), and (d).

(b) Underwater Junction Boxes and Other Underwater Enclosures. Junction boxes and other underwater enclosures shall be submersible and (1) be equipped with provisions for threaded conduit entries or compression glands or seals for cord entry; (2) be of copper, brass, or other approved corrosion-resistant material; (3) be filled with an approved potting compound to prevent the entry of moisture; and (4) be firmly attached to the supports or directly to the fountain surface and bonded as required. Where the junction box is supported only by the conduit, the conduit shall be of copper, brass, or other approved corrosion-resistant metal. Where the box is fed by nonmetallic conduit, it shall have additional supports and fasteners of copper, brass, or other approved corrosion-resistant material.

(FPN): See Section 370-23 for support of enclosures.

680-53. Bonding. All metal piping systems associated with the fountain shall be bonded to the equipment grounding conductor of the branch circuit supplying the fountain.

(FPN): See Section 250-95 for sizing of these conductors.

680-54. Grounding. The following equipment shall be grounded: (1) all electric equipment located within the fountain or within 5 feet (1.52 m) of the inside wall of the fountain; (2) all electric equipment associated with the recirculating system of the fountain; (3) panelboards that are not part of the service equipment and that supply any electric equipment associated with the fountain.

680-55. Methods of Grounding.

(a) Applied Provisions. The provisions of Section 680-25 shall apply, excluding paragraph (e).

(b) Supplied by a Flexible Cord. Electric equipment that is supplied by a flexible cord shall have all exposed noncurrent-carrying metal parts grounded by an insulated copper equipment grounding conductor that is an integral part of this cord. This grounding conductor shall be connected to a grounding terminal in the supply junction box, transformer enclosure, or other enclosure.

680-56. Cord- and Plug-Connected Equipment.

(a) Ground-Fault Circuit-Interrupter. All electric equipment, including power-supply cords, shall be protected by ground-fault circuit-interrupters.

(b) Cord Type. Flexible cord immersed in or exposed to water shall be of the hard service type as designated in Table 400-4 and shall be marked water-resistant.

(c) Sealing. The end of the flexible cord jacket and the flexible cord conductor termination within equipment shall be covered with, or encapsulated in, a suitable potting compound to prevent the entry of water into the equipment through the cord or its conductors. In addition, the ground connection within equipment shall be similarly treated to protect such connections from the deteriorating effect of water that may enter into the equipment.

(d) Terminations. Connections with flexible cord shall be permanent, except that grounding-type attachment plugs and receptacles shall be permitted to facilitate removal or disconnection for maintenance, repair, or storage of fixed or stationary equipment not located in any water-containing part of a fountain.

F. Therapeutic Pools and Tubs in Health Care Facilities

680-60. General. The provisions of Part F include therapeutic pools and tubs in health care facilities. See Section 517-2 for definition of health care facilities. Portable therapeutic appliances shall comply with Article 422.

680-61. Permanently Installed Therapeutic Pools. Therapeutic pools that are constructed in the ground, on the ground, or in a building in such a manner that the pool cannot be readily disassembled shall comply with Parts A and B of this article.

Exception: The limitations of Sections 680-6(b)(1) and (2) shall not apply where all lighting fixtures are of the totally enclosed type.

680-62. Therapeutic Tubs (Hydrotherapeutic Tanks). Therapeutic tubs, used for the submersion and treatment of patients, that are not easily moved from one place to another in normal use or that are fastened or otherwise secured at a specific location, including associated piping systems, shall conform to this part.

(a) Ground-Fault Circuit-Interrupter. A ground-fault circuit-interrupter shall protect all therapeutic equipment.

Exception: Portable therapeutic appliances shall comply with Section 250-45.

(b) Bonding. The following parts shall be bonded together:

(1) All metal fittings within or attached to the tub structure.

(2) Metal parts of electric equipment associated with the tub water circulating system, including pump motors.

(3) Metal-sheathed cables and raceways and metal piping that are within 5 feet (1.52 m) of the inside walls of the tub and not separated from the tub by a permanent barrier.

(4) All metal surfaces that are within 5 feet (1.52 m) of the inside walls of the tub and not separated from the tub area by a permanent barrier.

(5) Electrical devices and controls not associated with the therapeutic tubs shall be located a minimum of 5 feet (1.52 m) away from such units or be bonded to the therapeutic tub system.

(c) Methods of Bonding. All metal parts associated with the tub shall be bonded by any of the following methods: the interconnection of threaded metal piping and fittings; metal-to-metal mounting on a common frame or base; connections by suitable metal clamps; or by the provisions of a solid copper bonding jumper, insulated, covered, or bare, not smaller than No. 8.

(d) Grounding. The following equipment shall be grounded:

(1) All electric equipment located within 5 feet (1.52 m) of the inside wall of the tub.

(2) All electric equipment associated with the circulating system of the tub.

(e) Methods of Grounding.

(1) All electric equipment shall be grounded in accordance with Article 250 and connected by wiring methods of Chapter 3.

(2) Where equipment is connected with a flexible cord, the equipment grounding conductor shall be connected to a fixed metal part of the assembly.

(f) Receptacles. All receptacles within 5 feet (1.52 m) of a therapeutic tub shall be protected by a ground-fault circuit-interrupter.

(g) Lighting Fixtures. All lighting fixtures used in therapeutic tub areas shall be of the totally enclosed type.

G. Hydromassage Bathtubs

680-70. Protection. Hydromassage bathtubs and their associated electric components shall be protected by a ground-fault circuit-interrupter.

680-71. Other Electric Equipment. Lighting fixtures, switches, receptacles, and other electric equipment located in the same room, and not directly associated with a hydromassage bathtub, shall be installed in accordance with the requirements of Chapters 1 through 4 in this Code covering the installation of that equipment in bathrooms.

ARTICLE 685 — INTEGRATED ELECTRICAL SYSTEMS

A. General

685-1. Scope. This article covers integrated electrical systems, other than unit equipment, in which orderly shutdown is necessary to assure safe operation. An integrated electrical system as used in this article is a unitized segment of an industrial wiring system where all of the following conditions are met: (1) an orderly shutdown is required to minimize personnel hazard and equipment damage; (2) the conditions of maintenance and supervision assure that qualified persons will service the system; and (3) effective safeguards, acceptable to the authority having jurisdiction, are established and maintained.

685-2. Application of Other Articles. In other articles applying to particular cases of installation of conductors and equipment, there are orderly shutdown requirements that are in addition to those of this article or are modifications of them:

	Section
More than One Building or Other Structure	225-8
Ground-Fault Protection of Equipment	230-95, Exception No. 1
Protection of Conductors	240-3(a)
Electrical System Coordination	240-12
Ground-Fault Protection of Equipment	240-13, Exception No. 1
Grounding ac Systems of 50 to 1000 Volts	250-5(b), Exception No. 3
Orderly Shutdown	430-44
Disconnection	430-74, Exception Nos. 1 and 2
Disconnecting Means in Sight from Controller	430-102, Exception No. 2
Energy from More than One Source	430-113, Exception Nos. 1 and 2
Disconnecting Means	645-110, Exception
Point of Connection	705-12, Exception No. 1

B. Orderly Shutdown

685-10. Location of Overcurrent Devices in or on Premises. Location of overcurrent devices that are critical to integrated electrical systems shall be permitted to be accessible, with mounting heights allowed to assure security from operation by nonqualified personnel.

685-12. Direct-Current System Grounding. Two-wire direct-current circuits shall be permitted to be ungrounded.

685-14. Ungrounded Control Circuits. Where operational continuity is required, control circuits of 150 volts or less from separately derived systems shall be permitted to be ungrounded.

ARTICLE 690 — SOLAR PHOTOVOLTAIC SYSTEMS

A. General

690-1. Scope. The provisions of this article apply to solar photovoltaic electrical energy systems including the array circuit(s), power conditioning unit(s) and controller(s) for such systems. Solar photovoltaic systems covered by this article may be interactive with other electric power production sources or stand alone, with or without electrical energy storage such as batteries. These systems may have alternating- or direct-current output for utilization.

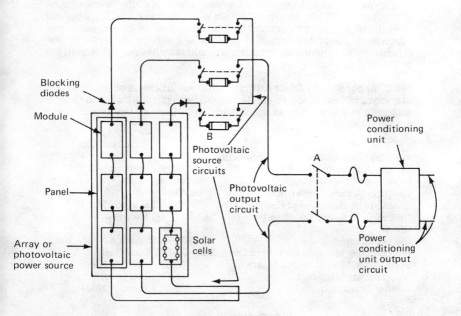

A: Disconnecting means required by Section 690-13.
B: Equipment permitted to be on the photovoltaic power source side of the photovoltaic power source disconnecting means, per Section 690-14, Exception No. 2. See Section 690-16.

Diagram 690-1. Solar Photovoltaic System (Simplified Circuit-grounding system not shown)

690-2. Definitions.

Array: A mechanically integrated assembly of modules or panels with a support structure and foundation, tracking, thermal control, and other components, as required, to form a direct-current power-producing unit.

Blocking Diode: A diode used to block reverse flow of current into a photovoltaic source circuit.

Interactive System: A solar photovoltaic system that operates in parallel with and may be designed to deliver power to another electric power production source connected to the same load. For the purpose of this definition, an energy storage subsystem of a solar photovoltaic system, such as a battery, is not another electric power production source.

Module: The smallest complete, environmentally protected assembly of solar cells, optics, and other components, exclusive of tracking, designed to generate direct-current power under sunlight.

Panel: A collection of modules mechanically fastened together, wired, and designed to provide a field-installable unit.

Photovoltaic Output Circuit: Circuit conductors between the photovoltaic source circuit(s) and the power conditioning unit or direct-current utilization equipment. See Diagram 690-1.

Photovoltaic Power Source: An array or aggregate of arrays that generates direct-current power at system voltage and current.

Photovoltaic Source Circuit: Conductors between modules and from modules to the common connection point(s) of the direct-current system. See Diagram 690-1.

Power Conditioning Unit: Equipment that is used to change voltage level or waveform, or both, of electrical energy. Commonly, a power conditioning unit is an inverter that changes a direct-current input to an alternating-current output.

Power Conditioning Unit Output Circuit: Conductors between the power conditioning unit and the connection to the service equipment or another electric power production source, such as a utility. See Diagram 690-1.

Solar Cell: The basic photovoltaic device that generates electricity when exposed to light.

Solar Photovoltaic System: The total components and subsystems that, in combination, convert solar energy into electrical energy suitable for connection to a utilization load.

Stand-Alone System: A solar photovoltaic system that supplies power independently but that may receive control power from another electric power production source.

690-3. Other Articles. Wherever the requirements of other articles of this Code and Article 690 differ, the requirements of Article 690 shall apply. Solar photovoltaic systems operating as interconnected power production sources shall be installed in accordance with the provisions of Article 705.

690-4. Installation.

(a) Photovoltaic System. A solar photovoltaic system shall be permitted to supply a building or other structure in addition to any service(s) of another electricity supply system(s).

(b) Conductors of Different Systems. Photovoltaic source circuits and photovoltaic output circuits shall not be contained in the same raceway, cable tray, cable, outlet box, junction box, or similar fitting as feeders or branch circuits of other systems.

Exception: Where the conductors of the different systems are separated by a partition or are connected together.

(c) Module Connection Arrangement. The connections to a module or panel shall be so arranged that removal of a module or panel from a photovoltaic source circuit does not interrupt a grounded conductor to another photovoltaic source circuit.

(d) Equipment. Inverters or motor generators shall be identified for use in solar photovoltaic systems.

690-5. Ground Fault Detection and Interruption. Roof-mounted photovoltaic arrays located on dwellings shall be provided with ground-fault protection to reduce fire hazard. The ground-fault protection circuit shall be capable of detecting a ground fault, interrupting the fault path, and disabling the array.

B. Circuit Requirements

690-7. Maximum Voltage.

(a) Voltage Rating. In a photovoltaic power source and its direct-current circuits, the voltage considered shall be the rated open-circuit voltage.

(b) Direct-Current Utilization Circuits. The voltage of direct-current utilization circuits shall conform with Section 210-6.

(c) Photovoltaic Source and Output Circuits. Photovoltaic source circuits and photovoltaic output circuits that do not include lampholders, fixtures, or receptacles shall be permitted up to 600 volts.

(d) Circuits Over 150 Volts to Ground. In one- and two-family dwellings, live parts in photovoltaic source circuits and photovoltaic output circuits over 150 volts to ground shall not be accessible while energized to other than qualified persons.

(FPN): See Section 110-17 for guarding of live parts, and Section 210-6 for voltage to ground and between conductors.

690-8. Circuit Sizing and Current.

(a) **Ampacity and Overcurrent Devices.** The ampacity of the conductors and the rating or setting of overcurrent devices in a circuit of a solar photovoltaic system shall not be less than 125 percent of the current computed in accordance with (b) below. The rating or setting of overcurrent devices shall be permitted in accordance with Section 240-3 (b) and (c). |

Exception: Circuits containing an assembly together with its overcurrent device(s) that is listed for continuous operation at 100 percent of its rating.

(b) **Computation of Circuit Current.** The current for the individual type of circuit shall be computed as follows:

(1) **Photovoltaic Source Circuits.** The sum of parallel module short-circuit current ratings.

(2) **Photovoltaic Output Circuit.** The photovoltaic power source short-circuit current rating. |

(3) **Power Conditioning Unit Output Circuit.** The power conditioning unit output current rating.

Exception: The current rating of a circuit without an overcurrent device shall be the short-circuit current, and it shall not exceed the ampacity of the circuit conductors.

(c) **Systems with Multiple DC Voltages.** For a photovoltaic power source having multiple output circuit voltages and employing a common-return conductor, the ampacity of the common-return conductor shall not be less than the sum of the ampere ratings of the overcurrent devices of the individual output circuits. |

690-9. Overcurrent Protection.

(a) **Circuits and Equipment.** Photovoltaic source circuit, photovoltaic output circuit, power conditioning unit output circuit, and storage battery circuit conductors and equipment shall be protected in accordance with the requirements of Article 240. Circuits connected to more than one electrical source shall have overcurrent devices so located as to provide overcurrent protection from all sources.

(FPN): Possible backfeed of current from any source of supply, including a supply through a power conditioning unit into the photovoltaic output circuit and photovoltaic source circuits, must be considered in determining whether adequate overcurrent protection from all sources is provided for conductors and modules.

(b) **Power Transformers.** Overcurrent protection for a transformer with a source(s) on each side shall be provided in accordance with Section 450-3 by considering first one side of the transformer, then the other side of the transformer, as the primary.

Exception: A power transformer with a current rating on the side connected toward the photovoltaic power source not less than the short-circuit output current rating of the power conditioning unit shall be permitted without overcurrent protection from that source.

(c) Photovoltaic Source Circuits. Branch-circuit or supplementary type overcurrent devices shall be permitted to provide overcurrent protection in photovoltaic source circuits. The overcurrent devices shall be accessible, but shall not be required to be readily accessible.

C. Disconnecting Means

690-13. All Conductors. Means shall be provided to disconnect all current-carrying conductors of a photovoltaic power source from all other conductors in a building or other structure.

Exception: Where a circuit grounding connection is not designed to be automatically interrupted as part of the ground-fault protection system required by Section 690-5, a switch or circuit breaker used as a disconnecting means shall not have a pole in the grounded conductor.

(FPN): It is intended that the grounded conductor be permitted to have a bolted or terminal disconnecting means to allow maintenance or troubleshooting by qualified personnel.

690-14. Additional Provisions. The provisions of Article 230, Part F, shall apply to the photovoltaic power source disconnecting means.

Exception No. 1: The disconnecting means shall not be required to be suitable as service equipment and shall be rated in accordance with Section 690-17.

Exception No. 2: Equipment such as photovoltaic source circuit isolating switches, overcurrent devices, and blocking diodes shall be permitted on the photovoltaic power source side of the photovoltaic power source disconnecting means.

690-15. Disconnection of Photovoltaic Equipment. Means shall be provided to disconnect equipment, such as a power conditioning unit, filter assembly, and the like, from all ungrounded conductors of all sources. If the equipment is energized from more than one source, the disconnecting means shall be grouped and identified.

690-16. Fuses. Disconnecting means shall be provided to disconnect a fuse from all sources of supply if the fuse is energized from both directions and is accessible to other than qualified persons. Such a fuse in a photovoltaic source circuit shall be capable of being disconnected independently of fuses in other photovoltaic source circuits.

690-17. Switch or Circuit Breaker. The disconnecting means for ungrounded conductors shall consist of a manually operable switch(es) or circuit breaker(s) (1) located where readily accessible, (2) externally operable without exposing the operator to contact with live parts, (3) plainly indicating whether in the open or closed position, and (4) having ratings not less than the load to be carried. Where all terminals of the disconnecting means may be energized in the open position, a warning sign shall be mounted on or adjacent to the disconnecting means. The sign shall be clearly legible and shall read substantially: WARNING — ELECTRIC SHOCK — DO NOT TOUCH — TERMINALS ENERGIZED IN OPEN POSITION.

Exception: A disconnecting means located on the direct-current side shall be permitted to have an interrupting rating less than the current-carrying rating where the system is designed so that the direct-current switch cannot be opened under load.

690-18. Disablement of an Array. Means shall be provided to disable an array or portions of an array.

(FPN): Photovoltaic modules are energized while exposed to light. Installation, replacement, or servicing of array components while a module(s) is irradiated may expose persons to electric shock.

D. Wiring Methods

690-31. Methods Permitted.

(a) **Wiring Systems.** All raceway and cable wiring methods included in this Code and other wiring systems and fittings specifically intended and identified for use on photovoltaic arrays shall be permitted. Where wiring devices with integral enclosures are used, sufficient length of cable shall be provided to facilitate replacement.

(b) **Single Conductor Cable.** Types SE, UF, and USE conductor cable shall be permitted in photovoltaic source circuits where installed in the same manner as a Type UF multiconductor cable in accordance with Article 339. Where exposed to direct rays of the sun, Type UF cable identified as sunlight-resistant or Type USE cable shall be used.

690-32. Component Interconnections. Fittings and connectors that are intended to be concealed at the time of on-site assembly, where listed for such use, shall be permitted for on-site interconnection of modules or other array components. Such fittings and connectors shall be equal to the wiring method employed in insulation, temperature rise and fault-current withstand, and shall be capable of resisting the effects of the environment in which they are used.

690-33. Connectors. The connectors permitted by Section 690-32 shall comply with (a) through (e) below.

(a) **Configuration.** The connectors shall be polarized and shall have a configuration that is noninterchangeable with receptacles in other electrical systems on the premises.

(b) **Guarding.** The connectors shall be constructed and installed so as to guard against inadvertent contact with live parts by persons.

(c) **Type.** The connectors shall be of the latching or locking type.

(d) **Grounding Member.** The grounding member shall be the first to make and the last to break contact with the mating connector.

(e) **Interruption of Circuit.** The connectors shall be capable of interrupting the circuit current without hazard to the operator.

690-34. Access to Boxes.   Junction, pull, and outlet boxes located behind modules or panels shall be installed so that the wiring contained in them can be rendered accessible directly or by displacement of a module(s) or panel(s) secured by removable fasteners and connected by a flexible wiring system.

E. Grounding

690-41. System Grounding.   For a photovoltaic power source, one conductor of a 2-wire system rated over 50 volts and a neutral conductor of a 3-wire system shall be solidly grounded.

Exception: Other methods that accomplish equivalent system protection and that utilize equipment listed and identified for the use shall be permitted.

(FPN):   See the first Fine Print Note under Section 250-1.

690-42. Point of System Grounding Connection.   The direct-current circuit grounding connection shall be made at any single point on the photovoltaic output circuit.

(FPN):   Locating the grounding connection point as close as practicable to the photovoltaic source will better protect the system from voltage surges due to lightning.

690-43. Size of Equipment Grounding Conductor.   The equipment grounding conductor shall be no smaller than the required size of the circuit conductors in systems where the available photovoltaic power source short-circuit current is less than twice the current rating of the overcurrent device. In other systems, the equipment grounding conductor shall be sized in accordance with Section 250-95.

690-44. Common Grounding Electrode.   Exposed noncurrent-carrying metal parts of equipment and conductor enclosures of a photovoltaic system shall be grounded to the grounding electrode that is used to ground the direct-current system. Two or more electrodes that are effectively bonded together shall be considered as a single electrode in this sense.

F. Marking

690-51. Modules.   Modules shall be marked with identification of terminals or leads as to polarity, maximum overcurrent device rating for module protection, and with rated: (1) open-circuit voltage, (2) operating voltage, (3) maximum permissible system voltage, (4) operating current, (5) short-circuit current, and (6) maximum power.

690-52. Photovoltaic Power Source.   A marking, specifying the photovoltaic power source rated: (1) operating current, (2) operating voltage, (3) open-circuit voltage, and (4) short-circuit current, shall be provided at an accessible location at the disconnecting means for the photovoltaic power source.

(FPN):   Reflecting systems used for irradiance enhancement may result in increased levels of output current and power.

G. Connection to Other Sources

690-61. Loss of System Voltage. The power output from a power conditioning unit in a solar photovoltaic system that is interactive with another electric system(s) shall be automatically disconnected from all ungrounded conductors in such other electric system(s) upon loss of voltage in that electric system(s) and shall not reconnect to that electric system(s) until its voltage is restored.

(FPN): For other interconnected electric power production sources, see Article 705.

A normally interactive solar photovoltaic system shall be permitted to operate as a stand-alone system to supply premises wiring.

690-62. Ampacity of Neutral Conductor. If a single-phase, 2-wire power conditioning unit output is connected to the neutral and one ungrounded conductor (only) of a 3-wire system or of a 3-phase, 4-wire wye-connected system, the maximum load connected between the neutral and any one ungrounded conductor plus the power conditioning unit output rating shall not exceed the ampacity of the neutral conductor.

690-63. Unbalanced Interconnections.

(a) Single-Phase. The output of a single-phase power conditioning unit shall not be connected to a 3-phase, 3- or 4-wire electrical service derived directly from a delta-connected transformer.

(b) Three-Phase. A 3-phase power conditioning unit shall be automatically disconnected from all ungrounded conductors of the interconnected system when one of the phases opens in either source.

Exception for (a) and (b): Where the interconnected system is designed so that significant unbalanced voltages will not result.

690-64. Point of Connection. The output of a power production source shall be connected as specified in (a) or (b) below.

(FPN): For the purposes of this section, a power production source is considered to be (1) the output of a power conditioning unit where connected to an alternating-current electric source; and (2) the photovoltaic output circuit where interactive with a direct-current electric source.

(a) Supply Side. To the supply side of the service disconnecting means as permitted in Section 230-82, Exception No. 6.

(b) Load Side. To the load side of the service disconnecting means of the other source(s), if all of the following conditions are met:

(1) Each source interconnection shall be made at a dedicated circuit breaker or fusible disconnecting means.

(2) The sum of the ampere ratings of overcurrent devices in circuits supplying power to a busbar or conductor shall not exceed the rating of the busbar or conductor.

Exception: For a dwelling unit, the sum of the ampere ratings of the overcurrent devices shall not exceed 120 percent of the rating of the busbar or conductor.

(3) The interconnection point shall be on the line side of all ground-fault protection equipment.

Exception: Connection shall be permitted to be made to the load side of ground-fault protection, provided that there is ground-fault protection for equipment from all ground-fault current sources.

(4) Equipment containing overcurrent devices in circuits supplying power to a busbar or conductor shall be marked to indicate the presence of all sources.

Exception: Equipment with power supplied from a single point of connection.

(5) Equipment such as circuit breakers, if back-fed, shall be identified for such operation.

H. Storage Batteries

690-71. Installation.

(a) General. Storage batteries in a solar photovoltaic system shall be installed in accordance with the provisions of Article 480.

Exception: As provided in Section 690-73.

(b) Dwellings.

(1) Storage batteries for dwellings shall have the cells connected so as to operate at less than 50 volts.

Exception: Where live parts are not accessible during routine battery maintenance, a battery system voltage in accordance with Section 690-7 shall be permitted.

(2) Live parts of battery systems for dwellings shall be guarded to prevent accidental contact by persons or objects, regardless of voltage or battery type.

(FPN):   Batteries in solar photovoltaic systems are subject to extensive charge-discharge cycles and typically require frequent maintenance, such as checking electrolyte and cleaning connections.

690-72. State of Charge.   Equipment shall be provided to control the state of charge of the battery. All adjusting means for control of the state of charge shall be accessible only to qualified persons.

Exception: Where the design of the photovoltaic power source is matched to the voltage rating and charge current requirements for the interconnected battery cells.

690-73. Grounding.   The interconnected battery cells shall be considered grounded where the photovoltaic power source is installed in accordance with Section 690-41, Exception.

Chapter 7. Special Conditions

ARTICLE 700 — EMERGENCY SYSTEMS

A. General

700-1. Scope. The provisions of this article apply to the electrical safety of the design, installation, operation, and maintenance of emergency systems consisting of circuits and equipment intended to supply, distribute, and control electricity for illumination or power, or both, to required facil- | ities when the normal electrical supply or system is interrupted.

Emergency systems are those systems legally required and classed as emergency by municipal, state, federal, or other codes, or by any governmental agency having jurisdiction. These systems are intended to automatically supply illumination or power, or both, to designated areas and equip- | ment in the event of failure of the normal supply or in the event of accident to elements of a system intended to supply, distribute, and control power and illumination essential for safety to human life.

(FPN No. 1):   For further information regarding wiring and installation of emergency systems in health care facilities, see Article 517.

(FPN No. 2):   For further information regarding performance and maintenance of emergency systems in health care facilities, see Health Care Facilities, NFPA 99-1990 (ANSI). |

(FPN No. 3):   Emergency systems are generally installed in places of assembly where artificial illumination is required for safe exiting and for panic control in buildings subject to occupancy by large numbers of persons, such as hotels, theaters, sports arenas, health care facilities, and similar institutions. Emergency systems may also provide power for such functions as ventilation where essential to maintain life, fire detection and alarm systems, elevators, fire pumps, public safety communication systems, industrial processes where current interruption would produce serious life safety or health hazards, and similar functions.

(FPN No. 4):   For specification of locations where emergency lighting is considered essential to life safety, see Life Safety Code, NFPA 101-1991 (ANSI). |

(FPN No. 5):   For further information regarding performance of emergency and standby power systems, see Emergency and Standby Power Systems, NFPA 110-1988 (ANSI).

700-2. Application of Other Articles.   Except as modified by this article, all applicable articles of this Code shall apply.

700-3. Equipment Approval.   All equipment shall be approved for use on emergency systems.

700-4. Tests and Maintenance.

(a) **Conduct or Witness Test.**   The authority having jurisdiction shall conduct or witness a test on the complete system upon installation and periodically afterward.

(b) Tested Periodically. Systems shall be tested periodically on a schedule acceptable to the authority having jurisdiction to assure their maintenance in proper operating condition.

(c) Battery Systems Maintenance. Where battery systems or unit equipments are involved, including batteries used for starting, control, or ignition in auxiliary engines, the authority having jurisdiction shall require periodic maintenance.

(d) Written Record. A written record shall be kept of such tests and maintenance.

(e) Testing Under Load. Means for testing all emergency lighting and power systems during maximum anticipated load conditions shall be provided.

700-5. Capacity.

(a) Capacity and Rating. An emergency system shall have adequate capacity and rating for all loads to be operated simultaneously. The emergency system equipment shall be suitable for the maximum available fault current at its terminals.

(b) Selective Load Pickup, Load Shedding, and Peak Load Shaving. The alternate power source shall be permitted to supply emergency, legally required standby, and optional standby system loads where automatic selective load pickup and load shedding is provided as needed to assure adequate power to (1) the emergency circuits; (2) the legally required standby circuits; and (3) the optional standby circuits, in that order of priority. The alternate power source shall be permitted to be used for peak load shaving, provided the above conditions are met.

(FPN): Peak load shaving operation may be acceptable for satisfying the test requirements of Section 700-4(b) where all conditions of Section 700-4 are met.

A portable or temporary alternate source shall be available whenever the emergency generator is out of service for major maintenance or repair.

700-6. Transfer Equipment.
Transfer equipment, including automatic transfer switches, shall be automatic and identified for emergency use and approved by the authority having jurisdiction. Transfer equipment shall be designed and installed to prevent the inadvertent interconnection of normal and emergency sources of supply in any operation of the transfer equipment. See Section 230-83.

Means shall be permitted to bypass and isolate the transfer equipment. Where bypass isolation switches are used, inadvertent parallel operation shall be avoided.

700-7. Signals.
Audible and visual signal devices shall be provided, where practicable, for the following purposes:

(a) Derangement. To indicate derangement of the emergency source.

(b) Carrying Load. To indicate that the battery is carrying load.

(c) Not Functioning. To indicate that the battery charger is not functioning.

(d) Ground Fault. To indicate a ground fault in solidly grounded wye emergency systems of more than 150 volts to ground and circuit protective devices rated 1000 amperes or more. The sensor for the ground-fault signal devices shall be located at, or ahead of, the main system disconnecting means for the emergency source, and the maximum setting of the signal devices shall be for a ground-fault current of 1200 amperes. Instructions on the course of action to be taken in event of indicated ground fault shall be located at or near the sensor location.

(FPN): For signals for generator sets, see Emergency and Standby Power Systems, NFPA 110-1988 (ANSI).

700-8. Signs.

(a) **Emergency Sources.** A sign shall be placed at the service entrance equipment indicating type and location of on-site emergency power sources.

(b) **Grounding.** Where the grounded circuit conductor connected to the emergency source is connected to a grounding electrode conductor at a location remote from the emergency source, there shall be a sign at the grounding location that shall identify all emergency and normal sources connected at that location.

B. Circuit Wiring

700-9. Wiring, Emergency System.

(a) **Identification.** All boxes, and enclosures (including transfer switches, generators, and power panels) for emergency circuits shall be permanently marked so they will be readily identified as a component of an emergency circuit or system.

(b) **Wiring.** Wiring from emergency source or emergency source distribution overcurrent protection to emergency loads shall be kept entirely independent of all other wiring and equipment and shall not enter the same raceway, cable, box, or cabinet with other wiring.

Exception No. 1: In transfer equipment enclosures.

Exception No. 2: In exit or emergency lighting fixtures supplied from two sources.

Exception No. 3: In a common junction box attached to exit or emergency lighting fixtures supplied from two sources.

Exception No. 4: Wiring of two or more emergency circuits supplied from the same source shall be permitted in the same raceway, cable, box, or cabinet.

Exception No. 5: In a common junction box attached to a unit equipment, and that contains only the branch circuit supplying the unit equipment and the emergency circuit supplied by the unit equipment.

Emergency wiring circuit(s) shall be designed and located to minimize the hazards that might cause failure due to flooding, fire, icing, vandalism, and other adverse conditions.

C. Sources of Power

700-12. General Requirements. Current supply shall be such that, in the event of failure of the normal supply to, or within, the building or group of buildings concerned, emergency lighting, emergency power, or both will be available within the time required for the application but not to exceed 10 seconds. The supply system for emergency purposes, in addition to the normal services to the building and meeting the general requirements of this section, shall be permitted to comprise one or more of the types of systems described in (a) through (e) below. Unit equipment in accordance with Section 700-12(f) shall satisfy the applicable requirements of this article.

In selecting an emergency source of power, consideration shall be given to the occupancy and the type of service to be rendered, whether of minimum duration, as for evacuation of a theater, or longer duration, as for supplying emergency power and lighting due to an indefinite period of current failure from trouble either inside or outside the building.

Equipment shall be designed and located to minimize the hazards that might cause complete failure due to flooding, fires, icing, and vandalism.

(FPN): Assignment of degree of reliability of the recognized emergency supply system depends upon the careful evaluation of the variables at each particular installation.

(a) Storage Battery. Storage batteries used as source of power for emergency systems shall be of suitable rating and capacity to supply and maintain the total load for a period of $1\frac{1}{2}$ hours minimum, without the voltage applied to the load falling below $87\frac{1}{2}$ percent of normal.

Batteries, whether of the acid or alkali type, shall be designed and constructed to meet the requirements of emergency service and shall be compatible with the charger for that particular installation.

For a sealed battery, the container shall not be required to be transparent. However, for the lead acid battery that requires water additions, transparent or translucent jars shall be furnished. Automotive-type batteries shall not be used.

An automatic battery charging means shall be provided.

(b) Generator Set.

(1) A generator set driven by a prime mover acceptable to the authority having jurisdiction and sized in accordance with Section 700-5. Means shall be provided for automatically starting the prime mover on failure of the normal service and for automatic transfer and operation of all required electrical circuits. A time-delay feature permitting a 15-minute setting shall be provided to avoid retransfer in case of short-time reestablishment of the normal source.

(2) Where internal combustion engines are used as the prime mover, an on-site fuel supply shall be provided with an on-premise fuel supply sufficient for not less than 2 hours full-demand operation of the system.

(3) Prime movers shall not be solely dependent upon a public utility gas system for their fuel supply or municipal water supply for their cooling systems. Means shall be provided for automatically transferring from one fuel supply to another where dual fuel supplies are used.

Exception: Where acceptable to the authority having jurisdiction, the use of other than on-site fuels shall be permitted where there is a low probability of a simultaneous failure of both the off-site fuel delivery system and power from the outside electrical utility company.

(4) Where a storage battery is used for control or signal power, or as the means of starting the prime mover, it shall be suitable for the purpose and shall be equipped with an automatic charging means independent of the generator set.

(5) Generator sets that require more than 10 seconds to develop power are acceptable, provided an auxiliary power supply will energize the emergency system until the generator can pick up the load.

(c) Uninterruptible Power Supplies. Uninterruptible power supplies used to provide power for emergency systems shall comply with the applicable provisions of Sections 700-12(a) and (b).

(d) Separate Service. Where acceptable to the authority having jurisdiction as suitable for use as an emergency source, a second service shall be permitted. This service shall be in accordance with Article 230, with separate service drop or lateral, widely separated electrically and physically from the normal service to minimize the possibility of simultaneous interruption of supply.

(e) Connection Ahead of Service Disconnecting Means. Where acceptable to the authority having jurisdiction as suitable for use as an emergency source, connection ahead of, but not within, the main service disconnecting means shall be permitted. The emergency service shall be sufficiently separated from the normal main service disconnecting means to prevent simultaneous interruption of supply through an occurrence within the building or groups of buildings served.

(FPN):   See Section 230-82 for equipment permitted on the supply side of a service disconnecting means.

(f) Unit Equipment. Individual unit equipment for emergency illumination shall consist of (1) a rechargeable battery; (2) a battery charging means; (3) provisions for one or more lamps mounted on the equipment, or shall be permitted to have terminals for remote lamps, or both; and (4) a relaying device arranged to energize the lamps automatically upon failure of the supply to the unit equipment. The batteries shall be of suitable rating and capacity to supply and maintain at not less than $87\frac{1}{2}$ percent of the nominal battery voltage for the total lamp load associated with the unit for a period of at least $1\frac{1}{2}$ hours, or the unit equipment shall supply and maintain not less than 60 percent of the initial emergency illumination for a period of at least $1\frac{1}{2}$ hours. Storage batteries, whether of the acid or alkali type, shall be designed and constructed to meet the requirements of emergency service.

Unit equipment shall be permanently fixed in place (i.e., not portable) and shall have all wiring to each unit installed in accordance with the requirements of any of the wiring methods in Chapter 3. Flexible cord- and plug-connection shall be permitted, provided that the cord does not exceed 3 feet (914 mm) in length. The branch circuit feeding the unit equipment shall be the same branch circuit as that serving the normal lighting in the area and connected ahead of any local switches. The branch circuit feeding unit equipment shall be clearly identified at the distribution panel. Emergency illumination fixtures that obtain power from a unit equipment and are not part of the unit equipment shall be wired to the unit equipment as required by Section 700-9 and by one of the wiring methods of Chapter 3.

Exception: In a separate and uninterrupted area supplied by a minimum of three normal lighting circuits, a separate branch circuit for unit equipment shall be permitted if it originates from the same panelboard as that of the normal lighting circuits and is provided with a lock-on feature.

D. Emergency System Circuits for Lighting and Power

700-15. Loads on Emergency Branch Circuits. No appliances and no lamps, other than those specified as required for emergency use, shall be supplied by emergency lighting circuits.

700-16. Emergency Illumination. Emergency illumination shall include all required means of egress lighting, illuminated exit signs, and all other lights specified as necessary to provide required illumination.

Emergency lighting systems shall be so designed and installed that the failure of any individual lighting element, such as the burning out of a light bulb, cannot leave in total darkness any space that requires emergency illumination.

Where high-intensity discharge lighting such as high- and low-pressure sodium, mercury vapor, and metal halide is used as the sole source of normal illumination, the emergency lighting system shall be required to operate until normal illumination has been restored.

Exception: Where alternative means have been taken to ensure that the emergency lighting illumination level is maintained.

700-17. Circuits for Emergency Lighting. Branch circuits that supply emergency lighting shall be installed to provide service from a source complying with Section 700-12 when the normal supply for lighting is interrupted. Such installations shall provide either one of the following: (1) an emergency lighting supply, independent of the general lighting supply, with provisions for automatically transferring the emergency lights upon the event of failure of the general lighting system supply, or (2) two or more separate and complete systems with independent power supply, each system providing sufficient current for emergency lighting purposes. Unless both systems are used for regular lighting purposes and are both kept lighted, means shall be provided for automatically energizing either system upon failure of the other. Either or both systems shall be permitted to be a part of the general lighting system of the protected occupancy if circuits supplying lights for emergency illumination are installed in accordance with other sections of this article.

700-18. Circuits for Emergency Power. For branch circuits that supply equipment classed as emergency, there shall be an emergency supply source to which the load will be transferred automatically upon the failure of the normal supply.

E. Control — Emergency Lighting Circuits

700-20. Switch Requirements. The switch or switches installed in emergency lighting circuits shall be so arranged that only authorized persons will have control of emergency lighting.

Exception No. 1: Where two or more single-throw switches are connected in parallel to control a single circuit, at least one of these switches shall be accessible only to authorized persons.

Exception No. 2: Additional switches that act only to put emergency lights into operation but not disconnect them are permissible.

Switches connected in series or 3- and 4-way switches shall not be used.

700-21. Switch Location.   All manual switches for controlling emergency circuits shall be in locations convenient to authorized persons responsible for their actuation. In places of assembly, such as theaters, a switch for controlling emergency lighting systems shall be located in the lobby or at a place conveniently accessible thereto.

In no case shall a control switch for emergency lighting in a theater, or motion-picture theater or place of assembly, be placed in a motion-picture projection booth or on a stage or platform.

Exception: Where multiple switches are provided, one such switch shall be permitted in such locations where so arranged that it can energize the circuit only, but it cannot deenergize the circuit.

700-22. Exterior Lights.   Those lights on the exterior of a building that are not required for illumination when there is sufficient daylight shall be permitted to be controlled by an automatic light-actuated device.

F. Overcurrent Protection

700-25. Accessibility.   The branch-circuit overcurrent devices in emergency circuits shall be accessible to authorized persons only.

(FPN):   Fuses and circuit breakers for emergency circuit overcurrent protection, where coordinated to ensure selective clearing of fault currents, increase overall reliability of the system.

700-26. Ground-Fault Protection of Equipment.   The alternate source for emergency systems shall not be required to have ground-fault protection of equipment with automatic disconnecting means. Ground-fault indication of the emergency source shall be provided per Section 700-7(d). |

ARTICLE 701 — LEGALLY REQUIRED STANDBY SYSTEMS

A. General

701-1. Scope.   The provisions of this article apply to the electrical safety of the design, installation, operation, and maintenance of legally required standby systems consisting of circuits and equipment intended to supply, distribute, and control electricity to required facilities for illumination or power, or both, when the normal electrical supply or system is interrupted. |

The systems covered by this article consist only of those that are permanently installed in their entirety, including the power source.

(FPN No. 1):   For additional information, see Health Care Facilities, NFPA 99-1990 (ANSI). |

(FPN No. 2):   For further information regarding performance of emergency and standby power systems, see Emergency and Standby Power Systems, NFPA 110-1988 (ANSI).

(FPN No. 3):   For further information, see Recommended Practice for Emergency and Standby Power Systems for Industrial and Commercial Applications, ANSI/IEEE 446-1987.

701-2. Legally Required Standby Systems. Legally required standby systems are those systems required and so classed as legally required standby by municipal, state, federal, or other codes or by any governmental agency having jurisdiction. These systems are intended to automatically supply power to selected loads (other than those classed as emergency systems) in the event of failure of the normal source.

(FPN):   Legally required standby systems are typically installed to serve loads, such as heating and refrigeration systems, communication systems, ventilation and smoke removal systems, sewerage disposal, lighting systems, and industrial processes, that, when stopped during any interruption of the normal electrical supply, could create hazards or hamper rescue or fire fighting operations.

701-3. Application of Other Articles. Except as modified by this article, all applicable articles of this Code shall apply.

701-4. Equipment Approval. All equipment shall be approved for the intended use.

701-5. Tests and Maintenance for Legally Required Standby Systems.

(a) Conduct or Witness Test. The authority having jurisdiction shall conduct or witness a test on the complete system upon installation.

(b) Tested Periodically. Systems shall be tested periodically on a schedule and in a manner acceptable to the authority having jurisdiction to assure their maintenance in proper operating condition.

(c) Battery Systems Maintenance. Where batteries are used for control, starting, or ignition of prime movers, the authority having jurisdiction shall require periodic maintenance.

(d) Written Record. A written record shall be kept on such tests and maintenance.

(e) Testing under Load. Means for testing legally required standby systems under load shall be provided.

701-6. Capacity and Rating. A legally required standby system shall have adequate capacity and rating for the supply of all equipment intended to be operated at one time. Legally required standby system equipment shall be suitable for the maximum available fault current at its terminals.

The alternate power source shall be permitted to supply legally required standby and optional standby system loads when automatic selective load pickup and load shedding is provided as needed to assure adequate power to the legally required standby circuits.

701-7. Transfer Equipment. Transfer equipment, including automatic transfer switches, shall be automatic and identified for standby use and approved by the authority having jurisdiction. Transfer equipment shall be designed and installed to prevent the inadvertent interconnection of normal and alternate sources of supply in any operation of the transfer equipment. See Section 230-83.

Means to bypass and isolate the transfer switch equipment shall be permitted. Where bypass isolation switches are used, inadvertent parallel operation shall be avoided.

701-8. Signals. Audible and visual signal devices shall be provided, where practicable, for the following purposes:

(a) Derangement. To indicate derangement of the standby source.

(b) Carrying Load. To indicate that the standby source is carrying load.

(c) Not Functioning. To indicate that the battery charger is not functioning.

(FPN): For signals for generator sets, see Emergency and Standby Power Systems, NFPA 110-1988 (ANSI).

701-9. Signs.

(a) Mandated Standby. A sign shall be placed at the service entrance | indicating type and location of on-site legally required standby power sources.

(b) Grounding. Where the grounded circuit conductor connected to the | emergency source is connected to a grounding electrode conductor at a location remote from the emergency source, there shall be a sign at the grounding location that shall identify all emergency and normal sources connected at that location.

B. Circuit Wiring

701-10. Wiring Legally Required Standby Systems. The legally required standby system wiring shall be permitted to occupy the same raceways, cables, boxes, and cabinets with other general wiring.

C. Sources of Power

701-11. Legally Required Standby Systems. Current supply shall be such that, in event of failure of the normal supply to, or within, the building or group of buildings concerned, legally required standby power will be available within the time required for the application but not to exceed 60 seconds. The supply system for legally required standby purposes, in addition to the normal services to the building, shall be permitted to comprise one or more of the types of systems described in (a) through (f) below. | Unit equipment in accordance with Section 701-11(f) shall satisfy the applicable requirements of this article.

In selecting a legally required standby source of power, consideration shall be given to the type of service to be rendered, whether of short-time duration or long duration.

Consideration shall be given to the location or design, or both, of all | equipment to minimize the hazards that might cause complete failure due to floods, fires, icing, and vandalism.

(FPN): Assignment of degree of reliability of the recognized legally required standby supply system depends upon the careful evaluation of the variables at each particular installation.

(a) Storage Battery. A storage battery of suitable rating and capacity to supply and maintain at not less than 87½ percent of system voltage the total load of the circuits supplying legally required standby power for a period of at least 1½ hours.

Batteries, whether of the acid or alkali type, shall be designed and constructed to meet the service requirements of emergency service and shall be compatible with the charger for that particular installation.

For a sealed battery, the container shall not be required to be transparent. However, for the lead acid battery which requires water additions, transparent or translucent jars shall be furnished. Automotive-type batteries shall not be used.

An automatic battery charging means shall be provided.

(b) Generator Set.

(1) A generator set driven by a prime mover acceptable to the authority having jurisdiction and sized in accordance with Section 701-6. Means shall be provided for automatically starting the prime mover on failure of the normal service and for automatic transfer and operation of all required electrical circuits. A time-delay feature permitting a 15-minute setting shall be provided to avoid retransfer in case of short-time reestablishment of the normal source.

(2) Where internal combustion engines are used as the prime mover, an on-site fuel supply shall be provided with an on-premise fuel supply sufficient for not less than 2 hours full-demand operation of the system.

(3) Prime movers shall not be solely dependent upon a public utility gas system for their fuel supply or municipal water supply for their cooling systems. Means shall be provided for automatically transferring one fuel supply to another where dual fuel supplies are used.

Exception: Where acceptable to the authority having jurisdiction, the use of other than on-site fuels shall be permitted where there is a low probability of a simultaneous failure of both the off-site fuel delivery system and power from the outside electrical utility company.

(4) Where a storage battery is used for control or signal power, or as the means of starting the prime mover, it shall be suitable for the purpose and shall be equipped with an automatic charging means independent of the generator set.

(c) Uninterruptible Power Supplies. Uninterruptible power supplies used to provide power for legally required standby systems shall comply with the applicable provisions of Sections 701-11(a) and (b).

(d) Separate Service. Where acceptable to the authority having jurisdiction, a second service shall be permitted. This service shall be in accordance with Article 230, with separate service drop or lateral widely separated electrically and physically from the normal service to minimize the possibility of simultaneous interruption of supply.

(e) Connection Ahead of Service Disconnecting Means. Where acceptable to the authority having jurisdiction, connections ahead of, but not within, the main service disconnecting means shall be permitted. The legally required standby service shall be sufficiently separated from the normal main service disconnecting means to prevent simultaneous interruption of supply through an occurrence within the building or groups of buildings served.

(FPN): See Section 230-82 for equipment permitted on the supply side of a service disconnecting means.

(f) Unit Equipment. Individual unit equipment for legally required standby illumination shall consist of (1) a rechargeable battery; (2) a battery charging means; (3) provisions for one or more lamps mounted on the equipment and shall be permitted to have terminals for remote lamps; and (4) a relaying device arranged to energize the lamps automatically upon failure of the supply to the unit equipment. The batteries shall be of suitable rating and capacity to supply and maintain at not less than 87½ percent of the nominal battery voltage for the total lamp load associated with the unit for a period of at least 1½ hours, or the unit equipment shall supply and maintain not less than 60 percent of the initial legally required standby illumination for a period of at least 1½ hours. Storage batteries, whether of the acid or alkali type, shall be designed and constructed to meet the requirements of emergency service.

Unit equipment shall be permanently fixed in place (i.e., not portable) and shall have all wiring to each unit installed in accordance with the requirements of any of the wiring methods in Chapter 3. Flexible cord- and plug-connection shall be permitted, provided that the cord does not exceed 3 feet (914 mm) in length. The branch circuit feeding the unit equipment shall be the same branch circuit as that serving the normal lighting in the area and connected ahead of any local switches. Legally required standby illumination fixtures that obtain power from a unit equipment and are not part of the unit equipment shall be wired to the unit equipment by one of the wiring methods of Chapter 3.

D. Overcurrent Protection

701-15. Accessibility.   The branch-circuit overcurrent devices in legally required standby circuits shall be accessible to authorized persons only.

701-17. Ground-Fault Protection of Equipment.   The alternate source for legally required standby systems shall not be required to have ground-fault protection of equipment.

ARTICLE 702 — OPTIONAL STANDBY SYSTEMS

A. General

702-1. Scope.   The provisions of this article apply to the installation and operation of optional standby systems.

The systems covered by this article consist only of those that are permanently installed in their entirety, including prime movers.

702-2. Optional Standby Systems.   Optional standby systems are intended to protect private business or property where life safety does not depend on the performance of the system. Optional standby systems are intended to supply on-site generated power to selected loads either automatically or manually.

(FPN):   Optional standby systems are typically installed to provide an alternate source of electric power for such facilities as industrial and commercial buildings, farms, and residences, and to serve loads such as heating and refrigeration systems, data processing and communications systems, and industrial processes that, when stopped during any power outage, could cause discomfort, serious interruption of the process, damage to the product or process, or the like.

702-3. Application of Other Articles.   Except as modified by this article, all applicable articles of this Code shall apply.

702-4. Equipment Approval.   All equipment shall be approved for the intended use.

702-5. Capacity and Rating.   An optional standby system shall have adequate capacity and rating for the supply of all equipment intended to be operated at one time. Optional standby system equipment shall be suitable for the maximum available fault current at its terminals.

(FPN):   The optional standby system rating need only be sufficient to carry loads selected by the user.

702-6. Transfer Equipment.   Transfer equipment shall be suitable for the intended use and so designed and installed as to prevent the inadvertent interconnection of normal and alternate sources of supply in any operation of the transfer equipment.

Transfer equipment, located on the load side of branch-circuit protection, shall be permitted to contain supplementary overcurrent protection having an interrupting rating sufficient for the available fault current that the generator can deliver.

702-7. Signals.   Audible and visual signal devices shall be provided, where practicable, for the following purposes:

(a) Derangement. To indicate derangement of the optional standby source.

(b) Carrying Load. To indicate that the optional standby source is carrying load.

702-8. Signs.

(a) Standby. A sign shall be placed at the service-entrance equipment indicating type and location of on-site optional standby power sources.

(b) Grounding. Where the grounded circuit conductor connected to the emergency source is connected to a grounding electrode conductor at a location remote from the emergency source, there shall be a sign at the grounding location that shall identify all emergency and normal sources connected at that location.

B. Circuit Wiring

702-9. Wiring Optional Standby Systems.   The optional standby system wiring shall be permitted to occupy the same raceways, cables, boxes, and cabinets with other general wiring.

ARTICLE 705 — INTERCONNECTED ELECTRIC POWER PRODUCTION SOURCES

705-1. Scope. This article covers installation of one or more electric power production sources operating in parallel with a primary source(s) of electricity.

(FPN): The primary source may be a utility supply, on-site electric power source(s), or other sources.

705-2. Definition. For purposes of this article, the following definition applies:

Interactive System: An electric power production system that is operating in parallel with and capable of delivering energy to an electric primary source supply system.

705-3. Other Articles. Interconnected electric power production sources shall comply with this article and also the applicable requirements of the following articles:

	Article
Generators	445
Solar Photovoltaic Systems	690
Emergency Systems	700
Legally Required Standby Systems	701
Optional Standby Systems	702

705-10. Directory. A permanent plaque or directory, denoting all electrical power sources on or in the premises, shall be installed at each service equipment location and at locations of all electric power production sources capable of being interconnected.

Exception: Installations with large numbers of power production sources shall be permitted to be designated by groups.

705-12. Point of Connection. The outputs of electric power production systems shall be interconnected at the premises service disconnecting means. See Section 230-82, Exception No. 6.

Exception No. 1: The outputs shall be permitted to be interconnected at a point or points elsewhere on the premises where the system qualifies as an integrated electric system and incorporates protective equipment in accordance with all applicable sections of Article 685.

Exception No. 2: The outputs shall be permitted to be interconnected at a point or points elsewhere on the premises where all of the following conditions are met:

a. The aggregate of nonutility sources of electricity has a capacity in excess of 100 kW, or the service is above 1000 volts;

b. The conditions of maintenance and supervision assure that qualified persons will service and operate the system; and

c. Safeguards and protective equipment are established and maintained.

705-14. Output Characteristics. The output of a generator or other electric power production source operating in parallel with an electric supply system shall be compatible with the voltage, wave shape, and frequency of the system to which it is connected.

(FPN): The term compatible does not necessarily mean matching the primary source wave shape.

705-16. Interrupting and Withstand Rating. Consideration shall be given to the contribution of fault currents from all interconnected power sources for the interrupting and withstand ratings of equipment on interactive systems.

705-20. Disconnecting Means, Sources. Means shall be provided to disconnect all ungrounded conductors of an electric power production source(s) from all other conductors. See Article 230.

705-21. Disconnecting Means, Equipment. Means shall be provided to disconnect equipment, such as inverters or transformers, associated with a power production source, from all ungrounded conductors of all sources of supply.

Exception: Equipment intended to be operated and maintained as an integral part of a power production source exceeding 1000 volts.

705-22. Disconnect Device. The disconnecting means for ungrounded conductors shall consist of a manually or power operable switch(es) or circuit breaker(s):

(1) Located where accessible;

(2) Externally operable without exposing the operator to contact with live parts and, if power operable, of a type that can be opened by hand in the event of a power supply failure;

(3) Plainly indicating whether in the open or closed position; and

(4) Having ratings not less than the load to be carried and the fault current to be interrupted.

For disconnect equipment energized from both sides, a marking shall be provided to indicate that all contacts of the disconnect equipment may be energized.

(FPN No. 1): In parallel generation systems, some equipment, including knife blade switches and fuses, may be energized from both directions. See Section 240-40.

(FPN No. 2): Interconnection to off-premises primary source may require a visibly verifiable disconnecting device.

705-30. Overcurrent Protection. Conductors shall be protected in accordance with Article 240. Equipment overcurrent protection shall be in accordance with the articles referenced in Article 240. Equipment and conductors connected to more than one electrical source shall have a sufficient number of overcurrent devices so located as to provide protection from all sources.

(1) Generators shall be protected in accordance with Section 445-4.

(2) Solar photovoltaic systems shall be protected in accordance with Article 690.

(3) Overcurrent protection for a transformer with a source(s) on each side shall be provided in accordance with Section 450-3 by considering first one side of the transformer, then the other side of the transformer, as the primary.

705-32. Ground-Fault Protection. Where ground-fault protection is used, the output of an interactive system shall be connected to the supply side of the ground-fault protection.

Exception: Connection shall be permitted to be made to the load side of ground-fault protection provided that there is ground-fault protection for equipment from all ground-fault current sources.

705-40. Loss of Primary Source. Upon loss of primary source, an electric power production source shall be automatically disconnected from all ungrounded conductors of the primary source and shall not be reconnected until the primary source is restored.

(FPN No. 1): Risks to personnel and equipment associated with the primary source may occur if an interactive electric power production source can operate as an island. Special detection methods may be required to determine that a primary source supply system outage has occurred, and whether there should be automatic disconnection. When the primary source supply system is restored, special detection methods may be required to limit exposure of power production sources to out-of-phase reconnection.

(FPN No. 2): Induction generating equipment on systems with significant capacitance may become self-excited upon loss of primary source and experience severe over-voltage as a result.

705-42. Unbalanced Interconnections. A 3-phase electric power production source shall be automatically disconnected from all ungrounded conductors of the interconnected systems when one of the phases of that source opens.

Exception: An electric power production source providing power for an emergency or legally required standby system.

705-43. Synchronous Generators. Synchronous generators in a parallel system shall be provided with the necessary equipment to establish and maintain a synchronous condition.

705-50. Grounding. Interconnected electric power production sources shall be grounded in accordance with Article 250.

Exception: For direct-current systems connected through an inverter directly to a grounded service, other methods that accomplish equivalent system protection and that utilize equipment listed and identified for the use shall be permitted.

ARTICLE 710 — OVER 600 VOLTS, NOMINAL GENERAL

A. General

710-1. Scope. This article covers the general requirements for all circuits and equipment operating at more than 600 volts, nominal.

710-2. Definition: For the purposes of this article, "high voltage" shall be defined as more than 600 volts, nominal.

710-3. Other Articles. Provisions applicable to specific types of installations are included in the following articles:

	Article
Busways	364
Cable Trays	318
Capacitors	460
Conductors for General Wiring	310
Definitions	100
Electric Signs and Outline Lighting	600
Flexible Cords and Cables	400
Grounding	250
Lighting Fixtures, Lampholders, Lamps, and Receptacles	410
Motors, Motor Circuits, and Controllers	430
Outlet, Device, Pull and Junction Boxes, Conduit Bodies, and Fittings	370
Outside Branch Circuits and Feeders	225
Overcurrent Protection	240
Requirements for Electrical Installations	110
Resistors and Reactors	470
Services	230
Surge Arresters	280
Transformers and Transformer Vaults	450
Wiring Methods	300

710-4. Wiring Methods.

(a) Aboveground Conductors. Aboveground conductors shall be installed in rigid metal conduit, in intermediate metal conduit, in rigid nonmetallic conduit, in cable trays, as busways, as cablebus, in other identified raceways, or as open runs of metal-clad cable suitable for the use and purpose.

In locations accessible to qualified persons only, open runs of Type MV cables, bare conductors, and bare busbars shall also be permitted.

(b) Underground Conductors. Underground conductors shall be identified for the voltage and conditions under which they are installed.

Direct burial cables shall comply with the provisions of Section 310-7.

Underground cables shall be permitted to be direct buried or installed in raceways identified for the use and shall meet the depth requirements of Table 710-4(b).

Nonshielded cables shall be installed in rigid metal conduit, in intermediate metal conduit, or in rigid nonmetallic conduit encased in not less than 3 inches (76 mm) of concrete.

Exception No. 1: Type MC cable with nonshielded conductor where the metallic sheath is grounded through an effective grounding path meeting the requirements of Section 250-51.

Exception No. 2: Lead-sheath cable with nonshielded conductor where the lead sheath is grounded through an effective grounding path meeting the requirements of Section 250-51.

Table 710-4(b). Minimum Cover Requirements
Cover is defined as the shortest distance in inches measured between a point on the top surface of any direct buried conductor, cable, conduit, or other raceway and the top surface of finished grade, concrete, or similar cover.

Circuit Voltage	Direct Buried Cables	Rigid Nonmetallic Conduit Approved for Direct Burial*	Rigid Metal Conduit and Intermediate Metal Conduit
Over 600-22kV	30	18	6
Over 22kV-40kV	36	24	6
Over 40kV	42	30	6

For SI units: one inch = 25.4 millimeters.
* Listed by a qualified testing agency as suitable for direct burial without encasement. All other nonmetallic systems shall require 2 inches (50.8 mm) of concrete or equivalent above conduit in addition to above depth.

Exception No. 1: Areas subject to vehicular traffic, such as thoroughfares or commercial parking areas, shall have a minimum cover of 24 inches (610 mm).

Exception No. 2: The minimum cover requirements for other than rigid metal conduit and intermediate metal conduit shall be permitted to be reduced 6 inches (152 mm) for each 2 inches (50.8 mm) of concrete or equivalent protection placed in the trench over the underground installation.

Exception No. 3: The minimum cover requirements shall not apply to conduits or other raceways that are located under a building or exterior concrete slab not less than 4 inches (102 mm) in thickness and extending not less than 6 inches (152 mm) beyond the underground installation. A warning ribbon or other effective means suitable for the conditions shall be placed above the underground installation.

Exception No. 4: Lesser depths shall be permitted where cables and conductors rise for terminations or splices or where access is otherwise required.

Exception No. 5: In airport runways, including adjacent defined areas where trespass is prohibited, cable shall be permitted to be buried not less than 18 inches (457 mm) deep and without raceways, concrete enclosement, or equivalent.

Exception No. 6: Raceways installed in solid rock shall be permitted to be buried at lesser depth where covered by 2 inches (50.8 mm) of concrete, which shall be permitted to extend to the rock surface.

(1) Protection from Damage. Conductors emerging from the ground shall be enclosed in approved raceway. Raceways installed on poles shall be of rigid metal conduit, intermediate metal conduit, PVC Schedule 80, or equivalent extending from the ground line up to a point 8 feet (2.44 m) above finished grade. Conductors entering a building shall be protected by an approved enclosure from the ground line to the point of entrance. Metallic enclosures shall be grounded.

(2) Splices. Direct burial cables shall be permitted to be spliced or tapped without the use of splice boxes, provided they are installed using materials suitable for the application. The taps and splices shall be watertight and protected from mechanical damage. Where cables are shielded, the shielding shall be continuous across the splice or tap.

(3) Backfill. Backfill containing large rock, paving materials, cinders, large or sharply angular substance, or corrosive materials shall not be placed in an excavation where materials can damage raceways, cables, or other substructures or prevent adequate compaction of fill or contribute to corrosion of raceways, cables, or other substructures.

Protection in the form of granular or selected material or suitable sleeves shall be provided to prevent physical damage to the raceway or cable.

(4) Raceway Seal. Where a raceway enters from an underground system, the end within the building shall be sealed with an identified compound so as to prevent the entrance of moisture or gases, or it shall be so arranged to prevent moisture from contacting live parts.

(c) Busbars. Busbars shall be permitted to be either copper or aluminum.

710-5. Braid-Covered Insulated Conductors — Open Installation. Open runs of braid-covered insulated conductors shall have a flame-retardant braid. If the conductors used do not have this protection, a flame-retardant saturant shall be applied to the braid covering after installation. This treated braid covering shall be stripped back a safe distance at conductor terminals, according to the operating voltage. This distance shall not be less than 1 inch (25.4 mm) for each kilovolt of the conductor-to-ground voltage of the circuit, where practicable.

710-6. Insulation Shielding. Metallic and semiconducting insulation shielding components of shielded cables shall be removed for a distance dependent on the circuit voltage and insulation. Stress reduction means shall be provided at all terminations of factory-applied shielding.

Metallic shielding components such as tapes, wires or braids, or combinations thereof and their associated conducting or semiconducting components shall be grounded.

710-7. Grounding. Wiring and equipment installations shall be grounded in accordance with the applicable provisions of Article 250.

710-8. Moisture or Mechanical Protection for Metal-Sheathed Cables. Where cable conductors emerge from a metal sheath and where protection against moisture or physical damage is necessary, the insulation of the conductors shall be protected by a cable sheath terminating device.

710-9. Protection of Service Equipment, Metal-Enclosed Power Switchgear, and Industrial Control Assemblies. Pipes or ducts foreign to the electrical installation that require periodic maintenance or whose malfunction would endanger the operation of the electrical system shall not be located in the vicinity of the service equipment, metal-enclosed power switchgear, or industrial control assemblies. Protection shall be provided where necessary to avoid damage from condensation leaks and breaks in such foreign systems. Piping and other facilities shall not be considered foreign if provided for fire protection of the electrical installation.

B. Equipment — General Provisions

710-11. Indoor Installations.   See Section 110-31(a).

710-12. Outdoor Installations.   See Section 110-31(b).

710-13. Metal-Enclosed Equipment.   See Section 110-31(c).

710-14. Oil-Filled Equipment.   Installation of electrical equipment, other than transformers, covered in Article 450, containing more than 10 gallons (37.85 L) of flammable oil per unit shall meet the requirements of Parts B and C of Article 450.

C. Equipment — Specific Provisions

(FPN):   See also references to specific types of installations in Section 710-3.

710-20. Overcurrent Protection.   Overcurrent protection shall be provided for each ungrounded conductor by one of the following:

(a) Overcurrent Relays and Current Transformers. Circuit breakers used for overcurrent protection of ac 3-phase circuits shall have a minimum of three overcurrent relays operated from three current transformers.

Exception No. 1: On 3-phase, 3-wire circuits, an overcurrent relay in the residual circuit of the current transformers shall be permitted to replace one of the phase relays.

Exception No. 2: An overcurrent relay, operated from a current transformer that links all phases of a 3-phase, 3-wire circuit, shall be permitted to replace the residual relay and one of the phase conductor current transformers. Where the neutral is not regrounded on the load side of the circuit as permitted in Section 250-152(b), the current transformer shall be permitted to link all three phase conductors and the grounded circuit conductor (neutral).

(b) Fuses. A fuse shall be connected in series with each ungrounded conductor.

710-21. Circuit-Interrupting Devices.

(a) Circuit Breakers.

(1) Indoor installations shall consist of metal-enclosed units or fire-resistant cell-mounted units.

Exception: Open mounting of circuit breakers shall be permitted in locations accessible to qualified persons only.

(2) Circuit breakers used to control oil-filled transformers shall be either located outside the transformer vault or be capable of operation from outside the vault.

(3) Oil circuit breakers shall be so arranged or located that adjacent readily combustible structures or materials are safeguarded in an approved manner.

(4) Circuit breakers shall have the following equipment or operating characteristics:

a. An accessible mechanical or other approved means for manual tripping, independent of control power.

b. Be release free (trip free).

c. If capable of being opened or closed manually while energized, the main contacts shall operate independently of the speed of the manual operation.

d. A mechanical position indicator at the circuit breaker to show the open or closed position of the main contacts.

e. A means of indicating the open and closed position of the breaker at the point(s) from which they may be operated.

f. A permanent and legible nameplate showing manufacturer's name or trademark, manufacturer's type or identification number, continuous current rating, interrupting rating in mVA or amperes, and maximum voltage rating. Modification of a circuit breaker affecting its rating(s) shall be accompanied by an appropriate change of nameplate information.

(5) The continuous current rating of a circuit breaker shall be not less than the maximum continuous current through the circuit breaker.

(6) The interrupting rating of a circuit breaker shall not be less than the maximum fault current the circuit breaker will be required to interrupt, including contributions from all connected sources of energy.

(7) The closing rating of a circuit breaker shall not be less than the maximum asymmetrical fault current into which the circuit breaker can be closed.

(8) The momentary rating of a circuit breaker shall not be less than the maximum asymmetrical fault current at the point of installation.

(9) The rated maximum voltage of a circuit breaker shall not be less than the maximum circuit voltage.

(b) Power Fuses and Fuseholders.

(1) Use. Where fuses are used to protect conductors and equipment, a fuse shall be placed in each ungrounded conductor. Two power fuses shall be permitted to be used in parallel to protect the same load, if both fuses have identical ratings, and both fuses are installed in an identified common mounting with electrical connections that will divide the current equally.

Power fuses of the vented type shall not be used indoors, underground, or in metal enclosures unless identified for the use.

(2) Interrupting Rating. The interrupting rating of power fuses shall not be less than the maximum fault current the fuse will be required to interrupt, including contributions from all connected sources of energy.

(3) Voltage Rating. The maximum voltage rating of power fuses shall not be less than the maximum circuit voltage. Fuses having a minimum recommended operating voltage shall not be applied below this voltage.

(4) Identification of Fuse Mountings and Fuse Units. Fuse mountings and fuse units shall have permanent and legible nameplates showing the manufacturer's type or designation, continuous current rating, interrupting current rating, and maximum voltage rating.

(5) Fuses. Fuses that expel flame in opening the circuit shall be so designed or arranged that they will function properly without hazard to persons or property.

(6) Fuseholders. Fuseholders shall be designed or installed so that they will be deenergized while replacing a fuse.

Exception: Fuse and fuseholder designed to permit fuse replacement by qualified persons using equipment designed for the purpose without deenergizing the fuseholder.

(7) High-Voltage Fuses. Metal-enclosed switchgear and substations that utilize high-voltage fuses shall be provided with a gang-operated disconnecting switch. Isolation of the fuses from the circuit shall be provided by either connecting a switch between the source and the fuses or providing roll-out switch and fuse-type of construction. The switch shall be of the load-interrupter type, unless mechanically or electrically interlocked with a load-interrupting device arranged to reduce the load to the interrupting capability of the switch.

Exception: More than one switch shall be permitted as the disconnecting means for one set of fuses where the switches are installed to provide connection to more than one set of supply conductors. The switches shall be mechanically or electrically interlocked to permit access to the fuses only when all switches are open. A conspicuous sign shall be placed at the fuses reading: "WARNING —FUSES MAY BE ENERGIZED FROM MORE THAN ONE SOURCE."

(c) Distribution Cutouts and Fuse Links — Expulsion Type.

(1) Installation. Cutouts shall be so located that they may be readily and safely operated and re-fused, and so that the exhaust of the fuses will not endanger persons. Distribution cutouts shall not be used indoors, underground, or in metal enclosures.

(2) Operation. Where fused cutouts are not suitable to interrupt the circuit manually while carrying full load, an approved means shall be installed to interrupt the entire load. Unless the fused cutouts are interlocked with the switch to prevent opening of the cutouts under load, a conspicuous sign shall be placed at such cutouts reading, "WARNING — DO NOT OPEN UNDER LOAD."

(3) Interrupting Rating. The interrupting rating of distribution cutouts shall not be less than the maximum fault current the cutout will be required to interrupt, including contributions from all connected sources of energy.

(4) Voltage Rating. The maximum voltage rating of cutouts shall not be less than the maximum circuit voltage.

(5) Identification. Distribution cutouts shall have on their body, door, or fuse tube a permanent and legible nameplate or identification showing the manufacturer's type or designation, continuous current rating, maximum voltage rating, and interrupting rating.

(6) Fuse Links. Fuse links shall have a permanent and legible identification showing continuous current rating and type.

(7) Structure Mounted Outdoors. The height of cutouts mounted outdoors on structures shall provide safe clearance between lowest energized parts (open or closed position) and standing surfaces, in accordance with Section 110-34(e).

(d) Oil-Filled Cutouts.

(1) Continuous Current Rating. The continuous current rating of oil-filled cutouts shall not be less than the maximum continuous current through the cutout.

(2) Interrupting Rating. The interrupting rating of oil-filled cutouts shall not be less than the maximum fault current the oil-filled cutout will be required to interrupt, including contributions from all connected sources of energy.

(3) Voltage Rating. The maximum voltage rating of oil-filled cutouts shall not be less than the maximum circuit voltage.

(4) Fault Closing Rating. Oil-filled cutouts shall have a fault closing rating not less than the maximum asymmetrical fault current that can occur at the cutout location, unless suitable interlocks or operating procedures preclude the possibility of closing into a fault.

(5) Identification. Oil-filled cutouts shall have a permanent and legible nameplate showing the rated continuous current, rated maximum voltage, and rated interrupting current.

(6) Fuse Links. Fuse links shall have a permanent and legible identification showing the rated continuous current.

(7) Location. Cutouts shall be so located that they will be readily and safely accessible for re-fusing, with the top of the cutout not over 5 feet (1.52 m) above the floor or platform.

(8) Enclosure. Suitable barriers or enclosures shall be provided to prevent contact with nonshielded cables or energized parts of oil-filled cutouts.

(e) Load Interrupters. Load-interrupter switches shall be permitted if suitable fuses or circuit breakers are used in conjunction with these devices to interrupt fault currents. Where these devices are used in combination, they shall be so coordinated electrically that they will safely withstand the effects of closing, carrying, or interrupting all possible currents up to the assigned maximum short-circuit rating.

Where more than one switch is installed with interconnected load terminals to provide for alternate connection to different supply conductors, each switch shall be provided with a conspicuous sign reading: "WARNING — SWITCH MAY BE ENERGIZED BY BACKFEED."

(1) Continuous Current Rating. The continous current rating of interrupter switches shall equal or exceed the maximum continuous current at the point of installation.

(2) Voltage Rating. The maximum voltage rating of interrupter switches shall equal or exceed the maximum circuit voltage.

(3) Identification. Interrupter switches shall have a permanent and legible nameplate including the following information: manufacturer's type or designation, continuous current rating, interrupting current rating, fault closing rating, maximum voltage rating.

(4) Switching of Conductors. The switching mechanism shall be arranged to be operated from a location where the operator is not exposed to energized parts and shall be arranged to open all ungrounded conductors of the circuit simultaneously with one operation. Switches shall be arranged to be locked in the open position. Metal-enclosed switches shall be operable from outside the enclosure.

(5) Stored Energy for Opening. The stored energy operator shall be permitted to be left in the uncharged position after the switch has been closed if a single movement of the operating handle charges the operator and opens the switch.

(6) Supply Terminals. Fused interrupter switches shall be so installed that all supply terminals shall be at the top of the switch enclosure.

Exception: Supply terminals shall not be required to be at the top of the switch enclosure if barriers are installed to prevent persons from accidentally contacting energized parts or dropping tools or fuses into energized parts.

710-22. Isolating Means. Means shall be provided to completely isolate an item of equipment. The use of isolating switches shall not be required where there are other ways of deenergizing the equipment for inspection and repairs, such as drawout-type metal-enclosed switchgear units and removable truck panels.

Isolating switches not interlocked with an approved circuit-interrupting device shall be provided with a sign warning against opening them under load.

A fuseholder and fuse, designed for the purpose, shall be permitted as an isolating switch.

710-23. Voltage Regulators. Proper switching sequence for regulators shall be assured by use of one of the following: (1) mechanically sequenced regulator bypass switch(es); (2) mechanical interlocks; or (3) switching procedure prominently displayed at the switching location.

710-24. Metal-Enclosed Power Switchgear and Industrial Control Assemblies.

(a) Scope. This section covers assemblies of metal-enclosed power switchgear and industrial control, including but not limited to switches, interrupting devices and their control, metering, protection and regulating equipment, where an integral part of the assembly, with associated interconnections and supporting structures. This section also includes metal-enclosed power switchgear assemblies that form a part of unit substations, power centers, or similar equipment.

(b) Arrangement of Devices in Assemblies. Arrangement of devices in assemblies shall be such that individual components can safely perform their intended function without adversely affecting the safe operation of other components in the assembly.

(c) Guarding of High-Voltage Energized Parts Within a Compartment. Where access for other than visual inspection is required to a compartment that contains energized high-voltage parts, barriers shall be provided as follows:

(1) To prevent accidental contact with energized parts.

Exception No. 1: Fuse and fuseholder designed to permit fuse replacement by qualified persons using equipment designed for the purpose without deenergizing the fuseholder.

Exception No. 2: Exposed live parts shall be permitted within the compartment where accessible to qualified persons only.

(2) To prevent tools or other equipment from being dropped on energized parts.

(d) Guarding of Low-Voltage Energized Parts Within a Compartment. Energized bare parts mounted on doors shall be guarded where the door must be opened for maintenance of equipment or removal of drawout equipment.

(e) Clearance for Cable Conductors Entering Enclosure. The unobstructed space opposite terminals or opposite raceways or cables entering a switchgear or control assembly shall be adequate for the type of conductor and method of termination.

(f) Accessibility of Energized Parts.

(1) Doors that would provide nonqualified persons access to high-voltage energized parts shall be locked.

(2) Low-voltage control equipment, relays, motors, and the like shall not be installed in compartments with exposed high-voltage energized parts or high-voltage wiring unless the access door or cover is interlocked with the high-voltage switch or disconnecting means to prevent the door or cover from being opened or removed unless the switch or disconnecting means is in its isolating position.

Exception No. 1: Instrument or control transformers connected to high voltage.

Exception No. 2: Space heaters.

(g) Grounding. Frames of switchgear and control assemblies shall be grounded.

(h) Grounding of Devices. Devices with metal cases and/or frames, such as instruments, relays, meters, and instrument and control transformers, located in or on switchgear or control, shall have the frame or case grounded.

(i) Door Stops and Cover Plates. External hinged doors or covers shall be provided with stops to hold them in the open position. Cover plates intended to be removed for inspection of energized parts or wiring shall be equipped with lifting handles and shall not exceed 12 square feet (1.11 sq m) in area or 60 pounds (27.22 kg) in weight, unless they are hinged and bolted or locked.

(j) Gas Discharge from Interrupting Devices. Gas discharged during operating of interrupting devices shall be so directed as not to endanger personnel.

(k) Inspection Windows. Windows intended for inspection of disconnecting switches or other devices shall be of suitable transparent material.

(l) Location of Devices. Control and instrument transfer switch handles or pushbuttons shall be in a readily accessible location at an elevation not over 78 inches (1.98 m).

Exception No. 1: Operating handles requiring more than 50 pounds (22.68 kg) of force shall not be higher than 66 inches (1.68 m) in either the open or closed position.

Exception No. 2: Operating handles for infrequently operated devices, such as drawout fuses, fused potential or control transformers and their primary disconnects, and bus transfer switches where they are otherwise safely operable and serviceable from a portable platform.

(m) Interlocks — Interrupter Switches. Interrupter switches equipped with stored energy mechanisms shall have mechanical interlocks to prevent access to the switch compartment unless the stored energy mechanism is in the discharged or blocked position.

(n) Stored Energy for Opening. The stored energy operator shall be permitted to be left in the uncharged position after the switch has been closed if a single movement of the operating handle charges the operator and opens the switch.

(o) Fused Interrupter Switches.

(1) Fused interrupter switches shall be so installed that all supply terminals shall be at the top of the switch enclosure.

Exception: Supply terminals shall not be required to be at the top of the switch enclosure if barriers are installed to prevent persons from accidentally contacting energized parts or dropping tools or fuses into energized parts.

(2) Where fuses can be energized by backfeed, a sign shall be placed on the enclosure door reading: "WARNING — FUSES MAY BE ENERGIZED BY BACKFEED."

(3) The switching mechanism shall be arranged to be operated from a location outside the enclosure where the operator is not exposed to energized parts and shall be arranged to open all ungrounded conductors of the circuit simultaneously with one operation. Switches shall be capable of being locked in the open position.

(p) Interlocks — Circuit Breakers.

(1) Circuit breakers equipped with stored energy mechanisms shall be designed to prevent the release of the stored energy unless the mechanism has been fully charged.

(2) Mechanical interlocks shall be provided in the housing to prevent the complete withdrawal of the circuit breaker from the housing when the stored energy mechanism is in the fully charged position.

Exception: Where a suitable device is provided that prevents the complete withdrawal of the circuit breaker unless the closing function is blocked.

D. Installations Accessible to Qualified Persons Only

710-31. Enclosure for Electrical Installations.　See Section 110-31.

710-32. Circuit Conductors.　Circuit conductors shall be permitted to be installed in raceways, in cable trays, as metal-clad cable, as bare wire, cable, and busbars, or as Type MV cables, or conductors as provided in Sections 710-4 through 710-6. Bare live conductors shall conform with Sections 710-33 and 710-34.

Insulators, together with their mounting and conductor attachments, where used as supports for wires, single-conductor cables, or busbars, shall be capable of safely withstanding the maximum magnetic forces that would prevail when two or more conductors of a circuit were subjected to short-circuit current.

Open runs of insulated wires and cables having a bare lead sheath or a braided outer covering shall be supported in a manner designed to prevent physical damage to the braid or sheath. Supports for lead-covered cables shall be designed to prevent electrolysis of the sheath.

710-33. Minimum Space Separation.　In field-fabricated installations, the minimum air separation between bare live conductors and between such conductors and adjacent grounded surfaces shall not be less than the values given in Table 710-33. These values shall not apply to interior portions or exterior terminals of equipment designed, manufactured, and tested in accordance with accepted national standards.

710-34. Work Space and Guarding.　See Section 110-34.

E. Mobile and Portable Equipment

710-41. General.

(a) Covered. The provisions of this part shall apply to installations and use of high-voltage power distribution and utilization equipment that is portable and/or mobile, such as substations and switch houses mounted on skids, trailers, or cars, mobile shovels, draglines, cranes, hoists, drills, dredges, compressors, pumps, conveyors, underground excavators, and the like.

Table 710-33.   Minimum Clearance of Live Parts*

Nominal Voltage Rating, kV	Impulse Withstand, B.I.L. kV		Minimum Clearance of Live Parts, in Inches			
			Phase-to-Phase		Phase-to-Ground	
	Indoors	Outdoors	Indoors	Outdoors	Indoors	Outdoors
2.4-4.16	60	95	4.5	7	3.0	6
7.2	75	95	5.5	7	4.0	6
13.8	95	110	7.5	12	5.0	7
14.4	110	110	9.0	12	6.5	7
23	125	150	10.5	15	7.5	10
34.5	150	150	12.5	15	9.5	10
	200	200	18.0	18	13.0	13
46		200		18		13
		250		21		17
69		250		21		17
		350		31		25
115		550		53		42
138		550		53		42
		650		63		50
161		650		63		50
		750		72		58
230		750		72		58
		900		89		71
		1050		105		83

For SI units: one inch = 25.4 millimeters.

* The values given are the minimum clearance for rigid parts and bare conductors under favorable service conditions. They shall be increased for conductor movement or under unfavorable service conditions or wherever space limitations permit. The selection of the associated impulse withstand voltage for a particular system voltage is determined by the characteristics of the surge protective equipment.

(b) Other Requirements. The requirements of this part shall be additional to, or amendatory of, those prescribed in Articles 100 through 725 of this Code. Special attention shall be paid to Article 250.

(c) Protection. Adequate enclosures and/or guarding shall be provided to protect portable and mobile equipment from physical damage.

(d) Disconnecting Means. Disconnecting means shall be installed for mobile and portable high-voltage equipment according to the requirements of Part H of Article 230 and shall disconnect all ungrounded conductors.

710-42. Overcurrent Protection. Motors driving single or multiple dc generators supplying a system operating on a cyclic load basis do not require running overcurrent protection, provided that the thermal rating of the ac drive motor cannot be exceeded under any operating condition. The branch-circuit protective device(s) shall provide short-circuit and locked-rotor protection, and shall be permitted to be external to the equipment.

710-43. Enclosures. All energized switching and control parts shall be enclosed in effectively grounded metal cabinets or enclosures. These cabinets or enclosures shall be marked "DANGER — HIGH VOLTAGE — KEEP OUT" and shall be locked so that only authorized and qualified persons can enter. Circuit breakers and protective equipment shall have the operating means projecting through the metal cabinet or enclosure so these units can be reset without opening locked doors. With doors closed, reasonable safe access for normal operation of these units shall be provided.

710-44. Collector Rings. The collector ring assemblies on revolving-type machines (shovels, draglines, etc.) shall be guarded to prevent accidental contact with energized parts by personnel on or off the machine.

710-45. Power Cable Connections to Mobile Machines. A metallic enclosure shall be provided on the mobile machine for enclosing the terminals of the power cable. The enclosure shall include provisions for a solid connection for the ground wire(s) terminal to effectively ground the machine frame. Ungrounded conductors shall be attached to insulators or terminated in approved high-voltage cable couplers (which include ground wire connectors) of proper voltage and ampere rating. The method of cable termination used shall prevent any strain or pull on the cable from stressing the electrical connections. The enclosure shall have provision for locking so only authorized and qualified persons may open it and shall be marked "DANGER — HIGH VOLTAGE — KEEP OUT."

710-46. High-Voltage Portable Cable for Main Power Supply. Flexible high-voltage cable supplying power to portable or mobile equipment shall comply with Article 250 and Article 400, Part C.

710-47. Grounding. Mobile equipment shall be grounded in accordance with Article 250.

F. Tunnel Installations

710-51. General.

(a) **Covered.** The provisions of this part shall apply to installation and use of high-voltage power distribution and utilization equipment that is portable and/or mobile, such as substations, trailers, or cars, mobile shovels, draglines, hoists, drills, dredges, compressors, pumps, conveyors, underground excavators, and the like.

(b) **Other Articles.** The requirements of this part shall be additional to, or amendatory of, those prescribed in Articles 100 through 710 of this Code. Special attention shall be paid to Article 250.

(c) **Protection Against Physical Damage.** Conductors and cables in tunnels shall be located above the tunnel floor and so placed or guarded to protect them from physical damage.

710-52. Overcurrent Protection. Motor-operated equipment shall be protected from overcurrent in accordance with Article 430. Transformers shall be protected from overcurrent in accordance with Article 450.

710-53. Conductors. High-voltage conductors in tunnels shall be installed in (1) metal conduit or other metal raceway; (2) Type MC cable; or (3) other approved multiconductor cable. Multiconductor portable cable shall be permitted to supply mobile equipment.

710-54. Bonding and Equipment Grounding Conductor.

(a) Grounded and Bonded. All noncurrent-carrying metal parts of electric equipment and all metal raceways and cable sheaths shall be effectively grounded and bonded to all metal pipes and rails at the portal and at intervals not exceeding 1000 feet (305 m) throughout the tunnel.

(b) Equipment Grounding Conductor. An equipment grounding conductor shall be run with circuit conductors inside the metal raceway or inside the multiconductor cable jacket. The equipment grounding conductor shall be permitted to be insulated or bare.

710-55. Transformers, Switches, and Electric Equipment. All transformers, switches, motor controllers, motors, rectifiers, and other equipment installed below ground shall be protected from physical damage by location or guarding.

710-56. Energized Parts. Bare terminals of transformers, switches, motor controllers, and other equipment shall be enclosed to prevent accidental contact with energized parts.

710-57. Ventilation System Controls. Electrical controls for the ventilation system shall be so arranged that the airflow can be reversed.

710-58. Disconnecting Means. A switching device meeting the requirements of Article 430 or 450 shall be installed at each transformer or motor location for disconnecting the transformer or motor. The switching device shall open all ungrounded conductors of a circuit simultaneously.

710-59. Enclosures. Enclosures for use in tunnels shall be dripproof, weatherproof, or submersible as required by the environmental conditions. Switch or contactor enclosures shall not be used as junction boxes or raceways for conductors feeding through or tapping off to other switches, unless special designs are used to provide adequate space for this purpose.

710-60. Grounding. Tunnel equipment shall be grounded in accordance with Article 250.

G. Electrode-type Boilers

710-70. General. The provisions of this part shall apply to boilers operating over 600 volts, nominal, in which heat is generated by the passage of current between electrodes through the liquid being heated.

710-71. Electric Supply System. Electrode-type boilers shall be supplied only from a 3-phase, 4-wire solidly grounded wye system, or from isolating transformers arranged to provide such a system. Control circuit voltages shall not exceed 150 volts, shall be supplied from a grounded system, and shall have the controls in the ungrounded conductor.

710-72. Branch Circuit Requirements.

(a) Rating. Each boiler shall be supplied from an individual branch circuit rated not less than 100 percent of the total load.

(b) Common-Trip Fault Interrupting Device. The circuit shall be protected by a 3-phase common-trip fault interrupting device, which shall be permitted to automatically reclose the circuit upon removal of an overload condition but shall not reclose after a fault condition.

(c) Phase Fault Protection. Phase fault protection shall be provided in each phase, consisting of a separate phase overcurrent relay connected to a separate current transformer in the phase.

(d) Ground Current Detection. Means shall be provided for detection of the sum of the neutral and ground currents and shall trip the circuit interrupting device if the sum of those currents exceeds the greater of 5 amperes or $7\frac{1}{2}$ percent of the boiler full-load current for 10 seconds or exceeds an instantaneous value of 25 percent of the boiler full-load current.

(e) Grounded Neutral Conductor. The grounded neutral conductor shall:

(1) Be connected to the pressure vessel containing the electrodes.

(2) Be insulated for not less than 600 volts.

(3) Have not less than the ampacity of the largest ungrounded branch-circuit conductor.

(4) Be installed in the same raceway or cable tray with the ungrounded conductors.

(5) Not be used for any other circuit.

710-73. Pressure and Temperature Limit Control. Each boiler shall be equipped with a means to limit the maximum temperature and/or pressure by directly or indirectly interrupting all current flow through the electrodes. Such means shall be in addition to the temperature and/or pressure regulating systems and pressure relief or safety valves.

710-74. Grounding. All exposed noncurrent-carrying metal parts of the boiler and associated exposed grounded structures or equipment shall be bonded to the pressure vessel or to the neutral conductor to which the vessel is connected, in accordance with Section 250-79, except the ampacity of the bonding jumper shall be not less than the ampacity of the neutral conductor.

ARTICLE 720 — CIRCUITS AND EQUIPMENT OPERATING AT LESS THAN 50 VOLTS

720-1. Scope. This article covers installations operating at less than 50 volts, direct current or alternating current.

Exception: As covered in Articles 551, 650, 669, 690, 725, and 760.

720-2. Hazardous (Classified) Locations. Installations coming within the scope of this article and installed in hazardous (classified) locations shall also comply with the appropriate provisions of Articles 500 through 517.

720-4. Conductors. Conductors shall not be smaller than No. 12 copper or equivalent. Conductors for appliance branch circuits supplying more than one appliance or appliance receptacle shall not be smaller than No. 10 copper or equivalent.

720-5. Lampholders. Standard lampholders having a rating of not less than 660 watts shall be used.

720-6. Receptacle Rating. Receptacles shall have a rating of not less than 15 amperes.

720-7. Receptacles Required. Receptacles of not less than 20-ampere rating shall be provided in kitchens, laundries, and other locations where portable appliances are likely to be used.

720-8. Overcurrent Protection. Overcurrent protection shall comply with Article 240.

720-9. Batteries. Installations of storage batteries shall comply with Article 480.

720-10. Grounding. Grounding shall comply with Sections 250-5(a) and 250-45.

ARTICLE 725 — CLASS 1, CLASS 2, AND CLASS 3 REMOTE-CONTROL, SIGNALING, AND POWER-LIMITED CIRCUITS

A. General

725-1. Scope. This article covers remote-control, signaling, and power-limited circuits that are not an integral part of a device or appliance.

(FPN): The circuits described herein are characterized by usage and electrical power limitations that differentiate them from electric light and power circuits and, therefore, alternative requirements to those of Chapters 1 through 4 are given with regard to minimum wire sizes, derating factors, overcurrent protection, insulation requirements, and wiring methods and materials.

725-2. Locations and Other Articles. Circuits and equipment shall comply with (a), (b), (c), (d), and (e) below.

(a) **Spread of Fire or Products of Combustion.** Section 300-21.

(b) **Ducts, Plenums, and Other Air-Handling Spaces.** Section 300-22, where installed in ducts or plenums or other space used for environmental air.

Exception to (b): As permitted in Section 725-53(a).

(c) **Hazardous (Classified) Locations.** Articles 500 through 516, and Article 517, Part D, where installed in hazardous (classified) locations.

(d) **Cable Trays.** Article 318, where installed in cable tray.

(e) **Motor Control Circuits.** Article 430, Part F, where tapped from the load side of the motor branch-circuit protective device(s) as specified in Section 430-72(a).

725-3. Classifications. A remote-control, signaling, or power-limited circuit is the portion of the wiring system between the load side of the overcurrent device or the power-limited supply and all connected equipment, and shall be Class 1, Class 2, or Class 3 as defined in (a) and (b) below.

(a) Class 1 Circuits. Circuits that comply with Part B of this article and in which the voltage and power limitations are in accordance with Section 725-11.

(b) Class 2 and Class 3 Circuits. Circuits that comply with Part C of this article and in which the voltage and power limitations are in accordance with Section 725-31.

(FPN): Due to their power limitations, both Class 2 and 3 circuits consider safety from a fire initiation standpoint. In addition, Class 2 circuits provide acceptable protection from electric shock. However, since Class 3 circuits permit higher allowable levels of voltage and current, additional safeguards are specified to provide protection against the electric shock hazard that could be encountered.

725-4. Safety-Control Equipment. Remote-control circuits to safety-control equipment shall be classified as Class 1 if the failure of the equipment to operate introduces a direct fire or life hazard. Room thermostats, water temperature regulating devices, and similar controls used in conjunction with electrically controlled household heating and air conditioning shall not be considered safety-control equipment.

725-5. Communications Cables. Class 1 circuits shall not be run in the same cable with communications circuits. Class 2 and Class 3 circuit conductors shall be permitted in the same cable with communications circuits, in which case the Class 2 and Class 3 circuits shall be classified as communications circuits and shall meet the requirements of Article 800. The cables shall be listed as communications cables or multipurpose cables.

Exception: Cables constructed of individually listed Class 2, Class 3, and communications cables under a common jacket shall not be required to be classified as communications cables. The fire resistance rating of the composite cable shall be determined by the performance of the composite cable.

725-6. Access to Electrical Equipment Behind Panels Designed to Allow Access. Access to equipment shall not be denied by an accumulation of wires and cables that prevents removal of panels, including suspended ceiling panels.

B. Class 1 Circuits

725-11. Power Limitations for Class 1 Circuits.

(a) Class 1 Power-Limited Circuits. These circuits shall be supplied from a source having a rated output of not more than 30 volts and 1000 volt-amperes. Power sources other than transformers shall be protected by overcurrent devices rated at not more than 167 percent of the volt-ampere rating of the source divided by the rated voltage. The overcurrent devices shall not be interchangeable with overcurrent devices of higher ratings. The overcurrent device shall be permitted to be an integral part of the power supply.

(1) Transformers. Transformers used to supply power-limited Class 1 circuits shall comply with Article 450.

(2) Other Power Sources. To comply with the 1000 volt-ampere limitation of Section 725-11(a), the maximum output of power sources other than transformers shall be limited to 2500 volt-amperes, and the product of the maximum current and maximum voltage shall not exceed 10,000 volt-amperes. These ratings shall be determined with any overcurrent-protective device bypassed.

(FPN): For definitions of V_{max}, I_{max}, and VA_{max}, see Note 1, Tables 725-31(a) and (b).

(b) Class 1 Remote-Control and Signaling Circuits. Class 1 remote-control and signaling circuits shall not exceed 600 volts; however, the power output of the source shall not be required to be limited.

725-12. Overcurrent Protection. Conductors No. 14 and larger shall be protected against overcurrent in accordance with Section 310-15. Derating factors shall not be applied. Overcurrent protection shall not exceed 7 amperes for No. 18 conductors and 10 amperes for No. 16.

Exception No. 1: Where other articles of this Code permit or require other overcurrent protection.

(FPN): For example, see Section 430-72 for motors, Section 610-53 for cranes and hoists, and Sections 517-74(b) and 660-9 for X-ray equipment.

Exception No. 2: Transformer Secondary Conductors. Class 1 circuit conductors supplied by the secondary of a single-phase transformer having only a 2-wire (single-voltage) secondary shall be permitted to be protected by overcurrent protection provided on the primary (supply) side of the transformer, provided this protection is in accordance with Section 450-3 and does not exceed the value determined by multiplying the secondary conductor ampacity by the secondary-to-primary transformer voltage ratio. Transformer secondary conductors other than 2-wire shall not be considered to be protected by the primary overcurrent protection.

Exception No. 3: Class 1 circuit conductors No. 14 and larger that are tapped from the load side of the overcurrent-protective device(s) of a controlled light and power circuit shall require only short-circuit and ground-fault protection and shall be permitted to be protected by the branch-circuit overcurrent-protective device(s) where the rating of the protective device(s) is not more than 300 percent of the ampacity of the Class 1 circuit conductor.

725-13. Location of Overcurrent Devices. Overcurrent devices shall be located at the point where the conductor to be protected receives its supply.

Exception No. 1: Where the overcurrent device protecting the larger conductor also protects the smaller conductor.

Exception No. 2: Where overcurrent protection is provided in accordance with Section 725-12, Exception No. 2.

725-14. Wiring Method. Installations of Class 1 circuits shall be in accordance with the appropriate articles in Chapter 3.

Exception No. 1: As provided in Sections 725-15 through 725-17.

Exception No. 2: Where other articles of this Code permit or require other methods.

725-15. Conductors of Different Circuits in Same Cable, Enclosure, or Raceway. Class 1 circuits shall be permitted to occupy the same cable, enclosure, or raceway without regard to whether the individual circuits are alternating current or direct current, provided all conductors are insulated for the maximum voltage of any conductor in the cable, enclosure, or raceway. Power supply and Class 1 circuit conductors shall be permitted in the same cable, enclosure, or raceway only where the equipment powered is functionally associated.

Exception No. 1: Where installed in factory- or field-assembled control centers.

Exception No. 2: Underground conductors in a manhole where one of the following conditions is met:

a. The power supply or Class 1 circuit conductors are in a metal-enclosed cable, or Type UF cable;

b. The conductors are permanently separated from the power-supply conductors by a continuous firmly fixed nonconductor, such as flexible tubing, in addition to the insulation on the wire;

c. The conductors are permanently and effectively separated from the power supply conductors and securely fastened to racks, insulators, or other approved supports.

725-16. Conductors.

(a) Sizes and Use. Conductors of sizes No. 18 and 16 shall be permitted to be used, provided they supply loads that do not exceed the ampacities given in Section 402-5 and are installed in a raceway, an approved enclosure, or a listed cable. Conductors larger than No. 16 shall not supply loads greater than the ampacities given in Section 310-15. Flexible cords shall comply with Article 400.

(b) Insulation. Insulation on conductors shall be suitable for 600 volts. Conductors larger than No. 16 shall comply with Article 310.

Conductors in sizes No. 18 and 16 shall be Type FFH-2, KF-2, KFF-2, PAF, PAFF, PF, PFF, PGF, PGFF, PTF, PTFF, RFH-2, RFHH-2, RFHH-3, SF-2, SFF-2, TF, TFF, TFFN, TFN, ZF, or ZFF.

Conductors with other types and thicknesses of insulation shall be permitted if listed for Class 1 circuit use.

725-17. Number of Conductors in Cable Trays and Raceway, and Derating.

(a) Class 1 Circuit Conductors. Where only Class 1 circuit conductors are in a raceway, the number of conductors shall be determined in accordance with Section 300-17. The derating factors given in Article 310, Note 8(a) of Notes to Ampacity Tables of 0 to 2000 Volts shall apply only if such conductors carry continuous loads in excess of 10 percent of the ampacity of each conductor.

(b) Power-Supply Conductors and Class 1 Circuit Conductors. Where power-supply conductors and Class 1 circuit conductors are permitted in a raceway in accordance with Section 725-15, the number of conductors shall be determined in accordance with Section 300-17. The derating factors given in Article 310, Note 8(a) of Notes to Ampacity Tables of 0 to 2000 Volts shall apply as follows:

(1) To all conductors where the Class 1 circuit conductors carry continuous loads in excess of 10 percent of the ampacity of each conductor and where the total number of conductors is more than three.

(2) To the power-supply conductors only, where the Class 1 circuit conductors do not carry continuous loads in excess of 10 percent of the ampacity of each conductor and where the number of power-supply conductors is more than three.

(c) Class 1 Circuit Conductors in Cable Trays. Where Class 1 circuit conductors are installed in cable trays, they shall comply with the provisions of Sections 318-9 through 318-11.

725-18. Physical Protection.   Where damage to remote-control circuits of safety control equipment would introduce a hazard, as covered in Section 725-4, all conductors of such remote-control circuits shall be installed in rigid metal conduit, intermediate metal conduit, rigid nonmetallic conduit, electrical metallic tubing, Type MI cable, Type MC cable, or be otherwise suitably protected from physical damage.

725-19. Circuits Extending Beyond One Building.   Class 1 circuits that extend aerially beyond one building shall also meet the requirements of Article 225.

725-20. Grounding.   Class 1 circuits and equipment shall be grounded in accordance with Article 250.

C. Class 2 and Class 3 Circuits

725-31. Power Limitations of Class 2 and Class 3 Circuits.   As specified in Table 725-31(a) for ac circuits and Table 725-31(b) for dc circuits, the power for Class 2 and Class 3 circuits shall be either inherently limited requiring no overcurrent protection or limited by a combination of a power source and overcurrent protection.

Notes for Tables 725-31(a) and (b)

Note 1.

V_{max}: Maximum output voltage regardless of load with rated input applied.

I_{max}: Maximum output current under any noncapacitive load, including short circuit, and with overcurrent protection bypassed, if used. When a transformer limits the output current, I_{max} limits apply after one minute of operation. Where a current-limiting impedance, listed for the purpose, or as part of a listed product, is used in combination with a nonpower-limited transformer or a stored energy source, e.g., storage battery, to limit the output current, I_{max} limits apply after 5 seconds.

VA_{max}: Maximum volt-ampere output after one minute of operation regardless of load and overcurrent protection bypassed, if used.

Current-limiting impedance shall not be bypassed when determining I_{max} and VA_{max}.

Note 2. For nonsinusoidal ac, V_{max} shall be not greater than 42.4 volts peak. Where wet contact (immersion not included) is likely to occur, Class 3 wiring methods shall be used, or V_{max} shall be not greater than: 15 volts for sinusoidal ac: 21.2 volts peak for nonsinusoidal ac.

Note 3. If the power source is a transformer, $(VA)_{max}$ is 350 or less where V_{max} is 15 or less.

Note 4. A dry cell battery shall be considered an inherently limited power source, provided the voltage is 30 volts or less and the capacity is equal to or less than that available from series connected No. 6 carbon zinc cells.

Note 5: For dc interrupted at a rate of 10 to 200 Hz, V_{max} shall not be greater than 24.8 volts. Where wet contact (immersion not included) is likely to occur, Class 3 wiring methods shall be used, or V_{max} shall not be greater than: 30 volts for continuous dc; 12.4 volts for dc that is interrupted at a rate of 10 to 200 Hz.

Table 725-31(a). Power Source Limitations for Alternating Current Class 2 and Class 3 Circuits

Circuit	Inherently Limited Power Source (Overcurrent protection not required)				Not Inherently Limited Power Source (Overcurrent protection required)			
	Class 2			Class 3	Class 2			Class 3
Circuit Voltage V_{max} (Volts) (Note 1)	0 through 20†	Over 20 and through 30†	Over 30 and through 150	Over 30 and through 100	0 through 20†	Over 20 and through 30†	Over 30 and through 100	Over 100 and through 150
Power Limitations $(VA)_{max}$ (Volt-Amps) (Note 1)	—	—	—	—	250 (Note 3)	250	250	N.A.
Current Limitations I_{max} (Amps) (Note 1)	8.0	8.0	0.005	$150/V_{max}$	$1000/V_{max}$	$1000/V_{max}$	$1000/V_{max}$	1.0
Maximum Overcurrent Protection (Amps)	—	—	—	—	5.0	$100/V_{max}$	$100/V_{max}$	1.0
Power Source Maximum Nameplate Ratings — VA (Volt-Amps)	$5.0 \times V_{max}$	100	$0.005 \times V_{max}$	100	$5.0 \times V_{max}$	100	100	100
Power Source Maximum Nameplate Ratings — Current (Amps)	5.0	$100/V_{max}$	0.005	$100/V_{max}$	5.0	$100/V_{max}$	$100/V_{max}$	$100/V_{max}$
Supply Conductors and Cables	See Section 725-37							
Circuit Conductors and Cables	See Sections 725-49 through 725-53							

† Voltage ranges shown are for sinusoidal ac in indoor locations or where wet contact is not likely to occur. For nonsinusoidal or wet contact conditions, see Note 2.

Table 725-31(b).   Power Source Limitations for Direct Current Class 2 and Class 3 Circuits

Circuit	Inherently Limited Power Source (Note 4) (Overcurrent protection not required)					Not Inherently Limited Power Source (Overcurrent protection required)			
	Class 2				Class 3	Class 2			Class 3
Circuit Voltage V_{max} (Volts) (Note 1)	0 through 20††	Over 20 and through 30††	Over 30 and through 60††	Over 60 and through 150	Over 60 and through 100	0 through 20††	Over 20 and through 60††	Over 60 and through 100	Over 100 and through 150
Power Limitations $(VA)_{max}$ (Volts-Amps) (Note 1)	—	—	—	—	—	250 (Note 3)	250	250	N.A.
Current Limitations I_{max} (Amps) (Note 1)	8.0	8.0	$150/V_{max}$	0.005	$150/V_{max}$	$1000/V_{max}$	$1000/V_{max}$	$1000/V_{max}$	1.0
Maximum Overcurrent Protection (Amps)	—	—	—	—	—	5.0	$100/V_{max}$	100	1.0
Power Source Maximum Nameplate Ratings — VA (Volt-Amps)	$5.0 \times V_{max}$	100	100	$0.005 \times V_{max}$	100	$5.0 \times V_{max}$	100	100	100
Power Source Maximum Nameplate Ratings — Current (Amps)	5.0	$100/V_{max}$	$100/V_{max}$	0.005	$100/V_{max}$	5.0	$100/V_{max}$	$100/V_{max}$	$100/V_{max}$
Supply Conductors and Cables	See Section 725-37								
Circuit Conductors and Cables	See Sections 725-49 through 725-53								

†† Voltage ranges shown are for continuous dc in indoor locations or where wet contact is not likely to occur. For interrupted dc or wet contact conditions, see Note 5.

725-32. Interconnection of Power Supplies. Class 2 or Class 3 power supplies shall not be paralleled or otherwise interconnected unless listed for such interconnection.

725-34. Marking. A Class 2 or Class 3 power supply unit shall be durably marked where plainly visible to indicate the class of supply and its electrical rating.

725-35. Overcurrent Protection. Where overcurrent protection is required, the overcurrent-protective devices shall not be interchangeable with devices of higher ratings. The overcurrent device shall be permitted as an integral part of the power supply.

725-36. Location of Overcurrent Devices. Overcurrent devices, where required, shall be located at the point where the conductor to be protected receives its supply.

725-37. Wiring Methods on Supply Side. Conductors and equipment on the supply side of overcurrent protection, transformers, or current-limiting devices shall be installed in accordance with the appropriate requirements of Chapter 3. Transformers or other devices supplied from electric light or power circuits shall be protected by an overcurrent device rated not over 20 amperes.

Exception: The input leads of a transformer or other power source supplying Class 2 and Class 3 circuits shall be permitted to be smaller than No. 14, but not smaller than No. 18 if they are not over 12 inches (305 mm) long and if they have insulation that complies with Section 725-16(b).

725-38. Wiring Methods and Materials on Load Side. Conductors on the load side of overcurrent protection, transformers, and current-limiting devices shall be insulated at not less than the requirements of Section 725-50 and shall be installed in accordance with Section 725-52.

725-42. Circuit Conductors Extending Beyond One Building. Where Class 2 or Class 3 circuit conductors that extend beyond one building and are so run as to be subject to accidental contact with electric light or power conductors operating at over 300 volts to ground, or are exposed to lightning on interbuilding circuits on the same premises, the requirements of the following shall also apply:

(1) Sections 800-10, 800-12, 800-13, 800-30, 800-31, 800-32, 800-33, and 800-40 for other than coaxial conductors, and

(2) Sections 820-10, 820-33, and 820-40 for coaxial conductors.

725-43. Grounding. Class 2 and Class 3 circuits and equipment shall be grounded in accordance with Article 250.

725-49. Fire Resistance of Cables Within Buildings. Single- and multi-conductor cables of Class 2 and Class 3 circuits, including PLTC cables, installed as wiring within buildings shall be listed as resistant to the spread of fire in accordance with Sections 725-50 and 725-51.

725-50. Listing, Marking, and Installation of Class 2, Class 3, and PLTC | Cables.

(a) Class 2 and Class 3 Cables. Class 2 and Class 3 cables installed as | wiring within buildings shall be listed as being suitable for the purpose, marked in accordance with Table 725-50, and installed in accordance with Section 725-38. The cable voltage ratings shall not be marked on the cable.

(FPN): Voltage markings on cables may be misinterpreted to suggest that the cables may be suitable for Class 1, electric light and power applications.

Exception No. 1: Voltage markings shall be permitted where the cable has multiple listings and voltage marking is required for one or more of the listings.

Exception No. 2: Cable substitutions as provided for in Section 725-53(g) shall be considered as suitable for the purpose and shall be permitted.

(b) PLTC. PLTC shall be marked in accordance with Section 310-11 and Table 725-50 and shall be listed as being suitable for use in cable trays.

Table 725-50.   Cable Markings

Cable Marking	Type	Reference
CL3P	Class 3 Plenum Cable	725-51(a) and 725-53(a)
CL2P	Class 2 Plenum Cable	725-51(a) and 725-53(a)
CL3R	Class 3 Riser Cable	725-51(b) and 725-53(b)
CL2R	Class 2 Riser Cable	725-51(b) and 725-53(b)
PLTC	Power-Limited Tray Cable	725-51(e) and 725-53(c) and (d)
CL3	Class 3 Cable	725-51(c), 725-53(b), Exception No. 2, and 725-53(e)
CL2	Class 2 Cable	725-51(c), 725-53(b), Exception No. 2, and 725-53(e)
CL3X	Class 3 Cable, Limited Use	725-51(d), 725-53(b), Exception No. 2, and 725-53(e), Exception Nos. 1, 2, and 3
CL2X	Class 2 Cable, Limited Use	725-51(d), 725-53(b), Exception No. 2, and 725-53(e), Exception Nos. 1, 2, and 3

(FPN No. 1):   Class 2 and Class 3 Cable types are listed in descending order of fire resistance rating, and Class 3 cables are listed above Class 2 cables, since Class 3 cables may substitute for Class 2 cables.

(FPN No. 2):   See the referenced sections for permitted uses.

725-51. Additional Listing Requirements.   Class 2 and Class 3 cables shall be listed in accordance with (a) through (f) below, and, where single conductors are used for Class 3 circuits, with (g) below:

(a) Types CL2P and CL3P.   Types CL2P and CL3P plenum cables shall be listed as being suitable for use in ducts, plenums, and other space used for environmental air and shall also be listed as having adequate fire-resistant and low-smoke-producing characteristics.

(FPN):   One method of defining low-smoke-producing cable is by establishing an acceptable value of the smoke produced when tested in accordance with Test for Fire and Smoke Characteristics of Wires and Cables, NFPA 262-1990 (ANSI) to a maximum peak optical density of 0.5 and a maxiumum average optical density of 0.15. Similarly, one method of defining fire-resistant cables is by establishing maximum allowable flame travel distance of 5 feet (1.52 m) when tested in accordance with the same test.

(b) Types CL2R and CL3R.   Types CL2R and CL3R riser cables shall be listed as being suitable for use in a vertical run in a shaft or from floor to floor and shall also be listed as having fire-resistant characteristics capable of preventing the carrying of fire from floor to floor.

(FPN):   One method of defining fire-resistant characteristics capable of preventing the carrying of fire from floor to floor is that the cables pass the requirements of Test for Flame Propagation Height of Electrical and Optical-Fiber Cable Installed Vertically in Shafts, ANSI/UL 1666-1986.

(c) Types CL2 and CL3.   Types CL2 and CL3 cables shall be listed as being suitable for general-purpose use, with the exception of risers, ducts, plenums, and other space used for environmental air, and shall also be listed as being resistant to the spread of fire.

(FPN):   One method of defining resistant to spread of fire is that the cables do not spread fire to the top of the tray in the "Vertical Tray Flame Test" in Reference Standard for Electrical Wires, Cables and Flexible Cords, ANSI/UL 1581-1985. Another method of defining resistant to the spread of fire is for the damage (char length) not to exceed 4 feet 11 inches (1.5 m) when performing the CSA "Vertical Flame Test — Cables in Cable Trays," as described in Test Methods for Electrical Wires and Cables, CSA C22.2 No. 0.3-M-1985.

(d) Types CL2X and CL3X.   Types CL2X and CL3X limited use cables shall be listed as being suitable for use in dwellings and for use in raceway and shall also be listed as being flame-retardant.

(FPN):   One method of determining that cable is flame-retardant is by testing the cable to the "VW-1 (Vertical-Wire) Flame Test" in the Reference Standard for Electrical Wires, Cables and Flexible Cords, ANSI/UL 1581-1985.

(e) Type PLTC.   Type PLTC nonmetallic-sheathed, power-limited tray cable shall be a factory assembly of two or more insulated conductors under a nonmetallic jacket. The insulated conductors shall be Nos. 22 through 12. The conductor material shall be copper (solid or stranded).

Insulation on conductors shall be suitable for 300 volts. The cable core shall be either (1) two or more parallel conductors; (2) one or more group assemblies of twisted or parallel conductors; or (3) a combination thereof. A metallic shield or a metallized foil shield with drain wire(s) shall be permitted to be applied either over the cable core, over groups of conductors, or both. The cable shall be listed as being resistant to the spread of fire. The outer jacket shall be a sunlight- and moisture-resistant nonmetallic material.

Exception: Where a smooth metallic sheath, welded and corrugated metallic sheath, or interlocking tape armor is applied over the nonmetallic jacket, an overall nonmetallic jacket shall not be required. On metallic-sheathed cable without an overall nonmetallic jacket, the information required in Section 310-11 shall be located on the nonmetallic jacket under the sheath.

(FPN):   One method of defining resistant to spread of fire is that the cables do not spread fire to the top of the tray in the "Vertical Tray Flame Test" in the Reference Standard for Electrical Wires, Cables and Flexible Cords, ANSI/UL 1581-1985.

Another method of defining resistant to the spread of fire is for the damage (char length) not to exceed 4 feet, 11 inches (1.5 m) when performing the CSA "Vertical Flame Test — Cables in Cable Trays," as described in Test Methods for Electrical Wires and Cables, CSA C22.2 No. 0.3-M-1985.

(f) Voltage Rating. Class 3 cables shall have a voltage rating of not less than 300 volts.

(g) Single Conductors. Class 3 single conductors shall not be smaller than No. 18 and shall be insulated in accordance with Section 725-16(b).

725-52. Installation of Conductors and Equipment.   Conductors and equipment on the load side of overcurrent protection, transformers, and current-limiting devices shall comply with (a) and (b) below.

(a) Separation from Electric Light, Power, Class 1, and Nonpower-Limited Fire Protective Signaling Circuit Conductors.

(1) Open Conductors. Conductors of Class 2 and Class 3 circuits shall be separated by at least 2 inches (50.8 mm) from conductors of any electric light, power, Class 1, or nonpower-limited fire protective signaling circuits.

Exception No. 1: Where either (1) all of the electric light, power, Class 1, and nonpower-limited fire protective signaling circuit conductors, or (2) all of the Class 2 and Class 3 circuit conductors are in raceway or in metal-sheathed, metal-clad, nonmetallic-sheathed, or Type UF cables.

Exception No. 2: Where the conductors are permanently separated from the conductors of the other circuits by a continuous and firmly fixed nonconductor, such as porcelain tubes or flexible tubing, in addition to the insulation on the wire.

(2) In Cable, Cable Trays, Enclosures, and Raceways. Conductors of Class 2 and Class 3 circuits shall not be placed in any cable, cable tray, compartment, enclosure, outlet box, raceway, or similar fitting with conductors of electric light, power, Class 1, and nonpower-limited fire protective signaling circuits.

Exception No. 1: Where the conductors of the different circuits are separated by a barrier. In enclosures, Class 2 or Class 3 circuits shall be permitted to be installed in a raceway within the enclosure to separate them from Class 1, electric light and power circuits.

Exception No. 2: Conductors in compartments, enclosures, outlet boxes, or similar fittings, where electric light, power, Class 1, or nonpower-limited fire protective signaling circuit conductors are introduced solely to connect to the equipment connected to Class 2 or Class 3 circuits to which the other conductors in the enclosure are connected. The electric light, power, Class 1, and nonpower-limited fire protective signaling circuit conductors shall be routed within the enclosure to maintain a minimum 0.25-inch (6.35-mm) separation from the conductors of Class 2 and Class 3 circuits.

Exception No. 3: Underground conductors in a manhole where one of the following conditions is met:

   a. The electric light, power, Class 1, and nonpower-limited fire protective signaling circuit conductors are in a metal-enclosed cable or Type UF cable;

   b. The conductors are permanently and effectively separated from the conductors of the other circuits by a continuous and firmly fixed nonconductor, such as flexible tubing, in addition to the insulation or covering on the wire;

   c. The conductors are permanently and effectively separated from conductors of the other circuits and securely fastened to racks, insulators, or other approved supports.

Exception No. 4: As permitted by Section 780-6(a) and installed in accordance with Article 780.

(3) In Hoistways. Class 2 or Class 3 circuit conductors shall be installed in rigid metal conduit, rigid nonmetallic conduit, intermediate metal conduit, or electrical metallic tubing in hoistways.

Exception: As provided for in Section 620-21, Exception Nos. 1 and 2 for elevators and similar equipment.

(4) In Shafts. Class 2 or Class 3 circuit conductors shall be separated by not less than 2 inches (50.8 mm) from electric light, power, Class 1, or nonpower-limited fire protective signaling circuit conductors run in the same shaft.

(b) Conductors of Different Circuits in Same Cable, Enclosure, or Raceway.

(1) Two or More Class 2 Circuits. Conductors of two or more Class 2 circuits shall be permitted within the same cable, enclosure, or raceway, provided all conductors in the cable, enclosure, or raceway are insulated for the maximum voltage of any conductor.

(2) Two or More Class 3 Circuits. Conductors of two or more Class 3 circuits shall be permitted within the same cable, enclosure, or raceway.

(3) Class 2 Circuits with Class 3 Circuits. Conductors of one or more Class 2 circuits shall be permitted within the same cable, enclosure, or raceway with conductors of Class 3 circuits, provided that the insulation of the Class 2 circuit conductors in the cable, enclosure, or raceway is at least that required for Class 3 circuits.

(4) Class 2 or Class 3 Circuits with Other Circuits. Jacketed cables of Class 2 or Class 3 circuits shall be permitted in the same enclosure or raceway with jacketed cables of any of the following:

a. Power-limited fire protective signaling systems in compliance with Article 760.

b. Nonconductive and conductive optical fiber cables in compliance with Article 770.

c. Communications circuits in compliance with Article 800.

d. Community antenna television and radio distribution systems in compliance with Article 820.

(c) Support of Conductors. Raceways shall not be used as a means of support for Class 2 or Class 3 circuit conductors.

Exception: Except as permitted by Section 300-11(b), Exception No. 2.

725-53. Applications of Listed Class 2, Class 3, and PLTC Cables. Class 2, Class 3, and PLTC cables shall comply with (a) through (g) below.

(a) Plenum. Cables installed in ducts, plenums, and other spaces used for environmental air shall be Type CL2P or CL3P.

Exception: Listed wires and cables installed in compliance with Section 300-22.

(b) Riser. Cables installed in vertical runs and penetrating more than one floor, or cables installed in vertical runs in a shaft, shall be Type CL2R or CL3R. Floor penetrations requiring Type CL2R or CL3R shall contain only cables suitable for riser or plenum use.

Exception No. 1: Other cables as covered in Table 725-53 and other listed wiring methods as covered in Chapter 3, where these are installed in metal raceways or located in a fireproof shaft having firestops at each floor.

Exception No. 2: Types CL2, CL3, CL2X, and CL3X cable in one- and two-family dwellings.

(FPN): See Section 300-21 for firestop requirements for floor penetrations.

(c) Cable Trays. Cables installed in cable trays shall be Type PLTC.

Exception: Conductors in PLTC cables used for Class 2 thermocouple circuits shall be permitted to be any of the materials used for thermocouple extension wire.

(d) Hazardous (Classified) Locations. Cables installed in hazardous (classified) locations shall be Type PLTC. Where the use of PLTC cable is permitted in Section 501-4(b), the cable shall be installed in cable trays, in raceways, supported by messenger wire, or directly buried where the cable is listed for this use.

Exception No. 1: For Class 2 circuits as permitted by Section 501-4(b), Exception.

Exception No. 2: Conductors in PLTC cables used for Class 2 thermocouple circuits shall be permitted to be any of the materials used for thermocouple extension wire.

(e) Other Wiring Within Buildings. Cables installed in building locations other than the locations covered in (a) through (d) above shall be Type CL2 or CL3.

Exception No. 1: Type CL2X or CL3X where installed in a raceway, or other wiring methods as covered in Chapter 3.

Exception No. 2: In nonconcealed spaces where the exposed length of cable does not exceed 10 feet (3.05 m).

Exception No. 3: Listed Type CL2X, Class 2 cables less than 0.25 inch (6.4 mm) in diameter and listed Type CL3X, Class 3 cables less than 0.25 inch (6.4 mm) in diameter installed in one- or two-family or multifamily dwellings.

(f) Cross-Connect Arrays. Type CL2 or CL3 wire or cable shall be used.

(g) Cable Substitutions. The substitutions for Class 2 and Class 3 cables listed in Table 725-53 and illustrated in Figure 725-53 shall be permitted.

Table 725-53.   Cable Substitutions

Cable Type	Permitted Substitutions
CL3P	MPP, CMP, FPLP
CL2P	MPP, CMP, FPLP, CL3P
CL3R	MPP, CMP, FPLP, CL3P, MPR, CMR, FPLR
CL2R	MPP, CMP, FPLP, CL3P, CL2P, MPR, CMR, FPLR, CL3R
CL3	MPP, CMP, FPLP, CL3P, MPR, CMR, FPLR, CL3R, MPG, MP, CMG, CM, FPL, PLTC
CL2	MPP, CMP, FPLP, CL3P, CL2P, MPR, CMR, FPLR, CL3R, CL2R MPG, MP, CMG, CM, FPL, PLTC, CL3
CL3X	MPP, CMP, FPLP, CL3P, MPR, CMR, FPLR, CL3R, MPG, MP, CMG, CM, FPL, PLTC, CL3, CMX
CL2X	MPP, CMP, FPLP, CL3P, CL2P, MPR, CMR, FPLR, CL3R, CL2R, MPG, MP, CMG, CM, FPL, PLTC, CL3, CL2, CMX, CL3X

(FPN): For information on FPLP, FPLR, and FPL cables, see Section 760-50. For information on MPP, MPR, MPG, MP, CMP, CMR, CMG, and CM cables, see Section 800-50.

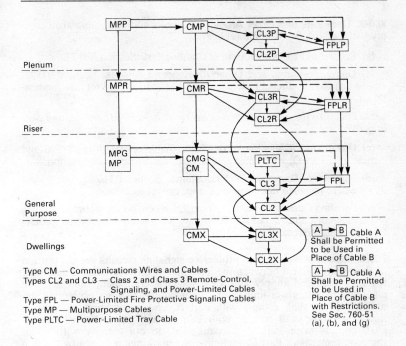

Plenum

Riser

General Purpose

Dwellings

Type CM — Communications Wires and Cables
Types CL2 and CL3 — Class 2 and Class 3 Remote-Control,
                    Signaling, and Power-Limited Cables
Type FPL — Power-Limited Fire Protective Signaling Cables
Type MP — Multipurpose Cables
Type PLTC — Power-Limited Tray Cable

A → B Cable A
Shall be Permitted
to be Used in
Place of Cable B

A → B Cable A
Shall be Permitted
to be Used in
Place of Cable B
with Restrictions.
See Sec. 760-51
(a), (b), and (g)

Figure 725-53. Cable Substitution Hierarchy

ARTICLE 760 — FIRE PROTECTIVE SIGNALING SYSTEMS

A. General

760-1. Scope. This article covers the installation of wiring and equipment of fire protective signaling systems operating at 600 volts, nominal, or less.

(FPN No. 1): Fire protective signaling systems include fire alarm, guard tour, sprinkler waterflow, and sprinkler supervisory systems. For further information on the installation and supervision requirements for fire protective signaling systems, refer to the following:

NFPA 71-1989 - Installation, Maintenance, and Use of Signaling Systems for Central Station Service.

NFPA 72-1990 - Installation, Maintenance, and Use of Protective Signaling Systems.

NFPA 72E-1990 - Automatic Fire Detectors.

NFPA 74-1989 - Installation, Maintenance, and Use of Household Fire Warning Equipment.

(FPN No. 2): Class 1, 2, and 3 circuits are defined in Article 725.

760-2. Locations and Other Articles. Circuits and equipment shall comply with (a), (b), (c), (d), and (e) below.

(a) Spread of Fire or Products of Combustion. Section 300-21.

(b) Ducts, Plenums, and Other Air-Handling Spaces. Section 300-22, where installed in ducts or plenums or other spaces used for environmental air.

Exception to (b): As permitted in Sections 760-17(e)(1) and (2) and Section 760-53(a).

(c) Hazardous (Classified) Locations. Articles 500 through 516 and Article 517, Part D, where installed in hazardous (classified) locations.

(d) Corrosive, Damp, or Wet Locations. Sections 110-11, 300-6, and 310-9 where installed in corrosive, damp, or wet locations.

(e) Building Control Circuits. Article 725 where building control circuits (e.g., elevator capture, fan shutdown) are associated with the fire protective signaling system.

760-3. Classifications. Fire protective signaling circuits are the portion of the wiring system between the load side of the overcurrent device or the power-limited supply and all connected equipment, and shall be classified as nonpower-limited or power-limited. All fire protective signaling circuits shall comply with Part A, and, in addition, nonpower-limited circuits shall comply with Part B, and power-limited circuits shall comply with Part C. Circuits that do not comply with all of the requirements of Part C, including the markings required by Section 760-22, are classified as nonpower-limited circuits and shall comply with all of the requirements of Part B.

760-4. Identification. Fire protective signaling circuits shall be identified at terminal and junction locations, in a manner that will prevent unintentional interference with the signaling circuit during testing and servicing.

760-5. Circuits Extending Beyond One Building. Fire protective signaling circuits that extend aerially beyond one building shall either meet the requirements of Article 800 and be classified as communications circuits, or shall meet the requirements of Article 225.

760-6. Grounding. Fire protective signaling circuits and equipment shall be grounded in accordance with Article 250.

Exception: DC power-limited fire protective signaling circuits having a maximum current of 0.030 amperes.

760-7. Access to Electrical Equipment Behind Panels Designed to Allow Access. Access to equipment shall not be denied by an accumulation of wires and cables that prevents removal of panels, including suspended ceiling panels.

B. Nonpower-Limited Fire Protective
Signaling Circuits

760-11. Power Limitations. The power supply of nonpower-limited fire protective signaling circuits shall comply with Chapters 1 through 4, and the output voltage shall not be more than 600 volts, nominal.

760-12. Overcurrent Protection. Conductors No. 14 and larger shall be protected against overcurrent in accordance with the values specified in Section 310-15, as applicable. Derating factors shall not be applied. Overcurrent protection shall not exceed 7 amperes for No. 18 conductors and 10 amperes for No. 16.

Exception: Where other articles of this Code permit or require other overcurrent protection.

760-13. Location of Overcurrent Devices. Overcurrent devices shall be located at the point where the conductor to be protected receives its supply.

Exception No. 1: Where the overcurrent device protecting the larger conductor also protects the smaller conductor.

Exception No. 2: Transformer secondary conductors. Nonpower-limited fire protective signaling circuit conductors supplied by the secondary of a single-phase transformer having only a 2-wire (single voltage) secondary shall be permitted to be protected by overcurrent protection provided by the primary (supply) side of the transformer, provided the protection is in accordance with Section 450-3 and does not exceed the value determined by multiplying the secondary conductor ampacity by the secondary-to-primary transformer voltage ratio. Transformer secondary conductors other than 2-wire shall not be considered to be protected by the primary overcurrent protection.

760-14. Wiring Method. Wiring installation shall be in accordance with the appropriate articles in Chapter 3.

Exception No. 1: As provided in Sections 760-15 through 760-18.

Exception No. 2: Where other articles of this Code require other methods.

760-15. Conductors of Different Circuits in Same Cable, Enclosure, or Raceway. Class 1 and nonpower-limited fire protective signaling circuits shall be permitted to occupy the same cable, enclosure, or raceway without regard to whether the individual circuits are alternating current or direct current, provided all conductors are insulated for the maximum voltage of any conductor in the enclosure or raceway. Power-supply and fire protective signaling circuit conductors shall be permitted in the same cable, enclosure, or raceway only where connected to the same equipment.

760-16. Copper Conductors.

(a) Types, Sizes, and Use. Only copper conductors shall be permitted to be used for fire protective signaling systems. Conductors of Nos. 18 and 16 shall be permitted to be used, provided they supply loads that do not exceed the ampacities given in Table 402-5 and are installed in a raceway or a listed cable. Conductors larger than No. 16 shall not supply loads greater than the ampacities given in Section 310-15, as applicable.

(b) Insulation. Insulation on conductors shall be suitable for 600 volts. Conductors larger than No. 16 shall comply with Article 310. Conductors in sizes No. 18 and 16 shall be Type KF-2, PF, PGF, RFH-2, RFHH-2, RFHH-3, SF-2, TF, TFN, or ZF. Conductors with other types and thickness of insulation shall be permitted if listed for nonpower-limited fire protective signaling circuit use.

(FPN):   For application provisions, see Table 402-3.

(c) Conductor Materials. Conductors shall be solid or bunch-tinned (bonded) stranded copper.

Exception No. 1: Stranded copper with a maximum of 7 strands for Nos. 16 and 18 shall be permitted.

Exception No. 2: Stranded copper with a maximum of 19 strands for Nos. 14 and larger shall be permitted.

Exception to (b) and (c): Wire Types PAF and PTF shall be permitted only for high-temperature applications between 90°C (194°F) and 250°C (482°F).

760-17. Multiconductor Nonpower-Limited Fire Protective Signaling Circuit Cables for Circuits Operating at 150 Volts or Less.   Multiconductor nonpower-limited fire protective signaling circuit cables that meet the requirements of (a) through (e) below shall be permitted to be used on fire protective signaling circuits operating at 150 volts or less.

(a) Fire Resistance of Cables Within Buildings. Multiconductor nonpower-limited fire protective signaling circuit cables installed as wiring within buildings shall be listed as being resistant to the spread of fire in accordance with sections (b) and (c) below.

(b) Listing and Markings. Multiconductor nonpower-limited fire protective signaling circuit cables in a building shall be listed as being suitable for the purpose, and cables shall be marked in accordance with Table 760-17(b).

Table 760-17(b).   Cable Markings

Cable Marking	Type	Reference
NPLFP	nonpower-limited fire protective signaling circuit plenum cable	Sections 760-17(c)(4) and 760-17(e)(2)
NPLFR	nonpower-limited fire protective signaling circuit riser cable	Sections 760-17(c)(5) and 760-17(e)(3)
NPLF	nonpower-limited fire protective signaling circuit cable	Sections 760-17(c)(6) and 760-17(e)(4)

(FPN No. 1):   See the referenced sections for listing requirements and permitted uses.

(FPN No. 2):   Cable types are listed in descending order of fire resistance rating.

(c) Listing Requirements. Nonpower-limited fire protective signaling circuit cables shall be listed in accordance with (1) through (3) below and, depending on the type, (4) through (6) below.

(1) Conductor Materials. Conductors shall be solid copper or bunch-tinned (bonded) stranded copper.

Exception No. 1: Stranded copper with a maximum of 7 strands for No. 16 and No. 18 shall be permitted.

Exception No. 2: Stranded copper with a maximum of 19 strands for Nos. 14 and larger shall be permitted.

(2) Sizes and Number. The cables shall have two or more No. 18 or larger conductors.

(3) Ratings. Each insulated conductor of the cable shall have a voltage rating of not less than 300 volts. The combination of the insulated conductors and the cable jacket shall have a voltage rating of not less than 600 volts.

(4) Type NPLFP. Type NPLFP nonpower-limited fire protective signaling plenum cable shall be listed as being suitable for use in other space used for environmental air as described in Section 300-22(c) and shall also be listed as having adequate fire-resistant and low-smoke-producing characteristics.

(FPN): One method of defining low-smoke-producing cable is by establishing an acceptable value of the smoke produced when tested in accordance with the Test for Fire and Smoke Characteristics of Wires and Cables, NFPA 262-1990 (ANSI) to a maximum peak optical density of 0.5 and a maximum average optical density of 0.15. Similarly, one method of defining fire-resistant cables is by establishing maximum allowable flame travel distance of 5 feet (1.52 m) when tested in accordance with the same test.

(5) Type NPLFR. Type NPLFR nonpower-limited fire protective signaling riser cable shall be listed as being suitable for use in a vertical run in a shaft or from floor to floor and shall also be listed as having fire-resistant characteristics capable of preventing the carrying of fire from floor to floor.

(FPN): One method of defining fire-resistant characteristics capable of preventing the carrying of fire from floor to floor is that the cables pass the Test for Flame Propagation Height of Electrical and Optical-Fiber Cables Installed Vertically in Shafts, ANSI/UL 1666-1986.

(6) Type NPLF. Type NPLF nonpower-limited fire protection signaling cable shall be listed as being suitable for general-purpose fire alarm use, with the exception of risers, ducts, plenums, and other space used for environmental air, and shall also be listed as being resistant to the spread of fire.

(FPN): One method of defining resistant to spread of fire is that the cables do not spread fire to the top of the tray in the "Vertical-Tray Flame Test" in the Reference Standard for Electrical Wires, Cables and Flexible Cords, ANSI/UL 1581-1985.
Another method of defining resistant to the spread of fire is for the damage (char length) not to exceed 4 feet 11 inches (1.5 m) when performing the CSA "Vertical Flame Test — Cables in Cable Trays," as described in Test Methods for Electrical Wires and Cables, CSA C22.2 No. 0.3-M-1985.

(d) Wiring Method. Multiconductor nonpower-limited fire protective signaling circuit cables described in Section 760-17(a) shall be installed as follows:

(1) In raceway or exposed on surface of ceiling and sidewalls or "fished" in concealed spaces. Where installed exposed, cables shall be adequately supported and terminated in approved fittings and installed in such a way that maximum protection against physical damage is afforded by building construction such as baseboards, door frames, ledges, etc. Where located within 7 feet (2.13 m) of the floor, cables shall be securely fastened in an approved manner at intervals of not more than 18 inches (457 mm).

(2) In metal raceway or rigid nonmetallic conduit where passing through a floor or wall to a height of 7 feet (2.13 m) above the floor unless adequate protection can be afforded by building construction such as detailed in (1) above, or unless an equivalent solid guard is provided.

(3) In rigid metal conduit, rigid nonmetallic conduit, intermediate metal conduit, or electrical metallic tubing where installed in hoistways.

Exception: As provided for in Section 620-21, Exception Nos. 1 and 2 for elevators and similar equipment.

(e) Applications of Listed Nonpower-Limited Fire Protective Signaling Circuit Cables. The use of nonpower-limited fire protective signaling circuit cables shall comply with (1) through (4) below.

(1) Ducts and Plenums. Multiconductor nonpower-limited fire protective signaling circuit cables shall not be installed exposed in ducts or plenums. See Section 300-22(b).

(2) Other Spaces Used for Environmental Air. Cables installed in other spaces used for environmental air shall be Type NPLFP.

Exception No. 1: Types NPLF and NPLFR cables installed in compliance with Section 300-22(c).

Exception No. 2: Other wiring methods in accordance with Section 300-22(c) and conductors in compliance with Section 760-16(c).

(3) Riser. Cables installed in vertical runs and penetrating more than one floor or cables installed in vertical runs in a shaft shall be Type NPLFR. Floor penetrations requiring Type NPLFR shall contain only cables suitable for riser or plenum use.

Exception No. 1: Type NPLF or other cables specified in Chapter 3 that are in compliance with Section 760-16(c) and encased in metal raceway.

Exception No. 2: Type NPLF cables located in a fireproof shaft having firestops at each floor.

(FPN): See Section 300-21 for firestop requirements for floor penetrations.

(4) Other Wiring Within Buildings. Cables installed in building locations other than the locations covered in (1), (2), and (3) above shall be Type NPLF.

Exception No. 1: Chapter 3 wiring methods with conductors in compliance with Section 760-16(c).

Exception No. 2: Type NPLFP or Type NPLFR cables shall be permitted.

760-18. Number of Conductors in Cable Trays and Raceways, and Derating.

(a) Nonpower-Limited Fire Protective Signaling Circuits and Class 1 Circuits. Where only nonpower-limited fire protective signaling circuit and Class 1 circuit conductors are in a raceway, the number of conductors shall be determined in accordance with Section 300-17. The derating factors given in Article 310, Note 8(a) of Notes to Ampacity Tables of 0 to 2000 Volts shall apply if such conductors carry continuous load in excess of 10 percent of the ampacity of each conductor.

(b) Power-Supply Conductors and Fire Protective Signaling Circuit Conductors. Where power-supply conductors and fire protective signaling circuit conductors are permitted in a raceway in accordance with Section 760-15, the number of conductors shall be determined in accordance with Section 300-17. The derating factors given in Article 310, Note 8(a) of Notes to Ampacity Tables of 0 to 2000 Volts shall apply as follows:

(1) To all conductors where the fire protective signaling circuit conductors carry continuous loads in excess of 10 percent of the ampacity of each conductor and where the total number of conductors is more than three.

(2) To the power-supply conductors only, where the fire protective signaling circuit conductors do not carry continuous loads in excess of 10 percent of the ampacity of each conductor and where the number of power-supply conductors is more than three.

(c) Cable Trays. Where fire protective signaling circuit conductors are installed in cable trays, they shall comply with Sections 318-9 through 318-11.

C. Power-Limited Fire Protective Signaling Circuits

760-21. Power Limitations. As specified in Table 760-21(a) for ac circuits and Table 760-21(b) for dc circuits, the power for power-limited fire protective signaling circuits shall be either inherently limited requiring no overcurrent protection or limited by a combination of a power source and overcurrent protection.

760-22. Circuit Marking. The equipment shall be durably marked where plainly visible to indicate each circuit that is a power-limited fire protective signaling circuit.

(FPN):   See Section 760-28(a), Exception No. 3 where a power-limited circuit is to be reclassified as a nonpower-limited circuit.

760-23. Overcurrent Protection. Where overcurrent protection is required, the overcurrent protective devices shall not be interchangeable with devices of higher ratings. The overcurrent device shall be permitted as an integral part of the power supply.

760-24. Location of Overcurrent Device. Overcurrent device, where required, shall be located at the point where the conductor to be protected receives its supply.

Table 760-21(a). Power Source Limitations for Alternating-Current Fire Protective Signaling Circuits

Power Source Maximum Nameplate Ratings	Inherently Limited Power Source (Overcurrent protection not required)			Not Inherently Limited Power Source (Overcurrent protection required)		
	0 through 20	Over 20 and through 30	Over 30 and through 100	0 through 20	Over 20 and through 100	Over 100 and through 150
Circuit Voltage V_{max} (Volts) (Note 1)	—	—	—	250 (Note 2)	250	N.A.
Power Limitations $(VA)_{max}$ (Volt-Amps) (Note 1)	—	—	—	$1000/V_{max}$	$1000/V_{max}$	1.0
Current Limitations I_{max} (Amps) (Note 1)	8.0	8.0	$150/V_{max}$	—	—	1.0
Maximum Over-current Protection (Amps)	—	—	—	5.0	$100/V_{max}$	1.0
VA (Volt-Amps)	$5.0 \times V_{max}$	100	100	$5.0 \times V_{max}$	100	100
Current (Amps)	5.0	$100/V_{max}$	$100/V_{max}$	5.0	$100/V_{max}$	$100/V_{max}$
Supply Conductors and Cables	See Section 760-25					
Circuit Conductors and Cables	See Sections 760-49 through 760-53					

Table 760-21(b). Power Source Limitations for Direct-Current Fire Protective Signaling Circuits

Power Source Maximum Nameplate Ratings	Inherently Limited Power Source (Overcurrent protection not required)				Not Inherently Limited Power Source (Overcurrent protection required)		
Circuit Voltage V_{max} (Volts) (Note 1)	0 through 20	Over 20 and through 30	Over 30 and through 100	Over 100 and through 250	0 through 20	Over 20 and through 100	Over 100 and through 150
Power Limitations $(VA)_{max}$ (Volt-Amps) (Note 1)	—	—	—	—	250 (Note 2)	250	N.A.
Current Limitations I_{max} (Amps) (Note 1)	8.0	8.0	$150/V_{max}$	0.030	$1000/V_{max}$	$1000/V_{max}$	1.0
Maximum Overcurrent Protection (Amps)	—	—	—	—	5.0	$100/V_{max}$	1.0
VA (Volt-Amps)	$5.0 \times V_{max}$	100	100	$0.030 \times V_{max}$	$5.0 \times V_{max}$	100	100
Current (Amps)	5.0	$100/V_{max}$	$100/V_{max}$	0.030	5.0	$100/V_{max}$	$100/V_{max}$
Supply Conductors and Cables	See Section 760-25						
Circuit Conductors and Cables	See Sections 760-49 through 760-53						

Notes for Tables 760-21(a) and (b)

Note 1. V_{max}: Maximum output voltage regardless of load with rated input applied.

I_{max}: Maximum output current under any noncapacitive load, including short circuit, and with overcurrent protection bypassed if used. When a transformer limits the output current, I_{max} limits apply after one minute of operation. Where a current-limiting impedance, listed for the purpose, is used in combination with a nonpower-limited transformer or a stored energy source, e.g., storage battery, to limit the output current, I_{max} limits apply after 5 seconds.

VA_{max}: Maximum volt-ampere output after one minute of operation regardless of load and overcurrent protection bypassed if used. Current-limiting impedance shall not be bypassed when determining I_{max} and VA_{max}.

Note 2. If the power source is a transformer, $(VA)_{max}$ is 350 or less when V_{max} is 15 or less.

760-25. Wiring Methods on Supply Side. Conductors and equipment on the supply side of overcurrent protection, transformers, or current-limiting devices shall be installed in accordance with the appropriate requirements of Part B and Chapter 3. Transformers or other devices supplied from power-supply conductors shall be protected by an overcurrent device rated not over 20 amperes.

Exception: The input leads of a transformer or other power source supplying power-limited fire protective signaling circuits shall be permitted to be smaller than No. 14, but not smaller than No. 18, if they are not over 12 inches (305 mm) long and if they have insulation that complies with Section 760-16(b).

760-28. Wiring Methods and Materials on Load Side. Circuits on the load side of overcurrent protection, transformers, and current-limiting devices shall be permitted to use wiring methods and materials in accordance with either (a) or (b) below.

(a) Nonpower-Limited Wiring Methods and Materials. The appropriate articles of Chapter 3, including Section 300-17, shall apply, and, in addition, conductors shall be copper, solid, bunch-tinned, or stranded with a maximum of 19 strands.

Exception No. 1: The derating factors given in Article 310, Note 8(a) of Notes to Ampacity Tables of 0 to 2000 Volts shall not apply.

Exception No. 2: Conductors and multiconductor cables described in and installed in accordance with Sections 760-16 and 760-17 shall be permitted.

Exception No. 3: Power-limited circuits shall be permitted to be reclassified and installed as nonpower-limited circuits if the markings required by Section 760-22 are eliminated (see Section 760-3) and the entire circuit is installed using the wiring methods and materials in accordance with Part B, Nonpower-Limited Fire Protective Signaling Circuits.

(b) Power-Limited Wiring Methods and Materials. Power-limited circuit conductors and cables described in Sections 760-49 through 760-51 shall be installed as follows:

(1) In raceway or exposed on the surface of ceiling and sidewalls or "fished" in concealed spaces. Where installed exposed, cables shall be adequately supported and terminated in approved fittings and installed in such a way that maximum protection against physical damage is afforded by building construction such as baseboards, door frames, ledges, etc. Raceways shall not be used as a means of support for power-limited fire protective signaling circuit conductors. Where located within 7 feet (2.13 m) of the floor, cables shall be securely fastened in an approved manner at intervals of not more than 18 inches (457 mm).

(2) In metal raceways or rigid nonmetallic conduit where passing through a floor or wall to a height of 7 feet (2.13 m) above the floor, unless adequate protection can be afforded by building construction such as detailed in (1) above, or unless an equivalent solid guard is provided.

(3) In rigid metal conduit, rigid nonmetallic conduit, intermediate metal conduit, or electrical metallic tubing where installed in hoistways.

Exception: As provided for in Section 620-21, Exception Nos. 1 and 2 for elevators and similar equipment.

760-49. Fire Resistance of Cables Within Buildings. Single and multiconductor power-limited fire protective signaling circuit cables installed as wiring within buildings shall be listed as being resistant to the spread of fire in accordance with Sections 760-50 and 760-51.

760-50. Listing, Marking, and Installation of Power-Limited Fire Protective Signaling Circuit Cables. Power-limited fire protective signaling circuit cables installed as wiring within buildings shall be listed as being suitable for the purpose, marked in accordance with Table 760-50, and installed in accordance with Section 760-52. The cable voltage rating shall not be marked on the cable.

(FPN): Voltage ratings on cables may be misinterpreted to suggest that the cables may be suitable for Class 1, electric light and power applications.

Exception No. 1: Voltage markings shall be permitted where the cable has multiple listings and voltage marking is required for one or more of the listings.

Exception No. 2: Cable substitutions as provided for in Section 760-53(d) shall be considered as suitable for the purpose and shall be permitted.

Table 760-50.   Cable Markings

Cable Marking	Type	Reference
FPLP	Power-limited fire protective signaling circuit plenum cable	Sections 760-51(d) and 760-53(a)
FPLR	Power-limited fire protective signaling circuit riser cable	Sections 760-51(e) and 760-53(b)
FPL	Power-limited fire protective signaling circuit cable	Sections 760-51(f) and 760-53(c)

(FPN No. 1):   See the referenced sections for listing requirements and permitted uses.

(FPN No. 2):   Cable types are listed in descending order of fire resistance rating.

760-51. Listing Requirements. Power-limited fire protective signaling cables shall be listed in accordance with (a) through (c) below and, depending on the type, (d) through (g) below.

(a) Conductor Materials. Conductors shall be solid copper or bunch-tinned (bonded) stranded copper.

Exception No. 1: Stranded copper with a maximum of 7 strands for Nos. 16 and 18 shall be permitted.

Exception No. 2: Stranded copper with a maximum of 19 strands for Nos. 14 and larger shall be permitted.

(b) Sizes and Number. The size and number of conductors in a cable shall meet the requirements of Table 760-51. Conductors of No. 26 shall only be permitted if spliced with a connector listed as suitable for No. 26 to No. 24 or larger conductors that are terminated on equipment or if the No. 26 conductors are terminated on equipment listed as suitable for No. 26 conductors.

Table 760-51. Minimum Size and Number of Conductors Required in Cables Used for Power-Limited Fire Protective Signaling Circuits

Size	Minimum Number of Conductors in Cable
26 AWG	10
24 AWG	6
22 AWG	4
19 AWG	2
16 AWG or larger	1

(c) Ratings. The cable shall have a voltage rating of not less than 300 volts.

(d) Type FPLP. Type FPLP power-limited fire alarm plenum cable shall be listed as being suitable for use in ducts, plenums, and other space used for environmental air and shall also be listed as having adequate fire-resistant and low-smoke-producing characteristics.

(FPN): One method of defining low-smoke-producing cable is by establishing an acceptable value of the smoke produced when tested in accordance with the Test for Fire and Smoke Characteristics of Wires and Cables, NFPA 262-1990 (ANSI) to a maximum peak optical density of 0.5 and a maximum average optical density of 0.15. Similarly, one method of defining fire-resistant cables is by establishing maximum allowable flame travel distance of 5 feet (1.52 m) when tested in accordance with the same test.

(e) Type FPLR. Type FPLR power-limited fire alarm riser cable shall be listed as being suitable for use in a vertical run in a shaft or from floor to floor and shall also be listed as having fire-resistant characteristics capable of preventing the carrying of fire from floor to floor.

(FPN): One method of defining fire-resistant characteristics capable of preventing the carrying of fire from floor to floor is that the cables pass the requirements of the Standard Test for Flame Propagation Height of Electrical and Optical-Fiber Cable Installed Vertically in Shafts, ANSI/UL 1666-1986.

(f) Type FPL. Type FPL power-limited fire alarm cable shall be listed as being suitable for general-purpose fire alarm use, with the exception of risers, ducts, plenums, and other space used for environmental air and shall also be listed as being resistant to the spread of fire.

(FPN): One method of defining resistant to the spread of fire is that the cables do not spread fire to the top of the tray in the "Vertical-Tray Flame Test" in the Reference Standard for Electrical Wires, Cables and Flexible Cords, ANSI/UL 1581-1985.

Another method of defining resistant to the spread of fire is for the damage (char length) not to exceed 4 feet 11 inches (1.5 m) when performing the CSA "Vertical Flame Test — Cables in Cable Trays," as described in Test Methods for Electrical Wires and Cables, CSA C22.2 No. 0.3-M-1985.

(g) Coaxial Cables. Coaxial cables shall have a minimum of No. 22 copper or 30 percent copper-covered steel center conductor, a voltage rating of not less than 300 volts, and shall be listed as a Type FPLP, FPLR, or FPL cable.

760-52. Installation of Conductors and Equipment. Conductors and cables on the load side of overcurrent protection, transformers, and current-limiting devices shall comply with (a) and (b) below.

(a) Separation from Electric Light, Power, Class 1, and Nonpower-Limited Fire Protective Signaling Circuit Conductors.

(1) Open Conductors. Power-limited circuit conductors shall be separated at least 2 inches (50.8 mm) from conductors of any electric light, power, Class 1, or nonpower-limited fire protective signaling circuit conductors.

Exception No. 1: Where the electric light, power, Class 1, or nonpower-limited fire protective signaling circuit conductors are in raceway or in metal-sheathed, metal-clad, nonmetallic-sheathed, or Type UF cables.

Exception No. 2: Where the power-limited circuit conductors are permanently separated from the conductors of the other circuits by a continuous and firmly fixed nonconductor, such as porcelain tubes or flexible tubing in addition to the insulation on the wire.

(2) In Cables, Compartments, Enclosures, Outlet Boxes, or Raceways. Power-limited circuit conductors shall not be placed in any cable, compartment, enclosure, outlet box, raceway, or similar fitting containing conductors of electric light, power, Class 1, or nonpower-limited fire protective signaling circuit conductors.

Exception No. 1: Where the conductors of the different circuits are separated by a barrier.

Exception No. 2: Conductors in compartments, enclosures, outlet boxes, or similar fittings where electric light, power, Class 1 circuit, or nonpower-limited circuit conductors are introduced solely to connect to the equipment connected to the power-limited fire protective signaling system or to other circuits controlled by the fire protective signaling system to which the other conductors in the enclosure are connected.

The electric light, power, Class 1, and nonpower-limited fire protective signaling circuit conductors shall be routed within the enclosure to maintain a minimum 0.25-inch (6.35-mm) separation from the conductors of power-limited fire protective signaling circuits.

(3) In Shafts. Power-limited circuit conductors shall be separated by not less than 2 inches (50.8 mm) from electric light, power, Class 1, or nonpower-limited fire protective signaling circuit conductors run in the same shaft.

Exception No. 1: Where the conductors of either the electric light, power, Class 1, the nonpower-limited fire protective signaling circuits, or the power-limited fire protective signaling circuits are encased in metal raceways.

Exception No. 2: Where the electric light, power, Class 1, or the nonpower-limited fire protective signaling circuit conductors are in a raceway or are in metal-sheathed, metal-clad, nonmetallic-sheathed, or Type UF cables.

(4) In Hoistways. Power-limited circuit conductors shall be installed in rigid metal conduit, intermediate metal conduit, or electrical metallic tubing in hoistways.

Exception: As provided for in Section 620-21, Exception Nos. 1 and 2 for elevators and similar equipment.

(b) Conductors of Different Power-Limited Fire Protective Signaling Circuits, Class 2, Class 3, and Communications Circuits in Same Cable, Enclosure, or Raceway.

(1) Cables and conductors of two or more power-limited fire protective signaling circuits, communications circuits, or Class 3 circuits shall be permitted in the same cable, enclosure, or raceway.

(2) Conductors of one or more Class 2 circuits shall be permitted within the same cable, enclosure, or raceway with conductors of power-limited fire protective signaling circuits, provided that the insulation of the Class 2 circuit conductors in the cable, enclosure, or raceway is at least that required by the power-limited fire protective signaling circuits.

760-53. Applications of Listed Power-Limited Fire Protective Signaling Circuit Cables. Power-limited fire protective signaling circuit cables shall comply with (a) through (c) or, where cable substitution is being made, (d) below.

(a) Plenum. Cables installed in ducts, plenums, and other spaces used for environmental air shall be Type FPLP.

Exception: Types FPLP, FPLR, and FPL cable installed in compliance with Section 300-22.

(b) Riser. Cables installed in vertical runs and penetrating more than one floor or cables installed in vertical runs in a shaft shall be Type FPLR. Floor penetrations requiring Type FPLR shall contain only cables suitable for riser or plenum use.

Exception No. 1: Where the cables are encased in metal raceway or are located in a fireproof shaft having firestops at each floor.

Exception No. 2: Type FPL in one- and two-family dwellings.

(FPN): See Section 300-21 for firestop requirements for floor penetrations.

(c) Other Wiring Within Buildings. Cables installed in building locations other than the locations covered in (a) and (b) above shall be Type FPL.

Exception No. 1: Where the cables are enclosed in raceway.

Exception No. 2: Cables specified in Chapter 3 that meet the requirements of Sections 760-51(a) and (b) and are installed in nonconcealed spaces where the exposed length of cable does not exceed 10 feet (3.05 m).

(d) Cable Substitutions. The substitutions for power-limited fire protective signaling circuit cables listed in Table 760-53 and illustrated in Figure 760-53 shall be permitted. Communications and Class 3 cables shall be permitted to substitute for power-limited fire protective signaling circuit cables only where the requirements of Sections 760-51(a), (b), and (c) for multiconductor cables and Section 760-51(g) for coaxial cables are satisfied.

Table 760-53. Cable Substitutions

Cable Type	Permitted Substitutions
FPLP	MPP, CMP, CL3P
FPLR	MPP, CMP, FPLP, CL3P, MPR, CMR, CL3R
FPL	MPP, CMP, FPLP, CL3P, MPR, CMR, CL3R, FPLR, MPG, MP, CMG, CM, PLTC, CL3

(FPN): For information on multipurpose cables (MPP, MRP, MPG, MP) and communications cables (CMP, CMR, CMG, CM), see Section 800-50. For information on Class 3 cables (CL3P, CL3R, CL3, PLTC), see Section 725-50.

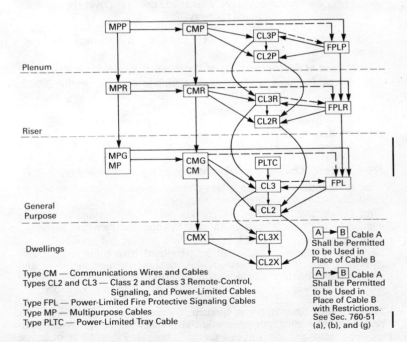

Type CM — Communications Wires and Cables
Types CL2 and CL3 — Class 2 and Class 3 Remote-Control,
                                Signaling, and Power-Limited Cables
Type FPL — Power-Limited Fire Protective Signaling Cables
Type MP — Multipurpose Cables
Type PLTC — Power-Limited Tray Cable

Figure 760-53. Cable Substitution Hierarchy

760-55. Current-Carrying Continuous Line-type Fire Detectors.

(a) Application. Listed continuous line-type fire detectors, including insulated copper tubing of pneumatically operated detectors, employed for both detection and carrying signaling currents shall be permitted to be used in circuits having power-limiting characteristics in accordance with Section 760-21.

(b) Insulation. Insulated continuous line-type fire detectors shall be listed as being resistant to the spread of fire in accordance with the requirements of Sections 760-49 and 760-50, have a voltage rating of not less than 300 volts, and the jacket compound shall have a high degree of abrasion resistance.

(c) Installation. Continuous line-type fire detectors shall be installed in accordance with Sections 760-22 through 760-28 and Section 760-52.

ARTICLE 770 — OPTICAL FIBER CABLES AND RACEWAYS

A. General

770-1. Scope. The provisions of this article apply to the installation of optical fiber cables and raceways. This article does not cover the construction of optical fiber cables and raceways.

770-2. Locations and Other Articles. Circuits and equipment shall comply with (a) and (b) below.

(a) Spread of Fire or Products of Combustion. See Section 300-21.

(b) Ducts, Plenums, and Other Air-Handling Spaces. Section 300-22, where installed in ducts or plenums or other space used for environmental air.

Exception to (b): As permitted in Section 770-53(a).

770-3. Optical Fiber Cables. Optical fiber cables transmit light for control, signaling, and communications through an optical fiber.

770-4. Types. Optical fiber cables can be grouped into three types.

(a) Nonconductive. These cables contain no metallic members and no other electrically conductive materials.

(b) Conductive. These cables contain noncurrent-carrying conductive members such as metallic strength members and metallic vapor barriers.

(c) Composite. These cables contain optical fibers and current-carrying electrical conductors, and shall be permitted to contain noncurrent-carrying conductive members such as metallic strength members and metallic vapor barriers. Composite optical fiber cables shall be classified as electrical cables in accordance with the type of electrical conductors.

770-5. Optical Fiber Raceway System. A raceway system designed for enclosing and routing only nonconductive optical fiber cables. Where optical fiber cables are installed in a raceway, the raceway shall be of a type permitted in Chapter 3 and installed in accordance with Chapter 3.

Exception: Listed optical fiber raceway.

(FPN): Plastic innerduct commonly used for underground or outside plant construction may not have appropriate fire safety characteristics for use as an optical fiber optic cable wiring method within buildings.

770-6. Cable Trays. Optical fiber cables of the types listed in Table 770-50 shall be permitted to be installed in cable trays.

(FPN): It is not the intent to require that these optical fiber cables be listed specifically for use in cable trays.

770-7. Access to Electrical Equipment Behind Panels Designed to Allow Access. Access to equipment shall not be denied by an accumulation of wires and cables that prevents removal of panels, including suspended ceiling panels.

B. Protection

770-33. Grounding of Entrance Cables. Where exposed to contact with electric light or power conductors, the noncurrent-carrying metallic members of optical fiber cables entering buildings shall be grounded as close to the point of entrance as practicable or shall be interrupted as close to the point of entrance as practicable by an insulating joint or equivalent device.

For purposes of this section, the point of entrance shall be considered to be at the point of emergence through an exterior wall, a concrete floor slab, or from a rigid metal conduit or an intermediate metal conduit grounded in accordance with Article 250.

C. Cables Within Buildings

770-49. Fire Resistance of Optical Fiber Cables. Optical fiber cables installed as wiring within buildings shall be listed as being resistant to the spread of fire in accordance with Sections 770-50 and 770-51.

770-50. Listing, Marking, and Installation of Optical Fiber Cables. Optical fiber cables in a building shall be listed as being suitable for the purpose, and cables shall be marked in accordance with Table 770-50.

Exception No. 1: Optical fiber cables shall not be required to be listed and marked where the length of cable within the building does not exceed 50 feet (15.2 m) and the cable enters the building from the outside and is terminated in an enclosure.

(FPN): Splice cases or terminal boxes, both metallic and plastic types, are typically used as enclosures for splicing or terminating optical fiber cables.

Exception No. 2: Conductive optical fiber cable shall not be required to be listed and marked where the cable enters the building from the outside and is run in rigid metal conduit or intermediate metal conduit and such conduits are grounded to an electrode in accordance with Section 800-40(b).

Exception No. 3: Nonconductive optical fiber cables shall not be required to be listed and marked where the cable enters the building from the outside and is run in raceway installed in compliance with Chapter 3.

Table 770-50.   Cable Markings

Cable Marking	Type	Reference
OFNP	Nonconductive optical fiber plenum cable	Sections 770-51(a) and 770-53(a)
OFCP	Conductive optical fiber plenum cable	Sections 770-51(a) and 770-53(a)
OFNR	Nonconductive optical fiber riser cable	Sections 770-51(b) and 770-53(b)
OFCR	Conductive optical fiber riser cable	Sections 770-51(b) and 770-53(b)
OFNG	Nonconductive optical fiber general-purpose cable	Sections 770-51(c) and 770-53(c)
OFCG	Conductive optical fiber general-purpose cable	Sections 770-51(c) and 770-53(c)
OFN	Nonconductive optical fiber general-purpose cable	Sections 770-51(d) and 770-53(c)
OFC	Conductive optical fiber general-purpose cable	Sections 770-51(d) and 770-53(c)

(FPN No. 1):   Cable types are listed in descending order of fire resistance rating. Within each fire resistance rating, nonconductive cable is listed first, since it may substitute for the conductive cable.

(FPN No. 2):   See the referenced sections for requirements and permitted uses.

770-51. Listing Requirements for Optical Fiber Cables and Raceways. Optical fiber cables shall be listed in accordance with (a) through (d) below, and optical fiber raceways shall be listed in accordance with (e) and (f) below.

(a) Types OFNP and OFCP. Types OFNP and OFCP nonconductive and conductive optical fiber plenum cables shall be listed as being suitable for use in ducts, plenums, and other space used for environmental air and shall also be listed as having adequate fire-resistant and low-smoke-producing characteristics.

(FPN):   One method of defining low-smoke-producing cables is by establishing an acceptable value of the smoke produced when tested in accordance with the Test for Fire and Smoke Characteristics of Wires and Cables, NFPA 262-1990 (ANSI) to a maximum peak optical density of 0.5 and a maximum average optical density of 0.15. Similarly, one method of defining fire-resistant cables is by defining maximum allowable flame travel distance of 5 feet (1.52 m) when tested in accordance with the same test.

(b) Types OFNR and OFCR. Types OFNR and OFCR nonconductive and conductive optical fiber riser cables shall be listed as being suitable for use in a vertical run in a shaft or from floor to floor and shall also be listed as having fire-resistant characteristics capable of preventing the carrying of fire from floor to floor.

(FPN):   One method of defining fire-resistant characteristics capable of preventing the carrying of fire from floor to floor is that the cables pass the requirements of the Standard Test for Flame Propagation Height of Electrical and Optical-Fiber Cable Installed Vertically in Shafts, ANSI/UL 1666-1986.

(c) Types OFNG and OFCG. Types OFNG and OFCG nonconductive and conductive general-purpose optical fiber cables shall be listed as being suitable for general-purpose use, with the exception of risers and plenums, and shall also be listed as being resistant to the spread of fire.

(FPN): One method of defining resistance to the spread of fire is for the damage (char length) not to exceed 4 feet 11 inches (1.5 m) when performing the "Vertical Flame Test — Cables in Cable Trays," as described in Test Methods for Electrical Wires and Cables, CSA C22.2 No. 0.3-M 1985.

(d) Types OFN and OFC. Types OFN and OFC nonconductive and conductive optical fiber cables shall be listed as being suitable for general-purpose use, with the exception of risers, plenums, and other space used for environmental air, and shall also be listed as being resistant to the spread of fire.

(FPN): One method of defining resistant to the spread of fire is that the cables do not spread fire to the top of the tray in the "Vertical-Tray Flame Test" in the Reference Standard for Electrical Wires, Cables and Flexible Cords, ANSI/UL 1581-1985.
Another method of defining resistant to the spread of fire is for the damage (char length) not to exceed 4 feet 11 inches (1.5 m) when performing the "Vertical Flame Test — Cables in Cable Trays," as described in Test Methods for Electrical Wires and Cables, CSA C22.2 No. 0.3-M 1985.

(e) Plenum Optical Fiber Raceway. Plenum optical fiber raceways shall be listed as having adequate fire-resistant and low-smoke-producing characteristics.

(f) Riser Optical Fiber Raceway. Riser optical fiber raceways shall be listed as having fire-resistant characteristics capable of preventing the carrying of fire from floor to floor.

770-52. Installation of Optical Fibers and Electrical Conductors.

(a) With Conductors for Electric Light, Power, or Class 1 Circuits. Optical fibers shall be permitted within the same composite cable for electric light, power, or Class 1 circuits operating at 600 volts or less only where the functions of the optical fibers and the electrical conductors are associated. Nonconductive optical fiber cables shall be permitted to occupy the same cable tray or raceway with conductors for electric light, power, or Class 1 circuits operating at 600 volts or less. Conductive optical fiber cables shall not be permitted to occupy the same cable tray or raceway with conductors for electric light, power, or Class 1 circuits. Composite optical fiber cables containing only current-carrying conductors for electric light, power, or Class 1 circuits rated 600 volts or less shall be permitted to occupy the same cabinet, cable tray, outlet box, panel, raceway, or other termination enclosure with conductors for electric light, power, or Class 1 circuits operating at 600 volts or less.

Nonconductive optical fiber cables shall not be permitted to occupy the same cabinet, outlet box, panel, or similar enclosure housing the electrical terminations of an electric light, power, or Class 1 circuit.

Exception No. 1: Occupancy of the same cabinet, outlet box, panel, or similar enclosure shall be permitted where nonconductive optical fiber cable is functionally associated with the electric light, power, or Class 1 circuit.

Exception No. 2: Occupancy of the same cabinet, outlet box, panel, or similar enclosure shall be permitted where nonconductive optical fiber cables are installed in factory- or field-assembled control centers.

Exception No. 3: In industrial establishments only, where conditions of maintenance and supervision assure that only qualified persons will service the installation, nonconductive optical fiber cables shall be permitted with circuits exceeding 600 volts.

Exception No. 4: In industrial establishments only, where conditions of maintenance and supervision assure that only qualified persons will service the installation, hybrid optical fiber cables shall be permitted to contain current-carrying conductors operating over 600 volts.

Installations in raceway shall comply with Section 300-17.

(b) With Other Conductors. Optical fibers shall be permitted in the same cable, and conductive and nonconductive optical fiber cables shall be permitted in the same cable tray, enclosure, or raceway with conductors of any of the following:

(1) Class 2 and Class 3 remote-control, signaling, and power-limited circuits in compliance with Article 725.

(2) Power-limited fire protective signaling systems in compliance with Article 760.

(3) Communications circuits in compliance with Article 800.

(4) Community antenna television and radio distribution systems in compliance with Article 820.

(c) Grounding. Noncurrent-carrying conductive members of optical fiber cables shall be grounded in accordance with Article 250.

770-53. Applications of Listed Optical Fiber Cables and Raceways. Nonconductive and conductive optical fiber cables shall comply with (a) through (d) below.

(a) Plenum. Cables installed in ducts, plenums, and other spaces used for environmental air shall be Type OFNP or OFCP.

Also, listed plenum optical fiber raceways shall be permitted to be installed in ducts and plenums as described in Section 300-22(b) and in other space used for environmental air as described in Section 300-22(c). Type OFNP cable shall be permitted to be installed in these raceways.

Exception: Types OFNR, OFCR, OFNG, OFN, OFCG, and OFC cables installed in compliance with Section 300-22.

(b) Riser. Cables installed in vertical runs and penetrating more than one floor or cables installed in vertical runs in a shaft shall be Types OFNR or OFCR. Floor penetrations requiring Types OFNR or OFCR shall contain only cables suitable for riser or plenum use.

Also, listed riser optical fiber raceways shall be permitted to be installed in vertical runs in a shaft or from floor to floor. Types OFNR and OFNP cables shall be permitted to be installed in these raceways.

Exception No. 1: Where Types OFNG, OFN, OFCG, and OFC cables are encased in metal raceway or are located in a fireproof shaft having fire-stops at each floor.

Exception No. 2: Type OFNG, OFN, OFCG, or OFC cable in one- and | two-family dwellings.

(FPN):   See Section 300-21 for firestop requirements for floor penetrations. |

(c) Other Wiring Within Buildings. Cables installed in building locations other than the locations covered in (a) and (b) above shall be Type OFNG, OFN, OFCG, or OFC.

(d) Cable Substitutions. The substitutions for optical fiber cables listed in Table 770-53 and illustrated in Figure 770-53 shall be permitted.

Table 770-53.   Cable Substitutions

Cable Type	Permitted Substitutions
OFNP	None
OFCP	OFNP
OFNR	OFNP
OFCR	OFNP, OFCP, OFNR
OFNG, OFN	OFNP, OFNR
OFCG, OFC	OFNP, OFCP, OFNR, OFCR, OFNG, OFN	

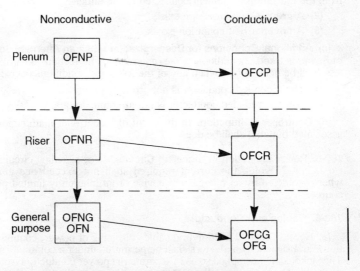

A ➔ B Cable A may be used in place of Cable B

Figure 770-53.   Cable Substitution Hierarchy

ARTICLE 780 — CLOSED-LOOP AND PROGRAMMED POWER DISTRIBUTION

780-1. Scope. The provisions of this article apply to premise power distribution systems jointly controlled by a signaling between the energy controlling equipment and utilization equipment.

780-2. General.

(a) Other Articles. Except as modified by the requirements of this article, all other applicable articles of this Code shall apply.

(b) Component Parts. All equipment and conductors shall be listed and identified.

780-3. Control. The control equipment and all power switching devices operated by the control equipment shall be listed and identified. The system shall operate such that:

(a) Characteristic Electrical Identification Required. Outlets shall not be energized unless the utilization equipment first exhibits a characteristic electrical identification.

(b) Conditions for Deenergization. Outlets shall be deenergized when any of the following conditions occur:

(1) A nominal-operation acknowledgement signal is not being received from the utilization equipment connected to the outlet.

(2) A ground-fault condition exists.

(3) An overcurrent condition exists.

(c) Additional Conditions for Deenergization When an Alternate Source of Power is Used. In addition to the requirements in Section 780-3(b), outlets shall be deenergized when any of the following conditions occur:

(1) The grounded conductor is not properly grounded.

(2) Any ungrounded conductor is not at nominal voltage.

(d) Controller Malfunction. In the event of a controller malfunction, all associated outlet(s) shall be deenergized.

780-5. Power Limitation in Signaling Circuits. For signaling circuits not exceeding 24 volts, the current required shall not exceed one ampere where protected by an overcurrent device or an inherently limited power source.

780-6. Cables and Conductors.

(a) Hybrid Cable. Listed hybrid cable consisting of power, communications, and signaling conductors shall be permitted under a common jacket. The jacket shall be applied so as to separate the power conductors from the communications and signaling conductors. An optional outer jacket shall be permitted to be applied. The individual conductors of a hybrid cable shall conform to the Code provisions applicable to their current, voltage, and insulation rating. The signaling conductors shall not be smaller than No. 24 copper.

(b) Cables and Conductors in the Same Cabinet, Panel, or Box. The power, communications, and signaling conductors of listed hybrid cable are permitted to occupy the same cabinet, panel, or outlet box (or similar enclosure housing the electrical terminations of electric light or power circuits) only if connectors specifically listed for hybrid cable are employed.

780-7. Noninterchangeability. Receptacles, cord connectors, and attachment plugs used on closed-loop power distribution systems shall be constructed so that they are not interchangeable with other receptacles, cord connectors, and attachment plugs.

Chapter 8.   Communications Systems

ARTICLE 800 — COMMUNICATIONS CIRCUITS

A. General

800-1. Scope.   This article covers telephone, telegraph (except radio), outside wiring for fire alarm and burglar alarm, and similar central station systems; and telephone systems not connected to a central station system but using similar types of equipment, methods of installation, and maintenance.

(FPN No. 1):   For further information for fire alarm, guard tour, sprinkler waterflow, and sprinkler supervisory systems, see Article 760.

(FPN No. 2):   For installation requirements of optical fiber cables, see Article 770.

800-2. Definitions.   See Article 100. For purposes of this article, the following additional definitions apply:

Cable: A cable is a factory assembly of two or more conductors having an overall covering.

Cable Sheath: A sheath is a covering over the conductor assembly that may include one or more metallic members, strength members, or jackets.

Point of Entrance: The point of entrance within a building is the point at which the wire or cable emerges from an external wall, from a concrete floor slab, or from a rigid metal conduit or an intermediate metal conduit grounded to an electrode in accordance with Section 800-40(b).

Wire: A wire is a factory assembly of one or more insulated conductors without an overall covering.

800-3. Hybrid Power and Communications Cables.   The provisions of Section 780-6 shall apply for listed hybrid power and communications cables in closed-loop and programmed power distribution.

(FPN):   See Section 800-51(i) for hybrid power and communications cable in other applications.

800-4. Equipment.   Equipment intended to be electrically connected to a telecommunications network shall be listed for the purpose.

(FPN):   One way to determine applicable requirements is to refer to the Standard for Telephone Equipment, UL 1459-1987, or Communication Circuit Accessories, UL 1863-1990.

Exception No. 1: For other than wires, cables, grounding conductors, and protectors, this listing requirement shall not apply to premises wiring components, such as plugs, jacks, connectors, terminal blocks, and cross-connect assemblies that were manufactured before October 1, 1990 and all other equipment manufactured before July 1, 1991.

Exception No. 2: This listing requirement shall not apply to test equipment intended for temporary connection to a telecommunications network by qualified persons during the course of installation, maintenance, or repair of telecommunications equipment or systems.

800-5. Access to Electrical Equipment Behind Panels Designed to Allow Access. Access to equipment shall not be denied by an accumulation of wires and cables that prevents removal of panels, including suspended ceiling panels.

800-6. Mechanical Execution of Work. Communication circuits and equipment shall be installed in a neat and workmanlike manner.

B. Conductors Outside and Entering Buildings

800-10. Overhead Communications Wires and Cables. Overhead communications wires and cables entering buildings shall comply with (a) and (b) below.

(a) On Poles and In-Span. Where communications wires and cables and electric light or power conductors are supported by the same pole or run in parallel in-span, the following conditions shall be met:

(1) Relative Location. Where practicable, the communications wires and cables shall be located below the electric light or power conductors.

(2) Attachment to Crossarms. Communications wires and cables shall not be attached to a crossarm that carries electric light or power conductors.

(3) Climbing Space. The climbing space through communications wires and cables shall comply with the requirements of Section 225-14(d).

(4) Clearance. Supply service drops of 0-750 volts running above and parallel to communications service drops shall be permitted to have a minimum separation of 12 inches (30.48 cm) at any point in the span, including the point of and at their attachment to the building, provided the non-grounded conductors are insulated and that a clearance of 40 inches (1.02 m) is maintained between the two services at the pole.

(b) Above Roofs. Communications wires and cables shall have a vertical clearance of not less than 8 feet (2.44 m) from all points of roofs above which they pass.

Exception No. 1: Auxiliary buildings, such as garages and the like.

Exception No. 2: A reduction in clearance above only the overhanging portion of the roof to not less than 18 inches (457 mm) shall be permitted if (1) not more than 4 feet (1.22 mm) of communications service-drop conductors pass above the roof overhang and (2) they are terminated at a through- or above-the-roof raceway or approved support.

Exception No. 3: Where the roof has a slope not less than 4 inches (102 mm) in 12 inches (305 mm), a reduction in clearance to not less than 3 feet (914 mm) shall be permitted.

800-11. Underground Circuits Entering Buildings. Underground communications wires and cables entering buildings shall comply with (a) and (b) below.

(a) With Electric Light or Power Conductors. Underground communications wires and cables in a raceway, handhole, or manhole containing electric light or power conductors shall be in a section separated from such conductors by means of brick, concrete, or tile partitions.

(b) Underground Block Distribution. Where the entire street circuit is run underground and the circuit within the block is so placed as to be free from likelihood of accidental contact with electric light or power circuits of over 300 volts to ground, the insulation requirements of Sections 800-12(a) and 800-12(c) shall not apply, insulating supports shall not be required for the conductors, and bushings shall not be required where the conductors enter the building.

800-12. Circuits Requiring Primary Protectors. Circuits that require primary protectors as provided in Section 800-30 shall comply with the following:

(a) Insulation, Wires, and Cables. Communications wires and cables without a metallic shield, running from the last outdoor support to the primary protector, shall be listed as being suitable for the purpose and having current-carrying capacity as specified in Section 800-30(a)(1)(b) or 800-30(a)(1)(c).

(b) On Buildings. Communications wires and cables in accordance with Section 800-12(a) shall be separated at least 4 inches (102 mm) from electric light or power conductors not in a raceway or cable, or be permanently separated from conductors of the other system by a continuous and firmly fixed nonconductor in addition to the insulation on the wires, such as porcelain tubes or flexible tubing. Communications wires and cables in accordance with Section 800-12(a) exposed to accidental contact with electric light and power conductors operating at over 300 volts to ground and attached to buildings shall be separated from woodwork by being supported on glass, porcelain, or other insulating material.

Exception: Separation from woodwork shall not be required where fuses are omitted as provided for in Section 800-31(a), or where conductors are used to extend circuits to a building from a cable having a grounded metal sheath.

(c) Entering Buildings. Where a primary protector is installed inside the building, the communications wires and cables shall enter the building either through a noncombustible, nonabsorbent insulating bushing, or through a metal raceway. The insulating bushing shall not be required where the entering communications wires and cables (1) are in metal-sheathed cable; (2) pass through masonry; (3) meet the requirements of Section 800-12(a) and fuses are omitted as provided in Section 800-31(a); or (4) meet the requirements of Section 800-12(a) above and are used to extend circuits to a building from a cable having a grounded metallic sheath. Raceways or bushings shall slope upward from the outside or, where this cannot be done, drip loops shall be formed in the communications wires and cables immediately before they enter the building.

Raceways shall be equipped with an approved service head. More than one communications wire and cable shall be permitted to enter through a single raceway or bushing. Conduits or other metal raceways located ahead of the primary protector shall be grounded.

800-13. Lightning Conductors. Where practicable, a separation of at least 6 feet (1.83 m) shall be maintained between open conductors of communications systems on buildings and lightning conductors.

C. Protection

800-30. Protective Devices.

(a) Application. A listed primary protector shall be provided on each circuit run partly or entirely in aerial wire or aerial cable not confined within a block. Also, a listed primary protector shall be provided on each circuit, aerial or underground, so located within the block containing the building served as to be exposed to accidental contact with electric light or power conductors operating at over 300 volts to ground. In addition, where there exists a lightning exposure, each interbuilding circuit on a premises shall be protected by a listed primary protector at each end of the interbuilding circuit. Installation of primary protectors shall also comply with Section 110-3(b).

(FPN No. 1):   The word "block" as used in this article means a square or portion of a city, town, or village enclosed by streets and including the alleys so enclosed, but not any street. The word "premises" as used in this article means the land and buildings of a user located on the user side of the utility-user network point of demarcation.

(FPN No. 2):   The word "exposed" as used in this article means that the circuit is in such a position that, in case of failure of supports or insulation, contact with another circuit may result.

(FPN No. 3):   On a circuit not exposed to accidental contact with power conductors, providing a listed primary protector in accordance with this article will help protect against other hazards, such as lightning and above-normal voltages induced by fault currents on power circuits in proximity to the communications circuit.

(FPN No. 4):   Interbuilding circuits are considered to have a lightning exposure unless one or more of the following conditions exist:

1. Circuits in large metropolitan areas where buildings are close together and sufficiently high to intercept lightning.

2. Interbuilding cable runs of 140 feet (42.7 m) or less, directly buried or in underground conduit, where a continuous metallic cable shield or a continuous metallic conduit containing the cable is bonded to each building grounding electrode system.

3. Areas having an average of five or fewer thunderstorm days per year and earth resistivity of less than 100 ohm-meters. Such areas are found along the Pacific coast.

(1) Fuseless Primary Protectors. Fuseless-type primary protectors shall be permitted under any of the following conditions:

a. Where conductors enter a building through a cable with grounded metallic sheath member(s) and if the conductors in the cable safely fuse on all currents greater than the current-carrying capacity of the primary protector and of the primary protector grounding conductor.

b. Where insulated conductors in accordance with Section 800-12(a) are used to extend circuits to a building from a cable with effectively grounded metallic sheath member(s) and if the conductors in the cable or cable stub, or the connections between the insulated conductors and the exposed plant, safely fuse on all currents greater than the current-carrying capacity of the primary protector, or the associated insulated conductors and of the primary protector grounding conductor.

c. Where insulated conductors in accordance with Section 800-12(a) or (b) are used to extend circuits to a building from other than a cable with metallic sheath member(s) if (1) the primary protector is listed for this purpose, and (2) the connections of the insulated conductors to the exposed plant or the conductors of the exposed plant safely fuse on all currents greater than the current-carrying capacity of the primary protector, or the associated insulated conductors and of the primary protector grounding conductor.

d. Where insulated conductors in accordance with Section 800-12(a) are used to extend circuits aerially to a building from an unexposed buried or underground circuit.

e. Where insulated conductors in accordance with Section 800-12(a) are used to extend circuits to a building from cable with effectively grounded metallic sheath member(s) and if (1) the combination of the primary protector and insulated conductors is listed for this purpose, and (2) the insulated conductors safely fuse on all currents greater than the current-carrying capacity of the primary protector and of the primary protector grounding conductor.

(FPN): Effectively grounded means intentionally connected to earth through a ground connection or connections of sufficiently low impedance and having sufficient current-carrying capacity to prevent the buildup of voltages that may result in undue hazard to connected equipment or to persons.

(2) Fused Primary Protectors. Where the requirements listed under items a through e above are not met, fused-type primary protectors shall be used. Fused-type primary protectors shall consist of an arrester connected between each line conductor and ground, a fuse in series with each line conductor, and an appropriate mounting arrangement. Primary protector terminals shall be marked to indicate line, instrument, and ground, as applicable.

(b) Location. The primary protector shall be located in, on, or immediately adjacent to the structure or building served and as close as practicable to the point at which the exposed conductors enter or attach.

For purposes of this section, the point at which the exposed conductors enter shall be considered to be the point of emergence through an exterior wall, a concrete floor slab, or from a rigid metal conduit or an intermediate metal conduit grounded to an electrode in accordance with Section 800-40(b).

For purposes of this section, primary protectors located at mobile home service equipment located in sight from and not more than 30 feet (9.14 m) from the exterior wall of the mobile home it serves, or at a mobile home disconnecting means grounded in accordance with Section 250-24 and located in sight from and not more than 30 feet (9.14 m) from the exterior wall of the mobile home it serves, shall be considered to meet the requirements of this section.

(FPN):   Selecting a primary protector location to achieve the shortest practicable primary protector grounding conductor will help limit potential differences between communications circuits and other metallic systems.

(c) Hazardous (Classified) Locations. The primary protector shall not be located in any hazardous (classified) location as defined in Article 500, nor in the vicinity of easily ignitible material.

Exception: As permitted in Sections 501-14, 502-14, and 503-12.

800-31. Primary Protector Requirements. The primary protector shall consist of an arrester connected between each line conductor and ground in an appropriate mounting. Primary protector terminals shall be marked to indicate line and ground as applicable.

(FPN):   One way to determine applicable requirements for a listed primary protector is to refer to the Standard for Protectors for Communications Circuits, ANSI/UL 497-1978.

800-32. Secondary Protector Requirements. Where a secondary protector is installed in series with the indoor communications wire and cable between the primary protector and the equipment, it shall be listed for the purpose. The secondary protector shall provide means to safely limit currents to less than the current-carrying capacity of listed indoor communications wire and cable, listed telephone set line cords, and listed communications terminal equipment having ports for external wire line communications circuits. Any overvoltage protection, arresters, or grounding connection shall be connected on the equipment terminals side of the secondary protector current-limiting means.

(FPN No. 1):   One way to determine applicable requirements for a listed secondary protector is to refer to the Standard for Secondary Protectors for Communications, UL497A-1990.

(FPN No. 2):   Secondary protectors on exposed circuits are not intended for use without primary protectors.

800-33. Cable Grounding. The metallic sheath of communications cables entering buildings shall be grounded as close as practicable to the point of entrance or shall be interrupted as close to the point of entrance as practicable by an insulating joint or equivalent device.

For purposes of this section, the point of entrance shall be considered to be at the point of emergence through an exterior wall, a concrete floor slab, or from a rigid metal conduit or an intermediate metal conduit grounded to an electrode in accordance with Section 800-40(b).

D. Grounding Methods

800-40. Cable and Primary Protector Grounding. The metallic member(s) of the cable sheath, where required to be grounded by Section 800-33, and primary protectors shall be grounded as specified in (a) through (d) below.

(a) Grounding Conductor.

(1) Insulation. The grounding conductor shall be insulated and shall be listed as suitable for the purpose.

(2) Material. The grounding conductor shall be copper or other corrosion-resistant conductive material, stranded or solid.

(3) Size. The grounding conductor shall not be smaller than No. 14.

(4) Run in Straight Line. The grounding conductor shall be run to the grounding electrode in as straight a line as practicable.

(5) Physical Damage. Where necessary, the grounding conductor shall be guarded from physical damage. Where the grounding conductor is run in a metal raceway, both ends of the raceway shall be bonded to the grounding conductor or the same terminal or electrode to which the grounding conductor is connected.

(b) Electrode. The grounding conductor shall be connected as follows:

(1) To the nearest accessible location on (1) the building or structure grounding electrode system as covered in Section 250-81, (2) the grounded interior metal water piping system as covered in Section 250-80(a), (3) the power service accessible means external to enclosures as covered in Section 250-71(b), (4) the metallic power service raceway, (5) the service equipment enclosure, (6) the grounding electrode conductor or the grounding electrode conductor metal enclosure, or (7) to the grounding conductor or the grounding electrode of a building or structure disconnecting means that is grounded to an electrode as covered in Section 250-24.

For purposes of this section, the mobile home service equipment or the mobile home disconnecting means, as described in Section 800-30(b), shall be considered accessible.

(2) If the building or structure served has no grounding means, as described in b(1), to any one of the individual electrodes described in Section 250-81; or

(3) If the building or structure served has no grounding means, as described in b(1) or b(2), to (1) an effectively grounded metal structure or (2) to a ground rod or pipe not less than 5 feet (1.52 m) in length and $\frac{1}{2}$ inch (12.7 mm) in diameter, driven, where practicable, into permanently damp earth and separated from lightning conductors as covered in Section 800-13 and at least 6 feet (1.83 m) from electrodes of other systems. Steam or hot water pipes or lightning-rod conductors shall not be employed as electrodes for protectors.

(c) Electrode Connection. Connections to grounding electrodes shall comply with Section 250-115. Connectors, clamps, fittings, or lugs used to attach grounding conductors and bonding jumpers to grounding electrodes or to each other that are to be concrete-encased or buried in the earth shall be suitable for its application.

(d) Bonding of Electrodes. A bonding jumper not smaller than No. 6 copper or equivalent shall be connected between the communications grounding electrode and power grounding electrode system at the building or structure served where separate electrodes are used. Bonding together of all separate electrodes shall be permitted.

Exception: At mobile homes as covered in Section 800-41.

(FPN No. 1): See Section 250-86 for use of lightning rods.

(FPN No. 2): Bonding together of all separate electrodes will limit potential differences between them and between their associated wiring systems.

800-41. Primary Protector Grounding and Bonding at Mobile Homes.

(a) Grounding. Where there is no mobile home service equipment located in sight from and not more than 30 feet (9.14 m) from the exterior wall of the mobile home it serves, or there is no mobile home disconnecting means grounded in accordance with Section 250-24 and located within sight from and not more than 30 feet (9.14 m) from the exterior wall of the mobile home it serves, the primary protector ground shall be in accordance with Section 800-40(b)(2) and (3).

(b) Bonding. The primary protector grounding terminal or grounding electrode shall be bonded to the metal frame or available grounding terminal of the mobile home with a copper grounding conductor not smaller than No. 12 under any of the following conditions:

(1) Where there is no mobile home service equipment or disconnecting means as in (a) above or;

(2) The mobile home is supplied by cord and plug.

E. Conductors Within Buildings

800-49. Fire Resistance of Communications Wires and Cables.
Communications wires and cables installed as wiring within a building shall be listed as being resistant to the spread of fire in accordance with Sections 800-50 and 800-51.

800-50. Listing, Marking, and Installation of Communications Wires and Cables.
Communications wires and cables installed as wiring within buildings shall be listed as being suitable for the purpose, marked in accordance with Table 800-50, and installed in accordance with Section 800-52. The cable voltage rating shall not be marked on the cable.

(FPN): Voltage markings on cables may be misinterpreted to suggest that the cables may be suitable for Class 1, electric light and power applications.

Exception No. 1: Voltage markings shall be permitted where the cable has multiple listings and voltage marking is required for one or more of the listings.

Exception No. 2: Listing and marking shall not be required where the cable enters the building from the outside and is continuously enclosed in rigid metal conduit or intermediate metal conduit and such conduits are grounded to an electrode in accordance with Section 800-40(b).

Exception No. 3: Listing and marking shall not be required where the length of cable within the building does not exceed 50 feet (15.2 m) and the cable enters the building from the outside and is terminated in an enclosure or on a listed primary protector.

(FPN No. 1): Splice cases or terminal boxes, both metallic and plastic types, are typically used as enclosures for splicing or terminating telephone cables.

(FPN No. 2): This exception limits the length of unlisted outside plant cable to 50 feet (15.2 m), while Section 800-30(b) requires that the primary protector shall be located as close as practicable to the point at which the cable enters the building. Hence, in installations requiring a primary protector, the outside plant cable may not be permitted to extend 50 feet (15.2 m) into the building if it is practicable to place the primary protector closer than 50 feet (15.2 m) to the entrance point.

Exception No. 4: Multipurpose cables shall be considered as being suitable for the purpose and shall be permitted to substitute for communications cables as provided for in Section 800-53(f).

Table 800-50.   Cable Markings

Cable Marking	Type	Reference
MPP	Multipurpose plenum cable	Sections 800-51(g) and 800-53(a)
CMP	Communications plenum cable	Sections 800-51(a) and 800-53(a)
MPR	Multipurpose riser cable	Sections 800-51(g) and 800-53(b)
CMR	Communications riser cable	Sections 800-51(b) and 800-53(b)
MPG	Multipurpose general-purpose cable	Sections 800-51(g) and 800-53(d)
CMG	Communications general-purpose cable	Sections 800-51(c) and 800-53(d)
MP	Multipurpose general-purpose cable	Sections 800-51(g) and 800-52(d)
CM	Communications general-purpose cable	Sections 800-51(d) and 800-53(d)
CMX	Communications cable, limited use	Sections 800-51(e) and 800-53(d), Exception Nos. 1, 2, 3, and 4
CMUC	Undercarpet communications wire and cable	Sections 800-51(f) and 800-53(d), Exception No. 5

(FPN No. 1):   Cable types are listed in descending order of fire resistance rating, and multipurpose cables are listed above communications cables because multipurpose cables may substitute for communications cables.

(FPN No. 2):   See the referenced sections for permitted uses.

800-51. Listing Requirements.   Communications wires and cables shall have a voltage rating of not less than 300 V and shall be listed in accordance with (a) through (i) below.

(FPN):   See Section 800-4 for listing requirement for equipment.

(a) Type CMP.   Type CMP communications plenum cable shall be listed as being suitable for use in ducts, plenums, and other space used for environmental air and shall also be listed as having adequate fire-resistant and low-smoke-producing characteristics.

(FPN):   One method of defining low-smoke-producing cables is by establishing an acceptable value of the smoke produced when tested in accordance with the Test for Fire and Smoke Characteristics of Wires and Cables, NFPA 262-1990 (ANSI) to a maximum peak optical density of 0.5 and a maximum average optical density of 0.15. Similarly, one method of defining fire-resistant cables is by establishing maximum allowable flame travel distance of 5 feet (1.52 m) when tested in accordance with the same test.

(b) Type CMR.   Type CMR communications riser cable shall be listed as being suitable for use in a vertical run in a shaft and shall also be listed as having fire-resistant characteristics capable of preventing the carrying of fire from floor to floor.

(FPN): One method of defining fire-resistant characteristics capable of preventing the carrying of fire from floor to floor is that the cables pass the requirements of the Standard Test for Flame Propagation Height of Electrical and Optical-Fiber Cable Installed Vertically in Shafts, ANSI/UL 1666-1986.

(c) Type CMG. Type CMG general-purpose communications cable shall be listed as being suitable for general-purpose communications use, with the exception of risers and plenums, and shall also be listed as being resistant to the spread of fire.

(FPN): One method of defining resistant to the spread of fire is for the damage (char length) not to exceed 4 feet 11 inches (1.5 m) when performing the "Vertical Flame Test — Cables in Cable Trays," as described in Test Methods for Electrical Wires and Cables, CSA C22.2 No. 0.3-M 1985.

(d) Type CM. Type CM communications cable shall be listed as being suitable for general purpose communications use, with the exception of risers and plenums, and shall also be listed as being resistant to the spread of fire.

(FPN): One method of defining resistant to the spread of fire is that the cables do not spread fire to the top of the tray in the "Vertical-Tray Flame Test" in the Reference Standard for Electrical Wires, Cables and Flexible Cords, ANSI/UL 1581-1985.
Another method of defining resistant to the spread of fire is for the damage (char length) not to exceed 4 feet 11 inches (1.5 m) when performing the "Vertical Flame Test—Cables in Cable Trays," as described in Test Methods for Electrical Wires and Cables, CSA C22.2 No. 0.3-M-1985.

(e) Type CMX. Type CMX limited use communications cable shall be listed as being suitable for use in dwellings and for use in raceway and shall also be listed as being flame-retardant.

(FPN): One method of determining that cable is flame-retardant is by testing the cable to the "VW-1 (Vertical-Wire) Flame Test" in the Reference Standard for Electrical Wires, Cables and Flexible Cords, ANSI/UL 1581-1985.

(f) Type CMUC Undercarpet Wire and Cable. Type CMUC undercarpet communications wire and cable shall be listed as being suitable for undercarpet use and shall also be listed as being flame-retardant.

(FPN): One method of determining that cable is flame-retardant is by testing the cable to the "VW-1 (Vertical-Wire) Flame Test" in the Reference Standard for Electrical Wires, Cables and Flexible Cords, ANSI/UL 1581-1985.

(g) Multipurpose (MP) Cables. Cables that meet the requirements for Types CMP, CMR, CMG, and CM and also satisfy the requirements of Sections 760-51(a) and (b) for multiconductor cables and Section 760-51(g) for coaxial cables shall be permitted to be listed and marked as Types MPP, MPR, MPG, and MP, respectively.

(h) Communications Wires. Communications wires, such as distributing frame wire and jumper wire, shall be listed as being resistant to the spread of fire.

(FPN): One method of defining resistant to the spread of fire is that the cables do not spread fire to the top of the tray in the "Vertical-Tray Flame Test" in the Reference Standard for Electrical Wires, Cables and Flexible Cords, ANSI/UL 1581-1985.

Another method of defining resistant to the spread of fire is for the damage (char length) not to exceed 4 feet 11 inches (1.5 m) when performing the "Vertical Flame Test—Cables in Cable Trays," as described in Test Methods for Electrical Wires and Cables, CSA C22.2 No. 0.3-M-1985.

(i) Hybrid Power and Communications Cable. Listed hybrid power and communications cable shall be permitted where the power cable is a listed Type NM conforming to the provisions of Article 336, and the communications cable is a listed Type CM, and the jackets on the listed NM and listed CM cables are rated for 600 volts minimum, and the hybrid cable is listed as being resistant to the spread of fire.

(FPN): One method of defining resistant to the spread of fire is that the cables do not spread fire to the top of the tray in the "Vertical-Tray Flame Test" in the Reference Standard for Electrical Wires, Cables and Flexible Cords, ANSI/UL 1581-1985.

Another method of defining resistant to the spread of fire is for the damage (char length) not to exceed 4 feet 11 inches (1.5 m) when performing the "Vertical Flame Test—Cables in Cable Trays," as described in Test Methods for Electrical Wires and Cables, CSA C22.2 No. 0.3-M-1985.

800-52. Installation of Communications Wires, Cables, and Equipment. Communications wires and cables from the protector to the equipment or, where no protector is required, communications wires and cables attached to the outside or inside of the building shall comply with (a) through (c) below.

(a) Separation from Other Conductors.

(1) Open Conductors. Communications wires and cables shall be separated at least 2 inches (50.8 mm) from conductors of any electric light or power circuits, Class 1, or nonpower-limited fire protective signaling circuits.

Exception No. 1: Where the electric light or power, Class 1, or nonpower-limited fire protective signaling circuit conductors are in a raceway or in metal-sheathed, metal-clad, nonmetallic-sheathed, Type AC, or Type UF cables.

Exception No. 2: Where the communications wires and cables are permanently separated from the conductors of the other circuit by a continuous and firmly fixed nonconductor, such as porcelain tubes or flexible tubing, in addition to the insulation on the wire.

(2) In Raceways, Boxes, and Cables.

a. Other Power-Limited Circuits. Communications cables shall be permitted in the same raceway or enclosure with cables of any of the following:

1. Class 2 and Class 3 remote-control, signaling, and power-limited circuits in compliance with Article 725.

2. Power-limited fire protective signaling systems in compliance with Article 760.

3. Nonconductive and conductive optical fiber cables in compliance with Article 770.

4. Community antenna television and radio distribution systems in compliance with Article 820.

b. Class 2 and Class 3 Circuits. Class 1 circuits shall not be run in the same cable with communications circuits. Class 2 and Class 3 circuit conductors shall be permitted in the same cable with communications circuits, in which case the Class 2 and Class 3 circuits shall be classified as communications circuits and shall meet the requirements of this article. The cables shall be listed as communications cables or multipurpose cables.

Exception: Cables constructed of individually listed Class 2, Class 3, and communications cables under a common jacket shall not be required to be classified as communications cable. The fire resistance rating of the composite cable shall be determined by the performance of the composite cable.

c. Electric Light or Power Circuits.

1. Communications conductors shall not be placed in any raceway, compartment, outlet box, junction box, or similar fitting with conductors of electric light or power circuits or Class 1 circuits.

Exception No. 1: Where all of the conductors of electric light or power, Class 1, or nonpower-limited fire protective signaling circuits are separated from all of the conductors of communications circuits by a barrier.

Exception No. 2: Electric light or power, Class 1, or nonpower-limited fire protective signaling circuit conductors in outlet boxes, junction boxes, or similar fittings or compartments where such conductors are introduced solely for power supply to communications equipment or for connection to remote-control equipment. The electric light or power, Class 1, or nonpower-limited fire protective signaling circuit conductors shall be routed within the enclosure to maintain a minimum of 0.25 inch (6.35 mm) separation from the communications circuit conductors.

2. In shafts. Communications wires and cables run in the same shaft with conductors of electric light or power, Class 1, or nonpower-limited fire protective signaling circuits shall be separated from light or power conductors, Class 1, or nonpower-limited fire protective signaling circuit conductors by not less than 2 inches (50.8 mm).

Exception No. 1: Where either (1) all of the conductors of the electric light or power, Class 1 or nonpower-limited fire protective signaling circuits, or (2) all of the conductors of communications circuits are encased in raceway.

Exception No. 2: Where the electric light or power, Class 1 or nonpower-limited fire protective signaling conductors are in a raceway, or in metal-sheathed, metal-clad, nonmetallic-sheathed, or Type UF cables.

(b) Spread of Fire or Products of Combustion. Installations in hollow spaces, vertical shafts, and ventilation or air-handling ducts shall be so made that the possible spread of fire or products of combustion will not be substantially increased. Openings around penetrations through fire resistance-rated walls, partitions, floors, or ceilings shall be firestopped using approved methods.

(c) Equipment in Other Space Used for Environmental Air. Section 300-22(c) shall apply.

(d) Cable Trays. Types MPP, MPR, MPG, and MP multipurpose cables and Types CMP, CMR, CMG, and CM communications cables shall be permitted to be installed in cable trays.

(e) Support of Conductors. Raceways shall not be used as a means of support for communications wires and cables.

800-53. Applications of Listed Communications Wires and Cables. Communications wires and cables shall comply with (a) through (f) below.

(a) Plenum. Cables installed in ducts, plenums, and other spaces used for environmental air shall be Type CMP.

Exception: Types CMP, CMR, CMG, CM, and CMX and communications wire installed in compliance with Section 300-22.

(b) Riser. Cables installed in vertical runs and penetrating more than one floor, or cables installed in vertical runs in a shaft, shall be Type CMR. Floor penetrations requiring Type CMR shall contain only cables suitable for riser or plenum use.

(FPN):   See Section 800-52(b) for firestop requirements for floor penetrations.

Exception No. 1: Where the listed cables are encased in metal raceway or are located in a fireproof shaft having firestops at each floor.

Exception No. 2: Types CM and CMX cable in one- and two-family dwellings.

(c) Distributing Frames and Cross-Connect Arrays. Communications wires shall be used in distributing frames and cross-connect arrays.

Exception: Types CMP, CMR, CMG, and CM are permitted to be used.

(d) Other Wiring Within Buildings. Cables installed in building locations other than the locations covered in (a), (b), and (c) above shall be Type CMG or Type CM.

Exception No. 1: Where listed communications wires and cables are enclosed in raceway.

Exception No. 2: Type CMX communications cable in nonconcealed spaces where the exposed length of cable does not exceed 10 feet (3.05 m).

Exception No. 3: Type CMX communications cables that are less than 0.25 inch (6.35 mm) in diameter and installed in one- or two-family dwellings.

Exception No. 4: Type CMX communications wires and cables that are less than 0.25 inch (6.35 mm) in diameter and installed in nonconcealed spaces in multifamily dwellings.

Exception No. 5: Type CMUC undercarpet communications wires and cables installed under carpet.

(e) Hybrid Power and Communications Cable. Hybrid power and communications cable listed in accordance with Section 800-51(i) shall be permitted to be installed in one-and two-family dwellings.

(f) Cable Substitutions. Substitutions for communications cables listed in Table 800-53 and illustrated in Figure 800-53 shall be permitted.

Table 800-53. Cable Substitutions

Cable Type	Permitted Substitutions
MPP	None
CMP	MPP
MPR	MPP
CMR	MPP, CMP, MPR
MPG, MP	MPP, MPR
CMG, CM	MPP, CMP, MPR, CMR, MPG, MP
CMX	MPP, CMP, MPR, CMR, MPG, MP, CMG, CM

(FPN): For the use of communications cable and multipurpose cable in place of Class 2 or Class 3 cables, see Section 725-53(g), and, for the use of communications cable and multipurpose cable in place of power-limited fire protective signaling cables, see Section 760-53(d). The permitted cable substitutions are illustrated in Figure 800-53.

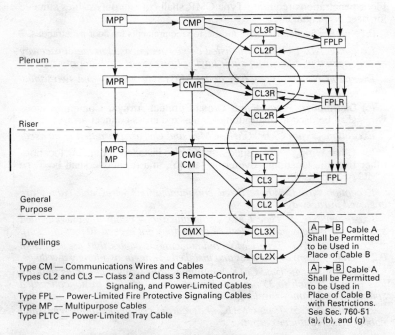

Type CM — Communications Wires and Cables
Types CL2 and CL3 — Class 2 and Class 3 Remote-Control,
    Signaling, and Power-Limited Cables
Type FPL — Power-Limited Fire Protective Signaling Cables
Type MP — Multipurpose Cables
Type PLTC — Power-Limited Tray Cable

[A] ➤ [B] Cable A
Shall be Permitted
to be Used in
Place of Cable B

[A] ⊷➤ [B] Cable A
Shall be Permitted
to be Used in
Place of Cable B
with Restrictions.
See Sec. 760-51
(a), (b), and (g)

Figure 800-53. Cable Substitution Hierarchy

ARTICLE 810 — RADIO AND TELEVISION EQUIPMENT

A. General

810-1. Scope. This article covers radio and television receiving equipment and amateur radio transmitting and receiving equipment, but not equipment and antennas used for coupling carrier current to power line conductors.

810-2. Other Articles. Wiring from the source of power to and between devices connected to the interior wiring system shall comply with Chapters 1 through 4 other than as modified by Sections 640-3, 640-4, and 640-5. Wiring for radio-frequency and audio-frequency equipment and loud speakers shall comply with Article 640. Where optical fiber is used Article 770 shall apply. Coaxial cable wiring for television receiving shall comply with Article 820.

810-3. Community Television Antenna. The antenna shall comply with this article. The distribution system shall comply with Article 820.

810-4. Radio Noise Suppressors. Radio interference eliminators, interference capacitors, or noise suppressors connected to power-supply leads shall be of a listed type. They shall not be exposed to physical damage.

810-5. Definitions. See Article 100.

B. Receiving Equipment — Antenna Systems

810-11. Material. Antennas and lead-in conductors shall be of hard-drawn copper, bronze, aluminum alloy, copper-clad steel, or other high-strength, corrosion-resistant material.

Exception: Soft-drawn or medium-drawn copper shall be permitted for lead-in conductors where the maximum span between points of support is less than 35 feet (10.67 m).

810-12. Supports. Outdoor antennas and lead-in conductors shall be securely supported. The antennas shall not be attached to the electric service mast. They shall not be attached to poles or similar structures carrying open electric light or power wires or trolley wires of over 250 volts between conductors. Insulators supporting the antenna conductors shall have sufficient mechanical strength to safely support the conductors. Lead-in conductors shall be securely attached to the antennas.

810-13. Avoidance of Contacts with Conductors of Other Systems. Outdoor antennas and lead-in conductors from an antenna to a building shall not cross over open conductors of electric light or power circuits and shall be kept well away from all such circuits so as to avoid the possibility of accidental contact. Where proximity to open electric light or power ser-vice conductors of less than 250 volts between conductors cannot be avoided, the installation shall be such as to provide a clearance of at least 2 feet (610 mm).

Where practicable, antenna conductors shall be so installed as not to cross under open electric light or power conductors.

810-14. Splices. Splices and joints in antenna spans shall be made mechanically secure with approved splicing devices or by such other means as will not appreciably weaken the conductors.

810-15. Grounding. Masts and metal structures supporting antennas shall be grounded in accordance with Section 810-21.

810-16. Size of Wire-Strung Antenna — Receiving Station.

(a) Size of Antenna Conductors. Outdoor antenna conductors for receiving stations shall be of a size not less than given in Table 810-16(a).

Table 810-16(a).
Size of Receiving-Station Outdoor Antenna Conductors

	Minimum Size of Conductors		
	When Maximum Open Span Length is		
Material	Less than 35 feet	35 feet to 150 feet	Over 150 feet
Aluminum alloy, hard-drawn copper	19	14	12
Copper-clad steel, bronze, or other high-strength material	20	17	14

For SI units: one foot = 0.3048 meter.

(b) Self-Supporting Antennas. Outdoor antennas, such as vertical rods, dishes, or dipole structures, shall be of corrosion-resistant materials and of strength suitable to withstand ice and wind loading conditions, and shall be located well away from overhead conductors of electric light and power cir-cuits of over 150 volts to ground, so as to avoid the possibility of the antenna or structure falling into or making accidental contact with such circuits.

810-17. Size of Lead-in — Receiving Station. Lead-in conductors from outside antennas for receiving stations shall, for various maximum open span lengths, be of such size as to have a tensile strength at least as great as that of the conductors for antennas as specified in Section 810-16. Where the lead-in consists of two or more conductors that are twisted together, are enclosed in the same covering, or are concentric, the conduc-tor size shall, for various maximum open span lengths, be such that the tensile strength of the combination will be at least as great as that of the conductors for antennas as specified in Section 810-16.

810-18. Clearances — Receiving Stations.

(a) Outside of Buildings. Lead-in conductors attached to buildings shall be so installed that they cannot swing closer than 2 feet (610 mm) to the conductors of circuits of 250 volts or less between conductors, or 10 feet (3.05 m) to the conductors of circuits of over 250 volts between conductors, except that in the case of circuits not over 150 volts between conductors, where all conductors involved are supported so as to ensure permanent separation, the clearance shall be permitted to be reduced but shall not be less than 4 inches (102 mm). The clearance between lead-in conductors and any conductor forming a part of a lightning-rod system shall not be less than 6 feet (1.83 m) unless the bonding referred to in Section 250-86 is accomplished. Underground conductors shall be separated at least 12 inches (305 mm) from conductors of any light or power circuits, or Class 1 circuits.

Exception: Where the electric light or power conductors, Class 1 conductors, or lead-in conductors are installed in raceways or metal cable armor.

(b) Antennas and Lead-ins — Indoors. Indoor antennas and indoor lead-ins shall not be run nearer than 2 inches (50.8 mm) to conductors of other wiring systems in the premises.

Exception No. 1: Where such other conductors are in metal raceways or cable armor.

Exception No. 2: Where permanently separated from such other conductors by a continuous and firmly fixed nonconductor, such as porcelain tubes or flexible tubing.

(c) In Boxes or Other Enclosures. Indoor antennas and indoor lead-ins shall be permitted to occupy the same box or enclosure with conductors of other wiring systems where separated from such other conductors by an effective permanently installed barrier.

810-19. Electric Supply Circuits Used in Lieu of Antenna — Receiving Stations.
Where an electric supply circuit is used in lieu of an antenna, the device by which the radio receiving set is connected to the supply circuit shall be listed.

810-20. Antenna Discharge Units — Receiving Stations.

(a) Where Required. Each conductor of a lead-in from an outdoor antenna shall be provided with a listed antenna discharge unit.

Exception: Where the lead-in conductors are enclosed in a continuous metallic shield that is either permanently and effectively grounded, or is protected by an antenna discharge unit.

(b) Location. Antenna discharge units shall be located outside the building or inside the building between the point of entrance of the lead-in and the radio set or transformers, and as near as practicable to the entrance of the conductors to the building. The antenna discharge unit shall not be located near combustible material nor in a hazardous (classified) location as defined in Article 500.

(c) Grounding. The antenna discharge unit shall be grounded in accordance with Section 810-21.

810-21. Grounding Conductors — Receiving Stations. Grounding conductors shall comply with (a) through (j) below.

(a) Material. The grounding conductor shall be of copper, aluminum, copper-clad steel, bronze, or similar corrosion-resistant material.

(b) Insulation. Insulation on grounding conductors shall not be required.

(c) Supports. The grounding conductors shall be securely fastened in place and shall be permitted to be directly attached to the surface wired over without the use of insulating supports.

Exception: Where proper support cannot be provided, the size of the grounding conductors shall be increased proportionately.

(d) Mechanical Protection. The grounding conductor shall be protected where exposed to physical damage, or the size of the grounding conductors shall be increased proportionately to compensate for the lack of protection.

(e) Run in Straight Line. The grounding conductor for an antenna mast or antenna discharge unit shall be run in as straight a line as practicable from the mast or discharge unit to the grounding electrode.

(f) Electrode. The grounding conductor shall be connected as follows:

(1) To the nearest accessible location on (1) the building or structure grounding electrode system as covered in Section 250-81, (2) the grounded interior metal water piping system as covered in Section 250-80(a), (3) the power service accessible means external to enclosures as covered in Section 250-71(b), (4) the metallic power service raceway, (5) the service equipment enclosure, or (6) the grounding electrode conductor or the grounding electrode conductor metal enclosures; or

(2) If the building or structure served has no grounding means, as described in (f)(1), to any one of the individual electrodes described in Section 250-81; or

(3) If the building or structure served has no grounding means, as described in (f)(1) or (f)(2), to (1) an effectively grounded metal structure or (2) to any of the individual electrodes described in Section 250-83.

(g) Inside or Outside Building. The grounding conductor shall be permitted to be run either inside or outside the building.

(h) Size. The grounding conductor shall not be smaller than No. 10 copper or No. 8 aluminum or No. 17 copper-clad steel or bronze.

(i) Common Ground. A single grounding conductor shall be permitted for both protective and operating purposes.

(j) Bonding of Electrodes. A bonding jumper not smaller than No. 6 copper or equivalent shall be connected between the radio and television equipment grounding electrode and the power grounding electrode system at the building or structure served where separate electrodes are used.

C. Amateur Transmitting and Receiving Stations — Antenna Systems

810-51. Other Sections. In addition to complying with Part C, antenna systems for amateur transmitting and receiving stations shall also comply with Sections 810-11 through 810-15.

810-52. Size of Antenna. Antenna conductors for transmitting and receiving stations shall be of a size not less than given in Table 810-52.

Table 810-52.
Size of Amateur Station Outdoor Antenna Conductors

	Minimum Size of Conductors	
	Where Maximum Open Span Length is	
Material	Less than 150 feet	Over 150 feet
Hard-drawn copper	14	10
Copper-clad steel, bronze, or other high-strength material	14	12

For SI units: one foot = 0.3048 meter.

810-53. Size of Lead-in Conductors. Lead-in conductors for transmitting stations shall, for various maximum span lengths, be of a size at least as great as that of conductors for antennas as specified in Section 810-52.

810-54. Clearance on Building. Antenna conductors for transmitting stations, attached to buildings, shall be firmly mounted at least 3 inches (76 mm) clear of the surface of the building on nonabsorbent insulating supports, such as treated pins or brackets equipped with insulators having not less than 3-inch (76-mm) creepage and airgap distances. Lead-in conductors attached to buildings shall also comply with these requirements.

Exception: Where the lead-in conductors are enclosed in a continuous metallic shield that is permanently and effectively grounded, they shall not be required to comply with these requirements. Where grounded, the metallic shield shall also be permitted to be used as a conductor.

810-55. Entrance to Building. Except where protected with a continuous metallic shield that is permanently and effectively grounded, lead-in conductors for transmitting stations shall enter buildings by one of the following methods: (1) through a rigid, noncombustible, nonabsorbent insulating tube or bushing; (2) through an opening provided for the purpose in which the entrance conductors are firmly secured so as to provide a clearance of at least 2 inches (50.8 mm); or (3) through a drilled window pane.

810-56. Protection Against Accidental Contact. Lead-in conductors to radio transmitters shall be so located or installed as to make accidental contact with them difficult.

810-57. Antenna Discharge Units — Transmitting Stations. Each conductor of a lead-in for outdoor antennas shall be provided with an antenna discharge unit or other suitable means that will drain static charges from the antenna system.

Exception No. 1: Where protected by a continuous metallic shield that is permanently and effectively grounded.

Exception No. 2: Where the antenna is permanently and effectively grounded.

810-58. Grounding Conductors — Amateur Transmitting and Receiving Stations. Grounding conductors shall comply with (a) through (c) below.

(a) Other Sections. All grounding conductors for amateur transmitting and receiving stations shall comply with Sections 810-21(a) through (j).

(b) Size of Protective Grounding Conductor. The protective grounding conductor for transmitting stations shall be as large as the lead-in, but not smaller than No. 10 copper, bronze, or copper-clad steel.

(c) Size of Operating Grounding Conductor. The operating grounding conductor for transmitting stations shall not be less than No. 14 copper or its equivalent.

D. Interior Installation — Transmitting Stations

810-70. Clearance from Other Conductors. All conductors inside the building shall be separated at least 4 inches (102 mm) from the conductors of any electric light, power, or signaling circuit.

Exception No. 1: As provided in Article 640.

Exception No. 2: Where separated from other conductors by raceway or some firmly fixed nonconductor, such as porcelain tubes or flexible tubing.

810-71. General. Transmitters shall comply with (a) through (c) below.

(a) Enclosing. The transmitter shall be enclosed in a metal frame or grille, or separated from the operating space by a barrier or other equivalent means, all metallic parts of which are effectively connected to ground.

(b) Grounding of Controls. All external metal handles and controls accessible to the operating personnel shall be effectively grounded.

(c) Interlocks on Doors. All access doors shall be provided with interlocks that will disconnect all voltages of over 350 volts between conductors when any access door is opened.

ARTICLE 820 — COMMUNITY ANTENNA TELEVISION AND RADIO DISTRIBUTION SYSTEMS

A. General

820-1. Scope. This article covers coaxial cable distribution of radio frequency signals typically employed in community antenna television (CATV) systems.

(FPN): Where the installation is other than coaxial, see Articles 770 and 800 as applicable.

820-2. Energy Limitations. The coaxial cable shall be permitted to deliver low-energy power to equipment directly associated with the radio frequency distribution system if the voltage is not over 60 volts and if the current supply is from a transformer or other device having energy-limiting characteristics.

820-3. Definitions. See Article 100. For the purposes of this Article, the following additional definition applies:

Point of Entrance. The point of entrance within a building is the point at which the cable emerges from an external wall, from a concrete floor slab, or from a rigid metal conduit or an intermediate metal conduit grounded to an electrode in accordance with Section 820-40(b).

820-5. Access to Electrical Equipment Behind Panels Designed to Allow Access. Access to equipment shall not be denied by an accumulation of wires and cables that prevents removal of panels, including suspended ceiling panels.

820-6. Mechanical Execution of Work. Community antenna television and radio distribution systems shall be installed in a neat and workmanlike manner.

B. Cables Outside and Entering Buildings

820-10. Outside Cables. Coaxial cables, prior to the point of grounding, as defined in Section 820-33, shall comply with (a) through (e) below.

(a) On Poles. Where practicable, conductors on poles shall be located below the electric light or power conductors and shall not be attached to a cross-arm that carries electric light or power conductors.

(b) Lead-in Clearance. Lead-in or aerial-drop cables from a pole or other support, including the point of initial attachment to a building or structure, shall be kept away from electric light or power circuits so as to avoid the possibility of accidental contact.

Exception: Where proximity to electric light or power service conductors cannot be avoided, the installation shall be such as to provide clearances of not less than 12 inches (305 mm) from light or power service drops.

(c) Above Roofs. Cables shall have a vertical clearance of not less than 8 feet (2.44 m) from all points of roofs above which they pass.

Exception No. 1: Auxiliary buildings such as garages and the like.

Exception No. 2: A reduction in clearance above only the overhanging portion of the roof to not less than 18 inches (457 mm) shall be permitted if (1) not more than 4 feet (1.22 m) of communication service drop conductors pass above the roof overhang, and (2) they are terminated at a through-the-roof raceway or support.

Exception No. 3: Where the roof has a slope not less than 4 inches (102 mm) in 12 inches (305 mm), a reduction in clearance to not less than 3 feet (914 mm) shall be permitted.

(d) Between Buildings. Cables extending between buildings and also the supports or attachment fixtures shall be acceptable for the purpose and shall have sufficient strength to withstand the loads to which they may be subjected.

Exception: Where a cable does not have sufficient strength to be self-supporting, it shall be attached to a supporting messenger cable that, together with the attachment fixtures or supports, shall be acceptable for the purpose and shall have sufficient strength to withstand the loads to which they may be subjected.

(e) On Buildings. Where attached to buildings, cables shall be securely fastened in such a manner that they will be separated from other conductors as follows:

(1) Electric Light or Power. The coaxial cable shall have a separation of at least 4 inches (102 mm) from electric light or power conductors not in raceway or cable, or be permanently separated from conductors of the other system by a continuous and firmly fixed nonconductor in addition to the insulation on the wires.

(2) Other Communications Systems. Coaxial cable shall be installed so that there will be no unnecessary interference in the maintenance of the separate systems. In no case shall the conductors, cables, messenger strand, or equipment of one system cause abrasion to the conductors, cable, messenger strand, or equipment of any other system.

(3) Lightning Conductors. Where practicable, a separation of at least 6 feet (1.83 m) shall be maintained between any coaxial cable and lightning conductors.

820-11. Entering Buildings.

(a) Underground Systems. Underground coaxial cables in a duct, pedestal, handhole, or manhole containing electric light or power conductors or Class 1 circuits shall be in a section permanently separated from such conductors by means of a suitable barrier.

(b) Direct Buried Cables and Raceways. Direct buried coaxial cable shall be separated at least 12 inches (305 mm) from conductors of any light or power or Class 1 circuit.

Exception No. 1: Where electric service conductors or coaxial cables are installed in raceways or have metal cable armor.

Exception No. 2: Where electric light or power branch-circuit or feeder conductors or Class 1 circuit conductors are installed in a raceway or in metal-sheathed, metal-clad, or Type UF or Type USE cables; or the coaxial cables have metal cable armor or are installed in a raceway.

C. Protection

820-33. Grounding of Outer Conductive Shield of a Coaxial Cable. Where coaxial cable is exposed to lightning or to accidental contact with lightning arrester conductors or power conductors operating at a voltage of over 300 volts to ground, the outer conductive shield of the coaxial cable shall be grounded at the building premises as close to the point of cable entry as practicable. For purposes of this section, the point at which the exposed cable enters shall be considered to be the point of emergence through an exterior wall, a concrete floor slab, or from a rigid or intermediate metal conduit grounded to an electrode in accordance with Section 820-40(b).

For purposes of this section, grounding located at mobile home service equipment located in sight from and no more than 30 feet (9.14 m) from the exterior wall of the mobile home it serves, or at a mobile home disconnecting means grounded in accordance with Section 250-24 and located in sight from and not more than 30 feet (9.14 m) from the exterior wall of the mobile home it serves, shall be considered to meet the requirements of this section.

(FPN): Selecting a grounding location to achieve the shortest practicable grounding conductor will help limit potential differences between CATV and other metallic systems.

(a) Shield Grounding. Where the outer conductive shield of a coaxial cable is grounded, no other protective devices shall be required.

(b) Shield Protection Devices. Grounding of a coaxial drop cable shield by means of a protective device that does not interrupt the grounding system within the premises shall be permitted.

D. Grounding Methods

820-40. Cable Grounding. Where required by Section 820-33, the shield of the coaxial cable shall be grounded as specified in (a) through (d) below.

(a) Grounding Conductor.

(1) Insulation. The grounding conductor shall be insulated and shall be listed as suitable for the purpose.

(2) Material. The grounding conductor shall be copper or other corrosion-resistant conductive material, stranded or solid.

(3) Size. The grounding conductor shall not be smaller than No. 14; it shall have a current-carrying capacity approximately equal to that of the outer conductor of the coaxial cable.

(4) Run in Straight Line. The grounding conductor shall be run to the grounding electrode in as straight a line as practicable.

(5) Physical Protection. Where subject to physical damage, the grounding conductor shall be adequately protected. Where the grounding conductor is run in a metal raceway, both ends of the raceway shall be bonded to the grounding conductor or the same terminal or electrode to which the grounding conductor is connected.

(b) Electrode. The grounding conductor shall be connected as follows:

(1) To the nearest accessible location on (1) the building or structure grounding electrode system as covered in Section 250-81, (2) the grounded interior metal water piping system as covered in Section 250-80(a), (3) the power service accessible means external to enclosures as covered in Section 250-71(b), (4) the metallic power service raceway, (5) the service equipment enclosure, (6) the grounding electrode conductor or the grounding electrode conductor metal enclosure, or (7) to the grounding conductor or to the grounding electrode of a building or structure disconnecting means that is grounded to an electrode as covered in Section 250-24; or

(2) If the building or structure served has no grounding means as described in (b)(1), to any one of the individual electrodes described in Section 250-81; or

(3) If the building or structure served has no grounding means as described in (b)(1) or (b)(2), to (1) an effectively grounded metal structure or (2) to any one of the individual electrodes described in Section 250-83.

(c) Electrode Connection. Connections to grounding electrodes shall comply with Section 250-115.

(d) Bonding of Electrodes. A bonding jumper not smaller than No. 6 copper or equivalent shall be connected between the antenna systems grounding electrode and the power grounding electrode system at the building or structure served where separate electrodes are used.

Exception: At mobile homes as covered in Section 820-42.

(FPN No. 1):   See Section 250-86 for use of lightning rods.

(FPN No. 2):   Bonding together of all separate electrodes will limit potential differences between them and between their associated wiring systems.

820-41. Equipment Grounding.   Unpowered equipment and enclosures or equipment powered by the coaxial cable shall be considered grounded where connected to the metallic cable shield.

820-42. Bonding and Grounding at Mobile Homes.

(a) Grounding. Where there is no mobile home service equipment located in sight from and not more than 30 feet (9.14 m) from the exterior wall of the mobile home it serves, or there is no mobile home disconnecting means grounded in accordance with Section 250-24 and located within sight from and not more than 30 feet (9.14 m) from the exterior wall of the mobile home it serves, the coaxial cable shield ground, or surge arrester ground, shall be in accordance with Sections 820-40(b) (2) and (3).

(b) Bonding. The coaxial cable shield grounding terminal, surge arrester grounding terminal, or grounding electrode shall be bonded to the metal frame or available grounding terminal of the mobile home with a copper grounding conductor not smaller than No. 12 under any of the following conditions:

(1) Where there is no mobile home service equipment or disconnecting means as in (a) above, or

(2) Where the mobile home is supplied by cord and plug.

E. Cables Within Buildings

820-49. Fire Resistance of CATV Cables.   Coaxial cables installed as wiring within buildings shall be listed as being resistant to the spread of fire in accordance with Sections 820-50 and 820-51.

820-50. Listing, Marking, and Installation of Coaxial Cables.   Coaxial cables in a building shall be listed as being suitable for the purpose, and cables shall be marked in accordance with Table 820-50. The cable voltage rating shall not be marked on the cable.

(FPN):   Voltage markings on cables may be misinterpreted to suggest that the cables may be suitable for Class 1, electric light and power applications.

Exception No. 1: Voltage markings shall be permitted where the cable has multiple listings and voltage marking is required for one or more of the listings.

Exception No. 2: Listing and marking shall not be required where the cable enters the building from the outside and is run in rigid metal conduit or intermediate metal conduit, and such conduits are grounded to an electrode in accordance with Section 820-40(b).

Exception No. 3: Listing and marking shall not be required where the length of cable within the building does not exceed 50 feet (15.2 m) and the cable enters the building from the outside and is terminated at a grounding block.

Table 820-50.   Cable Markings

Cable Marking	Type	Reference
CATVP	CATV plenum cable	Sections 820-51(a) and 820-53(a)
CATVR	CATV riser cable	Sections 820-51(b) and 820-53(b)
CATV	CATV cable	Sections 820-51(c) and 820-53(c)
CATVX	CATV cable, limited use	Sections 820-51(d) and 820-53(c), Exception Nos. 1, 2, and 3

(FPN No. 1):   Cable types are listed in descending order of fire-resistance rating.

(FPN No. 2):   See the referenced sections for listing requirements and permitted uses.

820-51. Additional Listing Requirements.   Cables shall be listed in accordance with (a) through (d) below.

(a) Type CATVP. Type CATVP community antenna television plenum cable shall be listed as being suitable for use in ducts, plenums, and other space used for environmental air and shall also be listed as having adequate fire-resistant and low-smoke-producing characteristics.

(FPN):   One method of defining low-smoke-producing cables is by establishing an acceptable value of the smoke produced when tested in accordance with the Test for Fire and Smoke Characteristics of Wires and Cables, NFPA 262-1990 (ANSI), to a maximum peak optical density of 0.5 and a maximum average optical density of 0.15. Similarly, one method of defining fire-resistant cables is by establishing maximum allowable flame travel distance of 5 feet (1.52 m) when tested in accordance with the same test.

(b) Type CATVR. Type CATVR community antenna television riser cable shall be listed as being suitable for use in a vertical run in a shaft or from floor to floor and shall also be listed as having fire-resistant characteristics capable of preventing the carrying of fire from floor to floor.

(FPN): One method of defining fire-resistant characteristics capable of preventing the carrying of fire from floor to floor is that the cables pass the requirements of the Standard Test for Flame Propagation Height of Electrical and Optical-Fiber Cable Installed Vertically in Shafts, ANSI/UL 1666-1986.

(c) Type CATV. Type CATV community antenna television cable shall be listed as being suitable for general purpose CATV use, with the exception of risers and plenums, and shall also be listed as being resistant to the spread of fire.

(FPN): One method of defining resistant to the spread of fire is that the cables do not spread fire to the top of the tray in the "Vertical-Tray Flame Test" in the Reference Standard for Electrical Wires, Cables and Flexible Cords, ANSI/UL 1581-1985.

Another method of defining resistant to the spread of fire is for the damage (char length) not to exceed 4 feet 11 inches (1.5 m) when performing the "Vertical Flame Test—Cables in Cable Trays," as described in Test Methods for Electrical Wires and Cables, CSA C22.2 No. 0.3-M-1985.

(d) Type CATVX. Type CATVX limited-use community antenna television cable shall be listed as being suitable for use in dwellings and for use in raceway and shall also be listed as being flame-retardant.

(FPN): One method of determining that cable is flame-retardant is by testing the cable to the "VW-1 (Vertical-Wire) Flame Test" in the Reference Standard for Electrical Wires, Cables and Flexible Cords, ANSI/UL 1581-1985.

820-52. Installation of Cables and Equipment. Beyond the point of grounding, as defined in Section 820-33, the cable installation shall comply with (a) through (e) below.

(a) Separation from Other Conductors.

(1) Open Conductors. Coaxial cable shall be separated at least 2 inches (50.8 mm) from conductors of any electric light or power circuits or Class 1 circuits.

Exception No. 1: Where the electric light or power or Class 1 or coaxial cable circuit conductors are in a raceway, or in metal-sheathed, metal-clad, nonmetallic-sheathed, or Type UF cables.

Exception No. 2: Where the conductors are permanently separated from the conductors of the other circuits by a continuous and firmly fixed nonconductor, such as porcelain tubes or flexible tubing, in addition to the insulation on the wire.

(2) In Enclosures and Raceways.

a. Other Power-Limited Circuits. Coaxial cables shall be permitted in the same raceway or enclosure with jacketed cables of any of the following:

    1. Class 2 and Class 3 remote-control, signaling, and power-limited circuits in compliance with Article 725.

    2. Power-limited fire protective signaling systems in compliance with Article 760.

    3. Communications circuits in compliance with Article 800.

    4. Optical fiber cables in compliance with Article 770.

b. Electric Light or Power Circuits. Coaxial cable shall not be placed in any raceway, compartment, outlet box, junction box, or other enclosures with conductors of electric light or power or Class 1 circuits.

Exception No. 1: Where the conductors of the different systems are separated by a permanent barrier.

Exception No. 2: Conductors in outlet boxes, junction boxes, or similar fittings or compartments where such conductors are introduced solely for power supply to the coaxial cable system distribution equipment or for power connection to remote-control equipment.

The electric light, power, and Class 1 circuit conductors and nonpower-limited fire protective signaling circuit conductors shall be routed within the enclosure to maintain a minimum 0.25-inch (6.35-mm) separation from the coaxial cable.

(3) In Shafts. Coaxial cables run in the same shaft with conductors of electric light or power shall be separated from electric light or power conductors by not less than 2 inches (50.8 mm).

Exception No. 1: Where the conductors of either system are encased in metal raceway.

Exception No. 2: Where the electric light or power conductors are in raceway, or in metal-sheathed, metal-clad, nonmetallic-sheathed, or Type UF cables.

(b) Spread of Fire or Products of Combustion. Installations in hollow spaces, vertical shafts, and ventilation or air-handling ducts shall be so made that the possible spread of fire or products of combustion will not be substantially increased. Openings around penetrations through fire-resistance-rated walls, partitions, floors, or ceilings shall be firestopped using approved methods.

(c) Equipment in Other Space Used for Environmental Air. Section 300-22(c) shall apply.

(d) Hybrid Power and Coaxial Cabling. The provisions of Section 780-6 shall apply for listed hybrid power and coaxial cabling in closed-loop and programmed power distribution.

(e) Support of Conductors. Raceways shall not be used as a means of support for coaxial cables.

820-53. Applications of Listed CATV Cables. CATV cables shall comply with (a) through (d) below.

(a) Plenum. Cables installed in ducts, plenums, and other spaces used for environmental air shall be Type CATVP.

Exception: Types CATVP, CATVR, CATV, and CATVX cables installed in compliance with Section 300-22.

(b) Riser. Cables installed in vertical runs and penetrating more than one floor, or cables installed in vertical runs in a shaft, shall be Type CATVR. Floor penetrations requiring Type CATVR shall contain only cables suitable for riser or plenum use.

Exception No. 1: Types CATV and CATVX cables encased in metal raceway or located in a fireproof shaft having firestops at each floor.

Exception No. 2: Types CATV and CATVX cables in one- and two-family dwellings.

(FPN): See Section 820-52(b) for the firestop requirements for floor penetrations.

(c) Other Wiring Within Buildings. Cables installed in building locations other than the locations covered in (a) and (b) above shall be Type CATV.

Exception No. 1: Type CATVX cable enclosed in raceway.

Exception No. 2: Type CATVX cable in nonconcealed spaces where the exposed length of cable does not exceed 10 feet (3.05 m).

Exception No. 3: Type CATVX cables that are less than .375 inch (9.52 mm) in diameter and installed in one- or two-family dwellings.

Exception No. 4: Type CATVX cables that are less than .375 inch (9.52 mm) in diameter and installed in nonconcealed spaces in multi-family dwellings.

(d) Cable Substitutions. The substitutions for community antenna television cables listed in Table 820-53 shall be permitted.

Table 820-53.   Coaxial Cable Substitutions

Cable Type	Permitted Substitutions
CATVP	MPP, CMP
CATVR	CATVP, MPP, CMP, MPR, CMR
CATV	CATVP, MPP, CMP, CATVR, MPR, CMR, MPG, MP, CMG, CM
CATVX	CATVP, MPP, CMP, CATVR, MPR, CMR, CATV, MPG, MP, CMG, CM

Chapter 9   Tables and Examples

A. Tables
Notes to Tables

1. Tables 3A, 3B, and 3C apply only to complete conduit or tubing systems and are not intended to apply to sections of conduit or tubing used to protect exposed wiring from physical damage.

2. Equipment grounding or bonding conductors, where installed, shall be included when calculating conduit or tubing fill. The actual dimensions of the equipment grounding or bonding conductor (insulated or bare) shall be used in the calculation.

3. Where conduit or tubing nipples having a maximum length not to exceed 24 inches (610 mm) are installed between boxes, cabinets, and similar enclosures, the nipple shall be permitted to be filled to 60 percent of its total cross-sectional area, and Article 310, Note 8(a) of Notes to Ampacity Tables of 0 to 2000 Volts need not apply to this condition.

4. For conductors not included in Chapter 9, such as multiconductor cables, the actual dimensions shall be used.

5. See Table 1 for allowable percentage of conduit or tubing fill.

(FPN):   Table 1 is based on common conditions of proper cabling and alignment of conductors where the length of the pull and the number of bends are within reasonable limits. It should be recognized that for certain conditions a larger size conduit or a lesser conduit fill should be considered.

Table 1.   Percent of Cross Section of Conduit and Tubing for Conductors
(See Table 2 for Fixture Wires)

Number of Conductors	1	2	3	4	Over 4
All conductor types except lead-covered	53	31	40	40	40
Lead-covered conductors	55	30	40	38	35

Note 1. See Tables 3A, 3B, and 3C for number of conductors all of the same size in trade sizes of conduit or tubing $\frac{1}{2}$ inch through 6 inch.

Note 2: See Table 5B for number of compact conductors all of the same size in trade sizes of conduit or tubing $\frac{1}{2}$ inch through 4 inch.

Note 3. For conductors larger than 750 kcmil or for combinations of conductors of different sizes, use Tables 4 through 8, Chapter 9, for dimensions of conductors, conduit, and tubing.

Note 4. Where the calculated conductors, all of the same size (total cross-sectional area including insulation), include a decimal fraction, the next higher whole number shall be used where this decimal is 0.8 or larger.

Note 5. Where bare conductors are permitted by other sections of this Code, the dimensions for bare conductors in Table 8 of Chapter 9 shall be permitted.

Note 6. A multiconductor cable of two or more conductors shall be treated as a single conductor cable for calculating percentage conduit fill area. For cables that have elliptical cross section, the cross-sectional area calculation shall be based on using the major diameter of the ellipse as a circle diameter.

Table 2. Maximum Number of Fixture Wires in Trade Sizes of Conduit or Tubing
(40 Percent Fill Based on Individual Diameters)

Conduit Trade Size (Inches) Wire Types	1/2					3/4					1					1¼					1½					2				
	18	16	14	12	10	18	16	14	12	10	18	16	14	12	10	18	16	14	12	10	18	16	14	12	10	18	16	14	12	10
PTF, PTFF, PGFF, PGF, PFF, PF, PAF, PAFF, ZF, ZFF	23	18	14			40	31	24			65	50	39			115	90	70			157	122	95			257	200	156		
TFFN, TFN	19	15				34	26				55	43				97	76				132	104				216	169			
SF-1	16					29					47					83					114					186				
SFF-1	15					26					43					76					104					169				
TF	11	10				20	18				32	30				57	53				79	72				129	118			
RFH-1	11					20					32					57					79					129				
TFF	11	10				20	17				32	27				56	49				77	66				126	109			
AF	11	9	7	4	3	19	16	12	7	5	31	26	20	11	8	55	46	36	19	15	75	63	49	27	20	123	104	81	44	34
SFF-2	9	7	6			16	12	10			27	20	17			47	36	30			65	49	42			106	81	68		
SF-2	9	8	6			16	14	11			27	23	18			47	40	32			65	55	43			106	90	71		
FFH-2	9	7				15	12				25	19				44	34				60	46				99	75			
RFH-2	7	5				12	10				20	16				36	28				49	38				80	62			
KF-1, KFF-1, KF-2, KFF-2	36	32	22	14	9	64	55	39	25	17	103	89	63	41	28	182	158	111	73	49	248	216	152	100	67	406	353	248	163	110

Table 3A. Maximum Number of Conductors in Trade Sizes of Conduit or Tubing
(Based on Table 1, Chapter 9)

Type Letters	Conductor Size AWG/kcmil	1/2	3/4	1	1 1/4	1 1/2	2	2 1/2	3	3 1/2	4	5	6
TW, XHHW (14 through 18) RH (14 + 12)	14	9	15	25	44	60	99	142	171	176	108	133	
	12	7	12	19	35	47	78	111	131	84			
	10	5	9	15	26	36	60	85	62				
	8	2	4	7	12	17	28	40					
RHW and RHH (without outer covering), RH (10 + 8), THW, THHW	14	6	10	16	29	40	65	93	143	192	163		
	12	4	8	13	24	32	53	76	117	157	85		
	10	4	6	11	19	26	43	61	95	127			
	8	1	3	5	10	13	22	32	49	66			
TW,	6	1	2	4	7	10	16	23	36	48	62	97	141
	4	1	1	3	5	7	12	17	27	36	47	73	106
THW,	3	1	1	2	4	6	10	15	23	31	40	63	91
	2	1	1	2	4	5	9	13	20	27	34	54	78
	1		1	1	3	4	6	9	14	19	25	39	57
FEPB (6 through 2), RHW and RHH (without outer covering)	1/0		1	1	2	3	5	8	12	16	21	33	49
	2/0		1	1	1	3	5	7	10	14	18	29	41
	3/0		1	1	1	2	4	6	9	12	15	24	35
	4/0			1	1	1	3	5	7	10	13	20	29
	250			1	1	1	2	4	6	8	10	16	23
	300			1	1	1	2	3	5	7	9	14	20
	350				1	1	1	3	4	6	8	12	18
	400				1	1	1	2	4	5	7	11	16
	500				1	1	1		3	4	6	9	14
RH, THHW	600				1	1	1	1	3	4	5	7	11
	700				1	1	1	1	2	3	4	7	10
	750				1	1	1		2	3	4	6	9

Note 1. This table is for concentric stranded conductors only. For cables with compact conductors, the dimensions in Table 5A shall be used.
Note 2. Conduit fill for conductors with a -2 suffix is the same as for those types without the suffix.

Table 3B. Maximum Number of Conductors in Trade Sizes of Conduit or Tubing
(Based on Table 1, Chapter 9)

Type Letters	Conductor Size AWG/kcmil	1/2	3/4	1	1¼	1½	2	2½	3	3½	4	5	6
THWN,	14	13	24	39	69	94	154	164	160	106	136	154	
	12	10	18	29	51	70	114	104	79				
	10	6	11	18	32	44	73	51					
	8	3	5	9	16	22	36						
THHN, FEP (14 through 2), FEPB (14 through 8), PFA (14 through 4/0), PFAH (14 through 4/0)	6	1	4	6	11	15	26	37	57	76	98	154	137
	4	1	2	4	7	9	16	22	35	47	60	94	116
	3	1	1	3	6	8	13	19	29	39	51	80	97
	2		1	3	5	7	11	16	25	33	43	67	72
	1		1	1	3	5	8	12	18	25	32	50	
Z (14 through 4/0) XHHW (4 through 500 kcmil)	1/0		1	1	3	4	7	10	15	21	27	42	61
	2/0		1	1	2	3	6	8	13	17	22	35	51
	3/0		1	1	1	3	5	7	11	14	18	29	42
	4/0		1	1	1	2	4	6	9	12	15	24	35
	250		1	1	1	1	3	4	7	10	12	20	28
	300		1	1	1	1	3	4	6	8	11	17	24
	350		1	1	1	1	2	3	5	7	9	15	21
	400				1	1	1	3	5	6	8	13	19
	500				1	1	1	2	4	5	7	11	16
	600				1	1	1	1	3	4	5	9	13
	700					1	1	1	3	4	5	8	11
	750					1	1	1	2	3	4	7	11
XHHW	6	1	3	5	9	13	21	30	47	63	81	128	185
	600				1	1	1	1	3	4	5	9	13
	700					1	1	1	3	4	5	7	11
	750					1	1	1	2	3	4	7	10

Note 1. This table is for concentric stranded conductors only. For cables with compact conductors, the dimensions in Table 5A shall be used.
Note 2. Conduit fill for conductors with a -2 suffix is the same as for those types without the suffix.

Table 3C. Maximum Number of Conductors in Trade Sizes of Conduit or Tubing
(Based on Table 1, Chapter 9)

Conductor Size AWG/kcmil	½	¾	1	1¼	1½	2	2½	3	3½	4	5	6
RHW,												
14	3	6	10	18	25	41	58	90	121	155		
12	3	5	9	15	21	35	50	77	103	132		
10	2	4	7	13	18	29	41	64	86	110		
8	1	2	4	7	9	16	22	35	47	60	94	137
RHH												
6	1	1	2	5	6	11	15	24	32	41	64	93
4	1	1	1	3	5	8	12	18	24	31	50	72
3	1	1	1	3	4	7	10	16	22	28	44	63
(with 2		1	1	3	4	6	9	14	19	24	38	56
outer 1		1	1	1	3	5	7	11	14	18	29	42
covering)												
1/0		1	1	1	2	4	6	9	12	16	25	37
2/0			1	1	1	3	5	8	11	14	22	32
3/0			1	1	1	3	4	7	9	12	19	28
4/0			1	1	1	2	4	6	8	10	16	24
250				1	1	1	3	5	6	8	13	19
300					1	1	3	4	5	7	11	17
350					1	1	2	4	5	6	10	15
400					1	1	1	3	4	6	9	14
500					1	1	1	3	4	5	8	11
600					1	1	1	2	3	4	6	9
700					1	1	1	1	3	3	6	8
750						1	1	1	3	3	5	8

Note 1. This table is for concentric stranded conductors only. For cables with compact conductors, the dimensions in Table 5A shall be used.
Note 2. Conduit fill for conductors with a -2 suffix is the same as for those types without the suffix.

Tables 4 through 8, Chapter 9. Tables 4 through 8 give the nominal size of conductors and conduit or tubing for use in computing size of conduit or tubing for various combinations of conductors. The dimensions represent average conditions only, and variations will be found in dimensions of conductors and conduits of different manufacture.

Table 4. Dimensions and Percent Area of Conduit and of Tubing

Areas of Conduit or Tubing for the Combinations of Wires Permitted in Table 1, Chapter 9.

Trade Size	Internal Diameter Inches	Total 100%	Area — Square Inches							
			Not Lead-Covered			Lead-Covered				
			2 Cond. 31%	Over 2 Cond. 40%	1 Cond. 53%	1 Cond. 55%	2 Cond. 30%	3 Cond. 40%	4 Cond. 38%	Over 4 Cond. 35%
½	.622	.30	.09	.12	.16	.17	.09	.12	.11	.11
¾	.824	.53	.16	.21	.28	.29	.16	.21	.20	.19
1	1.049	.86	.27	.34	.46	.47	.26	.34	.33	.30
1¼	1.380	1.50	.47	.60	.80	.83	.45	.60	.57	.53
1½	1.610	2.04	.63	.82	1.08	1.12	.61	.82	.78	.71
2	2.067	3.36	1.04	1.34	1.78	1.85	1.01	1.34	1.28	1.18
2½	2.469	4.79	1.48	1.92	2.54	2.63	1.44	1.92	1.82	1.68
3	3.068	7.38	2.26	2.95	3.91	4.06	2.21	2.95	2.80	2.58
3½	3.548	9.90	3.07	3.96	5.25	5.44	2.97	3.96	3.76	3.47
4	4.026	12.72	3.94	5.09	6.74	7.00	3.82	5.09	4.83	4.45
5	5.047	20.00	6.20	8.00	10.60	11.00	6.00	8.00	7.60	7.00
6	6.065	28.89	8.96	11.56	15.31	15.89	8.67	11.56	10.98	10.11

Table 5.   Dimensions of Rubber-Covered and Thermoplastic-Covered Conductors

Size AWG kcmil	Types RFH-2, RH, RHH,*** RHW-2, RHW,*** SF-2		Types TF, THW†, TW, THHW		Types TFN, THHN, THWN		Types FEP, FEPB, FEPW, TFE, PF, PFA, PFAH, PGF, PTF, Z, ZF, ZFF		Types XHHW, ZW††, XHHW-2		Types KF-1, KF-2, KFF-1, KFF-2	
	Approx. Diam. Inches	Approx. Area Sq. In.	Approx. Diam. Inches	Approx. Area Sq. In.	Approx. Diam. Inches	Approx. Area Sq. In.	Approx. Diam. Inches	Approx. Area Sq. In.	Approx. Diam. Inches	Approx. Area Sq. In.	Approx. Diam. Inches	Approx. Area Sq. In.
Col.1	Col.2	Col. 3	Col. 4	Col. 5	Col. 6	Col. 7	Col. 8	Col. 9	Col. 10	Col. 11	Col. 12	Col. 13
18	.146	.0167	.106	.0088	.089	.0062	.081 .105	.0052 .0087			.065	.0033
16	.158	.0196	.118	.0109	.100	.0079	.092	.0066			.070	.0038
14	30 mils .171	.0230	.131	.0135	.105	.0087	.105 .105	.0087 .0087			.083	.0054
14	45 mils .204*	.0327*										
14			.162†	.0206†			.121	.0115	.129	.0131		
12	30 mils .188	.0278	.148	.0172	.122	.0117	.121 .121	.0115 .0115			.102	.0082
12	45 mils .221*	.0384*										
12			.179†	.0252†			.142	.0158	.146	.0167		
10	.242	.0460	.168	.0222	.153	.0184	.142 .142	.0158 .0158	.166	.0216	.124	.0121
10			.199†	.0311†			.186	.0272				
8	.328	.0845	.245	.0471	.218	.0373	.206 .186	.0333 .0272	.241	.0456		
8			.276†	.0598†								
6	.397	.1238	.323	.0819	.257	.0519	.244	.0468	.282	.0625		
4	.452	.1605	.372	.1087	.328	.0845	.292	.0670	.328	.0845		
3	.481	.1817	.401	.1263	.356	.0995	.320	.0804	.356	.0995		
2	.513	.2067	.433	.1473	.388	.1182	.352	.0973	.388	.1182		
1	.588	.2715	.508	.2027	.450	.1590	.420	.1385	.450	.1590		
1/0	.629	.3107	.549	.2367	.491	.1893	.462	.1676	.491	.1893		
2/0	.675	.3578	.595	.2781	.537	.2265	.498	.1948	.537	.2265		
3/0	.727	.4151	.647	.3288	.588	.2715	.560	.2463	.588	.2715		
4/0	.785	.4840	.705	.3904	.646	.3278	.618	.3000	.646	.3278		

Table 5. (Continued)

Size AWG kcmil	Types RFH-2, RH, RHH, RHW-2, RHW, SF-2		Types TF, THW, TW, THHW		Types TFN, THHN, THWN		Types FEP, FEPB, FEPW, TFE, PF, PFA, PFAH, PGF, PTF, Z, ZF, ZFF		Types XHHW, ZW, XHHW-2	
	Approx. Diam. Inches	Approx. Area Sq. In.	Approx. Diam. Inches	Approx. Area Sq. In.	Approx. Diam. Inches	Approx. Area Sq. In.	Approx. Diam. Inches	Approx. Area Sq. In.	Approx. Diam. Inches	Approx. Area Sq. In.
Col.1	Col.2	Col. 3	Col. 4	Col. 5	Col. 6	Col. 7	Col. 8	Col. 9	Col. 10	Col. 11
250	.868	.5917	.788	.4877	.716	.4026	....	....	.716	.4026
300	.933	.6837	.843	.5581	.771	.4669	....	....	.771	.4669
350	.985	.7620	.895	.6291	.822	.5307	....	....	.822	.5307
400	1.032	.8365	.942	.6969	.869	.5931	....	....	.869	.5931
500	1.119	.9834	1.029	.8316	.955	.7163	....	....	.955	.7163
600	1.233	1.1940	1.143	1.0261	1.058	.8791	....	....	1.073	.9043
700	1.304	1.3355	1.214	1.1575	1.129	1.0011	....	....	1.145	1.0297
750	1.339	1.4082	1.249	1.2252	1.163	1.0623	....	....	1.180	1.0936
800	1.372	1.4784	1.282	1.2908	1.196	1.1234	....	....	1.210	1.1499
900	1.435	1.6173	1.345	1.4208	1.259	1.2449	....	....	1.270	1.2668
1000	1.494	1.7530	1.404	1.5482	1.317	1.3623	....	....	1.330	1.3893
1250	1.676	2.2062	1.577	1.9532	....	....	....	....	1.500	1.7671
1500	1.801	2.5475	1.702	2.2751	....	....	....	....	1.620	2.0612
1750	1.916	2.8832	1.817	2.5930	....	....	....	....	1.740	2.3779
2000	2.021	3.2079	1.922	2.9013	....	....	....	....	1.840	2.6590

* The dimensions of Types RHH and RHW with outer covering.
† Dimensions of THW in sizes No. 14 through No. 8. Nos. 6 THW and larger are the same dimension as TW.
*** Dimensions of RHH and RHW without outer coverings are the same as THW No. 18 through No. 8 and larger, stranded.
**** In Columns 8 and 9, the values shown for size Nos. 1 through No. 10, solid; Nos. 8 and larger, stranded. The right-hand values in Columns 8 and 9 are for FEPB, Z, ZF, and ZFF only.
†† No. 14 through No. 2.

Table 5A.　Compact Aluminum Building Wire Nominal Dimensions* and Areas

Bare Conductor			Types THW and THHW		Type THHN		Type XHHW		
Size AWG or kcmil	Number of Strands	Diam. Inches	Approx. Diam. Inches	Approx. Area Sq. In.	Approx. Diam. Inches	Approx. Area Sq. In.	Approx. Diam. Inches	Approx. Area Sq. Inches	Size AWG or kcmil
8	7	.134	.255	.0510	—	—	.224	.0394	8
6	7	.169	.290	.0660	.240	.0452	.260	.0530	6
4	7	.213	.335	.0881	.305	.0730	.305	.0730	4
2	7	.268	.390	.1194	.360	.1017	.360	.1017	2
1	19	.299	.465	.1698	.415	.1352	.415	.1352	1
1/0	19	.336	.500	.1963	.450	.1590	.450	.1590	1/0
2/0	19	.376	.545	.2332	.495	.1924	.490	.1885	2/0
3/0	19	.423	.590	.2733	.540	.2290	.540	.2290	3/0
4/0	19	.475	.645	.3267	.595	.2780	.590	.2733	4/0
250	37	.520	.725	.4128	.670	.3525	.660	.3421	250
300	37	.570	.775	.4717	.720	.4071	.715	.4015	300
350	37	.616	.820	.5281	.770	.4656	.760	.4536	350
400	37	.659	.865	.5876	.815	.5216	.800	.5026	400
500	37	.736	.940	.6939	.885	.6151	.880	.6082	500
600	61	.813	1.050	.8659	.985	.7620	.980	.7542	600
700	61	.877	1.110	.9676	1.050	.8659	1.050	.8659	700
750	61	.908	1.150	1.0386	1.075	.9076	1.090	.9331	750
1000	61	1.060	1.285	1.2968	1.255	1.2370	1.230	1.1882	1000

* Dimensions are from industry sources.

Table 5B. Maximum Number of Compact Conductors in Trade Sizes of Conduit or Tubing

Insulation Type

Conduit or Tubing Trade Size

Conductor Size AWG or kcmil	1 in. THW	1 in. THHN	1 in. XHHW	1¼ in. THW	1¼ in. THHN	1¼ in. THHW	1¼ in. XHHW	1½ in. THW	1½ in. THHN	1½ in. THHW	1½ in. XHHW	2 in. THW	2 in. THHN	2 in. THHW	2 in. XHHW	2½ in. THW	2½ in. THHN	2½ in. THHW	2½ in. XHHW	3 in. THW	3 in. THHN	3 in. THHW	3 in. XHHW	3½ in. THW	3½ in. THHN	3½ in. THHW	3½ in. XHHW	4 in. THW	4 in. THHN	4 in. THHW	4 in. XHHW
6	5	7	6	9	13	13	11	12	18	18	15	15	18	18																	
4	4	4	4	7	8	8	8	9	11	11	11	11	13	13																	
2	3	3	3	5	6	6	6	7	8	8	8	8	10	10																	
1				3	4	4	4	5	6	6	6					11	14	14	14												
1/0				3	3	3	3	4	5	5	5	7	8	8	8	9	12	12		12	15	15									
2/0					3	3	3	3	4	4	4	5	7	7	7	8	10	10	10	10	13	13									
3/0								3	3	3	3	5	6	6	6	7	8	8	8	9	10	10		12	14	14	14				
4/0									3	3	3	4	5	5	5	6	7	7	7	7	8	8									
250												3	4	4	4	4	5	5	5	7	8	8	8	9	11	11		10	12	12	
300												3	3	3	3	4	4	4	4	6	7	7	7	8	9	10		9	11	12	
350													3	3	3	3	4	4	4	5	6	6	6	7	8	8	9	8	9	11	
400																3	3	3	3	5	5	5	6	6	7	8	8	8	8	9	10
500																	3	3	3	4	4	4	5	5	6	6	7	7	8	8	8
600																				3	3	4	4	4	5	5	6	6	6	7	7
700																				3	3	3	4	4	4	5	5	5	5	6	6
750																							4	3	4	4	5	5	5	5	5
1000																							3	3	3	3	4	4	4	4	4

Definition: Compact stranding is the result of a manufacturing process where the standard conductor is compressed to the extent that the interstices (voids between strand wires) are virtually eliminated.

Table 6. Dimensions of Lead-Covered Conductors
Types RL, RHL, and THHW

Size AWG kcmil	Single Conductor		Two Conductor		Three Conductor	
	Diam. Inches	Area Sq. In.	Diam. Inches	Area Sq. In.	Diam. Inches	Area Sq. In.
14	.28	.062	.28 × .47	.115	.59	.273
12	.29	.066	.31 × .54	.146	.62	.301
10	.35	.096	.35 × .59	.180	.68	.363
8 sol.	.41	.132	.41 × .71	.255	.82	.528
8 str.	.43	.145	.43 × .75	.282	.86	.581
6	.49	.188	.49 × .86	.369	.97	.738
4	.55	.237	.54 × .96	.457	1.08	.916
2	.60	.283	.61 × 1.08	.578	1.21	1.146
1	.67	.352	.70 × 1.23	.756	1.38	1.49
1/0	.71	.396	.74 × 1.32	.859	1.47	1.70
2/0	.76	.454	.79 × 1.41	.980	1.57	1.94
3/0	.81	.515	.84 × 1.52	1.123	1.69	2.24
4/0	.87	.593	.90 × 1.64	1.302	1.85	2.68
250	.98	.754	....	....	2.02	3.20
300	1.04	.85	....	....	2.15	3.62
350	1.10	.95	....	....	2.26	4.02
400	1.14	1.02	....	....	2.40	4.52
500	1.23	1.18	....	....	2.59	5.28

The above cables are limited to straight runs or with nominal offsets equivalent to not more than two quarter bends.

Note: No. 14 through No. 10, solid conductors; No. 8, solid or stranded conductors; Nos. 6 and larger, stranded conductors.

Table 8. Conductor Properties

| Size AWG/ kcmil | Area Cir. Mills | Conductors | | | | DC Resistance at 75°C (167°F) | | |
| | | Stranding | | Overall | | Copper | | Aluminum |
		Quan- tity	Diam. In.	Diam. In.	Area In.²	Uncoated ohm/kFT	Coated ohm/kFT	ohm/ kFT
18	1620	1	—	0.040	0.001	7.77	8.08	12.8
18	1620	7	0.015	0.046	0.002	7.95	8.45	13.1
16	2580	1	—	0.051	0.002	4.89	5.08	8.05
16	2580	7	0.019	0.058	0.003	4.99	5.29	8.21
14	4110	1	—	0.064	0.003	3.07	3.19	5.06
14	4110	7	0.024	0.073	0.004	3.14	3.26	5.17
12	6530	1	—	0.081	0.005	1.93	2.01	3.18
12	6530	7	0.030	0.092	0.006	1.98	2.05	3.25
10	10380	1	—	0.102	0.008	1.21	1.26	2.00
10	10380	7	0.038	0.116	0.011	1.24	1.29	2.04
8	16510	1	—	0.128	0.013	0.764	0.786	1.26
8	16510	7	0.049	0.146	0.017	0.778	0.809	1.28
6	26240	7	0.061	0.184	0.027	0.491	0.510	0.808
4	41740	7	0.077	0.232	0.042	0.308	0.321	0.508
3	52620	7	0.087	0.260	0.053	0.245	0.254	0.403
2	66360	7	0.097	0.292	0.067	0.194	0.201	0.319
1	83690	19	0.066	0.332	0.087	0.154	0.160	0.253
1/0	105600	19	0.074	0.373	0.109	0.122	0.127	0.201
2/0	133100	19	0.084	0.419	0.138	0.0967	0.101	0.159
3/0	167800	19	0.094	0.470	0.173	0.0766	0.0797	0.126
4/0	211600	19	0.106	0.528	0.219	0.0608	0.0626	0.100
250	—	37	0.082	0.575	0.260	0.0515	0.0535	0.0847
300	—	37	0.090	0.630	0.312	0.0429	0.0446	0.0707
350	—	37	0.097	0.681	0.364	0.0367	0.0382	0.0605
400	—	37	0.104	0.728	0.416	0.0321	0.0331	0.0529
500	—	37	0.116	0.813	0.519	0.0258	0.0265	0.0424
600	—	61	0.099	0.893	0.626	0.0214	0.0223	0.0353
700	—	61	0.107	0.964	0.730	0.0184	0.0189	0.0303
750	—	61	0.111	0.998	0.782	0.0171	0.0176	0.0282
800	—	61	0.114	1.03	0.834	0.0161	0.0166	0.0265
900	—	61	0.122	1.09	0.940	0.0143	0.0147	0.0235
1000	—	61	0.128	1.15	1.04	0.0129	0.0132	0.0212
1250	—	91	0.117	1.29	1.30	0.0103	0.0106	0.0169
1500	—	91	0.128	1.41	1.57	0.00858	0.00883	0.0141
1750	—	127	0.117	1.52	1.83	0.00735	0.00756	0.0121
2000	—	127	0.126	1.63	2.09	0.00643	0.00662	0.0106

These resistance values are valid ONLY for the parameters as given. Using conductors having coated strands, different stranding type, and, especially, other temperatures changes the resistance.

Formula for temperature change: $R_2 = R_1 [1 + \alpha(T_2 - 75)]$ where: $\alpha_{cu} = 0.00323$, $\alpha_{AL} = 0.00330$.

Conductors with compact and compressed stranding have about 9 percent and 3 percent, respectively, smaller bare conductor diameters than those shown. See Table 5A for actual compact cable dimensions.

The IACS conductivities used: bare copper = 100%, aluminum = 61%.

Class B stranding is listed as well as solid for some sizes. Its overall diameter and area is that of its circumscribing circle.

(FPN): The construction information is per NEMA WC8-1976 (Rev 5=1980). The resistance is calculated per National Bureau of Standards Handbook 100, dated 1966, and Handbook 109, dated 1972.

Table 9. AC Resistance and Reactance for 600 V Cables, 3-Phase 60, Hz, 75°C (167°F) — Three Single Conductors in Conduit

Ohms to Neutral per 1000 feet

Size AWG/ kcmil	X_L (Reactance) for All Wires		AC Resistance for Uncoated Copper Wires			AC Resistance for Aluminum Wires			Effective Z at .85 PF for Uncoated Copper Wires			Effective Z at .85 PF for Aluminum Wires			Size AWG/ kcmil
	PVC, Al. Conduits	Steel Conduit	PVC Conduit	Al. Conduit	Steel Conduit	PVC Conduit	Al. Conduit	Steel Conduit	PVC Conduit	Al. Conduit	Steel Conduit	PVC Conduit	Al. Conduit	Steel Conduit	
14	.058	.073	3.1	3.1	3.1	—	—	—	2.7	2.7	2.7	—	—	—	14
12	.054	.068	2.0	2.0	2.0	3.2	3.2	3.2	1.7	1.7	1.7	2.8	2.8	2.8	12
10	.050	.063	1.2	1.2	1.2	2.0	2.0	2.0	1.1	1.1	1.1	1.8	1.8	1.8	10
8	.052	.065	0.78	0.78	0.78	1.3	1.3	1.3	0.69	0.69	0.70	1.1	1.1	1.1	8
6	.051	.064	0.49	0.49	0.49	0.81	0.81	0.81	0.44	0.45	0.45	0.71	0.72	0.72	6
4	.048	.060	0.31	0.31	0.31	0.51	0.51	0.51	0.29	0.29	0.30	0.46	0.46	0.46	4
3	.047	.059	0.25	0.25	0.25	0.40	0.41	0.40	0.23	0.24	0.24	0.37	0.37	0.37	3
2	.045	.057	0.19	0.20	0.20	0.32	0.32	0.32	0.19	0.19	0.20	0.30	0.30	0.30	2
1	.046	.057	0.15	0.16	0.16	0.25	0.26	0.25	0.16	0.16	0.16	0.24	0.24	0.25	1
1/0	.044	.055	0.12	0.13	0.12	0.20	0.21	0.20	0.13	0.13	0.13	0.19	0.20	0.20	1/0
2/0	.043	.054	0.10	0.10	0.10	0.16	0.16	0.16	0.11	0.11	0.11	0.16	0.16	0.16	2/0
3/0	.042	.052	0.077	0.082	0.079	0.13	0.13	0.13	0.088	0.092	0.094	0.13	0.13	0.14	3/0
4/0	.041	.051	0.062	0.067	0.063	0.10	0.11	0.10	0.074	0.078	0.080	0.11	0.11	0.11	4/0
250	.041	.052	0.052	0.057	0.054	0.085	0.090	0.086	0.066	0.070	0.073	0.094	0.098	0.10	250
300	.041	.051	0.044	0.049	0.045	0.071	0.076	0.072	0.059	0.063	0.065	0.082	0.086	0.088	300
350	.040	.050	0.038	0.043	0.039	0.061	0.066	0.063	0.053	0.058	0.060	0.073	0.077	0.080	350
400	.040	.049	0.033	0.038	0.035	0.054	0.059	0.055	0.049	0.053	0.056	0.066	0.071	0.073	400
500	.039	.048	0.027	0.032	0.029	0.043	0.048	0.045	0.043	0.048	0.050	0.057	0.061	0.064	500
600	.039	.048	0.023	0.028	0.025	0.036	0.041	0.038	0.040	0.044	0.047	0.051	0.055	0.058	600
750	.038	.048	0.019	0.024	0.021	0.029	0.034	0.031	0.036	0.040	0.043	0.045	0.049	0.052	750
1000	.037	.046	0.015	0.019	0.018	0.023	0.027	0.025	0.032	0.036	0.040	0.039	0.042	0.046	1000

Notes:

1. These values are based on the following constants: UL-type RHH wires with Class B stranding, in cradled configuration. Wire conductivitites are 100% IACS copper and 61% IACS aluminum, and aluminum conduit is 45% IACS. Capacitive reactance is ignored, since it is negligible at these voltages. These resistance values are valid only at 75°C (167°F) and for the parameters as given, but are representative for 600-volt wire types operating at 60 Hz. Multiplying current by effective impedance gives a good approximation for line-to-neutral voltage drop. Effective impedance values shown in this table are valid only at .85 power factor. For another circuit power factor (PF), effective impedance (Zc) can be calculated from R and X_L values given in this table as follows:

2. "Effective Z" is defined as R cos(theta) + X sin(theta), where "theta" is the power factor angle of the circuit.

$Zc = R \times PF + X_L \sin[\arccos(PF)]$.

Table 10.   Expansion Characteristics of PVC Rigid Nonmetallic Conduit
Coefficient of Thermal Expansion = 3.38×10^{-5} in./in./°F

Temperature Change in Degrees F	Length Change in Inches per 100 ft. of PVC Conduit	Temperature Change in Degrees F	Length Change in Inches per 100 ft. of PVC Conduit	Temperature Change in Degrees F	Length Change in Inches per 100 ft. of PVC Conduit	Temperature Change in Degrees F	Length Change in Inches per 100 ft. of PVC Conduit
5	0.2	55	2.2	105	4.2	155	6.3
10	0.4	60	2.4	110	4.5	160	6.5
15	0.6	65	2.6	115	4.7	165	6.7
20	0.8	70	2.8	120	4.9	170	6.9
25	1.0	75	3.0	125	5.1	175	7.1
30	1.2	80	3.2	130	5.3	180	7.3
35	1.4	85	3.4	135	5.5	185	7.5
40	1.6	90	3.6	140	5.7	190	7.7
45	1.8	95	3.8	145	5.9	195	7.9
50	2.0	100	4.1	150	6.1	200	8.1

B. Examples

Selection of Conductors. In the following examples, the results are generally expressed in amperes. To select conductor sizes, refer to the 0 to 2000 volt ampacity tables of Article 310 and the Notes that pertain to such tables.

Voltage. For uniform application of Articles 210, 215, and 220, a nominal voltage of 120, 120/240, 240, and 208Y/120 volts shall be used in computing the ampere load on the conductor.

Fractions of an Ampere. Except where the computations result in a major fraction of an ampere (0.5 or larger), such fractions shall be permitted to be dropped.

Ranges. For the computation of the range loads in these examples, Column A of Table 220-19 has been used. For optional methods, see Columns B and C of Table 220-19. Except where the computations result in a major fraction of a kilowatt (0.5 or larger), such fractions shall be permitted to be dropped.

SI Units: For SI units: one square foot = 0.093 square meter; one foot = 0.3048 meter.

Example No. 1(a). One-Family Dwelling

The dwelling has a floor area of 1500 sq. ft., exclusive of unoccupied cellar, unfinished attic, and open porches. Appliances are a 12-kW range and a 5.5 kW, 240-volt dryer. Assume range and dryer kW ratings equivalent to kVA ratings in accordance with Sections 220-18 and 220-19.

Computed Load [see Section 220-10(a)]:

General Lighting Load:

1500 sq. ft. at 3 volt-amperes per sq. ft = 4500 volt-amperes.

Minimum Number of Branch Circuits Required [see Section 220-4(b)]:

General Lighting Load:

4500 volt-amperes ÷ 120 volts = 37.5 A: This requires three 15 A 2-wire or two 20 A 2-wire circuits

Small Appliance Load: Two 2-wire 20 A circuits [see Section 220-3(b)]

Laundry Load: One 2-wire 20 A circuit [see Section 220-4(c)]

Minimum Size Feeder Required [see Section 220-10(a)]:

General Lighting	4500 volt-amperes
Small Appliance Load	3000 volt-amperes
Laundry	1500 volt-amperes
Total General Light and Small Appliance	9000 volt-amperes
3000 volt-amperes at 100%	3000 volt-amperes
9000 − 3000 = 6000 volt-amperes at 35%	2100 volt-amperes
Net General Lighting and Small Appliance Load	5100 volt-amperes
Range Load (see Table 220-19)	8000 volt-amperes
Dryer Load (see Table 220-18)	5500 volt-amperes
Total Load	18,600 volt-amperes

For 120/240-volt 3-wire single-phase service or feeder,

18,600 volt-amperes ÷ 240 volts = 77.5 A.

Net computed load exceeds 10kVA. Service conductors shall be 100 amperes [see Section 230-42(b)(2)].

Example No. 1(a) (Continued)

Neutral for Feeder and Service

Lighting and Small Appliance Load	5100 volt-amperes
Range Load 8000 volt-amperes at 70%	5600 volt-amperes
Dryer Load 5500 volt-amperes at 70%	3850 volt-amperes
Total	14,550 volt-amperes

14,550 volt-amperes ÷ 240 Volts = 60.6 amperes

Example No. 1(b). One-Family Dwelling

Same conditions as Example No. 1(a), plus addition of one 6-ampere 230-volt room air-conditioning unit and one 12-ampere 115-volt room air-conditioning unit,* one 8-ampere 115-volt rated disposal and one 10-ampere 120-volt rated dishwasher.* See Article 430 for general motors and Article 440, Part G, for air-conditioning equipment. Motors have nameplate ratings of 115 V and 230 V for use on 120 V and 240 V nominal voltage systems.

From previous Example No. 1(a), feeder current is 78 amperes (3-wire, 240 volts).

	Line A	Neutral	Line B
Amperes from Example No. 1(a)	78	61	78
One 230-V air conditioner	6	—	6
One 115-V air conditioner and 120-V dishwasher	12	12	10
One 115-V disposal	—	8	8
25% of largest motor (Section 430-24)	3	3	2
Amperes per line	99	84	104

* For feeder neutral, use largest of the two appliances for unbalance.

Example No. 2(a). Optional Calculation for One-Family Dwelling, Heating Larger than Air Conditioning

[See Section 220-30(a) and Table 220-30.]

Dwelling has a floor area of 1500 sq. ft., exclusive of unoccupied cellar, unfinished attic, and open porches. It has a 12-kW range, a 2.5-kW water heater, a 1.2-kW dishwasher, 9 kW of electric space heating installed in five rooms, a 5-kW clothes dryer, and a 6-ampere 230-volt room air-conditioning unit. Assume range, water heater, dishwasher, space heating, and clothes dryer kW ratings equivalent to kVA.

Air conditioner kVA is 6 × 230 ÷ 1000 = 1.38 kVA

1.38 kVA is less than the connected load of 9 kVA of space heating; therefore, the air conditioner load need not be included in the service calculation (see Section 220-21).

1500 sq. ft. at 3 volt-amperes	4500 volt-amperes
Two 20-ampere appliance outlet circuits at 1500 volt-amperes each	3000 volt-amperes
Laundry circuit	1500 volt-amperes
Range (at nameplate rating)	12,000 volt-amperes
Water heater	2500 volt-amperes
Dishwasher	1200 volt-amperes
Clothes dryer	5000 volt-amperes
	29,700 volt-amperes

Example No. 2(a) (Continued)

First 10 kVA of other load at 100% = 10,000 volt-amperes
Remainder of other load at

40% (19.7 kVA × .4)	=	7900 volt-amperes
Total of other load	=	17,900 volt-amperes
9 kVA of heat at 40% (9 × .4)	=	3600 volt-amperes
Total load	=	21,500 volt-amperes

Calculated load for service size
21.5 kVA = 21,500 volt-amperes
21,500 VA ÷ 240 volts = 90 amperes

Therefore, this dwelling shall be permitted to be served by a 100-ampere service.

Feeder Neutral Load, per Section 220-22:

1500 sq. ft. at 3 volt-amperes	4500 volt-amperes
Three 20-amp. circuits at 1500 volt-amperes	4500 volt-amperes
Total	9000 volt-amperes
3000 volt-amperes at 100%	3000 volt-amperes
9000 VA−3000 VA = 6000 volt-amperes at 35%	2100 volt-amperes
	5100 volt-amperes
Range — 8 kVA at 70%	5600 volt-amperes
Clothes dryer — 5 kVA at 70%	3500 volt-amperes
Dishwasher	1200 volt-amperes
Total	15,400 volt-amperes

15,400 volt-amperes ÷ 240 volts = 64.2 amperes

Example No. 2(b). Optional Calculation for One-Family Dwelling, Air Conditioning Larger than Heating

[See Section 220-30(a) and Table 220-30.]

Dwelling has a floor area of 1500 sq. ft., exclusive of unoccupied cellar, unfinished attic, and open porches. It has two 20-ampere small appliance circuits, one 20-ampere laundry circuit, two 4-kW wall-mounted ovens, one 5.1-kW counter-mounted cooking unit, a 4.5-kW water heater, a 1.2-kW dishwasher, a 5-kW combination clothes washer and dryer, six 7-ampere 230-volt room air-conditioning units, and a 1.5-kW permanently installed bathroom space heater. Assume wall-mounted ovens, counter-mounted cooking unit, water heater, dishwasher, and combination clothes washer and dryer kW ratings equivalent to kVA.

Air Conditioning kVA Calculation:

Total amperes 6 × 7 = 42.00 amperes
42 × 240 ÷ 1000 = 10.08 kVA of air-conditioned load; assume P.F. = 1.0

Load included at 100%:

Air Conditioning (see below)
Space heater (omit, see Section 220-21)

Other Load:

1500 sq. ft. at 3 volt-amperes	4500 volt-amperes
Two 20-amp. small appliance circuits at 1500 volt-amperes	3000 volt-amperes
Laundry circuit	1500 volt-amperes
Two ovens	8000 volt-amperes
One cooking unit	5100 volt-amperes

Example No. 2(b) (Continued)

Water heater	4500 volt-amperes
Dishwasher	1200 volt-amperes
Washer/dryer	5000 volt-amperes
Total other load	32,800 volt-amperes
1st 10 kVA at 100%	10,000 volt-amperes
Remainder at 40% (22.8 kVA × .4)	9120 volt-amperes
Total other load	19,120 volt-amperes
Other load	19,120 volt-amperes
Air conditioning	10,080 volt-amperes
	29,200 volt-amperes

29,200 volt-amperes ÷ 240 volts = 122 amperes (service rating)

Feeder Neutral Load, per Section 220-22:

(It is assumed that the two 4-kVA wall-mounted ovens are supplied by one branch circuit, the 5.1-kVA counter-mounted cooking unit by a separate circuit.)

1500 sq. ft. at 3 volt-amperes	4500 volt-amperes
Three 20-amp. circuits at 1500 volt-amperes	4500 volt-amperes
Total	9000 volt-amperes
3000 volt-amperes at 100%	3000 volt-amperes
9000 VA−3000 VA = 6000 volt-amperes at 35%	2100 volt-amperes
	5100 volt-amperes

Two 4-kVA ovens plus one 5.1-kVA cooking unit totals
13.1 kVA
Table 220-19 permits 55% demand factor
13.1 kVA × .55 = 7.2 kVA feeder capacity

7200 volt-amperes × 70% for neutral load	5040 volt-amperes
Clothes washer/dryer — 5 kVA × 70% for neutral load	3500 volt-amperes
Dishwasher	1200 volt-amperes
Total	14,840 volt-amperes

14,840 volt-amperes ÷ 240 volts = 61.83, use 62 amperes

**Example No. 2(c). Optional Calculation for
One-Family Dwelling with Heat Pump
Single-Phase, 240/120-Volt Service**

(See Section 220-30.)

Dwelling has a floor area of 2000 sq. ft., exclusive of unoccupied cellar, unfinished attic, and open porches. It has a 12-kW range, a 4.5-kW water heater, a 1.2-kW dishwasher, a 5-kW clothes dryer, and a 2½-ton (24-ampere) heat pump with 15 kW of back-up heat.

Heat pump kVA is 24 × 240 ÷ 1000 = 5.76 kVA. 5.76 kVA is less than 15 kVA of the back-up heat; therefore, the heat pump load need not be included in the service calculation. (See Table 220-30.)

2000 sq. ft. at 3 volt-amperes	6000 volt-amperes
Two 20-ampere appliance outlet circuits	
at 1500 volt-amperes each	3000 volt-amperes
Laundry circuit	1500 volt-amperes
Range (at nameplate rating)	12,000 volt-amperes
Water heater	4500 volt-amperes
Dishwasher	1200 volt-amperes
Clothes dryer	5000 volt-amperes
	33,200 volt-amperes

Example No. 2(c) (Continued)

First 10 kVA of other load at 100% = 10,000 volt-amperes

Remainder of other load at
40% (23,200 volt-amperes × 0.4) = 9,280 volt-amperes

Total of other load = 19,280 volt-amperes

Heat pump and supplementary heat*

240 × 24 = 5760 volt-amperes

15-kW electric heat: 5760 volt-amperes + 15,000 volt-amperes = 20,760 volt-amperes
or = 20.76 kVA

20.76 kVA at 65% = 13.49 kVA

*If supplementary heat is not on at same time as heat pump, heat pump kVA need not be added to total.

Totals:

Other load	19,280 volt-amperes
Heat pump and supplementary heat	13,490 volt-amperes
Total	32,770 volt-amperes

32.77 kVA ÷ 240 volts= 136.5 amperes

Therefore, this dwelling unit shall be permitted to be served by a 150-ampere service. |

Example No. 3. Store Building

A store 50 ft. by 60 ft., or 3000 sq. ft., has 30 ft. of show window. There are a total of 80 duplex receptacles. The service is 120/240-volt, single-phase (3-wire service). Actual connected lighting load: 8500 volt-amperes.

Computed Load (Section 220-10):
Noncontinuous Loads:

Receptacle Load (Section 220-13)

80 receptacles at 180 volt-amperes = 14,400 volt-amperes

10,000 VA at 100%	10,000 volt-amperes
(14,400 − 10,000) VA at 50%	2200 volt-amperes
Total	12,200 volt-amperes

Continuous Loads:

General Lighting Load:*

3000 sq. ft. at 3 volt-amperes per sq. ft	9000 volt-amperes
Show Window Lighting Load: 30 ft. at 200 volt-amperes per ft.	6000 volt-amperes
Outside sign circuit 1200 volt-amperes [Section 600-6(c)]	1200 volt-amperes
	16,200 volt-amperes
Total noncontinuous loads plus continuous loads	28,400 volt-amperes

Minimum Number of Branch Circuits Required:

General Lighting Load: Branch circuits need only be installed to supply the actual connected load [Section 220-4(d)].

8500 volt-amperes × 1.25 = 10,625 volt-amperes

10,625 volt-amperes ÷ 240 volts = 44 amperes for 3-wire, 120/240.

The lighting load shall be permitted to be served by 2-wire or 3-wire 15- or 20-ampere circuits with combined capacity equal to 44 amperes or greater for 3-wire circuits or 88 amperes or greater for 2-wire circuits. The feeder capacity as well as the number of branch-circuit positions available for lighting circuits in the panelboard must reflect the full calculated load of 9000 volt-amperes × 1.25 = 11,250 volt-amperes.

Example No. 3 (Continued)

Show Window:

6000 volt-amperes × 1.25 = 7500 volt-amperes

7500 volt-amperes ÷ 240 volts = 31 amperes for 3-wire, 120/240.

The show window lighting shall be permitted to be served by 2-wire or 3-wire circuits with a capacity equal to 31 amperes or greater for 3-wire circuits or 62 amperes or greater for 2-wire circuits.

Receptacles required by Section 210-62 are assumed to be included in the receptacle load above if these receptacles do not supply the show window lighting load.

Receptacles:

Receptacle Load: 14,400 volt-amperes ÷ 240 volts = 60 amperes for 3-wire, 120/240.

The receptacle load shall be permitted to be served by 2-wire or 3-wire circuits with a capacity equal to 60 amperes or greater for 3-wire circuits or 120 amperes or greater for 2-wire circuits.

Minimum Size Feeder (or Service) Overcurrent Protection
[Section 220-10(b) or Section 230-90(a)]:

Total noncontinuous loads	12,200 volt-amperes
Total continuous load @ 125%	
16,200 × 1.25	20,250 volt-amperes
Total load	32,450 volt-amperes

32,450 volt-amperes ÷ 240 volts = 135 amperes

The next higher standard size is 150 amperes. (Section 240-6).

Minimum Size Feeders (or Service Conductors) Required
[Sections 215-3, 220-10(b), 230-42(a)]:

For 120/240-volt, 3-wire system:

28,400 volt-amperes ÷ 240 volts = 118 amperes

Service or feeder conductor is No. 1 Cu per Table 310-16 (with 75°C terminations). The No. 1 Cu wire is protected by the 150-ampere overcurrent setting per Sections 215-3 [or Section 230-90(a), Exception No. 2] and 240-3(b).

In the example, 125 percent of the actual connected lighting load (8500 volt-amperes × 1.25 = 10,625 volt-amperes) is less than 125 percent of the load from Table 220-3(b), so the minimum lighting load from Table 220-3(b) is used in the calculation. Had the actual lighting load been greater than the value computed from Table 220-3(b), 125 percent of the actual connected lighting load would have been used.

Example No. 4(a). Multifamily Dwelling

Multifamily dwelling having 40 dwelling units.

Meters in two banks of 20 each and individual subfeeders to each dwelling unit.

One-half of the dwelling units are equipped with electric ranges not exceeding 12 kW each. Assume range kW rating equivalent to kVA rating in accordance with Section 220-19. Other half of ranges are gas ranges.

Area of each dwelling unit is 840 sq. ft.

Laundry facilities on premises available to all tenants. Add no circuit to individual dwelling unit. Add 1500 volt-amperes for each laundry circuit to house load and add to the example as a "house load."

Example No. 4(a) (Continued)

Computed Load for Each Dwelling Unit (Article 220):

General Lighting Load:

    840 sq. ft. at 3 volt-amperes per sq. ft ....   2520 volt-amperes

Special Appliance Load:

    Electric Range (Section 220-19) ....   8000 volt-amperes

Minimum Number of Branch Circuits Required for Each Dwelling Unit
(Section 220-4):

    General Lighting Load: 2520 volt-amperes ÷ 120 volts = 21 amperes or two 15-ampere, 2-wire circuits; or two 20-ampere, 2-wire circuits.

    Small Appliance Load: Two 2-wire circuits of No. 12 wire. [See Section 220-4(b).]

    Range Circuit: 8000 volt-amperes ÷ 240 = 33 amperes or a circuit of two No. 8 conductors and one No. 10 conductor as permitted by Section 220-22. (See Section 210-19.)

Minimum Size Subfeeder Required for Each Dwelling Unit
(Section 215-2):

Computed Load (Article 220):

General Lighting Load	2520 volt-amperes
Small Appliance Load: two 20-ampere circuits	3000 volt-amperes
Total Computed Load (without ranges)	5520 volt-amperes

Application of Demand Factor:

3000 volt-amperes at 100%	3000 volt-amperes
2520 volt-amperes at 35%	882 volt-amperes
Net Computed Load (without ranges)	3882 volt-amperes
Range Load	8000 volt-amperes
Net Computed Load (with ranges)	11,882 volt-amperes

Size of Each Subfeeder (see Section 215-3).

For 120/240-volt, 3-wire system (without ranges):

    Net Computed Load, 3882 volt-amperes ÷ 240 volts = 16.2 amperes.

For 120/240-volt, 3-wire system (with ranges):

    Net Computed Load, 11,882 volt-amperes ÷ 240 volts = 49.5 amperes.

Subfeeder Neutral:

Lighting and Small Appliance Load	3882 volt-amperes
Range Load: 8000 volt-amperes at 70% (see Section 220-22)	5600 volt-amperes
(Not included for apartments without electric range)	
Net Computed Load (neutral)	9482 volt-amperes

        9482 volt-amperes ÷ 240 volts = 39.5 amperes

Minimum Size Feeders Required from Service Equipment to Meter Bank
(For 20 Dwelling Units — 10 with Ranges):

Total Computed Load:

Lighting and Small Appliance Load:

20 × 5520 volt-amperes	110,400 volt-amperes

Application of Demand Factor:

3000 volt-amperes at 100%	3000 volt-amperes
107,400 volt-amperes at 35%	37,590 volt-amperes
Net Computed Lighting and Small Appliance Load	40,590 volt-amperes
Range Load, 10 ranges (less than 12 kVA, Col. A, Table 220-19)	25,000 volt-amperes
Net Computed Load (with ranges)	65,590 volt-amperes

Example No. 4(a) (Continued)

For 120/240-volt, 3-wire system:

Net Computed Load, 65,590 volt-amperes ÷ 240 volts = 273 amperes

Feeder Neutral:

Lighting and Small Appliance Load	40,590 volt-amperes
Range Load: 25,000 volt-amperes at 70% (see Section 220-22)	17,500 volt-amperes
Computed Load (neutral)	58,090 volt-amperes

58,090 volt-amperes ÷ 240 volts = 242 amperes

Further Demand Factor (Section 220-22):

200 amperes at 100%	200 amperes
42 amperes at 70%	29 amperes
Net Computed Load (neutral)	229 amperes

Minimum Size Main Feeder (or Service Conductors) Required (less house load). (For 40 Dwelling Units — 20 with Ranges):

Total Computed Load:

Lighting and Small Appliance Load: 40 × 5520 volt-amperes	220,800 volt-amperes

Application of Demand Factor:

3000 volt-amperes at 100%	3000 volt-amperes
117,000 volt-amperes at 35%	40,950 volt-amperes
100,800 volt-amperes at 25%	25,200 volt-amperes
Net Computed Lighting and Small Appliance Load	69,150 volt-amperes
Range Load, 20 ranges (less than 12 kVA, Col. A, Table 220-19)	35,000 volt-amperes
Net Computed Load	104,150 volt-amperes

For 120/240-volt, 3-wire system:

Net Computed Load, 104,150 volt-amperes ÷ 240 volts = 434 amperes.

Feeder Neutral:

Lighting and Small Appliance Load	69,150 volt-amperes
Range Load: 35,000 volt-amperes at 70% (see Section 220-22)	24,500 volt-amperes
Computed Load (neutral)	93,650 volt-amperes

93,650 volt-amperes ÷ 240 volts = 390 amperes.

Further Demand Factor (see Section 220-22):

200 amperes at 100%	200 amperes
190 amperes at 70%	133 amperes
Net Computed Load (neutral)	333 amperes

See Tables 310-16 through 310-31, Notes 8 and 10.

Example No. 4(b). Optional Calculation for Multifamily Dwelling

Multifamily dwelling equipped with electric cooking and space heating or air conditioning and having 40 dwelling units.

Meters in two banks of 20 each plus house metering and individual sub-feeders to each dwelling unit.

Each dwelling unit is equipped with an electric range of 8-kW nameplate rating, four 1.5-kW separately controlled 240-volt electric space heaters,

Example No. 4(b) (Continued)

and a 2.5-kW 240-volt electric water heater. Assume range, space heater, and water heater kW ratings equivalent to kVA.

A common laundry facility is available to all tenants [Section 210-52(f), Exception No. 1].

Area of each dwelling unit is 840 sq. ft.

Computed Load for Each Dwelling Unit (Article 220):

General Lighting Load:

840 sq. ft. at 3 volt-amperes per sq. ft	2520 volt-amperes
Electric Range	8000 volt-amperes
Electric Heat 6 kVA	6000 volt-amperes
(or air conditioning if larger)	
Electric Water Heater	2500 volt-amperes

Minimum Number of Branch Circuits Required for Each Dwelling Unit:

General Lighting Load: 2520 volt-amperes ÷ 120 volts = 21 amperes or two 15-ampere 2-wire circuits or two 20-ampere 2-wire circuits.

Small Appliance Loads: Two 2-wire circuits of No. 12 [see Section 220-4(b)].

Range circuit: 8000 volt-amperes × 80% ÷ 240 volts = 27 amperes on a circuit of three No. 10 conductors as permitted in Column C of Table 220-19.

Space Heating 6000 volt-amperes ÷ 240 volts = 25 amperes

No. of circuits (see Section 220-4).

Minimum Size Subfeeder Required for Each Dwelling Unit

(Section 215-2):

Computed Load (Article 220):

General Lighting Load	2520 volt-amperes
Small Appliance Load, two 20-ampere circuits	3000 volt-amperes
Total Computed Load	5520 volt-amperes
(without range and space heating)	

Application of Demand Factor:

3000 volt-amperes at 100%	3000 volt-amperes
2520 volt-amperes at 35%	882 volt-amperes
Net Computed Load	3882 volt-amperes
(without range and space heating)	
Range Load	6400 volt-amperes
Space Heating (Section 220-15)	6000 volt-amperes
Water Heater	2500 volt-amperes
Net Computed Load for individual dwelling unit	18,782 volt-amperes

For 120/240-volt, 3-wire system

Net Computed Load 18,782 volt-amperes ÷ 240 volts = 78 amperes

Subfeeder Neutral (Section 220-22)

Lighting and Small Appliance Load	3882 volt-amperes
Range Load: 6400 volt-amperes at 70%	
(see Section 220-22)	4480 volt-amperes
Space and Water Heating (no neutral)	
240 volt	0 volt-amperes
Net Computed Load (neutral)	8362 volt-amperes

8362 volt-amperes ÷ 240 volts = 35 amperes

Example No. 4(b) (Continued)

Minimum Size Feeder Required from Service Equipment to Meter Bank for 20 Dwelling Units:
Total Computed Load:

Lighting and Small Appliance Load 20 × 5520	110,400 volt-amperes
Water and Space Heating Load 20 × 8500	170,000 volt-amperes
Range Load 20 × 8000 volt-amperes	160,000 volt-amperes
Net Computed Load (20 dwelling units)	440,400 volt-amperes

Net Computed Load Using Optional Calculation (Table 220-32)

440,400 volt-amperes × .38	167,352 volt-amperes

167,352 volt-amperes ÷ 240 volts = 697 amperes

Minimum Size Main Feeder Required (less house load) for 40 Dwelling Units:
Total Computed Load:

Lighting and Small Appliance Load 40 × 5520	220,800 volt-amperes
Water and Space Heating 40 × 8500	340,000 volt-amperes
Range Load 40 × 8000 volt-amperes	320,000 volt-amperes
Net Computed Load (40 dwelling units)	880,800 volt-amperes

Net Computed Load Using Optional Calculation (Table 220-32)

880,800 volt-amperes × .28	246,624 volt-amperes

246,624 volt-amperes ÷ 240 volts = 1028 amperes

Feeder Neutral Load for Feeder from Service Equipment to Meter Bank for 20 Dwelling Units:
Lighting and Small Appliance Load

20 × 5520 volt-amperes	110,400 volt-amperes
First 3000 volt-amperes at 100%	3000 volt-amperes
107,400 volt-amperes at 35%	37,590 volt-amperes
Subtotal	40,590 volt-amperes
20 Ranges = 35,000 volt-amperes at 70%	24,500 volt-amperes
(See Table 220-19 and Section 220-22.)	
Total	65,090 volt-amperes

65,090 volt-amperes ÷ 240 volts = 271 amperes

Further Demand Factor (Section 220-22)

First 200 amperes at 100%	200 amperes
Balance: 71 amperes at 70%	50 amperes
Total	250 amperes

Feeder Neutral Load of Main Feeder (less house load) for 40 Dwelling Units:
Lighting and Small Appliance Load

40 × 5520 volt-amperes	220,800 volt-amperes
First 3000 volt-amperes at 100%	3000 volt-amperes
120,000 volt-amperes −3000 volt-amperes =	
117,000 volt-amperes at 35%	40,950 volt-amperes
220,800 volt-amperes −120,000 volt-amperes =	
100,800 volt-amperes at 25%	25,200 volt-amperes
Net Computed Lighting and Small Appliance Load	69,150 volt-amperes
40 Ranges = 55,000 volt-amperes at 70%	38,500 volt-amperes
(See Table 220-19 and Section 220-22)	
Total	107,650 volt-amperes

107,650 volt-amperes ÷ 240 volts = 449 amperes

Example No. 4(b) (Continued)

Further Demand Factor (Section 220-22)

First 200 amperes at 100%. .... 200 amperes

Balance: 249 amperes at 70% .... 174 amperes

Total .... 374 amperes

Example No. 5(a). Multifamily Dwelling Served at 208Y/120 Volts, Three Phase

All conditions and calculations the same as for Multifamily Dwelling [Example No. 4(a)] served at 120/240 volts, single phase except as follows: Service to each dwelling unit shall be two phase legs and neutral.

Minimum Number of Branch Circuits Required for Each Dwelling Unit (Section 220-4):

Range Circuit: 8000 volt-amperes ÷ 208 volts = 38 amperes or a circuit of two No. 8 conductors and one No. 10 conductor as permitted by Section 220-22.

Minimum Size Subfeeder Required for Each Dwelling Unit (Section 215-2):

For 120/208-volt, 3-wire system (without ranges)

Net Computed Load: 3882 volt-amperes ÷ 2 legs ÷ 120 volts/leg = 16.2 amperes

For 120/208-volt, 3-wire system (with ranges)

Net Computed Load: range, 8000 volt-amperes ÷ 208 volts = 38.5 amperes

Total load: range plus lighting = 54.7 amperes

Subfeeder neutral: range, 5600 volt-amperes ÷ 208 volts = 26.9 amperes

Total load: range plus lighting = 43.1 amperes

Minimum Size Feeders Required from Service Equipment to Meter Bank (For 20 Dwelling Units—10 with Ranges):

For 208Y/120-volt, 3-phase, 4-wire system

Ranges: Maximum number between any two phase legs = 4

2 × 4 = 8. Table 220-19 Demand = 23,000 volt-amperes.

Per phase demand: 23,000 volt-amperes ÷ 2 = 11,500 volt-amperes

Equivalent 3-phase load = 34,500 volt-amperes

Net Computed Load (total): 40,590 volt-amperes + 34,500 volt-amperes = 75,080 volt-amperes

75,080 volt-amperes ÷ (208)(1.732) = 208.4 amperes

Feeder Neutral Size

40,590 volt-amperes + 34,500 volt-amperes at 70% = 64,700 volt-amperes

Net Computed Neutral Load:

64,700 volt-amperes ÷ (208)(1.732) = 179.6 amperes

Minimum Size Main Feeder (less house load) (For 40 Dwelling Units — 20 with Ranges):

For 208Y/120-volt, 3-phase, 4-wire system

Ranges: Maximum number between any two phase legs = 7

2 × 7 = 14. Table 220-19 Demand = 29,000 volt-amperes.

Per phase demand: 29,000 ÷ 2 = 14,500 volt-amperes

Equivalent 3-phase load = 43,500 volt-amperes

Net Computed Load (total): 69,150 volt-amperes + 43,500 volt-amperes = 112,650 volt-amperes

112,650 volt-amperes ÷ (208)(1.732) = 312.7 amperes

Example No. 5(a) (Continued)

Main Feeder Neutral Size

69,150 volt-amperes + 43,500 volt-amperes at 70% = 99,600 volt-amperes

99,600 volt-amperes ÷ (208)(1.732) = 276.5 amperes

Further Demand Factor (Section 220-22)

200 amperes at 100%	200.0 amperes
76.5 amperes at 70%.	53.6 amperes
Net Computed Load	253.6 amperes

Example No. 5(b). Optional Calculation for Multifamily Dwelling Served at 208Y/120 Volts, Three Phase

All conditions and calculations the same as for Optional Calculation for Multifamily Dwelling [Example No. 4(b)] served at 120/240 volts, single phase except as follows:

Service to each dwelling unit shall be two phase legs and neutral.

Minimum Number of Branch Circuits Required for Each Dwelling Unit
(Section 220-4):

Range Circuit: 8000 volt-amperes at 80% ÷ 208 volts = 30.7 amperes or a circuit of two No. 8 conductors and one No. 10 conductor as permitted by Section 220-22.

Space Heating: 6000 volt-amperes ÷ 208 volts = 28.8 amperes. Two 20-ampere, 2-pole circuits required, No. 12 conductor.

Minimum Size Subfeeder Required for Each Dwelling Unit.

Computed Load (120/208-volt, 3-wire circuit)

Net Computed Load: 18,782 volt-amperes ÷ 208 volts = 90.3 amperes

Net Computed Load. Lighting (line to neutral)

3882 volt-amperes ÷ 2 legs ÷ 120 volts/leg	16.2 amperes
Line to line:	
14,900 volt-amperes ÷ 208 volts	71.6 amperes
Total load:	87.8 amperes

Minimum Size Feeder Required for Service Equipment to Meter Bank
(For 20 dwelling units):

Net Computed Load: 167,352 volt-amperes ÷ (208)(1.732) = 464.9 amperes
Feeder Neutral Load:

65,080 volt-amperes ÷ (208)(1.732) = 180.67 amperes

Minimum Size Main Feeder Required (less house load):
(For 40 dwelling units):

Net Computed Load: 246,624 volt-amperes ÷ (208)(1.732) = 685.1 amperes
Main Feeder Neutral Load:

107,650 volt-amperes ÷ (208)(1.732) = 298.8 amperes

Further Demand Factor (Section 220-22)

200 amperes at 100%	200.0 amperes
98.8 amperes at 70%	69.2 amperes
Net Computed Load	269.2 amperes

Example No. 6. Maximum Demand for Range Loads

Table 220-19, Column A applies to ranges not over 12 kW. The application of Note 1 to ranges over 12 kW (and not over 27 kW) and Note 2 to ranges over 8¾ kW (and not over 27 kW) is illustrated in the following examples:

A. Ranges all the same rating (Table 220-19, Note 1).

Assume 24 ranges, each rated 16 kW.

From Column A, the maximum demand for 24 ranges of 12 kW rating is 39 kW.

16 kW exceeds 12 kW by 4.

5% × 4 = 20% (5% increase for each kW in excess of 12).

39 kW × 20% = 7.8 kW increase.

39 + 7.8 = 46.8 kW: value to be used in selection of feeders.

| B. Ranges of unequal rating (Table 220-19, Note 2).

Assume 5 ranges, each rated 11 kW.

2 ranges, each rated 12 kW.

20 ranges, each rated 13.5 kW.

3 ranges, each rated 18 kW.

5 × 12 = 60 use 12 kW for range rated less than 12.

2 × 12 = 24

20 × 13.5 = 270

3 × 18 = 54

30   408 kW

408 ÷ 30 = 13.6 kW (average to be used for computation).

From Column A, the demand for 30 ranges of 12 kW rating is 15 + 30 = 45 kW.

13.6 exceeds 12 by 1.6 (use 2).

5% × 2 = 10% (5% increase for each kW in excess of 12).

45 kW × 10% = 4.5 kW increase.

45 + 4.5 = 49.5 kW: value to be used in selection of feeders.

Example No. 8. Motors, Conductors, Overload, and Short-Circuit and Ground-Fault Protection

(See Sections 240-6, 430-6, 430-7, 430-22, 430-23, 430-24, 430-32, 430-34, 430-52, and 430-62 and Tables 430-150 and 430-152.)

Determine the conductor size, the motor overload protection, the branch-circuit short-circuit and ground-fault protection and the feeder protection, for one 25-horsepower squirrel-cage induction motor (full voltage starting nameplate current 31.6 amperes, service factor 1.15, Code letter F), and two 30-horsepower wound-rotor induction motors (nameplate primary current 36.4 amperes, nameplate secondary current 65 amperes, 40°C rise), on a 460-volt, 3-phase, 60-Hertz supply.

Conductor Loads

The full-load current value used to determine the ampacity of conductors for the 25-horsepower motor is 34 amperes [Section 430-6(a) and Table 430-150].

Example No. 8 (Continued)

A full-load current of 34 amperes × 1.25 = 42.5 amperes (Section 430-22). The full-load current value used to determine the ampacity of primary conductors for each 30-horsepower motor is 40 amperes [Section 430-6(a) and Table 430-150]. A full-load primary current of 40 amperes × 1.25 = 50 amperes (Section 430-22). A full-load secondary current of 65 amperes × 1.25 = 81.25 amperes [Section 430-23(a)]. The feeder ampacity will be 125 percent of 40 plus 40 plus 34, or 124 amperes (Section 430-24).

Overload and Short-Circuit and Ground-Fault Protection

Overload. Where protected by a separate overload device, the 25-horsepower motor, with nameplate current of 31.6 amperes, shall have overload protection of not over 39.5 amperes [Sections 430-6(a) and 430-32(a)(1)]. Where protected by a separate overload device, the 30-horsepower motor, with nameplate current of 36.4 amperes, shall have overload protection of not over 45.5 amperes [Sections 430-6(a) and 430-32(a)(1)]. If the overload protection is not sufficient to start the motor or to carry the load, it shall be permitted to be increased according to Section 430-34. For a motor marked "thermally protected," overload protection is provided by the thermal protector [see Sections 430-7(a)(12) and 430-32(a)(2)].

Branch-Circuit Short-Circuit and Ground-Fault. The branch circuit of the 25-horsepower motor shall have branch-circuit short-circuit and ground-fault protection of not over 300 percent of a nontime-delay fuse (Table 430-152) or 3.00 × 34 = 102 amperes. The next smaller standard size fuse is 100 amperes (Section 240-6). Since a 100-ampere fuse is adequate to carry the load, Section 430-52(a), Exception No. 1 does not apply. If a time-delay fuse is to be used, see Section 430-52(a), Exception No. 2b. Where the 100-ampere fuse will not allow the motor to run, the value for a nontime-delay fuse shall be permitted to be increased to the next larger standard size, or 110 amperes. If these fuses are not sufficient to start the motor, the value for a nontime-delay fuse shall be permitted to be increased to 400 percent [See Section 430-52(a), Exception No. 2a.]

Feeder Circuit. The maximum rating of the feeder short-circuit and ground-fault protection device is based on the sum of the largest branch-circuit protective device (100-ampere fuse) plus the sum of the full-load currents of the other motors, or 100 + 40 + 40 = 180 amperes. The nearest standard fuse that does not exceed this value is 175 amperes [Section 430-62(a)].

Example No. 9. Feeder Size Determination for Generator Field Control.

Determining conductor ampacity for a group of six elevators on a single feeder.

(See Sections 620-15, 620-13, 430-26, and 430-24 and Tables 620-15 and 430-150.)

Determine the conductor ampacity for a feeder supplying the six elevators. The system is generator field control. The ac drive motor horsepower rating of the largest MG set for one elevator is 40 horsepower, and the remaining elevators each have a 30-horsepower ac drive motor rating for their MG sets. Each controller draws 10 amperes to operate relays, power supplies, and the dooroperator. The line power supply is 460 volts, 3-phase, 60 Hz ac. The MG sets are rated continuous.

Example No. 9 (Continued)

Conductor ampacity is determined as follows:

(a) Per Table 430-150, the full-load current for a 30-horsepower ac motor at 460 volts is 40 amps; for a 40-horsepower ac motor at 460 volts, 52 amps.

(b) Per Section 430-24, Exception No. 1, determine the ampere rating for each motor used for other than continuous duty from Table 430-22(a), Exception.

(c) Per Table 430-22(a), Exception, for intermittent duty using a continuous rated motor, the percentage of nameplate current rating to be used is 140%.

(d) For the 30-horsepower ac drive motors, 1.4 × 40 amps = 56 amps. For the 40-horsepower ac drive motors, 1.4 × 52 amps = 73 amps.

(e) Per Section 430-24, the total conductor ampacity is:
1.25 × largest motor current + the sum of the remaining motor currents = (1.25 × 73) + (5 × 56) = 371 amps.

(f) Per Section 620-15 and Table 620-15, the conductor (feeder) ampacity shall be permitted to be reduced by the use of a demand factor. For six elevators, the demand factor is .79. The feeder motor ampacity is, therefore, .79 × 371 amps = 293 amps.

(g) The total feeder ampacity is the sum of the motor diverse current and the controller (constant) current. I total = 293 + (6 × 10) = 353 amps.

(h) This total ampacity shall be permitted to be used to select the wire size.

Example No. 10. Determining Conductor Ampacity for a Group of Six Elevators on a Single Feeder Using Adjustable Speed Drive.

(See Sections 620-14, 620-15, and 430-26 and Table 620-15.)

Determine the conductor ampacity for a feeder supplying six identical elevators. The system is adjustable speed SCR dc drive. The transformers are external to the drive cabinet. Each drive has an ac nameplate rating of 204 amps. Each transformer is rated 85 kVA with a 300V secondary; efficiency is 90%. The ac line voltage is 460 volts, 3-phase, 60 Hz.

Conductor ampacity is determined as follows:

(a) Per 620-14(b), calculate the nameplate current rating of the transformer:

$$I = \frac{85 \times 1000}{\sqrt{3} \times 460 \times 0.9} = 119 \text{ amps}$$

(b) Per Section 620-14(b), calculate the drive input current reflected to the transformer primary:

I = 204 A × 300 V/460 V = 133 amps

(c) Per Section 620-14(b), the larger current is 133 amps.

(d) For six elevators, the total conductor ampacity is 6 × 133 amps = 798 amps.

(e) Per Section 620-15 and Table 620-15, the conductor (feeder) ampacity shall be permitted to be reduced by the use of a demand factor.

For six elevators, the demand factor is .79. The feeder ampacity is, therefore, .79 × 798 amperes = 630 amperes.

(f) This ampacity shall be permitted to be used to select the wire size subject to Section 430-26.

Appendix A

Extracts

(See Section 90-3, Paragraph 4)

A-500-3, FPN No. 13	Classification of Gases, Vapors, and Dusts for Electrical Equipment in Hazardous (Classified) Locations, NFPA 497M-1991	Chapter 3, Class II Materials— Dusts
A-500-3, FPN No. 14	Classification of Gases, Vapors, and Dusts for Electrical Equipment in Hazardous (Classified) Locations, NFPA 497M-1991	Chapter 3, Class II Materials— Dusts
A-514-2	Automotive and Marine Service Station Code, NFPA 30A-1990	Section 7-3
A-Table 514-2	Automotive and Marine Service Station Code, NFPA 30A-1990	Table 7
A-Fig. 514-2	Automotive and Marine Service Station Code, NFPA 30A-1990	Figure 7-1
A-Fig. 514-5(b)	Automotive and Marine Service Station Code, NFPA 30A-1990	Section 9-5.4
A-Fig. 514-5(c)	Automotive and Marine Service Station Code, NFPA 30A-1990	Section 9-5.3
A-515-2	Flammable and Combustible Liquids Code, NFPA 30-1990	Paragraph 5-3.5.3
A-Table 515-2	Flammable and Combustible Liquids Code, NFPA 30-1990	Table 5-3.5.3
A-516-2(a)(1)	Standard for Spray Application Using Flammable and Combustible Materials, NFPA 33-1989	Section 1-2 Spray Area
A-516-2(a)(2)	Standard for Spray Application Using Flammable and Combustible Materials, NFPA 33-1989	Section 1-2 Spray Area

A-516-2(a)(3)	Standard for Spray Application Using Flammable and Combustible Materials, NFPA 33-1989	Section 1-2 Spray Area
A-516-2(a)(4)	Standard for Dipping and Coating Processes Using Flammable or Combustible Liquids, NFPA 34-1989	Section 1-2, Vapor Source, and Paragraph 6-2.2
A-516-2(a)(5)	Standard for Dipping and Coating Processes Using Flammable or Combustible Liquids, NFPA 34-1989	Paragraph 6-2.1
A-516-2(a)(6)	Standard for Dipping and Coating Processes Using Flammable or Combustible Liquids, NFPA 34-1989	Figure 6-1(b)
A-516-2(b)(1)	Standard for Spray Application Using Flammable and Combustible Materials, NFPA 33-1989	Paragraph 4-7.1
A-516, Fig. 1	Standard for Spray Application Using Flammable and Combustible Materials, NFPA 33-1989	Figure 1
A-516-2(b)(2)	Standard for Spray Application Using Flammable and Combustible Materials, NFPA 33-1989	Paragraph 4-7.2
A-516, Fig. 2	Standard for Spray Application Using Flammable and Combustible Materials, NFPA 33-1989	Figure 2
A-516-2(b)(3)	Standard for Spray Application Using Flammable and Combustible Materials, NFPA 33-1989	Paragraph 4-7.3
A-516-2(b)(4)	Standard for Spray Application Using Flammable and Combustible Materials, NFPA 33-1989	Paragraph 4-7.4
A-516, Fig. 3	Standard for Spray Application Using Flammable and Combustible Materials, NFPA 33-1989	Figure 3

A-516-2(b)(5)	Standard for Dipping and Coating Processes Using Flammable or Combustible Liquids, NFPA 34-1989	Paragraphs 6-2.3 and 6-2.4
A-516-2(b)(6)	Standard for Dipping and Coating Processes Using Flammable or Combustible Liquids, NFPA 34-1989	Section 6-3
A-516, Fig. 4	Standard for Dipping and Coating Processes Using Flammable or Combustible Liquids, NFPA 34-1989	Figure 6-1(a)
A-516-3(b)	Standard for Spray Application Using Flammable and Combustible Materials, NFPA 33-1989	Section 4-5
A-516-3(d)	Standard for Spray Application Using Flammable and CombustibleMaterials, NFPA 33-1989	Section 4-9
A-516-4	Standard for Spray Application Using Flammable and Combustible Materials, NFPA 33-1989	Chapter 9
A-516-5	Standard for Spray Application Using Flammable and Combustible Materials, NFPA 33-1989	Chapter 10
A-516-6	Standard for Spray Application Using Flammable and Combustible Materials, NFPA 33-1989	Chapter 13
A-517-10, Exception No. 3	Standard for Health Care Facilities, NFPA 99-1990	Section 16-2, Exception; Paragraph 17-3.3.2, Exception
A-517-15	Standard for Health Care Facilities, NFPA 99-1990	3-5.2.1.6
A-517-25	Standard for Health Care Facilities, NFPA 99-1990	Paragraphs 3-2.2.1, 3-4.2.3.2, 3-4.2.4.1
A-517-30(b)(1)	Standard for Health Care Facilities, NFPA 99-1990	Paragraph 3-4.2.2.1

A-517-30(b)(2)	Standard for Health Care Facilities, NFPA 99-1990	Paragraph 3-4.2.2.1
A-517-30(b)(3)	Standard for Health Care Facilities, NFPA 99-1990	Paragraph 3-4.2.2.1
A-517-30(b)(4)	Standard for Health Care Facilities, NFPA 99-1990	Paragraph 3-4.2.2.1
A-517-31	Standard for Health Care Facilities, NFPA 99-1990	Paragraphs 3-4.2.2.2(a), 3-5.1.2.1(a)
A-517-32	Standard for Health Care Facilities, NFPA 99-1990	Paragraph 3-4.2.2.2(b)
A-517-33(a)	Standard for Health Care Facilities, NFPA 99-1990	Paragraph 3-4.2.2.2(c)
A-517-34	Standard for Health Care Facilities, NFPA 99-1990	Paragraph 3-4.2.2.3(b)
A-517-34(a)	Standard for Health Care Facilities, NFPA 99-1990	Paragraph 3-4.2.2.3(c)
A-517-34(b)	Standard for Health Care Facilities, NFPA 99-1990	Paragraph 3-4.2.2.3(d)
A-517-35(a)	Standard for Health Care Facilities, NFPA 99-1990	Paragraph 3-3.2.1.2
A-517-35(b)	Standard for Health Care Facilities, NFPA 99-1990	Paragraph 3-3.2.1.3
A-517-41(a)	Standard for Health Care Facilities, NFPA 99-1990	Paragraph 3-4.2.3.2
A-517-41(b)	Standard for Health Care Facilities, NFPA 99-1990	Paragraph 3-4.2.3.2
A-517-42	Standard for Health Care Facilities, NFPA 99-1990	Paragraphs 3-4.2.3.3, 3-5.1.2.1(b)
A-517-43	Standard for Health Care Facilities, NFPA 99-1990	Paragraph 3-4.2.3.4
A-517-44(a)	Standard for Health Care Facilities, NFPA 99-1990	Paragraph 3-3.2.1.2
A-517-44(b)	Standard for Health Care Facilities, NFPA 99-1990	Paragraphs 3-3.2.1.3, 16-3.3.2.1, 17-3.3.2.1
A-517-50(a)	Standard for Health Care Facilities, NFPA 99-1990	Paragraphs 14-3.3.2, 15-3.3.2, 18-3.3.2

A-517-50(b)	Standard for Health Care Facilities, NFPA 99-1990	Paragraph 3-4.2.4.2	
A-517-50(c)(1)	Standard for Health Care Facilities, NFPA 99-1990	Paragraph 3-4.2.4.3	
A-517-50(c)(2)	Standard for Health Care Facilities, NFPA 99-1990	Paragraph 3-5.1.2.1(c)	
A-517-50(c)(3)	Standard for Health Care Facilities, NFPA 99-1990	Paragraph 3-5.1.2.1(c)	
A-517-60(a)(1)	Standard for Health Care Facilities, NFPA 99-1990	Section 2-2, Hazardous Area in Flammable Anesthetizing Location	
A-517-61(a)(1)	Standard for Health Care Facilities, NFPA 99-1990	Paragraph 12-4.1.3.2	
A-517-61(a)(3)	Standard for Health Care Facilities, NFPA 99-1990	Paragraphs 3-4.1.2.1(e)(2) 7-5.1.2.4, 7-5.1.2.5	
A-517-64	Standard for Health Care Facilities, NFPA 99-1990	Paragraph 7-5.1.2.3	
A-517-160(a)(2)	Standard for Health Care Facilities, NFPA 99-1990	Paragraph 3-4.3.1.2	
A-517-160(b)	Standard for Health Care Facilities, NFPA 99-1990	Paragraph 3-4.3.3	

Appendix B

This appendix is not part of the requirements of this Code, but is included for information purposes only.

B-310-15(b)(1). Formula Application Information. This appendix provides application information for ampacities calculated under engineering supervision.

B-310-15(b)(2). Typical Applications Covered by Tables. Typical ampacities for conductors rated 0 through 2000 volts are shown in Tables B-310-1 through B-310-10. Underground electrical duct bank configurations, as detailed in Figures B-310-3, B-310-4, and B-310-5, are utilized for conductors rated 0 through 5000 volts. In Figures B-310-2 through B-310-5, where adjacent duct banks are used, a separation of 5 feet (1.52 m) between the centerlines of the closest ducts in each bank or 4 feet (1.22 m) between the extremities of the concrete envelopes is sufficient to prevent derating of the conductors due to mutual heating. These ampacities were calculated as detailed in the basic ampacity paper, "The Calculation of the Temperature Rise and Load Capability of Cable Systems," by J. H. Neher and M. H. McGrath, AIEE Paper 57-660. For additional information concerning the application of these ampacities, see Power Cable Ampacities, IEEE/ICEA Standard S-135/P-46-426.

Typical values of thermal resistivity (Rho) are as follows:

Polyethylene (PE) = 450
Polyvinyl Chloride PVC = 650
Rubber and Rubber-like = 500
Paper Insulation = 550
Average Soil (90 percent of USA) = 90

Concrete = 55
Very Dry Soil (rocky or sandy) = 120
Damp Soil
   (coastal areas, high water table) = 60

Thermal resistivity, as used in this appendix, refers to the heat transfer capability through a substance by conduction. It is the reciprocal of thermal conductivity and is normally expressed in the units °C-cm/watt. For additional information on determining soil thermal resistivity (Rho), see IEEE *Guide for Soil Thermal Resistivity Measurements*, ANSI/IEEE Standard 442-1981.

B-310-15(b)(3). Criteria Modifications.

Where values of load factor and Rho are known for a particular electrical duct bank installation and they are different from those shown in a specific table or figure, the ampacities shown in the table or figure can be modified by the application of factors derived from the use of Figure B-310-1.

Where two different ampacities apply to adjacent portions of a circuit, the higher ampacity can be used beyond the point of transition, a distance equal to 10 feet (3.05 m) or 10 percent of the circuit length figured at the higher ampacity, whichever is less.

Where the burial depth of direct burial or electrical duct bank circuits are modified from the values shown in a figure or table, ampacities can be modified as shown in (1) and (2), which follow.

(1) Where burial depths are increased in part(s) of an electrical duct run to avoid underground obstructions, no decrease in ampacity of the conductors is needed, provided the total length of parts of the duct run increased in depth to avoid obstructions is less than 25 percent of the total run length.

(2) Where burial depths are deeper than shown in a specific underground ampacity table or figure, an ampacity derating factor of 6 percent per increased foot (305 mm) of depth for all values of Rho can be utilized. No rating change is needed where the burial depth is decreased.

B-310-15(b)(4). Electrical Ducts. The term electrical duct(s) is defined in Section 310-15(d).

B-310-15(b)(5). Tables B-310-6 and B-310-7.

(1) To obtain the ampacity of cables installed in two electrical ducts in one horizontal row with 7.5-inch (191-mm) center-to-center spacing between electrical ducts, similar to Detail 1, Figure B-310-2, multiply the ampacity shown for one duct in Tables B-310-6 and B-310-7 by 0.88.

(2) To obtain the ampacity of cables installed in four electrical ducts in one horizontal row with 7.5-inch (191-mm) center-to-center spacing between electrical ducts, similar to Detail 2, Figure B-310-2, multiply the ampacity shown for three electrical ducts in Tables B-310-6 and B-310-7 by 0.94.

B-310-15(b)(6). Electrical Ducts Utilized in Figure B-310-2. If spacing between electrical ducts, as shown in Figure B-310-2, is less than specified in Figure B-310-2, where electrical ducts enter equipment enclosures from underground, the ampacity of conductors contained within such electrical ducts need not be reduced.

B-310-15(b)(7). Examples Showing Use of Figure B-310-1 for Electrical Duct Bank Ampacity Modifications. Figure B-310-1 is used for interpolation or extrapolation for values of Rho and load factor for cables installed in electrical ducts. The upper family of curves shows the variation in ampacity and Rho at unity load factor in terms of I_1, the ampacity for Rho = 60, and 50 percent load factor. Each curve is designated for a particular ratio I_2/I_1, where I_2 is the ampacity at Rho = 120 and 100 percent load factor.

The lower family of curves shows the relationship between Rho and load factor that will give substantially the same ampacity as the indicated value of Rho at 100 percent load factor.

As an example, to find the ampacity of a 500 kcmil copper cable circuit for six electrical ducts as shown in Table B-310-5: At the Rho = 60, LF = 50, I_1 = 583; for Rho = 120 and LF = 100, I_2 = 400. The ratio I_2/I_1 = 0.686. Locate Rho = 90 at bottom of chart and follow the 90 Rho line to the intersection with 100 percent load factor where the equivalent Rho = 90. Then follow the 90 Rho line to I_2/I_1 ratio of 0.686 where F = 0.74. The desired ampacity = 0.74 × 583 = 431, which agrees with the table for Rho = 90, LF = 100.

To determine the ampacity for the same circuit where Rho = 80 and LF = 75, using Figure B-310-1, the equivalent Rho = 43, F = 0.855, and the desired ampacity = 0.855 × 583 = 498 amperes. Values for using Figure B-310-1 are found in the electrical duct bank ampacity tables of this appendix.

Where the load factor is less than 100 percent and can be verified by measurement or calculation, the ampacity of electrical duct bank installations can be modified as shown. Different values of Rho can be accommodated in the same manner.

Table B-310-1. Ampacities of Two or Three Insulated Conductors, Rated 0 through 2000 Volts, within an Overall Covering (Multiconductor Cable), in Raceway in Free Air Based on Ambient Air Temperature of 30°C (86°F)

Size	Temperature Rating of Conductor. See Table 310-13.						Size
	60°C (140°F)	75°C (167°F)	90°C (194°F)	60°C (140°F)	75°C (167°F)	90°C (194°F)	
AWG kcmil	TYPES TW, UF	TYPES RH, RHW, THHW, THW, THWN, XHHW, ZW	TYPES THHN, THHW, THW-2, THWN-2, RHH, RWH-2, USE-2, XHHW XHHW-2, ZW-2	TYPE TW	TYPES RH, RHW, THHW, THW, THWN, XHHW	TYPES THHN, THHW, THW-2, THWN-2, RWH-2, USE-2, XHHW, XHHW-2, ZW-2	AWG kcmil
	COPPER			ALUMINUM OR COPPER-CLAD ALUMINUM			
14	16†	18†	21†	. . . .	. . . .	. . . .	14
12	20†	24†	27†	16†	18†	21†	12
10	27†	33†	36†	21†	25†	28†	10
8	36	43	48	28	33	37	8
6	48	58	65	38	45	51	6
4	66	79	89	51	61	69	4
3	76	90	102	59	70	79	3
2	88	105	119	69	83	93	2
1	102	121	137	80	95	106	1
1/0	121	145	163	94	113	127	1/0
2/0	138	166	186	108	129	146	2/0
3/0	158	189	214	124	147	167	3/0
4/0	187	223	253	147	176	197	4/0
250	205	245	276	160	192	217	250
300	234	281	317	185	221	250	300
350	255	305	345	202	242	273	350
400	274	328	371	218	261	295	400
500	315	378	427	254	303	342	500
600	343	413	468	279	335	378	600
700	376	452	514	310	371	420	700
750	387	466	529	321	384	435	750
800	397	479	543	331	397	450	800
900	415	500	570	350	421	477	900
1000	448	542	617	382	460	521	1000
Ambient Temp. °C	For ambient temperatures other than 30°C (86°F), multiply the ampacities shown above by the appropriate factor shown below.						Ambient Temp. °F
21-25	1.08	1.05	1.04	1.08	1.05	1.04	70-77
26-30	1.00	1.00	1.00	1.00	1.00	1.00	79-86
31-35	.91	.94	.96	.91	.94	.96	88-95
36-40	.82	.88	.91	.82	.88	.91	97-104
41-45	.71	.82	.87	.71	.82	.87	106-113
46-50	.58	.75	.82	.58	.75	.82	115-122
51-55	.41	.67	.76	.41	.67	.76	124-131
56-60	. . . .	.58	.71	. . . .	.58	.71	133-140
61-70	. . . .	.33	.58	. . . .	.33	.58	142-158
71-80	. . . .	. . . .	.41	. . . .	. . . .	.41	160-176

† Unless otherwise specifically permitted elsewhere in this Code, the overcurrent protection for conductor types marked with an obelsik (†) shall not exceed 15 amperes for No. 14, 20 amperes for No. 12, and 30 amperes for No. 10 copper; or 15 amperes for No. 12 and 25 amperes for No. 10 aluminum and copper-clad aluminum.

Table B-310-2. Ampacities of Two or Three Single Insulated Conductors, Rated 0 through 2000 Volts, Supported on a Messenger, Based on Ambient Air Temperature of 40°C (104°F)

Size	Temperature Rating of Conductor. See Table 310-13.				Size
	75°C (167°F)	90°C (194°F)	75°C (167°F)	90°C (194°F)	
AWG kcmil	TYPES RH, RHW, THW, THWN, XHHW, ZW	TYPES THHN, THHW, THW-2, THWN-2, RHH, RWH-2, USE-2, XHHW, XHHW-2, ZW-2	TYPES RH, RHW, THW, THWN, THHW, XHHW	TYPES THHN, THHW, RHW-2, XHHW, RHH, XHHW, THW-2, THWN-2, USE-2, ZW-2	AWG kcmil
	COPPER		ALUMINUM OR COPPER-CLAD ALUMINUM		
8	57	66	44	51	8
6	76	89	59	69	6
4	101	117	78	91	4
3	118	138	92	107	3
2	135	158	106	123	2
1	158	185	123	144	1
1/0	183	214	143	167	1/0
2/0	212	247	165	193	2/0
3/0	245	287	192	224	3/0
4/0	287	335	224	262	4/0
250	320	374	251	292	250
300	359	419	282	328	300
350	397	464	312	364	350
400	430	503	339	395	400
500	496	580	392	458	500
600	553	647	440	514	600
700	610	714	488	570	700
750	638	747	512	598	750
800	660	773	532	622	800
900	704	826	572	669	900
1000	748	879	612	716	1000
Ambient Temp. °C	For ambient temperatures other than 40°C (104°F), multiply the ampacities shown above by the appropriate factor shown below.				Ambient Temp. °F
21-25	1.20	1.14	1.20	1.14	70-77
26-30	1.13	1.10	1.13	1.10	79-86
31-35	1.07	1.05	1.07	1.05	88-95
36-40	1.00	1.00	1.00	1.00	97-104
41-45	.93	.95	.93	.95	106-113
46-50	.85	.89	.85	.89	115-122
51-55	.76	.84	.76	.84	124-131
56-60	.65	.77	.65	.77	133-140
61-70	.38	.63	.38	.63	142-158
71-80	. . . .	.45	. . . .	.45	160-176

Table B-310-3. Ampacities of Multiconductor Cables with not more than Three Insulated Conductors, Rated 0 through 2000 Volts, in Free Air. Based on Ambient Air Temperature of 40°C (104°F)
(For TC, MC, MI, UF, USE, and SNM cables)

Size	Temperature Rating of Conductor. See Table 310-13.								Size
AWG	60°C (140°F)	75°C (167°F)	85°C (185°F)	90°C (194°F)	60°C (140°F)	75°C (167°F)	85°C (185°F)	90°C (194°F)	AWG
kcmil	COPPER				ALUMINUM OR COPPER-CLAD ALUMINUM				kcmil
18				11†					18
16				16†					16
14	18†	21†	24†	25†					14
12	21†	28†	30†	32†	18†	21†	24†	25†	12
10	28†	36†	41†	43†	21†	28†	30†	32†	10
8	39	50	56	59	30	39	44	46	8
6	52	68	75	79	41	53	59	61	6
4	69	89	100	104	54	70	78	81	4
3	81	104	116	121	63	81	91	95	3
2	92	118	132	138	72	92	103	108	2
1	107	138	154	161	84	108	120	126	1
1/0	124	160	178	186	97	125	139	145	1/0
2/0	143	184	206	215	111	144	160	168	2/0
3/0	165	213	238	249	129	166	185	194	3/0
4/0	190	245	274	287	149	192	214	224	4/0
250	212	274	305	320	166	214	239	250	250
300	237	306	341	357	186	240	268	280	300
350	261	337	377	394	205	265	296	309	350
400	281	363	406	425	222	287	317	334	400
500	321	416	465	487	255	330	368	385	500
600	354	459	513	538	284	368	410	429	600
700	387	502	562	589	306	405	462	473	700
750	404	523	586	615	328	424	473	495	750
800	415	539	604	633	339	439	490	513	800
900	438	570	639	670	362	469	514	548	900
1000	461	601	674	707	385	499	558	584	1000

Ambient Temp. °C	For ambient temperatures other than 30°C (86°F), multiply the ampacities shown above by the appropriate factor shown below.								Ambient Temp. °F
21-25	1.32	1.20	1.15	1.14	1.32	1.20	1.15	1.14	70-77
26-30	1.22	1.13	1.11	1.10	1.22	1.13	1.11	1.10	79-86
31-35	1.12	1.07	1.05	1.05	1.12	1.07	1.05	1.05	88-95
36-40	1.00	1.00	1.00	1.00	1.00	1.00	1.00	1.00	97-104
41-45	.87	.93	.94	.95	.87	.93	.94	.95	106-113
46-50	.71	.85	.88	.89	.71	.85	.88	.89	115-122
51-55	.50	.76	.82	.84	.50	.76	.82	.84	124-131
56-60	....	.65	.75	.77	....	.65	.75	.77	133-140
61-70	....	.38	.58	.63	....	.38	.58	.63	142-158
71-80	....	....	.33	.44	....	....	.33	.44	160-176

† Unless otherwise specifically permitted elsewhere in this Code, the overcurrent protection for conductor types marked with an obelisk (†) shall not exceed 15 amperes for No. 14, 20 amperes for No. 12, and 30 amperes for No. 10 copper; or 15 amperes for No. 12 and 25 amperes for No. 10 aluminum and copper-clad aluminum.

Table B-310-4. Ampacities for Bare or Covered Conductors Based on 40°C (104°F) Ambient, 80°C (176°F) Total Conductor Temperature, 2 Feet (610 mm) per Second Wind Velocity

Bare Copper Conductors		Covered Copper Conductors	
AWG kcmil	AMPS	AWG kcmil	AMPS
8	98	8	103
6	124	6	130
4	155	4	163
2	209	2	219
1/0	282	1/0	297
2/0	329	2/0	344
3/0	382	3/0	401
4/0	444	4/0	466
250	494	250	519
300	556	300	584
500	773	500	812
750	1000	750	1050
1000	1193	1000	1253

Bare AAC Aluminum Conductors		Covered AAC Aluminum Conductors	
AWG kcmil	AMPS	AWG kcmil	AMP
8	76	8	80
6	96	6	101
4	121	4	127
2	163	2	171
1/0	220	1/0	231
2/0	255	2/0	268
3/0	297	3/0	312
4/0	346	4/0	364
266.8	403	266.8	423
336.4	468	336.4	492
397.5	522	397.5	548
477.0	588	477.0	617
556.5	650	556.5	682
636.0	709	636.0	744
795.0	819	795.0	860
954.0	920		
1033.5	968	1033.5	1017
1272	1103	1272	1201
1590	1267	1590	1381
2000	1454	2000	1527

Table B-310-5. Ampacities of Single Insulated Conductors, Rated 0 through 2000 Volts, in Nonmagnetic Underground Electrical Ducts (One Conductor per Electrical Duct) Based on Ambient Earth Temperature of 20°C (68°F), Electrical Duct Arrangement per Figure B-310-2, Conductor Temperature 75°C (167°F)

COPPER

Size kcmil	3 Electrical Ducts (Fig. B-310-2 Detail 2) TYPES RHW, THHW, THW, THWN, XHHW, USE			6 Electrical Ducts (Fig. B-310-2 Detail 3) TYPES RHW, THHW, THW, THWN, XHHW, USE			9 Electrical Ducts (Fig. B-310-2 Detail 4) TYPES RHW, THHW, THW, THWN, XHHW, USE		
	RHO 60 LF 50	RHO 90 LF 100	RHO 120 LF 100	RHO 60 LF 50	RHO 90 LF 100	RHO 120 LF 100	RHO 60 LF 50	RHO 90 LF 100	RHO 120 LF 100
250	410	344	327	386	295	275	369	270	252
350	503	418	396	472	355	330	446	322	299
500	624	511	484	583	431	400	545	387	360
750	794	640	603	736	534	494	674	469	434
1000	936	745	700	864	617	570	776	533	493
1250	1055	832	781	970	686	632	854	581	536
1500	1160	907	849	1063	744	685	918	619	571
1750	1250	970	907	1142	793	729	975	651	599
2000	1332	1027	959	1213	836	768	1030	683	628

ALUMINUM OR COPPER-CLAD ALUMINUM

Size kcmil	3 Electrical Ducts (Fig. B-310-2 Detail 2) TYPES RHW, THHW, THW, THWN, XHHW, USE			6 Electrical Ducts (Fig. B-310-2 Detail 3) TYPES RHW, THHW, THW, THWN, XHHW, USE			9 Electrical Ducts (Fig. B-310-2 Detail 4) TYPES RHW, THHW, THW, THWN, XHHW, USE		
	RHO 60 LF 50	RHO 90 LF 100	RHO 120 LF 100	RHO 60 LF 50	RHO 90 LF 100	RHO 120 LF 100	RHO 60 LF 50	RHO 90 LF 100	RHO 120 LF 100
250	320	269	256	302	230	214	288	211	197
350	393	327	310	369	277	258	350	252	235
500	489	401	379	457	337	313	430	305	284
750	626	505	475	581	421	389	538	375	347
1000	744	593	557	687	491	453	629	432	399
1250	848	668	627	779	551	508	703	478	441
1500	941	736	689	863	604	556	767	517	477
1750	1026	796	745	937	651	598	823	550	507
2000	1103	850	794	1005	693	636	877	581	535

For ambient temperatures other than 20°C (68°F), multiply the ampacities shown above by the appropriate factor shown below.

Ambient Temp. °C	Factor	Ambient Temp. °F
6-10	1.09	43-50
11-15	1.04	52-59
16-20	1.00	61-68
21-25	.95	70-77
26-30	.90	79-86

Table B-310-6. Ampacities of Three Insulated Conductors, Rated 0 through 2000 Volts, within an Overall Covering (Three-Conductor Cable) in Underground Electrical Ducts (One Cable per Electrical Duct) Based on Ambient Earth Temperature of 20°C (68°F), Electrical Duct Arrangement per Figure B-310-2, Conductor Temperature 75°C (167°F)

COPPER

Size AWG/kcmil	1 Electrical Duct (Detail 1) RHO 60 LF 50	RHO 90 LF 100	RHO 120 LF 100	3 Electrical Ducts (Detail 2) RHO 60 LF 50	RHO 90 LF 100	RHO 120 LF 100	6 Electrical Ducts (Detail 3) RHO 60 LF 50	RHO 90 LF 100	RHO 120 LF 100
8	58	54	53	56	48	46	53	42	39
6	77	71	69	74	63	60	70	54	51
4	101	93	91	96	81	77	91	69	65
2	132	121	118	126	105	100	119	89	83
1	154	140	136	146	121	114	137	102	95
1/0	177	160	156	168	137	130	157	116	107
2/0	203	183	178	192	156	147	179	131	121
3/0	233	210	204	221	178	158	205	148	137
4/0	268	240	232	253	202	190	234	168	155
250	297	265	256	280	222	209	258	184	169
350	363	321	310	340	267	250	312	219	202
500	444	389	375	414	320	299	377	261	240
750	552	478	459	511	388	362	462	314	288
1000	628	539	518	579	435	405	522	351	321

For ambient temperatures other than 20°C (68°F), multiply the ampacities shown above by the appropriate factor shown below.

Ambient Temp. °C	Factor
6-10	1.09
11-15	1.04
16-20	1.00
21-25	.95
26-30	.90

ALUMINUM OR COPPER-CLAD ALUMINUM

Size AWG/kcmil	1 Electrical Duct (Detail 1) RHO 60 LF 50	RHO 90 LF 100	RHO 120 LF 100	3 Electrical Ducts (Detail 2) RHO 60 LF 50	RHO 90 LF 100	RHO 120 LF 100	6 Electrical Ducts (Detail 3) RHO 60 LF 50	RHO 90 LF 100	RHO 120 LF 100
8	45	42	41	43	37	36	41	32	30
6	60	55	54	57	49	47	54	42	39
4	78	72	71	75	63	60	71	54	51
2	103	94	92	98	83	78	92	70	65
1	120	109	106	114	94	89	107	79	74
1/0	138	125	122	131	107	101	122	90	84
2/0	158	143	139	150	122	115	140	102	95
3/0	182	164	159	172	139	131	160	116	107
4/0	209	187	182	198	158	149	183	131	121
250	233	207	201	219	174	163	205	144	132
350	285	252	244	267	209	196	245	172	158
500	352	308	297	328	254	237	299	207	190
750	446	386	372	413	314	293	374	254	233
1000	521	447	430	480	361	336	433	291	266

For ambient temperatures other than 20°C (68°F), multiply the ampacities shown above by the appropriate factor shown below.

Ambient Temp. °F	Factor
43-50	1.09
52-59	1.04
61-68	1.00
70-77	.95
79-86	.90

Table B-310-7. Ampacities of Three Single Insulated Conductors, Rated 0 through 2000 Volts, in Underground Electrical Ducts (Three Conductors per Electrical Duct) Based on Ambient Earth Temperature of 20°C (68°F), Electrical Duct Arrangement per Figure B-310-2, Conductor Temperature 75°C (167°F)

COPPER — TYPES RHW, THHW, THW, THWN, XHHW, USE

Size (AWG kcmil)	1 Electrical Duct (Detail 1) RHO 60 / LF 50	RHO 90 / LF 100	RHO 120 / LF 100	3 Electrical Ducts (Detail 2) RHO 60 / LF 50	RHO 90 / LF 100	RHO 120 / LF 100	6 Electrical Ducts (Detail 3) RHO 60 / LF 50	RHO 90 / LF 100	RHO 120 / LF 100
8	63	58	57	61	51	49	57	44	41
6	84	77	75	80	67	63	75	56	53
4	111	100	98	105	86	81	98	73	67
3	129	116	113	122	99	94	113	83	77
2	147	132	128	139	112	106	130	93	86
1	171	153	148	161	128	121	149	106	98
1/0	197	175	169	185	146	137	170	121	111
2/0	226	200	193	212	166	156	194	136	126
3/0	260	228	220	243	189	177	222	154	142
4/0	301	263	253	280	215	201	255	175	161
250	334	290	279	310	236	220	281	192	176
300	373	321	308	344	260	242	310	210	192
350	409	351	337	377	283	264	340	228	209
400	442	376	361	394	302	280	368	243	223
500	503	427	409	460	341	316	412	273	249
600	552	468	447	511	371	343	457	296	270
700	602	509	486	553	402	371	492	319	291
750	632	529	505	574	417	385	509	330	301
800	654	544	520	597	428	395	527	338	308
900	692	575	549	628	450	415	554	355	323
1000	730	605	576	659	472	435	581	372	338

ALUMINUM OR COPPER-CLAD ALUMINUM — TYPES RHW, THHW, THW, THWN, XHHW, USE

Size (AWG kcmil)	1 Electrical Duct (Detail 1) RHO 60 / LF 50	RHO 90 / LF 100	RHO 120 / LF 100	3 Electrical Ducts (Detail 2) RHO 60 / LF 50	RHO 90 / LF 100	RHO 120 / LF 100	6 Electrical Ducts (Detail 3) RHO 60 / LF 50	RHO 90 / LF 100	RHO 120 / LF 100
8	49	45	44	47	40	38	45	34	32
6	66	60	58	63	52	49	59	44	41
4	86	78	76	79	67	63	77	57	52
3	101	91	89	92	77	73	84	65	60
2	115	103	100	108	87	82	101	73	67
1	133	119	115	126	100	94	116	83	77
1/0	153	136	132	144	114	107	133	94	87
2/0	176	156	151	165	130	121	151	106	98
3/0	203	178	172	189	147	138	173	121	111
4/0	235	205	198	219	168	157	199	137	126
250	261	227	218	242	185	172	220	150	137
300	293	252	242	272	204	190	245	165	151
350	321	276	265	296	222	207	266	179	164
400	349	297	284	321	238	220	288	191	174
500	397	338	323	364	270	250	326	216	197
600	446	373	356	408	296	274	365	236	215
700	488	408	389	443	321	297	394	255	232
750	508	425	405	461	334	309	409	265	241
800	530	439	418	481	344	318	427	273	247
900	563	466	444	510	365	337	450	288	261
1000	597	494	471	538	385	355	475	304	276

For ambient temperatures other than 20°C (68°F), multiply the ampacities shown above by the appropriate factor shown below.

Ambient Temp. °C	Factor	Ambient Temp. °F	Factor
6–10	1.09	43–50	1.09
11–15	1.04	52–59	1.04
16–20	1.00	61–68	1.00
21–25	.95	70–77	.95
26–30	.90	79–86	.90

Table B-310-8. Ampacities of Two or Three Insulated Conductors, Rated 0 through 2000 Volts, Cabled within an Overall (Two-or-Three Conductor) Covering, Directly Buried in Earth Based on Ambient Earth Temperature of 20°C (68°F), Arrangement per Figure B-310-2, 100 Percent Load Factor, Thermal Resistance (RHO) of 90

Size	1 Cable (Fig. B-310-2 Detail 5)		2 Cables (Fig. B-310-2 Detail 6)		1 Cable (Fig. B-310-2 Detail 5)		2 Cables (Fig. B-310-2 Detail 6)		Size
	60°C (140°F)	75°C (167°F)	60°C (140°F)	75°C (167°F)	60°C (140°F)	75°C (167°F)	60°C (140°F)	75°C (167°F)	
	TYPES		TYPES		TYPES		TYPES		
AWG kcmil	UF	RHW, THHW, THW, THWN, XHHW, USE	UF	RHW, THHW, THW, THWN, XHHW, USE	UF	RHW, THHW, THW, THWN, XHHW, USE	UF	RHW, THHW, THW, THWN, XHHW, USE	AWG kcmil
	COPPER				ALUMINUM OR COPPER-CLAD ALUMINUM				
8	64	75	60	70	51	59	47	55	8
6	85	100	81	95	68	75	60	70	6
4	107	125	100	117	83	97	78	91	4
2	137	161	128	150	107	126	110	117	2
1	155	182	145	170	121	142	113	132	1
1/0	177	208	165	193	138	162	129	151	1/0
2/0	201	236	188	220	157	184	146	171	2/0
3/0	229	269	213	250	179	210	166	195	3/0
4/0	259	304	241	282	203	238	188	220	4/0
250		333		308		261		241	250
350		401		370		315		290	350
500		481		442		381		350	500
750		585		535		473		433	750
1000		657		600		545		497	1000

Ambient Temp. °C	For ambient temperatures other than 20°C (68°F), multiply the ampacities shown above by the appropriate factor shown below.								Ambient Temp. °F
6-10	1.12	1.09	1.12	1.09	1.12	1.09	1.12	1.09	43-50
11-15	1.06	1.04	1.06	1.04	1.06	1.04	1.06	1.04	52-59
16-20	1.00	1.00	1.00	1.00	1.00	1.00	1.00	1.00	61-68
21-25	.94	.95	.94	.95	.94	.95	.94	.95	70-77
26-30	.87	.90	.87	.90	.87	.90	.87	.90	79-86

For ampacities of UF cable in underground electrical ducts, multiply the ampacities shown in the table by 0.74.

**Table B-310-9. Ampacities of Three Triplexed Single Insulated
Conductors, Rated 0 through 2000 Volts, Directly Buried in Earth
Based on Ambient Earth Temperature of 20°C (68°F),
Arrangement per Figure B-310-2, 100 Percent Load Factor,
Thermal Resistance (RHO) of 90**

Size	See Fig. B-310-2 Detail 7		See Fig. B-310-2 Detail 8		See Fig. B-310-2 Detail 7		See Fig. B-310-2 Detail 8		Size
	60°C (140°F)	75°C (167°F)	60°C (140°F)	75°C (167°F)	60°C (140°F)	75°C (167°F)	60°C (140°F)	75°C (167°F)	
AWG kcmil	TYPES		TYPES		TYPES		TYPES		AWG kcmil
	UF	USE	UF	USE	UF	USE	UF	USE	
	COPPER				ALUMINUM OR COPPER-CLAD ALUMINUM				
8	72	84	66	77	55	65	51	60	8
6	91	107	84	99	72	84	66	77	6
4	119	139	109	128	92	108	85	100	4
2	153	179	140	164	119	139	109	128	2
1	173	203	159	186	135	158	124	145	1
1/0	197	231	181	212	154	180	141	165	1/0
2/0	223	262	205	240	175	205	159	187	2/0
3/0	254	298	232	272	199	233	181	212	3/0
4/0	289	339	263	308	226	265	206	241	4/0
250		370		336		289		263	250
350		445		403		349		316	350
500		536		483		424		382	500
750		654		587		525		471	750
1000		744		665		608		544	1000

Ambient Temp. °C	For ambient temperatures other than 20°C (68°F), multiply the ampacities shown above by the appropriate factor shown below.								Ambient Temp. °F
6-10	1.12	1.09	1.12	1.09	1.12	1.09	1.12	1.09	43-50
11-15	1.06	1.04	1.06	1.04	1.06	1.04	1.06	1.04	52-59
16-20	1.00	1.00	1.00	1.00	1.00	1.00	1.00	1.00	61-68
21-25	.94	.95	.94	.95	.94	.95	.94	.95	70-77
26-30	.87	.90	.87	.90	.87	.90	.87	.90	79-86

Table B-310-10. Ampacities of Three Single Insulated Conductors, Rated 0 through 2000 Volts, Directly Buried in Earth Based on Ambient Earth Temperature of 20°C (68°F), Arrangement per Figure B-310-2, 100 Percent Load Factor, Thermal Resistance (RHO) of 90

Size	See Fig. B-310-2 Detail 9		See Fig. B-310-2 Detail 10		See Fig. B-310-2 Detail 9		See Fig. B-310-2 Detail 10		Size
	60°C (140°F)	75°C (167°F)	60°C (140°F)	75°C (167°F)	60°C (140°F)	75°C (167°F)	60°C (140°F)	75°C (167°F)	
AWG kcmil	TYPES		TYPES		TYPES		TYPES		AWG kcmil
	UF	USE	UF	USE	UF	USE	UF	USE	
	COPPER				ALUMINUM OR COPPER-CLAD ALUMINUM				
8	84	98	78	92	66	77	61	72	8
6	107	126	101	118	84	98	78	92	6
4	139	163	130	152	108	127	101	118	4
2	178	209	165	194	139	163	129	151	2
1	201	236	187	219	157	184	146	171	1
1/0	230	270	212	249	179	210	165	194	1/0
2/0	261	306	241	283	204	239	188	220	2/0
3/0	297	348	274	321	232	272	213	250	3/0
4/0	336	394	309	362	262	307	241	283	4/0
250		429		394		335		308	250
350		516		474		403		370	350
500		626		572		490		448	500
750		767		700		605		552	750
1000		887		808		706		642	1000
1250		979		891		787		716	1250
1500		1063		965		862		783	1500
1750		1133		1027		930		843	1750
2000		1195		1082		990		897	2000

Ambient Temp. °C	For ambient temperatures other than 20°C (68°F), multiply the ampacities shown above by the appropriate factor shown below.								Ambient Temp. °F
6-10	1.12	1.09	1.12	1.09	1.12	1.09	1.12	1.09	43-50
11-15	1.06	1.04	1.06	1.04	1.06	1.04	1.06	1.04	52-59
16-20	1.00	1.00	1.00	1.00	1.00	1.00	1.00	1.00	61-68
21-25	.94	.95	.94	.95	.94	.95	.94	.95	70-77
26-30	.87	.90	.87	.90	.87	.90	.87	.90	79-86

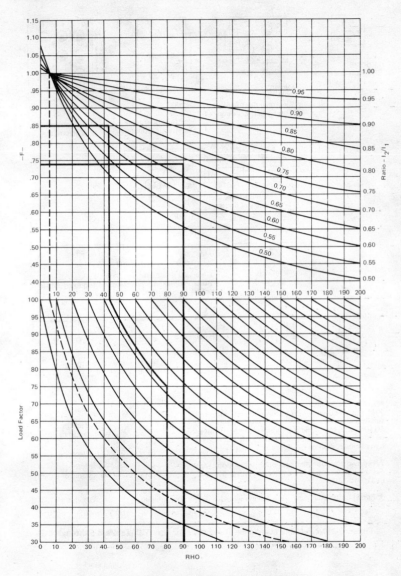

Figure B-310-1. Interpolation Chart for Cables in a
Duct Bank I_1 = Ampacity for Rho - 60, 50 LF;
I_2 = Ampacity for Rho-120, 100 LF;
Desired Ampacity = F X I_1

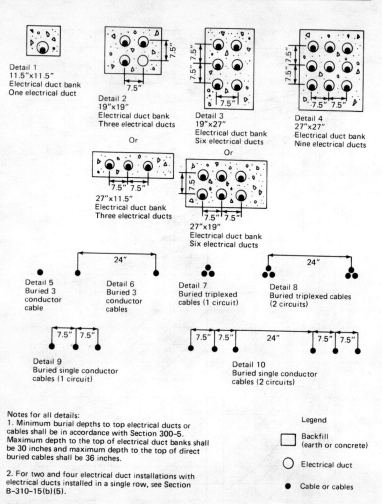

Notes for all details:

1. Minimum burial depths to top electrical ducts or cables shall be in accordance with Section 300-5. Maximum depth to the top of electrical duct banks shall be 30 inches and maximum depth to the top of direct buried cables shall be 36 inches.

2. For two and four electrical duct installations with electrical ducts installed in a single row, see Section B-310-15(b)(5).

3. For SI units: one inch = 25.4 millimeters; one foot = 305 millimeters.

Legend

☐ Backfill (earth or concrete)

◯ Electrical duct

● Cable or cables

Figure B-310-2.   Cable Installation Dimensions for Use With Tables B-5 through B-10

Figure B-310-3. Ampacities of Single Insulated Conductors Rated 0 through 5000 Volts in Underground Electrical Ducts (Three Conductors per Electrical Duct) Nine Single-Conductor Cables per Phase Based on Ambient Earth Temperature of 20°C (68°F), Conductor Temperature 75°C (167°F)

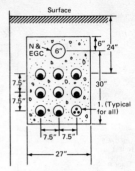

Design Criteria

Neutral and Electrical Grounding
Conductor (EGC) Duct = 6 inch
Phase Ducts = 3 to 5 inch
Conductor Material = Copper
Number of Cables per Duct = 3
Number of Cables per Phase = 9
Rho Concrete = Rho Earth − 5
Rho PVC Duct = 650
Rho Cable Insulation = 500
Rho Cable Jacket = 650

NOTES:

Neutral configuration per Section 300-5(i), Exception No. 2 for isolated phase installations in nonmagnetic ducts.

Phasing is A, B, C in rows or columns. Where magnetic electrical ducts are used, conductors are installed A, B, C per electrical duct with the neutral and all equipment grounding conductors in the same electrical duct. In this case, the 6-inch trade size neutral duct is eliminated.

Maximum harmonic loading on the neutral conductor cannot exceed 50 percent of the phase current for the ampacities shown in the table.

Metallic shields of Type MV-75 cable shall be grounded at one point only where using A, B, C phasing in rows or columns.

For SI units: one inch = 25.4 millimeters.

Size	Types RHW, THHW, THW, THWN, XHHW, USE, OR MV-75			Size
kcmil	Total Per Phase Ampere Rating			kcmil
	RHO EARTH 60 LF 50	RHO EARTH 90 LF 100	RHO EARTH 120 LF 100	
250	2340 (260A/Cable)	1530 (170A/Cable)	1395 (155A/Cable)	250
350	2790 (310A/Cable)	1800 (200A/Cable)	1665 (185A/Cable)	350
500	3375 (375A/Cable)	2160 (240A/Cable)	1980 (220A/Cable)	500

Ambient Temp. °C	For ambient temperatures other than 20°C (68°F), multiply the ampacities shown above by the appropriate factor shown below.				Ambient Temp. °F
6-10	1.09	1.09	1.09	1.09	43-50
11-15	1.04	1.04	1.04	1.04	52-59
16-20	1.00	1.00	1.00	1.00	61-68
21-25	.95	.95	.95	.95	70-77
26-30	.90	.90	.90	.90	79-86

Figure B-310-4.   Ampacities of Single Insulated Conductors Rated 0 through 5000 Volts in Nonmagnetic Underground Electrical Ducts (One Conductor per Electrical Duct) Four Single-Conductor Cables per Phase Based on Ambient Earth Temperature of 20°C (68°F), Conductor Temperature 75°C (167°F)

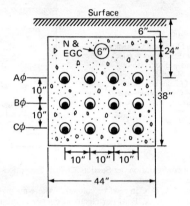

Design Criteria

Neutral and Electrical
Grounding Conductor Duct = 6 inch
Phase Ducts = 3 inch (min.)
Conductor Material = Copper
Number of Cables per Duct =1
Number of Cables per Phase = 4
Rho Concrete = Rho Earth − 5
Rho PVC Duct = 650
Rho Cable Insulation = 500
Rho Cable Jacket = 650

NOTES:
Neutral configuration per Section 300-5(i), Exception No. 2.
Maximum harmonic loading on the neutral conductor cannot exceed 50 percent of the phase current for the ampacities shown in the table.
Metallic shields of Type MV-75 cable shall be grounded at one point only.
For SI units, one inch = 25.4 millimeters.

Size	Types			Size
	RHW, THHW, THW, THWN, XHHW, USE, OR MV-75			
kcmil	Total Per Phase Ampere Rating			kcmil
	RHO EARTH 60 LF 50	RHO EARTH 90 LF 100	RHO EARTH 120 LF 100	
750	2520 (705A/Cable)	1860 (465A/Cable)	1680 (420A/Cable)	750
1000	3300 (825A/Cable)	2140 (535A/Cable)	1920 (480A/Cable)	1000
1250	3700 (925A/Cable)	2380 (595A/Cable)	2120 (530A/Cable)	1250
1500	4060 (1015A/Cable)	2580 (645A/Cable)	2300 (575A/Cable)	1500
1750	4360 (1090A/Cable)	2740 (685A/Cable)	2460 (615A/Cable)	1750

Ambient Temp. °C	For ambient temperatures other than 20°C (68°F), multiply the ampacities shown above by the appropriate factor shown below.					Ambient Temp. °F
6-10	1.09	1.09	1.09	1.09	1.09	43-50
11-15	1.04	1.04	1.04	1.04	1.04	52-59
16-20	1.00	1.00	1.00	1.00	1.00	61-68
21-25	.95	.95	.95	.95	.95	70-77
26-30	.90	.90	.90	.90	.90	79-86

Figure B-310-5. Ampacities of Single Insulated Conductors Rated 0 through 5000 Volts in Nonmagnetic Underground Electrical Ducts (One Conductor per Electrical Duct) Five Single-Conductor Cables per Phase Based on Ambient Earth Temperature of 20°C (68°F), Conductor Temperature 75°C (167°F)

Design Criteria

Neutral and Electrical Grounding
Conductor (EGC) Duct = 6 inch
Phase Ducts = 3 inch (min.)
Conductor Material = Copper
Number of Cables per Duct = 1
Number of Cables per Phase = 5
Rho Concrete = Rho Earth − 5
Rho PVC Duct = 650
Rho Cable Insulation = 500
Rho Cable Jacket = 650

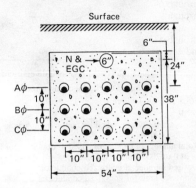

NOTES:
Neutral configuration per Section 300-5(i), Exception No. 2.
Maximum harmonic loading on the neutral conductor cannot exceed 50 percent of the phase current for the ampacities shown in the table.
Metallic shields of Type MV-75 cable shall be grounded at one point only.
For SI units: one inch = 25.4 millimeters.

Size	Types RHW, THHW, THW, THWN, XHHW, USE, OR MV-75			Size
kcmil	Total Per Phase Ampere Rating			kcmil
	RHO EARTH 60 LF 50	RHO EARTH 90 LF 100	RHO EARTH 120 LF 100	
2000	5575 (1115A/Cable)	3375 (675A/Cable)	3000 (600A/Cable)	2000

Ambient Temp. °C	For ambient temperatures other than 20°C (68°F), multiply the ampacities shown above by the appropriate factor shown below.					Ambient Temp. °F
6-10	1.09	1.09	1.09	1.09	1.09	43-50
11-15	1.04	1.04	1.04	1.04	1.04	52-59
16-20	1.00	1.00	1.00	1.00	1.00	61-68
21-25	.95	.95	.95	.95	.95	70-77
26-30	.90	.90	.90	.90	.90	79-86

Table B-310-11. Adjustment Factors for More than Three Current-Carrying Conductors in a Raceway or Cable with Load Diversity

Number of Current-Carrying Conductors	Percent of Values in Tables as Adjusted for Ambient Temperature if Necessary
4 through 6	80
7 through 9	70
10 through 24*	70
25 through 42*	60
43 and above*	50

*These factors include the effects of a load diversity of 50 percent.

(FPN): This table is based on the following formula:

$$A_2 = \frac{\sqrt{0.5\,N}}{E} \times (A_1) \text{ where}$$

A_1 = Table ampacity multiplied by factor from Article 310, Notes to Ampacity Tables of 0 to 2000 Volts, Note·8(a).

N = Total number of conductors used to obtain factor from Article 310, Notes to Ampacity Tables of 0 to 2000 Volts, Note 8(a).

E = Desired number of energized conductors

A_2 = Ampacity limit for energized conductors

Index

Blocking diode, solar photovoltaic systems
Definition, 690-2

Boatyards, see Marinas and boatyards

Bodies, conduit, see Conduit bodies

Boilers
Electrode-type, 424-H
Resistance-type, 424-G

Bonding, 250-G
Definition, Art. 100-A
Grounding-type receptacles, 250-74
Hazardous (classified) locations, see
   Hazardous (classified) locations
Intrinsically safe systems, 504-60
Loosely jointed raceways, 250-77
Other enclosures, 250-75
Outside raceway, 250-79(f)
Over 250 volts, 250-76
Piping systems, 250-80
Receptacles, 250-74
Service equipment, 250-71, 250-72
Swimming pools, 680-22

Bonding jumpers, see Jumpers, bonding

Bored holes through studs, joists,
300-4(a)(1), 545-4(b)

Bowling alleys, Art. 518
Emergency lighting system, Art. 700

Boxes (outlet, device, pull, and junction), Art. 370; see also Hazardous (classified) locations
Accessibility, 370-29
Concealed work, 370-24
Conductors, number in box, 370-16
   Entering boxes, conduit bodies or fittings, 370-17
Construction specifications, 370-C
Covers, 370-25, 370-28(c), 370-41
Cutout, see Cabinets and cutout boxes
Damp locations, 370-15
Depth, 370-24
Exposed surface extensions, 370-22
Floor, for receptacles, 370-27(b)
Grounding, 250-E, 250-114, 300-9
Insulating, see Boxes, nonmetallic
Junction, pull, see Junction boxes
Lighting outlets, 370-27(a), 410-C, 410-16
Metal
   Construction, 370-40
   Grounding, 370-14
   Installation, 370-B
Nonmetallic, 324-13, 336-17, 370-13,
   370-17(c), 370-43
Over 600 volts, 370-D
Portable, in theaters, 520-62
Remote control, signal circuits,
   725-11(b), 725-38(a)(2)
Repairing plaster around, 370-11
Required location, 300-15
Round, 370-12
Secured supports, 300-11, 370-23
Snap switches over 300 volts, not
   ganged, 380-8(b)
Unused openings, closed, 110-12(a),
   370-18

Vertical raceway runs, 300-19
Wall or ceiling, 370-20
Wet locations, 370-15

Branch circuits, Art. 210, Art. 220
Air conditioners, 440-D, 440-G
Appliances, 422-B and C
   Definition, Art. 100-A
Busways as branch circuits, 364-12 thru
   364-14
Calculation of loads, 220-3, Chap. 9 Part B
Car lighting, 620-22
Color code, 210-5
Conductors, minimum ampacity and size,
   210-19
Critical
   Definition, 517-3
Definition, Art. 100-A
Fixed electric space heating equipment,
   424-3
General, 210-A
General purpose
   Definition, Art. 100-A
Health care facilities, 517-18(a)
Individual
   Definition, Art. 100-A
   Overcurrent protection, 210-20
   Permissible loads, 210-23
   Required, 520-53(g), 600-6(b), 605-8(b),
     620-22, 710-72
Infrared lamps, 422-15
Isolated power systems, 517-160(c)
Maximum loads, 210-22
Mobile homes, 550-7
Motor, on individual branch circuit,
   430-B
Multiwire, 210-4, 501-18, 502-18
   Definition, Art. 100-A
Outside, Art. 225
Overcurrent protection, 210-20, 240-3
Patient bed location, 517-19(a)
Permissible loads, 210-23
Ratings, 210-B
Recreational vehicles, 551-42
Requirements for, Table 210-24
Selection current, 440-4(c)
   Definition, 440-2
Small appliance, 220-4(b)
Stage or set, 530-22
Taps from, 210-19(c), 210-24, 240-3
   Ex. 5, 240-4 Ex. 1 and 2
Through fixtures, 410-11, 410-31
Two or more outlets on, 210-23
Voltage drop, 210-19(a) FPN
Voltage limitations, 210-6
X-ray equipment, 517-E, 660-4, 660-6(b)

Branch-circuit selection current
Definition, 440-2

Building component
Definition, 545-3

Building system
Definition, 545-3

Building wire, see Conductors

Buildings
Definition, Art. 100-A
First floor of, 336-4(a)

Service-entrance, Types SE and USE, see Service-entrance cable
Shielded nonmetallic, Type SNM, see Shielded nonmetallic cable
Splices in boxes, 300-15
Stage, 530-18(a)
Through studs, joists, rafters, 300-4
Underground, 230-C, 300-5, 310-7, 710-3(b)
Underground feeder and branch circuit Type UF, see Underground feeder and branch circuit cable

Calculations, Chap. 9, Part B; see also Loads

Camping trailer
Definition, 551-2

Canopies
Boxes and fittings, 370-15
Lighting fixtures
Conductors, space for, 410-10
Cover
At boxes, 410-12
Combustible finishes, covering required between canopy and box, 410-13
Live parts exposed, 410-3

Capacitors, Art. 460; see also Hazardous (classified) locations
Conductors, 460-8
Connections at services, 230-82
Discharge of stored energy, 460-6
Draining charge, 460-6
Induction and dielectric heating, 665-24
Over 600 volts, 460-B
X-ray equipment, 660-36

Capacity, interrupting, see Interrupting capacity

Caps, see Attachment plugs

Cartridge fuses, 240-F
Disconnection, 240-40

CATV systems, see Community antenna television and radio distribution systems

Ceiling fans, 680-6, 680-41(b)
Support of, 370-27(c), 422-18

Cell
Cellular concrete floor raceways
Definition, 358-2
Electrolytic
Definition, 688-2
Sealed, storage batteries
Definition, 480-2
Solar
Definition, 690-2

Cell line, electrolytic cells
Attachments and auxiliary equipment
Definition, 668-2
Definition, 668-2

Cellars, see Basements

Cellular concrete floor raceways, Art. 358
Connection to cabinets and other enclosures, 358-6
Definitions, 358-2

Discontinued outlets, 358-13
Header ducts, 358-5
Inserts, 358-9
Junction boxes, 358-7
Markers, 358-8
Number of conductors, 358-11
Size of conductors, 358-10
Uses not permitted, 358-4

Cellular metal floor raceways, Art. 356
Connection to cabinets and extension from cells, 356-11
Construction, 356-12
Definitions, 356-1
Discontinued outlets, 356-7
Inserts, 356-10
Junction boxes, 356-9
Markers, 356-8
Number of conductors, 356-5
Size of conductors, 356-4
Uses not permitted, 356-2

Chair lifts, see Elevators, dumbwaiters, escalators, moving walks, wheelchair lifts, and stairway chair lifts

Churches, Art. 518

Cinder fill
Intermediate or rigid metal conduits and electrical metallic tubing, in or under, 345-3(c), 346-3, 348-1
Rigid nonmetallic conduit, 347-2

Circuit breakers, Art. 240; see also Hazardous (classified) locations
Accessibility and grouping, 380-8
Circuits over 600 volts, 710-21
Definition, Art. 100-A
Disconnection of grounded circuits, 380-2(b), 514-5
Enclosures, 380-3
General, 110-9, 240-A
Overcurrent protection, 230-208, 240-A
Generators, 445-4
Motors, 430-52, 430-58, 430-110
Transformers, 450-3
Panelboards, 384-B
Rating, motor branch circuits, 430-58
Rating, nonadjustable trip, 240-3 Ex. 4, 240-83(c)
Service overcurrent protection, 230-90, 230-91
Services, disconnecting means, 230-70, 230-208
Switches, use as, 240-83(d), 380-11, 410-81(b)
Wet locations, in, 380-4

Circuit directory, panelboards, 110-22

Circuit-interrupters, ground-fault, see Ground-fault circuit-interrupters

Circuits
Anesthetizing locations, 517-63
Bonding jumper
Definition, Art. 100-A
Branch, see Branch circuits
Burglar alarm, see Remote-control signaling and power limited circuits